T0212071

Lecture Notes in Computer Science   11166

Commenced Publication in 1973
Founding and Former Series Editors:
Gerhard Goos, Juris Hartmanis, and Jan van Leeuwen

More information about this series at http://www.springer.com/series/7409

embers, reviewers, session chairs, student volunteers, and supporters.
ons are much appreciated.

Richang Hong
Wen-Huang Cheng
Toshihiko Yamasaki
Meng Wang
Chong-Wah Ngo

Richang Hong · Wen-Huang Cheng
Toshihiko Yamasaki · Meng Wang
Chong-Wah Ngo (Eds.)

Advances in Multimedia Information Processing – PCM 2018

19th Pacific-Rim Conference on Multimedia
Hefei, China, September 21–22, 2018
Proceedings, Part III

 Springer

Editors
Richang Hong
Hefei University of Technology
Hefei
China

Wen-Huang Cheng
National Chiao Tung University
Hsinchu
Taiwan

Toshihiko Yamasaki
University of Tokyo
Tokyo
Japan

Meng Wang
Hefei University of Technology
Hefei
China

Chong-Wah Ngo
City University of Hong Kong
Hong Kong
Hong Kong, China

ISSN 0302-9743   ISSN 1611-3349   (electronic)
Lecture Notes in Computer Science
ISBN 978-3-030-00763-8   ISBN 978-3-030-00764-5   (eBook)
https://doi.org/10.1007/978-3-030-00764-5

Library of Congress Control Number: 2018954671

LNCS Sublibrary: SL3 – Information Systems and Applications, incl. Internet/Web, and HCI

This Springer imprint is published by the registered company Springer Nature Switzerland AG
The registered company address is: Gewerbestrasse 11, 6330 Cham, Switzerland

Preface

all committee m
Their contributi

September 201

The 19th Pacific-Rim Conference on Multimed
China, during September 21–22, 2018, and host
nology (HFUT). PCM is a major annual intern
researchers and practitioners across academia and
entific achievements and industrial innovations in

It is a great honor for HFUT to host PCM 20
multimedia conferences, in Hefei, China. Hefei Un
the capital of Anhui province, is one of the key
Ministry of Education, China. Recently its multime
more and more attentions from local and international
the capital city of Anhui Province, and is located in
Yangtze and Huaihe rivers. Well known both as a hist
Kingdoms Period and the hometown of Lord Bao, Hefe
than 2000 years. In modern times, as an important bas
China, Hefei is the first and sole Science and Techno
China, and a member city of WTA (World Technopoli
PCM 2018 is a memorable experience for all participant

PCM 2018 featured a comprehensive program. We rec
main conference by authors from more than ten countries.
a large number of high-quality papers in multimedia c
signal processing and communications, and multimedia ap
thank our Technical Program Committee with 178 memb
reviewing papers and providing valuable feedback to the aut
submissions, the program chairs decided to accept 209 regula
at least three reviews per submission. In total, 30 papers were
sessions, while 20 of them were accepted. The volumes of th
contain all the regular and special session papers.

We are also heavily indebted to many individuals for their s
We wish to acknowledge and express our deepest appreciation
Wang and Chong-Wah Ngo; program chairs, Richang Hong,
Toshihiko Yamasaki; organizing chairs, Xueliang Liu, Yun Tie
publicity chairs, Jingdong Wang, Min Xu, Wei-Ta Chu and
chairs, Zhengjun Zha and Liqiang Nie. Without their efforts a
2018 would not have become a reality. Moreover, we want to
Springer, Anhui Association for Artificial Intelligence, Shandong
Institute, Kuaishou Co. Ltd., and Zhongke Leinao Co. Ltd. Finall

Organization

General Chairs

Meng Wang   Hefei University of Technology, China
Chong-Wah Ngo   City University of Hong Kong, Hong Kong, China

Technical Program Chairs

Richang Hong   Hefei University of Technology, China
Wen-Huang Cheng   National Chiao Tung University, Taiwan, China
Toshihiko Yamasaki   University of Tokyo, Japan

Organizing Chairs

Xueliang Liu   Hefei University of Technology, China
Yun Tie   Zhengzhou University, China
Hanwang Zhang   Nanyang Technological University, Singapore

Publicity Chairs

Jingdong Wang   Microsoft Research Asia, China
Min Xu   University of Technology Sydney, Australia
Wei-Ta Chu   National Chung Cheng University, Taiwan, China
Yi Yu   National Institute of Informatics, Japan

Special Session Chairs

Zhengjun Zha   University of Science and Technology of China, China
Liqiang Nie   Shandong University, China

Contents – Part III

Poster Papers

MFDCNN: A Multimodal Fusion DCNN Framework for Object Detection and Segmentation

Feng Zhou[1]([✉])[iD], Yong Hu[1,2], and Xukun Shen[1,2][iD]

[1] State Key Laboratory of Virtual Reality Technology and Systems,
Beihang University, Beijing, China
{zhoufeng,huyong,xkshen}@buaa.edu.cn

[2] School of New Media Art and Design, Beihang University, Beijing, China

Abstract. In this paper, we study the problem of object detection and segmentation in the cluttered indoor scenes based on RGB-D data. The main issues about object detection and segmentation in the indoor scenes are coming from serious obstruction, inconspicuous classes, and confusion categories. To solve these problems, we propose a multimodal fusion deep convolutional neural network (MFDCNN) framework for object detection and segmentation, which can boost the performance effectively at two levels whilst keeping the framework end-to-end training. Towards the object detection, we adopt a multimodal region proposal network to solve the problem of object-level detection, towards the semantic segmentation, we utilize a multimodal fully convolutional network to provide the class labels to which each pixel belongs. We focus on learning object detection and segmentation simultaneous, we propose a novel loss function to combine these two kind networks together. Under this framework, we focus on cluttered indoor scenes with challenging settings and evaluate the performance of our MFDCNN on the NYU-Depth V2 dataset. Our MFDCNN achieves state-of-the-art performance on the object detection task and earns the comparable state-of-the-art performance on the task of semantic segmentation.

Keywords: RGB-D Perception · Semantic segmentation
Convolutional neural network · Object detection

1 Introduction

Object detection and semantic segmentation of indoor scene are the fundamental problems in computer vision, which can benefit a variety of intelligent applications such as domestic robots, Simultaneous localization and mapping (SLAM), content-based image retrieval. However, totally understanding of a scene, especially cluttered indoor scenes remains a challenging issue.

Marking out a tight bounding box to each object, and assigning a category label to all the pixels in the image in the bounding box is crucial to the

© Springer Nature Switzerland AG 2018
R. Hong et al. (Eds.): PCM 2018, LNCS 11166, pp. 3–13, 2018.
https://doi.org/10.1007/978-3-030-00764-5_1

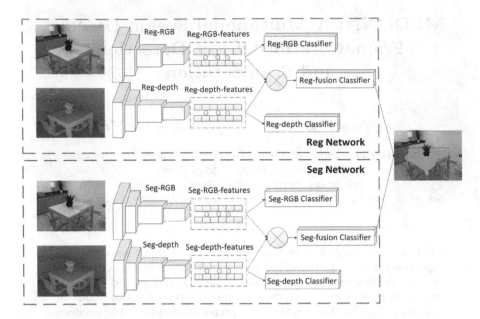

Fig. 1. Overview of our MFDCNN learning framework for object detection and segmentation. The training of the framework mainly contains four steps: (1) Training the $Reg - RGB$ and $Reg - depth$ models over the respective training data (indicated as purple box) each is for object detection; (2) Training the $Seg - RGB$ and $Seg - depth$ (indicated as red box) each is for semantic segmentation; (3) Discarding classify layer in Seg and Reg, replacing them with a new fusion classify layer; (4) Combining them together via novel loss function to obtain object detection and segmentation. Best viewed in color.

understanding of the scene. In this paper, we have designed and implemented a framework (Fig. 1) based on the deep convolutional neural network for cluttered indoor scene understanding. The overall architecture is based on the current very successful networks for object detection and segmentation.

The whole pipeline of our MFDCNN framework is visualized as Fig. 1 which contains two essential parts, object detection, and semantic segmentation. Object detection part provides the bounding box of an object instance, and semantic segmentation offers a per-pixel segmentation. For clarity, in the following the two-stream line object detection network is called Reg network (purple box) in which the network for RGB is called $Reg - RGB$, and the other network for depth is called $Reg-depth$, and the other two-stream line semantic segmentation network is called Seg network (red box) in which the network for RGB is called $Seg - RGB$, and the other network for depth is called $Seg - depth$. The detail of these two parts is described as following.

(1) Reg network: As described in Fig. 1, Reg network is composed of two substreamlines. It takes a color and depth image pair as inputs. In this system, we convert the depth modality image into HHA (horizontal disparity, height

above ground, and the angle with gravity for the local surface normals of each pixel in the depth image) [8]. In the next, we do not distinguish disparity, depth, and HHA. We use RPN to generate 2.5D region candidates by computing features on the depth and color image for use in [5]. Using the *Reg* model, we classify the 2.5D region proposals to object level categories.

(2) *Seg* network: As described in Fig. 1, the *Seg* network contains $Seg - RGB$ and $Seg - depth$, these two sub-streamlines which trained by RGB and depth image pairs as above. $Seg - RGB$ and $Seg - depth$ take each corresponding modal images as training data. Each *Seg* produces a densely score map via the combination between a series atrous convolution and a bilinear interpolation stage, and then obtains $Seg - fusion\ Classify$ via the combining.

In this paper, we make three contributions, which are shown to have practical merit through experiments. First, we propose MFDCNN for object detection and segmentation. Second, Our *Reg* network achieves state-of-the-art performance on NYU-Depth V2 dataset, and *Seg* network earns the comparable state-of-the-art performance. Third, we propose a novel loss function to combine these two networks together. Experimental results show that our MFDCNN learning framework improves the performance of object detection and segmentation.

In the following section of this paper, we will discuss the detail of our framework. In the multimodal fusion architecture section, we will talk about the algorithms of *Reg* network and *Seg* network, and in network training and inference section, we will give the details of our parameters and loss constraint. In the experiment section, we will give the results and evaluation of our framework.

2   Related Work

The approach proposed in this paper is related to a large body of work on both DCNN for object detection and segmentation as well as the multimodal deep learning framework. We will briefly describe the similarities and differences between our work and other existing approaches with a highlight on recent literature.

There is large body of works having focused on object detection in the last decades, and most of them are based considerably on the use of a handcrafted feature. ConvNets is not a brand new notion, which has been widely used in 1990s [13], but it lost its crown because of the popularity of SVM and some other reasons. In 2012 Krizheysky [12] rekindled researchers' interest. Based on the success of the region based convolutional neural networks, Grishick [6] repurposed the deep classification network onto the tasks of object detection, it made the result of object detection soar up. The output of object detection is bounding boxes which surround the object. After this work, so far there have been many object detection tasks that have been proposed. The general trend of these tasks is fast, accurate and able to recognize a wide variety of objects [5,15,17–20]. These works answered the question about locating object, but they are not suitable for solving which part belongs to the located object.

Through DCNNs we can learn tens of thousands feature easily and efficiently. Before [16], the network of semantic segmentation based on ConvNets is not trained end-to-end. Fully convolutional networks (FCNs) [1–3,16,23] make pixel-level semantic segmentation feasible efficiently and easily, and they exceed the previous methods. Meantime it provides the dense output without any pre- and post-processing. It offers a more nuanced angle to view the scene, but only pixel-level annotation cannot tell the intraclass difference over the same category.

There have been many works about RGB-D images object detection [9,10], semantic segmentation [7,8] and instance segmentation [21]. [8] proposed a new way, a geocentric embedding, to encode each pixel in the depth map by height above ground, horizontal disparity and angle with gravity. Adding this new encoding way into R-CNN framework, [8] proposed a new way to learn rich features from RGB-D images for object detection and segmentation. Our network architecture shares similarities with theirs in the fine tune stage, while is different in the stage of classification and segmentation. Lenz [14] claimed that there were many different network architectures for fusing multimodel (depth and RGB) images, such as early fusion, late fusion and the combining early and late fusion. The author concluded that late fusion and the combined approach perform best for the fusion problem. Late fusion is simple and effective for our task. So in this work, we use the late fusion way to train our model for object detection and segmentation.

3   Multimodal Fusion DCNN

3.1   Multimodal Fusion Architecture

Totally speaking, there are two kinds of methods to do object detection and segmentation. The first one is to associate a pixel mask to each detected bounding box, and the second one is to separate this job into two steps, object detection, and semantic segmentation, through this way, we can get the pixel label of all object instances. In this paper, we use the second method to combine "where" and "which" together to achieve our work. From the framework visualized in Fig. 1, our framework MFDCNN can be divided into two parts - Reg network and Seg network - to discuss. The Reg network, which gives the location of the object instance by bounding boxes, and Seg network, which gives the pixel-level segmentation results for objects belonging to the same category.

In the Reg network, a new fusion classification layer - $Reg - fusion\ Classifier$ is merged into the $Reg - RGB$ and $Reg - depth$ network where the last $Classific - ation$ layer is discarded. Then the last $F(4096)$ in $Reg - RGB$ and $Reg - depth$ is concatenated and merged and sent to the last $Reg - fusion\ Classifier$ layer. In the seg network, we discard the last $Classification$ layer in the $Seg - RGB$ and $Seg - depth$ network, afterward concatenate and merge the features extracted from the $ASPP$ (6,12,18,24) layer and then send them into the newly added $Seg - fusion\ Classifier$ layer. Then the results obtained by the above networks are overlaid according to an extended

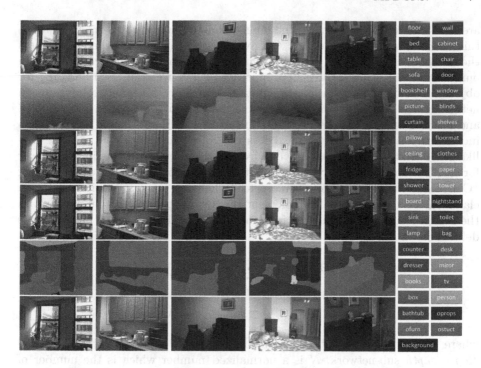

Fig. 2. The output of our MFDCNN. Here we showed the object detection and segmentation result, from the first row to the last, they are RGB images, depth images, the ground truth of object detection and segmentation of an instance, the result of the *Reg* network, the result of our *Seg* network, and object detection and segmentation result. Best viewed in color.

constraint. Through this operation, we can optimize object detection and segmentation simultaneously.

In short, we combine the region proposals from object detection with the score map gotten from semantic segmentation to get object detection and segmentation simultaneously.

3.2 Network Training and Inference

In our MFDCNN framework, each stream is initialized by a pre-trained CNN on the ImageNet database, the root causes of the use pre-trained model to initialize our model is that training on a smaller database is easy to overfitting. To be specific,

Training. Let $D = \{(I^1, D^1, y^1, l^1), ...(I^n, D^n, y^n, l^n)\}$ be the training data of our MFDCNN learning framework with n pairwise RGB-D images, where I^i and D^i denote the RGB and depth modalities of the i-th examples separately. The label y^i represents coordinates of the predicted bounding boxes. Suppose there are k objects in (I^i/D^i) image, the label y^i has $4k$ attributes corresponding to this image. The label $l^i \in \{0, 1, ..., C\}$ gives the per-pixel label, where C denotes the number of the category, the label 0 means background in this work. We train our multimodal using a three-step approach. First, we train Reg and Seg for object detection and semantic segmentation separately. Each network is a fusion network, in which RGB and depth image pair are firstly trained individually, and fused at last. In the Reg, we finetune on the RGB and depth training data individually, and use the extended loss of [20] to minimize the negative log-likelihood. In the fusion stage, we discard softmax weights in classification layer for each stream in Reg over each modality and concatenate them together using "CONCAT" operation. And then we feed them to the last additional fusion classification softmax layer to implement the fusion object detection network. In the fusion Reg network, the loss function for an RGB and depth image pair is defined as:

$$L_{Reg}\left\{I^i_{t_j}, D^i_{t_j}, y^i_{t_{j*}}\right\} = \sum_j (\lambda \frac{1}{N} P^*_j L_{Reg}(I^i_{t_j}, y^i_{t_{j*}}) \\ + (1 - \lambda)\frac{1}{N} P^*_j L_{Reg}(D^i_{t_j}, y^i_{t_{j*}}))$$

(1)

where λ is a constant coefficient to balance the loss between $Reg - RGB$ and $Reg - depth$ sub-network. N is a normalized number which is the number of anchor locations (around 6200). $I^i_{t_j}$ and $D^i_{t_j}$ denote two coordinates vectors of the predicted bounding boxes in RGB and depth images respectively. j is the index of an anchor in this pair of images, The value of p_j is 1 or 0, the detail about the loss, please refer to [20]. In the fusion Seg network, the loss function is defined as follow:

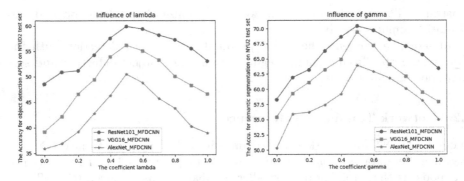

Fig. 3. The effect of the coefficient number on the NYU-Depth V2 dataset of object detection and semantic segmentation. The left gives the results of the object detection with different λ. And the right gives the results of the semantic segmentation with different γ.

$$L_{Seg}\left\{I^i, D^i, l^i\right\} - \sum_k (\gamma L_{Seg}(W_{I_k}^C g_{I_k}^C(I_k^i; \theta^C), l_k^i)$$
$$+ (1-\gamma)L_{Seg}(W_{D_k}^C g_{D_k}^C(D_k^i; \theta^C), l_k^i)) \tag{2}$$

Here, the per-pixel loss on RGB and depth modality are weighted by γ the balancing parameter. $g_I^C(I^i; \theta^C)$ denotes the representation extracted at the bilinear interpolation stage -with parameters θ^C- when applied to an RGB modality image I^i. k denotes the pixel of the image, the detail about the loss, please refer to [1].

In a word, in this work, we solve

$$\min \sum_{i=1}^N (\alpha L_{Seg}\left\{I^i, D^i, l^i\right\} + \beta L_{Reg}\left\{I_{t_j}^i, D_{t_j}^i, y_{t_{j*}}^i\right\}) \tag{3}$$

where, α and β are two constant coefficients to balance the loss between Reg network and Seg network, $L_{Seg}(I^i, D^i, l^i)$ is the loss of sub network - Seg, and $L_{Reg}(I_{t_j}^i, D_{t_j}^i, y_{t_{j*}}^i)$ is the loss of Reg.

Inference. Given an unseen RGB-D image, we utilize the final RGB-D model to predict the bounding box and the corresponding per-pixel classification to each object. In the experiment section, we compare the state-of-the-art with our method, the result shows the method proposed in this paper is better in object detection and segmentation.

4 Experiment

We design and implement our experiment on the NYU-Depth V2 dataset, which contains 1449 densely labeled pairs of aligned RGB and depth images. In the experiment, we use standard splits to train and test our model. There are 795 images in the training dataset, and the remaining 654 images belong to the testing dataset. The dataset comes with dense semantic segmentation annotation and has 894 classes, in the experiment, we map these 894 classes into 40 major furniture categories. We enclose each semantic segmentation class with a tight rectangle bounding box as the ground truth of each instance and study the task of object detection. For the sake of comparison to the state-of-the-art, we map all the 40 major furniture categories into all categories studied by [22]. In this paper, we train models on NYU-Depth V2 dataset and afterward extend the trained model into RGB-D modal using the proposed loss.

Experimental Setup. All experiments are performed using the publicly available Caffe framework [11]. As described previously we use [20] and [1] as the basis for our MFDCNN. Our MFDCNN contains two sub-networks - Reg and Seg (fusion model). Each is initialized with weights and biases of the all weighted layers from the trained $Reg - RGB/Reg - depth$ and $Seg - RGB/Seg - depth$

Table 1. Control experiments for object detection AP (%) on NYUD2 test set: we compare our performance against several state-of-the-art methods. And also, we give the results of ablation study.

method	modality	RGB Arch.	D Arch.	bath tub	bed	book shelf	box	chair	counter	desk	door	dresser	garbage bin
Fast R-CNN [5]	RGB	VGG	-	37.4	69.1	47.0	2.9	44.4	48.6	11.5	28.7	43.1	33.6
Gupta et al. [8]	RGB-D	AlexNet	AlexNet	36.4	70.8	35.1	3.6	47.3	46.8	14.9	23.3	38.6	43.9
Gupta et al. [9]	RGB-D	VGG	AlexNet	50.6	81.0	52.6	5.4	53.0	56.1	20.9	34.6	57.9	46.2
MFDCNN	RGB-D	AlexNet	AlexNet	51.2	82.2	53.1	5.9	55.0	59.1	22.1	35.8	59.1	47.1
MFDCNN	RGB-D	AlexNet	VGG	51.7	82.3	54.3	6.3	55.0	59.2	22.8	36.3	59.6	48.3
MFDCNN	RGB-D	VGG	VGG	57.5	85.4	57.1	7.7	58.1	65.4	28.7	39.3	59.9	51.1
MFDCNN	RGB-D	ResNet101	VGG	59.8	86.1	59.2	7.9	59.0	67.1	30.1	41.8	64.6	56.1
MFDCNN	RGB-D	ResNet101	ResNet101	61.2	87.4	62.7	8.3	59.9	68.2	32.8	43.3	66.4	57.5

method	modality	RGB Arch.	D Arch.	lamp	monitor	nightstand	pillow	sink	sofa	table	TV	toilet	mean
Fast R-CNN [5]	RGB	VGG	-	32.9	50.9	32.6	34.4	39.0	50.3	24.5	44.1	61.5	38.8
Gupta et al. [8]	RGB-D	AlexNet	AlexNet	37.6	52.7	40.7	42.4	43.5	51.6	22.0	38.0	47.7	38.8
Gupta et al. [9]	RGB-D	AlexNet	VGG	42.5	62.9	54.7	49.1	50.0	65.9	31.9	50.1	68.0	49.1
MFDCNN	RGB-D	AlexNet	AlexNet	44.1	66.1	54.8	49.9	50.9	66.2	34.3	53.2	69.7	50.5
MFDCNN	RGB-D	AlexNet	VGG	44.2	67.2	55.1	51.2	51.0	67.1	35.1	54.1	69.9	50.8
MFDCNN	RGB-D	VGG	VGG	55.5	74.1	64.9	55.8	54.6	75.8	43.3	56.4	76.9	56.2
MFDCNN	RGB-D	ResNet101	VGG	56.2	75.9	67.7	57.3	57.0	77.4	46.8	58.8	77.1	58.2
MFDCNN	RGB-D	ResNet101	ResNet101	57.9	77.1	69.9	59.3	59.0	79.1	48.2	59.8	79.4	59.9

(single model) except for the last classification layer. In the stage of RGB and depth network training, we initialize the $Reg - RGB$ and $Reg - depth$ using the pre-trained model released in [20], meantime, initialize the $Seg - RGB$ and $Seg - depth$ via the pre-trained model released in [1]. In the stage of training single model, the parameters of the learning rate start from 0.001 and are reduced by 10 times every 10k iterations, and the mini-batch size is set to 50. In the stage of fusion model training, the mini-batch size is set to 16, the initial learning rate base starts from 1e-4, multiplying the learning rate by 0.1 at every 5k iterations. Experimentally verified, α and β are fixed to 1, and λ and γ are set to 0.5 visualised in Fig. 3 in our MFDCNN learning framework.

Overall Performance. Tables 1 and 2 present the object detection and semantic segmentation results using Reg and Seg. In Table 1, we compare our Reg to the other different object detection methods, we find that the better result can be obtained using deeper and better network architecture. In here, we use the PASCAL VOC box detection average precision (mAP) to measure our fusion model. Our Reg network obtains the best object detection result. In Table 2, we use the pixel accuracy, mean accuracy, mean IU and frequency weighted IU to measure our experiment. Our Seg network get the best semantic segmentation

Table 2. Control experiments for semantic segmentation of 40 class on NYU2 test set. Comparison of our Seg with other state-of-the-arts, the last two rows shows our model is plauseble. ∗ denotes we retrain their codes on the NYU-Depth v2 dataset.

Method	Pixel acc.	Mean acc.	Mean IU	f.w. IU
Gupta [8]	60.3	—	28.6	47.0
FCN-32s [16] RGB	60.0	42.2	29.2	43.9
FCN-32s [16] RGB-HHA	61.5	42.4	30.5	45.5
FCN-32s [16] HHA	57.1	35.2	24.2	40.4
Deng [4]	63.8	—	31.5	48.5
Long [16]	65.4	46.1	34.0	49.5
PSPNet∗ [23]	67.4	55.9	40.2	55.7
DeepLab V3∗ [2]	69.9	57.1	41.8	57.2
MFDCNN (VGG)	69.4	56.2	40.7	55.9
MFDCNN (ResNet101)	**70.4**	**57.6**	**42.4**	**57.7**

result compare to the other state-of-the-art, we can get global accuracy of 70.4%. Figure 2 shows the object segmentation result using our MFDCNN. Through the results, we find that our framework is easy to handle the very structurally scene like a wall or ceiling, but is difficult to deal with the less structurally thin scene like picture or mirror.

5   Conclusion

In this paper, we propose a novel MFDCNN framework for indoor object detection and segmentation with RGB-D data. Our framework consists of two sub-networks that learn feature representation from both object detection and segmentation. According to the proposed extended constraint, the two sub-networks -*Reg* and *Seg*- are combined together to achieve object detection and segmentation simultaneously. Extensive experiments on the recent NYU-Depth V2 scene benchmark demonstrate that our MFDCNN achieves significant performance improvements compared to recent state-of-the-art methods. The results from our MFDCNN show that the proposed framework provides useful information for understanding the cluttered indoor scenes.

Acknowledgements. This work was supported in part by National Science and Technology Fund (61811530330) and MSRA CCRP Funding "FY16-RES-THEME-039". The authors thank all the anonymous reviewers for their very helpful comments and suggestions to improve this paper.

References

1. Chen, L.C., Papandreou, G., Kokkinos, I., Murphy, K., Yuille, A.L.: DeepLab: semantic image segmentation with deep convolutional nets, atrous convolution, and fully connected CRFs. IEEE Trans. Pattern Anal. Mach. Intell. **40**(4), 834–848 (2018)
2. Chen, L.C., Papandreou, G., Schroff, F., Adam, H.: Rethinking atrous convolution for semantic image segmentation. arXiv preprint arXiv:1706.05587 (2017)
3. Chen, L.C., Zhu, Y., Papandreou, G., Schroff, F., Adam, H.: Encoder-decoder with atrous separable convolution for semantic image segmentation. arXiv preprint arXiv:1802.02611 (2018)
4. Deng, Z., Todorovic, S., Jan Latecki, L.: Semantic segmentation of RGBD images with mutex constraints. In: Proceedings of the IEEE International Conference on Computer Vision, pp. 1733–1741 (2015)
5. Girshick, R.: Fast R-CNN. In: Proceedings of the IEEE International Conference on Computer Vision, pp. 1440–1448 (2015)
6. Girshick, R., Donahue, J., Darrell, T., Malik, J.: Rich feature hierarchies for accurate object detection and semantic segmentation. In: Proceedings of the IEEE Conference on Computer Vision and Pattern Recognition, pp. 580–587 (2014)
7. Gupta, S., Arbelaez, P., Malik, J.: Perceptual organization and recognition of indoor scenes from RGB-D images. In: 2013 IEEE Conference on Computer Vision and Pattern Recognition (CVPR), pp. 564–571. IEEE (2013)
8. Gupta, S., Girshick, R., Arbeláez, P., Malik, J.: Learning rich features from RGB-D images for object detection and segmentation. In: Fleet, D., Pajdla, T., Schiele, B., Tuytelaars, T. (eds.) ECCV 2014. LNCS, vol. 8695, pp. 345–360. Springer, Cham (2014). https://doi.org/10.1007/978-3-319-10584-0_23
9. Gupta, S., Hoffman, J., Malik, J.: Cross modal distillation for supervision transfer. In: 2016 IEEE Conference on Computer Vision and Pattern Recognition (CVPR), pp. 2827–2836. IEEE (2016)
10. Hariharan, B., Arbeláez, P., Girshick, R., Malik, J.: Simultaneous detection and segmentation. In: Fleet, D., Pajdla, T., Schiele, B., Tuytelaars, T. (eds.) ECCV 2014. LNCS, vol. 8695, pp. 297–312. Springer, Cham (2014). https://doi.org/10.1007/978-3-319-10584-0_20
11. Jia, Y., et al.: Caffe: convolutional architecture for fast feature embedding. In: Proceedings of the 22nd ACM International Conference on Multimedia, pp. 675–678. ACM (2014)
12. Krizhevsky, A., Sutskever, I., Hinton, G.E.: Imagenet classification with deep convolutional neural networks. In: Advances in neural information processing systems, pp. 1097–1105 (2012)
13. Lecun, Y., Bottou, L., Bengio, Y., Haffner, P.: Gradient-based learning applied to document recognition. Proc. IEEE **86**(11), 2278–2324 (1998)
14. Lenz, I., Lee, H., Saxena, A.: Deep learning for detecting robotic grasps. Int. J. Robot. Res. **34**(4–5), 705–724 (2015)
15. Liu, W., et al.: SSD: single shot multibox detector. In: Leibe, B., Matas, J., Sebe, N., Welling, M. (eds.) ECCV 2016. LNCS, vol. 9905, pp. 21–37. Springer, Cham (2016). https://doi.org/10.1007/978-3-319-46448-0_2
16. Long, J., Shelhamer, E., Darrell, T.: Fully convolutional networks for semantic segmentation. In: Proceedings of the IEEE Conference on Computer Vision and Pattern Recognition, pp. 3431–3440 (2015)

17. Redmon, J., Divvala, S., Girshick, R., Farhadi, A.: You only look once: unified, real-time object detection. In: Proceedings of the IEEE Conference on Computer Vision and Pattern Recognition, pp. 779–788 (2016)
18. Redmon, J., Farhadi, A.: YOLO9000: better, faster, stronger. In: 2017 IEEE Conference on Computer Vision and Pattern Recognition (CVPR), pp. 6517–6525. IEEE (2017)
19. Redmon, J., Farhadi, A.: YOLOv3: an incremental improvement. arXiv preprint arXiv:1804.02767 (2018)
20. Ren, S., He, K., Girshick, R., Sun, J.: Faster R-CNN: towards real-time object detection with region proposal networks. In: IEEE Transactions on Pattern Analysis and Machine Intelligence, p. 1 (2015)
21. Silberman, N., Sontag, D., Fergus, R.: Instance segmentation of indoor scenes using a coverage loss. In: Fleet, D., Pajdla, T., Schiele, B., Tuytelaars, T. (eds.) ECCV 2014. LNCS, vol. 8689, pp. 616–631. Springer, Cham (2014). https://doi.org/10.1007/978-3-319-10590-1_40
22. Ye, E.S., Malik, J.: Object detection in RGB-D indoor scenes. Technical report, University of California, Berkeley (2013)
23. Zhao, H., Shi, J., Qi, X., Wang, X., Jia, J.: Pyramid scene parsing network. In: Proceedings of IEEE Conference on Computer Vision and Pattern Recognition (CVPR) (2017)

Mixup-Based Acoustic Scene Classification Using Multi-channel Convolutional Neural Network

Kele Xu[1,2], Dawei Feng[1], Haibo Mi[1], Boqing Zhu[1], Dezhi Wang[3(✉)],
Lilun Zhang[3], Hengxing Cai[4], and Shuwen Liu[5]

[1] Science and Technology on Parallel and Distributed Laboratory,
National University of Defense Technology, Changsha, China
[2] College of Information Communication, National University of Defense Technology,
Wuhan, China
[3] College of Meteorology and Oceanography,
National University of Defense Technology, Changsha, China
wang_dezhi@hotmail.com
[4] College of Engineering, Sun Yat-Sen University, Guangzhou, China
[5] College of Computer Science, Nanjing University of Technology, Nanjing, China

Abstract. Audio scene classification, the problem of predicting class labels of audio scenes, has drawn lots of attention during the last several years. However, it remains challenging and falls short of accuracy and efficiency. Recently, Convolutional Neural Network (CNN)-based methods have achieved better performance with comparison to the traditional methods. Nevertheless, conventional single channel CNN may fail to consider the fact that additional cues may be embedded in the multi-channel recordings. In this paper, we explore the use of Multi-channel CNN for the classification task, which aims to extract features from different channels in an end-to-end manner. We conduct the evaluation compared with the conventional CNN and traditional Gaussian Mixture Model-based methods. Moreover, to improve the classification accuracy further, this paper explores the using of mixup method. In brief, mixup trains the neural network on linear combinations of pairs of the representation of audio scene examples and their labels. By employing the mixup approach for data augmentation, the novel model can provide higher prediction accuracy and robustness in contrast with previous models, while the generalization error can also be reduced on the evaluation data.

1 Introduction

Acoustic scene classification (ASC) refers to the identification of the environment in which the audios have been acquired, which associates a semantic label to each audio. In 1997, Sawhney proposed the first method to address the ASC problem in an MIT technical report [1]. A set of classes, including "people", "voices", "subway", "traffic" is recorded. An overall classification accuracy of 68% was

R. Hong et al. (Eds.): PCM 2018, LNCS 11166, pp. 14–23, 2018.
https://doi.org/10.1007/978-3-030-00764-5_2

obtained based on the recurrent neural networks and the K-nearest neighbor criterion. Indeed, the recognition of environments has become an important application in the field of machine listening, and ASC enables devices to make sense of their environments. The potential applications of ASC seem evident in several fields, such as security surveillance and context-aware services.

In order to solve the problem of lacking common benchmarking datasets, the first Detection and Classification of Acoustic Scenes and Events (DCASE) 2013 challenge [2] was organized by the IEEE Audio and Acoustic Signal Processing (AASP) Technical Committee. Many audio processing techniques have been proposed during the past years. The applications of deep learning in the ASC have witnessed a dramatic increase during last five years, especially the convolutional neural network (CNN). Compared to the traditional method, which commonly involves training a Gaussian Mixture Model (GMM) on the frame-level features such as Mel-Frequency Cepstral Coefficients (MFCCs) [3], CNN-based methods can achieve better performance. However, most of the previous attempts aimed to apply the deep learning method by using one single channel (or just the average between the left and right channels) [4]. A robust Audio Scene classification model should be able to capture temporal patterns at different channels as additional cues may be embedded in the multi-channel recordings [5]. In this paper, we explore the use of multi-channel CNN for the ASC task, which achieves better accuracy with comparison to the standard CNN.

On the other hand, the deep neural network architectures have a large number of parameters, and they are prone to overfitting. The easiest and most widely used approach to reduce overfitting is to employ larger datasets. As an alternative, data augmentation method can be used to improve the performance of neural network by artificially enlarging the dataset using label-preserving transformation. However, only a few attempts have been made for the data augmentation for audio scene classification.

In this paper, we explore the use of mixup-based method for data augmentation [6], with the goal to obtain superior accuracy and robustness. In brief, mixup constructs virtual training examples, and the neural network can be trained by using the linear combinations of pairs of the representation of examples and their labels.

Theoretically, mixup extends the training distribution by incorporating the prior knowledge that linear interpolations of audio feature vectors should lead to linear interpolations of the associated targets [6]. Mixup can be implemented in a few lines of code, and induces the minimal computation overhead. Despite its simplicity, mixup allows a performance improvement using the DCASE 2017 audio scene classification dataset.

The paper is organized as follows. Section 2 discusses the relationship between our method and prior work, while, the multi-channel CNN classification method is presented in Sect. 3. Section 4 describes the mixup method, and the experimental results are given in Sect. 5. Section 6 gives the conclusion of this paper.

2   Related Work

Scene classification (detection) has been explored by computer vision using different techniques, and dramatic progress has been made during last two decades. However, compared to the progress of scene classification using image (or video), audio-based approaches have been under-explored, and the state-of-the-art audio-based techniques are not able to achieve the comparable performance to its image/video counterpart. In fact, audios can sometimes be more descriptive than videos/images, especially when it comes to the description of an event.

Recently, due to the release of the relatively larger labeled data, there has been a plethora of efforts have been made for the audio scene classification task [7,8]. In brief, the main contributions can be divided into three parts: the representation of the audio signal (or handcrafted feature design) [9–11]; more sophisticated shallow-architecture classifiers [12–14] and the applications of deep learning in ASC task [15,16].

Indeed, deep learning has witnessed dramatic progress during the last decade and achieved success in several different fields, such as, image classification [16], speech recognition [17], natural language processing [18] and so on. Although, there are some attempts, which employ CNN as the tool to solve the ASC task, most of them tried to solve the problem within the context of using the monaural signals. In [11], the author proposed to concatenate different channels, resulting in a one-channel file with longer duration. This kind of method employed the one-channel CNN architecture. In [19], the author proposed to use all-convolutional neural network and masked global pooling for the ASC task. However, only left-hand channel was employed for the classification task. Here, we argue that additional cues may be embedded in the binaural recordings [11]. The combination of information in multi-channels may lead to advanced feature representations for the classification.

On the other hand, the trend of deep neural network' architecture is to become deeper and wider, and millions of parameters need to be trained. To improve the generalization ability of neural networks, plenty of regularization approaches have been used, which include: batch normalization, dropout, etc. When there is only limited training data available, data augmentation using preserving transformation is a widely-used technique for the neural network training to improve the robustness. Although following the same concept of improving the prediction invariance of deep neural network, the data augmentation in audio scene classification is different from the image classification tasks, and the traditional augmentation, such as rotation, flipping, distorting and deformation cannot be applied directly. The procedure is dataset-dependent and requires the use of expert knowledge [6].

In this paper, we explore the use of mixup data augmentation approach, which was proposed in [6]. In brief, the new samples are created by mixing two inputs of the neural network with a ratio, and the labels of the samples are similar to the between-class label. Normally, the ratio ranges from 0 to 1. Using the DCASE 2017 audio scene classification dataset, improved performance has been observed after employing mixup approach.

3   Multi Channel Convolutional Neural Network

Due to its ability of automatic learning complex feature representations, CNNs have achieved great success. CNN has the potential to identify the various salient patterns of the audio signals. In more detail, the processing units in the lower layers can obtain the local feature of the signals, while the higher layers can extract the features of high-level representations.

The input for a CNN architecture can be the raw audio signal or the spatial frequency representation of the raw signal (for example: MFCCs, Short time Fourier transform, spectrograms). In our experiments, we employ the widely used feature representation: Mel-filter bank features of the audio signal segments as the input for the CNN. However, it is not complicated to extend our framework for other kinds of input.

Unlike the attempts which aim to maintain the one-channel CNN architecture [11], we extract features in terms of three different channels. The three different channels are: left channel, right channel, the mean between the left and right channels. The Mel-filter bank features of different channels will be concatenated as a multi-channel image, which results in training a system in an end-to-end manner. Note that, the Mel-filter bank features configuration was kept the same for each single channel during our experiments. In our experiment, Mel-filter bank features is calculated for each channel. We employ the first half of the symmetric Hann window as the window function with a window size of 25 ms and a hop size of 25 ms.

The input of the network is three-channels Mel-filter bank features with size $3 \times 128 \times 128$, where 3 represents the number of channels, 128×128 denotes the size of Mel-filter bank features for single channel. The input sizes are kept the same during the experiments.

There are numerous variants of CNN architectures in the literature. However, their basic components are very similar. Since the starting with LeNet-5 [20], convolutional neural networks have typically standard structure-stacked convolutional layers (optionally followed by batch normalization and max-pooling) are followed by fully-connected layers.

In this paper, we followed the VGG-style [21] networks and Xception [22] networks due to its relatively high accuracy and simplicity. The main contribution of VGG net is to increase the depth using an architecture with very small (3×3) convolution filters.

While VGG achieves an impressive accuracy on the image classification task, its deployment on even the most modest-sized GPUs is a problem because of huge computational requirements, both in terms of memory and time. It becomes inefficient due to large width of convolutional layers.

As the-state-of-the-art model in Inception model group, Xception architecture employs the depthwise separable convolution operation to replace the regular Inception modules, which has an excellent performance on a larger image classification dataset like ImageNet, and becomes a cornerstone of convolutional neural network architecture design. Another change that Xception model made, was to replace the fully-connected layers at the end with a simple global average pooling which averages out the channel values across the 2D feature map, after the last convolutional layer. This drastically reduces the total number of parameters. This can be understood from VGGNet, where fully connected layers contain about 90% of parameters.

The only changes we made to VGG were to the final layer (using the global average layer) as well as the use of batch normalization instead of Local Response Normalization (LRN). The parameters of the CNN model are optimized with stochastic gradient descent. The cross-entropy was selected as the objective function. Moreover, an L2 weight decay penalty of 0.002 was employed in our model. To train the CNN, we used Keras with tensorflow backend, which can fully utilize GPU resource. CUDA and cuDNN were also used to accelerate the system.

It is worthwhile to note that each layer consists of many convolutions or pooling operators. The convolutional filters can be interpreted as the filter-banks learning. For the activation layer, the rectified linear unit is used to introduce the non-linearity into a neural network. The last layer is the probability output layer, that converts the output vector of the fully connected layer to a vector of probabilities, which sum up to 1, each probability corresponding to one class. The probabilities can be used to predict the scene label of the audio segment.

For the final prediction of an input instance, there are many widely used approaches to perform the final prediction, for example, maximum probability, median probability, average probability and majority votes. In this paper, for the evaluation of the CNN-based method, we use the maximum probability to obtain the label.

4   Mixup for Data Augmentation

We evaluate the multi-channel CNN on the TUT sound events detection 2017 database [7]. The database consists of stereo recordings which were collected using 44.1 kHz sampling rate and 24-bit resolution. The recordings came from 15 various acoustic scenes, which have distinct recording locations, for example: office, train, forest path. For different locations, 3–5 min long audio was recorded. And the audio files were split into 30-s segments. The acoustic scene classes considered in this task were: bus, cafe/restaurant, car, city center, forest path, grocery store, home, lakeside beach, library, metro station, office, residential area, train, tram, and urban park.

Currently, most publicly available ASC datasets have limited sizes [3,7]. The disadvantage of small dataset is that the model is prone to overfitting. In the DCASE 2017 audio scene classification task, it is found that the generalization gap is big, and the accuracy difference between development dataset and evaluation dataset ranges from 4% to 30% by using different approaches. The ability to generalization is a research topic for the deep neural network. To improve the generalization ability of deep neural network, especially the CNN, a plethora approaches have been proposed, such as dropout [23], batch normalization [24]. Data augmentation is another explicit form of regularization, which is also widely used in the deep neural network. In more detail, for the deep CNN, random cropping and random flipping are two most popular data augmentation approaches. However, these methods cannot be applied to ASC directly. Recently, it is found that Generative adversarial network can be used for ASC data augmentation, and impressive performance have been obtained on the task [25]. Indeed, the data augmentation is under-explored in previous ASC study.

In this paper, we explore the use of mixup data augmentation. In more detail, virtual training examples can be constructed by using the following formula:

$$x = \alpha \times x_i + (1 - \alpha) \times x_j \tag{1}$$

$$y = \alpha \times y_i + (1 - \alpha) \times y_j \tag{2}$$

Where (x_i, y_i) and (x_j, y_j) are two examples random selected from the training data of the DCASE 2017 ASC task. α is the mixed ratio. In our experiments, $\alpha \in [0, 1]$. A mixup example is given in Fig. 1, and the α is set as 0.2 for the example. In the figure, two labeled audio scenes are selected randomly, while a new training sample is constructed by weighted average between two given samples.

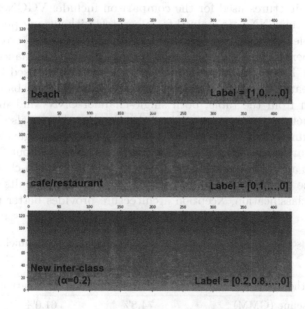

Fig. 1. An example for mixup data augmentation for audio scene classification.

Despite its simplicity, the mixup data augmentation methods have provided state-of-the-art performance in many datasets, which include the CIFAR-10, CIFAR-100, and ImageNet-2012 image classification datasets. Similar to create inter-class, mixup increases the robustness of deep CNN when the samples contains corrupted labeled ones. In the following section, we will demonstrate that mixup data augmentation can also improve the ASC performance.

5   Experimental Results

As the DCASE 2017 scene classification dataset provides cross-validation splits, we follow the 4 fold cross-validation splits. We used the same experiment settings from development set for the evaluation set. In the development stage, the results are evaluated in terms of average accuracy for 4 folds. The performance of evaluation data is also given in this section.

In our experiments, we made two sets of comparison: the performance comparison between the single channel CNN and multi-channel CNN; the performance comparison between Multi-channel CNN with mixup data augmentation and Multi-channel CNN without mixup data augmentation.

5.1   Single/Multi-channel CNN

The first set of experience aims to evaluate single-channel and multi-channel based audio scene classification using convolutonal neural network. As aforementioned, the architectures used for the comparison include: VGGNet and Xception, and all of the CNNs are trained from scratch without any pre-trained initialization. Table 1 presents the validation results for the 4 fold cross-validation as well as the performance on the evaluation data. The performance of baseline is also given in Table 1. In more detail, the baseline system used here consists of 60 MFCC features and a Gaussian mixture model (GMM) based classifier. As can be seen from the table, both single-channel CNNs and multi-channel CNN showed better performance against the baseline. It can also be observed that multi-channel CNN performance is better than the single-channel CNN using different architectures, and the accuracy is increased about 2% to 3%. It may imply that additional features can be extracted from multi-channels, which can improve the accuracy for the ASC task. Similar to the results obtained on the ImageNet classification, Xception architecture provides better performance

Table 1. Audio scene classification accuracy using single/multi channel convolutional neural network

Method	Cross-validation	Evaluation
Baseline (GMM)	74.8%	61.0%
Single-channel using VGGNet	82.4%	67.8%
Multi-channel using VGGNet	84.7%	71.5%
Single-channel using Xception	83.3%	72.1%
Multi-channel using Xception	85.4%	74.5%

than VGGNet. The reason is that Xception architecture can take multi-scale information into account as different kernels size are learned in the model. On the other hand, the accuracy difference between the development dataset and evaluation dataset) is significant, which ranges from 10.9% to 14.2% using different approaches. This may indicate that the trained CNN models are prone to overfitting during the training procedure.

5.2   Multi-channel CNN With/Without Mixup

The second set of experiments aimed to show that mixup-base data augmentation can improve the performance of ASC. Moreover, it is also shown that mixup can reduce the generalization gap.

Table 2. Audio scene classification accuracy using mixup data augmentation

Method	α	Cross-validation	Evaluation
Multi-channel VGGNet	0	84.7%	71.5%
Multi-channel VGGNet	0.2	85.2%	73.4%
Multi-channel VGGNet	0.5	86.9%	73.2%
Multi-channel VGGNet	0.8	85.8%	72.1%
Multi-channel Xception	0	85.4%	74.5%
Multi-channel Xception	0.2	86.7%	75.6%
Multi-channel Xception	0.5	87.2%	76.7%
Multi-channel Xception	0.8	86.9%	74.8%

The performance of different approaches, which employing the mixup data augmentation, is given in Table 2, and different ratios are used for the mixup. Due to the computation resource constraint, only three ratios are used in our experiments (when $\alpha = 0$, the mixup approach is not employed). As can be seen from the table: without mixup data augmentation, the cross-validation accuracy of multi-channel VGGNet is 84.7% and the accuracy of evaluation data is 71.5%, while the accuracy of evaluation data ranges from 72.1% to 73.2% if mixup data augmentation is employed. For multi-channel CNN using Xception architecture, the accuracy was also improved using the mixup data augmentation, which demonstrates that mixup approach is effective despite its simplicity. In our experimental results, mixup with ratio 0.5 provides superior performance.

6   Conclusion

In this paper, we have presented the multi-channel convolutional neural network-based method for the multi-class acoustic scene classification. To summarize, the contributions of this paper are twofold: firstly, we present a multi-channel

CNN architecture for the classification task. Secondly, we explore the mixup data augmentation method, and experiments demonstrated that by employing the mixup dataset augmentation, the classification can be improved, and the generalization error can also be reduced. To the best knowledge of the authors, this is the first attempt of employing mixup for the audio scene classification task. For future work, we will investigate the CNN architecture to utilize multi-scale information embedded in the audio signal, thus improving the classification accuracy. The mixup approach is also needed to be fully explored. Presently, the mixup processing is relied on the log mel spectrum of the audio signal. We did not observe significant improvement by mixup of the raw audio signal directly. Moreover, mixup may also be useful for audio event tagging and detection.

Acknowledgments. This work is sponsored by the Scientific Research Project of NUDT (No. ZK17-03-31).

References

1. Sawhney, N., Schmandt, C.: Nomadic radio: speech and audio interaction for contextual messaging in nomadic environments. ACM Trans. Comput. Interact. **7**(3), 353–383 (2000)
2. Stowell, D., Giannoulis, D., Benetos, E., Lagrange, M., Plumbley, M.D.: Detection and classification of acoustic scenes and events. IEEE Trans. Multimed. **17**(10), 1733–1746 (2015)
3. Marchi, E., Tonelli, D., Xu, X., Ringeval, F., Deng, J., Squartini, S.: The up system for the 2016 DCASE challenge using deep recurrent neural network and multiscale kernel subspace learning. In: IEEE AASP Challenge on Detection and Classification of Acoustic Scenes and Events (DCASE) (2016)
4. Adavanne, S.: Sound event detection in multichannel audio using spatial and harmonic features. In: IEEE AASP Challenge on Detection and Classification of Acoustic Scenes and Events (DCASE) (2016)
5. Elizalde, B., Lei, H., Friedland, G., Peters, N.: An i-vector based approach for audio scene detection. In: IEEE AASP Challenge on Detection and Classification of Acoustic Scenes and Events (DCASE) (2013)
6. Zhang, H., Cisse, M., Dauphin, Y.N., Lopez-Paz, D.: mixup: Beyond Empirical Risk Minimization. In: International Conference on Learning Representations (2018)
7. Mesaros, A., et al.: DCASE 2017 challenge setup: tasks, datasets and baseline system. In: IEEE AASP Challenge on Detection and Classification of Acoustic Scenes and Events (DCASE) (2017)
8. Mesaros, A., Heittola, T., Virtanen, T.: Metrics for Polyphonic sound event detection. Appl. Sci. **6**(6), 162 (2016)
9. Rakotomamonjy, A., Gasso, G.: Histogram of gradients of time-frequency representations for audio scene classification. IEEE/ACM Trans. Audio Speech Lang. Process. (TASLP) **23**(1), 142–153 (2015)
10. Salamon, J., Bello, J.P.: Unsupervised feature learning for urban sound classification. In: IEEE International Conference on Acoustics, Speech and Signal Processing, pp. 171–175 (2015)

11. Eghbal-Zadeh, H.: CP-JKU submissions for DCASE-2016: a hybrid approach using binaural i-vectors and deep convolutional neural networks. In: IEEE AASP Challenge on Detection and Classification of Acoustic Scenes and Events (DCASE) (2016)

12. Bisot, V., Serizel, R., Essid, S., Richard, G.: Supervised nonnegative matrix factorization for acoustic scene classification. In: IEEE AASP Challenge on Detection and Classification of Acoustic Scenes and Events (DCASE) (2016)

13. Schröder, J., Anemüller, J., Goetze, S.: Performance comparison of GMM, HMM and DNN based approaches for acoustic event detection within task 3 of the DCASE 2016 challenge. In: IEEE AASP Challenge on Detection and Classification of Acoustic Scenes and Events (DCASE) (2016)

14. Phan, H., Hertel, L., Maass, M., Koch, P., Mertins, A.: Label tree embeddings for acoustic scene classification. In: IEEE AASP Challenge on Detection and Classification of Acoustic Scenes and Events (DCASE) (2016)

15. Han, Y., Lee, K.: Acoustic scene classification using convolutional neural network and multiple-width frequency-delta data augmentation. In: IEEE AASP Challenge on Detection and Classification of Acoustic Scenes and Events (DCASE) (2016)

16. Xu, Y., Kong, Q., Huang, Q., Wang, W., Plumbley, M.D.: Convolutional gated recurrent neural network incorporating spatial features for audio tagging. arXiv:1702.07787 (2017)

17. Krizhevsky, A., Sutskever, I., Hinton, G.E.: ImageNet classification with deep convolutional neural networks. In: Advances in Neural Information Processing Systems, pp. 1–9 (2012)

18. Hinton, G., et al.: Deep neural networks for acoustic modeling in speech recognition. IEEE Signal Process. Mag. **29**, 82–97 (2012)

19. Hertel, L., Phan, H., Mertins, A.: Classifying variable-length audio files with all-convolutional networks and masked global pooling. arXiv:1607.02857 (2016)

20. LeCun, Y., Bottou, L., Bengio, Y., Haffner, P.: Gradient-based learning applied to document recognition. Proc. IEEE **86**(10), 2278–2324 (1998)

21. Simonyan, K., Zisserman, A.: Very deep convolutional networks for large-scale image recognition. arXiv preprint arXiv:1409.1556 (2014)

22. Chollet, F.: Xception: deep learning with depthwise separable convolutions. In: Proceedings of the IEEE Conference on Computer Vision and Pattern Recognition (2017)

23. Srivastava, N., Hinton, G., Krizhevsky, A., Sutskever, I., Salakhutdinov, R.: Dropout: a simple way to prevent neural networks from overfitting. J. Mach. Learn. Res. **15**(1), 1929–1958 (2014)

24. Ioffe, S., Szegedy, C.: Batch normalization: accelerating deep network training by reducing internal covariate shift. In: International Conference on Machine Learning (2015)

25. Mun, S., Park, S., Han D.K., Ko, H.: Generative adversarial network based acoustic scene training set augmentation and selection using SVM hyper-plane. In: IEEE AASP Challenge on Detection and Classification of Acoustic Scenes and Events (DCASE) (2017)

Multimodal Fusion for Traditional Chinese Painting Generation

Sanbi Luo, Si Liu, Jizhong Han, and Tao Guo[(⊠)]

Institute of Information Engineering, Chinese Academy of Sciences,
Beijing 100093, China
{luosanbi,liusi,hanjizhong,guotao}@iie.ac.cn

Abstract. Creativity is a fundamental feature of human intelligence, and a challenge for artificial intelligence (AI). In recent years, AI has gained tremendous development in many single tasks with single models, such as classification, detection and parsing. As the development continued, AI has been increasingly used for more complex tasks, multitasking for example, and then research in multimodal fusion naturally became a new trend. In this paper, we propose a multimodal fusion framework and system to generate traditional Chinese paintings. We select suitable existing networks for different elements generation in this oldest continuous artistic traditions artwork, and finally fusion these networks and elements to create a complete new painting. Meanwhile, we propose a divide-and-conquer strategy to generate large images with limited GPU resources. In our end-to-end system, a large image becomes a traditional Chinese painting in minutes automatically. It shows that our multimodal fusion framework works well and AI methods has good performance in traditional Chinese painting creation.

Keywords: Multimodal fusion · Image style · GAN
Traditional Chinese painting

1 Introduction

Boden said in 1998 [1], "Creativity is a fundamental feature of human intelligence, and a challenge for AI." Nearly 20 years later, AI has made astonishing development, especially in recent years, the big bang of AI has been fueled by deep learning (DL). Neural networks, such as RNN and GAN, can generate poetry and images. In this paper, we plan to use DL to generate traditional Chinese paintings, which is a multi-tasking challenge that requires the fusion of many DL networks and technologies.

Traditional Chinese painting is one of the oldest continuous artistic traditions in the world. As show in Fig. 1, it usually consists of four parts: painting, blank space, poem and calligraphy, and seal [3]. Painting is the most important and essential part in traditional Chinese painting. It has a unique style, which emphasis on spirit rather than realism. Blank space is a philosophical concept in traditional Chinese painting. It is given a lot of meaning. Here we only consider it as a form of composition. There are usually poems and seals in it. The poems are usually an expression and a supplement of a traditional Chinese painting. Calligraphy is understood in China as Chinese art of

© Springer Nature Switzerland AG 2018
R. Hong et al. (Eds.): PCM 2018, LNCS 11166, pp. 24–34, 2018.
https://doi.org/10.1007/978-3-030-00764-5_3

writing a good hand with the brush or the study of the rules and techniques of this art. Seals always represent the author, or owners.

To generate a traditional Chinese painting means at least has to face following challenges:

- Style transfer a painting from a image
- Find and generate the blank space (optional)
- Generate poems about the painting
- Generate the seal of the author
- Transfer poems and seals to a Chinese calligraphy style
- Fusion all elements to create a traditonal Chinese painting

In this paper, we chose the latest and most suitable networks, such as YOLO, Mask RCNN, Neural Style Transfer network, Cycle GAN, pix2pixHD and LSTM, to be part of our fusion system. We select the key parts of these networks and integrate them into a fusion one and finally used for generating traditional Chinese paintings.

Fig. 1. A traditional Chinese painting. It always has four elements: painting, blank space, poem and calligraphy, and seal

2 Related Work

Many works can generate images. They can be roughly divided into three categories: GANs [4] and GANs-based networks [5–8], RNN-bested networks [9–11], and some other methods [12–14, 16, 17]. These methods can generate hand-written numbers, human faces, indoor and outdoor scenes, and something else.

There are also many works that can generate stylized-images. Cycle-GAN [8] is an approach for learning to translate an image from a source domain X to a target domain Y in the absence of paired examples, such as to translate a horse to a zebra. In [12], authors use neural representations to separate and recombine content and style of arbitrary images, providing a neural algorithm for the creation of artistic images, which is based on VGG [15] network. However, these networks are always limited to GPU's performance and cannot translate large images directly with limited GPU resources. With a 12G Video Random Access Memory (VRAM) GPU, these methods always hard to generate more than 1024 * 1024 pixel size images. To generate larger images, more GPU VRAM is needed. If one wants to translate a big image as 10240 * 10240 pixel size, 1200G VRAM GPU may be necessary. GPUs are always very expensive, and it is often a bottleneck in image processing.

Scene parsing, or recognizing and segmenting objects in an image, is one of the key problems in computer vision. It can recognize and segment objects in an image, such as the sky and mountains. It also can be used in identifying and segmenting blank space and painting in traditional Chinese paintings. A great deal of works has gained remarkable achievements, such as FCN [22] and DeepLab [2]. ADE20K [18] collects precise dense annotation of scenes, objects, parts of objects with a large and open vocabulary. It works well in image parsing and provides a good code implementation in GitHub.

Object detection is a core problem in computer vision too. There are many methods available, such as Deformable parts models (DPM), a sliding window approach to object detection [25], or R-CNN and its variants [24], using region proposals instead of sliding windows to find objects in images. In [19], authors introduced YOLO, a real-time detection system. This model is simple to construct and can be trained directly on full images, and it is one of the fastest object detector at present.

The success of recurrent neural network (RNN) models in complex tasks like machine translation and audio synthesis has inspired immense interest in learning from sequence data. In [26], the author proposed a character-level Language model to generate sentences in English. In [20], the author trained an RNN character-level language model on Chinese poetry datasets, and it shows that RNN-based models can generate Chinese poetry well.

Multimodal fusion learning has been extensively studied in recent years, such as Visual Question Answering (VQA) [27]. As single neural models for single tasks become efficient and practical, Multimodal fusion naturally receives attention and development when AI faces more complex challenges.

3   Traditional Chinese Painting Generation

Figure 2 shows the framework in our system how to generate traditional Chinese paintings by fusion network. With the input of a content image and a style image, our system generates four elements through four branches, and finally merges all branches to generate a traditional Chinese painting. Branch 1 is used to generate poetry and calligraphic styles. It usually requires CNN, RNN and GAN networks. Branch 2 is optional. It is used to generate blank space in traditional Chinese paintings. We can use parsing networks to erase some non-critical parts in picture, such as the sky, and in traditional Chinese painting the sky is always blank. Branch 3 is used to generate stylized large paintings. The famous Neural Style Transfer network and GAN nets could be used for reference. Branch 4 is used to make seals, and a stylized network is required.

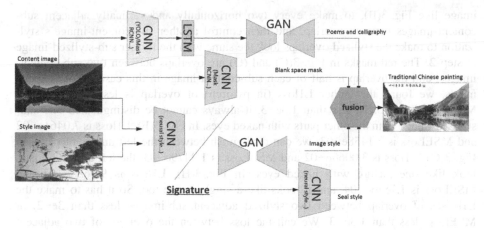

Fig. 2. The multimodal fusion framework of our system in traditional Chinese painting generation

The characteristic of our fusion framework is that it focuses on the selection and integration of sub-networks. Therefore, on the one hand we can choose and integrate the latest and best single network into our framework to generate traditional Chinese painting elements, on the other hand we must find a balance between elements fusion and networks fusion, and it brings us challenges and innovation opportunities in fusion research. Just like human brains integrate a variety of information to make a decision, fusion will be one of the trends in the future of AI industry. In this version of our system, we used the key parts of existed YOLO, Mask RCNN, Neural Style Transfer network, Cycle GAN and LSTM-based network to form the multimodal fusion framework. We have integrated all the networks in branch 1and branch 4, and we fuse all the elements generated by 4 branches to draw a painting.

3.1 A Divide-and-Conquer Strategy for Large Image Style Transfer

In branch 3, the original Neural Style Transfer Nets is hard to generate a picture larger than 1024 * 1024 pixel size with 12G GPU. Pix2pixHD [23] can generate 1024 * 2048 pixel size pictures with 12G GPU, but it requires thousands of big paintings for training, and so many big paintings are very hard to collect. Therefore, we propose a divide-and-conquer strategy for large image style transfer, which can generate large paintings with a single content and a single style image. There are three steps to generate style paintings: divide, stylize, and merge.

Step 1: split larger images into smaller ones. When dividing an original large content image into sub-content-images, we let the adjacent sub-content-images have overlaps. These overlaps will be used in step 2 to guide the stylization of adjacent sub-content-images.

As shown in Fig. 3, if we divide an image like Fig. 3(A) without overlaps, the stylized image merged by sub-stylized-images will probably like Fig. 3(E). It does not look like a single image but a puzzle with four pieces. Therefore, we try to divide the image like Fig. 3(B), to make every two horizontally and vertically adjacent sub-content-images have an overlap, and then control another sub-content-image's stylization to make the stylized overlaps look the same with the former sub-stylized-image in step 2. The red masks in Fig. 3(C) and (D) are overlaps between two sub-content-images, and the overlap is half of each sub-content-image in this case. In our experiments, we found that either L1loss (in pytorch) of overlap is less than $3e-2$ or MSELoss (in pytorch) less than $1.3e-3$, it always cannot be distinguished one sub-stylized-image from the other parts with naked eyes. In Fig. 3(F), L1loss is $7.0461e-02$ and MSELoss is $7.1458e-03$, we can distinguish between the left and right parts. In Fig. 3(G), L1loss is $3.0060e-02$ and MSELoss is $1.3479e-03$, the left and right parts look like one picture with naked eyes. In Fig. 3(H), L1loss is $9.9688e-03$ and MSELoss is $1.7576e-04$, and it looks like a single image too. So it has to make the L1losses of overlap between two stylized adjacent sub-images less than $3e-2$, or MSELoss less than $1.3e-3$. We call the loss between the overlaps of two adjacent images similar1 loss.

Fig. 3. Image division with overlap. (Color figure online)

Fig. 4. A problem in sub-image transfer. A is a sub-content-image, B is style image, C is a sub-stylized-image that the problem has occurred, and D is the image that a sub-stylized-image should be.

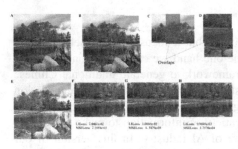

Fig. 5. Sub-content-images stylization one by one.

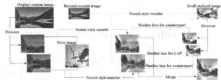

Fig. 6. Two similar losses for image transfer.

Step 2: transfer sub-content-images one by one. Our strategy is that at first a large image is divided into sub-content-images, and then they are translated sub-content-images in a certain order gradually. As shown in Fig. 5, we firstly translate sub-content-image 1 to a sub-stylized-image according to the style image. Then translate sub-content-image 2 depends on the style image and sub-stylized-image 1, and so on until the end of the first line. Sub-content-image 6 should depend on the style image and sub-stylized-image 1. Sub-content-image 7 has to depend on the style image, sub-stylized-image 6 and sub-stylized-image 2. It seems like the generate order of

PixelRNN [16]. The difference is that PixelRNN is for pixels, and our strategy is for sub-content-images.

As mentioned in step 1, if we divide an image into sub-content-images using Neural Style Transfer network [12] without overlaps, the merging stylized image may look like a puzzle with many parts. Therefore we use divide-and-conquer strategy with overlaps to deal with such a problem. Also, there is another problem that if the sub-content-image is very monotonous, a whole piece of blue sky or calm water for instance, the sub-stylized-images will not be monotonous but tend to the content of the style image. It is determined by the structure of Neural Style Transfer networks [12]. As shown in Fig. 4, Fig. 4(A) is a sub-content-image, Fig. 4(B) is style image, Fig. 4 (C) is a sub-stylized-image that the problem has happened, and Fig. 4(D) is the image that a sub-stylized-image should be. To solve this problem, we added a new loss between the sub-stylizing-image and the corresponding area of the stylized image, which is a small one resized from the original large content image. We call this loss as similar2 loss.

As illustrated in Fig. 6, we use two losses to control sub-content-image stylization. The one is to improve the similarity of overlaps between adjacent sub-stylized-images, and the other is to eliminate the problem caused by local sub-content-image stylization. Resized content image is minified from the original content image, a small one (under 1024 * 1024 pixel size) that can be stylized by a 12G VRAM GPU at a time.

Let p and a be the original content image and the style image, f and c be the overlap of the former stylized image and the corresponding area of the minified and stylized image, and x be the image that is generated. The loss function that should be minimized is:

$$L_{total}(p, a, f, c, x) = \alpha L_{content}(p, x) + \beta L_{style}(a, x) + \gamma L_{similar1}(f, x) + \delta L_{similar2}(c, x) \quad (1)$$

α, β, γ and δ are the weighting factors for content, style, similar1 and similar2 loss respectively.

Fig. 7. Generate large stylized image (1920 * 1080 pixel size).

Step 3: merge the stylized small images into a big picture. When sub-content-images style translation is finished, it is time to merge sub-stylized-images to a big one. Figure 7 shows the process of picture integration. The merged styled-image consists of four sub-stylized-images, show in Fig. 7(B), and each part is 960 * 540 pixel size and divided from the 1024 * 800 pixel size sub-stylized-image separately, show in Fig. 7 (A). The experiment uses a 12G VRAM GPU, which can only generate 1024 * 1024 pixel resolution images at most by using Neural Style Transfer Nets directly. It proves that our divide-and-conquer strategy with neural style translate networks could transfer 1920 * 1080 pixel size image by a 12G VRAM GPU, which can only transfer a 1024 * 1024 pixel size image using Neural Style Transfer network directly.

3.2 Find and Generate the Blank Space

In [18], the authors introduces a new densely annotated dataset with the instances of stuff, objects, and parts, covering a diverse set of visual concepts in scenes. A generic network design was proposed to parse scenes into stuff, objects, and object parts in a cascade. We use these image semantic segmentation achievements to find and generate the blank space in stylized images, showing in upper parts of paintings in Fig. 9.

3.3 Generate Poetry and Stylize into Chinese Calligraphy Style

In [19], the author introduces a real-time detection system YOLO. It can recognize the objects in the content image, such as sky, person, mountain, house, water, and so on. Meanwhile, we found that RNN-based poetry generate model [20] is very suitable for our system to generate Chinese poems. We combined these two networks to form a fusion network to generate poetics from pictures, and combined a trained GAN-based neural network [21] to stylize the text. In Fig. 8(A), branch 1 of our framework generated Chinese poems for the picture of the first row and first column in Fig. 9. It means the mountain is moist with water surrounding, and the house and yard are full of romance. It sounds objective and poetic.

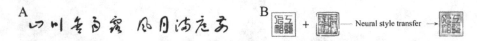

Fig. 8. A Chinese calligraphy style poem and a seal stylization.

3.4 Generate and Stylize the Seal

Seals in traditional Chinese paintings represent the author or owners. They use Small Seal Script. Small Seal Script cannot be transferred from the other fonts for its special structure, so we use TrueType fonts directly. However, you can use Neural Style Transfer network to stylize the seal. As shown in Fig. 8(B), the first seal on the left is generated by TrueType font of Small Seal Script, and it means "artificial intelligence". The rightmost one is stylized by a real seal image.

3.5   Create a Complete Traditional Chinese Painting

After four elements of traditional Chinese painting have been generated, we can integrate these four elements to generate a large traditional Chinese painting, as shown in Fig. 9.

Fig. 9. Traditional Chinese paintings generated by our system. The first column is landscape photos, and the second column is traditional Chinese paintings created by our system

4   Results and Limitation

Figure 9 shows some results of our system. The first column is landscape photos, and the second column is traditional Chinese paintings created by our system, which contain painting, blank space, poem and calligraphy, and seal. The first row is a

1920 * 1080 pixel size. The second is 3824 * 2144 pixel size. The third and fourth rows are all 2600 * 916 pixel size. In 12G VRAM GPU environment, we divide the content image into suitable size sub-content-images, generally no more than 1024 * 1024 pixel size. It cost 10–20 min. These successful cases with divide-and-conquer strategy show that our system works well.

Table 1 shows that compared to the current mainstream methods, our method can also generate poem, seal and blank space instead of just generating big size stylized images. Although our system could generate larger images than that Neural Style Transfer network can do with limited GPU resources. Due to the divide-and-conquer strategy, the generation speed depends on both the speed of Neural Style Transfer network and the size of the picture. Meanwhile, we use semantic segmentation to generate blank space and use RNN character-level language model and GAN-based networks to generate poem and calligraphy. Naturally the quality of the generated traditional Chinese paintings also depends on the development of these technologies. In the aspect of multimodal fusion, we also operate in elements fusion and networks fusion together. These are the limitations of our system.

Table 1. Comparison of painting generation methods.

Heading level	Image style	Big size	Poem	Seal	Blank space
Neural style [12]	✓				
Cycle GAN [8]	✓				
Pix2pixHD [23]	✓	✓			
Our system	✓	✓	✓	✓	✓

5   Conclusion and Future Work

In this paper, we propose a multimodal fusion framework and system to generate traditional Chinese paintings. We select suitable existing networks for different elements generation in this oldest continuous artistic traditions artwork, and finally fusion these networks and elements to create a new painting. Meanwhile, we propose a divide-and-conquer strategy to generate large images with limited GPU resources. In our end-to-end system, a large image becomes a traditional Chinese painting in minutes automatically. It shows that our solution works effectively and AI methods has good performance in traditional Chinese painting creation. We believe that just like human brains integrate a variety of information to make a decision, fusion network research will be one of the trends in the future of AI industry. Next we will continue to focus on multimodal fusion research.

Acknowledgments. This work was supported by the National Natural Science Foundation of China (Nos. U1536203, 61572493), IIE project (No. Y6Z0021102, No. Y7Z0241102) and CCF-Tencent Open Research Fund.

References

1. Boden, M.A.: Creativity and artificial intelligence. Artif. Intell. **103**(1/2), 347–356 (1998)
2. Chen, L.-C., Papandreou, G., Kokkinos, I., Murphy, K., Yuille, A.L.: Semantic image segmentation with deep convolutional nets and fully connected CRFs. In: ICLR (2015)
3. http://oilpaintingfactory.com/traditional-Chinese-painting.html
4. Goodfellow, I., et al.: Generative adversarial nets. In: NIPS (2014)
5. Gauthier, J.: Conditional generative adversarial nets for convolutional face generation. In: Class Project for Stanford CS231N: Convolutional Neural Networks for Visual Recognition, Winter semester (2014)
6. Radford, A., Metz, L., Chintala, S.: Unsupervised representation learning with deep convolutional generative adversarial networks. arXiv preprint arXiv:1511.06434 (2015)
7. Johnson, J., Alahi, A., Fei-Fei, L.: Perceptual losses for real-time style transfer and super-resolution. In: Leibe, B., Matas, J., Sebe, N., Welling, M. (eds.) ECCV 2016. LNCS, vol. 9906, pp. 694–711. Springer, Cham (2016). https://doi.org/10.1007/978-3-319-46475-6_43
8. Zhu, J.-Y., Park, T., Isola, P., Efros, A.A.: Unpaired image-to-image translation using cycle-consistent adversarial networks. arXiv preprint arXiv:1703.10593 (2017)
9. Gregor, K., Danihelka, I., Graves, A., Wierstra, D.D.: A recurrent neural network for image generation. arXiv preprint arXiv:1502.04623 (2015)
10. van den Oord, A., Kalchbrenner, N., Kavukcuoglu, K.: Pixel recurrent neural networks. arXiv preprint arXiv:1601.06759 (2016)
11. Yang, J., Reed, S., Yang, M.-H., Lee, H.: Weakly-supervised disentangling with recurrent transformations for 3D view synthesis. In: NIPS (2015)
12. Gatys, L.A., Ecker, A.S., Bethge, M.: A neural algorithm of artistic style. arXiv preprint arXiv:1508.06576 (2015)
13. Tieleman, T.: Optimizing neural networks that generate images. Ph.D. thesis, University of Toronto (2014)
14. Dosovitskiy, A., Springenberg, J., Tatarchenko, M., Brox, T.: Learning to generate chairs, tables and cars with convolutional networks. IEEE Trans. Pattern Anal. Mach. Intell. **PP**(99), 1 (2016). https://doi.org/10.1109/TPAMI.2016.2567384
15. Simonyan, K., Zisserman, A.: Very deep convolutional networks for large-scale image recognition. CoRR, abs/1409.1556 (2014)
16. van den Oord, A., et al.: Conditional image generation with PixelCNN decoders. CoRR, abs/1606.05328 (2016)
17. Reed, S., Akata, Z., Yan, X., Logeswaran, L., Schiele, B., Lee, H.: Generative adversarial text to image synthesis. arXiv preprint arXiv:1605.05396 (2016)
18. Zhou, B., Zhao, H., Puig, X., Fidler, S., Barriuso, A., Torralba, A.: Scene paring through ADE20K dataset. In: Proceedings of CVPR (2017)
19. Redmon, J., Farhadi, A.: YOLO9000: better, faster, stronger. In: CVPR (2017)
20. https://github.com/justdark/pytorch-poetry-gen
21. https://github.com/kaonashi-tyc/zi2zi
22. Long, J., Shelhamer, E., Darrell, T.: Fully convolutional networks for semantic segmentation. In: CVPR (2015)
23. Wang, T.-C., Liu, M.-Y., Zhu, J.-Y., Tao, A., Kautz, J., Catanzaro, B.: High-resolution image synthesis and semantic manipulation with conditional GANs. arXiv preprint arXiv:1711.11585
24. Dai, J., Li, Y., He, K., Sun, J.: R-FCN: object detection via region-based fully convolutional networks. In: NIPS (2016)

25. Felzenszwalb, P.F., Girshick, R.B., McAllester, D., Ramanan, D.: Object detection with discriminatively trained part based models. IEEE Trans. Pattern Anal. Mach. Intell. **32**(9), 1627–1645 (2010)
26. Karpathy, A.: The unreasonable effectiveness of recurrent neural networks. Andrej Karpathy Blog (2015). http://karpathy.github.io
27. Antol, S., et al.: VQA: visual question answering. In: ICCV (2015)

Optimal Feature Selection for Saliency Seed Propagation in Low Contrast Images

Nan Mu[1], Xin Xu[1,2(✉)], and Xiaolong Zhang[1,2]

[1] School of Computer Science and Technology, Wuhan University of Science and Technology, Wuhan 430065, China
xuxin0336@163.com
[2] Hubei Province Key Laboratory of Intelligent Information Processing and Real-Time Industrial System, Wuhan University of Science and Technology, Wuhan 430065, China

Abstract. Salient object detection can substantially facilitate a wide range of applications. Although significant improvements have been made in recent years, low contrast image still pose great challenges to current methods due to its low signal to noise ratio property. In this paper, an optimal feature selection based saliency seed propagation method is presented to detect salient objects in low contrast images. The key idea of the proposed approach is to hierarchically refine the saliency map guided by adaptively selecting the optimal features in low contrast images recursively. Multiscale superpixel segmentation is firstly utilized to suppress background interference. Then, the initial saliency map can be generated via global contrast and spatial relationship. Local and global fitness are finally utilized to optimize the resulting saliency maps. Extensive experimental evaluations on four datasets demonstrate that the proposed model outperforms 15 state-of-the-art methods in terms of efficiency and accuracy.

Keywords: Salient object detection · Multiscale superpixel · Optimal feature
Seed propagation · Low contrast image

1 Introduction

Aiming to mimic human visual system which can effortlessly pick out the most relevant objects from the scene, salient object detection can substantially facilitate a wide range of applications, such as image segmentation [1], image retrieval [2], image compression [3], wireless network node allocation [4–6], and etc.

By calculating pixel/region uniqueness in either low-level or high-level cues, existing saliency models can be classified into two categories. (1) Bottom-up methods are generally unsupervised and are based on local or/and global contrasts. One typical example of this type was proposed by Itti et al. [7], which is based on the center-surround differences of multiple feature maps. Following this model, various methods were proposed in recent years. For example, Goferman et al. [8] implemented local and global contrasts to compute the saliency values of image patches. Cheng et al. [9] performed saliency computation based on the histogram and region contrasts. Xu et al. [10] utilized the contrast and spatial distribution to estimate the image saliency. Kim

© Springer Nature Switzerland AG 2018
R. Hong et al. (Eds.): PCM 2018, LNCS 11166, pp. 35–45, 2018.
https://doi.org/10.1007/978-3-030-00764-5_4

et al. [11] estimated the global saliency and local saliency by high-dimensional color transform and regression, respectively. Hu et al. [12] constructed saliency map by using compactness hypothesis of color and texture features. Huang and Zhang [13] proposed a minimum directional contrast based saliency model. Typically, these bottom-up methods encounter difficulties in handling images of messy background, and struggle to find the real salient object whenever the image contrast is relatively low. (2) Top-down methods mainly rely on supervised learning to guide the target acquisition. Xu et al. [14] utilized the *support vector machine* (SVM) training to generate the superpixel-level saliency map. Qu et al. [15] designed a deep learning based saliency model by fusing the superpixel-based Laplacian propagation and the trained *convolutional neural network* (CNN). Luo et al. [16] put forward a simplified CNN which combines the local and global features. Mu et al. [17] presented a covariance based CNN model to learn the image saliency. Peng et al. [18] proposed a structured matrix decomposition method based on the low-rank and sparse matrixes. Huang et al. [19] treated the saliency detection as the multiple instances learning task. Wang et al. [20] utilized the edge preserving and multi-scale contextual neural network to generate the saliency map. These top-down methods suffer from high computational complexity, the current deep neural network models are extremely time-consuming and their performance are relatively weak in determining precise localization.

Although a number of bottom-up and top-down saliency models have been proposed, most of them are designed for salient object detection in daytime scenes. These models may face great challenges in low lighting scenario, due to the lack of well-defined features to represent saliency information in low contrast images. The most likely reasons may lie in as follows: (1) current hand-crafted features can hardly evaluate objectness in an image; (2) current high-level features face great challenges in detecting precise object boundaries, which may be easily blurred by multiple levels of convolutional and pooling layers in CNN model.

This paper presents an optimal feature selection based saliency seed propagation method in low contrast images. Several hand-crafted features are selected to hierarchically refine the saliency map generated from the high-level cue recursively. The overview of this method is shown in Fig. 1. The optimal features are firstly selected to provide robust representation towards low contrast and/or normal images. Then, two cost functions are implemented to iteratively optimize the foreground and background seeds. In order to evaluate the performance, our model is compared with 15 state-of-the-art models on four datasets.

In summary, this work offers three major contributions: (1) an effective feature selection method is introduced to provide robust representation of saliency information in low contrast images, (2) two cost functions are provided to refine initial saliency seeds recursively, (3) a nighttime image dataset[1] is built to verify the effectiveness of the proposed model.

---

[1] The nighttime image (NI) dataset of this paper can be downloaded from https://drive.google.com/open?id=0BwVQK2zsuAQwX2hXbnc3ZVMzejQ.

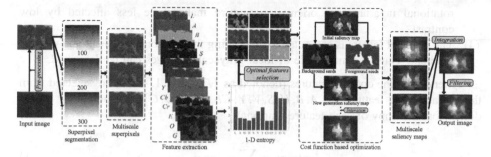

Fig. 1. Overview of the proposed model to detect salient object in low contrast image.

2 The Proposed Saliency Model

The proposed optimal feature selection based saliency seed propagation method is introduced in this section.

2.1 Multiscale Superpixel Segmentation

To preserve object structure information and make use of the mid-level information of the original image, the *simple linear iterative clustering* (SLIC) algorithm [21] is utilized to segment the image into superpixels (denoted as $\{s_i\}$, $i = 1, \ldots, N$). This processing can improve the model's efficiency by regarding superpixel as a processing unit. Since the accuracy of detection results are highly depended on the number of superpixels, the proposed model captures superpixels of three different scales, where N is set to 100, 200, and 300, respectively.

2.2 Optimal Feature Selection

Given an image, several low-level visual features are extracted including nine color features, the texture feature, the orientation feature, and the gradient feature. Since the validity of these visual features varies according to different contrasts of input images, we only select nine optimal ones from them, the adaptive selection method is based on the information entropy of the feature maps. The feature extraction procedures are described as follows.

(1) The RGB color space of the original image is firstly normalized to eliminate the influence of light and shadow. Then, the image is converted to LAB, HSV, and YCbCr color spaces to extract nine color features.

(2) The proposed method utilizes 2-dimensional entropy of the image to represent its texture features. The changes of texture features can be determined by variations in entropy, which also has strong resistance against noise and geometric deformation.

(3) The orientation feature is acquired by performing Gabor filter of different directions $\theta \in \{0°, 45°, 90°, 135°\}$ on grayscale image. The global property and

rotational invariance of orientation feature make it be less affected by low contrast.

(4) The gradient feature is computed by averaging the horizontal gradient and vertical gradient. Thus, the magnitude of local grayscale changes can be described.

Afterward, nine optimal features (denoted as $\{F_k\}$, $k = 1, \cdots, 9$) are selected from the extracted 12 features $\{L, A, B, H, S, V, Y, Cb, Cr, E, O, G\}$ by computing the 1-dimensional entropy of each feature map via:

$$entropy = \sum_{I=0}^{255} p_I \log p_I, \tag{1}$$

where p_I denotes the proportion of pixels of which the gray values are I. As a feature statistical form, the average information contained in aggregation characteristics of image grayscale distribution can be reflected by image entropy. The larger the entropy value of the feature map is, the higher the validity of the feature will be. In this paper, nine features are selected which can ensure the well description of image information.

2.3   Initial Saliency Map Generation

The global region contrast and the spatial relationship are used to compute the saliency value of each superpixel via:

$$Sal(s_i) = \left(\sum_{j=1, j\neq i}^{N} \frac{\sqrt{(F_k(s_i) - F_k(s_j))^2}}{1 + pos(s_i, s_j)} \right) \times c(s_i), \tag{2}$$

$$c(s_i) = \exp\left(-\frac{(x_i - x')^2}{2v_x^2} - \frac{(y_i - y')^2}{2v_y^2} \right), \tag{3}$$

where $pos(s_i, s_j)$ denotes the distance between s_i and s_j. $c(s_i)$ measures the spatial distance between coordinate (x_i, y_i) and image center (x', y'). v_x and v_y are variables decided by the horizontal and vertical information of the image.

2.4   Saliency Map Optimization

The initial saliency map (denoted as $Smap_k$, $k = 0$) is obtained by computing the saliency of each superpixel s_i, then it is segmented into non-salient and salient regions using the Otsu's thresholding [22]. The non-salient and salient regions can be regarded as background seeds (denoted as BS) and foreground seeds (denoted as FS) of the original image, respectively. Because the greater the difference between a superpixel and the background is, the higher the saliency value of the superpixel is. On the contrary, the larger the difference with foreground is, the lower the saliency value will be. Then, the saliency value of superpixel s_i can be recomputed based on BS and FS via:

$$Sal_{BS}(s_i) = \sum_{s_j \in BS, j \neq i} \sqrt{(F_k(s_i) - F_k(s_j))^2}/(1 + pos(s_i, s_j)), \tag{4}$$

$$Sal_{FS}(s_i) = \sum_{s_j \in FS, j \neq i} 1/(\sqrt{(F_k(s_i) - F_k(s_j))^2} + pos(s_i, s_j)), \tag{5}$$

$$Sal(s_i) = \left(1 - \exp\left(-\frac{Sal_{BS}(s_i) + Sal_{FS}(s_i)}{2}\right)\right) \times c(s_i). \tag{6}$$

Then, a new saliency map $Smap_k$, $k = 1$ of the first iteration optimization is generated. Next, the Otsu's method is reused to acquire new BS and FS, the saliency map of next generation $Smap_{k+1}$ can be obtained by Eqs. (4–6). Here, two cost functions are defined to determine whether the iteration has reached the end condition or not.

$$\text{minimize} \begin{cases} f_1(k) = (Smap_k - Smap_{k-1})^2, \\ f_2(k) = \sum_{i=1}^{N} \sum_{j=1}^{N} \frac{(Sal(s_i) - Sal(s_j))^2}{1 + pos(s_i, s_j)}, \end{cases} \tag{7}$$

where $k \geq 1$, $s_i, s_j \in Smap_k$, $1 \leq i, j \leq N$.

The function $f_1(k)$ measures the global fitness, which means that the smaller the difference between saliency map of the new generation and the previous generation is, the objective tends to be more accurate. The function $f_2(k)$ measures the local fitness, which means that the smaller the change between a superpixel and its neighbouring superpixels is, the better saliency value of each decision variable can get. By minimizing $f_1(k)$ and $f_2(k)$, the optimal superpixel-level saliency map can be generated.

3 Experiments

Extensive experiments are conducted on four datasets to evaluate the performance of the proposed model against 15 state-of-the-art saliency models. The four datasets include three public image datasets and a low contrast image dataset built by us: (1) the MSRA dataset [23] contains 10000 natural images which have simple background and high contrast; (2) the SOD dataset [24] contains images of multiple objects and complex background; (3) the DUT-OMRON dataset [25] contains relatively complex and challenging images; (4) our NI dataset contains 200 low contrast images captured in the evening with a stand camera. The resolution of each image is 640 × 480, and the manually labeled ground truths are also provided.

The 15 saliency models include: IT [7], SR [26], FT [27], NP [28], CA [8], IS [29], LR [30], PD [31], MR [25], SO [32], BL [33], GP [34], SC [35], SMD [18], and MIL [19]. All of the experiments are executed using MATLAB on an Intel i5-5250 CPU (1.6 GHz) PC with 8 GB RAM.

To evaluate the performance, seven criteria are utilized, including the *precision-recall* (PR) curve, the *true positive rates* and *false positive rates* (TPRs-FPRs) curve, the *area under the curve* (AUC) score, the *mean absolute error* (MAE) score, the *weighted F-measure* (WF) score, the *overlapping ratio* (OR) score, and the average execution time per image (in second).

By varying the threshold over the generated saliency map, different precisions, recalls, TPRs, and FPRs can be computed by comparing the obtained different binary images with the ground-truth. The PR curve and the TPRs-FPRs curve can be created by plotting these ratios. The AUC score is the percentage of area under the TPRs-FPRs curve, which gives an intuitive indication of how well the saliency map can predict the true salient objects. The MAE score is computed as the average absolute difference between the obtained saliency map and the ground-truth. The smaller the value is, the higher the similarity is. The F-measure score is defined as the weighted harmonic mean of precision and recall, and the WF score is computed by introducing a weighting function to the detection errors [36]. The OR score is computed as the ratio of over-lapping salient pixels between the binary saliency map and the ground-truth.

The quantitative detection performances of the proposed saliency model against the other 15 models on the four datasets are shown in Figs. 2, 3, and Table 1. The best three results in Table 1 are highlighted in red, blue and green fonts, respectively. The up-arrow ↑ indicates the lager the value is, the better the performance can be. While the down-arrow ↓ denotes the opposite meaning. The results show that the proposed model ranks the first or second on the three public image datasets in most cases and achieves

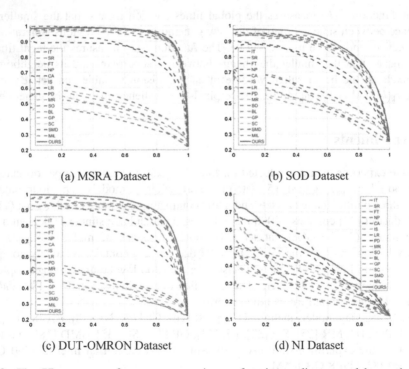

(a) MSRA Dataset

(b) SOD Dataset

(c) DUT-OMRON Dataset

(d) NI Dataset

Fig. 2. The PR curves performance comparisons of various saliency models on the four datasets. (Color figure online)

the best performance on the low contrast image dataset in a relatively short time consumption.

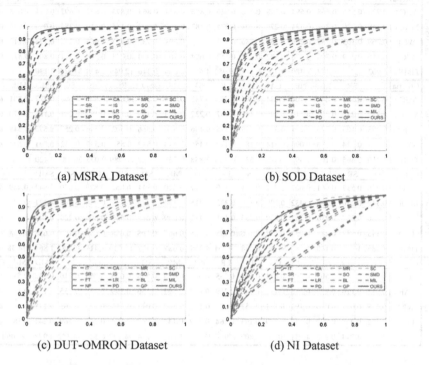

(a) MSRA Dataset

(b) SOD Dataset

(c) DUT-OMRON Dataset

(d) NI Dataset

Fig. 3. The TPRs-FPRs curves performance comparisons of various saliency models on the four datasets. (Color figure online)

On MSRA and DUT-OMRON datasets ((a) and (c) of Figs. 2, 3, and Table 1), the proposed model obtains the best performance on TPRs-FPRs curve, PR curve and AUC score, while SO model achieves the best MAE and WF scores, and MIL model achieves the best OR scores. This is because SO model utilized the boundary connectivity and global optimization to improve the robustness, and MIL model proposed a multiple instance learning strategy to improve the accuracy. These two models took advantage of background measures, which are effective in detecting the salient object of complex background. Although the MAE, WF and OR values of our model are slightly lower than the SO and MIL models, our scores are more competitive than the others. The average time consuming of MIL model is more than 100 s per image, which is highly inefficient.

On SOD dataset (Figs. 2(b), 3(b), and Table 1(b)), the proposed model achieves the best performance on TPRs-FPRs, PR, AUC, WF, and OR. In terms of MAE score, our model achieves the second best, which only has a small difference (0.002) to the best result of SO model.

Table 1. Quantitative results of various saliency models using five criteria on the four datasets.

	Criteria	IT	SR	FT	NP	CA	IS	LR	PD	MR	SO	BL	GP	SC	SMD	MIL	OURS
(a) MSRA	AUC↑	0.726	0.556	0.668	0.694	0.758	0.738	0.846	0.856	0.854	0.864	**0.873**	0.872	0.692	0.871	**0.875**	*0.879*
	MAE↓	0.348	0.220	0.266	0.421	0.270	0.304	0.224	0.187	0.107	*0.093*	0.155	0.110	0.284	**0.096**	0.096	**0.095**
	WF↑	0.268	0.103	0.289	0.269	0.332	0.286	0.433	0.477	0.694	*0.732*	0.577	0.685	0.245	**0.725**	0.713	**0.727**
	OR↑	0.155	0.249	0.300	0.070	0.332	0.257	0.552	0.604	0.769	**0.804**	0.754	0.786	0.211	0.802	*0.818*	**0.807**
	TIME↓	1.902	1.797	0.172	1.480	78.612	0.231	29.952	4.462	0.684	0.316	12.949	1.428	31.568	2.130	101.54	7.437
(b) SOD	AUC↑	0.738	0.554	0.644	0.704	0.734	0.704	0.782	0.782	0.765	0.773	**0.818**	**0.806**	0.697	0.792	0.801	*0.829*
	MAE↓	0.332	0.256	0.267	0.410	0.287	0.349	0.257	0.227	0.173	*0.163*	0.198	0.184	0.291	**0.166**	0.170	**0.165**
	WF↑	0.298	0.103	0.275	0.294	0.330	0.278	0.375	0.406	0.550	**0.566**	0.502	0.543	0.257	**0.573**	0.549	*0.574*
	OR↑	0.188	0.224	0.296	0.096	0.314	0.180	0.409	0.455	0.544	0.582	0.552	0.537	0.232	**0.591**	**0.593**	*0.594*
	TIME↓	2.450	2.465	0.162	1.397	81.028	0.533	39.654	5.888	0.768	0.527	20.489	2.023	30.181	2.420	139.16	9.967
(c) DUT-OMRON	AUC↑	0.726	0.533	0.661	0.667	0.701	0.661	0.830	0.817	0.820	0.844	**0.852**	0.829	0.670	0.844	**0.849**	*0.866*
	MAE↓	0.349	0.230	0.268	0.432	0.290	0.346	0.232	0.215	0.134	*0.114*	0.177	0.144	0.288	**0.121**	0.125	**0.119**
	WF↑	0.266	0.087	0.276	0.257	0.289	0.255	0.411	0.413	0.609	*0.665*	0.530	0.604	0.229	**0.644**	0.627	**0.656**
	OR↑	0.143	0.181	0.283	0.054	0.257	0.166	0.495	0.465	0.667	**0.705**	0.658	0.640	0.191	0.698	*0.708*	**0.699**
	TIME↓	1.986	1.962	0.140	1.385	54.163	0.238	30.361	4.305	0.687	0.437	18.530	1.411	35.257	1.861	114.38	8.749
(d) NI	AUC↑	0.672	0.512	0.507	0.708	0.614	0.643	**0.755**	0.612	0.673	0.538	0.625	**0.753**	0.710	0.642	0.584	*0.793*
	MAE↓	0.241	**0.116**	0.125	0.218	0.131	0.186	0.219	**0.129**	0.361	0.136	0.399	0.278	0.155	0.158	0.175	*0.105*
	WF↑	0.139	0.027	0.036	0.198	0.143	0.157	0.211	0.172	0.167	0.074	0.140	**0.223**	0.165	**0.218**	0.155	*0.230*
	OR↑	0.162	0.098	0.097	0.230	0.230	0.193	**0.264**	0.219	0.143	0.062	0.104	0.248	**0.260**	0.241	0.184	*0.296*
	TIME↓	4.279	4.744	0.293	4.941	107.13	0.807	179.62	13.639	1.387	1.274	94.297	6.520	34.162	4.978	191.06	13.202

On NI dataset (Figs. 2(d), 3(d), and Table 1(d)), the proposed model is superior, as it obtains the best results on these criteria with a relatively low time consumption.

The qualitative comparisons of saliency maps of the 16 saliency models on the four datasets are presented in Fig. 4, which show that the proposed model can accurately extract the true salient object in complex and low contrast images[2]. The IT, NP, IS, and SC models are susceptible to the influence of background noises. The SR and FT models cannot accurately locate the salient objects. The CA and PD models are unable to reflect the internal structure of the salient objects. The subjective performance of LR, MR, BL, and GP models are greatly influenced by the low contrast background. The SO, SMD, and MIL models cannot robustly detect the salient objects under low contrast environments. The proposed model has the ability to achieve the state-of-the-art performance in low contrast images.

---

[2] More detected saliency maps of various models on four datasets can be downloaded from https:// drive.google.com/open?id=0BwVQK2zsuAQwSENvVVR1NUJzVGc.

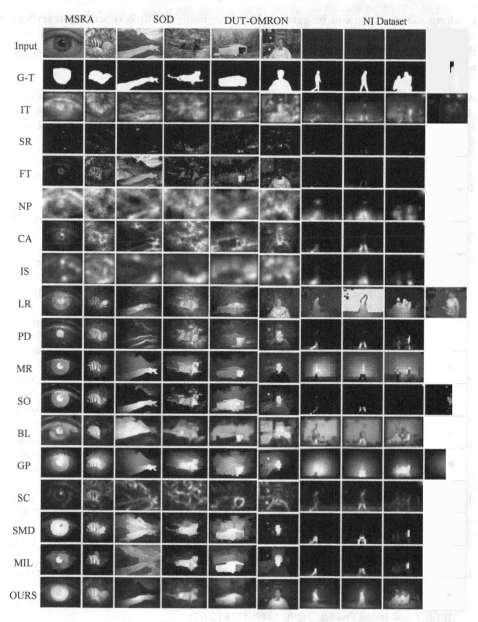

Fig. 4. Visual comparisons of saliency maps obtained by different models on four datasets

4   Conclusions

In this paper, we have constructed an optimal feature selection based saliency seed propagation method, which carries out saliency calculation via refining foreground and background seeds recursively. Guided by the optimal features and saliency seeds, the resulting saliency map can be generated by integrating multiple superpixel-level saliency maps at three scales. Experimental results show that the proposed model achieves the best performance over the 15 state-of-the-art models on three public datasets and a nighttime image dataset.

Acknowledgment. This work was supported by the Natural Science Foundation of China (61602349, 61440016, and 61273225), Hubei Chengguang Talented Youth Development Foundation (2015B22), and the Educational Research Project from the Educational Commission of Hubei Province (2016234).

References

1. Wang, L., Hua, G., Sukthankar, R., Xue, J., Niu, Z., Zheng, N.: Video object discovery and co-segmentation with extremely weak supervision. IEEE Trans. Pattern Anal. Mach. Intell. **39**(10), 2074–2088 (2017)
2. Yang, X., Qian, X., Xue, Y.: Scalable mobile image retrieval by exploring contextual saliency. IEEE Trans. Image Process. **24**(6), 1709–1721 (2015)
3. Li, S., Xu, M., Ren, Y., Wang, Z.: Closed-form optimization on saliency-guided image compression for HEVC-MSP. IEEE Trans. Multimed. **99**, 1–16 (2017)
4. Gao, L., Wang, X., Xu, Y., Zhang, Q.: Spectrum trading in cognitive radio networks: a contract-theoretic modeling approach. IEEE J. Sel. Areas Commun. **29**(4), 843–855 (2011)
5. Wang, X., Huang, W., Wang, S., Zhang, J., Hu, C.: Delay and capacity tradeoff analysis for motioncast. IEEE/ACM Trans. Netw. (TON) **19**(5), 1354–1367 (2011)
6. Gao, L., Xu, Y., Wang, X.: MAP: multiauctioneer progressive auction for dynamic spectrum access. IEEE Trans. Mobile Comput. **10**(8), 1144–1161 (2011)
7. Itti, L., Koch, C., Niebur, E.: A model of saliency-based visual attention for rapid scene analysis. IEEE Trans. Pattern Anal. Mach. Intell. **20**(11), 1254–1259 (1998)
8. Goferman, S., Zelnik-Manor, L., Tal, A.: Context-aware saliency detection. IEEE Trans. Pattern Anal. Mach. Intell. **34**(10), 1915–1926 (2012)
9. Cheng, M.-M., Mitra, N.J., Huang, X., Torr, P.H.S., Hu, S.-M.: Global contrast based salient region detection. IEEE Trans. Pattern Anal. Mach. Intell. **37**(3), 569–582 (2015)
10. Xu, X., Mu, N., Chen, L., Zhang, X.: Hierarchical salient object detection model using contrast based saliency and color spatial distribution. Multimed. Tools Appl. **75**(5), 2667–2679 (2016)
11. Kim, J., Han, D., Tai, Y.-W., Kim, J.: Salient region detection via high-dimensional color transform and local spatial support. IEEE Trans. Image Process. **25**(1), 9–23 (2016)
12. Hu, P., Wang, W., Zhang, C., Lu, K.: Detecting salient objects via color and texture compactness hypotheses. IEEE Trans. Image Process. **25**(10), 4653–4664 (2016)
13. Huang, X., Zhang, Y.-J.: 300-FPS salient object detection via minimum directional contrast. IEEE Trans. Image Process. **26**(9), 4243–4254 (2017)
14. Xu, X., Mu, N., Zhang, H., Fu, X.: Salient object detection from distinctive features in low contrast images. In: IEEE International Conference on Image Processing, pp. 3126–3130 (2015)
15. Qu, L., He, S., Zhang, J., Tian, J., Tang, Y., Yang, Q.: RGBD salient object detection via deep fusion. IEEE Trans. Image Process. **26**(5), 2274–2285 (2017)
16. Luo, Z., Mishra, A., Achkar, A., Eichel, J., Li, S., Jodoin, P.-M.: Non-local deep features for salient object detection. In: IEEE Conference on Computer Vision and Pattern Recognition, pp. 6609–6617 (2017)

17. Mu, N., Xu, X., Zhang, X., Zhang, H.: Salient object detection using a covariance-based CNN model for low-contrast images. Neural Comput. Appl. **29**(8), 181–192 (2018)
18. Peng, H., Li, B., Ling, H., Hu, W., Xiong, W., Maybank, S.J.: Salient object detection via structured matrix decomposition. IEEE Trans. Pattern Anal. Mach. Intell. **39**(4), 818–832 (2017)
19. Huang, F., Qi, J., Lu, H., Zhang, L., Ruan, X.: Salient object detection via multiple instance learning. IEEE Trans. Image Process. **26**(4), 1911–1922 (2017)
20. Wang, X., Ma, H., Chen, X., You, S.: Edge preserving and multi-scale contextual neural network for salient object detection. IEEE Trans. Image Process. **27**(1), 121–134 (2018)
21. Achanta, R., Shaji, A., Smith, K., Lucchi, A., Fua, P., Susstrunk, S.: SLIC superpixels compared to state-of-the-art superpixel methods. IEEE Trans. Pattern Anal. Mach. Intell. **34** (11), 2274–2282 (2012)
22. Otsu, N.: A threshold selection method from gray-level histograms. IEEE Trans. Syst. Man Cybern. **9**(1), 62–66 (1979)
23. Liu, T., Sun, J., Zheng, N.-N., Tang, X., Shum, H.-Y.: Learning to detect a salient object. In: IEEE Conference on Computer Vision and Pattern Recognition, pp. 1–8 (2007)
24. Movahedi, V., Elder, J. H.: Design and perceptual validation of performance measures for salient object segmentation, In: IEEE Computer Society Conference on Computer Vision and Pattern Recognition Workshops, pp. 49–56 (2010)
25. Yang, C., Zhang, L., Lu, H., Ruan, X., Yang, M.-H.: Saliency detection via graph-based manifold ranking. In: IEEE Conference on Computer Vision and Pattern Recognition, pp. 3166–3173 (2013)
26. Hou, X., Zhang, L.: Saliency detection: a spectral residual approach. In: IEEE Conference on Computer Vision and Pattern Recognition, pp. 1–8 (2007)
27. Achanta, R., Hemami, S., Estrada, F., Susstrunk, S.: Frequency-tuned salient region detection. In: IEEE Conference on Computer Vision and Pattern Recognition, pp. 1597–1604 (2009)
28. Murray, N., Vanrell, M., Otazu, X., Parraga, C.A.: Saliency estimation using a non-parametric low-level vision model. In: IEEE Computer Conference on Computer Vision and Pattern Recognition, pp. 433–440 (2011)
29. Hou, X., Harel, J., Koch, C.: Image signature: highlighting sparse salient regions. IEEE Trans. Pattern Anal. Mach. Intell. **34**(1), 194–201 (2012)
30. Shen, X., Wu, Y.: A unified approach to salient object detection via low rank matrix recovery. In: IEEE Conference on Computer Vision and Pattern Recognition, pp. 853–860 (2012)
31. Margolin, R., Zelnik-Manor L., Tal, A.: What makes a patch distinct? In: IEEE Conference on Computer Vision and Pattern Recognition, pp. 1139–1146 (2013)
32. Zhu, W., Liang, S., Wei, Y., Sun, J.: Saliency optimization from robust background detection. In: IEEE Conference on Computer Vision and Pattern Recognition, pp. 2814–2821 (2014)
33. Tong, N., Lu, H., Yang, M.: Salient object detection via bootstrap learning. In: IEEE Conference on Computer Vision and Pattern Recognition, pp. 1884–1892 (2015)
34. Jiang, P., Vasconcelos, N., Peng, J.: Generic promotion of diffusion-based salient object detection. In: IEEE International Conference on Computer Vision, pp. 217–225 (2015)
35. Zhang, J., Wang, M., Zhang, S., Li, X., Wu, X.: Spatiochromatic context modeling for color saliency analysis. IEEE Trans. Neural Netw. Learn. Syst. **27**(6), 1177–1189 (2016)
36. Margolin, R., Zelnik-Manor, L., Tal, A.: How to evaluate foreground maps? In: IEEE Conference on Computer Vision and Pattern Recognition, pp. 248–255 (2014)

New Fusion Based Enhancement for Text Detection in Night Video Footage

Chao Zhang[1], Palaiahnakote Shivakumara[2], Minglong Xue[1],
Liping Zhu[3], Tong Lu[1(✉)], and Umapada Pal[4]

[1] National Key Lab for Novel Software Technology, Nanjing University,
Nanjing, China
zhangchao_nju@126.com, xueml@smail.nju.edu.cn,
lutong@nju.edu.cn
[2] Faculty of Computer Science and Information Technology,
University of Malaya, Kuala Lumpur, Malaysia
shiva@um.edu.my
[3] School of Information Management, Nanjing University, Nanjing, China
chemzlp@163.com
[4] Computer Vision and Pattern Recognition Unit, Indian Statistical Institute,
Kolkata, India
umapada@isical.ac.in

Abstract. Text Detection in night video footage is hard due to low contrast and low resolution caused by distance variations between camera and ground under poor light. In this paper, we propose a new fusion based enhancement method for text detection especially in night video footage. The proposed method integrates the merits of color space and frequency based enhanced methods for sharpening low contrast details. Specifically, for each enhanced image, the proposed method derives weighted mean for the pixels values to widen the gap between high contrast (texts) and low contrast (background) pixels. The weighed means are further modified as dynamic weights with respect to enhanced images. These weights are convolved with pixel values of respective enhanced images to produce fused images. The proposed fusion based enhancement method is tested on images collected from night video footage to demonstrate the effectiveness of the method. For the output of each enhancement method including the proposed method, text detection rates are computed to show that the proposed enhancement method outperforms the existing enhancement methods.

Keywords: Night video · Enhancement · Color space · Frequency domain
Image fusion · Text detection

1 Introduction

As noticed literature on text detection and recognition in natural scene images and video, researchers are developing new methods to tackle several challenges such as low contrast, uneven illumination effect, multiple scripts or orientations, complex background, and font or font size variations [1, 2]. This shows that text detection and

© Springer Nature Switzerland AG 2018
R. Hong et al. (Eds.): PCM 2018, LNCS 11166, pp. 46–56, 2018.
https://doi.org/10.1007/978-3-030-00764-5_5

recognition is important for real time emerging applications, such as surveillance and navigation apart from image understanding for retrieval [3, 6]. But it is noticed from literature that none of the methods addressed the issue of text detection in night video footage, where texts suffer from poor quality due to poor light conditions including other challenges mentioned above. One such illustration can be seen in Fig. 1, where (a) is an input image captured at night, in which texts are even invisible from our naked eyes, (b) gives the result of the existing enhancement method [7] which explores color space, (c) shows the result of one more state-of-art existing enhancement method [8], which explores frequency domain, and (d) gives the result of the proposed method. Note that to test the effectiveness of text detection performances for enhancement methods, we run the latest method [6] that uses powerful deep learning for text detection in natural scene images. It can be seen that the text detection method fails to detect the texts in the input image, and misses the texts from the results of both the existing enhancement method-1 and method-2. However, the same method [6] detects texts properly for the result of the proposed method. This shows that text detection performance is poor for night images, which can be improved for enhanced images. This is also true that understanding night video footage is essential for surveillance and monitoring applications especially for the purpose of forensic investigations in crime cases. Therefore, we focus on image enhancement for improving text detection performance for night video images in this work.

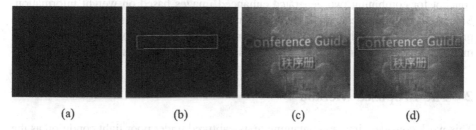

(a)   (b)   (c)   (d)

Fig. 1. Text detection performances prior to enhancement and after enhancement, where (a) is the input image, (b) shows the enhanced image from color space, (c) gives the enhanced image from frequency domain, and (d) shows the enhanced result from the proposed method.

There are methods for general image enhancement and text image enhancement through fusion concept. For example, Sharma et al. [9] proposed contrast enhancement using pixel based image fusion in wavelet domain. The method explores the combination of low and high pass filters. Lee et al. [10] proposed an edge enhancement algorithm for low dose X ray fluoroscopic imaging. This method uses the same filters as mentioned in above in gradient domain. Jiang et al. [7] proposed night video enhancement using improved dark channel priors. The method uses inversion pixel operation in color domain. Rui et al. [8] proposed a medical X-ray image enhancement method based on TV-Homomorphic filter. It has the ability to balance bright and fine pixels in images. Maurya et al. [11] proposed a social spider optimized image fusion approach for contrast enhancement and brightness preservation. The method produces one high sharp image and one more with high peak signals for each input image. Then it fuses the two images to enhance fine details. From the above reviews, it is noticed that none of the methods considers images with texts for enhancement. In other words, the objective of the above methods is to enhance the content in general images.

Recently, we can see a few methods for text image enhancement. Pei et al. [12] proposed multi-orientation scene text detection with multi-information fusion. The method explores convolutional neural networks for detecting low contrast texts in natural scene images. The focus of the method is to enhance text information in natural scene images but not texts in night images. Roy et al. [13] proposed a fractional Poisson enhancement model for text detection and recognition in video frames. However, the target of the work is to enhance images affected by Laplacian operation but not those captured at night. Overall, none of the methods addresses the issue of text enhancement in night images for improving the performance of text detection.

Hence, in this work, we focus on the enhancement of images captured under poor light to improve the performance of text detection. Inspired by the method [11], where two different domains for enhancing low contrast images are used, we exploit the same idea for generating enhanced images using color space and frequency information. When there are texts in images, one can expect high contrast information, and meanwhile the values of such text pixels often have almost the uniform color values. Based on this observation, we propose new criteria for weight derivation for each pixel in enhanced images. Additionally, motivated by the method [14] where fusion is introduced for medical image enhancement, we propose to explore the same fusion concept for combining two generated enhanced images based on weight information. Since our weights are derived from fusion operation based on text properties, the proposed method well enhances text information by suppressing background information. This is our contribution, which is different from the existing methods.

2   The Proposed Method

This work considers images containing texts captured under poor light condition as the input for enhancement. As noticed from the literature, color space and frequency domain are the main concepts for image enhancement. This is because color is sensitive to human perception, while frequency coefficients are sensitive to tiny changes at pixel level. To take the advantage of both the domains, we propose to use color space for generating one enhanced image, and frequency domain for generating another enhanced image for each input image. Then to integrate both the enhanced images, we propose a new fusion method, which derives weights based on text properties to produce the final fused image, resulting in text enhancement. The block diagram of the proposed method can be seen in Fig. 2.

2.1   Enhanced Images Using Color Space and Frequency Domain

As mentioned in the previous section, we consider the idea presented in [7] for obtaining enhanced images using color space. The method treats night video images as foggy or haze removal. As a result, the method proposes a degradation model, which requires estimating the global atmospheric light and medium transmission using an

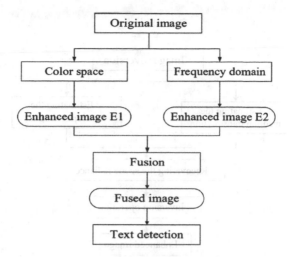

Fig. 2. The block diagram of the propose method.

improving dark channel prior. The institution to estimate the global atmospheric light is that the pixels that correspond to atmospheric light should be located in a large smooth area and during a period of time, the atmospheric light has no changes. In the same way, the main basis for estimating medium transmission using an improving dark channel prior is that the pixels on the same object should have the same or similar depth values. By knowing the transmission map and the global atmospheric light, the method can obtain the inversion of the enhanced image. The steps can be seen in Fig. 3. The effect of the enhancement method can be seen in Fig. 4(a), where we can see image details are enhanced compared to the input image in Fig. 1(a). However, since the method considers image enhancement as degradation, the method alone is insufficient to enhance text details in night video footage.

We thus propose another concept to enhance image details based on TV-Homomorphic filter [8]. It has the ability to reduce low frequency and increase high frequency information simultaneously, which results in reducing illumination changes and sharpening edge pixels. It works based on the incident and reflected light model. For the purpose of filtering, the method uses Fourier transform as the filter, which gives a filtered image for each input image. The total variation model has been used in homomorphic filter as a new transfer function. The total variation model is widely used for image restoration and noise removal. Since homomorphic filter uses the total variation model, it has the ability to adjust brightness and details of enhancement. The effect of the homomorphic filter is shown in Fig. 4(b), where one can see the image is brighter than the result shown in Fig. 4(a). It is also observed from Fig. 4(b) that the method enhances image details but not only text pixels. Overall, we can conclude that method-1 focuses on enhancing foreground information, while method-2 focuses on enhancing background information. Therefore, text pixels that have low contrast and high contrast are enhanced in separate images. To combine both low contrast and high contrast text pixels, we propose a new fusion method in the next section.

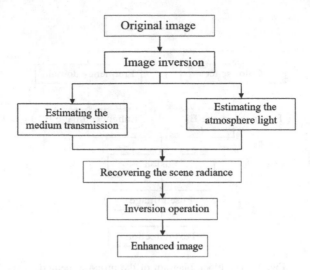

Fig. 3. The method for obtaining enhanced image-1 using color space.

| (a) | (b) | (c) |

Fig. 4. Fusing two enhanced images for generating a better enhanced image, where (a) is the enhanced image from color space, (b) is the enhanced image from frequency domain and (c) is the enhanced image from the proposed method.

2.2 Fusion Criteria for Text Enhancement

As mentioned in the Proposed Method Section, for each pixel in the two enhanced images given by the methods presented in the previous section, the proposed method computes mean as defined in Eq. (1). It is true that when there is an edge that represents text pixels, there will be high contrast values compared to its background. Therefore, the mean of such pixels gives high values. If there are no text pixels, the mean gives low values because of low contrast background information. To extract such difference, the proposed method multiplies the mean values with the current pixel values as defined in Eq. (2), which we call weights for each pixel in the two enhanced images. This operation increases text pixel values and decreases non-text values simultaneously. However, sometimes, arbitrary contrast variations in an image cause problem for identifying the gap between text and non-text pixels. To alleviate this problem, we recalculate weights based on the fact that an image can contain three type of values, namely, high values which represent edges (text), low values which represent background, and middle values which usually represent text pixels affected by arbitrary contrast variations. Therefore, we recalculate weights as if a weight is greater than 0.6 in the values (text pixels) of the range, 0 to 1 must be multiplied by 0.8. If the weight is less than 0.6 and greater than 0.2 (affected values), the values are multiplied by 0.5 in the range of 0 and 1 values. The weight recalculation is defined in Eq. (3). The main objective of recalculating weights is to classify the pixels which represent texts as white and the pixels which represent non-texts as black. The values are determined empirically. The recalculated weights for each pixel in the enhanced images are then used for obtaining the fused image as defined in Eq. (4). The effect of the proposed fusion can be seen in Fig. 4(c), where one can see text pixels are enhanced compared to the images in Fig. 4(a) and (b). This is the advantage of the proposed work. The above steps are formulated mathematically as follows.

We define the grayscale image of the enhanced image from color space as $f^1(x, y)$ and enhanced image from frequency domain as $f^2(x, y)$. For two enhanced images, we calculate the local mean value of the $(i, j)^{th}$ pixel over an $n \times n$ window as $m^1_{(i,j)}$, $m^2_{(i,j)}$ using the Eq. (1):

$$m^k_{(i,j)} = \frac{1}{n * n} \sum_{x=0}^{n-1} \sum_{y=0}^{n-1} f^k(x, y) \tag{1}$$

where k denotes the index of the enhanced images and $k \in \{1, 2\}$, while n denotes the size of the window and is set as 7 according to experiments. We compute the corresponding weight using Eq. (2):

$$w^k_{(i,j)} = \frac{m^k_{(i,j)}}{\sum_{l=1}^{l=p} m^l_{(i,j)}} \tag{2}$$

where $w^k_{(i,j)}$ denotes the weight at position (i, j) of the image, p is the number of the enhanced images and is set as 2, $k \in \{1, 2\}$. Then we apply the following Eq. (3) to change weight values:

$$w^k_{(i,j)} = \begin{cases} w^k_{(i,j)} \times \mu_1 & t_1 \leq w^k_{(i,j)} < 1 \\ w^k_{(i,j)} \times \mu_2 & t_2 \leq w^k_{(i,j)} < t_1 \\ w^k_{(i,j)} & 0 \leq w^k_{(i,j)} < t_2 \end{cases} \tag{3}$$

where $k \in \{1, 2\}$, $w^k_{(i,j)}$ is the weight at position (i, j) of the enhanced image. $\mu_1 = 0.8$, $\mu_2 = 0.5$, $t_1 = 0.6$, $t_2 = 0.2$ and these values are determined empirically.

$$F(i, j) = \sum_{q=1}^{n} w^q_{(i,j)} \times f^q(i, j) \tag{4}$$

where F denotes the fused image, $f(i,j)$ is the gray pixel value at position (i,j) in the enhanced image, while n denotes the number of enhance images, which is set as 2.

3 Experimental Results

As there is no standard dataset available for evaluating the proposed method on night video footage, we collect our own data by capturing images at night, which includes 500 images under different poor light conditions. The dataset comprises images of Chinese and English captured by mobile phones in dark scenes. Besides, it also includes images of book covers, daily necessities, and boxes containing texts. To test the effectiveness of the proposed method, we collect low contrast images from standard ICDAR 2015 video [15] and YVT video, which provide 500 images. In total, we consider 1000 images for experimentation and evaluation of the proposed and existing methods.

To show the superiority to existing methods, we implement three state of the art existing enhancement methods, namely, Jiang et al.'s method [7] which uses color space and a degradation model for enhancing general images, Rui et al.'s method [8] which uses a TV-Homomorphic filter for enhancing details in images, and Roy et al.'s method [13] which is developed for enhancing text information in images affected by Laplacian operation. The reason to consider the above three methods is that Jiang et al.'s method is the state-of-the-art method for enhancing details in night video images as the proposed work, Rui et al.'s method is the state-of-the-art method that enhances image details in low contrast images, while Roy et al.'s method [13] is the state-of-the-art method that focusses on enhancement of text pixels affected by video noises as the proposed work.

To validate the result of enhancement given by the proposed and existing methods, we implement two well-known text detection methods to run on input images and the result of enhancement images, that is, Zhou et al.'s method [5] which proposes an efficient and accurate scene text detector based on deep learning and a large number of features, and Shi et al.'s method [4] which detects oriented texts in natural scene images based on deep learning and linking segments. The reason to consider these two methods is that the former one is good for low contrast and low resolution images, while the latter is good for images affected by blur and uneven illuminations. In addition, both the methods used deep learning tools for achieving their results. The results are obtained for prior to enhancement and after enhancement to show that the proposed enhancement method helps in improving text detection performance compared to prior to enhancement.

To measure the performance of the proposed method, we use standard measures, namely, Recall (R), Precision (P) and F-Measure (F) as defined in Eqs. (5)–(7), where True Positive (TP) means the number of items labeled correctly and belong to the positive class, True Negative (TN) means the number of items labeled correctly and belong to the negative class, False Positive (FP) and False Negative (FN) means the number of items labeled incorrectly in positive class and negative class, respectively.

$$\text{Precision} = \frac{TP}{TP + FP} \tag{5}$$

$$\text{Recall} = \frac{TP}{TP + FN} \tag{6}$$

$$\text{F-Measure} = \frac{2 * Precision * Recall}{Precision + Recall} \tag{7}$$

Qualitative results of the proposed and existing enhancement methods are shown in Fig. 5, where we can see that text detection methods detect texts properly for the output of the proposed enhancement method compared to the results of the other three existing enhancement methods. It is also observed from Fig. 5 that proposed fusion is better than the existing methods in terms of fine details of texts. This shows that the proposed enhancement is better than the existing enhancement methods. The reason of the existing methods for poor results is that the existing methods are developed with specific objectives but not text enhancement in night images. In addition, the way the proposed method integrates the advantage of enhancement results is something new and contributes for enhancing text pixels in night images.

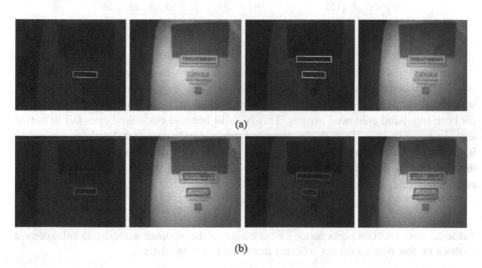

Fig. 5. Text detection performances for the enhanced results of proposed and existing methods where (a) is text detection results of EAST [5] and (b) is text detection results of SegLink [4]. The enhancement images in (a) and (b) are obtained by color space method, frequency domain method, Roy et al. [13] and the proposed method, respectively.

Quantitative results of the text detection methods for the output of the proposed and existing enhancement methods on our dataset are reported in Table 1. Table 1 shows that the text detection performance for input images is lower than that of enhanced images. When we compare the text detection performance on the output of the proposed method with the text detection performance on input images, the text detection performance for the proposed enhanced images is improved significantly especially on F-measure. Similarly, the text detection results for the output of the proposed and existing enhancement methods on low contrast dataset are reported in Table 2, where the same conclusions can be drawn as Table 1. It is observed from Tables 1 and 2 that when we compare the text detection performances of the existing enhancement methods with the text detection performance of the proposed method, text detection performance of proposed enhancement is better than those of the existing enhancement methods. This shows that the proposed enhancement is better than the existing enhancement, and hence we can conclude that text detection performance improves significantly for the output of the proposed enhancement method.

Table 1. Text detection performance for the output of proposed and existing enhancement methods on our dataset.

Methods	SegLink [4]			EAST [5]		
	P	R	F	P	R	F
Input (prior to enhancement)	43.6	16.0	23.4	73.4	18.6	29.7
Enhancement-1 [7]	41.5	21.2	28.1	69.8	38.2	49.4
Enhancement-2 [8]	43.2	33.0	37.4	69.6	54.8	61.3
Roy's et al. [13]	41.4	31.4	35.7	70.8	50.7	59.1
Proposed method	**46.5**	**36.6**	**41.0**	**73.7**	**57.1**	**64.3**

Sometimes, for the images captured under full dark as shown in Fig. 6, the proposed enhancement method does not work well due to the limitation of automatic weight calculation. As a result, text detection method [6] does not detect texts properly for both input and enhanced images. This is valid because our naked eyes fail to notice texts in input images. Though the proposed method enhances text details compared to input images, it is insufficient to improve text detection performance. This shows that there is a scope for improvement in future. To overcome this issue, it is necessary to consider context information for text detection in such full dark images.

Table 2. Text detection performance for the output of the proposed and existing enhancement methods on low contrast dataset collected from standard video dataset.

Methods	SegLink [4]			EAST [5]		
	P	R	F	P	R	F
Input (prior to enhancement)	70.3	71.9	71.1	74.5	75.4	74.9
Enhancement-1 [7]	69.4	**76.5**	72.8	77.2	74.2	75.7
Enhancement-2 [8]	74.5	68.4	71.3	78.0	72.7	75.3
Roy's et al. [13]	69.5	73.8	71.6	73.2	**77.0**	75.0
Proposed method	**79.1**	71.2	**74.9**	**80.1**	74.5	**77.2**

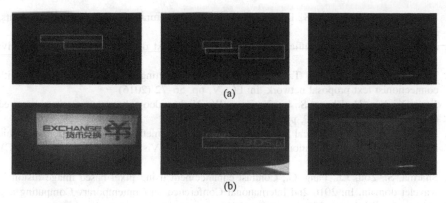

(a)

(b)

Fig. 6. Limitation of the proposed method. Text detection results are given by the CTPN [6]. (a) Gives input images of different situations, and (b) shows fused images.

4 Conclusion and Future Work

In this work, we have proposed a new method for enhancing texts in night video footage to improve text detection performance. The proposed method generates enhanced images based on color and frequency domains respectively to take the advantage of respective domains. We propose a new fusion criterion for integrating the advantages of color domain and frequency domain, which results in text enhancement in night images. Experimental results on the proposed and existing enhancement methods show that the proposed method is better than the existing enhancement methods. Additionally, text detection performance of after enhancement is better than prior to enhancement. However, when an image contains non-uniform blur or variations in degree of illumination effect, the performance is still poor. This would be our near future work.

Acknowledgment. The work described in this paper was supported by the Natural Science Foundation of China under Grant No. 61672273 and No. 61272218, the Science Foundation for Distinguished Young Scholars of Jiangsu under Grant No. BK20160021, and Scientific Foundation of State Grid Corporation of China (Research on Ice-wind Disaster Feature Recognition and Prediction by Few-shot Machine Learning in Transmission Lines).

References

1. Ye, Q., Doermann, D.: Text detection and recognition in imagery: a survey. IEEE Trans. Pattern Anal. Mach. Intell. **37**(7), 1480–1500 (2015)
2. Yin, X.C., Zuo, Z.Y., Tian, S., Liu, C.L.: Text detection, tracking and recognition in video: a comprehensive survey. IEEE Trans. Image Process. **25**(6), 2752–2773 (2016)
3. Tian, S., Yin, X.C., Su, Y., Hao, H.W.: A unified framework for tracking based text detection and recognition from web videos. IEEE Trans. Pattern Anal. Mach. Intell. **40**(3), 542–554 (2018)

4. Shi, B., Bai, X., Belongie, S.: Detecting oriented text in natural images by linking segments. In: Proceedings CVPR, vol. 3 (2017)
5. Zhou, X., et al.: East: an efficient and accurate scene text detector. arXiv preprint arXiv: 1704.03155 (2017)
6. Tian, Z., Huang, W., He, T., He, P., Qiao, Y.: Detecting text in natural image with connectionist text proposal network. In: ECCV, pp. 56–72 (2016)
7. Jiang, X., Yao, H., Zhang, S., Lu, X., Zeng, W.: Night video enhancement using improved dark channel prior. In: ICIP, pp. 553–557. IEEE (2013)
8. Rui, W., Guoyu, W.: Medical X-ray image enhancement method based on tvhomomorphic filter. In: 2017 2nd International Conference on Image, Vision and Computing (ICIVC), pp. 315–318. IEEE (2017)
9. Sharma, S., Zou, J.J., Fang, G.: Contrast enhancement using pixel based image fusion in wavelet domain. In: 2016 2nd International Conference on Contemporary Computing and Informatics (IC3I), pp. 285–290. IEEE (2016)
10. Lee, M.S., Park, C.H., Kang, M.G.: Edge enhancement algorithm for low-dose X-ray fluoroscopic imaging. Comput. Methods Programs Biomed. **152**, 45–52 (2017)
11. Maurya, L., Mahapatra, P.K., Kumar, A.: A social spider optimized image fusion approach for contrast enhancement and brightness preservation. Appl. Soft Comput. **52**, 575–592 (2017)
12. Pei, W.Y., Yang, C., Kau, L.J., Yin, X.C.: Multi-orientation scene text detection with multi-information fusion. In: 2016 23rd International Conference on Pattern Recognition (ICPR), pp. 657–662. IEEE (2016)
13. Roy, S., Shivakumara, P., Jalab, H.A., Ibrahim, R.W., Pal, U., Lu, T.: Fractional poisson enhancement model for text detection and recognition in video frames. Pattern Recogn. **52**, 433–447 (2016)
14. Xu, X., Wang, Y., Chen, S.: Medical image fusion using discrete fractional wavelet transform. Biomed. Signal Process. Control **27**, 103–111 (2016)
15. Karatzas, D., et al.: ICDAR 2015 competition on robust reading. In: 2015 13th International Conference on Document Analysis and Recognition (ICDAR), pp. 1156–1160. IEEE (2015)

Deformable Feature Pyramid Network for Ship Recognition

Yao Ding, Yichen Zhang, Yanyun Qu[✉], and Cuihua Li

Computer Science Department, Xiamen University, Xiamen, China
dingyao_0221@163.com, chinesexzyc@gmail.com,
quyanyun@gmail.com, chli@xmu.edu.cn

Abstract. Ship recognition under complex sea environment and weather condition is a challenging task because the ship appearances change greatly, especially under large geometric transformation. The original feature pyramid network (FPN) [4] does not achieve good performance if it is implemented on ship detection directly, because it uses the fixed geometric structures in their building modules. In this paper, a deformable feature pyramid network is designed for ship recognition. The contributions are three folds: (1) We change the fixed geometric structure model of the original feature pyramid network to deformable geometric structure model and use the dilated convolution [12] instead of the original convolution. Correspondingly, deformable position-sensitive RoI pooling is used instead of the fixed geometric RoI pooling in the RoI-wise subnetwork. (2) The focal loss function [6] replaces the original mixed cross-entropy loss function. (3) Decay-NMS, a new post-processing method, is designed in this paper to improve the detection accuracy. The experimental results demonstrate the effectiveness and efficiency of our model.

Keywords: Ship recognition · Deformable feature pyramid network
Dilated convolution · Deformable position-sensitive RoI pooling
Focal loss · Decay-NMS

1 Introduction

Ships are very important targets in the sea and automatic ship recognition is helpful for port monitoring, sea transportation and military action in the sea. However, the research on ship recognition is still in primary stage. In order to seek the solution of ship recognition, Chinese Computer Federation organized the Big Data and Computing Intelligence (BDCI) Challenge [14] on ship recognition under complex sea environment and weather condition in 2017.

Ship recognition is a challenging task until now. Ship appearances vary largely with the changes of illuminations, viewpoints, and scales. Especially, in the BDCI challenge, training and testing images are generated through Google Earth screenshots combined with simulation algorithms. They not only include the good weather conditions, but also include extreme weather conditions such as overcast and fog. Ship appearances change very largely in scale and geometric transformation. Some ships occupy several hundreds of pixels while some ships only occupy dozens of pixels.

© Springer Nature Switzerland AG 2018
R. Hong et al. (Eds.): PCM 2018, LNCS 11166, pp. 57–66, 2018.
https://doi.org/10.1007/978-3-030-00764-5_6

Moreover, different from the objects in ImageNet [17], which usually show an object in a regular geometric shape, ships in BDCI are of a large geometric transformation and in a long oriented bar. Ship images shot from the air are vulnerable to the complex sea environments and weather conditions, thus the ships may look blurring, and are occluded by clouds, and have low-contrast due to fog. Though the sponsor of BDCI gave the ship dataset, there are many flaws in the original labels. We relabel all the training data and build a well-labelled ship dataset.

With the development of deep learning, deep neural network is applied in more and more vision tasks in the field of computer vision. There are many promising deep neural network for object detection and recognition, such as faster R-CNN [3], SSD [7], and Yolo [8] and so on. However, the original convolution layers in deep convolutional network fix the geometric structure of the convolution kernel, thus they cannot exactly extract the features of the ship area. If they are applied directly to ship recognition, they do not achieve good performance due to the great changes of geometric transformations and scales. In order to solve ship recognition, deformable convolution and dilated convolution are introduced to the feature pyramid network (FPN). We design a deformable feature pyramid network for ship recognition.

The contributions of our approach are three folds: (1) In the backbone network, we change the fixed geometric structure model of the original feature pyramid network to deformable geometric structure model and use the dilated convolution [12] instead of the original convolution. Correspondingly, deformable position-sensitive RoI pooling is used instead of the fixed geometric RoI pooling in the RoI-wise subnetwork. Deformable convolution greatly increases the feature extraction capability of the network and dilated convolution is helpful for detecting the small ships. (2) The focal loss function [6] and online hard example mining (OHEM) [10] replaces the original mixed cross-entropy loss function. (3) Decay-NMS, a new post-processing method, is designed in this paper to improve the detection accuracy.

In the following, we describe the implementation details of the proposed method in Sect. 2. Section 3 evaluates the proposed method on BDCI ship dataset. Finally, conclusions are given in Sect. 4.

2 The Proposed Method

The architecture of the deformable FPN is shown in Fig. 1. ResNet-101 [5] pre-trained on the ImageNet1k classification set [17] is used as the backbone network. The original FPN consists of bottom-up pathway and top-down pathway. The bottom-up pathway is the feedforward computation of the backbone network, while the top-down pathway and lateral connections combine the high-resolution shallow features with low-resolution deep features. We introduce the deformable convolution layer, deformable position-sensitive RoI pooling, the focal loss function and OHEM, and the Decay-NMS to the original FPN.

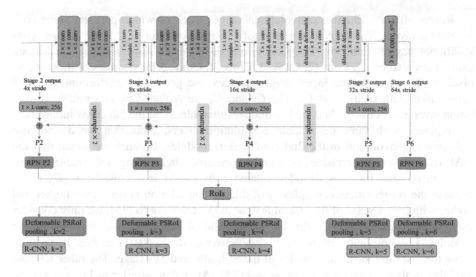

Fig. 1. Our deformable FPN model.

Deformable Convolution [11] in the Backbone Network. Because the ship objects are mostly in an oriented long bar, and the ratio between the length and width of a cargo ship is 10:1, the uniform sampling points of the regular rectangular convolution kernel cannot effectively extract the distinct features of the ship area. In order to make the sampling points more concentrated in the target area of the ship, we replace the convolution kernel in the fixed geometric structure with deformable convolution in some convolution layers of the residual modules. As shown in Fig. 2, the former extract the features of the ship area together with the background in the left subfigure, while the latter can extract the features of the concentrated ship area in the right subfigure.

Fig. 2. The regular rectangular convolution (left) and the deformable convolution [11] (right).

Dilated Convolution [12]. Many ships shot from the air in a long distance become small targets. For example, the yacht usually becomes very small if the shooting distance is far. Because the dilated convolution can effectively increase the size of the receptive field under the premise of maintaining the spatial resolution of the feature map, we use it to improve the detection accuracy on small objects.

Resnet-101 contains five convolutional stages, each of which contains some convolutional layers. The activation output of the last convolutional layer in each convolutional stage is used for FPN. Let the output of the ith stage's last residual block Conv_i denoted by C_i. Conv_2, Conv_3, Conv_4, Conv_5 have strides of 4, 8, 16, 32 pixels with respect to the input image. Besides, we get C_6 by performing a 3×3 convolution with stride 2 on C_5. The C_3 and C_4 are replaced by deformable convolution layer, respectively. And the last three residual blocks in the 5th convolution stage are replaced with three deformable convolution layers. Note that we do not use deformable convolution in the 2nd convolution modules, because we found that the mAP (mean average precision) is slightly decreased after adding deformable convolution in the 2nd modules. Besides, dilated convolution is not used for $\{C_3, C_4\}$, because the computational complexity of dilated convolution is becoming higher and higher with the increase of the size of feature map. Referring to [15], we only consider using dilated convolution in the last three residual blocks of the 5th convolution modules. For simplicity, in the backbone network (ResNet-101) in Fig. 1, we only show the last three residual modules of the 3rd, 4th, and 5th stage. The other residual modules in these stages are the same as in [5]. After that, similar to [4], we get the feature pyramids $\{P_2, P_3, P_4, P_5, P_6\}$ corresponding to $\{C_2, C_3, C_4, C_5, C_6\}$ that are of the same spatial sizes.

Deformable PSRoI Pooling in the RoI-Wise Subnetwork. As shown in Fig. 1. We replace RoI pooling with deformable position-sensitive RoI (PSRoI) pooling to enable adaptive partial positioning of ships with different shapes. It is worth noting that deformable PSRoI pooling combines deformable RoI pooling [11] and position-sensitive RoI pooling [16]. Therefore, as shown in Fig. 3, when performing deformable PSRoI pooling, the split k × k bins are not only well concentrated in the target area, but also added a location-sensitive score map, which increases the accuracy of the ship's object detection. Besides, in our model, deformable PSRoI pooling is followed by two 1024-dimension fully-connected layers, and then two sibling 1×1 convolutions are added for classification and regression. In order to ensure the detection efficiency, these fully-connected layers and 1×1 convolutions are parameter-shared.

Fig. 3. The original RoI pooling (left) and the Deformable PSRoI pooling (right).

The Focal Loss Function [6] and OHEM [10]. In loss function design, we not only consider the imbalance of positive and negative samples, but also consider the degree of difficulties in classification. Thus, we replace the original mixed cross-entropy loss function in the RoI-Wise subnetwork with the OHEM and focal loss respectively.

For the purpose of simplicity, we discuss the loss function of the two-class problem, but in fact it can be applied to multi-class problems. Based on OHEM, the loss function in our model can be expressed as,

$$L = \lambda(y, \hat{y}) \cdot L_{ce} \tag{1}$$

where $\hat{y}$ is the predicted value given by the network, $y \in \{0, 1\}$ is the corresponding real label value. And $\lambda(y, \hat{y})$ is defined as:

$$\lambda(y, \hat{y}) = \begin{cases} 0 & when(y = 1 \ and \ \hat{y} > 0.6) \ or \ (y = 0 \ and \ \hat{y} < 0.6) \\ 1 & otherwise \end{cases} \tag{2}$$

L_{ce} is defined as:

$$L_{ce} = -y\log\hat{y} - (1 - y)\log(1 - \hat{y}) = \begin{cases} -\log\hat{y} & if \ y = 1 \\ -\log(1 - \hat{y}) & if \ y = 0 \end{cases} \tag{3}$$

When the real label of a certain RoI is 1 and the predicted confidence value given by the network is greater than 0.6, or the real label is 0 and predicted value is less than 0.6, the RoI is considered as an easily separable sample. Similar to OHEM, its loss value is set to 0, which makes the network pay more attention to difficult samples with inaccurate predictions. However, the experimental results show that, the network with this kind of loss function is extremely difficult to converge, and the reason is analyzable. The above modification of the loss function is a mandatory truncation of the loss value, resulting in a non-conductible point. Therefore, $\lambda(y, \hat{y})$ will not have any effect on the updating of network weights. That is to say, the network cannot get any feedback from these easy-to-divide samples.

For the above issues, the paper integrates focal loss into the model. In fact, focal loss is equivalent to the "softened" version of (2). It can be replaced with (4) to obtain a two-class version of focal loss:

$$\lambda(y, \hat{y}) = \begin{cases} \alpha(1 - \hat{y})^\gamma & if \ y = 1 \\ (1 - \alpha)\hat{y}^\gamma & if \ y = 0 \end{cases} \tag{4}$$

Extending (4) to a multi-class version in [6], the focal loss function can be described as,

$$FL(p_t) = -\alpha(1 - p_t)^\gamma \log(p_t) \tag{5}$$

p_t is defined as:

$$p_t = \begin{cases} \hat{y} & if \ y = 1 \\ 1 - \hat{y} & otherwise \end{cases} \tag{6}$$

In (5), for the $1 - p_t$, we know that when a sample is easily misclassified, it corresponds to a difficult sample. In this case, p_t will be very small, and $1 - p_t$

approaches 1, which means focal loss is similar to the original loss function. When a sample can be easily classified correctly, then it corresponds to a simple sample. In this case, p_t is relatively large (close to 1) and $1 - p_t$ approximates 0, which means that this kind of samples will have a smaller contribution to the loss value. Different from (1) and OHEM, the contribution of simple samples to the loss value is small but should not be 0, thus, the weight parameters of these samples can be updated through the forward and back propagation of a network. The hyper-parameter γ, named as the adjustment coefficient, is used to adjust $1 - p_t$ and we usually set it to 2. When $\gamma = 0$, the focal loss degenerates into the original mixed cross-entropy loss function λ. Another hyper-parameter α is to weight the proportion of positive and negative samples and we empirically set $\alpha = 0.5$. Thus, the Eq. (5) can not only adjust the weights of positive and negative samples but also adjust the weights of difficult and simple samples to achieve better results.

Decay-NMS. Based on the Soft-NMS in [22], we design Decay-NMS. The confidence score S_i can be expressed as:

$$S_i = \begin{cases} S_i & IoU(M, b_i) < N_t \\ S_i(1 - IoU(M, b_i))e^{-IoU(M, b_i)} & IoU(M, b_i) \geq N_t \end{cases} \tag{7}$$

In (7), Decay-NMS attenuate the confidence scores of the neighboring detection boxes with the increase of the IoU score rather than set it directly to 0 when the IoU between two detection boxes M and b_i is greater than the threshold. In other words, when the IoU score between two detection boxes is larger, it has great chances to make a mistake, so its corresponding confidence score should be attenuated. When the IoU score is small, its confidence score of a candidate object is not be substantially affected.

3   Experimental Results

3.1   Data Augmentation

There are many flaws in the labelled ship dataset given by BDCI which let down the quality of the dataset, such as unbalanced labels (unbalanced number of labels between different categories), wrong labels, missed labels, empty labels (labels with no objects), repeatable labels (on the same target ship), labels on the incomplete ship bodies and so on. For the unbalanced labels, we implement histogram equalization and Laplacian enhancement on an image as the preprocessing. We correct the wrong labels and delete the empty and repeatable labels. For the missed labels, we annotate the missed ships in the images manually. For a ship with 80% of the ship body appearing in an image, we label it, otherwise, we do not label it.

The final data set contains 25,548 images with 75801 labels in total. Ship images are divided into three categories: cargo ship, cruise ship and yacht. The details of the ship dataset are shown in Table 1. We use 90% of the data set (22,621 images) for training and 10% (2827 images) for testing. The ship detector is trained on a GPU computer with i7-6800 K and NVIDIA GeForce GTX 1080.

Table 1. The ship dataset.

Category name	Cargo ship	Yacht	Cruise ship	
Number of samples	19856	13101	9707	25448
Number of labels	41591	23030	11180	75801

We treat the AP (at a IoU threshold of 0.5 and 0.7) used in the PASCAL object detection challenge as the criteria to evaluate the ship detection performance. Similar as most supervised detection works, mean Average Precision (mAP) score [1, 2] is used as the final evaluation criteria. In our experiments, we use ResNet101 pre-trained on ImageNet as the backbone network. And all the default settings and hyper-parameters are equal to their counterpart used in the original paper [4].

Table 2. The effects of deformable and dilated convolution and deformable PSRoI pooling.

Models	Category & mAP	AP@0.5	AP@0.7
FPN	cargo ship	93.95	85.31
	yacht	92.98	65.47
	cruise ship	96.10	91.41
	mAP	94.34	80.82
Deformable FPN	cargo ship	**94.74**	**87.27**
	yacht	**93.06**	**67.91**
	cruise ship	96.01	**91.72**
	mAP	**94.60**	**82.30**
Deformable FPN + OHEM	cargo ship	95.27	87.22
	yacht	93.95	66.73
	cruise ship	96.57	91.79
	mAP	95.26	81.91
Deformable FPN + focal loss	cargo ship	**95.49**	**87.82**
	yacht	**94.17**	67.33
	cruise ship	**96.79**	**92.39**
	mAP	**95.48**	**82.51**
Deformable FPN + Decay NMS	cargo ship	**95.24**	**87.61**
	yacht	**93.60**	**68.12**
	cruise ship	**96.64**	**92.18**
	mAP	**95.16**	**82.64**
Deformable FPN + Decay NMS + focal loss	cargo ship	**95.72**	**87.93**
	yacht	**94.40**	**67.44**
	cruise ship	**97.02**	**92.50**
	mAP	**95.71**	**82.62**

3.2   Ablation Study

Next, we investigate the effectiveness of the major changes in the proposed approach: deformable and dilated convolution, deformable PSRoI pooling, OHEM and focal loss. Considering the trade-off between GPU memory capacity and image size, images for training and test is normalized with the size of 704 * 704. The detection results are shown in Table 2.

Table 2 shows that deformable and dilated convolution with the deformable PSRoI pooling can make better detection performance with the mAP gain of 0.26% and 1.48% in AP@0.5 and AP@0.7 respectively. Furthermore, OHEM and focal loss with the Deformable FPN can improve APs and mAP with different degrees except the AP of yacht which is abnormal in the Deformable FPN, For OHEM and focal loss, their effects on improving the performance is almost equivalent. However, compared to OHEM, using focal loss saves nearly 4 to 5 h in training. The faster training speed means that we can upgrade the algorithm several times, and the faster test speed is more conducive to practical applications. Finally, after combining all our changes together, the mAP increases by 1.37% and 1.8% in AP@0.5 and AP@0.7 respectively. The Fig. 4 shows that the detection accuracy of this algorithm is higher than that of the original FPN under high recall. It further demonstrates the robustness of our algorithms. Figure 5 shows some of the experimental results. We can see that in different weather conditions, our model has excellent performance.

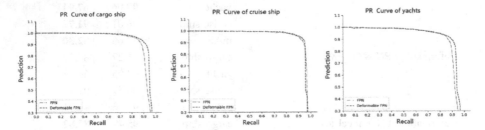

Fig. 4. The PR Curve of three categories in FPN and our model.

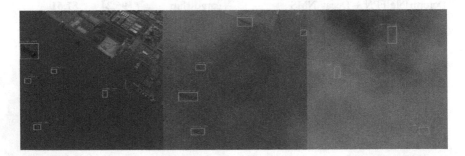

Fig. 5. Some experimental results in three types of weather. From left to right are the detection results with good weather, the fog day, and the cloud day respectively.

4 Conclusions

In this paper, we propose a deformable FPN for ship recognition. We introduce deformable convolution and dilated convolution to the backbone of Resnet-101 of the original FPN. Moreover, in the RoI-wise subnetwork, the fixed geometric RoI pooling is replaced by the deformable PSRoI pooling. For further improving the ship recognition, the focal loss and OHEM are used in the optimal problem in order to trade off the unbalance between the positive and negative samples as well as the degree of difficulties in classification. At last, Decay-NMS is designed to further improve the detection accuracy. The experimental results show that our model has strong robustness and excellent detection performance under different weather conditions.

References

1. Everingham, M., Gool, L., Williams, C.K., Winn, J., Zisserman, A.: The pascal visual object classes (VOC) challenge. Int. J. Comput. Vis. **88**(2), 303–338 (2010)
2. Lin, T.-Y., et al.: Microsoft COCO: common objects in context. In: Fleet, D., Pajdla, T., Schiele, B., Tuytelaars, T. (eds.) ECCV 2014. LNCS, vol. 8693, pp. 740–755. Springer, Cham (2014). https://doi.org/10.1007/978-3-319-10602-1_48
3. Ren, S., He, K., Girshick, R., Sun, J.: Faster R-CNN: towards real-time object detection with region proposal networks. IEEE Trans. Pattern Anal. Mach. Intell. **39**(6), 1137–1149 (2017)
4. Lin, T.Y., Dollár, P., Girshick, R., He, K., Hariharan, B., Belongie, S.: Feature pyramid networks for object detection. arXiv:1612.03144 (2016)
5. He, K., Zhang, X., Ren, S., Sun, J.: Deep residual learning for image recognition. In: Proceedings of the IEEE Conference on Computer Vision and Pattern Recognition (CVPR), Las Vegas, pp. 770–778 (2016)
6. Lin, T.Y., Goyal, P., Girshick, R., He, K., Dollár, P.: Focal loss for dense object detection. In: IEEE International Conference on Computer Vision (ICCV), Venice, pp. 2999–3007 (2017)
7. Liu, W., Anguelov, D., et al.: SSD: single shot multibox detector. In: Leibe, B., Matas, J., Sebe, N., Welling, M. (eds.) ECCV 2016. LNCS, vol. 9905, pp. 21–37. Springer, Cham (2016). https://doi.org/10.1007/978-3-319-46448-0_2
8. Redmon, J., Divvala, S., Girshick, R., Farhadi, A.: You only look once: unified, real-time object detection. In: Proceedings of the IEEE Conference on Computer Vision and Pattern Recognition (CVPR), pp. 779–788 (2016)
9. Neubeck, A., Gool, L.V.: Efficient non-maximum suppression. In: 18th International Conference on Pattern Recognition (ICPR 2006), Hong Kong, pp. 850–855 (2006)
10. Shrivastava, A., Gupta, A., Girshick, R.: Training region-based object detectors with online hard example mining. In: 2016 IEEE Conference on Computer Vision and Pattern Recognition (CVPR), Las Vegas, pp. 761–769 (2016)
11. Dai, J., Qi, H., Xiong, Y., Li, Y., Zhang, G., Hu, H., et al.: Deformable convolutional networks. In: IEEE International Conference on Computer Vision (ICCV), Venice, pp. 764–773 (2017)
12. Yu, F., Koltun, V.: Multi-scale context aggregation by dilated convolutions. In: International Conference on Learning Representations (2016)

13. Bodla, N., Singh, B., Chellappa, R., Davis, L.S.: Soft-NMS—improving object detection with one line of code. In: IEEE International Conference on Computer Vision (ICCV), Venice, pp. 5562–5570 (2017)
14. http://www.datafountain.cn/projects/2017CCF/
15. Yu, F., Koltun, V., Funkhouser, T.: Dilated residual networks. In: IEEE Conference on Computer Vision and Pattern Recognition (CVPR), Honolulu, pp. 636–644 (2017)
16. Li, Y., He, K., Sun, J., et al.: R-FCN: Object detection via region-based fully convolutional networks. In: Advances in Neural Information Processing Systems, pp. 379–387 (2016)
17. Russakovsky, O., et al.: ImageNet large scale visual recognition challenge. Int. J. Comput. Vis. **115**(3), 211–252 (2015)

Learning Hierarchical Context for Action Recognition in Still Images

Haisheng Zhu[1], Jian-Fang Hu[1,2,3(✉)], and Wei-Shi Zheng[1,3]

[1] School of Data and Computer Science, Sun Yat-sen University, Guangzhou, China
hyesunzhu@outlook.com, hujf5@mail.sysu.edu.cn, wszheng@ieee.org
[2] Guangdong Province Key Laboratory of Computational Science, Guangzhou, China
[3] Key Laboratory of Machine Intelligence and Advanced Computing, MOE,
Guangzhou, China

Abstract. Recognizing actions from still images is challenging due to the lack of movement information. Most of the existing works intend to characterize actions by the interaction context between each pair of features extracted from local image regions, which often fails to capture the complex structures in actions. Different from these works, in this paper we propose to divide the human body into a set of increasingly finer body parts, forming our hierarchical composition. To model the interaction patterns among these body parts, we further develop the hierarchical propagation network. By propagating information bottom-up in the composition, our model efficiently mines the interaction among local body parts, and integrates those discriminative context cues hierarchically into a compact action representation. Our experiments on the HICO and VOC 2012 Action datasets demonstrate the efficiency of our method for characterizing static actions.

Keywords: Action recognition · Hierarchical context
Human-object interaction

1 Introduction

Recognizing actions from still images is an important computer vision task with many applications in the field of image caption [1], image analysis [19], collective activity recognition [11], human-computer interactions [16], etc. Indeed it is very challenging to infer actions from observed still images without any motion cues which are found to be useful in the recognition of actions.

Many works believed that actions can be inferred by only observing the actor's poses. [13,14,17,20,22,24,25,28] proposed methods to model human's appearances and poses, generating more robust features. The main focus of these methods is to leverage useful foreground information, and build an effective connection between human poses and corresponding action categories. However, different actions may be performed with similar poses, and these approaches ignored the texture context in their modeling, which made them less applicable to recognize complex actions.

© Springer Nature Switzerland AG 2018
R. Hong et al. (Eds.): PCM 2018, LNCS 11166, pp. 67–77, 2018.
https://doi.org/10.1007/978-3-030-00764-5_7

(a) using computer   (b) playing instru-   (c) phoning   (d) playing tennis
ment

Fig. 1. Example images of different actions. As shown in (a) and (b), the local body parts could contain some informative cues for characterizing actions. (c) illustrates that some actions can be recognized from only the region around a joint (the hand). (d) shows that for some actions can be identified by observing higher level body parts (the arms and even the upper body, etc). Thus, methods focusing on only a single level of body parts are not enough. Hierarchical methods combining different levels are needed.

Recent researches show that some local static contextual cues can be explored for characterizing actions. Many researchers make great efforts on mining such local cues [2,12,15,26,27,29]. The most popular way is to treat different human body parts as separate action components, extract useful action cues from these components and finally combine them by concatenation. This proves to be more effective than methods focusing only on human poses, as the local regions around each body part could include some important action contexts, such as objects and hand gestures, etc.

However, it remains a problem how to combine these local cues. Those methods mentioned above intend to treat the contextual cues gained from different local regions (body parts for example) independently, or simply explore interactions of pair-wise regions. This is not the best for characterizing human actions, because an action is a complex system where contextual cues actively interact with each other (see Fig. 1). In this paper, we propose to explicitly explore the interactions among local parts for action reasoning. Specifically, we divide the upper body into increasingly finer body parts, and then propose a hierarchical propagation network to integrate the information gained in each body part. In our model, information contained in local parts are coupled and propagated bottom-up, from the smaller parts to the larger parts, forming a hierarchical structure. At the top of this hierarchical structure, cues are integrated and produce a compact action representation. As a result, discriminative cues from local body parts are more efficiently mined and integrated, compared to other existing models.

Our contribution is three-fold: (1) A hierarchical part composition with increasingly finer body parts for capturing local action context; (2) A hierarchical propagation network for integrating the cues gained in the body parts in a recursive manner; (3) Extensive experiments on the HICO [3] and VOC Action [5] datasets for demonstrating the efficiency of the proposed method.

2   Related Work

In this section, we briefly review the development still image action recognition. We mainly focus on methods that explores hierarchical modeling for human action recognition and developed models based on individual human body part.

Hierarchical Action Recognition. Recent works found that human actions can be factorized into action units, and thus proposed hierarchical structures to represent an action. [2] extracted informative action cues from the regions of human body and manipulated objects. [15] aimed to infer actions using the information depicted in different human body parts. [27] found that objects and body parts can serve as mutual context to each other, and proposed a random field model to encode that context information. [26] proposed a new image representation, "grouplet", to encode the visual features and their spatial configurations, yielding the structured information of an image. [29] realized that local contextual cues of different body parts characterize different action units, which can be used to constitute the whole action. However, it needs the users to define the atomic actions of each body part, which make it less applicable in practice.

Human Pose Based Methods. Originally, images for action recognition only contain a single person performing specific actions. At this stage, an amount of prior work exploited the key role that human poses play in the image. [20] proposed "poselet activation vector" to represent human poses, which was robust to occlusion. [24] utilized linear programming relaxation to compute the distance of image pairs, matched by the shape of human. [25] treated poses as latent variables. Some other works leveraged discriminative templates [22,28], color [17], or exemplars [13].

Another line of works leveraged successful object detectors such as [4,7,18], and analyzed the relationship between detected humans and objects. For example, [8] utilized an object detector to detect humans and objects. Using the human poses as guidance, their method then located the most likely object and perform recognition on that human object pair. These methods are more object oriented, while ours focuses more on the human, and incorporates the object implicitly. Hopefully our framework can be easily extended with object detection mechanisms.

3   Approach

In this paper, we aim to mine informative action contexts from a set of increasingly finer body parts. A graphic illustration of our model is presented in Figs. 2 and 3. In the following, we first describe how to construct human body parts in a hierarchical manner and then introduce our model for learning hierarchical action context.

3.1   Hierarchical Part Composition

Here, we define the hierarchical part composition as a set of increasingly finer body parts collected in a pyramid manner. It is constructed by iteratively split

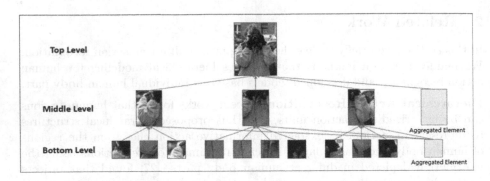

Fig. 2. An example for the construction of our hierarchical part composition.

the body parts into smaller parts (e.g. upper body can be split into arms, the head, and the torso etc.). Let's denote the composition as $A = \{H_i\}, i = 1, 2, \ldots, I$, where H_i indicates the set of body parts for the i-th level in the composition and I is the total number of levels. Specifically, H_1 have the largest body parts, which serves as a root node of the whole hierarchical composition. In this paper, the part in the first level is the upper body. Each node (body part) in H_i is split into smaller parts, forming the nodes of H_{i+1}. And nodes in H_I contains the smallest body parts. An example of our hierarchical part composition can been seen in Fig. 2. As shown, each level captures action context from different scales and perspectives.

The advantage of using hierarchical part composition for action modeling is three-fold. First, the part composition can provide a comprehensive description about the action, from coarse to fine and local to global perspectives. Second, the body parts in the composition could also capture some useful context like manipulated objects or scene. Third, the complex interactions among different parts can be easily modeled based on the composision.

3.2   Hierarchical Propagation Network

Here we describe in detail how information is propagated within the composition A. Instead of directly combining the cues by concatenation, we propose a hierarchical propagation model to generate a compact feature representation. As illustrated in Fig. 3 (middle part), information from nodes in H_I is propagated to their corresponding parent nodes in H_{I-1}, forming the integrated information at those parent nodes. Again this integrated information is propagated to those parent nodes in H_{I-2}. Recursively, information in A is propagated to the node in H_1, the root node of A, forming the final feature representation.

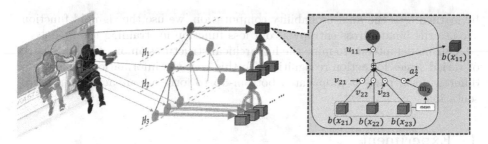

Fig. 3. A graphical illustration of our proposed hierarchical propagation network. The left part is an example of body part division. The middle part shows the formation of the hierarchical part composition (where $I = 3$) and the procedure of information propagation in the network. The right part is the concrete mathematical explanation which computes the integrated information $b(x_{11})$.

Specifically, denote the j-th node in H_i as $\mathbf{x}_{ij}$, the information gained at it is modeled as (see Fig. 3 for an example, where $i = 1$ and $j = 1$):

$$b(\mathbf{x}_{ij}) = \mathbf{u}_{ij} \odot \mathbf{x}_{ij} + \sum_{k \in S_{ij}} \mathbf{v}_{i+1,k} \odot b(\mathbf{x}_{i+1,k}) + \mathbf{a}_{i+1}^{j} \odot \mathbf{m}_{i+1},$$

$$\mathbf{m}_{i+1} = \frac{1}{K} \sum_{k} b(\mathbf{x}_{i+1,k}), \qquad i = 1, ..., I - 1,$$

(1)

where $\mathbf{x}_{i+1,k}$ is the k-th node in H_{i+1}. As described before, information of $\mathbf{x}_{i+1,k}$ only propagates to its parent node. This is enforced by S_{ij}, where $k \in S_{ij}$ indicates that $\mathbf{x}_{i+1,k}$ is the child node of $\mathbf{x}_{ij}$. $\mathbf{m}_{i+1}$ represents the information aggregated from H_{i+1} and K is the number of nodes in H_{i+1}. The use of $\mathbf{m}_{i+1}$ enables our model to capture some minor relationship among different body structures (lefthand and righthand, etc). $\mathbf{u}_{ij}$, $\mathbf{v}_{i+1,k}$ and $\mathbf{a}_{i+1}^{j}$ are model parameters used to fuse information of the current node, information propagated from its children nodes and information aggregated from H_{i+1}, respectively, which would be learned from training images. $\odot$ represents element-wise multiplication.

For the bottom level (i.e., the I-th level), $b(\mathbf{x}_{Ij}) = \mathbf{u}_{Ij} \odot \mathbf{x}_{Ij}$. Then, all the $b(\mathbf{x}_{ij})$ are defined in a recursive manner. It integrates information from the current node $\mathbf{x}_{ij}$ and the nodes from the lower level that are related to $\mathbf{x}_{ij}$. Thus, $b(\mathbf{x}_{11})$ encodes all the contextual information from all the nodes in the hierarchical composition. It forms a compact representation for actions in still images.

Finally, we feed the compact representation $b(\mathbf{x}_{11})$ into a fully connected layer along with additional information to obtain a final recognition score S:

$$S = \Theta^{T}(\mathbf{w}_{\alpha} \odot b(\mathbf{x}_{11}) + \mathbf{w}_{\beta} \odot \mathbf{f}_{\mathbf{I}}),$$

(2)

where $\mathbf{w}_{\alpha}$ and $\mathbf{w}_{\beta}$ are model parameters to be learned. $\mathbf{f}_{\mathbf{I}}$ is the feature of the whole image. Θ is the parameter of the last *fully connected layer*, which would be learned. For the final probability computation, we use the sigmoid function. We use the binary cross entropy as our loss function for training the model.

Our model intends to mine the hierarchical structures among the body parts of varied scales for action recognition. To this end, the information gained from different local parts is propagated bottom-up to form a compact action representation.

4   Experiment

We evaluated our methods on two still image action datasets: HICO [3] and PASCAL VOC Action dataset [5]. In the following, we'll briefly introduce the implementation details and then describe our experimental results.

4.1   Implementation Details

In our experiment, we divide the human body into 3 levels, which means $I = 3$. For H_1, we choose the upper body as the only node. Here, we don't use the lower part since most images don't contain the lower body parts. For H_2, we divide the upper body into 3 parts, the trunk (including the head, the neck and the torso), the lefthand and the righthand. For H_3, we further split the body parts in H_2 to 10 skeleton joints, which correspond to the body parts obtained by the pose estimator in [6]. To extract image patches for body parts in H_1 and H_2, we crop the rectangle region that surrounds the corresponding joints, extended by 10 pixels in each direction to contain more context. To extract image patches for the joints in H_3, we crop a 64×64 region around the joint. All these patches are resized to 224×224. We train a ResNet-50 [10] for each body part with the corresponding patches. All the networks are pretrained on ImageNet [21] and then finetuned for each part. We use SGD and set the learning rate as 10^{-4}, decreased by a factor of 0.1 every 10 epochs, for the parameters of ResNet-50. *The results output by the last FC layers are used as the feature representations, which are C-dim prediction score vectors (C is the number of classes).*

The CNN features of the body parts are further fed to our hierarchical propagation network, so that each node x_{ij} in the propagation network represents the feature extracted from its corresponding part. We use SGD to update the model parameters of the propagation network, where the initial learning rate is set as 10^{-4}, decreased by 0.1 every 10 epochs.

4.2   Results on HICO Dataset

The HICO dataset is a large scale dateset for the research of action recognition in still images, which contains 38116 training images and 9658 testing samples from 600 action classes. This dataset is challenging as each of the actions involves a kind of interaction between human and certain objects. And an image could corresponds to more than one action classes.

Table 1. Comparison results on the HICO dataset.

Method	mAP
HOCNN [3]	4.9
SVM (V) [3]	17.7
SVM (O) [3]	19.4
SVM (VO) [3]	18.1
R*CNN [9]	28.5
Scene-RCNN [9]	29.0
ResNet-whole	27.8
Ours	**29.3**

Table 2. The influence of the number of levels in our hierarchical part composition.

Method	mAP
ResNet-torso-joint-1	7.3
ResNet-lefthand-joint-1	8.0
ResNet-trunk	10.7
ResNet-lefthand	8.1
ResNet-body	17.6
Ours (2 levels)	**28.9**
Ours (3 levels)	**29.3**

The results on the HICO dataset are presented in Table 1. We also report the result obtained by the baseline using ResNet-50 network trained on the whole images, which is denoted as *ResNet-whole* in the table. As can be seen, our approach outperforms all of the competitors and obtains a mAP of 29.3 on this set. Compared with the system developed in [3], which intends to finetune the DNN on either the verbs (denoted as **V**), the objects (denoted as **O**) or directly on the human-object interaction (denoted as **VO**), our system even has a gain of 10.6%. This demonstrates that exploring the massive complex interactions among local parts using our propagation network can benefit the recognition. We also compare our model with the one proposed in [9] (Denoted as **R*CNN** and **Scene-RCNN** in Table 1). These methods compute the average sum over the action scores of the human bounding box and an Region of Interest (RoI). In the R*CNN version this RoI is explicitly discovered whereas in the Scene-RCNN version this RoI is the whole image. As for our system, we focus on local body parts as our region of interest (ROI), and mine the hierarchical structures between them. Our system outperforms **R*CNN** and **Scene-RCNN** by 0.8% and 0.3%, respectively. We conclude that combination of our part composition and hierarchical propagation model can mine more informative action cues.

Table 3. Comparison results on the PASCAL VOC Action 2012 val set.

Method	mAP
RCNN [9]	84.9
Scene-RCNN [9]	85.7
R*CNN [9]	**87.9**
Semantic Parts [29]	84
ResNet-bb	85.5
Ours	86.4

Table 4. APs for the classes in the VOC Action 2012 val set.

Classes	Jumping	Phoning	Playing Instrument	Reading	Riding Bike
R*CNN	88.9	79.9	95.1	82.2	96.1
Ours	87.9	75.5	91.3	75.4	**97.3**

Classes	Riding Horse	Running	Taking Photo	Using Computer	Walking
R*CNN	97.8	87.9	85.3	94.0	71.5
Ours	**99.1**	**90.6**	82.8	87.7	**73.2**

We also study the influence of the number of levels in our hierarchical part composition (see Table 2). The second block of Table 2 presents the results of using different numbers of hierarchical levels. We see that using more levels yields better performance. From this we observe that, low-level local cues are also important for action recognition, since they're better at capturing rich details. On the other hand, local cues of different levels are of different granularity, and our hierarchical propagation network is capable of discovering the interactions among the parts of different levels.

4.3   Results on VOC 2012 Dataset

The PASCAL VOC Action dataset consists of 10 different actions, i.e. *Jumping, Phoning, Playing Instrument, Reading, Riding Bike, Riding Horse, Running, Taking Photo, Using Computer* and *Walking*. This set contains some samples where the actors of interest are not performing any of the actions specified above, which are annotated as action *Other*. Bounding boxes indicating the actors of interest are manually annotated in this set, and the image patches corresponding to the bounding boxes are treated as action samples for training and testing. Therefore, this set has a total of 6278 samples for the model training (train + val) and 6283 for testing. The mAP is computed for evaluation.

Table 3 presents comparison results on the VOC Action validation set. As shown, our system can obtain a better performance than Semantic Parts [29], which even need the users to additionally define the atomic actions of each body part. By this we claim that, action cues in local body parts are related in a hierarchical manner, and mining structures among them is beneficial. We also observe that our model performs worse than the R*CNN [9] on this set. We suspect that his is because the VOC dataset is relatively small, which is not enough for our hierarchical propagation network to fit. Further, the R*CNN is more object oriented, which uses Selective Search [23] and ROI pooling [7] for object capturing. In contrast, our method intends to implicitly capture the object information from the local region around each body part. However, by directly comparing the per-class APs in Table 4, we can find that our system obtains better results for the actions where human are performing discriminative and cyclic actions. For example, *Riding Bike* and *Walking*. This may be because in such actions, body parts include more discriminative local action cues, and these cues have stronger implication for the whole action. This observation supports our claim that action cues are propagated bottom-up to form the action representation. Whereas for other action categories, local regions of background or objects are more meaningful, and thus the object detection mechanism of R*CNN is more capable at recognizing them.

5 Conclusion

In this paper, we propose to divide the human body into a hierarchical part composition, which consists of gradually finer body parts. Further, we define the hierarchical propagation network to intergrate local contextual cues recursively. Compared with methods that treat local cues parallel and independent, our model allows local cues to actively interact with each other, so that information is coupled and propagated bottom-up. Our experiments on the HICO dataset and VOC 2012 Action dataset demonstrate that our model can capture action context more efficiently and is able to boost performance on action recognition.

Acknowledgements. This work was supported by the NSFC (No. 61702567, 61522115, 61472456, 61628212).

References

1. Bernardi, R., et al.: Automatic description generation from images: a survey of models, datasets, and evaluation measures. J. Artif. Intell. Res. **55**, 409–442 (2016)
2. Chao, Y.W., Liu, Y., Liu, X., Zeng, H., Deng, J.: Learning to detect human-object interactions (2017)
3. Chao, Y.W., Wang, Z., He, Y., Wang, J., Deng, J.: HICO: a benchmark for recognizing human-object interactions in images. In: 2015 IEEE International Conference on Computer Vision (ICCV) (2015)
4. Dai, J., Li, Y., He, K., Sun, J.: R-FCN: Object detection via region-based fully convolutional networks. In: Neural Information Processing Systems (2016)
5. Everingham, M., Van Gool, L., Williams, C.K., Winn, J., Zisserman, A.: The pascal visual object classes (VOC) challenge. Int. J. Comput. Vis. **88**(2), 303–338 (2010)
6. Fang, H., Xie, S., Tai, Y.W., Lu, C.: RMPE: regional multi-person pose estimation. arXiv preprint arXiv:1612.00137 (2016)

7. Girshick, R.B.: Fast R-CNN. In: 2015 IEEE International Conference on Computer Vision (ICCV) (2015)
8. Gkioxari, G., Girshick, R., Dollár, P., He, K.: Detecting and recognizing human-object interactions. arXiv preprint arXiv:1704.07333 (2017)
9. Gkioxari, G., Girshick, R., Malik, J.: Contextual action recognition with R* CNN. In: Proceedings of the IEEE International Conference on Computer Vision, pp. 1080–1088 (2015)
10. He, K., Zhang, X., Ren, S., Sun, J.: Deep residual learning for image recognition. In: Proceedings of the IEEE Conference on Computer Vision and Pattern Recognition, pp. 770–778 (2016)
11. Herath, S., Harandi, M., Porikli, F.: Going deeper into action recognition: a survey. Image Vis. Comput. **60**, 4–21 (2017)
12. Hu, J., Zheng, W., Lai, J., Zhang, J.: Jointly learning heterogeneous features for RGB-D activity recognition. IEEE Trans. Pattern Anal. Mach. Intell. **39**(11), 2186–2200 (2017)
13. Hu, J.F., Zheng, W.S., Lai, J., Gong, S., Xiang, T.: Recognising human-object interaction via exemplar based modelling. In: 2013 IEEE International Conference on Computer Vision (2013)
14. Hu, J.F., Zheng, W.S., Lai, J., Gong, S., Xiang, T.: Exemplar-based recognition of human-object interactions. IEEE Trans. Circuits Syst. Video Technol. **26**(4), 647–660 (2016)
15. Ikizler, N., Cinbis, R.G., Pehlivan, S., Duygulu, P.: Recognizing actions from still images. In: 2008 19th International Conference on Pattern Recognition (2008)
16. Jaimes, A., Sebe, N.: Multimodal human-computer interaction: a survey. Computer Vis. Image Understand. **108**(1–2), 116–134 (2007)
17. Khan, F.S., Anwer, R.M., van de Weijer, J., Bagdanov, A.D., Lopez, A.M., Felsberg, M.: Coloring action recognition in still images. Int. J. Comput. Vis. **105**(3), 205–221 (2013)
18. Liu, W., et al.: SSD: single shot multibox detector. In: Leibe, B., Matas, J., Sebe, N., Welling, M. (eds.) ECCV 2016. LNCS, vol. 9905, pp. 21–37. Springer, Cham (2016). https://doi.org/10.1007/978-3-319-46448-0_2
19. Lu, D., Weng, Q.: A survey of image classification methods and techniques for improving classification performance. Int. J. Remote Sens. **28**(5), 823–870 (2007)
20. Maji, S., Bourdev, L.D., Malik, J.: Action recognition from a distributed representation of pose and appearance. In: CVPR 2011 (2011)
21. Russakovsky, O., et al.: Imagenet large scale visual recognition challenge. Int. J. Comput. Vis. **115**(3), 211–252 (2015)
22. Sharma, G., Jurie, F., Schmid, C.: Expanded parts model for human attribute and action recognition in still images. In: 2013 IEEE Conference on Computer Vision and Pattern Recognition (2013)
23. Uijlings, J.R., Van De Sande, K.E., Gevers, T., Smeulders, A.W.: Selective search for object recognition. Int. J. Comput. Vis. **104**(2), 154–171 (2013)
24. Wang, Y., Jiang, H., Drew, M.S., Li, Z.N., Mori, G.: Unsupervised discovery of action classes. In: 2006 IEEE Computer Society Conference on Computer Vision and Pattern Recognition (CVPR 2006) (2006)
25. Yang, W., Wang, Y., Mori, G.: Recognizing human actions from still images with latent poses. In: 2010 IEEE Computer Society Conference on Computer Vision and Pattern Recognition (2010)
26. Yao, B., Fei-Fei, L.: Grouplet: a structured image representation for recognizing human and object interactions. In: 2010 IEEE Computer Society Conference on Computer Vision and Pattern Recognition (2010)

27. Yao, B., Fei-Fei, L.: Modeling mutual context of object and human pose in human-object interaction activities. In: 2010 IEEE Computer Society Conference on Computer Vision and Pattern Recognition (2010)
28. Yao, B., Khosla, A., Fei-Fei, L.: Combining randomization and discrimination for fine-grained image categorization. In: CVPR 2011 (2011)
29. Zhao, Z., Ma, H., You, S.: Single image action recognition using semantic body part actions. arXiv preprint arXiv:1612.04520 (2016)

iMakeup: Makeup Instructional Video Dataset for Fine-Grained Dense Video Captioning

Xiaozhu Lin, Qin Jin[✉], Shizhe Chen, Yuqing Song, and Yida Zhao

Multimedia Computing Lab, School of Information, Renmin University of China, Beijing, China
{linxz,qjin,cszhe1,syuqing,zyiday}@ruc.edu.cn

Abstract. Generating natural language descriptions for short videos has made fast progress due to the advances in deep learning and various annotated datasets. However, automatic analysis, understanding and learning from long videos remain very challenging and request more exploration. To support investigation for this challenge, we introduce a large-scale makeup instructional video dataset named iMakeup. This dataset contains 2000 videos which covers 50 popular topics for makeup, amounting to 256 h, with 12,823 annotated clips in total. This dataset contains both visual and auditory modalities with a large coverage and diversity in the specific makeup domain. We further extend existing long video understanding techniques to present the feasibility of our dataset, showing the results of baseline video segmentation and caption models. We expect this dataset to support research works in various problems such as video segmentation, video dense captioning, object detection and tracking, action tracking, learning for fashion, etc.

Keywords: Large-scale dataset · Makeup · Dense video captioning

1 Introduction

Automatically describing images or videos with natural language sentences has received significant attention in recent years [1–3]. The increasing availability of large-scale image or video datasets [4,6,9] is one of the key supporting factors to the rapid progress on the challenging captioning problems. For video captioning, most of the works have focused on generating a single caption sentence for short videos [1,9]. However, using a single sentence cannot well recognize or articulate numerous details within long videos, like user-uploaded instructional videos of complex tasks (how to tie a tie, how to prepare a dish, etc.) on the internet. We humans learn from this kind of videos relying on these details. Hence, challenging tasks such as dense video captioning [2,3,10], which that aim to simultaneously describe all detected contents within a long video with multiple natural language sentences, have attracted increasing attentions.

© Springer Nature Switzerland AG 2018
R. Hong et al. (Eds.): PCM 2018, LNCS 11166, pp. 78–88, 2018.
https://doi.org/10.1007/978-3-030-00764-5_8

Given that few large-scale long video datasets are available for this task, we collect a large-scale instructional video dataset in the specific makeup domain, which is named iMakeup. Makeup tutorials are popular on commercial website such as Youtube which people rely on to learn how to do makeup. In such a tutorial video, the makeup artist or vlogger is always in the viewfinder and the camera is focusing on her/his face. Also, makeup sometimes requires very small, precise movements, which makes detection and tracking fine-grained actions challenging. Makeup involves explicit steps and different cosmetics used in each step, which makes it intriguing to investigate automatic techniques for makeup procedure learning, fine-grained object detection, and dense captioning tasks.

We collect the iMakeup dataset from Youtube, which consists of 2000 makeup tutorial videos from top 50 popular makeup topics, amounting to 256 h in total. Each video is segmented into multiple clips with temporal boundary and caption annotations according to the makeup procedures, resulting to 12,823 clips in total. To the best of our knowledge, this dataset is the first large-scale long video dataset in makeup domain with both temporal boundaries and manual annotations for video segments.

For long videos, one single sentence suffers from describing all procedural and logical information, while this is the crucial part to understand instructional videos. Hence, captioning of this kind of video requires combination of several modalities, like video segmentation, procedure analysis, concept detection, etc. We define our dense captioning task into three steps as following: (1) We first segment a complex long video into several short video clips, with one specific action in each clip; (2) We then generate the natural language sentences of video clips; (3) Finally we connect all captions into a long paragraph narration as the description of the whole video.

We then conduct several baseline experiments to validate the availability of our dataset. We first adopt segment-based convolutional networks (SCNNs) to conduct temporal segmentation on long videos. We also conduct the experiments on action proposal leveraging sliding window. Next, we generate captions of video clips based on commonly-used encoder-decoder framework. The encoder extracts multi-modal features, encoding to fixed-length vectors. While the decoder applies simple recurrent neural network (RNN).

2   Related Work

Video Datasets. Early works in computer vision have closely tied connection with large-scale image datasets, like ImageNet [4], MSCOCO [13], etc. Having access to such datasets enabled the research community to improve deep neural models to achieve significant effects. Recently, with increasing attention on video domain, researchers have proposed a large number of video datasets, like UCF101 [14], Kinetics [15] and Youtube-8M [16] for classification with videos crawled from the Internet. Datasets like MSVD [17], MSR-VTT [9] contain multiple topics and natural language captions. ActivityNet [18] and MomentsInTime [19] can help researchers to conduct activity detection tasks.

Datasets mentioned above suffer from limited video duration. For untrimmed long videos, there are YouCook [5], YouCookII [20], TACos [21], etc. related to cooking and food. Rohrbach et al. using classic HD movies as raw videos to collect a large-scale annotated video dataset MPII-MD [6]. We can find that long video datasets for fine-grained dense captioning mostly consist of logical procedures and distinguished activities. Hence, we believe makeup instructional videos has significant leverage for dense captioning.

Video Segmentation. A substantial topic related to fine-grained video captioning is video segmentation. It considers the task of segmenting a long video into continuous segments corresponding to a series of sequential procedures. A simple method is to determine break point using absolute difference of pixels color intensities between two adjacent frames [22]. Obviously, this way suffers from segmenting videos which consist of contiguous frames with rare visual changes. Thus, researchers constantly addressed a large number of methods like color histogram comparisons [24], extracting edge [25] and motion features [26], etc.

Recently, Huang et al. [27] leverage frame-wise visual similarities to perform match and alignment of actions and temporal boundaries. Kuehne et al. [28] adopt HMM to segment long videos, using hidden action states to learn likelihood of image features. Besides, Shou et al. [11] proposed a segment-based 3D convolutional network named Segment CNN (SCNN). Zhou et al. [10] suggested a segment-level recurrent network to train relationship between video clips and generate segmentation according to the sequence of procedures.

Video Captioning. Another line of work significantly impressing dense video captioning is short video captioning. Short video here means video clips which only contain one specific action each. For this kind of video captioning, Venugopalan et al. [1] proposed a baseline method through mean pooling of frame-wise visual features based on image captioning. Recently, researchers proposed a great many method, like attention mechanism [2], hierarchical networks [28], multi-modal fusion [29], etc.

3 iMakeup Dataset

3.1 Collection and Annotation

We used category "makeup" on WikiHow [7] to obtain the most popular queries that the internet users used in makeup domain. We then discarded repetitive or extraneous queries, which leads to 50 popular queries in makeup domain, shown in Fig. 1. With each query, we crawled YouTube and obtained the top 40 videos. These makeup videos are instructional videos with distinguishable text descriptions, like narrations, comparing with common videos. Each video contains 2–20 procedure steps. We therefore target at creating annotations of temporal boundaries for each step and text descriptions of the procedure for each step. An annotated example is shown in Fig. 2. We recruit a dozen young ladies to be the annotators, provided with raw videos, a brief task description, a makeup vocabulary, and several annotation examples. For each raw video, they are asked to segment the whole video into clips according to the makeup procedure and annotate the start and end time of each clip. For each clip, annotators are also required to write an English sentence to describe the procedure. Hence, for each step in a makeup instructional video, we generate its unique time boundaries and caption. Resembling most instructional video datasets (e.g. YouCook [5], MPII-MD [6], etc.), We annotate each video with one annotator due to time and resource constraints. And we make more captions available for future search.

1	How to apply makeup to dry skin	26	how to do korean "k-pop" style makeup
2	how to cover up a bruise	27	how to take day makeup into night
3	how to cover vitiligo patches with makeup	28	how to apply makeup for a dinner meeting
4	how to apply bb cream	29	how to apply makeup for a glamour photography shoot
5	how to apply makeup for dark indian skin	30	how to apply makeup for a morning meeting
6	how to apply makeup for middle eastern skin tones	31	how to apply makeup for a professional setting
7	how to apply makeup for dark skin (girls)	32	how to apply makeup for prom
8	how to do makeup on a copper toned redhead	33	how to choose the best bridal makeup artist
9	how to sculpt your face with makeup	34	how to do a St. Patrick's Day makeup look
10	how to apply dark eyeshadow	35	how to do rave makeup
11	how to apply eyelid primer	36	how to look pretty in pink for Valentine's Day
12	how to apply eye makeup (for women over 50)	37	how to apply makeup in 10 minutes
13	how to apply subtle eyeshadow	38	how to avoid carrying too much makeup
14	how to get pink lips at home	39	how to bake your makeup
15	how to make lipstick matte	40	how to care for your permanent makeup procedure
16	how to use makeup to make nose look smaller	41	how to choose and start wearing makeup
17	how to use makeup to make eyes appear larger	42	how to choose a makeup bag
18	how to apply bright makeup	43	how to compare makeup brands
19	how to apply goth barbie makeup	44	how to have great makeup on a budget
20	how to choose fall makeup colors	45	how to contour without looking fake
21	how to apply makeup to look more masculine	46	how to clean an eyelash curler
22	how to apply matte makeup	47	how to get dark makeup without over doing it
23	how to apply strobing makeup	48	how to keep makeup from melting or fading
24	how to create a no makeup makeup look	49	how to make makeup last all day
25	how to do a makeup parody	50	how to remove makeup

Fig. 1. Top 50 popular makeup topics

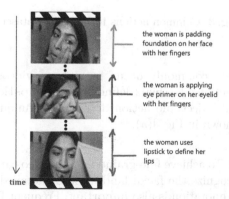

Fig. 2. An annotation example of iMakeup dataset.

3.2   Dataset Statistics

The dataset contains 2000 makeup instructional videos from 50 most popular makeup topics, with 40 videos for each topic. The total video length is about 256 h with an average duration of 7.68 min per video. Each video is manually segmented into multiple clips with temporal boundary and caption annotations according to the makeup procedures. There are 12,823 annotated clips in total. All video clips are temporally localized and described in complete English sentences. The average length of annotated sentences is 11.29 words. The total vocabulary size is around 2183 words.

Actions: The most frequent action word used in captions is "apply". Some specific actions like "pad", "dab", "brush", and "define" occur in less videos. Since the distribution of action vocabulary is quite biased, we then consider "Verb+Object" pairs as fine-grained actions in subsequent work. Common actions are shown in Fig. 3.

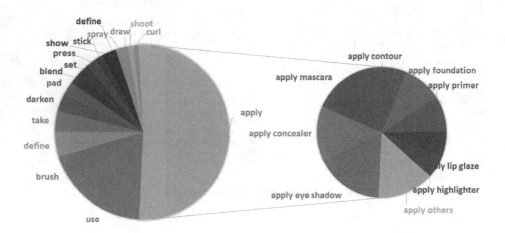

Fig. 3. Common actions in iMakeup dataset.

Cosmetics: They are commonly occur in makeup videos as action objects (apply **mascara**) or action adverbial (define lips using **lipstick**). They pose challenges for fine-grained object and action detection techniques. The commonly-used cosmetics are shown in Fig. 4(a).

Facial Landmarks: To achieve fine-grained dense video captioning, the models should be able to recognize the facial landmark for detailed description. Hence the facial landmark annotation is also important. Frequent facial landmarks are shown in Fig. 4(b).

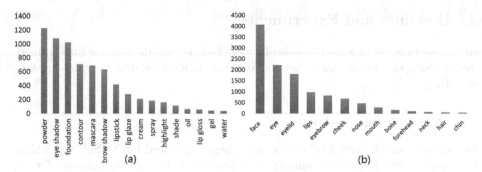

Fig. 4. Common cosmetics and facial landmarks in iMakeup dataset. (a) The cosmetics, (b) the facial landmarks.

Cosmetic Applicators: Appropriate cosmetic applicators are essential for perfect application or blending of various cosmetics. Hence we emphasize this part in annotation, as well. Frequently occured applicators are "brush", "beauty blender", "sponge", "puff", etc.

Cosmetic Brands: A small portion of video annotations mentioned the cosmetic brands. For example, we can find that some cosmetics like "Estee Lauder Double Wear Foundation", "Maybelline Eraser Concealer", "Nyx Setting Spray", etc. are highly in common use. With more annotations, these can help create a knowledge base for future makeup products and facial style recommendation.

Table 1. Comparisons of large-scale video datasets. We collect their duration by hour. FAnn. is short for Fine-grained Annotation.

Name	Duration	Domain	Videos	FAnn.
YouCook [5]	2.3	Cooking	88	No
MPII-MD [6]	73.6	Movie	94	Yes
TACoS [8]	-	Cooking	123	No
YouCookII [10]	176	Cooking	2000	Yes
iMakeup	256	Makeup	2000	Yes

3.3   Comparison

We compare our dataset with several popular large-scale video datasets in Table 1. iMakeup is a brand-new domain-specific large-scale long video dataset with detailed annotations, which can support tasks of learning complicated information or intelligence from long videos, such as temporal action proposal, dense video captioning, and video summarization, etc.

4   Baselines and Experiments

In this section, we explore the feasibility of the iMakeup dataset for fine-grained dense video captioning using several baseline models related to video understanding.

4.1   Video Segmentation

We segment untrimmed long videos using Segment-based CNN (SCNN), which is a multi-stage framework aiming to determine temporal boundaries of crucial actions. SCNN leverages three 3D convolutional networks [30], which are proposal network, classification network and localization network. These three networks use the same architecture, while the last fully connected layer sets different length for different tasks.

We use C3D model pre-trained on Sports1M [31] and test on iMakeup dataset. And we use traditional Jaccard [32] and mean Intersection over Union (mIoU) to evaluate the segmentation performance. Jaccard measures the intersection over prediction between proposals and ground truth. While mIoU considers the intersection over union. Table 2 shows the results of evaluation metrics on temporal segmentation. A segmentation sample is shown in Fig. 5, suggesting that our dataset iMakeup is useful for video segmentation.

Table 2. Results on video temporal segmentation.

Dataset	Model	Jaccard	mIoU
iMakeup	SCNN	45.2	26.3

Fig. 5. Samples of video temporal segmentation.

4.2   Action Proposal

We apply sliding window to generate action proposal in long videos, and present the results on ActivityNet [18] and iMakeup dataset.

We first conduct clustering on all ground truth proposal in training videos with k-means, resulting general length of action proposals. Next, we consider these lengths as the sizes of sliding window, sliding on videos in validation set

Table 3. Results on video action proposal.

# of cluster	ActivityNet				iMakeup			
	Proposal number/video	IoU @0.3	IoU @0.5	IoU @0.7	Proposal number/video	IoU @0.3	IoU @0.5	IoU @0.7
10	73	0.96	0.91	0.66	231	0.99	0.97	0.68
20	202	0.99	0.97	0.84	1147	1.0	1.0	0.93

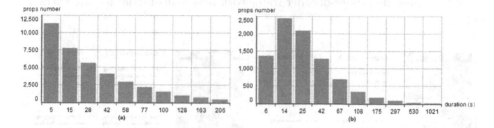

Fig. 6. ActivityNet (a) and iMakeup (b) proposal length distribution.

with 1/2 overlapping. And temporal clips covered in each window generates a proposal.

We set the numer of clusters to 10 and 20. And results of baseline model on ActivitiNet and iMakeup are shown in Table 3. From Fig. 6, we can find that the distribution of proposal lengths of videos in iMakeup dataset is more uniform, resulting that clustering on it converges faster. Hence the baseline model outperforms on iMakeup.

4.3   Video Captioning

The basic fine-grained dense video captioning pipeline includes video segmentation and short video captioning. Hence, we conduct video captioning on our dataset based on commonly-used encoder-decoder framework. We extract visual and auditory features for feature encoding of fixed length, and decode the feature encoding to generate natural language output through a LSTM-RNN network.

In video captioning section, we split the dataset to 65%:5%:30% for training, validation and testing according to makeup topics. Then we extract iResnet.v2 [33] and MFCC [34] (Bag Of Words) for training. We adopt standard metrics like Bleu [35] for evaluating caption proposals. Table 4 compares the video caption performance with two different training setups. We can see that makeup instructional dataset works for video captioning tasks. And the performance of multi-modal features increases comparing to using visual features merely.

We demonstrate qualitative results with videos from iMakeup test set in Fig. 7. We connect captions of clips in a video as a paragraph to describe the whole video. We can find that the baseline method propose most aligned captions, while several objects are missed. For example, the proposal is "concealer",

Table 4. Results on video captioning.

Features	Bleu1	Bleu2	Bleu3	Bleu4	METEOR
iResnet.v2	59.58	49.53	39.70	29. 90	27.40
iResnet.v2+mfccbow	61.86	51.87	41.99	31.63	27.50

yet the ground-truth is "contour". In future works, we would like to extend our work to improved encoder-decoder framework and multi-modal feature extracting.

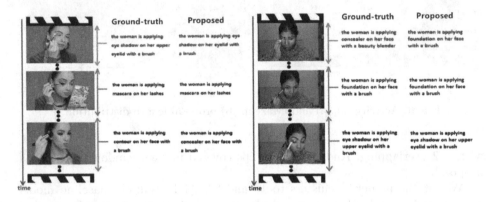

Fig. 7. Qualitative results of video captioning.

5   Conclusion

We propose a large-scale video dataset named iMakeup, containing 2000 videos that are equally distributed over 50 topics, addressing makeup instructional videos uploaded by users on Youtube. The total duration of this dataset is about 256 h, containing about 12,823 video clips in total which are segmented based on makeup procedures. We describe the collection and annotation process of our dataset, analyze the scale, the text statistics and diversity in comparison with other video datasets for similar problems. We then present the results of our baseline video segmentation and caption models on this dataset. The iMakeup dataset contains information from both visual and auditory modalities with a large coverage and diversity of content. We expect that it can be used in an extensive range of problems, like video temporal segmentation, object detection, fine-grained dense video captioning, etc.

Acknowledgment. This work is supported by National Natural Science Foundation of China under Grant No. 61772535 and National Key Research and Development Plan under Grant No. 2016YFB1001202. We also appreciate the support from the National Demonstration Center for Experimental Education of Information Technology and Management (Renmin University of China).

References

1. Venugopalan, S., et al.: Translating Videos to Natural Language Using Deep Recurrent Neural Networks. Computer Science (2014)
2. Yao, L., Torabi, A., et al.: Describing videos by exploiting temporal structure. In: IEEE International Conference on Computer Vision, pp. 4507–4515 (2015)
3. Krishna, R., et al.: Dense-captioning events in videos. In: IEEE International Conference on Computer Vision, p. 6 (2017)
4. Russakovsky, O., et al.: Imagenet large scale visual recognition challenge. Int. J. Comput. Vis. **115**(3), 211–252 (2015)
5. Das, P., et al.: A thousand frames in just a few words: lingual description of videos through latent topics and sparse object stitching. In: IEEE Conference on Computer Vision and Pattern Recognition (2013)
6. Rohrbach, A., et al.: A dataset for movie description. In: IEEE Conference on Computer Vision and Pattern Recognition (CVPR) (2015)
7. http://www.wikihow.com
8. Regneri, M., et al.: Transactions of the Association for Computational Linguistics (TACL), Grounding Action Descriptions in Videos, vol. 1, pp. 25–36 (2013)
9. Xu, J., et al.: MSR-VTT: a large video description dataset for bridging video and language. In: IEEE International Conference on Computer Vision and Pattern Recognition (CVPR) (2016)
10. Zhou, L., et al.: End-to-End Dense Video Captioning with Masked Transformer. arXiv preprint arXiv:1804.00819 (2018)
11. Shou, Z., et al.: Temporal Action Localization in Untrimmed Videos via Multi-stage CNNs, pp. 1049–1058 (2016)
12. Hochreiter, S.: Longshort-term memory. Neural Comput. **9**(8), 1735–1780 (1997)
13. Lin, T.-Y., et al.: Microsoft COCO: common objects in context. In: Fleet, D., Pajdla, T., Schiele, B., Tuytelaars, T. (eds.) ECCV 2014. LNCS, vol. 8693, pp. 740–755. Springer, Cham (2014). https://doi.org/10.1007/978-3-319-10602-1_48
14. Soomro, K., et al.: UCF101: a dataset of 101 human actions classes from videos in the wild. arXiv preprint arXiv:1212.0402 (2012)
15. Kay, W., et al.: The kinetics human action video dataset. arXiv preprint arXiv:1705.06950 (2017)
16. Abu-El-Haija, S., et al.: YouTube-8M: a large-scale video classification benchmark. arXiv preprint arXiv:1609.08675 (2016)
17. Chen, D.L., Dolan, W.B.: Collecting highly parallel data for paraphrase evaluation. In: ACL, pp. 190–200 (2011)
18. Heilbron, F.C., Escorcia, V., Ghanem, B., Niebles, J.C.: ActivityNet: a large-scale video benchmark for human activity understanding. In: IEEE Conference on Computer Vision and Pattern Recognition, pp. 961–970 (2015)
19. Monfort, M., et al.: Moments in time dataset: one million videos for event understanding (2018)
20. Zhou, L., Xu, C., Corso, J.J.: Towards automatic learning of procedures from web instructional videos. In: AAAI (2018)
21. Regneri, M.: Grounding action descriptions in videos. Trans. Assoc. Comput. Linguist. **1**, 25–36 (2013)
22. Zhang, H.J.: Automatic partitioning of full-motion video. Multimed. Syst. **1**, 10–28 (1993)
23. Lienhart, R., Pfeiffer, S., Effelsberg, W.: Video abstracting. Commun. ACM, 1–12 (1997)

24. Yuan, J.: A formal study of shot boundary detection. IEEE Trans. Circuits Syst. Video Tech. **17**, 168–186 (2007)
25. Zabih, R., Miller, J., Mai, K.: A feature-based algorithm for detecting and classifying scene breaks. ACM Multimed. **95**, 189–200 (1995)
26. Porter, S.V., et al.: Video cut detection using frequency domain correlation. In: 15th International Conference on Pattern Recognition, pp. 413–416 (2000)
27. Huang, D.-A., Fei-Fei, L., Niebles, J.C.: Connectionist temporal modeling for weakly supervised action labeling. In: Leibe, B., Matas, J., Sebe, N., Welling, M. (eds.) ECCV 2016. LNCS, vol. 9908, pp. 137–153. Springer, Cham (2016). https://doi.org/10.1007/978-3-319-46493-0_9
28. Kuehne, H., et al.: Weakly supervised learning of actions from transcripts. CVIU **163**, 78–89 (2010)
29. Jin, Q., Chen, J., Chen, S., et al.: Describing videos using multi-modal fusion. In: ACM on Multimedia Conference, pp. 1087–1091. ACM (2016)
30. Tran, D., Bourdev, L., et al.: Learning spatiotemporal features with 3D convolutional networks. In: Proceedings of the IEEE International Conference on Computer Vision, pp. 4489–4497 (2015)
31. Karpathy, A., et al.: Large-scale video classification with convolutional neural networks. In: CVPR (2014)
32. Bojanowski, P., et al.: Weakly supervised action labeling in videos under ordering constraints. In: Fleet, D., Pajdla, T., Schiele, B., Tuytelaars, T. (eds.) ECCV 2014. LNCS, vol. 8693, pp. 628–643. Springer, Cham (2014). https://doi.org/10.1007/978-3-319-10602-1_41
33. Szegedy, C., et al.: Inception-v4, inception-ResNet and the impact of residual connections on learning (2016)
34. Davis, S., Mermelstein, P.: Comparison of parametric representations for monosyllabic word recognition in continuously spoken sentences. IEEE Trans. Acoust. Speech Signal Process. **28**(4), 357366 (1980)
35. Papineni, K., Roukos, S., Ward, T., Zhu, W.-J.: BLEU: a method for automatic evaluation of machine translation. In: ACL, pp. 311–318 (2002)

Embedded Temporal Visualization of Collaboration Networks

Li Zhang[1]([✉]), Ming Jing[2], and Yongli Zhou[3]

[1] School of Information, Qilu University of Technology
(Shandong Academy of Sciences), Jinan, China
lizhang@qlu.edu.cn
[2] Department of Computer Science and Technology, Shandong University,
Jinan, China
jingming@sdu.edu.cn
[3] Telchina Smart Industry Group Co., Ltd, Jinan 250001, Shandong, China
24077161@qq.com

Abstract. Literature data are often visualized as collaboration networks to show the connection between researchers. However, the static networks barely transfer much information when the dataset including temporal variable. In this paper, we propose an embedded network visualization to display the temporal patterns hiding in the data and to avoid occlusion by intelligent filters. We proposed a graph with rich edges to draw the temporal feature in the data. An integrated interface is developed to demonstrate the usability of our approach with case studies on IEEE Vis publications dataset.

Keywords: Collaboration networks · Temporal visualization
Bibliographic analysis

1 Introduction

A large number of scientific publications have been presented these decades. To analyze and discover the feature hiding in the data is an important task to understand the data quickly and directly. Many online databases, such as DBLP, are available for researcher to survey [1].

The academic social networks are typically based on co-authorship and co-citation relationship among researchers and publications. Some popular variables such as subject, author and citation, are very useful to understand the situation [2, 3]. When the relationship between entities gets intricate, the network would become huge and overlapping [4]. The use of technologies such as clustering and threshold filtering, would simplify the final view to a great extent. Besides, with the help of coloring and shading, network view displays patterns intuitively. However, it becomes more difficult when the dataset contains temporal information. Compared to other unsteady dataset with space and temporal information in scientific computing, academic social network always along with lots of lines or curves to represent the relationship between entities, which always be a mess [5, 6].

© Springer Nature Switzerland AG 2018
R. Hong et al. (Eds.): PCM 2018, LNCS 11166, pp. 89–98, 2018.
https://doi.org/10.1007/978-3-030-00764-5_9

In this paper, we address the problem of finding a way to embed the temporal information into networks. A novel context preserving visualization technique is presented to integrate temporal displays into networks. Our method utilizes the empty line spaces or the strokes of points (presented as objects of authors or papers) on the network. An algorithm of carving is developed to keep the graph balance when the user interacts with interface. To avoid occlusion and reduce the overlapping, some interacts technologies are implemented such as keyword filter. We demonstrate our method with case studies on real bibliographic data

The major contents of this paper are summarized as below. First, we summarize the related work on time-oriented information visualization and bibliographic analysis technologies. Then we describe the framework and details of our proposed method in Sects. 3 and 4. And give the case study and evaluation to show the visualization results in Sects. 5 and 6. At last, we draw conclusion based on our discussion and outline of future work.

2 Related Work

2.1 Time-Oriented Data Visualization

Time-oriented data visualization requires dedicated visualization tools. A large variety of time-oriented visualization techniques have been proposed, many of which are useful for domain-specific datasets [7]. For examples, History Flow [8] is used to visualize text edit history and is very effectively at conveying changes over time and revealing typical patterns. Bach [9] et al. proposed a general approach, named time curves, for visualizing patterns of evolution in temporal data by folding the timeline. Cui et al. [10] proposed TextFlow, a seamless integration of visualization and topic mining techniques, complemented with timeline view for helping users interactively inspect and refine the mining results.

Other techniques are more generic and can be applied to a wider range of datasets, many of which are derived from space-time cube representation [11], such as small multiples, time-flattened views, animations and 3D space-time cubes.

2.2 The Bibliographic Analysis

Over the last decade, some library datasets have been used to do bibliographic analysis, such as DBLP, IEEE trans [12] or IEEE conferences data [13]. Xu et al. [14] analysis the CNKI series database to study the coauthor cooperation relationship. From these dataset, research topics, co-author, co-citation [15] and collaboration relationships are mined and visualized on the screen. A deep learning model is applied by Wang et al. [2] to represent coauthor network features for relationships identification, which has better performance compared with other state-of-the-art methods. And Daud et al. [16] apply machine learning techniques to predict features in the co-author networks. Billah et al. [17] designed and trained a SVM classifier to identify and predict the emerging researchers. Sándor [18] involved the temporal data into bibliographic coupling to discover cognitive structure of research areas. While the dataset includes complex information, the network-like visualization always displays the mess result. Some network simplify algorithms [19, 20] are proposed to make the view clearer. Daud et al.

[21] apply the Latent Dirichlet Allocation algorithm, one of the topic modeling techniques to capture the group level structures and temporal trends in the academic social network. Meanwhile, interaction technologies are very useful to explore the interested features among the plenty of objects. Jiang [22] designed a simple effective interface and a hierarchical topic model for mining cross-domain research topic. Varlamis et al. [23] proposed a representation model for visualizing bibliographic databases as graphs and suggest potential synergies between researchers. Besides the basic analysis, the temporal information in the dataset gets more attentions. Chen [24] designed and implemented CiteSpace II, a java application provided a modeling and visualization process to detect and visualize emerging trends and transient patterns in scientific literature. Alsakran et al. [25] proposed STREAMIT, an interactive visualization system that enables users to explore streaming-in text documents without prior knowledge of the data. The Science of Science (Sci2) Tool is a modular toolset specifically designed for the study of science, supports the temporal, geospatial, topical, and network analysis and visualization of scholarly datasets at the micro (individual), meso (local), and macro (global) levels.

3 Proposed Technique

3.1 System Overview

Figure 1 shows the pipeline of our system that contains three major components. The data component load and pre-process data on demand. The visualization component computes the position and size of each object. According to the task analysis, we classify the objects by the topic and field then group them by some specific rules. Concerning their complex relationship, apply lines bundling algorithm and community detection technology to reduce the number of the links. Then, visualize all groups as force-directed network with focused elements. After that, embed the temporal information into

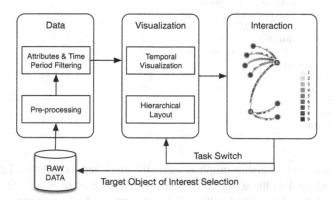

Fig. 1. System overview. The data component deals with pre-processing and filter data according to user's configuration. The visualization module supports the temporal display technology. The interaction part provides user interactions to help them observing the result more efficiently.

the links or objects. In the last component, user can select any elements they are interested in and the relative information will display automatically for further analysis.

3.2 Graph with Rich Edges

The first important decision to make is the meaning of edges. We treat the first author as the key author and the other co-authors as contributors to the first author. Given a published paper data with n authors, a new directed connection information will be added from the other $n - 1$ author nodes to the first author. Combing the published year attribution of the paper, the edges information between these nodes will only increase the current year connection amount as 1. Along the edges, number of connections are sequenced year by year. To represent the complex information between author nodes, we propose a graph with Rich Edges.

Input: database of papers D, keywords k, empty graph $G(V,RE)$
Output: graph with Rich Edges $G(V,RE)$

Algorithm 1 The graph with Rich Edges creation algorithm

1: **procedure** GRAPHRICHEDGESCREATION$(D, k, G(V, RE))$ ▷ database of papers D, keywords
    k, empty graph G(V,RE),set of authors V, rich edges RE
2:    $CP = \emptyset$
3:    **for** each paper $p \in D$ **do**
4:      **if** $k \in p.keywords \lor k \in p.abstract$ **then**
5:        $CP = CP \cup \{p\}$
6:      **end if**
7:    **end for**
8:    $ry = maxyear(CP) - minyear(CP) + 1$ ▷ ry: Range of Year
9:    $RE.seq.length = ry$
10:   **for** each paper $p \in CP$ **do**
11:      **for** each author $a \in p.authors$ **do**
12:        **for** each author $b \in p.authors$ **do**
13:          **if** $a \neq b$ **then**
14:            $V = V \cup \{a, b\}$ ▷ V: without duplicate values
15:            $e = E(a, b)$
16:            **if** $E(a, b) \not\subset RE$ **then**
17:              $e.RS = 0, e.AR = [ry]$
18:              $RE = RE \cup \{e\}$
19:            **end if**
20:            $e.RS+ = 1, e[p.year]+ = 1$
21:          **end if**
22:        **end for**
23:      **end for**
24:   **end for**
25:   Return G
26: **end procedure**

The resulting co-authorship graph is formally modelled as follows. Let the graph $G = (V, RE)$, where V is the set of authors and a rich edge as $re = \{v1, v2, sum, seq\} \in RE$ represents that author $v1$ and author $v2$ are in a paper. And sum represents the total amount of connections in the range of year. The seq is a sequence of collaborate frequency each year.

3.3   Algorithmic Description

The algorithmic description of our method given a publication database D is presented in Algorithm 1. As the pseudocode shows, the first step (lines 3 to 7) is to filter all the papers by given keywords by scanning the abstract and keywords fields. Multiple keywords are supported in our method as conditional $\vee$ (*OR*). If keyword list is empty, none papers would be excluded. Then the candidate paper data *CP* is created. The second step (lines 8 to 9) is to set the length of *seq* as range of year in *CP*. This is also the number of walls on lines while rendering the graph. The third step of our method is the main step to create co-authorship graph G with Rich Edges (line 10 to 24). For each paper p in database *CP*, a set of Rich Edges is added (or updated) to the co-authorship graph.

4   Temporal Display Design

4.1   Time Direction and Temporal Data Visualization

Embedding temporal displays on two dimension networks could disturb the readability if some problems have not been solved reasonably. Time direction of temporal displays on the links is one of them.

We observe that the main problem here is to find a design, which has an easy understanding style like arrow and a simple link; so, we introduce directed curves instead of lines as shown in Fig. 2.

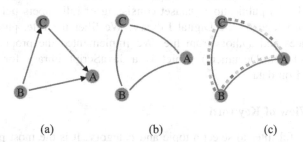

(a)   (b)   (c)

Fig. 2. Time direction designs using arrows (a), and directed curves (b). (c) is temporal display embedded in links in proposed method.

Directed links, as Fig. 2(a) shows, use arrows to point the direction, which is clear and simple. However, arrows would drown in massive links and nodes when the dataset is complex. Then it is hard to find out the direction quickly. Therefore, we remove the arrows and take place of directed links as clock-wise curves, as shown in Fig. 2(b).

The temporal display is another key design in our proposed method [26]. The ideal design should be clear, simply and intuitive to the temporal features. We combine the color wall and directed curve to display the temporal feature. Here, each rectangle represents a value on a time spot and the color represents the amount of this value.

Color is a popular element to use. We draw the information rectangles as saturation low to high representing the value from small to big.

4.2   Radius of Nodes and Height of Color Walls

Like other co-authorship graph, the size of node represents the importance of the author in our method. According to the graph $G(V,RE)$ what we mentioned in Sect. 3.3, the total amount of papers, the variable *sum*, is related to the contribution of the author. Given a $G(V,RE)$, sum_max and sum_min is the maximum and minimum of total papers of one author respectively, s is the saliency and b is the base radius which can be configured by user. Then the radius of node is defined as follows:

$$ri = b + s * \frac{sum_{max} - sum_i}{sum_{max} - sum_{min}}$$

Because the temporal display is on the lines of network, the height of color wall dominates the readability of result. First, the height should be enough to hold the temporal information. On the other hand, oversize color wall will increase the probability of overlapping. We set the default height automatically as the radius of smaller node on this link.

5   Results

In this section, we demonstrate the usage and effectiveness of our technique by applying to IEEE Vis publications dataset consisting of full papers published during 1990 to 2015, provided by the Digital Library. We filter the title, publication year, abstract, references and authors from file. We implemented the proposed technique with D3 (Data-driven Document), which is a JavaScript library for manipulating documents based on data.

5.1   Case 1: View of Keyword

Keyword is a useful filter to select a topic and category. It is the most popular search parameter to find relative literatures. Figure 3 shows a visualization result when a user input keyword "temporal, spatial" and set the paper amount threshold as 12. The data containing only "temporal or spatial" appears in the main window of the view. We can observe 2 larger groups on the left side and some other smaller groups. The 2 larger groups are a litter difficult to understand the inner feature however the key nodes of the group are much easier to identify: node Daniel (top) and node Eduard (bottom). And Daniel's publication almost published in past few years (red rectangles). For other smaller group, the collaboration relationship almost happened lately similarly. That means the published papers relative to "temporal or spatial" had got more attention recently.

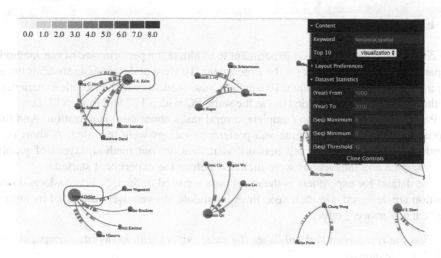

Fig. 3. Example of keyword "Temporal or Spatial". (Color figure online)

5.2 Case 2: User Interaction

Interaction is a necessary part of visualization interface. Besides the configuration discussed in Sect. 4.2, we also provide other operations. Select object to check out details is a basic function we offer.

Another operation is dragging nodes to change the relative location of networks. As Fig. 4 shown, the two images are the same visualization results. Figure 4(a) is the original images but there are many links overlapping which makes it hard to understand.

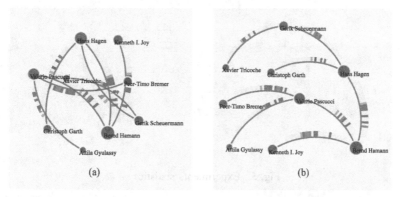

(a)   (b)

Fig. 4. Change the locations of objects by dragging nodes. (a) is the original visualization and (b) is the interaction result.

6 Evaluation

We conducted a controlled user experiment to evaluate the performance of our method. 20 participants were recruited for the experiment. 10 were undergraduate students unfamiliar with network visualization, 10 were graduate students who had a little experience. All the experiments were carried out in the same PC with a 17" widescreen LCD.

Participants were asked to complete several tasks about our visualization. And the response accuracy, response time and preference ratings were recorded. A short oral introduction from authors about network visualization, our method, layout of graph, user interface and interactions were included before the experiment started.

The dataset for experiment is the exact data we used in Sect. 5. Five tasks and one question are designed. On each task, three candidate answers were provided including one "cannot answer" option.

T1: Find the researcher that publishes the most papers with keywords "temporal" or "flow" in the dataset.
T2: Find the new rising researchers that work at "visualization" in the year 2006-2016 in the larger research group of the dataset.
T3: Find the researchers that co-authors the most papers with Wolfgang Aigner in the year 2012–2016.
T4: Estimate the papers amount of Kwan-Liu Ma in the year 2014–2016 in the dataset.
T5: Find the time slots when Huamin Qu co-authors with Weiwei Cui frequently.

We collected results and observed that for all task the response accuracy reached 100% for every participant that means the design can be understanding rightly. So, we focused on the analysis of the response time. Figure 5 shows the statistics for the experiments.

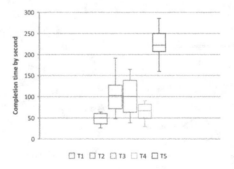

Fig. 5. Experiments statistics

T1 is a traditional task to use keyword which is easy to finish. It is used to examine the performance of filters and to study the interface. T2 to T5 are temporal task combing the network to evaluate the temporal design of our method. T2 is the simplest task and the T5 is the most difficult task. Results show that the response time of T1 is short reasonably. But response time of T2 gets much longer. We observed that the participants took much time to read the temporal design. Once they got familiar with the interface, complete time reduced even facing the more complex task T3 and T4. Task T5 is an integrated task which takes more time to work out.

From the analysis, we can summarize that on both non-temporal and temporal tasks, our design gains a high accuracy, and shorter task response time. We caution that our results are only on our own method and we haven't compared with various co-author network visualization tools.

7 Conclusion and Future Work

We presented a temporal visualization technique of collaboration networks of publication papers. The main goal of this method is to embed the temporal features into the collaboration networks links. We discussed the designs of time direction and temporal display to ensure that the temporal visualization result is meaningful and easy to understand. We applied the technology to a real dataset to demonstrate the efficiency.

As a future work, we would like to research other temporal displays and conduct user evaluation. Also, more complex temporal relationship would be embedded into the networks.

Acknowledgment. This material is based upon work supported by Shandong Province National Science Foundation, China (ZR2017LF006).

References

1. Horak, Z., Kudelka, M., Snasel, V., et al.: Forcoa.NET: an interactive tool for exploring the significance of authorship networks in DBLP data. In: 2011 International Conference on Computational Aspects of Social Networks (CASoN 2011), pp. 261–266. IEEE (2011)
2. Wang, W., Liu, J., Yu, S., et al.: Mining advisor–advisee relationships in scholarly big data. In: JCDL 2016, pp. 209–210. ACM Press, New York (2016)
3. Chang, Y.-W., Huang, M.-H., Lin, C.-W.: Evolution of research subjects in library and information science based on keyword, bibliographical coupling, and co-citation analyses. Scientometrics **105**, 2071–2087 (2015). https://doi.org/10.1007/s11192-015-1762-8
4. Ishida, R., Takahashi, S., Wu, H.-Y.: Interactively uncluttering node overlaps for network visualization. In: 2015 19th International Conference on Information Visualisation (iV), pp. 200–205. IEEE (2015)
5. Fulda, J., Brehmel, M., Munzner, T.: TimeLineCurator: interactive authoring of visual timelines from unstructured text. IEEE Trans. Vis. Comput. Graph. **22**, 300–309 (2015). https://doi.org/10.1109/TVCG.2015.2467531
6. Nakazawa, R., Itoh, T., Saito, T.: A visualization of research papers based on the topics and citation network. In: 2015 19th International Conference on Information Visualisation (iV), pp. 283–289. IEEE (2015)
7. Aigner, W., Miksch, S., Schumann, H., Tominski, C.: Visualization of Time-Oriented Data, pp. 1–296. Springer, London (2011). https://doi.org/10.1007/978-0-85729-079-3
8. Viégas, F.B., Wattenberg, M., Dave, K.: Studying cooperation and conflict between authors with history flow visualizations. In: CHI 2004, pp. 575–582. ACM Press, New York (2004)

9. Bach, B., Shi, C., Heulot, N., et al.: Time curves: folding time to visualize patterns of temporal evolution in data. IEEE Trans. Vis. Comput. Graph. **22**, 559–568 (2015). https://doi.org/10.1109/TVCG.2015.2467851

10. Cui, W., Liu, S., Tan, L., et al.: TextFlow: towards better understanding of evolving topics in text. IEEE Trans. Vis. Comput. Graph. **17**, 2412–2421 (2011). https://doi.org/10.1109/TVCG.2011.239

11. Bach, B., Dragicevic, P., Archambault, D., et al.: A review of temporal data visualizations based on space-time cube operations. In: Borgo, R., Maciejewski, R., Viola, I. (eds.) EuroVis, pp. 23–41 (2014)

12. Xu, X., Wang, W., Liu, Y., et al.: A bibliographic analysis and collaboration patterns of IEEE transactions on intelligent transportation systems between 2000 and 2015. IEEE Trans. Intell. Transp. Syst. **17**, 2238–2247 (2016). https://doi.org/10.1109/TITS.2016.2519038

13. Isenberg, P., Heimerl, F., Koch, S., et al.: Vispubdata.org: a metadata collection about IEEE visualization (VIS) publications. IEEE Trans. Vis. Comput. Graph. **23**, 2199–2206 (2016). https://doi.org/10.1109/TVCG.2016.2615308

14. Xueyou, X., Weiwei, J., Meng, T., et al.: Author cooperation relationship in digital publishing based on social network analysis. In: 2015 12th International Conference on Fuzzy Systems and Knowledge Discovery (FSKD), pp. 1631–1635. IEEE (2015)

15. Zhang, J., Chen, C., Li, J.: Visualizing the intellectual structure with paper-reference matrices. IEEE Trans. Vis. Comput. Graph. **15**, 1153–1160 (2009). https://doi.org/10.1109/TVCG.2009.202

16. Daud, A., Ahmad, M., Malik, M.S.I., Che, D.: Using machine learning techniques for rising star prediction in co-author network. Scientometrics **102**, 1687–1711 (2015). https://doi.org/10.1007/s11192-014-1455-8

17. Billah, S.M., Gauch, S.: Social network analysis for predicting emerging researchers. In: 7th International Conference on Knowledge Discovery and Information Retrieval, pp. 27–35. SCITEPRESS—Science and and Technology Publications (2015)

18. Soós, S.: Age-sensitive bibliographic coupling reflecting the history of science: the case of the species problem. Scientometrics **98**, 23–51 (2013). https://doi.org/10.1007/s11192-013-1080-y

19. Shi, L., Wang, C., Wen, Z., et al.: 1.5D egocentric dynamic network visualization. IEEE Trans. Vis. Comput. Graph. **21**, 624–637 (2016). https://doi.org/10.1109/TVCG.2014.2383380

20. Qingsong, L., Hu, Y., Shi L., et al.: EgoNetCloud: event-based egocentric dynamic network visualization. In: 2015 IEEE Conference on Visual Analytics Science and Technology (VAST), pp. 65–72. IEEE (2015)

21. Daud, A.: Group level temporal academic social network mining through topic models, pp. 1–105 (2010)

22. Jiang, X., Zhang, J.: A text visualization method for cross-domain research topic mining. J. Vis. **19**, 561–576 (2015). https://doi.org/10.1007/s12650-015-0323-9

23. Varlamis, I., Tsatsaronis, G.: Visualizing bibliographic databases as graphs and mining potential research synergies. In: 2011 International Conference on Advances in Social Networks Analysis and Mining (ASONAM 2011), pp. 53–60. IEEE (2011)

24. Chen, C.: CiteSpace II: detecting and visualizing emerging trends and transient patterns in scientific literature. J. Am. Soc. Inf. Sci. **57**, 359–377 (2006). https://doi.org/10.1002/asi.20317

25. Alsakran, J., Chen, Y., Zhao, Y., et al.: STREAMIT: dynamic visualization and interactive exploration of text streams. In: 2011 IEEE Pacific Visualization Symposium (PacificVis), pp. 131–138. IEEE (2011)

26. Jing, M., Xueqing, L., Yupeng, H.: Interactive Temporal Visualization of Collaboration Networks. Springer, New York (2018). https://doi.org/10.1007/978-3-319-77383-4_70

Particle Swarm Programming-Based Interactive Content-Based Image Retrieval

Xiao-Hui Yang[1], Chen-Xi Tian[1], Fei-Ya Lv[2], Jing Zhang[2(✉)],
and Zheng-Jun Zha[3]

[1] Data Analysis Technology Lab, Institute of Applied Mathematics,
School of Mathematics and Statistics, Henan University, Kaifeng, China
[2] School of Automation, HangZhou DianZi University, Hangzhou, China
jingzhang@hdu.edu.cn
[3] Department of Automation, University of Science and Technology of China,
Hefei, China

Abstract. Particle structure in particle swarm optimization (PSO) is fixed in initialization and may result in premature or slow convergence. To tackle this problem, an improved PSO approach called particle swarm programming (PSP) is presented. PSP forms flexible nonlinear distribution representation of particles by introducing hierarchical tree structure into PSO. Furthermore, PSP is introduced in relevance feedback (RF) process of interactive content-based image retrieval (CBIR) by constructing a nonlinear updated query vector. Tests with five benchmark functions demonstrate that PSP can indeed increase diversity of initial particles, enhance search power and improve convergence over PSO. Extensive experiments on Corel-1000 and Catch-256 datasets show that the proposed PSP-based CBIR technique outperforms other linear or recent RF methods proposed for CBIR.

Keywords: Hierarchical tree structure · Particle swarm programming
Nonlinear distribution representation · Nonlinear query vector
Content-based image retrieval

1 Introduction

As an important branch of optimization techniques, swarm intelligence algorithms [1] have been proved efficient and become popular tools in solving real problems [2, 3].

PSO is an evolutionary algorithm inspired by simplified social system [1]. PSO performs optimization by updating personal optimal particle (pbest) and global optimal particle (gbest). However, the existing PSO approaches [4] have some problems: (1) focus on movement regulations of particles rather than representation of particles themselves. (2) structures of particles are all linear and search space of PSO is fixed, which do not support dynamic variability [5]. (3) relevance feedback (RF) is based on fixed linear scheme rather than flexible updated scheme.

Apart from studies of PSO itself, extensive application is another inspiration of its development [6]. PSO-based image retrieval is an active and challenging research area [6–9]. PSO converts the RF into an optimization algorithm, which means the true

© Springer Nature Switzerland AG 2018
R. Hong et al. (Eds.): PCM 2018, LNCS 11166, pp. 99–111, 2018.
https://doi.org/10.1007/978-3-030-00764-5_10

retrieval intentions of users can be viewed as an optimization to query function. Query point movement (QPM) [10] is a widely used RF technique by constructing a linear query vector. PSO-CBIR [7] retrieved the most similar images through its effective space exploration mechanism. Evolutionary PSO-RF [8] derived a feature re-weighting process with the swarm moving towards the goal of minimizing a fitness function according to users' feedback information. [19] proposed a dynamically weight method to make up this shortage. In general, different feature descriptors are combined statically in retrieval process [11]. If they can be combined more flexible, visual perceptions of different users can be characterized better. Some other intelligence optimization techniques have been used directly in CBIR [11–14]. GP-RF [13] changed the current situation about most of the CBIRs through performing index by a set of fixed and pre-specific parameters. However, it is time-consuming for a large number of individuals because of its genetic operations. To tackle these problems, a more powerful representation of particles is urgent.

In this paper, we mainly focus on two issues. (1) An improved PSO algorithm, particle swarm programming (PSP) is presented for providing more flexible and efficient particle representation. (2) A PSP-based interactive CBIR technology is proposed by introducing PSP in RF process to construct a nonlinear updated query vector (NUQV). This work differs from [6, 7, 9–11] as follows: (1) PSP provides flexible and nonlinear distribution representation, which is similar to deep learning [15] and deep random forest [16]. PSP can improve particle's internal structure dynamically. By generating particles based on a hierarchical tree of dynamically varying sizes and shapes, PSP makes up deficiency of PSO in representing particles by fixed-length strings. PSP increases diversity of particles and enlarges search space. Moreover, PSP improves search ability and convergence of PSO. (2) A PSP-based CBIR system is constructed by updating nonlinear query vector. A NUQV is proposed for showing relations between positives or negatives indicated by users.

This paper is organized as follows. In Sect. 2, PSP will be introduced containing its structure and convergence. Section 3 focuses on details in the PSP-based RF framework for CBIR. In Sect. 4, experiments validate the effectiveness of the proposed algorithm. Finally, conclusions are discussed in Sect. 5.

2 Particle Swarm Programming

Swarm intelligent optimization algorithm mainly has population initialization and optimization. It is well known that the choice of initial population has a great influence on optimization results. Therefore, an initialization method based on nonhierarchical tree structure is discussed as follows.

Before describing PSP, PSO is briefly reviewed. In order to control the global and the local exploitation validly, Shi and Eberhart introduced a concept of inertia weight PSO and developed a modified PSO [17], which is the basement of the PSP.

PSO is initialized with a population of random solutions, the individuals in D-dimensional space can be represented as,

$$X_i^{PSO} = S_f = \{f_i | f_i \in D, i = 1, 2, \cdots\}, \tag{1}$$

For the function to be optimized, PSO represents individuals by fixed-length string and linear combination. Hence, PSO may fall into local minimum and less diversity.

To increase diversity and enlarge search space, an improved PSO called PSP is represented by introducing the LISP S-expression [5]. PSP starts with an initial population, which consists of randomly generated individuals composed of functions and terminals that generated by nonlinear hierarchical tree structures. So the size and shape of a group can change dynamically during optimization process. Therefore, PSP has more flexible search ability and larger search space. Each individuals X_i^{PSP} contain leaf nodes S_f and branch nodes $S_{op} = \{+, -, \div, \times (dot), ./, \log, \sin, \cos, \ldots\}$,

$$X_i^{PSP} = S_{op}(S_f). \tag{2}$$

PSP assigns the initial velocity to particles by using nonlinear combination of random particles as initial particles. Suppose $f_{pos,i(j,k,l)}$ $(i = j = k = l = 1, 2, \cdots)$ are vectors who belong to S_f, we choose $S_f = f_{pos,i(j,k,l)}$ and $S_{op} = \{+, \times, . \times (dot)\}$. And the "ramped half-and-half [5]" is adopted to produce a wide variety of particles $X_i^{PSP,1}$ and $X_i^{PSP,2}$, where $i = 1, 2, \cdots$. Two examples are showed as follows.

$$X_i^{PSP,1} = S_{op}(S_f) = [f_{pos,i} \times f_{pos,j}] \cdot \times (dot)(f_{pos,k} + f_{pos,l}). \tag{3}$$

$$X_i^{PSP,2} = S_{op}(S_f) = (f_{pos,i} + f_{pos,j}). \times (dot)f_{pos,k}. \tag{4}$$

The maximum number of available nodes in tree structure is determined up to the depth of the tree defined before learning process [1].

In optimization, suppose the initial population has n particles, the location for the i-th particle is $X_i^{PSP} = (X_{i,1}, \ldots, X_{i,d})$, and the optimal position in the history of flight is $P_i = (P_{i,1}, \ldots, P_{i,D})$. Denote P_j as the best optimal location in $P_j(j = 1, \ldots, n)$ for the j-th particle, $V_i^{PSP} = (V_{i,1}, \ldots, V_{i,D})$ as the velocity for the i-th particle. Then velocity and position of each particle update as follows.

$$\begin{aligned} V_{i,d}^{PSP}(t+1) = w \times V_{i,d}^{PSP}(t) + c_1 \times rand(1) \times [p_{i,d} - X_{i,d}^{PSP}(t)] \\ + c_2 \times rand(1) \times [p_{g,d} - X_{g,d}^{PSP}(t)] \end{aligned} \tag{5}$$

$$X_{i,d}^{PSP}(t+1) = X_{i,d}^{PSP}(t) + V_{i,d}^{PSP}(t+1), 1 \leq i \leq n, 1 \leq d \leq D. \tag{6}$$

The acceleration constants c_1 and c_2 in Eq. (5) represent weights of the stochastic acceleration terms that pull each particle toward pbest and gbest positions. Thus, c_1 and c_2 are always set to $2.0.rand(1)$ is a random value between $[0, 1]$. w is an inertia factor for suiting wide range exploration when it is large, or suiting small scale exploration when it is small. $[-Xmaxd, Xmaxd]$ is variation range of the location for d-th dimension while $[-Vmaxd, Vmaxd]$ is for velocity. If position and velocity are out of range in iteration process, boundaries will be chosen instead. $X_i^{PSP}(0) = (S_f \odot S_{op})_i$ is thus the only parameter to be adjusted. It is often set at about 10–20% of the dynamic range of the variable on each dimension. Similar to PSO, PSP also faces the parameters selection. For more details please refer to [18].

Convergence is critical to an optimization algorithm. PSP maintains the advantages of the original PSO but does not contain genetic operation. At the same time, PSP has the evolutionary programming operator similar to GP [5], and updates the size and shape of hierarchical tree structure dynamically. Therefore, apart from maintaining the convergence of PSO, PSP has better performance because of its flexible nonlinear distribution structure. Due to the combination of the nonlinear hierarchical tree of the particles, search space of PSP is larger than PSO. PSP leads to faster global convergence and improves the local optimum.

3   The Interactive CBIR Based on PSP

PSP can be introduced into RF process in CBIR to further verify its performance. Based on user feedback information, a NUQV, as an improvement of QPM, is constructed based on both positive and negative examples.

3.1   Relevance Feedback Based on Linear Query Vector

When dimension of feature is low, one can easily retrieve images by QPM. QPM fulfills RF based on linear query vector consists of positive and negative images.

$$\vec{q}_{opt} = \alpha \vec{q}_0 + \beta \sum_{I \in pos} f_I - \gamma \sum_{J \in neg} f_J, \tag{7}$$

where parameters α, β, γ in Rocchio formula [10] are static and fixed empirically.

However, different user may have different visual perceptions of an image. In [9], the query vector has been improved by Eq. (8).

$$\vec{q} = \alpha \vec{q_0} + \frac{1}{rel} \sum_{I \in pos} \beta_I f_I - \frac{1}{irr} \sum_{J \in neg} \gamma_J f_J = \alpha \vec{q_0} + T_1 - T_2, \tag{8}$$

where $\vec{q}_{opt}$ is the updated retrieval vector, and $\vec{q}_0$ is the primary one. $pos = \{I_{pos,1}, I_{pos,2}, \cdots, I_{pos,m}\}$ is a set of positive images indicated by user, and rel is the number of this set. Similarly, $neg = \{I_{neg,1}, I_{neg,2}, \cdots, I_{neg,n}\}$ is a set of negative images, and irr is the number of this set. α, β_I, γ_J are previous query vector weights, positive vector weight set and negative vector weight set, respectively. T_1 is combination function between positive images, T_2 is that of negative ones. Apparently T_1, T_2 are linear functions.

3.2   Relevance Feedback Based on Nonlinear Query Vector

When dimension of image feature is high, a stationary combination may not characterize the complex retrieval request properly. Intuitively, linear functions can be replaced by nonlinear functions. If the relationship between positive and negative images can be represented by hierarchy tree structures, the corresponding retrieval accuracy can be increased.

In RF process, users identify positive and negative images for the first L images output. Define *pos* and *neg* as terminals of two hierarchy trees in PSP, some appropriate function operators can be selected as internal-node. Thus two nonlinear functions ϕ and φ are obtained to show the relations of positive images and negative ones correspondingly.

With the query vector in Eq. (9), a NUQV is described by

$$\vec{q}_{opt} = \vec{q}_0 + \phi(\vec{f}_{pos,1}, \vec{f}_{pos,2}, \cdots, \vec{f}_{pos,m}) - \varphi(\vec{f}_{neg,1}, \vec{f}_{neg,2}, \cdots, \vec{f}_{neg,n}). \tag{9}$$

If only positive examples are considered, the special case of Eq. (10) is as follows.

$$\vec{q}_{opt} = \vec{q}_0 + \phi(\vec{f}_{pos,1}, \vec{f}_{pos,2}, \cdots, \vec{f}_{pos,m}), \tag{10}$$

In Eqs. (9) and (10), $\phi(\vec{f}_{pos,1}, \vec{f}_{pos,2}, \cdots, \vec{f}_{pos,m})$ is a positive example set of $pos = \{I_{pos,1}, I_{pos,2}, \cdots, I_{pos,m}\}$ as leaf nodes of the hierarchical tree. $\varphi(\vec{f}_{neg,1}, \vec{f}_{neg,2}, \cdots, \vec{f}_{neg,n})$ are those of a negative example set $neg = \{I_{neg,1}, I_{neg,2}, \cdots, I_{neg,n}\}$. ϕ and φ are constructed by the way of "ramped half-and-half". It can be seen that the NUQV is much closer to the users' real inquiry intention. Because the NUQV technique enhances the impact of positives while weaken that of negatives using nonlinear forms. NUVQ can be used to update the position and velocity adaptively.

As the computer program emerges from PSP is a consequence of fitness, the highest fitness value is assigned to the best individual competition works by comparing the fitness f_{δ_i} to choose the local optimum and global optimum. The evaluation function $f(rk_{\delta_i})$ is defined similar to [19].

$$f(rk_{\delta_i}) = \sum_{l=1}^{L} r(rk_{\delta_i}[l]) \times \frac{1}{A} \times \left(\frac{A-1}{A}\right)^{i-1}, \tag{11}$$

where $A = 2$ and $rk_{\delta_i}[l]$ is the l-th image in ranking rk_{δ_i}. The training set is re-arranged and determined by the distance: $d(t_{i_1}, r_n) \leq d(t_{i_2}, r_n) \leq \cdots \cdots \leq d(t_{i_{NT}}, r_n)$.

$$r(rk_{\delta_i}[l]) = \begin{cases} 1, & rk_{\delta_i}[l] \in positive, \\ 0, & rk_{\delta_i}[l] \in negative. \end{cases} \tag{12}$$

The performance of each particle is determined by the fitness function. The smaller the fitness function value is, the better the solution is, and then the better the particle performance is. The individual particle of the optimal solution is,

$$p_i(t+1) = \begin{cases} p_i(t), & f(x_i(t+1)) \geq f(p_i(t)) \\ x_i(t+1), & f(x_i(t+1)) < f(p_i(t)) \end{cases}. \tag{13}$$

The formula for calculating the global optimal solution is,

$$f(p_g(t)) = \min\{f(p_0(t)), f(p_1(t)), \cdots, f(p_N(t))\}, \tag{14}$$

where N is the population size, the updated NUQVs are obtained by the PSP.

The flow chart of the interactive image retrieval based on PSP is shown in Fig. 1.

4   Experiments and Discussions

In this subsection, extensive experiments are conducted to verify the effectiveness of PSP on interactive CBIR by testing the nonlinear query vector, robustness and computational cost.

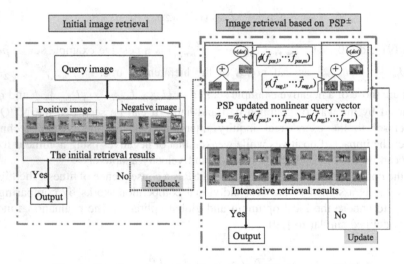

Fig. 1. The flow chart of the interactive CBIR based on PSP

4.1   Search Space of PSP

For comparing the search space between PSO and PSP, the "ramped half-and-half" generative method is adopted to produce initial particles. Figure 2(a) shows that the difference of the initialization mode between PSO and PSP. The comparison of search spaces between PSO and PSP is described by a motivating example corresponding to Eqs. (3) and (4) and shown in Fig. 2(b). The subimages are three-dimensional distributions of initial particles, and the first column is from PSO, the second to the fourth are from PSP. The red surface is the control boundary of PSO. The location distribution of the blue dots embodies the search space of particles. Figure 2(b) demonstrates the search space of PSP is larger than that of PSO.

Comparison of convergence for Sphere's function [20] is shown in Fig. 3, which shows that PSP is superior to PSO.

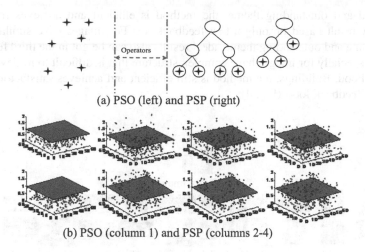

(a) PSO (left) and PSP (right)

(b) PSO (column 1) and PSP (columns 2-4)

Fig. 2. Examples of initial particle structures (a) and search spaces (b) between PSO and PSP.

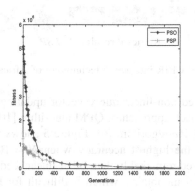

Fig. 3. Comparison of convergence for Sphere's function.

4.2 Interactive CBIR Based on PSP

In this subsection, we demonstrate the performance of PSP on interactive CBIR. For simplicity, color histogram feature and Euclidean distance are chosen. We chose $F=\{+,. \times (dot)\}$ as the internal-nodes of the hierarchy tree in the optimization process. The population size is selected as 60, the number of iterations is 5 and the training set is 35. The inertia factor w is 0.95, the acceleration factors are set as $c_1 = c_2 = 2$. It is worth mentioning that all the predefined parameters in this paper are obtained by experiments similar with [8]. Average precision (AP) is used to evaluate the performance of the compared methods.

Firstly, an overall analysis will be given based on intrinsic different characteristics of images. Corel 1000 dataset, for example, can be divided into three types. All images are tested and some images are randomly selected to exhibit. (1) For images with single background and outstanding theme, the method is efficient and achieves relatively satisfactory result based on only a first feedback. (2) For images have similar color, simple outline and outstanding theme, ideal results also can be got in the third feedback loop. (3) Especially for images have complex structure that is difficult to retrieval, such as Indian, Food, Buildings, the method is still efficient and achieves satisfactory result in the first feedback loop (Fig. 4).

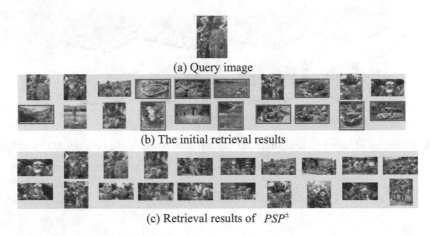

(a) Query image

(b) The initial retrieval results

(c) Retrieval results of $PSP^{\pm}$

Fig. 4. Results for PSP-based CBIR interactive technology of Africa (the first feedback loop).

The proposed PSP-based non-linear query vector updating technique is compared with following linear feedback approaches, QPM algorithm [10], PSO-based algorithm [9] and Evolutionary PSO_RF algorithm [6]. Figure 5 shows that $AP(20)$ of the proposed algorithm achieves the highest accuracy when $N = 10, 20, \ldots, 100$. The performance of PSP is improved remarkably than other linear feedback methods. Figure 6 gives a retrieval example of mountains, which is difficult for PSO-based method.

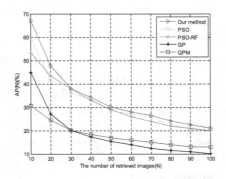

Fig. 5. AP(N) (%) of five methods, N = 10:10:100.

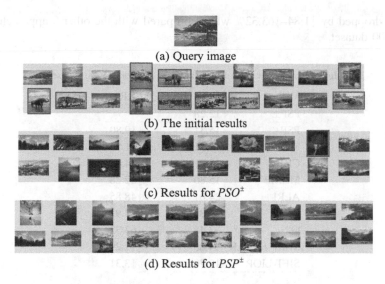

(a) Query image

(b) The initial results

(c) Results for $PSO^\pm$

(d) Results for $PSP^\pm$

Fig. 6. Comparison of retrieval results of PSO and PSP.

Table 1 enumerates the retrieval error rates and error reduction rate (ERR) [21], which denoted by a notion ↓, to characterize the proportion of the errors reduced by switching the other methods to $PSP^\pm$. For instance, the last one in Table 1 shows the error reduces from 32.67 to 7.32% by switching QPM to $PSP^\pm$, where the ERR is ↓77.59% [(32.67–7.32)/32.67]. This suggests that 77.59% retrieval errors can be avoided by using $PSP^\pm$ instead of QPM. For all the six methods, the $PSP^\pm$ error rate drops from ↓32.78 to ↓77.59%, which is significant.

Table 1. Error rates (%) of six methods on Corel-1000 dataset.

Methods	PSP±	PSP+	PSO±	PSO+	PSO_RF	QPM
Error rate	7.32	10.89	10.87	12.34	11.61	32.67
ERR	–	↓32.78	↓32.65	↓40.68	↓36.95	↓77.59

As an online learning technique, RF has been shown to provide dramatic performance boost in CBIR to mitigate semantic gap [22, 23]. In order to show the performance of PSP-based feedback technique, we compare it with other recent RF techniques proposed for CBIR. These methods are color difference histogram (CDH)-based algorithm [24], GP algorithm [11], and other states-of-the-art methods such as ALRI, MLRIP, MLRI-based feedback algorithms [25–27], SURF-FREAK-based visual words fusion algorithm [28] and SIFT-LIOP-based visual words fusion algorithm [29]. Compared results on Corel-1000 dataset are shown in Table 2. One can see that the error reduction between $PSP^\pm$ and the other methods is also remarkable. Our frameworks have the best retrieval precision. Moreover, $PSP^\pm$ is superior to PSP^+ in considering both positive and negative images. By ERR, one can see that the error rate of $PSP^{\pm}$ dropped by $\downarrow 1.84$–$\downarrow 63.32\%$ when compared with the other 9 approaches on Corel-1000 dataset.

Table 2. Error rates (%) of ten methods on Corel-1000 dataset.

Method	Error rate (%)	ERR
PSP±	12.28	–
PSP+	15.31	↓19.80
MLRI	12.51	↓1.84
CDH	12.91	↓4.88
MLRIP	15.80	↓22.28
ALRI	24.01	↓48.85
GP±	29.89	↓58.92
GP+	33.48	↓63.32
SURF-FREAK	14.00	↓12.29
SIFT-LIOP	12.70	↓3.31

This subsection aims to test the robustness of PSP-based interactive CBIR scheme. Figure 7(a) illustrates the changes in ranking the target image when one increases the significance of image alterations. When the size of filter window and rotate angle is increased to a certain degree, the rank order of target image still remains a strong

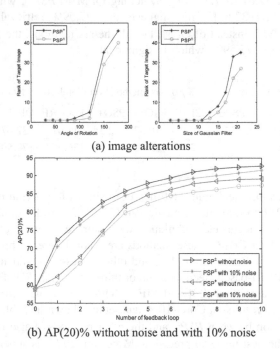

(a) image alterations

(b) AP(20)% without noise and with 10% noise

Fig. 7. Robustness of the PSP-based CBIR system.

position. To further demonstrate the robustness of the system, similar to [17], retrieval performance without noise and with noise is tested from 1 to 5 feedback loop. Figure 7 (b) gives retrieval results on 10% noise. It indicates that although the system with noise performs a little worse than the system without noise, the difference can be kept as the number of feedback loop increases. Experiments show that the proposed PSP-based framework is robust to noise.

All experiments in this paper have been implemented on an AMD Phenom Quad-Core 9750 PC (2.4 GHz, 4 GB RAM) 32Bit Windows XP operating system in MATLAB environment. For Corel-1000, average execution time at initial query stage is 0.0055 s, and in each feedback loop is 0.1891 s for PSP^+ method, 0.1954 s for $PSP^\pm$, which is lower than 0.4108 s for ALRI method, 0.9093 s for MLRIR and 0.9357 s for MLRI method. Considering the classic advice on response time which states that 10 s is the limit for keeping the user's attention focused on the task, the proposed PSP-based methods are competitive in both retrieval accuracy and computational complexity.

5 Conclusions

An improved PSO named PSP based on a nonlinear hierarchical tree initialization is introduced in this paper. PSP provides a flexible nonlinear representation of initial particles and reaches better convergence. In interactive CBIR, a PSP-based NUQV replaces linear query update vectors and solves the problem that linear model with low accuracy may not satisfy complex searching. Experiments show that the PSP-based interactive CBIR improves the retrieval performance, and can detect more positive images with better sorts via easy programming and simple computing. PSP belongs to memory search while every search procedure follows the best solution currently found. PSP is extremely simple and seems to be effective for optimizing a wide range of applications. It could be used in function optimization and data mining problems. It will be interesting to study PSP-based large-scale web image retrieval.

Acknowledgments. We would like to thank James Z. Wang et al., G. Griffin, et al. for their test datasets. This work was supported in part by NSF of China (61751304, 61622211, 61472392 and 61620106009), Key Project of Science and Technology of the Education Department Henan Province (14A120009), and NSF of Henan Province (162300410061) and Project of Emerging Interdisciplinary of Henan University (xxjc20170003).

References

1. Kennedy, J., Eberhart, R.C.: Particle swarm optimization. In: Proceedings of IEEE International Conference on Neural Network, pp. 1942–1948(1995)
2. Li, Y., Jiao, L., Shang, R., Stolkin, R.: Dynamic-context cooperative quantum-behaved particle swarm optimization based on multilevel thresholding applied to medical image segmentation. Inform. Sci. **294**, 408–422 (2015)
3. Zhang, Y.D., Wang, S., Ji, G.: A comprehensive survey on particle swarm optimization algorithm and its applications. Math. Probl. Eng. **1**, 1–38 (2015)

4. Gao, H., Xu, W., Sun, J., Tang, Y.: Multilevel thresholding for image segmentation through an improved quantum-behaved particle swarm algorithm. IEEE Trans. Instrum. Meas. **59**(4), 934–946 (2010)
5. Koza, J.R.: Genetic Programming: On the Programming of Computers by Means of Natural Selection. MIT Press, Cambridge (1992)
6. Broilo, M., Natale, D., Francesco, G.B.: A stochastic approach to image retrieval using relevance feedback and particle swarm optimization. IEEE Trans. Multimed. **12**(4), 267–277 (2010)
7. Broilo, M., Rocca, P., Natale, F.G.B.D.: Content-based image retrieval by a semi-supervised particle swarm optimization. In: Proceedings of MMSP, pp. 666–671 (2008)
8. Wei, K.P., Lu, T.W., Bi, W., Sheng, H.H.: A kind of feedback image retrieval algorithm based on PSO, wavelet and sub-block sorting thought. In: Proceedings of ICFCC, pp. 796–801 (2010)
9. Cai, L.J., Yang, X.H., Li, S.C., Li, D.F.: Relevance feedback based on particle swarm optimization for image retrieval. In: Proceedings of ITSE, pp. 749–756 (2012)
10. Baeza-Yates, R.A., Riberio-Neto, B.: Modern Information Retrieval, pp. 305–306. Mcgraw-Hill, New York (1999)
11. Ferreira, C.D., Asntos, J.A., Torres, R.S., Goncalves, M.A., Rezende, R.C., Fan, W.: Relevance feedback based on genetic programming for image retrieval. Pattern Recognit. **32**(1), 27–37 (2011)
12. Younus, Z.S., Mohamad, D., Saba, T., et al.: Content-based image retrieval using PSO and k-means clustering algorithm. Arab. J. Geosci. **8**(8), 6211–6224 (2015)
13. Koza, J. R.: Hierarchical genetic algorithms operating on populations of computer programs. In: Proceedings of IJCAI, pp. 768–774 (1989)
14. Yang, Y., Nie, F.P., Xu, D., Luo, J.B.: A multimedia retrieval framework based on semi-supervised ranking and relevance feedback. IEEE Trans. Pattern Anal. **34**(4), 723–742 (2012)
15. Zhang, Q., Yang, L., Chen, Z., Li, P.: A survey on deep learning for big data. Inf. Fusion **42**, 146–157 (2018)
16. Bai, S.: Growing Random Forest on Deep Convolutional Neural Networks for Scene Categorization, vol. 71. Pergamon Press, Oxford (2017)
17. He, X.F., King, O., Ma, W.Y., Li, M.J., Zhang, H.J.: Learning a semantic space from user's relevance feedback for image retrieval. IEEE Trans. Circuits Syst. Video **13**(1), 39–48 (2003)
18. Shi, Y.H., Eberhart, R.C.: Parameter selection in particle swarm optimization. In: Proceedings of 7th ICEP VII, pp. 591–600 (1998)
19. Fan, W., Fox, E.A., Pathak, P., Wu, H.: The effects of fitness functions on genetic programming-based ranking discovery for web search. J. Am. Soc. Inf. Sci. Technol. **55**(7), 628–636 (2004)
20. Xi, M.L., Jun, S., Xu, W.B.: An improved quantum-behaved particle swarm optimization algorithm with weighted mean best position. Appl. Math. Comput. **205**(2), 751–759 (2008)
21. Deng, W.H., Hu, J.N., Guo, J.: Extended SRC: undersampled face recognition via intraclass variant dictionary. IEEE Trans. Pattern Anal. **34**(9), 1864–1870 (2012)
22. Rui, Y., Thomas, H.: Optimizing learning in image retrieval. In: Proceedings of CVPR, pp. 1236–1243 (2000)
23. Wu, J., Shen, H., Li, Y.D., Xiao, Z.B., Lu, M., Wang, C.L.: Learning a hybrid similarity measure for image retrieval. Pattern Recognit. **46**(11), 2927–2939 (2013)
24. Liu, G.H., Yang, J.Y.: Content-based image retrieval using color difference histogram. Pattern Recognit. **46**(1), 188–198 (2013)

25. Zhang, B., Lin, F., Ma, W.Y., Zhang, H.J.: A novel region-based image retrieval method using relevance feedback. In: Proceedings of ACM MM, pp. 28–31 (2001)
26. Zhang, L., Shum, H., Shao, L.: Discriminative semantic subspace analysis for relevance feedback. IEEE Trans. Signal Process. **25**(3), 1275–1287 (2016)
27. Yang, X.H., Lv, F.Y., Cai, L.J., Li, D.F.: Adaptive learning region importance for region-based image retrieval. IET Comput. Vis. **9**(3), 1–10 (2015)
28. Jabeen, S., Mehmood, Z., Mahmood, T., Saba, T., Rehman, A., Mahmood, M.T.: An effective content-based image retrieval technique for image visuals representation based on the bag-of-visual-words model. Plos One **13**(4), e0194526 (2018)
29. Yousuf, M., et al.: A novel technique based on visual words fusion analysis of sparse features for effective content-based image retrieval. Math. Probl. Eng. **1**, 13 (2018)

Video Clip Growth: A General Algorithm for Multi-view Video Summarization

Gang Pan[1], Xingming Qu[1(✉)], Liangfu Lv[1], Shuai Guo[2], and Di Sun[1,3]

[1] Tianjin University, Tianjin 300350, China
{pangang,quxingming}@tju.edu.cn
[2] Hefei University of Technology, Hefei 230009, China
[3] Tianjin University of Science and Technology, Tianjin 300222, China

Abstract. Plenty of multi-view video processing tasks such as video abstract, key-frame extraction and camera selection focus on presenting to audiences the most significant information in a certain period of time. Basically, the main idea of these techniques is to show audiences the video segments or frames that have the highest spatio-temporal significances. However, existing approaches are not enough to deal with these tasks with a general framework. In this paper, we develop a novel bottom-up algorithm called *video clip growth* that generates multi-view video abstract through an accurate frames adding process, which allows users to customize the length of the video summaries. This approach firstly uses an energy function to evaluate each frame's importance from both time and space dimension. Then video clips and frames are gradually selected according to their energy rank, until reaching the target length. Besides, our algorithm can also extend to several multi-view video processing tasks. The experimental results on the Lobby and Office dataset have demonstrated the effectiveness of our algorithm.

Keywords: Multi-view video · Video summarization
Key-frame extraction · Camera selection

1 Introduction

With the development of multimedia industry, multi-view camera systems that can record a group of videos of scenes at the same time have become increasingly popular. Generally, in order to better present videos recorded by multi-view camera systems to audiences, further processing is always necessary. For instance, to select the best viewing angle from a number of cameras, to grasp the most important information and to extract the key frames. However, these processes often require highly specialized persons to handle and are time-consuming.

This work was supported by Tianjin Philosophy and Social Science Planning Program under grant TJSR15-008, National Social Science Foundation under grant 15XMZ057.

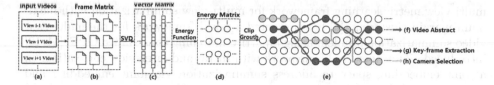

Fig. 1. Illustration of the multi-view videos summarization architecture and the applications of video clip growth algorithm. (a) the raw multi-view video. (b) all the frames are extracted to form a frame matrix. (c) dimensionality reduction is used for converting frames into vectors. (d) using energy function to evaluate the importance of each frame(now is vector). (e) how to select appropriate frames by different tasks.

In fact, the basic purpose of these multi-view video processing tasks is to select video segments or frames that have the most spatio-temporal significances. Therefore, these tasks can be attributed to two steps: first to measure the significances of video frames and then select the appropriate video frame according to the different needs of users.

Based on the above ideas, we propose a novel bottom-up approach called *video clip growth* for video summarization. A video clip represents a spatio-temporally coherent frame sequence, which is initialized with a constant length. Next, an energy function is introduced to measure each frame's importance, we called frame's energy. Based on frame's energy, the video clip's average energy can then be calculated. Finally, the video clip growth algorithm is used to select the appropriate video clips by their average energy to generate video abstract with a specific length. Moreover, this algorithm can be easily extended to other multi-view summarization tasks such as key-frame extraction and camera selection.

2   Related Work

Several kinds of multi-video summarization methods have been proposed at present: The method of keyframes [3–5,9] represents the original video using a small set of selected frames; the method of storyboard [2,7] shows the content of the original video using a specially arranged keyframe layout to tell the story; the method of video skimming [1,2,8,13] generates a short video highlight of the original video.

In addition, multi-view video summarization has received considerable attentions in recent years. Generally, the research of multi-video summarization can be roughly divided into two categories. The first category is to summarize a group of videos after all videos are recorded. [2] presented a multi-view video summarization technique using random walks applied on spatio-temporal shot graphs. [9] presented another method for abstracting multi-key frames from video datasets using Support Vector Machine (SVM) and rough sets. [13] enhanced summarization in a way that takes objects identifying into consideration. [14] introduced a multi-view metric learning framework for multi-view summarization with maximum margin clustering. [6] has presented a novel framework for the multi-view video summarization problem using bipartite matching constrained OPF. [11,12] introduced an unsupervised framework with intra- and inter-view correlations in a joint embedding space to address summarization via joint embedding and sparse representative selection.

Although the aforementioned works have successfully generated kinds of multi-view summarization tasks, to our best knowledge, they have ignored the potential connections between different tasks. In this paper, we focus on a novel *video clip growth* algorithm that can generate multi-view video abstract through a sustainable growth way. Moreover, this algorithm can be easily extended to other multi-view summarization tasks.

Table 1. Table of symbols

Symbol	Meaning
VC	Video clip
L_{VC}	Initial length of VC, least period of shot cut
N	VC's neighbour
E_N	Neighbour's energy
S	Selected video clip set
T_L	Length of video abstract defined by users
C	Candidate video clip set
E_{ave}	Average energy of VC
$CurrentL$	Total length of all VCs in S
$\Theta(VC)$	Weight coefficient of VC

3 The Algorithm

3.1 Pre-processing

In order to reduce the amount of computation, a pre-processing step is presented in this section to convert frames into vectors.

To begin with, frames from each video (Fig. 1(a)) are extracted and then combined into a frame matrix $\{X(i,j)\}$, where $X(i,j)$ is the jth frame from camera i (Fig. 1(b)).

Dimensionality Reduction: Several types of dimensionality reduction methods such as Principal Component Analysis (PCA), color quantization, Singular Value Decomposition (SVD), etc., are all feasible. SVD is adopted in this paper.

$$X(i,j) = U\Sigma V^T \tag{1}$$

where U, V^T are the real left and right singular vector matrices respectively and Σ is the real $\lambda \times \lambda$ diagonal matrix.

Next, the first λ left singular vectors will be picked out and reshaped to a column vector V, which can be used as the feature of a frame. In this paper, V is a 720D vector. After dimensionality reduction, the vector matrix is obtained $\{V(i,j)\}$ (Fig. 1(c)), where $V(i,j)$ is the corresponding vector from $X(i,j)$.

3.2   Energy Function

The proposed energy function used to measure the importance of the frame vector is defined as the sum of the Time-dependent energy and the Space-dependent energy. In addition, an optional active pixel ratio is introduced to compensate the influence due to the different distances from cameras to the object. By integrating these three parts, the Energy function is used to convert the vector matrix $\{V\}$ to Energy Matrix $\{E\}$. Each vector will be converted to a specific value (the energy of this vector) (Fig. 1(d)). Energy function is defined as follows:

$$E(i,j) = \alpha * EH(i,j) + \beta * EV(i,j) + \gamma * Active(i,j) \tag{2}$$

where $EH(i,j)$, $EV(i,j)$ are respectively the Time-dependent energy and the Space-dependent energy of vector $V(i,j)$. $Active(i,j)$ is the percentage of the active pixels of frame compared to its neighbour frame. α, β, γ are hyperparameters controlling the importance of the three parts, respectively. The above items will be discussed in detail below:

Time-Dependent Energy: Since the frame sequence is continuous, we define the Time-dependent Energy to represent each frame's energy in the horizontal direction. Specifically, if there is a great difference between the content before and after a frame, this frame tends to have higher Time-dependent Energy, vice versa. To compute the Time-dependent Energy, a sliding window with length L is created to collect vectors in the vector matrix. L can be adjusted according to different type of video.

$$\omega_1 = (V_{(i,j-l)}, V_{(i,j-l+1)}, \cdots, V_{(i,j-1)})$$
$$\omega_2 = (V_{(i,j+1)}, V_{(i,j+2)}, \cdots, V_{(i,j+l)})$$

where $l = L/2$, and ω_1, ω_2 are the sample sets before and after vector $V(i,j)$ respectively. Then we calculate the mean vector of the two sample sets by:

$$\bar{\omega}_i = \frac{1}{l} \sum_{V \in \omega_i} V \qquad i = 1, 2 \tag{3}$$

This leads to the following Time-dependent Energy function:

$$EH(i,j) = \|\bar{\omega}_1 - \bar{\omega}_2\| \tag{4}$$

Where $\|..\|$ operation represents Euclidian norm in this paper.

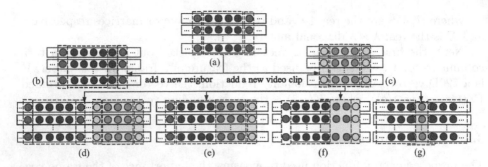

Fig. 2. Video clip growth cases. Blue: video clips in selected set. Cambridge blue: video clips in candidate set. Red and yellow: neighbours of video clips. (a) first we only have one VC and we have two growth cases (b) and (c). (e),(f) and (g) shows 3 cases of overlap when constantly adding new VCs. (d) is normal condition. (Color figure online)

Space-Dependent Energy: Space-dependent Energy $EV(i,j)$ is defined to represent a frame's significance compared to frames from other cameras at the same moment. We calculate the difference between each vector and the average of all the vectors in the same column. If there is a great difference, this vector tends to have great Space-dependent Energy. The average vector of each column is expressed as

$$\bar{m}_j = \frac{1}{n} \sum_{V(i,j) \in V_j} V(i,j) \qquad i = 1, 2 \dots n \tag{5}$$

where n is the number of cameras and V_j is the vectors in j^{th} column. The Space-dependent Energy function is defined as:

$$EV(i,j) = \|V(i,j) - \bar{m}_j\| \tag{6}$$

Active Pixel Ratio: The intention of the introduction of active pixel is to judge frame energy in a fair way. Adding active pixel term compensates the influence due to the different distances from cameras to the object. The higher the active pixels ratio of a frame, the greater the energy this frame has. A pixel whose change exceeds a threshold σ in two adjacent frames is called active pixel. The amount of active pixel divided by the total pixel is the active pixel ratio. We use $\sigma = 3$.

$$Active(i,j) = \frac{\sum_{a,b} I(X(i,j)(a,b) - X(i,j+1)(a,b) > \sigma)}{a * b} \tag{7}$$

In experiments, we believe that the $EH, EV, Active$ are equally important. Therefore, in Eq. (2), we set $\alpha = 1, \beta = \frac{\sum EH}{\sum EV}, \gamma = \frac{\sum EH}{\sum Active}$ to make them have the same magnitude.

Algorithm 1. Video clip growth method

Input: Energy Matrix E

Output: non-overlapping selected set S

1 **Initialize** Target video length T_L, minimum window length L_{VC};
2 Using sliding window that moves $L_{VC}/2$ each time to generate VC candidate set C;
3 Sort all VC in C by their total energy value in descending order ;
4 add the first VC from C into S ;
5 **while** $CurrentL < T_L$ **do**
6   pick out N with the highest E_N ;
7   compute corresponding E_{ave};
8   **if** $(E_N * \Theta < E_{ave})\&\&(CurrentL + L_{VC} < T_L)$ **then**
9   add the next VC from C into S;
10   **else**
11   add N and into corresponding VC ;
12   update Θ;
13   check and merge overlapping VCs in S;
14 **return** S;

3.3   Video Clip Growth Method

In this section, the idea of video clip VC is proposed. VC is a collection of frames from different cameras over the same time period (Fig. 2(a)). VC expresses the overall information of the entire camera system for a period of time. The initial length L_{VC} of VC is fixed and each VC has its own left and right "neighbour" N (Fig. 2, red and yellow dots).

The idea of the proposed algorithm is to generate a bunch of VCs. These VCs can be considered as "Candidate". Higher energy VCs will be added into a video clip selected set S. Then repeatedly adding candidate VCs or frames into S to achieve target video abstract length T_L defined by users.

Video Clip Generation: First create a sampling window to represent VC with size $n * L_{VC}$ and the stride is $L_{VC}/2$. Every time the window slides on $\{E\}$, a new VC is generated. Save each VC and sort them by their average energy in descending order to form a VC candidate set C. Take the first VC in C as the initial S and then we need to make VC in S grow bigger.

Video Clip Growth: Compare all Ns of VCs in S and pick out the N with the highest energy E_N (i.e. one neighbour has been selected). Then compare E_N and the average energy E_{ave} of it's corresponding VC. For example, in (Fig. 2(a)), assume right neighbour has been picked out. Compare right red dots' average energy and all blue dots' average energy. The motivation of this comparison is to make sure that the most important N has been added into its VC. If there is no neighbour whose E_N is bigger than E_{ave}, a new VC needs to be added. (i.e. if E_N is less than the E_{ave} and the current length $CurrentL$ of all VCs in S plus L_{VC} is less than T_L, we add the next VC from C into S (Fig. 2(c)),

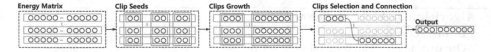

Fig. 3. Flowchart of multi-view videos abstract process based on the proposed algorithm. (a) is the Energy Matrix whose elements represent a frame's importance. (b) is the candidate VC generated by sampling set. (c) video clips growth method is performed. (d) how to generate video abstract.

otherwise add N into its VC (Fig. 2(b))). The pseudo code of the *video clip growth* algorithm is shown in Algorithm 1.

Weight Coefficient: Since T_L can be achieved by combining many shorter VCs or several longer VCs, we introduce a weight coefficient $\Theta(VC)$ to determine the number and lengths of VCs according to the users' preferences. It is worth mentioning that each VC has its own $\Theta(VC)$. $\Theta(VC)$ has a negative correlation with its VC's length. Because the N with the highest E_N has been picked out as indicated in the previous section now we multiply a Θ to E_N and then redo the comparison between E_N and E_{ave}. Each time we add a N, Θ needs to be updated.

For example, if users prefer shorter VC, the Θ can be expressed as a monotonically decreasing function related to the length of VC, as long as the domain is 0 to positive infinity and the range is 0 to 1. Thus, the product of the E_N and its Θ is more likely to be smaller than E_{ave} and then more VCs tend to be selected. In this paper, we used

$$\theta(VC) = e^{-\delta length(VC)} \tag{8}$$

where $length(VC)$ is current length of the video clip; δ is a constant, and we set $\delta = 0.1$ in our experiments.

Merge Video Clips: When constantly adding new VCs, it will inevitably encounter the situation that overlaps happen between the new VC and previous VCs. If so, merge those overlapped VCs into a new one. Through continuous checking and merging, we can guarantee that all the VCs in S are non-overlapping.

There are three conflict cases when merging video clips:

Case.1 New VC is adjacent with one previous VC (Fig. 2(e)).

Case.2 New VC overlaps with one previous VC (Fig. 2(f)).

Case.3 New VC overlaps with two previous VCs (Fig. 2(g)).

4   Applications on Multi-view Video Processing

Multiple multi-view video processing tasks can be completed easily by our *video clip growth* method. This will be explained in details as follows.

Video Abstract: Multi-View Video abstract, simplifies the long video into a brief one. This task can be easily achieved by using *video clip growth* method to generate VCs and choosing the rows with the highest energy in each VC (Fig. 3(d)). In this way, the camera that captured the most important information can be always selected out. After combining all the selected rows chronologically, this task is completed.

Key Frame Extraction: The key-frame is the mostly representative image or images of a video. Key-frame extraction is a technique on how to use the fewest video frames to reflect the main content of the video. Suppose we have already got several important VCs, higher energy frames in these VCs should be more important as well. To obtain these important VCs, set the Θ to a small value so that more VCs tend to be selected. Let users define the number of key-frames K. Picking out the top K VCs in energy values and finding out their highest energy frames respectively, the user gets K key-frames. The selected keyframes won't be close to each other in time dimension because all the VCs are non-overlapping.

Camera Selection: Camera Selection is to choose the most appropriate perspective according to the characteristics(action, direction) of the object when processing multi-view video. This task can be considered as making a 100% length video summarization. Our method is that first to find rows that with highest energies in each VC and the camera number C_Ns will be identified according to the rows' numbers. After this, merge overlapped VCs.

Three merge situation which are similar to "Merge Video clips" in Sect. 3.3 are discussed here:

Case.1 New VC is adjacent with one previous VC. Merge them and the C_N of each part remains constant.

Case.2 New VC overlaps with one previous VC. Merge them and the C_N of merged VC is consistent with the C_N from higher energy VC.

Case.3 New VC overlaps with two previous VCs. Merge them and only the C_N of non-overlapping part is changed to the C_N from higher energy previous VC. Finally, different video segments from different cameras are seamed together chronologically to make the output video.

5   Experimental Results

Dateset: Experiments have been conducted using a publicly available dataset given in $[2,10]^{1,2}$: LOBBY dataset captured with 3 cameras in a large lobby area and OFFICE dataset which contains more crowded scenes with richer activities. For the proposed method, input videos should be synchronized and with stable frame rates.

Results: To provide a quantitative comparison, we employ Precision, Recall and F1-measure from [2]. Our results were then compared with results based on the

---

[1] http://www.sdspeople.fudan.edu.cn/fuyanwei/summarization/.

[2] http://media.ee.ntu.edu.tw/research/summarization/.

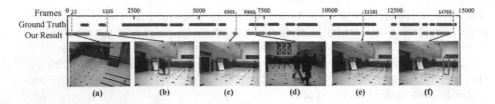

Fig. 4. The comparison between our result and groundtruth. (a) Our method mistakenly labeling camera moving as event. (b) Deviation of groundtruth: this event is still happening. (c) Our method mistakenly labeling trivial motion as event. (d) Our method never fails to detect drastic motion. (e) Mislabel in groundtruth, nothing happened here. (f) Our method tends to miss last few frames because they do not have EH.

methods in [2], including two user attention methods, Graph-based method, and Multi-view method. The comparative results in Table 2 show that our method has a better recall rate but a lower precision rate. Three aspects will be detailed for explaining the comparative results.

Generally, for parameter selection, if the content of the video changes drastically, small L should be used, vice versa. We use $L = 30$ in this paper. The selection of L_{VC} needs some tradeoffs. For instance, if the L_{VC} is too large, it may miss some details. If it is too small, the calculation time will be slightly longer. $L_{VC} = 20$ is used in this paper. In order to maintain the objective, we assume that $EH, EV, Active$ are equally important. So in the last of Sect. 3.2, α, β, γ have been set to the same magnitude. If the content of the video changes drastically, α can be appropriately increased. If the distances from the cameras to the object are similar, $Active$ is not necessary.

In addition, we tested our algorithm at frame level. Because a novel feature of the algorithm is that it can generate a specific length of video abstract, we make our video abstract the same number of frames as groundtruth. To visualize our results, a dot at time axis indicating a frame from a detected event is kept at that moment. Therefore, an event will be presented as a line segment. Figure 4 shows the comparison between our results and the groundtruth. In addition, we also conduct experiment using BL-7F dataset and Office dataset from [10]. Video abstracts are shown in supplementary material. Following discusses will help to better understand the results in Fig. 4:

The Sensitivity: The use of sliding window makes our method very sensitive to the movement of camera angle or the movement of object, even very small movement can be detected as an event. For example, camera angle was moving from the first frame to the 300th frame in $View2$ (Fig. 4(a)). From the 6363th frame to the 6450th frame in $View1$ (Fig. 4(c)), two girls were wandering at the left door and their actions were not obvious at all. Mistakenly labeling these small movement as events makes the number of False positive increase.

Deviation of Groundtruth: The groundtruth of lobby dataset was artificially labeled and different people defined the beginning and end of an event differently.

For instance, the 1220th frame to the 1335th frame in $View1$ (Fig. 4(b))was labeled as event "two men walks across the lobby towards the gate, and a man walks into the lobby." Actually, it is reasonable to prolong this event because the third man was still moving in the 1335th frame. However, groundtruth believed this event had ended.

Table 2. Performance comparison with previous methods for office and lobby dataset.

Method	Office					Lobby				
	Length of sum-mary(s)	Number of events	P(%)	R(%)	F1(%)	Length of sum-mary(s)	Number of events	P(%)	R(%)	F1(%)
User attention method1	40	10	100	38	55.07	184	31	95	72.1	81.98
User attention method2	55	12	100	46	63.01	179	30	100	69.8	82.21
Graph-based	64	7	100	26	41.27	201	25	100	58	73.42
Multi-view method	55	16	100	61	75.78	176	33	100	76.7	86.81
Ours	55	21	88.78	80.77	84.58	176	41	82.12	97.62	89.2

Limitation: The limitation of our proposed method lies in the fact that the using of sliding windows makes the last few frames can not have Time-dependent energy and this leads to the situation that they always have less energy than other frames and will be seldomly selected (Fig. 4(f)). A simple solution is to use the average Time-dependent energy to fill them. The second drawback of the sliding window is that our method never fails to capture motion (Fig. 4(d)), although some of them are "not important". Like motions in (Fig. 4(c)), their visual perception are not obvious and they are not considered as an event in groundtruth.

6 Conclusion and Future Work

In this paper, a novel bottom-up method for multi-view video summarization is proposed. An energy function is also introduced for evaluating each frame's importance. By extending *video clip growth* algorithm, we have easily completed several multi-view video processing tasks such as video abstract, keyframe extraction and camera selection. We show the effectiveness of our algorithm through rigorous experimentation on public datasets. The future work is to develop dynamic programming method for selecting video clips from candidate set.

References

1. Almeida, J., Leite, N.J., Torres, R.D.S.: Vison: video summarization for online applications. Pattern Recognit. Lett. **33**(4), 397–409 (2012)
2. Fu, Y., Guo, Y., Zhu, Y., Liu, F., Song, C., Zhou, Z.H.: Multi-view video summarization. IEEE Trans. Multimed. **12**(7), 717–729 (2010)
3. Guan, G., Wang, Z., Yu, K., Mei, S., He, M., Feng, D.: Video summarization with global and local features. In: IEEE International Conference on Multimedia and Expo Workshops, pp. 570–575 (2012)
4. Ioannidis, A.I., Chasanis, V.T., Likas, A.C.: Key-frame extraction using weighted multi-view convex mixture models and spectral clustering. In: International Conference on Pattern Recognition, pp. 3463–3468 (2014)
5. Khosla, A., Hamid, R., Lin, C.J., Sundaresan, N.: Large-scale video summarization using web-image priors. In: Computer Vision and Pattern Recognition, pp. 2698–2705 (2013)
6. Kuanar, S.K., Ranga, K.B., Chowdhury, A.S.: Multi-view video summarization using bipartite matching constrained optimum-path forest clustering. IEEE Trans. Multimed. **17**(8), 1166–1173 (2015)
7. Lee, K.M.: A unified framework for event summarization and rare event detection. IEEE Trans. Pattern Anal. Mach. Intell. **37**(9), 1737–1750 (2015)
8. De Leo, C., Manjunath, B.S.: Multicamera video summarization from optimal reconstruction. In: Koch, R., Huang, F. (eds.) ACCV 2010. LNCS, vol. 6468, pp. 94–103. Springer, Heidelberg (2011). https://doi.org/10.1007/978-3-642-22822-3_10
9. Li, P., Guo, Y., Sun, H.: Multi-keyframe abstraction from videos. In: IEEE International Conference on Image Processing, pp. 2473–2476 (2011)
10. Ou, S.H., Lee, C.H., Somayazulu, V.S., Chen, Y.K., Chien, S.Y.: On-line multi-view video summarization for wireless video sensor network. IEEE J. Sel. Top. Signal Process. **9**(1), 165–179 (2015)
11. Panda, R., Chowdhury, A.R.: Multi-view surveillance video summarization via joint embedding and sparse optimization. IEEE Trans. Multimed. **PP**(99), 1–1 (2017)
12. Panda, R., Das, A., Roy-Chowdhury, A.K.: Embedded sparse coding for summarizing multi-view videos. In: IEEE International Conference on Image Processing, pp. 191–195 (2016)
13. Wang, F., Ngo, C.W.: Summarizing rushes videos by motion, object, and event understanding. IEEE Trans. Multimed. **14**(1), 76–87 (2012)
14. Wang, L., Fang, X., Guo, Y., Fu, Y.: Multi-view metric learning for multi-view video summarization. In: International Conference on Cyberworlds, pp. 179–182 (2016)

Cross-Media Retrieval via Deep Semantic Canonical Correlation Analysis and Logistic Regression

Hong Zhang[1,2(✉)] and Liangmeng Xia[1,2]

[1] College of Computer Science and Technology,
Wuhan University of Science and Technology, Wuhan 430065, China
zhanghong_wust@163.com
[2] Intelligent Information Processing and Real-Time Industrial Systems Hubei
province Key Laboratory, Wuhan University of Science and Technology,
Wuhan 430065, China

Abstract. With the rapid growth of multimedia content on the Internet, multimedia retrieval has been extensively studied for decades. Inspired by the progress of deep neural network (DNN) in single-media retrieval, the researchers have applied the DNN to cross-media retrieval. Most of the existing methods cast low-level features of cross-media data onto a unified feature space. However, most of these feature spaces usually do not have explicit semantics which results in an unsatisfactory retrieval result. By considering the above issue, cross-media joint representation by hierarchical learning with DSCCA-LR is proposed in this paper which is able to explore jointly the correlation and semantic information. It combines Deep Canonical Correlation Analysis (DCCA) and Logistic Regression (LR) to learn two types of semantic features of of heterogeneous data and concatenate them to form common features for cross modal retrieval. Extensive experiments on two datasets show that the proposed approach achieves superior retrieval accuracy comparing to several state-of-the-art cross-media retrieval methods.

Keywords: Cross-media retrieval · Deep canonical correlation analysis
Logistic Regression

1 Introduction

Over the last decade there has been a massive explosion of multimedia content on the Internet. Such as in a web page, images and texts which have similar meaning are often utilized together to describe the same thing. Retrieval between different media forms is increasingly important. In the beginning, much research efforts devoted to the single-media retrieval where the retrieval results and user query are of the same media type, such as text retrieval [1, 2], image retrieval [3, 4], video retrieval [5, 6] and audio retrieval [7, 8]. But those retrieval methods do not meet the requirements when people need to use text to retrieve the images. To handle this problem, cross-media retrieval has been proposed where user's query and the retrieval results are of different media type.

© Springer Nature Switzerland AG 2018
R. Hong et al. (Eds.): PCM 2018, LNCS 11166, pp. 123–133, 2018.
https://doi.org/10.1007/978-3-030-00764-5_12

The most challenge problem in cross-media retrieval is how to exploit the correlation between low-level features of multimedia objects and high-level concepts. The unified feature representation based approaches is widely utilized for cross-media retrieval. Firstly, it tries to obtain a unified feature representation then it projects the objects of different modalities into the common space, lastly it computes the similarities between two multimedia objects such as Canonical Correlation Analysis (CCA) [9], Cross-modal Factor Analysis (CFA) [10] and deep Canonical Correlation Analysis (DCCA) [11]. However, the unified feature space obtained by these approaches usually has no explicit semantic meanings, which might ignore the hints contained in the original media content so it's not capable to fully measure the similarities among different media types. Hardoon proposed a Semantic Matching (SM) approach to address the cross-modal retrieval problem which is to represent data of different modalities at a higher level of abstraction [12]. But it takes only the semantic meanings into account, so the performance of this method is not so famous.

By considering the above issue, a new approach to cross-media retrieval via Deep Semantic Canonical Correlation Analysis and Logistic Regression (DSCCA-LR) is proposed in this paper which is able to explore jointly the correlation and semantic information. Our approach consists of three steps. Firstly, an improved version of Deep Canonical Correlation Analysis named DSCCA is utilized to learn the maximally correlated subspace. Deep Canonical Correlation Analysis (DCCA) is widely used for modeling multimedia data [13–15] whose loss function is to maximize the Pearson correlation coefficient between two sets of heterogeneous data. The improvement of DSCCA is that the semantic information is added in the loss function. Secondly, we utilize Logistic Regression (LR) to learn pure semantic space where each dimension is a semantic concept and each point is a weight vector over these concepts. By these two steps, both DSCCA features and LR features are extracted. Finally, we concatenate these two features and multi-class logistic regression is utilized ones more. By this step, we get the ultimate common features which are the final retrieval features.

Extensive experiments on two datasets show that the proposed approach achieves best search accuracy comparing to state-of-the-art cross-media retrieval methods. The remainder of the paper is organized as follows. We firstly review some related work on cross-media retrieval in Sect. 2. Secondly, we present DSCCA-LR method together with some analysis in Sect. 3. In the Sect. 4, we show the experiments on the two widely used cross-media datasets. Finally, we conclude this paper in Sect. 5.

2   Related Work

Canonical correlation analysis (CCA) is a natural possible solution to analyze the correlation between two multivariate random vectors. Through CCA, we could learn the subspace that maximizes the correlation between two sets of heterogeneous data. Reference [16] applied CCA to localize visual events associated with sound sources.

Bases on CCA, many extensions [17, 18] are applied to cross-media retrieval. Rasi-wasia et al. [19] proposed semantic correlation matching (SCM) approach, where the multiclass logistic regression is applied to the maximally correlated feature represen-tations obtained by CCA, to produce an isomorphic semantic space for cross-media retrieval. Hwang and Grauman [20] proposed an unsupervised learning method based on kernel CCA to discover the relationship between texts and images. Cross-media factor analysis (CFA) evaluates the association between two types of media. In the transformed domain, CFA minimizes the Frobenius norm between pairwise data, and performs better than CCA.

In recent years, since deep learning has shown superiority in image classification [4, 21] and image content representation, it has also been widely used in cross-media research to learn uniform representation. Andrew et al. [22] introduced a deep CCA model, which can be viewed as a nonlinear ectension of the linear CCA, to learn complex nonlinear transformations of two modalities of the data. Wang et al. [13] proposed an effective mapping mechanism, which can capture both intramodal and intermodal semantic relationships of multimodal data from heterogeneous sources, based on the stacked auto-encoders deep model. Wei et al. [23] proposed a deep semantic matching (deep-SM) method that uses the convolutional neural network and fully connected network to map images and texts into their label vectors, achieving state-of-the-art accuracy. The cross-media multiple deep network (CMDN) is a hier-archical structure with multiple deep network and can simultaneously preserve intra-media and inter-media information to further improve the retrieval accuracy.

Although there are significant research efforts on uniform representation learning for cross-media analysis tasks, a large gap still exists between these methods and user expectations. This is caused by the fact that existing schemes still have not achieved a satisfactory performance; i.e., their accuracies are far from acceptable. Therefore, we still need to investi-gate better uniform representation methods for cross-media research.

3   DSCCA-LR Model

As shown in Fig. 1, our DSCCA-LR model can be divided into three stages: in the first stage, we use DSCCA to learn a subspace which maximizes the weighted sum of Pearson correlation and the Euclidean distance. Pearson correlation is describing the difference between the two sets of heterogeneous samples, and Euclidean distance is the difference between the samples under the same media. Then we map the original feature to the subspace to get DSCCA features. In the second stage, we compute posterior probability distributions through multi-class logistic regression. In the last stage, we connect DSCCA features and LR features together and multi-class logistic regression is utilized ones more. By this step, we get the ultimate common feature.

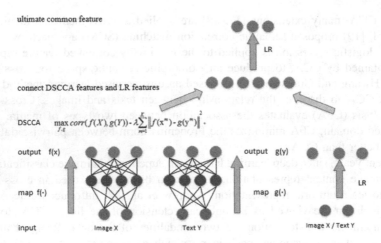

Fig. 1. The over view of our DSCCA-LR model

3.1 DSCCA Features Learning

First, we use two DNNs f and g to extract nonlinear features for images and texts. The weighted sum of Pearson correlation of these two features and the Euclidean distance of the same type under various medias are maximized. In outher words, we optimize the following objective:

$$\max_{W_f, W_g, U, V} \frac{1}{N} \mathrm{tr}(U^T f(X) g(Y)^T V) - \lambda \sum_{m=1}^{K} \| f(x^m) - g(y^m) \|^2$$

$$\text{s.t.} \quad U^T (\frac{1}{N} f(x) f(x)^T + r_x I) U = I, \qquad (1)$$

$$U^T (\frac{1}{N} g(x) g(x)^T + r_y I) V = I,$$

$$u_i^T f(X) g(Y)^T v_j = 0, \text{for } i \neq j,$$

Where $U = [u_1, \ldots, u_L]$ and $V = [v_1, \ldots, v_L]$ are the CCA directions that project the DNN outputs, $(r_x, r_y) > 0$ are the regularization parameters for sample convariance estimation [24] and $\lambda > 0$ is a trade-off parameter. In DSCCA, $f(\bullet)$ is the projection mapping. x^m and y^m represent for image and text objects with the same labels.

One intuition for CCA-based objectives is that, while it may be difficult to accurately reconstruct one view from the other view, it may be easier, and may be sufficient, to learn a predictor of a function (or subspace) of the second view. In addition, it should be helpful for the learned dimensions within each view to be uncorrelated so that they provide complementary information.

The DSCCA objective couples all training samples through the whitening constraints, so stochastic gradient descent (SGD) cannot be applied in a standard way. DSCCA can still be optimized efficiently as long as the gradient is estimated using a sufficiently large minibatch with gradient formulas because a large minibatch contains enough information for estimating the covariances.

3.2   LR Features Learning

We obtain posterior probability distributions for each media type. The input features I and T are original features. Before introducing the logical regression model, we first introduce the sigmoid function, which is in the form of:

$$g(x) = \frac{1}{1 + e^{-x}} \tag{2}$$

The corresponding function curve is as shown in Fig. 2. As is shown in Fig. 2, the sigmoid function is a s-shaped curve, and its value is between 0 and 1, and the value of a function far away from 0 will quickly close to 0 or 1. The assumptions made by the logistic regression model are follows:

$$p(y = 1|x; \theta) = g(\theta^T x) = \frac{1}{1 + e^{-\theta^T x}} \tag{3}$$

The $g(x)$ here is the sigmoid function mentioned above, and the corresponding decision function is:

$$\begin{aligned} \hat{y} = 1, & \quad \text{if } p(y = 1|x) > 0.5 \quad \hat{y} = 1, \quad \text{if } p(y = 1|x) > 0.5 \\ \hat{y} = 0, & \quad \text{if } p(y = 1|x) < 0.5 \quad \hat{y} = 0, \quad \text{if } p(y = 1|x) < 0.5 \end{aligned} \tag{4}$$

We use the maximum likelihood estimation to find the parameters. In the logistic regression model, the likelihood can be expressed as:

$$L(\theta) = \prod (y|x; \theta) = \prod g(\theta^T x)^y (1 - g(\theta^T x))^{1-y} \tag{5}$$

Logarithmic likelihood can be obtained by taking logarithms.

$$l(\theta) = \sum y \log g(\theta^T x) + (1 - y) \log(1 - g(\theta^T x)) \tag{6}$$

We can get the average log loss on the whole set of data.

$$J(\theta) = -\frac{1}{N} l(\theta) \tag{7}$$

Gradient descent, also known as the fastest gradient descent, is an iterative solution method. By adjusting every parameter in the direction of the fastest change of the objective function, it approximates the optimal value. The gradient method is used to calculate the gradient of the loss function.

$$\frac{\partial J}{\partial \theta} = -\frac{1}{N}\sum_{i=1}^{N}(y_i - \hat{y}_i)x_i + \lambda\theta \tag{8}$$

After the semantic features of images and texts are obtained by logistic regression.

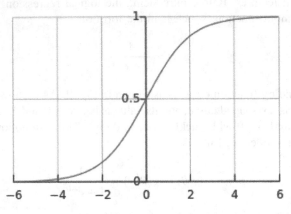

Fig. 2. The sigmoid curve

We ultilize the oroginal features and labels to train two different logistic regression models for images and texts. The outputs of the image/text model are LR features.

3.3   Ultimate Common Features Learning

Finally, we concatenate DSCCA features and LR features then multi-class logistic regression is utilized ones more. By this step, we get the ultimate common features which are the final retrieval features and can be used directly for cross media retrieval.

4   Experiment

In this section, we evaluate the performance of our approach by comparing with several existing approaches in the literature for cross media retrieval.

4.1   Data Sets

Wikipedia dataset is chosen from a collection of 2700 "feature articles", which are selected and reviewed by Wikipedia's editors since 2009. These articles are categorized into 29 classes, and the Wikipedia dataset consists of the 10 most populated ones. Each article is attached with one or more images from Wikimedia Commons. Based on its section headings, each article is split into several section. The final corpus contains a total of 2866 documents and is randomly splited into a training set of 2173 documents and a testing set of 693 documents. They are all text-image pairs and labeled by a vocabulary of 10 semantic classes.

XMedia dataset consists of 5000 texts, 5000 images, 500 videos, 1000 audios and 500 3D models. All the media data are crawled from the Internet websites, including Wikipedia, Youtube, Flickr, freesound, findsound, 3D Warehouse and Princeton 3D Model Search Engine. This dataset is organized into 20 categories with 600 media data per category and is randomly split into a training set of 9600 media data and a testing set of 2400 media data.

4.2   Evaluation Metrics

The final performance is evaluated using the mean average precision (MAP) since it is widely adopted to evaluate the performance of ranked retrieval results. Given one query and the N top-ranked retrieved results, the average precision (AP) can be calculated by (9):

$$AP = \frac{1}{R} \sum_{i=1}^{N} \frac{R_i}{i} \times rel_i \tag{9}$$

where R is the number of relevant results, R_i is the number of the relevant results among the top i results, and rel_i denotes the relevance of the rank i result ($rel_i = 1$ if relevant and 0 otherwise). Further, we can obtain the MAP by averaging AP of all queries.

4.3   Compared Methds

For comparison purpose, we adopt 4 state-of-the-art cross-media retrieval methods, namely CCA, SM, DCCA, JRL. The CCA and SM are the classical baselines. DCCA is the DNN-based cross-media retrieval method proposed recently. JRL is the state-of-the-art method based on linear projection.

CCA (Canonical correlation analysis): Through CCA we could learn the subspace that is able to maximize the correlation between two sets of heterogeneous data. It is widely used for cross-media relationship analysis.

SM (Semantic Matching): SM represents the image as well as text at a higher level of abstraction, so that there are natural correspondences between the text and image spaces.

DCCA (Deep canonical correlation analysis): DCCA learns simultaneously learns two deep nonlinear map-pings of two views that are maximally correlated.

JRL (Joint representation learning): Learns the joint representation for different media types, which is able to jointly exploit the pairwise correlation and semantic information in a unified optimization framework.

4.4   Comparisons and Results

First, we compare the proposed DSCCA-LR with four algorithms, including CCA, Semantic Matching (SM) and DCCA and JRL which are implemented with the codes provided by their authors respectively. The experimental results of these approaches are shown in Table 1. Compared with the state-of-the-art methods on cross-media retrieval tasks, our proposed DSCCA-LR improves the average MAP from 0.255 to 0.22. The accuracy of SM is close to CCA. DCCA improve the accuracy since the method use deep neural network. The increase of JRL is due to it takes both correlation analysis and semantic abstraction into consideration.

Table 1. Comparison results of map performance on different datasets

DataSet	Method	Task		
		Image → Text	Text → Image	Average
Wikipedia dataset	CCA	0.226	0.120	0.173
	SM	0.219	0.160	0.190
	DCCA	0.241	0.194	0.217
	JRL	0.290	0.219	0.255
	DSCCA-LR	**0.310**	**0.234**	**0.272**
XMedia dataset	CCA	0.100	0.118	0.065
	SM	0.120	0.071	0.095
	DCCA	0.100	0.098	0.990
	JRL	0.147	0.138	0.154
	DSCCA-LR	**0.215**	**0.210**	**0.213**

Then Figs. 3 and 4 show the MAP scores achieved per category on Wikipedia dataset by all approaches. For most categories, the MAP of DSCCA-LR outperform those of compared methods.

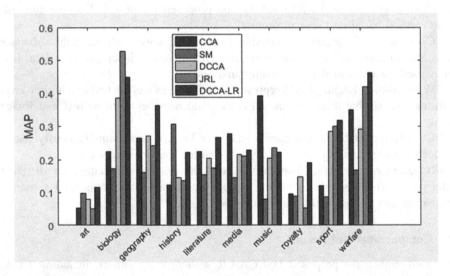

Fig. 3. MAP performance of Image → Text retrieval for each category on Wikipedia dataset.

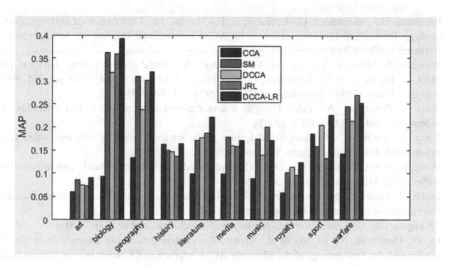

Fig. 4. MAP performance of Text → Image retrieval for each category on Wikipedia dataset.

From the experiment result, it can be seen that the proposed approach achieves the best results among these approaches, demonstrating the effectiveness of the proposed approach.

5   Conclusion

In this paper, we have proposed a DSCCA-LR algorithm to explore jointly the correlation and semantic information. Extracting related features and semantic features using DSCCA and LR respectively. This solves the shortcomings of most existing cross-media retrieval methods, which model the pairwise correlation or semantic information separately. Extensive experiments on cross-media retrieval show the effectiveness of our proposed approach as compared with the state-of-the-art methods. In the future, we intend to model more kinds of correlations among different media, such as the hyperlink on the Internet, on the other hand, we plan to apply DSCCA-LR to more applications.

Acknowledgments. This work described in this paper was supported by National Natural Science Foundation of China (No. 61373109).

References

1. Moffat, A., Zobel, J.: Self-indexing inverted files for fast text retrieval. ACM Trans. Inf. Syst. (TOIS) **14**(4), 349–379 (1996)
2. Jaderberg, M., Simonyan, K., Vedaldi, A., et al.: Reading text in the wild with convolutional neural networks. Int. J. Comput. Vis. **116**(1), 1–20 (2016)

3. Escalante, H.J., H´ernadez, C. A., Sucar, L. E., Montes, M.: Late fusion of heterogeneous methods for multimedia image retrieval. In: ACM International Conference on Multimedia Retrieval (ACM-ICMR), pp. 172–179 (2008)
4. Babenko, A., Slesarev, A., Chigorin, A., Lempitsky, V.: Neural codes for image retrieval. In: Fleet, D., Pajdla, T., S, B., Tuytelaars, T. (eds.) ECCV 2014. LNCS, vol. 8689, pp. 584–599. Springer, Cham (2014). https://doi.org/10.1007/978-3-319-10590-1_38
5. Sivic, J., Zisserman, A.: Video Google: a text retrieval approach to object matching in videos. In: Null, p. 1470. IEEE (2003)
6. Yang, H., Meinel, C.: Content based lecture video retrieval using speech and video text information. IEEE Trans. Learn. Technol. 7(2), 142–154 (2014)
7. Wold, E., Blum, T., Keislar, D., et al.: Content-based classification, search, and retrieval of audio. IEEE Multimed. 3(3), 27–36 (1996)
8. Ewert, S., Pardo, B., Müller, M., et al.: Score-informed source separation for musical audio recordings: an overview. IEEE Signal Process. Mag. 31(3), 116–124 (2014)
9. Thompson, B.: Canonical correlation analysis. Encycl. Stat. Behav. Sci. (2005)
10. Li, D., Dimitrova, N., Li, M., et al.: Multimedia content processing through cross-modal association. In: Proceedings of the Eleventh ACM International Conference on Multimedia, pp. 604–611. ACM (2003)
11. Zebende, G.F.: DCCA cross-correlation coefficient: quantifying level of cross-correlation. Phys. A: Stat. Mech. Appl. 390(4), 614–618 (2011)
12. Hardoon, D.R., Saunders, C., Szedmak, S., Shawe-Taylor, J.: A correlation approach for automatic image annotation. In: Li, X., Zaïane, O.R., Li, Z. (eds.) ADMA 2006. LNCS (LNAI), vol. 4093, pp. 681–692. Springer, Heidelberg (2006). https://doi.org/10.1007/11811305_75
13. Wang, W., Arora, R., Livescu, K., et al.: On deep multi-view representation learning. In: Proceedings of the 32nd International Conference on Machine Learning (ICML-15), pp. 1083–1092 (2015)
14. Yang, E., Deng, C., Liu, W., et al.: Pairwise relationship guided deep hashing for cross-modal retrieval. In: AAAI, pp. 1618–1625 (2017)
15. Qi, J., Huang, X., Peng, Y.: Cross-media similarity metric learning with unified deep networks. Multimed. Tools Appl. 76, 1–19 (2017)
16. Lessard, N., Paré, M., Lepore, F., et al.: Early-blind human subjects localize sound sources better than sighted subjects. Nature 395(6699), 278–280 (1998)
17. Vert, J.P., Kanehisa, M.: Graph-driven feature extraction from microarray data using diffusion kernels and kernel CCA. In: Advances in Neural Information Processing Systems, pp. 1449–1405 (2003)
18. Hardoon, D.R., Shawe-Taylor, J.: KCCA for different level precision in content-based image retrieval. In: Proceedings of Third International Workshop on Content-Based Multimedia Indexing, IRISA, Rennes, France (2003)
19. Rasiwasia, N., Costa Pereira, J., Coviello, E., et al.: A new approach to cross-modal multimedia retrieval. In: Proceedings of the 18th ACM International Conference on Multimedia, pp. 251–260. ACM (2010)
20. Hwang, S.J., Sha, F., Grauman, K.: Sharing features between objects and their attributes. In: 2011 IEEE Conference on Computer Vision and Pattern Recognition (CVPR), pp. 1761–1768. IEEE (2011)

21. Shin, H.C., Roth, H.R., Gao, M., et al.: Deep convolutional neural networks for computer-aided detection: CNN architectures, dataset characteristics and transfer learning. IEEE Trans. Med. Imaging **35**(5), 1285–1298 (2016)
22. Andrew, G., Arora, R., Bilmes, J., et al.: Deep canonical correlation analysis. In: International Conference on Machine Learning, pp. 1247–1255 (2013)
23. Willsey, A.J., Sanders, S.J., Li, M., et al.: Coexpression networks implicate human midfetal deep cortical projection neurons in the pathogenesis of autism. Cell **155**(5), 997–1007 (2013)
24. Hardoon, D.R., Szedmak, S., Shawe-Taylor, J.: Canonical correlation analysis: an overview with application to learning methods. Neural Comput. **16**(12), 2639–2664 (2004)

3D Global Trajectory and Multi-view Local Motion Combined Player Action Recognition in Volleyball Analysis

Yang Liu$^{(\boxtimes)}$, Shuyi Huang, Xina Cheng, and Takeshi Ikenaga

Graduate School of Information, Production, Systems, Waseda University,
Kitakyushu, Japan
leon@fuji.waseda.jp

Abstract. Volleyball video analysis is important for developing applications such as player evaluation system or tactic analysis system. Among its different topics, player action recognition serves as an elementary building brick for understanding player's behavior. Most conventional player action recognition methods have limits in real volleyball game due to the occlusion and intra-class variation problems. This paper proposes a 3D global trajectory and multi-view local motion combined volleyball player action recognition method. 3D global trajectory extracts global motion feature through 3D trajectories, which hides the unstable and incomplete 2D motion feature caused by the above problems. Multi-view local motion gets detailed local motion feature of arms and legs in multiple viewpoints and removes clutter features caused by occlusion problem. Through the combination, global 3D feature and local motion feature mutually promote each other and the actions are recognized well. Experiments are conducted on game videos from the Semifinal and Final Game of 2014 Japan Inter High School Games of Men's Volleyball in Tokyo Metropolitan Gymnasium. The experiments show the combing result accuracy achieves 98.39%, 95.50%, 96.86%, 96.98% for spike, block, receive, toss respectively and improve 11.33% averagely than the sing-view local motion based result.

Keywords: Player action recognition · Volleyball analysis
Occlusion · Intra-class variation

1 Introduction

Sports video analysis attracts lots of attentions due to the ballooning of sports video data. Among various kinds of sports videos, volleyball game analysis becomes an attractive research target due to its representative condition, which consists of multiple players and complex background. Current exiting volleyball game systems such as Data Volley require to manually input the game data, which is heavy cost of human labor. Therefore, automatic volleyball game analysis system is eager to develop. Among various kinds of topics in volleyball

© Springer Nature Switzerland AG 2018
R. Hong et al. (Eds.): PCM 2018, LNCS 11166, pp. 134–144, 2018.
https://doi.org/10.1007/978-3-030-00764-5_13

analysis, player action recognition becomes an essential problem since it serves as an elementary building brick for lots of applications, like player evaluation system and highlights extraction system. Therefore, developing a robust volleyball player action recognition system is an urgent task for volleyball game analysis.

There are mainly 2 problems: occlusion problem and intra-class variation problem. Firstly, the occlusion problem not only includes the self-occlusion, which means the partial occlusion among body parts due to the view limitation, but also the occlusion problem caused by frequent interaction with other team players. The occlusion problem results in difficulty in obtaining players' complete and stable motion features. Secondly, the intra-class variation problem, which means the diversity inside one action category, increases the difficulty in recognition. For example, player performs the same type action with different deformed poses, shape and speed according to different situations.

With the same target of action recognition in sports, Miyamori [1] has proposed a prototype player action recognition system in tennis, which is based on the silhouette transition. However, this method is sensitive to action deformation, because it heavily depends on the target appearance. So this method is less robust in volleyball due to the occlusion and intra-class variation. Ozturk [2] also proposed a player action recognition based on the skeleton model. However, this method cannot work well in occlusion situation since the skeleton model is obtained from depth maps, which are very noisy if serious occlusion occurs. With the same target of volleyball video analysis, Chen [3] proposed a basic action detection method based on the transitions of ball. However, there are various action categories in volleyball game. These actions share similar ball trajectory transition. So only the transition of ball trajectory is not enough. Kubota [4] proposed a volleyball player action recognition method based on local motion feature. While due to the occlusion and intra-class variation problems in real game, this method is less robust in real game volleyball player action recognition.

To make a robust volleyball player action recognition system, this paper proposes a 3D global trajectory and multi-view local motion combined method. The 3D global trajectory means to describe the global motion features in 3D space. These features represent the action with stable global feature while hiding the mess 2D information caused by intra-class variation and occlusion problems. Multi-view local motion describes moving details of players arms and legs in multiple viewpoints. A noise reduction method is proposed to remove the noise caused by occlusion and solves the intra-class variation problem through a multi-view processing. After combing global motion and local motion, the actions are recognized well.

The rest part of this paper is organized as follows. Section 2 presents the entire volleyball player action recognition method and details. Section 3 presents experiment results and analysis. Final conclusion is drawn in Sect. 4.

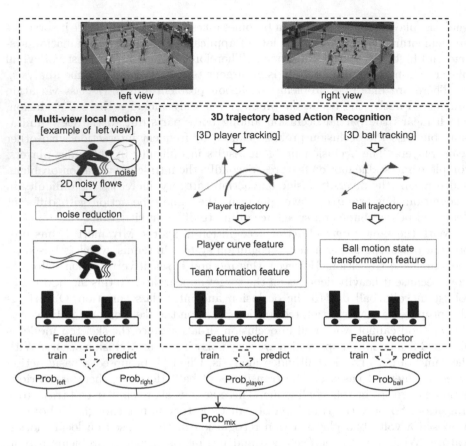

Fig. 1. Framework of global trajectory and multi-view local motion combined method

2 Proposal

2.1 Combing 3D Global Trajectory and Multi-view Local Motion

This proposal combines the 3D global trajectory and multi-view local motion. 3D global trajectory part aims to represent the action through the players' and ball's trajectories in 3D space. 3D trajectory represents the action with stable global feature as well as ignores the 2D mess and inconstant features caused by occlusion and intra-class variation. However, only 3D global trajectory, detailed features of motion are lost. Multi-view local motion part extracts details of arms and legs in multiple viewpoints. But due to the occlusion and intra-class variation, local motion feature is unstable and incomplete. Through the combination, these two parts remedy mutually and improve the performance.

The whole framework is depicted in Fig. 1. The framework consists of two parts: 3D global trajectory and multi-view local motion. A machine learning model, random forest [5] is selected to do training and prediction separately. For the 3D global trajectory and 2 directional local motion parts, we can get four probability value: $prob_{player}$, $prob_{ball}$, $prob_{2D-left}$ and $prob_{2D-right}$. The combination are conducted in following steps:

Step 1: Sort two multi-view recognition result $prob_{2D-left}$ and $prob_{2D-right}$. Pick the biggest value as $prob_{2D}$.

Step 2: Calculate the final probability $prob_{final}$ by assigning 3D and 2D results different weights according to the action category:

$$prob_{final} = w_1 * prob_{player} + w_2 * prob_{ball} + w_3 * prob_{2D}. \tag{1}$$

2.2 3D Global Trajectory

This proposal utilizes 3D trajectories of players and ball to get global motion features. Compared with 2D image based methods, global motion has 2 merits. Firstly, through multi-players tracking, it solves the occlusion problem caused by interaction between team members. Secondly, due to the utilization of 3D information, this proposal is robust to intra-class variation caused by appearance change. The 3D trajectory remains stable global motion feature and spatial relationship, no matter whether action deformation happens and from which direction the action is observed.

To get the player trajectory and ball trajectory automatically, a multiple players tracking algorithm [6] and a 3D ball tracking algorithm [7] are employed. The player curve feature and the team formation feature are extracted through the players' trajectory. Besides, ball motion state transformation feature is extracted through the ball's trajectory.

Firstly, we define the coordinate system as Fig. 2 shows. The origin of coordinates is located at the center of the court. Define the 3D space as P and to present multiple targets in one team, the jersey number $c \in C$ is used. C is the set of all jersey numbers. Then, the 3D position of player c at time k, X_k^c is defined as:

$$X_k^c = [x_k, y_k, z_k]^T, c \in C, (x_k, y_k, z_k) \in P, \tag{2}$$

given a l-length 3D trajectory of the player c, it is represented with a series of 3D points as follow.

$$Traj^c = (X_1^c, X_2^c \cdots X_i^c \cdots X_l^c), i \in 1 \cdots l, c \in C. \tag{3}$$

Also we define the 3D position of ball at time k, X_k^B and a m-length 3D trajectory of ball as:

$$X_k^B = (x_{ball_k}, y_{ball_k}, z_{ball_k})^T, (x_{ball_k}, y_{ball_k}, z_{ball_k}) \in P, \tag{4}$$

$$Traj^{Ball} = (X_1^B, X_2^B \cdots X_i^B \cdots X_l^B), i \in 1 \cdots m. \tag{5}$$

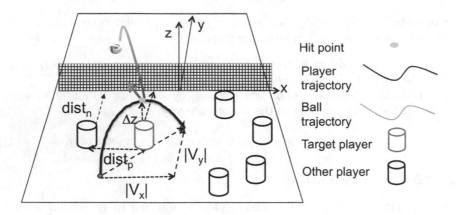

Fig. 2. Concept of global trajectory feature.

Player Curve Feature. This proposal utilizes 3D trajectories of players to get global motion features. These features indicate the player motion curve during the action. As Fig. 2 describes, for the 3D player trajectory, this proposal extracts 4 features including: (1) the maximum jump height Δz, (2) distance to the net at the hit point y_{hit}, (3) the absolute velocity towards the net $|v_y|$ and (4) absolute velocity along the net $|v_x|$. These 4 higher-level features describe target player's $X^T = [x^T, y^T, z^T]$ global motion feature on the court.

These four features are extracted and explained as follows:

The maximum jump height of target player during the trajectory period are extracted as Eq. (6) shows.

$$\Delta z = max\left(z_1^T, z_2^T \cdots z_l^T\right) - min\left(z_1^T, z_2^T \cdots z_l^T\right), \tag{6}$$

for the velocity $|v_y|$, we use linear least square fitting method to approximate the coordinate change Δy according to time t during l length trajectory:

$$|v| = \left| \frac{\Sigma t_i y_i - \Sigma t_i \Sigma y_i / l}{\Sigma t_i^2 - (\Sigma t_i)^2 / l} \right|. \tag{7}$$

To detect the distance to the net at the hit point y_{hit}, the hit point should be detected first. Based on the fact that distance between ball and target player becomes smallest at the hit point.

$$if \ dist_i = min_{i\epsilon\{1\cdots l\}}\left(dist\left(X_i^B, X_i^T\right)\right)$$

$$X_{hit}^T = X_i^T, \tag{8}$$

then the y_{hit} of target player can be obtained as:

$$y_{hit} = y_{hit}^T, hit \ \epsilon \ 1 \cdots l. \tag{9}$$

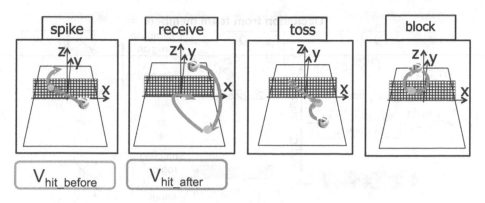

Fig. 3. Typical ball motion models of actions

Team Formation Feature. But only the target player curve information is directly influenced by the appearance of target player. So to lighten the personal appearance influence, make the algorithm more robust for target player change, we proposed the team formation feature.

To present the team formation difference between different action, we extract 5 features including: (1) extreme distance $dist_{p-min}$ and $dist_{p-max}$ between target player and other players; (2) extreme distance $dist_{n-min}$ and $dist_{n-max}$ between net and other players; (3) maximum velocity among other players $|v_{max}|$.

Firstly, for the $dist_{p-min}/dist_{p-max}$, action like block, normally there are two or three players doing the block, so the extreme distance $dist_{p-min}$ is small, while in receive, players stand far to each other, so the $dist_{p-max}$ is large. Secondly, for the $dist_{n-min}/dist_{n-max}$, spike and block usually happen near the net, so the distance to the net $dist_{n-min}$ is relatively small but for receive normally happen far to net, the $dist_{n-max}$ is large. Finally for the $|v_{max}|$, usually after toss the next action is spike, so there must be one or two spiker run towards to the net and jump to spike, so the $|v_{max}|$ is large. But for receive, there is no obviously motion of other players. As the definition before, the extreme distance between ball and player can be easily got according to the trajectory:

$$dist_{min/max} = min/max_{m \in C \setminus \{c\}} \left(min/max_{i \in \{1 \dots l\}} \left(dist \left(X_i^c, X_i^m \right) \right) \right), \qquad (10)$$

and for the velocity, we use linear least square fitting method to approximate the coordinate change according to time t, as Eq. (7) shows.

Ball Motion State Transformation Feature. As the Fig. 3 shows, this proposal extracts ball motion feature around hit-point based on the ball trajectory. Features including the hit point position X_{hit}^B, the ball velocity before the hit v_{hit_before} and ball velocity after the hit v_{hit_after}. Since different actions have different target, so the results of actions are different. As the Fig. 3 shows, obviously for each action the hit point position is different from each other and the ball motion state transformation models varies according to the actions.

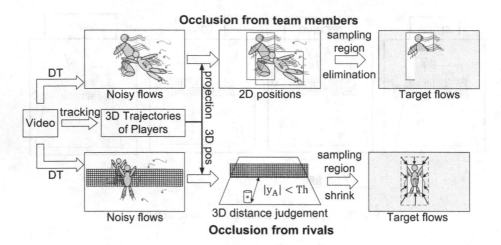

Fig. 4. Concept of player position based noise reduction in DT algorithm

As we defined before, the ball trajectory is described as Eq. (4). For the velocity v_{hit_before} and v_{hit_after}, the difference represent how the ball is acted by the action. Through Eqs. (8), (9), we get the hitpoint h during time $\{1 \cdots m\}$. Then, we use linear least square fitting method to approximate the 3D coordinate change according to time t. Make X direction as an example:

$$v_{hit\_before}^x = \frac{\Sigma t_i x_i - \Sigma t_i \Sigma x_i / k}{\Sigma t_i^2 - (\Sigma t_i)^2 / k}, t \epsilon \{1 \cdots h\}, \tag{11}$$

$$v_{hit\_after}^x = \frac{\Sigma t_i x_i - \Sigma t_i \Sigma x_i / l}{\Sigma t_i^2 - (\Sigma t_i)^2 / l}, t \epsilon \{h \cdots m\}, \tag{12}$$

the other 2 directional velocity changes can be obtained in the same way. Finally we get $v_{hit\_before} = \left[v_{hit\_before}^x, v_{hit\_before}^y, v_{hit\_before}^z \right]$ and $v_{hit\_after} = \left[v_{hit\_after}^x, v_{hit\_after}^y, v_{hit\_after}^z \right]$.

2.3   Multi-view Local Motion

For the local motion part, we employ the dense trajectory [8] method to extract local motion feature. We propose a noise reduction process to remove the noise caused by occlusion problem.

Figure 4 shows the concept of this propose. As Eqs. (13), (14) shows, if the distance between target player and the net is larger than threshold, only trajectories located inside the 2D projected region of the target player and meanwhile not inside 2D regions of other players will be selected. In the second situation, where the 3D distance to the net is less than a threshold, a parameter w is given

Table 1. Data sets of Actions

		Spike	Block	Toss	Receive
Sequence_A	Set1	25	32	20	39
	Set2	28	35	22	49
	Total	53	67	42	88
Sequence_B	Set3	29	36	22	43
	Set4	28	39	24	48
	Total	57	75	46	91

to shrink the 2D region area of the target. This propose shrinks the feature sampling regions of the target player to reduce the overlapping area with rivals and team members.

Here, P_l^c is the new selected feature point of player. $r\left(X_k^c \rightarrow I_k^i\right)$ means the projection from 3D space to 2D region. $|y\left(X_k^c\right)|$ is the distance of player to the net.

$$if\ |y\left(X_k^c\right)| > Th$$

$$P_l^c \epsilon\ r\left(X_k^c \rightarrow I_k^i\right) \&\& P_l^c \notin \bigcup_{b \epsilon C_k \backslash \{c\}} r\left(X_k^b \rightarrow I_k^i\right), \tag{13}$$

else

$$P_l^c \epsilon\ wr\left(X_k^c \rightarrow I_k^i\right) \&\& P_l^c \notin \bigcup_{b \epsilon C_k \backslash \{c\}} r\left(X_k^b \rightarrow I_k^i\right). \tag{14}$$

3   Experimental Results

3.1   Experimental Data and Environment

The experiment is based on multi-view videos of Semifinal Game and Final Game of 2014 Japan Inter High School Games of Men's Volleyball in Tokyo Metropolitan Gymnasium. The cameras are set at each corner of the court. Video resolution is 1920*1080, frame rate is 60 frames per second, and shutter speed of the camera is 1000 frames per second. The latter parameter prevents motion blur in video sequence. For the parameter in Eq. (1), for the spike and receive, the $w1 = 0.6, w2 = 0.25, w3 = 0.15$, for toss and block, the $w1 = 0.35, w2 = 0.4, w3 = 0.25$. The experiments are conducted by programming with c++ language and OpenCV 2.4.10 on a machine of 3.60 GHz CPU and 8 GB RAM.

The semifinal game and final game are named as Sequence_A and Sequence_B, each game consists of 2 sets. The amount of each action clips in sets are described in Table 1. Three classical evaluation criteria defined in [9], the accuracy, recall and precision are used to evaluate the final result.

Table 2. Learn Sequence_A and test Sequence_B

	Criteria	Conventional (best view)	3D trajectory (player)	3D trajectory (ball)	Combination (3D + multi-view 2D)
Spike	Accuracy	80.89%	98.47%	95.06%	99.24%
	Recall	39.74%	94.11%	82.35%	96.80%
	Precision	77.50%	97.96%	91.30%	100%
Block	Accuracy	88.05%	92.02%	93.92%	94.68%
	Recall	53.85%	73.33%	82.67%	85.33%
	Precision	71.79%	98.21%	95.38%	96.58%
Receive	Accuracy	84.64%	98.10%	94.68%	96.58%
	Recall	66.67%	95.60%	97.80%	96.70%
	Precision	81.58%	98.86%	88.12%	93.62%
Toss	Accuracy	84.30%	94.68%	96.58%	98.48%
	Recall	44.07%	78.26%	82.61%	91.30%
	Precision	66.67%	90.00%	97.44%	100%

3.2   Evaluation Result and Discussion

The experiments are conducted in 2 models, Training Sequence_A, testing Sequence_B and training Sequence_B, testing Sequence_A. Since the data sets are from different games, so the result can also show the method robustness for target team change.

In Tables 2 and 3, these two models experimental results are described. Since conventional work [4] didn't consider multi-view processing, we test conventional work on 2 directional viewpoints and pick up the best result for comparison. We can find that the recall of each action is very low. The occlusion problem and the intra-class variation of one action category make local motion features not distinguishable and lack of stability in the training step. Those both 2 reasons cause the low ability to extract relevant information of all kinds of actions.

For the 3D part, thanks to the stability of 3D global feature, we achieve a much better result than local motion, especially for recall. That means the 3D global feature has high ability to extract stable and relevant information in spite of seriously mess information in 2D frames.

For the combing result, we get a better performance. Comparing with 3D results, we got a better precision result. That means that the noise reduction processing reduce mess information and multi-view processing contribute to detect negative actions correctly. Comparing with best viewpoints 2D results, the recall results are improved a lot. This is mainly due to the 3D trajectories represent the global information of action while hiding the unrepresentative local motion. It solves the serious occlusion problem and intra-class variation problem.

Table 3. Learn Sequence_B and test Sequence_A

	Criteria	Conventional (best view)	3D trajectory (player)	3D trajectory (ball)	Combination (3D + multi-view 2D)
Spike	Accuracy	87.79%	96.72%	94.67%	97.54%
	Recall	56.14%	84.91%	77.36%	88.68%
	Precision	94.42%	100%	97.62%	100%
Block	Accuracy	84.64%	95.90%	95.90%	96.31%
	Recall	66.67%	95.52%	94.03%	95.52%
	Precision	81.58%	90.14%	91.30%	91.43%
Receive	Accuracy	86.01%	96.72%	97.13%	97.13%
	Recall	68.82%	94.32%	95.45%	95.46%
	Precision	84.21%	96.51%	96.55%	96.55%
Toss	Accuracy	88.47%	93.03%	90.98%	95.48%
	Recall	45.83%	85.77%	87.22%	90.40%
	Precision	73.33%	100%	85.00%	100%

4 Conclusion

In this paper, we propose a 3D trajectory and multi-view local motion combined method for volleyball player action recognition. 3D global trajectory represents the global motion while multi-view local motion extracts the motion details. By combing 3D global motion and multi-view local motion, we got a good performance in volleyball player action recognition. Experimental results show that the average accuracy achieve 96.93% with a 11.33% improvement. And the average recall achieve 92.53%, 37.36% improvement than conventional work.

Furthermore, we expert to generalize our method to other sports. For sports with obvious 3D trajectory feature, combing the 3D trajectory and 2D motion may give inspiration to solve problems for action recognition in other sports.

Acknowledgment. This work was supported by KAKENHI (16K13006) and Waseda University Grant for Special Research Projects (2018K-302).

References

1. Miyamori, H., Iisaku, S.: Video annotation for content-based retrieval using human behavior analysis and domain knowledge. In: IEEE International Conference on Automatic Face and Gesture Recognition, pp. 320–325 (2000)
2. Ozturk, B., Scahin, P.D.: Recogntion of tennis actions using a depth camera. In: IEEE Signal Processing and Communications Applications Conference (SIU), pp. 1–4, May 2017
3. Chen, H.-T., Chen, H.-S., Lee, S.-Y.: Physics-based ball tracking in volleyball videos with its applications to set type recognition and action detection. In: IEEE International Conference on Acoustics, Speech and Signal Processing, pp. 1097–1100, April 2007

4. Kubota, E., Honda, M., Ikenaga, T.: Action detection of volleyball using features based on clustering of body trajectories. Technical report of IIEEJ, vol. 272, Feburary 2015. (in Japanese)
5. Liaw, A., Wiener, M.: Classification and regression by randomForest. R News **2**, 18–22 (2002)
6. Huang, S.Y., Zhuang, X.Z., Ikoma, N., Honda, M., Ikenaga, T.: Particle filter with least square fitting prediction and spatial relationship based multi-view elimination for 3D volleyball players tracking. In: 12th IEEE Colloquium on Signal Processing & its Applications (CSPA), pp. 28–31, March 2016
7. Cheng, X.N., Ikoma, N., Honda, M., Ikenaga, T.: Anti-occlusion observation model and automatic recovery for multi-view ball tracking in sports analysis. In: 41st IEEE International Conference on Acoustics, Speech and Signal Processing (ICASSP 2016), March 2016
8. Wang, H., Kláser, A., Schmid, C., Liu, C.L.: Action recognition by dense trajectories. In: Proceedings of IEEE Conference on Computer Vision and Pattern Recognition (CVPR), pp. 3169–3176, June 2011
9. Fawcett, T.: An introduction to ROC analysis. Pattern Recogn. Lett. **27**(8), 861–874 (2006)

Underwater Image Enhancement by the Combination of Dehazing and Color Correction

Wenhao Zhang, Ge Li$^{(\boxtimes)}$, and Zhenqiang Ying

Peking University Shenzhen Graduate School, Lishui Road 2199, Nanshan District, Shenzhen 518055, Guangdong Province, China
1601214037@sz.pku.edu.cn, geli@ece.pku.edu.cn, zqying@pku.edu.cn

Abstract. Underwater image processing is crucial for many practical applications in the ocean filed, which is not a trivial thing since the environment of underwater is often complicated and short of light. The major difficulty is that a captured image is fuzzy, under-exposed and often has the color cast due to the fact that the light is absorbed and scattered. To overcome those difficulties, we propose a new underwater image enhancement method, which is composed of two successive vital processings: the dehazing and color correction. Firstly, considering the characteristic of light propagation under water, we propose a new dehazing algorithm to restore the visibility of degraded underwater images based on the dark channel prior, through building up the relationship of the transmission rates among three color channels. Then, to further improve the image quality, we adopt an effective color correction method on the obtained haze-free underwater images. We conduct extensive experiments under measure tests of both subjective and objective, and the results show that our method is superior to several existing approaches.

Keywords: Underwater image enhancement · Image dehazing
Color correction

1 Introduction

Underwater image processing is a very useful technique for ocean applications, such as aquatic robot inspection [21], underwater rescue and salvage tasks. Although the research in this area has been conducted for several years, the performance is still unsatisfied due to severe underwater environment. The major difficulty is that the captured image is fuzzy, under-exposure and color cast due to the absorption and scattering of the light [5,21]. Specifically, due to severe light absorption, the captured image is under-exposed. Meanwhile, since different wavelengths of light have different absorbing characteristic [5], the captured image has a more severe color cast than that filmed on the land. As illustrated in Fig. 4, a longer wavelength will experience worse attenuation along the travel path. The blue and green color travel further in the water due to the shorter wavelengths. This is the reason why underwater images are dominated by blue

© Springer Nature Switzerland AG 2018
R. Hong et al. (Eds.): PCM 2018, LNCS 11166, pp. 145–155, 2018.
https://doi.org/10.1007/978-3-030-00764-5_14

Fig. 1. **Top Row:** three raw underwater images. **Bottom Row:** the enhanced results by our method

or green color [21]. In addition, the light is scattered by the suspended particles in water which results in the fuzzy of image [18] (Fig. 1).

Several efforts have been made to tackle those issues. Some works attempt to enhance underwater image via the means adopted for generic image enhancement. Jaspers [14] proposed Histogram Equalization (HE). Based on HE, Zuiderveld proposed Contrast Limited Adaptive Histogram Equalization (CLAHE), Deng et al. proposed Generalized Unsharp Masking (GUM) to improve the sharpness of image, and Fu [8] proposed a Probability-Based method (PB) from the perspective of statistical. HE method can improve the contrast simply and well. However, HE method can't improve the contrast in proportion to the distance from the object to the camera, so it may lead in artifacts when processing underwater image.

In recent years, some dehazing methods have been proposed. They can greatly remove the scattering effects caused by the fog. Fattal [6] employed a method based on the fact that the surface shading and transmission functions are statistically uncorrelated in local to remove the fog. He et al. [15] proposed the dark channel prior that haze-free images have at least one color channel with a very low intensity, therefore one can estimate the transmission map with the dark channel prior. A few dehazing methods have made improvements based on the dark channel prior [16]. Directly applying the dehazing methods for underwater images can also make some improvement [7]. However, the dehazing methods applied for the terrestrial environment only consider the scattering and ignore the fact that underwater exists the color cast if adopted for underwater image enhancement straightway. Therefore, they can't provide satisfactory color correction for the reduction in the red channel caused by the absorption.

Among the research works for underwater image enhancement, there are some methods specially designed to solve the underwater imaging problems. Carlevaris-Bianco [3] exploited the strong difference in attenuation rates under water among the three color channels to estimate the depth of the scene. Chiang and Chen [4] exploited the dark channel prior to remove the scattering effects and exploited different wavelength attenuation rates to compensate the degraded color, which can effectively restore image color balance and remove haze-like effect. However, it has large computing overhead. Ancuti and Ancuti [2] made a confusion with two methods, the white balance and the bilateral filter, to considerably improve the visibility range. However, distant region in the image is not considering in [2]. Fu et al. [9] used the retinex-based model to decompose the reflectance and illumination from a single underwater image, then made the enhancement independently. However, the sharpness of result image was distorted. Wen et al. [22] proposed a new optical model to enhance the single underwater image. However, some assumptions made in the Wen's method are not suitable for many underwater situations.

According to the previous research, we attempt to propose a new underwater image enhancement method, which can get better visual quality improvement than existing methods and can be suitable for most underwater degraded images. Firstly, we propose an new method for underwater image dehazing based on the characteristic of light propagation under water and the dark channel prior. The proposed dehazing algorithm can build up the relationship of the transmission rates among three color channels to restore the visibility of degraded underwater images. Besides, an effective color enhancement method is adopted to solve the color cast of the obtained haze-free underwater images.

In summary, the major contributions of this paper are two-folds:

(1) A new underwater image dehazing algorithm is proposed, which based on the characteristic of light propagation under water and the dark channel prior.
(2) The relationship of the transmission rates among three color channels is built up to acquire better image quality enhancement.

The rest of the paper is organized as follows. In Sect. 2, the proposed methods are presented; In Sect. 3, the experimental results are evaluated and discussed. In Sect. 4, conclusions are given.

2   Proposed Method

The proposed method mainly includes underwater image dehazing and color correction. A flowchart of the proposed method is shown in Fig. 2.

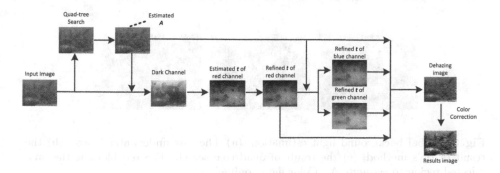

Fig. 2. Flowchart of the proposed new method.

2.1   Underwater Image Dehazing

The radiance of image captured under water can be modeled as two components: the radiance reflected from the observed scene and the atmospheric light from the ambient environment. Mathematically, this can be written as [15]:

$$I^c(x) = t^c(x)J^c(x) + (1 - t^c(x))A^c, c \in \{r, g, b\}. \tag{1}$$

where x is one point in the scene, c is color channel, I is the observed intensity, J is the real scene radiance, A is the global background light, and t is the medium transmission rate describing the portion of the light reflected from the observed scene that is not scattered and reaches the camera.

The goal of haze removal is to recover J, A, and t from I. So, we estimate the global background light A, the transmission rates of three color channels t, and the real scene radiance J.

Global Background Light Estimation. In an underwater image, the global background light A, in Eq. (1) is often assumed as the brightest color. However, the assumption is not suitable in a scene where objects are brighter than the global background light [15]. To robustly estimate A, we exploit the fact that the variance of the pixel values is lower, when the haze is thicker [16].

In this paper, we adopt the quad-tree search method to seek for the global background light. As shown in Fig. 3(c). First, we divide the image into four regions with the same size. Then, we calculate the average of the pixels subtracted by the standard deviation of the pixels within each region. We pick the region that has the highest values as the candidate region for the next iteration. The minimum of the image's size is set as 10×10. When the size is less or equal the set value, the search gets finished. Then, we propose a new method to calculate the global background light by computing the average value of the pixels within the final region.

As shown in Fig. 3, our method can seek the proper location of A lead to an accurate value of background light compare to He's method [15], which averaged top 0.1 lightest pixels of the image, filtered by the minimum operation in three channels (RGB), to calculate A. Therefore He's method often mistakes the white object as global background light, as illustrated in Fig. 3(b). Our estimation

(a)   (b)   (c)

Fig. 3. Global background light estimation. (a) The raw underwater image, (b) the result of He's method, (c) the result of quad-tree search. The red block is the final selected region to estimate A. (Color figure online)

of the global background light reflects the effects of absorption on underwater imaging, i.e. the global background light is blue or green, not white.

Estimation of Transmission Rate. For the haze image captured in the terrestrial environment, the haze can be regarded as a mask covering on the real natural image. As described in Eq. (1), the haze image is the composition of the real image and the atmospheric light according to the variance of t. So we also can regard the underwater image as the composition of the real image and the atmospheric light in the water.

According to He's dark channel prior [15], the transmission rates of underwater image can be described as follows:

$$min_c(\tilde{t}^c(x)) = 1 - min_c \left\{ \frac{\min\limits_{y \in \Omega(x)} (I^c(y))}{A^c} \right\}. \tag{2}$$

where $\Omega(x)$ denotes one patch, $\tilde{t}(x)$ denotes the estimate value of transmission rate. Meanwhile, according to the Fig. 4, the t of red channel is the minimum among three channels, we can approximately regard the transmission rate of red channel as the transmission rate of full image. So we can get that:

$$\tilde{t}^r(x) = min_c(\tilde{t}^c(x)) \tag{3}$$

Caused by block-wise calculation, $\tilde{t}^r$ will exist the block effect. We can use the guided filter [13] to refine the transmission rate image of red color.

The transmission rate is based on the Lambert-Beer law for transparent objects, which states that light traveling through a transparent material will be attenuated exponentially [12]:

$$t^c(x) = \exp(-\beta^c d(x)), \tag{4}$$

where $d(x)$ is the distance from observed scene to camera, β is the attenuation coefficient. According to Li's research [17], the global background light is directly proportional to the ratio of scattering coefficient and attenuation coefficient:

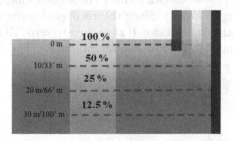

Fig. 4. Different wavelengths of light are attenuated at different rates in water [4]. (Color figure online)

$$A^c(\infty) \propto \frac{b_{sc}^c}{\beta^c},\tag{5}$$

where $A^c(\infty)$ means the air light from the infinity distance to the camera (i.e. the global background light, which has been estimated at above). And b_{sc} is the scattering coefficient. Meanwhile, Gould et al. [1] discovered that the scattering coefficient has a linear relationship with wavelength as follows:

$$b_{sc}^c = (-0.00113\lambda_c + 1.62517)\tau(\lambda_c),\tag{6}$$

where λ_c is the wavelength of channel c. The wavelength of red, green and blue channel are 620 nm, 540 nm, 450 nm, respectively. $\tau(\lambda_c)$ can simply be regarded as a constant for three channels. By putting Eqs. (5) and (6) into Eq. (4), we can build the relationship of transmission rates among three channels as follows:

$$\tilde{t}^g(x) = (\tilde{t}^r(x))^{\frac{\beta^g}{\beta^r}},\tag{7}$$

$$\tilde{t}^b(x) = (\tilde{t}^r(x))^{\frac{\beta^b}{\beta^r}}.\tag{8}$$

where $\tilde{t}^g$ and $\tilde{t}^b$ denote the estimated values. The colormap of transmission rates of three channels is shown in Fig. 5. Obviously to see that three channels' transmission rates have much difference. We can't simply think that three channels have the same transmission rate, liking in [3].

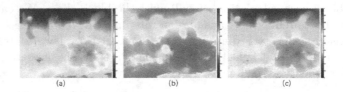

(a)   (b)   (c)

Fig. 5. The colormap of transmission rates. From left to right: red channel, green channel, blue channel. (Color figure online)

Recovering the Scene Radiance. With the estimated global background light and transmission rates, we can recover the scene radiance of three channels respectively, according to Eq. (1). Since the recovered scene radiance term $J(x)$ may be infinite when the transmission $t(x)$ is close to zero. To avoid this difficulty, we restrict the transmission $t(x)$ to a lower bound t_0. The final scene radiance $J(x)$ is recovered by:

$$J^c(x) = A^c + \frac{(I^c(x) - A^c)}{max(\tilde{t}^c(x), t_0)},\tag{9}$$

where t_0 is set to 0.1 [15].

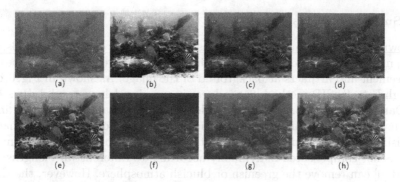

Fig. 6. Subjective measurement: (a) raw underwater images, (b) HE-Lab method, (c) SID method, (d) DCP method, (e) Fu method, (f) new optical method, (g) our dehazed results, (h) our color correction results. (Color figure online)

2.2 Underwater Image Color Correction

Since the blue and green light travel the longer distance through water for the shorter wavelengths, most underwater images appear greenish or bluish. To address the color cast, a color correction method based on gray world and white patch is adopted [11]. The method is a simplified model of the human visual system: the enhancement process is consistent with perception. Meanwhile, the method may be seen as a smoothed and localized modification of uniform histogram equalization.

For a color image, the following operation is performed independently on the RGB channels:

$$R(x) = \sum_{y \in \Omega \setminus x} \frac{s_\alpha(I(x) - I(y))}{\|x - y\|}, x \in \Omega \qquad (10)$$

where I denotes one channel in a color image with domain Ω, $\Omega \setminus x$ denotes $y \in \Omega : y \neq x$, $\|x - y\|$ denotes Euclidean distance.

$$s_\alpha(t) = \min\{\max\{\alpha t, -1\}, 1\}, \alpha \geq 1 \qquad (11)$$

In the second stage, the enhanced channel is computed by stretching R to [0, 1] as

$$L(x) = \frac{R(x) - \min R}{\max R - \min R}. \qquad (12)$$

3 Experimental Results and Discussion

3.1 Dataset

Due to the lack of underwater image datasets, we create a new dataset. The images are collected from other researchers' experiments[1], including some classical underwater images. The full image dataset has nearly 50 images.

---

[1] https://sites.google.com/site/kyutech8luhuimin/underwater_image_datasets.

3.2   Subjective Measurement

As shown in Fig. 6, we make comparisons among several existing methods. HE-Lab [3] method can improve the contrast in Fig. 6(b). However, it may cause over-enhancement and artifacts at background of the image. As shown in Fig. 6(c–d), although Fattal's SID method [6] produces impressive dehaze results, He's DCP [15] (Dark Channel Prior) method is able to handle distant objects even in the heavy haze image, they don't consider the characteristic of underwater imaging. The visibility, color, and details are not good enough because the attenuated energy is not compensated individually according to different wavelengths. Fu's method [9] can remove the greenish or blueish atmosphere. However, the sharpness of image will be degraded as displayed in Fig. 6(e). Wen's method [22] proposed a new optical model to enhance the single underwater image. However, the method will lead in artifacted halo and distortion. As shown in Fig. 6(f), the visual effect of image is bad. As shown in Fig. 6(g–h), our underwater enhancement algorithm can effectively remove the haze-like effect in image and correct the color cast.

3.3   Objective Measurement

Apart from subjective measurement, we make objective comparisons using different evaluation criterion. NIQE [19] is one of the methods to quantize the distortion degree of natural images. CNI is an evaluation criterion for natural color image enhancement, which is based on human visual system. Recently, UCIQE [23] has been proposed by Yang Miao for quantifying the nonuniform color cast, blurring, and low contrast of underwater images. Higher UCIQE values indicate the image has better balance among the chroma, saturation, and contrast. Meanwhile, Panetta proposed UIQM [20] to measure the quality of underwater image. UIQM includes three components and can be described as follows:

$$UIQM = c_1 \times UICM + c_2 \times UISM + c_3 \times UIConM, \qquad (13)$$

where UICM is underwater image colorfulness measure, UISM is underwater image sharpness measure, UIConM is underwater image contrast measure. They are linearly combined together with the weight [20]: $c_1 = 0.0282$, $c_2 = 0.2953$, $c_3 = 3.5753$.

In the Table 1, the values in bold represent the best results. As shown in Table 1, our enhancement method stands out among the compared methods in terms of NIQE, CNI, UCIQE and UIQM values. The highest NIQE and CNI scores indicate that our method can effectively restore the degraded image and preserve the natural appearance. The best UCIQE and UIQM values mean that our enhancement method performance better than other methods, when processing underwater image. As shown in Table 1, the values of CNI, UCIQE and UIQM for SID method and DCP method are all the lowest. It shows that dehazing only methods are not suitable for underwater image enhancement.

Table 1. Comparison of objective measurements

Method	NIQE	CNI	UCIQE	UIQM
HE-Lab	4.422	0.605	0.589	1.397
SID	5.240	0.539	0.541	1.336
DCP	4.848	0.572	0.500	1.452
Fu	4.187	0.611	0.598	1.625
New_optical	4.372	0.626	0.560	1.589
Ours	**4.053**	**0.688**	**0.606**	**1.708**

Additionally, we further test the proposed method on three different datasets, which come from Lu's personal website (See footnote 1), in order to demonstrate the effectiveness and robustness of the proposed method. These datasets are collected from researcher's papers: Ancuti dataset [2], CB dataset [3], Galdran dataset [10]. In Table 2, the values in bold represent the best results. We can see that our method can score the highest values of UIQM and UCIQE, which shows the robustness of our method.

Table 2. Comparison of objective measurements on different datasets

	Ancuti		CB		Galdran	
	uciqe	uiqm	uciqe	uiqm	uciqe	uiqm
HE-Lab	0.57	1.36	0.65	1.82	0.58	1.37
SID	0.48	1.31	0.66	1.67	0.53	1.31
DCP	0.46	1.52	0.64	1.88	0.49	1.44
Fu	0.58	1.59	0.65	1.86	0.60	1.66
New_optical	0.49	1.63	0.66	1.93	0.56	1.58
Ours	**0.60**	**1.79**	**0.68**	**2.14**	**0.61**	**1.73**

4   Conclusion

In this paper, we propose a new underwater image enhancement method that can resolve the problems caused by the light absorption and particle scattering existed in the environment under water. The method improves the visual quality of images and corrects the distorted color cast of raw images. The proposed method is consisted of two components: an effective dehazing algorithm specially designed for underwater image and a general color correction algorithm. First, based on the knowledge of the dark channel prior, the optical properties of underwater imaging and the relationship of transmission rates among three color channels, we can effectively remove the haze-like effect presenting on the underwater image. Second, by adopting color correction method, we can greatly adjust the color distribution of an underwater image to look like be more natural. Experiment results demonstrate that our method is superior than some existing methods in terms of genuine color, natural appearance, and visibility. Besides, our method is robust among different underwater image datasets.

Acknowledgements. This work was supported by the Project of National Engineering Laboratory for Video Technology -Shenzhen Division, Shenzhen Key Laboratory for Intelligent Multimedia and Virtual Reality under Grant ZDSYS201703031405467, and the Shenzhen Municipal Development and Reform Commission (Disciplinary Development Program for Data Science and Intelligent Computing) under Grant 1230233753.

References

1. Spectral dependence of the scattering coefficient in case: 1 and case 2 waters. Appl. Opt. **38**(12), 2377–2383 (1999). https://doi.org/10.1364/ao.38.002377. http://www.ncbi.nlm.nih.gov/pubmed/18319803
2. Ancuti, C., Ancuti, C.O., Haber, T., Bekaert, P.: Enhancing underwater images and videos by fusion. In: Proceedings of the IEEE Computer Society Conference on Computer Vision and Pattern Recognition, pp. 81–88 (2012). https://doi.org/10.1109/CVPR.2012.6247661
3. Carlevaris-Bianco, N., Mohan, A., Eustice, R.M.: Initial results in underwater single image dehazing. In: Oceans 2010, pp. 1–8. IEEE (2010)
4. Chiang, J.Y., Chen, Y.C.: Underwater image enhancement by wavelength compensation and dehazing. IEEE Trans. Image Process. **21**(4), 1756–1769 (2012)
5. Duntley, S.Q.: Light in the sea. J. Opt. Soc. Am. **53**(2), 214–233 (1963)
6. Fattal, R.: Single image dehazing. ACM Trans. Graph. **27**(3), 1 (2008). https://doi.org/10.1145/1360612.1360671
7. Fattal, R.: Dehazing using color-lines. ACM Trans. Graph. (TOG) **34**(1), 13 (2014)
8. Fu, X., Liao, Y., Zeng, D., Huang, Y., Zhang, X.P., Ding, X.: A probabilistic method for image enhancement with simultaneous illumination and reflectance estimation. IEEE Trans. Image Process. **24**(12), 4965 (2015). A Publication of the IEEE Signal Processing Society
9. Fu, X., Zhuang, P., Huang, Y., Liao, Y., Zhang, X.P., Ding, X.: A retinex-based enhancing approach for single underwater image. In: 2014 IEEE International Conference on Image Processing (ICIP), pp. 4572–4576. IEEE (2014)
10. Galdran, A., Pardo, D., Picón, A., Alvarez-Gila, A.: Automatic red-channel underwater image restoration. J. Vis. Commun. Image Represent. **26**, 132–145 (2015)
11. Getreuer, P.: Automatic color enhancement (ACE) and its fast implementation. Image Process. Line **2**, 266–277 (2012)
12. Gordon, H.R.: Can the lambert-beer law be applied to the diffuse attenuation coefficient of ocean water? Limnol. Oceanogr. **34**(8), 1389–1409 (1989)
13. He, K., Sun, J., Tang, X.: Guided image filtering. In: Daniilidis, K., Maragos, P., Paragios, N. (eds.) ECCV 2010. LNCS, vol. 6311, pp. 1–14. Springer, Heidelberg (2010). https://doi.org/10.1007/978-3-642-15549-9_1
14. Jaspers, C.A.M.: Histogram equalization (2004)
15. He, K., Sun, J., Tang, X.: Single image haze removal using dark channel prior. IEEE Trans. Pattern Anal. Mach. Intell. **33**(12), 2341–2353 (2010)
16. Kim, J.H., Jang, W.D., Sim, J.Y., Kim, C.S.: Optimized contrast enhancement for real-time image and video dehazing. J. Vis. Commun. Image Represent. **24**(3), 410–425 (2013)

17. Li, C.Y., Guo, J.C., Cong, R.M., Pang, Y.W., Wang, B.: Underwater image enhancement by dehazing with minimum information loss and histogram distribution prior. IEEE Trans. Image Process. **25**(12), 5664–5677 (2016). https://doi.org/10.1109/TIP.2016.2612882

18. Lu, H., Li, Y., Zhang, L., Serikawa, S.: Contrast enhancement for images in turbid water. J. Opt. Soci. Am. A Opt. Image Sci. Vis. **32**(5), 886 (2015)

19. Mittal, A., Soundararajan, R., Bovik, A.C.: Making a "completely blind" image quality analyzer. IEEE Signal Process. Lett. **20**(3), 209–212 (2013)

20. Panetta, K., Gao, C., Agaian, S.: Human-visual-system-inspired underwater image quality measures. IEEE J. Ocean. Eng. **41**(3), 541–551 (2016). https://doi.org/10.1109/JOE.2015.2469915

21. Torres-Méndez, L.A., Dudek, G.: Color correction of underwater images for aquatic robot inspection. In: Rangarajan, A., Vemuri, B., Yuille, A.L. (eds.) EMMCVPR 2005. LNCS, vol. 3757, pp. 60–73. Springer, Heidelberg (2005). https://doi.org/10.1007/11585978_5

22. Wen, H.: Single underwater image enhancement with a new optical model. In: ISCAS, pp. 753–756 (2013)

23. Yang, M., Sowmya, A.: An underwater color image quality evaluation metric. IEEE Trans. Image Process. **24**(12), 6062–6071 (2015)

A Novel No-Reference QoE Assessment
Model for Frame Freezing
of Mobile Video

Bing Wang[1]([✉]), Qiang Peng[1], Xiao Wu[1], Eric Wang[2], and Wei Xiang[2]

[1] Southwest Jiaotong University, Chengdu, China
iceice_wang@outlook.com
[2] James Cook University, Cairns, Australia

Abstract. In this paper, a novel no-reference (NR) Quality of Experience (QoE) assessment model for frame freezing of mobile video is proposed. Four source video sequences with smooth motion intensity which were extracted from LIVE mobile database have been used to create different types of test sequences. Two subjective experiments are conducted with these distorted sequences, and the Differential Mean Opinion Scores (DMOS) are obtained. Then a QoE model is proposed based on the experimental results. This model can quantitatively measure the perceptual quality of users' experience when they are watching the frame freezing videos. Due to the lack of publicly available datasets, we establish a new database of mobile videos with frame freezing distortion based on the LIVE mobile database. The proposed model is compared with three other QoE assessment metrics on the new database, and the result shows the proposed model has a better performance than others.

Keywords: Mobile video · Video quality assessment
Quality of experience · Frame freezing

1 Introduction

QoE measures the overall perceived quality of video delivered from subjective users' experience and objective system performance. To improve the quality of mobile video services, researchers from academics and industry service providers have focused on developing QoE models to predict overall user-perceived quality for optimizing quality provision.

Frame freezing is a common video temporal distortion in video applications due to both packet losses and delays in the real-time mobile video transmission system and significantly affects the video fluency. Therefore, the assessment of frame freezing can be used as a measurement of QoE because frame freezing is one of the biggest influencing factors of the users' experience. There are several related works for QoE metrics based on frame freezing [1–6]. In [1], the author proposed a method for detecting drop video frames which are divided into drop (multi-frame duration) and dip (one frame duration). Then other scholars used

© Springer Nature Switzerland AG 2018
R. Hong et al. (Eds.): PCM 2018, LNCS 11166, pp. 156–165, 2018.
https://doi.org/10.1007/978-3-030-00764-5_15

the same method to detect frame freezing position [2]. This method can be applied to different resolutions and variable frame rate video sequences, but it neglected the different content of each video. A full-reference (FR) video assessment based on frame freezing was proposed in [3], but the performance was not satisfactory. In [4], an NR metric for measuring the quality of frame freezing was presented. However, this method not considered the impact of video content for final evaluation scores. In [5], a novel NR metric based on neural network for frame freezing was proposed. This method required a lot of data to be trained, and the processing time was huge. According to the decoder buffer overflow, a new frame freezing assessment method was proposed in [6]. All of the above QoE models based on frame freezing are not oriented to mobile devices. The existing QoE models are limited to apply to mobile video.

To solve the shortcoming of previous works, we firstly proposed a mobile video oriented frame freezing QoE model. This model can quantitatively measure the value of QoE when users are watching frame freezing video sequences. Then we establish a new frame freezing video database to compare the proposed metric with other three methods, and the results show the proposed model has the best performance. The remainder of the paper is organized as follows: in Sect. 2, we introduce two subjective experiments based on frame freezing and establish our QoE model. The proposed frame freeze database and comparative experiment are given in Sect. 3. This is followed by conclusion in Sect. 4.

2 QoE Model

In this section, we conduct two subjective experiments to establish our QoE models based on frame freezing. The settings and procedures of the study are described in following sections.

2.1 QoE Model of Single Frame Freezing Event

We choose 10 source video sequences, (a) *fc (Friend Drinking Coke)*, (b) *sd: (Two Swan Dunking)*, (c) *rb: (Runners Skinny Guy)*, (d) *ss: (Students Across Street)*, (e) *bf: (Bulldozer With Fence)*, (f) *po: (Panning Under Oak)*, (g) *la: (Landing Airplane)*, (h) *dv: (Springs Pool Diving)*, (i)*tk: (Trail Pink Kid)*, (j) *hc: (Harmonicat)*, from the LIVE mobile database [7].

The same type of frame freezing occurred at the position with different motion intensity has different effects on the users' QoE. Hence, (1) is used to compute motion intensity of the 10 source video sequences.

$$MI = \frac{1}{N} \times \sum_{K=1}^{M} \sqrt{mv(k)_x^2 + mv(k)_y^2} \tag{1}$$

MI is the motion intensity of Region of Interest (ROI) in source sequences, N is the number of the macroblock in a frame and $(mv(k)_x, mv(k)_y)$ is the motion vector of the kth macroblock. $(mv(k)_x, mv(k)_y)$ equals to $(0,0)$ when the macroblock belongs to the background region. Motion estimation for the macroblock in ROI is used to obtain the motion vector of ROI, and the motion vector of background is set to 0. Based on the (1), the motion intensity figures of 10 source sequences are gained, and shown in Fig. 1. Four source video sequences (fc, ss, bf and dv) are selected because their motion intensity are gentle ($\Delta MI \approx 1$).

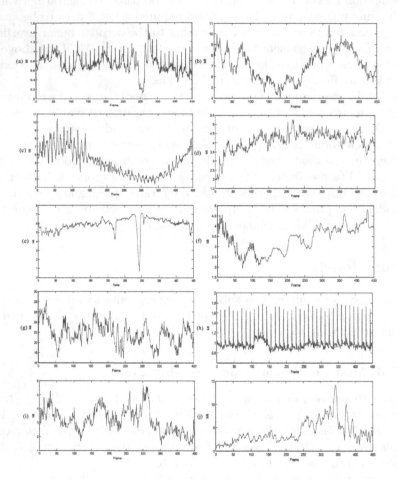

Fig. 1. The motion intensity of the 10 source sequences. (a) fc sequence, (b) sd sequence, (c) rb sequence, (d) ss sequence, (e) bf sequence, (f) po sequence, (g) la sequence, (h) dv sequence, (i) tk sequence, (j) hc sequence.

Each source video produces 10 frame freezing sequences which only have one frame freezing event. The duration of single frame freezing event are 0.1 s, 0.2 s, 0.4 s, 0.9 s, 1.6 s, 2.2 s, 3.1 s, 4 s, 5 s and 6 s for the distorted sequences respectively. Single frame freezing event is generated randomly in the position where the motion intensity is smooth.

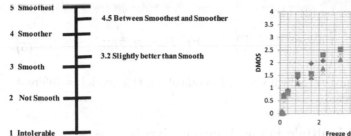

Fig. 2. The improved five-level scale.   **Fig. 3.** The relationship between DMOS and the duration of frame freeze.

The subjective experiments are designed in agreement with the recommendation of ITU-R BT.500-13 [8]. The number of participants for the subjective experiment is 22, and all aged from 18 to 35 years old. All participants are volunteers. The handheld mobile device is IPad mini2 which is used for displaying video sequences.

We use the five-level scale in the subjective experiments. The five-level scale in BT.500 is divided into excellent, good, fair, poor, bad. The frame freezing mainly affects the fluency of video display, so the excellent, good, fair, poor and bad was transformed into smoothest, smoother, smooth, not smooth and intolerable in our model. The improved five-level scale model is shown in Fig. 2. The score from 1 to 5 points can be accurate to one decimal place.

Our subjective experiments use the Subjective Assessment Methodology for Video Quality (SAMVIQ) described in ITU-R BT.1788 [9], and the DMOS values of frame freezing sequences are obtained. Experimental results are shown in Fig. 3. The DMOS values of the four videos are rising with the increase of the duration of single frame freezing and tend to a limit value of DMOS. When freeze duration is less than 0.1 s, people can hardly feel slowness in dv and fc due to their weak motion intensity, bf and ss are contrary. Therefore, motion intensity has an important impact on user's QoE of frame freezing. Then a new QoE model is defined as (2).

$$Dmos_p = a \times e^{(-b \times (ft-c))} + d \tag{2}$$

where $Dmos_p$ is the prediction of perceptual quality, ft is the duration of one frame freezing event. a, b, c, d are model parameters, and a, b, d are defined empirically by (3), c is obtained by (4). M_{seq} is the average motion intensity of the whole sequence. MI is the predicted value of motion intensity of the corresponding position of the frame freezing, and fps is the frame rate.

$$\begin{cases} a = -4.0, b = -4.0, d = 4.0, M_{seq} > 1.5 \\ a = -3.3, b = -3.3, d = 3.3, M_{seq} \leq 1.5 \end{cases} \tag{3}$$

$$c = \frac{1}{MI \times fps} \tag{4}$$

Then, Dmosp is derived by (2)–(4).

$$Dmos_p = \begin{cases} a \times e^{(b \times (ft - \frac{1}{MI \times fps}))} + d, \; ft > \frac{1}{MI \times fps} \\ 0, \hspace{4.5cm} other \end{cases} \tag{5}$$

Note that the training set which used for establishing this model is different from the test set in Sect. 3.

2.2 QoE Model of Multiple Frame Freezing Events

In the traditional subjective experiments of multiple frame freezing events, the frequency of frame freezing events was ignored. Hence, we conduct a subjective experiment mainly about the relationship between the frequency of frame freezing and QoE. Four sequences (fc, ss, bf and dv) are selected, and each of them produces 12 frame freezing sequences, as shown in Table 1. FT is the freeze duration, FTI is the inter-freeze distance.

Table 1. Multiple frames freezing distortion types.

Sympol	Distortion type
DT1	Occur in 3 s, $FT = 0.067$ s, $FTI = 0.1$ s, 5times
DT2	Occur in 3 s, $FT = 0.067$ s, $FTI = 0.1$ s, 5times
DT3	Occur in 3 s, $FT = 0.067$ s, $FTI = 0.1$ s, 5times
DT4	Occur in 3 s, $FT = 0.067$ s, $FTI = 0.1$ s, 5times
DT5	Occur in 3 s, $FT = 0.067$ s, $FTI = 0.1$ s, 5times
DT6	Occur in 3 s, $FT = 0.067$ s, $FTI = 0.1$ s, 5times
DT7	Occur in 3 s, $FT = 0.067$ s, $FTI = 0.1$ s, 5times
DT8	Occur in 3 s, $FT = 0.067$ s, $FTI = 0.1$ s, 5times
DT9	Occur in 3 s, $FT = 0.067$ s, $FTI = 0.1$ s, 5times
DT10	Occur in 3 s, $FT = 0.067$ s, $FTI = 0.1$ s, 5times
DT11	Occur in 3 s, $FT = 0.067$ s, $FTI = 0.1$ s, 5times
DT12	Occur in 3 s, $FT = 0.067$ s, $FTI = 0.1$ s, 5times

The experimental procedure is the same as the subjective experiment of single frame freezing event. The results are shown in Figs. 4 and 5.

Hence, the inter-freeze distance between the two adjacent frames has little impact on user's QoE, the impact of each frame freezing on the subjective experience is relatively independent. The pooling strategy of QoE is defined as (6), where $ft(i)$ is the ith frame freezing event, d, a and b are obtained by (6). $c(i)$ is the value of parameter c, obtained by (4). N is the sum of the frame freezing events in an entire sequence, QoE_{Freeze} is the value of QoE based on frame freezing.

$$QoE_{Freeze} = a \times \prod_{i=1}^{N} e^{(-b \times (ft(i) - c(i)))} + d \tag{6}$$

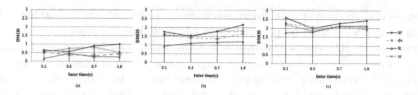

Fig. 4. The relationship between DMOS and the inter-freeze distance at different freeze duration. (a) single freeze duration is 0.0067 s, (b) single freeze duration is 0.2 s, (c) single freeze duration is 0.4s.

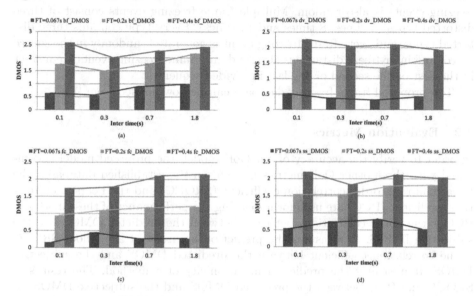

Fig. 5. The relationship between DMOS and the inter-freeze distance of different sequences. (a) bf sequence, (b) dv sequence, (c) fc sequence, (d) ss sequence

3   Database and Experiment Results

3.1   The Database for Frame Freezing Video

The existing databases for frame freezing are VQEGHD dataset [10] and LIVE mobile database [7]. VQEGHD dataset contains 7 source videos, 28 frame freezing distortion sequences (packet loss). LIVE mobile database contains 10 source videos, 40 frame freezing sequences (packet loss and packet delay). In these databases, the number of frame freezing sequences is small and frame freezing sequences are all multiple frames freezing events. So in this paper, we establish a new frame freezing database based on the LIVE mobile database, which contains single frame freezing events, multiple frames freezing events and all sequences caused by packet loss.

Six sequences (hc, la, po, rb, sd, tk) are selected as source video sequences from LIVE mobile database. Our database consists of 30 single frame freezing sequences and 78 multi-frame freezing sequences (each sequence contains 6 distortion sequences caused by packet loss in LIVE mobile database).

Research in [11] shows the frame freezing event can be perceived by the humans' eyes when the duration of a single frame freezing event exceeds 0.2 s, so a single frame freezing event in our database lasts at least 0.2 s. We randomly generate one frame freezing event in a different position of the source video. Single frame freezing includes five distorted types, and the duration of the frame freezing are 0.2 s, 0.5 s, 1 s, 2 s and 3.2 s respectively. Then we randomly generate three frame freezing events in the source video, and the distance between frame freezing events is also random. Multiple frame freezing events consist of three distorted types (5 times, 8 times and 12 times). Every type contains three distorted types, and every frame freezing event is generated randomly in the source video, and inter-freeze distance is also random. Frame freezing events are not set in the first or the second or the last in a video sequence.

The proposed frame freezing database can be downloaded from [13].

3.2   Evaluation Metrics

In order to verify the accuracy of our QoE model, the proposed model is compared with three other QoE metrics in [3–5] on the established database. The Spearman rank order correlation coefficient (SROCC) and the Pearson correlation coefficient (PCC) are used for measuring the performance of these methods. PCC is the linear correlation coefficient between the predicted DMOS and the subjective DMOS. It measures the prediction accuracy of a method. SROCC is the correlation coefficient between the predicted DMOS and the subjective DMOS. It measures the prediction monotonicity of a method. The results of SROCC and PCC between the predicted DMOS and the subjective DMOS for different QoE quality assessment methods are shown in Tables 2 and 3.

Table 2. The SROCC of the different methods.

Method	SROCC			Vs. Proposed
	Single	Multi	All	
VQM_VFD [3]	0.5072	0.2747	0.6508	−30.97%
Hyun_Thu [4]	0.9607	0.1156	0.6335	−32.70%
Xue [5] (whole set)	\	0.9203	0.9203	−4.02%
Xue [5] (training set)	\	0.9550	0.9550	\
Xue [5] (testing set)	\	0.8601	0.8601	−10.04%
Proposed	**0.9421**	**0.8852**	**0.9605**	\

Table 3. The PCC of the different methods.

Method	PCC			Vs. proposed
	Single	Multi	All	
VQM_VFD [3]	0.3882	0.2197	0.5080	−44.55%
Hyun_Thu [4]	0.9566	0.4042	0.7866	−16.69%
Xue [5] (whole set)	\	0.9493	0.9493	−0.42%
Xue [5] (training set)	\	0.9847	0.9847	\
Xue [5] (testing set)	\	0.8865	0.8865	−6.7%
Proposed	**0.9621**	**0.8247**	**0.9535**	\

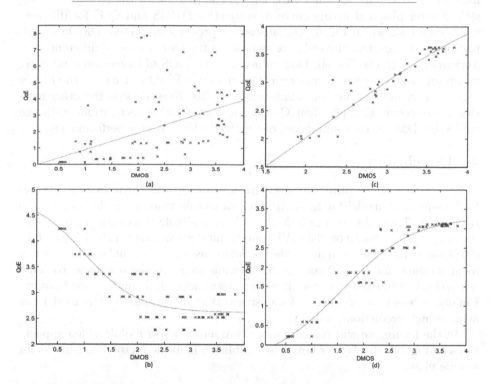

Fig. 6. Scatter plot and fitting curve for different metrics. (a) the metric in [3], (b) the metric in [4], (c) the metric in [5] and (d) the proposed metric.

3.3 Performance Comparison

The Video Quality Metric (VQM) software tools [12] provide standardized methods to measure the perceived video quality of digital video systems. The method in [3] is a full-reference video quality assessment method, but the performance is not satisfactory. For the proposed model, the SROCC is increased about 31% and the PCC is increased about 45% compared with [3]. The metric in [4] neglects the impact of video content. When the frame freezing occurs in the same position for different distortion sequences, it gets the same score. The performance of [4] has a worse performance for the multiple frame freezing. Since the method in [5] does not apply to the single frame freezing distortion, the data obtained by [5] in Tables 2 and 3 are only the values of the multiple frames freezing events. This method uses a trained neural network which needs a large number of data to train the network nodes. It splits the whole dataset into a training set and a testing set with 80/20 proportion. In Tables 2 and 3, Xue [5] (whole set) are the coefficients of all frame freezing distortion sequences, Xue [5] (training set) are the coefficients of the training set and Xue [5] (testing set) are the coefficients of the testing set. Obviously, Xue [5] (testing set) are more valuable. Our model has increased SROCC by 10% and PCC by 7% compared with the Xue [5] (testing set). Scatter plot and fitting curve of subjective DMOS and QoE for different methods are shown in Fig. 6. The abscissa represents subjective DMOS of distortion sequences, and the ordinate represents the QoE values of different QoE assessment methods. For [4], QoE values are the MOS of distortion sequences, which are negative correlations with the subjective DMOS. It means the higher QoE value represents the better subjective quality. Nevertheless, the other metrics are opposite with [4], their QoE values are positive correlations with the subjective DMOS. By comparison, our metric achieve better performance.

4   Conclusion

Based on the DMOS values obtained from the subjective experiments, a novel QoE assessment model for frame freezing of mobile video is firstly proposed in this paper. Then, due to the lack of publicly available data sets, a new frame freezing database based on the LIVE mobile database was established. This video database is used for comparing the performance of perceptual quality assessment methods for frame freezing. We provide access to the database, so other researchers will be able to reproduce the experiments and validate our findings. Finally, extensive experiments demonstrate that the proposed QoE model can achieve high prediction accuracy.

In the future, we will consider more characteristics of mobile video applications and mobile display terminals to establish a multiple-factors QoE model for mobile video.

Acknowledgement. This work was supported in part by the National Natural Science Foundation of China (Grant No: 61772436), Sichuan Science and Technology Innovation Seedling Fund (2017RZ0015), and the Fundamental Research Funds for the Central Universities.

We extend a special acknowledgement to the participants for their contributions to the subjective experiments.

References

1. Wolf, S., Pinson, M.: A no reference (Nr) and reduced reference (Rr) metric for detecting dropped video frames. In: 4th International Workshop on Video Processing and Quality Metrics for Consumer Electronics (VPQM), p. 3. (2009)

2. Borer, S.: A model of jerkiness for temporal impairments in video transmission. In: 2nd International Workshop on Video Processing and Quality Metrics for Consumer Electronics (QoMEX), pp. 218–223. IEEE, Trondheim (2010)
3. Wolf, S., Pinson, M. H.: Video quality model for variable frame delay (VQM-VFD). U.S. Department of Commerce, National Telecommunications and Information Administration NTIA Technical Memorandum TM-11-482 (2011)
4. Huynh-Thu, Q., Ghanbari, M.: No-reference temporal quality metric for video impaired by frame freezing artefacts. In: 16th International Conference on Image Processing (ICIP), pp. 2221–2224. IEEE, Cairo (2009)
5. Xue, Y., Erkin, B., Wang, Y.: A novel no-reference video quality metric for evaluating temporal jerkiness due to frame freezing. IEEE Trans. Multimed. **17**(1), 134–139 (2015)
6. Rodriguez, D.Z., Abrahao, J., Begazo, D.C.: Quality metric to assess video streaming service over TCP considering temporal location of pauses. IEEE Trans. Consum. Electron. **58**(3), 985–992 (2012)
7. Moorthy, A.K., Choi, L.K., Bovik, A.C.: Video quality assessment on mobile devices: subjective, behavioral and objective studies. IEEE J. Sel. Top. Signal Process. **6**(6), 652–671 (2012)
8. BT Series: Methodology for the subjective assessment of the quality of television pictures. Recommendation ITU-R BT (2012)
9. ITU, I.: Methodology for the subjective assessment of video quality in multimedia applications. Rapport Technique, International Telecommunication Union (2007)
10. Webster, A., Speranza, F.: Report on the validation of video quality models for high definition video content. The Video Quality Experts (2010)
11. Pastrana-Vidal, R. R., Gicquel, J. C., Colomes, C.: Sporadic frame dropping impact on quality perception. In: Human Vision and Electronic Imaging IX, pp. 182–194 (2004)
12. Pinson, M.H., Wolf, S.: A new standardized method for objectively measuring video quality. IEEE Trans. Broadcast. **50**(3), 312–322 (2004)
13. Wang, B., Peng, Q., Wu, X.: The proposed frame freezing database, Baidu Cloud. https://pan.baidu.com/s/1g10OvAjZSlgAld0WUgjDmQ. Accessed 25 May 2018

Saliency Detection Based on Deep Learning and Graph Cut

Hu Lu[1(✉)] [iD], Yuqing Song[1], Jun Sun[2], and Xingpei Xu[1]

[1] School of Computer Science and Communication Engineering,
Jiangsu University, Zhenjiang, China
luhu@ujs.edu.cn
[2] School of Electrical and Information Engineering, Jiangsu University,
Zhenjiang, China

Abstract. In this paper, we propose a new saliency detection method based on deep learning detection models and graph-cut partitioning methods. First, we present a full convolution deep neural network saliency detection model. By training a binary classification, full convolution, deep neural network, the image is divided into foreground and background. Given the initial image, the output is the coarse-grained saliency map with foreground and background. Second, we present a saliency detection method based on graph community partitioning. Using the superpixels algorithm, the initial image is transformed from pixel level to region level, and the super-pixel region similarity matrix is constructed. The similarity matrix is automatically divided into areas by the community-partitioning algorithm, and the segmentation map is obtained. The saliency values of each region are computed by the Gauss kernel function. Lastly, we combine the high-level global saliency map of deep learning with the low-level local saliency map of graph partitioning cut, and realize the saliency detection task of the original image. To illustrate the effectiveness of the proposed method, we also compare the method with the seven popular saliency detection algorithms and evaluate them on four public image databases. A large number of experimental results demonstrate that this algorithm has obvious superiority in comparison with existing algorithms.

Keywords: Deep learning · Saliency detection · Superpixels · Graph cut
Convolutional neural network

1 Introduction

In recent years, saliency detection has been used to reduce computational complexity, as an important step in the field of computer vision. This attracted a large amount research attention and has a wide range of applications, such as object tracking and recognition, image compression, image and video retrieval, and video event detection. From the human perception point of view, the characteristics of the image can be represented by three different levels. Low-level visual features include image color, edge, and texture, among other features. Intermediate features often comprise object information, shape, and spatial information. In contrast, advanced features have connections. With object information, segmentation, and intrinsic semantic interactions

© Springer Nature Switzerland AG 2018
R. Hong et al. (Eds.): PCM 2018, LNCS 11166, pp. 166–177, 2018.
https://doi.org/10.1007/978-3-030-00764-5_16

between objects. The substantial saliency test is related to the three levels of the above. Currently, saliency detection is divided into two methods: top-down [1] and bottom-up [2]. The bottom-up approach is from the region's color, texture, and many underlying characteristics that can effectively detect the details of the information. However, it cannot detect global information. On the contrary, the top-down model is for the training sample representative features, which can detect fixed size and target category. However, the top-down model of the test results may be rough, lacking detailed information. Although there are many good algorithms in this field, there are still challenges to develop a simple and effective target detection algorithm.

Deep learning has a good characteristic of learning ability, which is widely used in traditional classification and target recognition. We propose a deep saliency model based on a fully convolutional neural network. The model can effectively encode the saliency information of the image (i.e. obtain the advanced feature from the low-level feature combination). The weight of the network structure can effectively reduce the complexity of the network model structure, and reduce the number of weights. Therefore, in order to avoid the feature extraction and data reconstruction process in traditional identification methods.

In this paper, we propose a deep convolution, neural network saliency, target detection model. On the one hand, it is used to solve the bottom-up model that exists in inaccurate test results and the characteristics of the constraints. On the other hand, it can solve the top-down model with the details of the missing problem. This paper also introduces the traditional method of graph cut, which is used for Slic super-pixels and the Qcut algorithm, to realize the community division of the super-pixel region network through the center a priori, to obtain the local saliency value. The main contributions of this paper are as follows.

First, we improve the full convolution neural network model, which is based on the VGG model. We obtain the experimental model of this paper through data training. The input of the model is the original image, and the output is a binary map of the foreground and background. In principle, the deep saliency detection model is used to capture the advanced features of the image, so as to obtain prior saliency information. The model can obtain the global information of the image well. However, it cannot obtain the details of the image information.

Second, we propose the community detection of the graph cut from the bottom-up for saliency detection, to refine the image of the border. In this paper, we combine the Slic superpixels algorithm and community detection, and then use the Gaussian kernel to calculate the saliency value of each region to obtain the local saliency information. The deep saliency model is more concerned with the global shape information.

Finally, the above two models are combined and unified into a new saliency detection framework, where only the original image is provided, and then used to generate the final saliency detection.

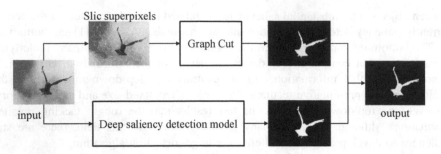

Fig. 1. Schematic diagram of the saliency target detection algorithm based on deep learning. First, SLIC and graph cut are used to construct the prior saliency of the image, and then combined with the deep saliency model to obtain the final saliency.

2 Related Work

Currently, saliency detection has been widely studied, using many bottom-up saliency models. Itti [10] proposed a saliency model based on neural networks. This model combines the color, brightness, and direction characteristics, and the three kinds of features of the central one peripheral mechanism operator to find the difference, which can detect the contrast is obvious saliency area. Yang et al. constructed a top-down saliency model combing the joint conditional random field (CRF) and distinguishing dictionary learning. They then established a complex energy function that achieved saliency detection based on task and purpose by minimizing energy functions. However, it could only roughly detect the target location, Shen and Wu [3] constructed a unified model of saliency detection. The model is based on the low-rank matrix theory, combined with the underlying characteristics and high-level vivid information. Yang et al. [4] used the tree model to construct a multi-scale saliency target detection algorithm. Jiang et al. proposed a saliency detection model based on Markov chains, which analyzed the saliency and non-saliency regions from the perspective of the background and the foreground. Zhu et al. proposed a method based on the color histogram to calculate saliency value, by calculating the Euclidean distance between a single pixel and other pixels. Achanta et al. [9] obtained the saliency value of the saliency model. First, it calculates the mean value of each layer of the lab color space of the image, and then subtracted the mean of the corresponding layers by the lab value of the corresponding layer. Hou [16] proposed a method based on the spatial frequency domain, to obtain the saliency region of the entire graph.

The methods based on deep learning have been widely used recently. Li et al. [13] proposed the Mdf algorithm, which used CNN to study the image features and establish an efficient visual saliency model by learning the multi-scale features. In order to study this model, a full connection layer is introduced in the first layer of CNN, to extract the characteristics of images on three different scales. The results obtained in the previous step are optimized to improve the spatial consistency of the test results. Finally, the different scales were calculated, the saliency graphs were fused to further improve the detection effect. Zhang combined with the convolution neural network model of multi-scale and migration learning, to identify the saliency region. Chen used deep convolution neural networks for target detection of satellite imagery.

3 The Proposed Method

The algorithm flow of this paper is illustrated in Fig. 1, which is divided into two parts: (1) The saliency detection model based on deep convolution neural network [6], and (2) The saliency of transcoding and community division. We obtain the deep full convolution network model by training. For the image of any input model, the characteristics of the image are extracted, and used to obtain the global information of the image, thus completing the contents of (1). For (2), we transform the image from the pixel level to the region level through the Slic superpixels algorithm [5]. Next, the lab color feature of the image is extracted, and the image is segmented by the Qcut algorithm. The saliency values of different regions are calculated to obtain prior saliency information.

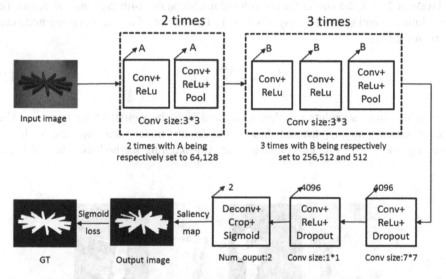

Fig. 2. Architecture for a full convolution neural network for training. The model has a 15-layer convolution layer, which is the input image, network model, saliency graph output, and groundtruth.

3.1 Deep Learning Model

The network model structure of this paper in Fig. 2 has 15 convolution layers, and each convolution layer has an activation function for ReLu [7]. In addition, the pooling method of the pooling layer is sampled at the maximum value. As shown in Fig. 2, we refer to the VGG [8] model of the network structure in the top 13 layers of the model, to better adjust the image spatial correlation. In the model, we only use convolution layers without using full connections. Because of the full convolution operation's ability to share convolution when the image is extracted and reduce the redundancy of the information, the convolution of the network model is simple, efficient.

The deep full convolution neural network model has multiple convolutions and down-sampling layers. Nevertheless, the operations of the two are certainly different. While the former focuses on the extraction of features, the latter focuses on the calculation of features. The output of the convolution node is expressed as:

$$a_n^l = f\left(\sum\nolimits_{\forall m}\left(a_m^{l-1} * k_{m,n}^l\right) + b_n^l\right) \tag{1}$$

Among them, a_n^l and a_n^{l-1} are the feature graph of the current layer and the feature map of the previous layer, respectively. $k_{m,n}^l$ represents the mth feature graph from the previous layer to the current nth, $f(x) = 1/[1 + \exp(-x)]$ is the neuron activation function, and b_n^l represents the neuron bias. It is the response of the convolution kernel to the convolution kernel of the upper layer, and different convolution cores can extract different features. As illustrated in Fig. 3, the size of the convoluted nuclei in the first 13 layers is 3×3, the size of the convoluted nuclei on the 14th layer is 7×7, and the convolution kernel of the 15th layer is 1×1. The output of the subsampling node can be expressed as:

$$a_n^l = f\left(k_n^l \times \frac{1}{s^2}\sum\nolimits_{s \times s} a_n^{l-1} + b_n^l\right) \tag{2}$$

$s \times s$ is the down-sampling template scale, and k_n^l is the weight of the template. The general down-sampling method has a maximum under-sampling, mean down-sampling, and random down-sampling. The sub-sampling method used in this paper

Fig. 3. The figure is based on the deep convolution of the neural network model test results. The left image is the original, and the right is a saliency detection.

is sampled at maximum. In this paper, we use stochastic gradient descent (SGD) to minimize the regularization coefficient.

As illustrated in Fig. 3, the left side images are the original, and the right sides images are the resulting binary map through the deep full convolution neural network model in this paper. From Fig. 3, we can easily see that the deep convolution neural network model often has a fuzzy boundary, such as the human shoulder, as well as the thigh, the athlete's hand, and petals bifurcation.

In order to earn better local information of the image, this paper presents a saliency detection model based on graph cut, which is used to refine the boundary of an image. The model divides the image through the super-pixels and the Qcut algorithm, and obtains the saliency value of the cutting area through the center a priori, combining the two.

3.2   Saliency Detection Model of Graph Cut

Since widely used in various fields of computer vision, super-pixels can effectively capture the structure of the image information. A simple linear iterative clustering (Slic) method in this paper is used to segment and obtain a certain number of super-pixels, which transform the image from pixel level to region level. Following this, we use the lab color feature to represent an image. The reason why lab color is used in this paper is that we find it can be a good description of the region's saliency when performing saliency target detection.

Specifically, we first provide an image through the Slic superpixels algorithm. Next, we obtain the N superpixels regions named $\{x_i\}_{i=1}^{N}$. We extract the lab color feature of each superpixels region, and construct the adjacency matrix of the super-pixels. The affinity matrix named $W = (w_{ij})_{N \times N}$ of the superpixels region can be expressed as:

$$w_{ij} = \begin{cases} S(x_i, x_j) & \text{if } x_i \text{ and } x_j \text{ are adjacent} \\ 0 & \text{Otherwise} \end{cases} \tag{3}$$

$S(x_i, x_j)$ is used to compute the similarity through RBF named $S(x_i, x_j) = exp\left(-\frac{1}{\rho}\|x_i - x_j\|_2^2\right)$. We use the Qcut algorithm to cluster the image. Since the Qcut algorithm can be used to automatically divide the number of associations, we assume that the number of associations is n. Therefore, the a priori saliency detection model is implemented by the following formula:

$$f(c_i) = g(c_i) \times \sum_{i \in n} d_k(c_i, n_j) \tag{4}$$

$d_k(c_i, n_j)$ represents the Euclidean distance between the lab color features corresponding to the regions c_i and n_j. In addition, $g(c_i)$ is the central a priori weight, which is obtained by the Qcut algorithm to divide the center of the center with the center of the normalized image. According to the above formula, the saliency value of each region can be obtained and a saliency graph of the pixel level is obtained, by assigning a saliency value for each region to all the pixels in the region. Finally, as shown in Fig. 4, the Gaussian filter is used to smooth the saliency graph.

Fig. 4. The figure is based on the deep convolution of the neural network model test results. The left image is the original, and the right is a saliency detection.

3.3 The Combination of the Deep Saliency Diagram and Image Segmentation of the Saliency Graph

This paper presents the bottom-up and top-down models, which have complementary features, as shown in Fig. 5 in the fusion process of this algorithm. In the end, we obtain the goal saliency graph of this paper from the deep saliency model, which tends to detect the global information of the image and the saliency pattern generated. The graph-cut method tends to detect image detail information. At the same time, it's easier to capture the local structure information of the image, due to using the contrast-based method. For any given image, the saliency of the calculated image is divided into the following steps: (1) The saliency detection model is based on the deep convolution neural network to obtain the foreground background referred to as S1; (2) The image a priori information S2 is obtained by the Slic and Qcut cut algorithm; and (3) We combine the two models with the saliency graph to produce a smooth saliency graph. The calculation is:

Fig. 5. Example of algorithm fusion. From left to right are the input image, the graph method obtained by the saliency graph, the deep learning method to obtain a saliency graph, the smooth saliency graph, and ground-truth.

$$S = \frac{S1 + S2}{2} \qquad (5)$$

3.4 Result

In order to illustrate the effectiveness of the proposed method, we selected four datasets to test our algorithm. Every dataset contained thousands of images. Typically, these datasets have been pretty fair to evaluate the performance of various algorithms. We compared these with seven saliency detection algorithms, such as Mdf [13], AC [14], FT [9], GBVS [15], Itti [10], LC [17], and SR [16].

3.5 Datasets

We collect four different datasets, which contain approximately 10,000 images, all providing ground-truth. These datasets are widely used to evaluate the performance of various saliency detection methods. We present a detailed description for these databases:

ASD [9]: 1000 images, including the pixel level of ground-truth, where the image is selected from MSRA-10000 [12]. Most of the image is single, and the background is relatively simple.

THUR [10]: 6232 images. The dataset is divided into five categories: "butterfly", "coffee", "dog", "giraffe", and "plane", and contains the pixel level of ground truth.

ECSSD [11]: 1000 images. The dataset is part of a BSD dataset and the PASCAL VOC data, and includes the pixel-level ground truth.

Model Training Process
In this paper, the number of super-pixels N = 100 and RBF is $\rho = 0.1$. We used the VGG-16 model hierarchy in the front 13 layers of the entire model. We migrated the VGG-16 semantic segmentation model for our weight model. As shown in Fig. 3, the front 3 layers of this paper were still the convolution layer, which were different from the full connection layer of the VGG-16 model. The activation function used in the deconvolution layer is the Sigmoid function.

We select approximately 10,000 images for our training set. The source of the training set includes the five databases in MSRA, ASD, THUR, ECSSD, and DUT-RMRON. The test set is part of the image in the MSRA. In the training process, the training set and is a one-to-one relationship. In the setting section of the parameter, the learning rate of the training model is 10^{-8}, and the training is performed using the SGD (gradient descent). We set the number of iterations to 10000×100, and finally obtain a saliency detection model.

Quantitative Evaluation
This paper evaluates all algorithms using the precision and recall rate (P-R) curves. The process of calculating the P-R curve is also called fixed-threshold segmentation. First, the saliency map is measured to [0, 255]. The saliency map is segmented according to the threshold value. With reference to the truth map, we calculate the average precision and recall rate of images in the test database, then obtain 256 groups of P-R values. At

the same time, we use the F-measure method to further evaluate the quality of the saliency graphs. The process of calculating the F-measure is also used in calculating the average precision and recall rate of all images in the test database. The original image is divided into superpixels, then in the ultra-pixel units, the entire image of the average gray value of the double as the adaptive threshold for segmentation. In other words, the average gray value is greater than the threshold value of the ultra-pixel as the foreground; otherwise, as the background. We obtain the average accuracy and recall rate of all images in the test image library, and obtain the F-measure value, using the following formula:

$$F_\eta = \frac{(1+\eta^2) \times Precision \times Recall}{\eta^2 \times Precision + Recall} \tag{6}$$

In this paper, we set $\eta^2 = 0.3$.

Experimental Analysis

In this paper, the algorithm is compared with Mdf [13], AC [14], FT [9], GBVS [15], Itti [10], LC [17], and SR [16]. We selected several images from the datasets. The saliency detection results are shown in Fig. 6. It can be seen from Fig. 6 that the coincidence of the saliency region detection and manual detection is the best. Furthermore, satisfying the pleasure of human eye vision is consistent with the visual point of attention. In addition, the saliency graph generated by this algorithm can accurately

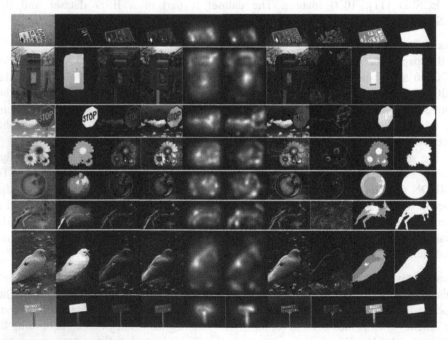

Fig. 6. The experimental results in ASD, ECSSD, DUT-OMRON, and THUR datasets. From left to right, these are followed by Mdf, AC, FT, GBVS, Itti, LC, SR, Ours, and ground-truth.

show the saliency target accurately, and the background noise is less. Although the saliency region is relatively small or large, the saliency region has great similarity to the background. Using the center a priori, this algorithm for the saliency target in other locations of the detection effect is also very good. Because the deep convolution neural network saliency model is good at capturing the global information of the image, the graph-cut method is more focused on capturing detailed information. Therefore, this paper can detect the multi-target image after the combination of the two models, and better display the image of the saliency information (Figs. 7 and 8).

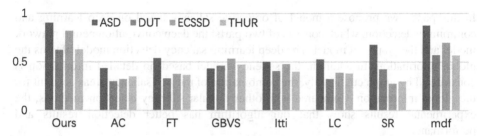

Fig. 7. The F-measure values of the algorithm and other existing algorithms on four datasets

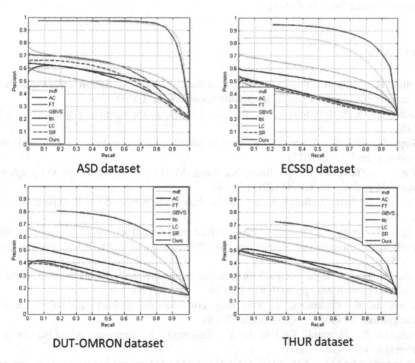

Fig. 8. The P-R curves of this algorithm and seven existing algorithms on ASD, ECSSD, DUT-OMRON, and THUR datasets.

P-R curves and F-measure histograms on four datasets are not difficult to determine that the other three datasets are not as good as the proposed algorithm, except that the P-R curve of the Mdf algorithm of the ASD dataset is closer to the upper-right corner. Meanwhile, with the increase in the abscissa, the mean of P is the decreasing trend. Even when R is relatively large, the precision P is relatively large. Finally, compared to other algorithms, the value of the F-measure is better from the bar chart to see comprehensive indicators.

4   Conclusion

In this paper, we propose a model of objective saliency based on deep learning and community detection, which consists of two parts: the deep convolution neural network model and the graph cut model. The deep learning saliency detection model obtains the global information. In addition, the saliency model based on detailed information is constructed by graph cut. Finally, the combination of the two saliency areas account for the global information. Compared with other popular saliency detection methods, the experimental results show that our algorithm has better detection results and performance.

Acknowledgments. This study was supported by the National Natural Science Foundation of China (Project No. 61572239). Scientific Research Foundation for Advanced Talents of Jiangsu University (Project No. 14JDG040).

References

1. Jiang, H., Wang, J., Yuan, Z., Wu, Y., Zheng, N., Li, S.: Salient object detection: a discriminative regional feature integration approach. In: CVPR (2013)
2. Goferman, S., Zelnik-Manor, L., Tal, A.: Context-aware saliency detection. In: CVPR (2010)
3. Shen, X., Wu, Y.: A unified approach to salient object detection via low rank matrix recovery. In: CVPR (2012)
4. Yang, C., Zhang, L., Lu, H., Ruan, X., Yang, M.-H.: Saliency detection via graph-based manifold ranking. In: CVPR (2013)
5. Achanta, R., Shaji, A., Smith, K., Lucchi, A., Fua, P., Susstrunk, S.: SLIC superpixels. EPFL (2010)
6. Eigen, D., Fergus, R.: Predicting depth, surface normals and semantic labels with a common multi-scale convolutional architecture. In: Proceedings of IEEE ICCV, pp. 1850–1858 (2015)
7. Nair, V., Hinton, G.E.: Rectified linear units improve restricted Boltzmann machines. In: Proceedings of ICML, pp. 807–814 (2010)
8. Simonyan, K., Zisserman, A.: Very deep convolutional networks for large-scale image recognition. In: Proceedings of ICLR (2015)
9. Achanta, R., Hemami, S., Estrada, F., Susstrunk, S.: Frequency-tuned salient region detection. In: CVPR (2009)
10. Cheng, M.-M., Mitra, N.J., Huang, X., Hu, S.-M.: Salientshape: group saliency in image collections. Vis. Comput. **30**(4), 443–453 (2014)

11. Yan, Q., Xu, L., Shi, J., Jia, J.: Hierarchical saliency detection. In: Proceedings of IEEE Conference CVPR, pp. 1155–1162 (2013)
12. Liu, T., Sun, J., Zheng, N.-N., Tang, X., Shum, H.-Y.: Learning to detect a salient object. In: CVPR (2007)
13. Li, G., Yu, Y.: Visual saliency based on multiscale deep features. In: IEEE Conference on Computer Vision and Pattern Recognition, pp. 5455–5463. IEEE Press, Boston, USA (2015)
14. Achanta, R., Estrada, F., Wils, P., Süsstrunk, S.: Salient region detection and segmentation. In: Gasteratos, A., Vincze, M., Tsotsos, J.K. (eds.) ICVS 2008. LNCS, vol. 5008, pp. 66–75. Springer, Heidelberg (2008). https://doi.org/10.1007/978-3-540-79547-6_7
15. Harel, J., Koch, C., Perona, P.: Graph-based visual saliency. Adv. Neural. Inf. Process. Syst. **19**, 545–552 (2007)
16. Hou, X., Zhang, L.: Saliency detection: a spectral residual approach. In: CVPR (2007)
17. Zhu, W., Liang, S., Wei, Y., Sun, J.: Saliency optimization from robust background detection. In: CVPR (2014)

Rethinking Fusion Baselines for Multi-modal Human Action Recognition

Hongda Jiang, Yanghao Li, Sijie Song, and Jiaying Liu[✉]

Institute of Computer Science and Technology, Peking University, Beijing, China
{jianghd,lyttonhao,ssj940920,liujiaying}@pku.edu.cn

Abstract. In this paper we study fusion baselines for multi-modal action recognition. Our work explores different strategies for multiple stream fusion. First, we consider the early fusion which fuses the different modal inputs by directly stacking them along the channel dimension. Second, we analyze the late fusion scheme of fusing the scores from different modal streams. Then, the middle fusion scheme in different aggregation stages is explored. Besides, a modal transformation module is developed to adaptively exploit the complementary information from various modal data. We give comprehensive analysis of fusion schemes described above through experimental results and hope our work could benefit the community in multi-modal action recognition.

Keywords: Fusion · Multi-modality · Action recognition

1 Introduction

With the rapid development of deep learning, there has been tremendous progress in computer vision [9,11,13]. The two-stream network [17] makes remarkable contribution for action recognition, by fusing the results of spatial and temporal streams, and achieves good performance on popular action recognition benchmarks. However, it still remains confusing whether combining different modalities on the final results as two-stream is the best choice. Recently, Spatiotemporal Multiplier Networks [5] investigate the middle connections in the two-stream architecture. The connections in their work are straightforward and their method only considers RGB and optical flow data. With the development of depth cameras, depth and other modal data become more available, which are able to provide clues for action recognition from other perspectives [14]. However, there is a lack of exploration of generic multi-modal fusion schemes. Thus, in our paper, we rethink various fusion schemes and design sufficient experiments to give some insights in the multi-modal fusion.

We conduct a general research on the fusion baselines for multi-modal human action recognition. We consider aggregating modalities in different levels, *i.e.*,

This work was supported by National Natural Science Foundation of China under contract No. 61772043 and CCF-Tencent Open Research Fund.

R. Hong et al. (Eds.): PCM 2018, LNCS 11166, pp. 178–187, 2018.
https://doi.org/10.1007/978-3-030-00764-5_17

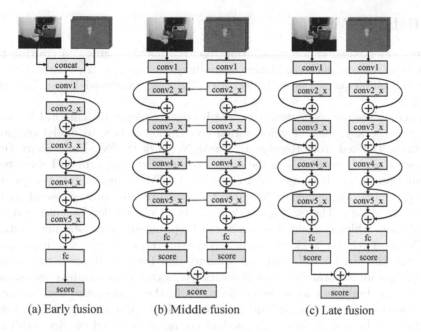

Fig. 1. The architectures of different multi-modal fusion schemes with ResNet101 as backbone (From left to right, early fusion, middle fusion and late fusion). Note that the input modalities are not limited in the above two modalities. We will explore more different modalities in the following.

early fusion, middle fusion and late fusion. In the early fusion, we stack different modalies together as a single input which is the most straightforward and easiest way to fuse different modalities. Next, we explore the late fusion scheme which directly fuses the softmax scores of different modal streams. Finally, the middle fusion is presented which combines multi-modal data in the feature level. We systematically explore various stages of middle fusion and propose a modal transformation module in adaptive middle fusion. Our networks for different fusion methods are based on ResNet101 [7] and the architectures of our different fusion schemes are illustrated in Fig. 1.

Our contributions are listed as follows:

- We conduct a thorough investigation on various fusion baselines which aggregate multiple modalities in different levels.
- We adopt deep ablation analysis of different fusion stages for middle fusion method. Sufficient experiments are conducted and discussed to compare different fusion methods with multiple modalities.
- We further propose a novel modal transformation module for middle fusion method, leading to a more efficient model combination over existing simple middle fusion methods.

2 Related Work

Pioneer video-based action recognition research mainly focuses on crafting features from videos, such as Motion Boundary Histograms [2], Histograms Of Flow [12], subsequent Dense Trajectories [22], and Improved Dense Trajectories [23].

Since videos are sequential data which contain plenty of temporal information [10], the key point of video action recognition is how to model temporal dynamics. Derived from Recurrent Neural Network (RNN), Long Short-Term Memory (LSTM) network has the ability to capture long-term and short-term information [6,20]. It is natural to employ LSTM to model these sequential data [3,15,19]. Meanwhile, CNN architectures are also proved useful in the video-related task. The C3D network [8,21] is widely applied because of its 3D convolution, which can simultaneously catch spatial and temporal information.

The most relative work to ours is two-stream [17] structure, which parallelly processes the spatial and temporal streams and fuses their prediction scores. Lately, Spatiotemporal Residual Networks [4] extend the original two-stream approach by building middle connections. To further understand how the interaction works, Spatiotemporal Multiplier Networks [5] provide a systematic investigation on the middle fusion in residual connections based on ResNet50 and ResNet152 [7]. In contrast to previous efforts which only consider RGB and optical flow, we take a step further and conduct a comprehensive exploration on various multi-modal fusion schemes, taking RGB, optical flow and depth information into account. Besides, a modal transformation module is proposed to achieve a more efficient modal combination.

3 Exploring Different Multi-modal Fusion Schemes

In this section, we investigate different fusion schemes for action recognition. We fuse the multi-modal data from input level, score level and feature level, respectively, which corresponds to early fusion, late fusion and mid fusion.

We use ResNet101 [7] as our backbone. ResNet101 is a fully convolutional architecture, with a chain of residual units. Each residual unit consists of closely linked 1×1 and 3×3 convolutions and is equipped with additive skip connections. In the end of the ResNet101 there is an average pooling and a fully connected layer. Fusions take place at different parts of the networks.

3.1 Early Fusion

Early fusion suggests that we fuse the multiple modalities on the input by stacking them along the channel dimension. Assume there are M kinds of modalities for action recognition and the input for the i-th modality is $\mathbf{I}_i (i = 0, ..., M-1)$. Note that $\mathbf{I}_i$ for different modalities should have the same spatial resolution but may differ in the number of channels. We concatenate the different modal data as:

$$\mathbf{I} = \mathbf{I}_0 \oplus \mathbf{I}_1 \oplus \mathbf{I}_2 \oplus ... \oplus \mathbf{I}_i \oplus ... \oplus \mathbf{I}_{M-2} \oplus \mathbf{I}_{M-1}, \tag{1}$$

where $\mathbf{I}$ is the final input for early fusion, $\oplus$ indicates concatenation along channels. Early fusion is the most straightforward and comprehensive method to combine multiple modalities. However, the multi-modal data get mixed in a low-level manner. It might be hard for the network to extract discriminative features for action recognition.

3.2 Late Fusion

Late fusion fuses the scores of different modal streams. Similar to two-stream, the softmax scores from multiple streams are combined together by average fusion. Supposing M modalities are adopted to classify N actions and the score of the i-th modality is $\mathbf{p}_i = (p_{i,0}, p_{i,1}, ..., p_{i,N-1})$. The final score $\mathbf{y} = (y_0, y_1, ..., y_{N-1})$ and the prediction label z can be given as below:

$$y_j = \frac{1}{M} \sum_{k=0}^{M-1} p_{k,j}, \tag{2}$$

$$z = \operatorname*{argmax}_{j}(y_j), \tag{3}$$

where y_j indicates the score of the j-th action class ($j = 0, 1, ..., N - 1$).

3.3 Multiple Middle Fusion

Direct and Adaptive Connections. For simplicity, we explain the middle fusion with two modalities, while more modalities could be introduced for middle fusion. As illustrated in Fig. 1, we design our middle fusion networks by building adaptive connections between different streams. A detailed schematic is proposed as (b) in Fig. 2. $W_{l,c}^s$ indicates the weights of the c-th convolution layer in the l-th residual unit and s, t represent different streams. We formulize the proposed method as:

$$\widehat{x}_l^t = g(x_l^t), \tag{4}$$

$$\widehat{x}_{l+1}^s = x_l^s + f(x_l^s \cdot \widehat{x}_l^t), \tag{5}$$

where x_l^s, x_l^t are features from the l-th layer of the two streams, respectively, and $f(\cdot)$ is the residual function. In addition, we insert a general transformation, $g(\cdot)$, into each connection which transforms the features adaptively for a more sufficient fusion. As a special case, the Spatiotemporal Multiplier Networks [5] directly fuse two modalities in which the $g(\cdot)$ is an identity transformation and we call it direct connection in Fig. 2. The final results are obtained by averaging the scores of two streams.

The middle fusion is more complicated when extending to multiple modalities since the middle connections is directional. The number of schemes increases when including new modalities and we design our connections according to the involved modalities. An example will be given in Sect. 4.

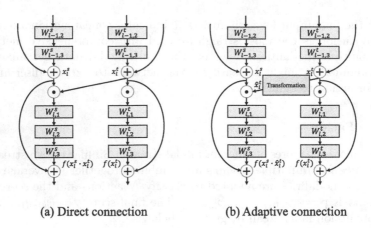

(a) Direct connection   (b) Adaptive connection

Fig. 2. An illustration of middle fusion. (a) Direct connection represents identity transformation connection and (b) adaptive connection corresponds to variable transformation connection.

4   Experiments

4.1   Dataset and Settings

In this section, we evaluate the performance of the above methods with NTU RGB+D Dataset [16] (NTU). NTU is a multi-modal dataset which is captured indoor by Microsoft Kinect v2 cameras concurrently. It consists of 56,880 action samples containing aligned RGB videos, depth map sequences, 3D skeleton data and infrared videos. We adopt RGB, optical flow and depth videos for our multi-modal action recognition. We split the videos into 40,400 training samples and 16,480 testing samples following the cross-subject rule in [16].

Implementation Details. We use ResNet101 [7] as our backbone network structure and follow the same training schemes in Temporal Segment Networks (TSN) [24] architecture. During training, the input images are first resized to 256×340 and then cropped with the specific width and height which are randomly chosen from {256, 224, 192, 168}, followed by resizing to 224×224. The cropping is performed on the four corners or the center of images. Since the input channels of different modalities may differ, we follow the initialization method in [24] to use models pretrained on ImageNet for all modalities and then modify the weights in the first convolution layer. In the training process, we randomly select 3 segments for each video where each segment contains one RGB image, one depth image or stacked optical flow frames. To speed up the testing process, we average the results of 3 segments for each video to obtain the final results. Note that when comparing to the state-of-the-art methods, we average the results of 25 segments for a fair comparison.

For middle fusion, we proceed in two steps during training. We first independently train each modality and then insert our adaptive connections among different streams. Finally, the connected networks with different modalities are optimized jointly with cross entropy losses. We propose the adaptive connection in Sect. 3. In the experiments, we apply a 1×1 convolution layer as the adaptive transformation which is expected to lead to a more efficient fusion.

Table 1 shows the classification results with a single modality. This demonstrates the importance of motion information in action recognition. The performance from different modalities also reveals that different modalities may have complementary information and the fusion could achieve better results.

Table 1. Results with a single modality on the NTU dataset in accuracy (%).

	RGB	Flow	Depth
Acc. (%)	83.87	**92.02**	85.23

4.2 Middle Fusion Stage Exploration

Since ResNet101 has four stages of convolution blocks, we conduct an exploration experiment to investigate where to append the middle connections. For each stage, ResNet101 has multiple residual units and we link the second residual unit for middle connection. We first compare the results of appending one middle connection in each stage, respectively, and then apply four connections for all stages. Here we apply the experiment based on the direct connections.

The results are in Table 2. With the fusion stage changing from 1 to 4, the result is continuously increasing and we achieve the best performance when appending all connections. This indicates that our middle fusion method could benefit from different levels of feature fusions. Therefore, in the following experiments, we apply connections in all four stages.

Table 2. Results for different stages of connections in middle fusion with direct connections on the NTU dataset.

Stage1	Stage2	Stage3	Stage4	Acc. (%)
√	-	-	-	93.83
-	√	-	-	93.93
-	-	√	-	94.07
-	-	-	√	94.29
√	√	√	√	**94.37**

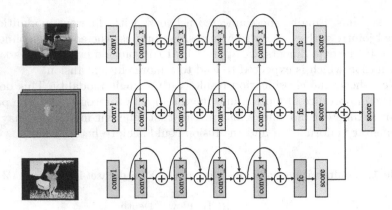

Fig. 3. Connections for middle fusion with RGB, optical flow and depth data.

4.3 Comparisons for Different Fusion Methods

Fusion of Multiple Modalities. It is easy to extend early fusion and late fusion to more modalities. However, the connections are directional for middle fusion. For two modalities, we follow the connection direction in [5] to construct connections from optical flow to RGB. For three modalities, we append adaptive connections from optical flow to RGB and depth at the same time, as illustrated in Fig. 3.

We conduct fusion experiments among all the combinations and fusion methods above. The results are summarized in Table 3. From the results, we can find that:

1. The results of early fusion are even worse for those with single modal data. For example, the combination of RGB and optical flow by early fusion is inferior to optical flow. We attribute this to the high complexity of the early fusion input. It is hard for the network to extract discriminative features for the task of action recognition.
2. Comparing the direct connection and the adaptive connection for middle fusion, the latter has better performance among all the combinations. This demonstrates that our proposed model is able to achieve a more sufficient fusion by transforming the features adaptively.
3. Experimental results show that late fusion performs better than direct connections. It is mainly because the middle fusion network with direct connections suffers from overfitting. Nevertheless, our middle fusion by adaptive connections achieve better performance than late fusion, illustrating the effectiveness of our modal transformation module.
4. The combination of depth and optical flow is even superior to the three modalities. We ascribe the confusing result to the considerable noise as we only select 3 segments for the test. Once we average the results of 25 segments, our model achieves the best performance with the three modalities in Table 4.

Table 3. Results of different fusion methods with multi-modal data on the NTU dataset in accuracy (%).

RGB	Flow	Depth	Early	Late	Direct connections	Adaptive connections
√	√	-	89.77	94.45	94.37	**94.62**
√	-	√	80.63	87.99	87.16	**88.14**
-	√	√	89.29	94.98	94.47	**95.03**
√	√	√	82.49	94.98	94.53	**94.98**

4.4 Comparisons with the State-of-the-Art Methods

We compare our method with current state-of-the-art models. For a fair comparison, we mark the modalities employed in each method. Table 4 shows that our method outperforms other methods using the same or fewer modalities. With the three modalities, our model achieves the best result.

Table 4. Comparisons with state-of-the-art methods on the NTU dataset in accuracy (%).

Methods	Skeleton	RGB	Flow	Depth	Acc. (%)
STA-LSTM [18]	√	-	-	-	73.4
VA-LSTM [25]	√	-	-	-	79.4
P-CNN [1]	-	√	√	-	53.8
TSN (BN-Inception) [24]	-	√	√	-	88.5
Chained MT [26]	√	√	√	-	80.8
Late fusion	-	√	√	-	94.8
Late fusion	-	√	√	√	95.2
Adaptive connections	-	√	√	-	**95.2**
Adaptive connections	-	√	√	√	**95.5**

5 Conclusion

In this paper, we investigate different fusion baselines for multi-modal action recognition. We explore early fusion, middle fusion and late fusion, respectively, which aggregate the multi-modal data from different levels. A modal transformation module is proposed to help effectively utilize the complementary information and improve the action recognition results. Analysis shows that our modal transformation module with adaptive connections has the best performance among all the fusion methods. We hope the insights from this work could encourage further research in multi-modal action recognition.

References

1. Chéron, G., Laptev, I., Schmid, C.: P-CNN: pose-based CNN features for action recognition. In: Proceedings of the IEEE International Conference on Computer Vision, pp. 3218–3226 (2015)
2. Dalal, N., Triggs, B., Schmid, C.: Human detection using oriented histograms of flow and appearance. In: Proceedings of European Conference on Computer Vision, pp. 428–441 (2006)
3. Donahue, J., et al.: Long-term recurrent convolutional networks for visual recognition and description. In: Proceedings of IEEE International Conference on Computer Vision and Pattern Recognition, pp. 2625–2634 (2015)
4. Feichtenhofer, C., Pinz, A., Wildes, R.: Spatiotemporal residual networks for video action recognition. In: Proceedings of Advances in Neural Information Processing Systems, pp. 3468–3476 (2016)
5. Feichtenhofer, C., Pinz, A., Wildes, R.P.: Spatiotemporal multiplier networks for video action recognition. In: Proceedings of IEEE International Conference on Computer Vision and Pattern Recognition, pp. 7445–7454 (2017)
6. Gers, F.A., Schmidhuber, J., Cummins, F.: Learning to forget: continual prediction with LSTM. Neural Comput. **12**(10), 2451–2471 (2000)
7. He, K., Zhang, X., Ren, S., Sun, J.: Deep residual learning for image recognition. In: Proceedings of IEEE International Conference on Computer Vision and Pattern Recognition, pp. 770–778 (2016)
8. Ji, S., Xu, W., Yang, M., Yu, K.: 3D convolutional neural networks for human action recognition. IEEE Trans. Pattern Anal. Mach. Intell. **35**(1), 221–231 (2013)
9. Jia, Y., et al.: Caffe: convolutional architecture for fast feature embedding. In: Proceedings of ACM International Conference on Multimedia, pp. 675–678 (2014)
10. Kang, S.B., Uyttendaele, M., Winder, S., Szeliski, R.: High dynamic range video. ACM Trans. Graph. **22**, 319–325 (2003)
11. Krizhevsky, A., Sutskever, I., Hinton, G.E.: Imagenet classification with deep convolutional neural networks. In: Proceedings of Advances in Neural Information Processing Systems, pp. 1097–1105 (2012)
12. Laptev, I., Marszalek, M., Schmid, C., Rozenfeld, B.: Learning realistic human actions from movies. In: Proceedings of IEEE International Conference on Computer Vision and Pattern Recognition, pp. 1–8 (2008)
13. LeCun, Y., Bengio, Y., Hinton, G.: Deep learning. Nature **521**(7553), 436 (2015)
14. Liu, J., Li, Y., Song, S., Xing, J., Lan, C., Zeng, W.: Multi-modality multi-task recurrent neural network for online action detection. IEEE Trans. Circ. Syst. Video Technol. (2018)
15. Liu, J., Shahroudy, A., Xu, D., Wang, G.: Spatio-temporal LSTM with trust gates for 3D human action recognition. In: Leibe, B., Matas, J., Sebe, N., Welling, M. (eds.) ECCV 2016. LNCS, vol. 9907, pp. 816–833. Springer, Cham (2016). https://doi.org/10.1007/978-3-319-46487-9_50
16. Shahroudy, A., Liu, J., Ng, T.T., Wang, G.: NTU RGB+D: a large scale dataset for 3D human activity analysis. In: Proceedings of IEEE International Conference on Computer Vision and Pattern Recognition, pp. 1010–1019 (2016)
17. Simonyan, K., Zisserman, A.: Two-stream convolutional networks for action recognition in videos. In: Proceedings of Advances in Neural Information Processing Systems, pp. 568–576 (2014)
18. Song, S., Lan, C., Xing, J., Zeng, W., Liu, J.: An end-to-end spatio-temporal attention model for human action recognition from skeleton data. In: AAAI, vol. 1, p. 7 (2017)

19. Song, S., Lan, C., Xing, J., Zeng, W., Liu, J.: Spatio-temporal attention-based LSTM networks for 3D action recognition and detection. IEEE Trans. Image Process. **27**(7), 3459–3471 (2018)
20. Srivastava, N., Mansimov, E., Salakhudinov, R.: Unsupervised learning of video representations using LSTMs. In: Proceedings of International Conference on Machine Learning, pp. 843–852 (2015)
21. Tran, D., Bourdev, L., Fergus, R., Torresani, L., Paluri, M.: Learning spatiotemporal features with 3D convolutional networks. In: Proceedings of IEEE International Conference on Computer Vision, pp. 4489–4497 (2015)
22. Wang, H., Kläser, A., Schmid, C., Liu, C.L.: Action recognition by dense trajectories. In: Proceedings of IEEE International Conference on Computer Vision and Pattern Recognition, pp. 3169–3176 (2011)
23. Wang, H., Schmid, C.: Action recognition with improved trajectories. In: Proceedings of IEEE International Conference on Computer Vision, pp. 3551–3558 (2013)
24. Wang, L., et al.: Temporal segment networks: towards good practices for deep action recognition. In: Proceedings of European Conference on Computer Vision, pp. 20–36 (2016)
25. Zhang, P., Lan, C., Xing, J., Zeng, W., Xue, J., Zheng, N.: View adaptive recurrent neural networks for high performance human action recognition from skeleton data. In: Proceedings of IEEE International Conference on Computer Vision and Pattern Recognition, pp. 2117–2126 (2017)
26. Zolfaghari, M., Oliveira, G.L., Sedaghat, N., Brox, T.: Chained multi-stream networks exploiting pose, motion, and appearance for action classification and detection. In: Proceedings of IEEE International Conference on Computer Vision, pp. 2923–2932 (2017)

A DCT-JND Profile for Disorderly Concealment Effect

Hongkui Wang[1], Li Yu[1(✉)], Tiansong Li[1], Mengting Fan[2],
and Haibing Yin[3]

[1] School of Electronic Information and Communications,
Huazhong University of Science and Technology, Wuhan, China
{hkwang, hustlyu}@hust.edu.cn
[2] College of Information Engineering,
China Jiliang University, Hangzhou, China
[3] School of Communication Engineering, Hangzhou Dianzi University,
Hangzhou, China

Abstract. Just noticeable distortion (JND) refers to the smallest visibility threshold of the human visual system (HVS). The existing JND profiles always overestimate the visibility threshold in orderly region and underestimate that of the disorderly region. In order to obtain a more accurate DCT-JND profile, a novel block-level disorder metric is proposed and disorderly concealment effect is taken into account in the DCT-JND model. Specifically, an improved perceptive Local Binary Patterns (LBP) algorithm is proposed to evaluate the disorder of each block in this paper. Since the visual acuity is insensitive to the disorder stimulus, a disorderly concealment effect factor is defined as the function of block disorder and background disorder in this paper. The factor is used to adjust the conventional JND threshold appropriately. The experimental result shows that the proposed JND model tolerates much more distortion with the same perceptual quality compared with the existing JND profiles.

Keywords: Disorderly concealment effect · Block-level disorder metric
DCT-JND model

1 Introduction

The human visual system perceives the pixel change above a certain visibility threshold. The minimal visibility threshold is called the just noticeable distortion. Exposing the perceived characteristics of HVS, the JND profile has been adopted in many image and video processing fields such as image/video compression [1], quality assessment [2], scene enhance [3], etc.

In general, the JND profile is divided into pixel-based and subband-based (e.g., DCT, DFT) JND models. Because most of image and video compression schemes are performed in DCT domain, the DCT-JND model attracts many researchers' attention. In 1192, Ahumada and Peterson [4] proposed the first DCT-JND model which gives the JND threshold for each DCT component. The threshold is determined by the spatial contrast sensitivity function (CSF) which is used to describe the band-pass property of

R. Hong et al. (Eds.): PCM 2018, LNCS 11166, pp. 188–199, 2018.
https://doi.org/10.1007/978-3-030-00764-5_18

sensitivity of HVS in spatial frequency. DCTune [5] is an improved model proposed by Watson, in which the luminance adaptation (LA) effect and the contrast masking (CM) effect had been incorporated. The LA effect indicates that the visual sensitivity is low in the dark and light regions. As for the CM effect, it is referred to as the reduction in the visibility of one visual component in the presence of another. In conventional DCT-JND model, the edge pixel density calculated by the Canny operator is employed to evaluate the contrast intensity of each block [6]. By calculating the standard deviation of pixels and the power spectrum density of each frequency, Bae recently proposed the structural contrast index (SCI) metric to evaluate the contrast intensity of DCT blocks [7]. Although it was reported that the SCI metric describes the block contrast accurately and effectively, the contrast intensity of edge regions calculated by SCI metric is always higher relatively. In fact, the existing JND profiles mainly consider the LA and CM effects, which always overestimates the visibility threshold of the edge region and underestimates that of the texture region [8].

In order to overcome the aforementioned limitation of JND models, some literatures had made meaningful explorations to improve the accuracy of JND models. Wu et al. [9] divides the image into orderly image and disorderly image with an autoregressive model. The disorderly concealment effect was taken into account in pixel-JND model at the first time. Disorderly concealment effect reveals that HVS is sensitive to the orderly stimulus but tries to avoid uncertainties caused by disorderly information. Recently, the pattern masking based on both CM effect and structural uncertainty was introduced in [8]. Due to the masking is determined by both luminance contrast and pattern complexity, it is limited in regular pattern and is strong in irregular pattern. It was reported that the JND model considering pattern masking is more consistent with the HVS than other JND models. However, these models are constructed in pixel domain and can't be applied in DCT domain directly.

As an extension of Wu's model [8], a disorderly concealed effect factor is proposed to adjust the conventional DCT-JND threshold appropriately. Since the visual acuity is insensitive to disorderly stimulus, the factor is defined as a function of the block disorder and the background disorder. The disorder of each block is evaluated by an improved perceptive LBP algorithm in this paper. Experimental results confirm that the proposed model tolerates much more distortion with the same perceptual quality compared with the existing JND profiles.

2   The Proposed DCT-JND Model

2.1   Conventional DCT-JND Model

The conventional DCT-JND model is expressed as a product of three modulation factors: J_{base} is applied to measure the JND threshold for each spatial frequency component. F_{LA} and F_{CM} are LA and CM effect factors.

$$J(N, \omega, n, i, j, \tau) = J_{base}(N, \omega) \cdot F_{LA}(n) \cdot F_{CM}(\omega, \tau) \tag{1}$$

Here, n is the block index and (i,j) is the DCT coefficient index. Ordinarily, human eyes are sensitive to the horizontal and vertical frequency and insensitive to the diagonal components. Considering the spatial summation effect [10] and oblique effect [10], the J_{base} of each DCT coefficient is modified as

$$J_{base}(i,j) = N \cdot F_s \cdot \exp(c\omega_{i,j})/(a+b\omega_{i,j})/[(r+(1-r)\cdot\cos^2\varphi_{i,j})] \quad (2)$$

where N is DCT normalization factor for an N × N DCT block. $\omega_{i,j}$ is the corresponding frequency of (i, j)-th subband in DCT block. F_s is spatial summation effect factor and takes the value of 0.125 [7]. The parameters are set a = 1.33, b = 0.11, c = 0.18, r = 0.6 according to Wei's psychophysical experiments [10]. In term $r + (1 -r)\cdot\cos^2\varphi_{i,j}$ accounts for the oblique effect and $\varphi_{i,j}$ equals to $\arcsin(2\cdot\omega_{i,0}\cdot\omega_{0,j}/\omega^2_{i,j})$. Due to the SCI metric has better performance in contrast measurement, we adopt Bae's CM effect factor [7] in our DCT-JND model. The LA and CM factors are computed by formula (3) and (4).

$$F_{LA}(n) = \begin{cases} (60 - I(n))/150 + 1 & I \leq 60 \\ 1 & 60 < I \leq 170 \\ (I(n) - 170)/425 + 1 & I > 170 \end{cases} \quad (3)$$

$$F_{CM} = f(\omega) \cdot \tau \cdot \varepsilon + 1 \quad (4)$$

where $I(n)$ is the average intensity value of the n-th block. $f(\omega)$ is the gain function of spatial frequency. ε is an invariant constant which is equal to 0.2. τ is the contrast intensity of the DCT block which is equal to C_t^α/K_t^β. C_t represents the contrast intensity that can be calculated by the Parseval's theorem. The K_t is the Kurtosis of normalized power spectral density of DCT coefficients. α, β are the paraments that are empirically set to 1.4 and 0.7.

2.2   Considerations for the Improvement of the DCT-JND Model

The relationship between the contrast intensity (τ) and edge pixel density (ρ) of each block is shown in Fig. 1. The block is divided into three types (i.e., plain, edge and texture) according to the edge pixel density [9]. As shown in Fig. 1, compared with other regions, the contrast intensity in edge region is higher relatively. Nevertheless, noise is easily observed in the smooth and edge areas. Therefore, the conventional DCT-JND model always overestimates the visibility threshold of the edge region and underestimates that of the texture region [8].

Recent research on human perception indicates that the HVS possesses an internal generative mechanism (IGM) for visual signal processing. The IGM theory suggests the visual acuity is sensitive to the orderly region and is insensitive to the disorderly region [11]. For applying the perceived characteristic of HVS effectively, the disorderly concealed effect should be properly incorporated into the DCT-JND model. How to efficiently incorporate the disorderly concealed effect into DCT-JND model is an open problem. In this paper, a disorderly concealed effect factor is designed to adjust the conventional DCT-JND threshold. Since the visual acuity is insensitive to the

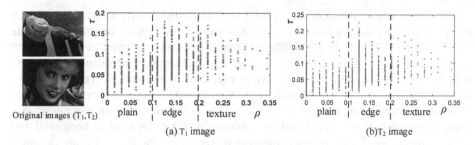

Fig. 1. The relationship between contrast intensity (τ) and the edge pixels density (ρ).

disorderly stimulus, the factor (F_{DC}) can be defined as a function of the block disorder (ξ_{block}) and the background disorder (ξ_{back}) based on the psychophysical experiments. In DCT domain, the block is the image processing unit. Therefore, a block-level disorder metric is designed to evaluate the disorder of the block. Therefore, the total DCT-JND can be written by

$$J(N, \omega, n, i, j, \tau) = J_{base}(N, \omega) \cdot F_{LA}(n) \cdot F_{CM}(\omega, \tau) \cdot F_{DC}(\xi_{back}, \xi_{back}) \qquad (5)$$

3   The Block-Level Disorder Metric

The visual acuity is sensitive to the orderly stimulus and is insensitive to the disorderly stimulus [11]. In order to evaluate the disorder intensity of the stimulus, a block-level disorder metric is introduced in this section.

Ojala et al. [12] analyzed the spatial relationship among pixels and introduced a classic LBP algorithm to describe the orderly information of each pixel. In particular, for an N × N block, the orderly information of the intermediate pixel (x_c) is represented by the relationship between the x_c and the adjacent pixels x_i. The block size N is always set as 3 by considering the accuracy and computational complexity. The LBP value of x_c is expressed as

$$LBP(x_c) = \sum_{i=1}^{N} t(x_i - x_c) \times 2^{i-1} \quad and \quad t(x_i - x_c) = \begin{cases} 1, x_i \geq x_c \\ 0, x_i < x_c \end{cases} \qquad (6)$$

where t is the sign of differences between the pixel x_c and x_i. Wu et al. [8] considered the LA effect and proposed a perceptive LBP algorithm based on a luminance adaptative threshold. The luminance adaptive threshold is expressed by formula (7). $I(x_c)$ is the average background luminance intensity of pixel x_c.

$$LA(x_c) = \begin{cases} 17 \times (1 - \sqrt{I(x_c)/127}), & if \ I(x_c) \leq 127 \\ 3/128 \times (I(x_c) - 127) + 3, & else \end{cases} \qquad (7)$$

However, Wu's LBP algorithm is used to evaluate the orderly information of each pixel and can't be used in block-level disorder evaluation directly. In order to estimate the disorder of the block with different dimensions, an improved perceptive LBP algorithm is proposed. For the sake of simplicity, the example of calculation process with N = 8 is shown in Fig. 2. As shown in Fig. 2(a), four directions (i.e., D_0, D_1, D_2, D_3) are considered in an N × N block (e.g., N = 8) and the pixels which are applied to calculate the orderly information are defined. Due to the dimension of block is always even, there are two row pixels and two column pixels to be used in horizontal and vertical directions. Therefore, six computing patterns (i.e., P_0, P_1, ... P_5) are designed in four directions. There are two patterns in horizontal and vertical directions respectively. For the convenience of description, the pixels in each pattern are defined x_0 to x_7 as Fig. 2(b). The sign of differences of each pattern is expressed as

$$
t_{ik} = \begin{cases} 1, & x_{ik} - x_{ik-1} \geq LA(x_{ik-1}) \\ 0, & |x_{ik} - x_{ik-1}| < LA(x_{ik-1}) \\ -1, & x_{ik} - x_{ik-1} \leq -LA(x_{ik-1}) \end{cases} \tag{8}
$$

where k is the index of the pattern (i.e., k = 0, 1, 2, ... 5). i is the pixel index in each pattern. The LBP value of each pattern is calculated by

$$
\xi_k = \sum_{i=0}^{7} |t_{ik}| \times 2^{q(ik)} \tag{9}
$$

$2^{q(ik)}$ is the weight function. Human eyes pay more attention to the center of the stimulus because of the center prior principle. The visual weight increases when the distance between the current t_{ik} and the center one decreases in each pattern. The disorder intensity is related to the number of continuous unequal t_{ik} in each pattern.

$$
q(i) = \begin{cases} 0, & i = 0 \, or \, i = 7 \, or \, t_i = t_{i-1} \\ q_{i-1} + 1, & t_i \neq_{i-1} \, and \, i < 4 \\ q_{i+1} + 1, & t_i \neq t_{i+1} \, and \, i \geq 4 \end{cases} \tag{10}
$$

$q(i)$ represents the number of continuous unequal t_i. The underlying idea of Eq. (10) is that the larger of $q(i)$, the disorder is much stronger. An example of t_{ik} is shown in Fig. 2(c). At last, for the sake of discussion without losing its generality, the block-level disorder is calculated by formula (11). w_k is the weight of k-th pattern. Considering the oblique effect, the weights in horizontal diagonal direction (i.e., w_2, w_5) are equal to 1.67. The weights in other directions (i.e., w_0, w_1, w_3, w_4, w_6) are equal to 1.0 [10]. ξ_{max} is the normalization factor and is set to 151. Figure 3 shows the block-level disorder intensity of images. Higher brightness means a larger disorder value. It is obvious that the block in disorderly region has high disorder value in Fig. 3.

$$
\xi_{block}(n) = \sum_{k=0}^{5} \xi_i \cdot w_k \bigg/ \xi_{max} \tag{11}
$$

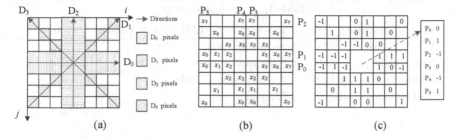

Fig. 2. The calculation process of the improved perceptive LBP algorithm: (a) four directions; (b) six computing patterns; (c) the sign of differences.

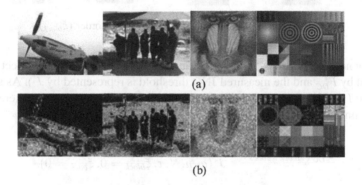

Fig. 3. The block-level disorder: (a) the original images; (b) the disorder map.

4   The Disorderly Concealment Effect Factor

The sensitivity of HVS is related to the block disorder (ξ_{block}) and the background disorder (ξ_{back}). The background disorder is defined as the average disorder intensity of the background. In order to measure the disorderly concealment effect of block and background to human visual acuity, we perform the psychophysical experiment for DCT-JND measurement. The test conditions for our psychophysical experiments are summarized in Table 1. For the background disorder of test stimulus, 5 test background disorder values are set to 0, 0.2, 0.4, 0.6 and 0.8. In only considering the disorderly concealment effect process, the LA and CM effects should be eliminated. According to the analysis in Sect. 2, The LA and CM effects of the block are related on the average intensity ($I(n)$) and the contrast intensity (τ) [7]. Therefore, the test patches are designed as shown in Fig. 4. μ is the normalized value of $I(n)$ and is set to 0.5 approximately. The contrast intensity (τ) is 0.25 approximately.

The test example is shown in Fig. 5. As shown in Fig. 5(a), the test patch and test background are set in the center of the real background. Figure 5(b) represents the test image in DCT domain and shows the noise injected into R_1. The amplitude of noise in R_1 increases until testers perceive the resulting distortion called the JND measured value. Each measured JND value is the value that 50% of the testers start to perceive

Table 1. Experimental setup

Display information	SONY KD-55X8000E (LED 55 inch, Full-HD resolution)
Viewing distance	1.75 times Screen height ($\approx$ 1.2 m)
Test patch (R_1)	8 $\times$ 8 pixels (square blocks)
Test background (R_2)	24 $\times$ 24 pixels (square blocks)
Real background (R_3)	256 $\times$ 256 pixels

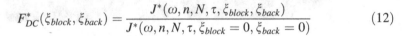

ξ_{block}=0.02 μ=0.482 τ=0.248 ξ_{block}=0.21 μ=0.486 τ=0.252 ξ_{block}=0.41 μ=0.501 τ=0.249 ξ_{block}=0.61 μ=0.499 τ=0.251 ξ_{block}=0.80 μ=0.50 τ=0.254

Fig. 4. The test patches with different block disorder (ξ_{block}).

the corresponding distortions [7]. The measured disorderly concealment effect factor is represented by F_{DC}^* and the measured JND threshold is represented by J^*). As shown in Eq. (5), the F_{DC}^* is defined as a multiplier in our total DCT-JND model. Therefore, the F_{DC}^* is obtained by

$$F_{DC}^*(\xi_{block}, \xi_{back}) = \frac{J^*(\omega, n, N, \tau, \xi_{block}, \xi_{back})}{J^*(\omega, n, N, \tau, \xi_{block} = 0, \xi_{back} = 0)} \tag{12}$$

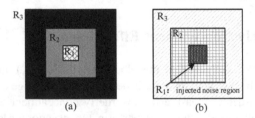

(a)   (b)

Fig. 5. A test image example: (a) the test patch and test background in the center of the real background; (b) the injected noise at the test patch in DCT domain.

It is known that the visual sensitivity to distortions decreases as the disorder intensity increases. In JND modelling, this indicates that the disorderly concealment effect factor (F_{DC}) can be modeled as a monotonically increasing function for the block disorder (ξ_{block}) and background disorder (ξ_{back}). The test results are shown in Fig. 6. It is observed that the measured disorderly concealment effect (F_{DC}^*) values increase approximately in an exponential manner as the block disorder for the given background disorder in Fig. 6(a). In Fig. 6(b), the measured (F_{DC}^*) values increase approximately in a linear manner as the background disorder for the given block disorder. The relationship between the measured F_{DC}^* values and the different combinations of block disorder and background disorder is shown in Fig. 6(c).

Based on the observations in Figs. 6(a, b), we projected the F_{DC}^* values onto the two parameter spaces of the block disorder and the background disorder. The linear model is adopted to model the F_{DC}.

$$F_{DC} = \Psi(\xi_{back}) \cdot \Phi(\xi_{block}) + \eta \quad (13)$$

$\Phi(\xi_{block})$ is an exponential function to represent that disorderly concealment effect factor (F_{DC}) increases as the block irregularity increases. $\Psi(\xi_{back})$ is the weight of the exponential function and is modeled as a function of background disorder. η is the basis and is set to -0.25, empirically. The factor is constructed based on the least mean square as shown in Fig. 6(d). $\Phi(\xi_{block})$ and $\Psi(\xi_{back})$ are expressed with $\eta_1 = 20$, $\eta_2 = 40$, $\eta_3 = 0.8$, $\eta_4 = 0.4$, $\eta_5 = 1$.

$$\Phi(\xi_{block}) = \eta_1^{\xi_{block}/\eta_2} + \eta_3 \quad (14)$$

$$\Psi(\xi_{back}) = \eta_4 \cdot \xi_{back} + \eta_5 \quad (15)$$

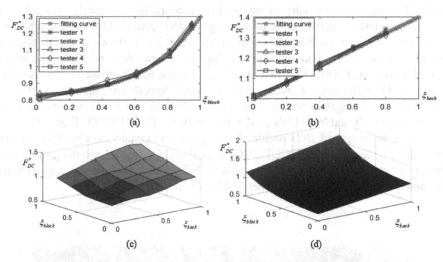

(a)   (b)

(c)   (d)

Fig. 6. The disorderly concealment effect factor model: (a) F_{DC}^* versus ξ_{block} in the case of $\xi_{back} = 0$; (b) F_{DC}^* versus ξ_{back} in the case of $\xi_{block} = 0$; (c) the measured F_{DC}^* values with different combinations of ξ_{block} and ξ_{back}; (d) the proposed F_{DC} model with ξ_{block} and ξ_{back}.

5 Experimental Result

To demonstrate the effectiveness of the proposed DCT-JND profile, the ability of hiding distortion of our DCT-JND model is verified. The disorderly concealment effect factor is incorporated into our DCT-JND model and is used to adjust the conventional DCT-JND threshold. Noise is added to each DCT coefficient in an image according to [7] such as

$$C'(n,i,j) = C(n,i,j) + S_r(n,i,j) \times J(n,i,j) \tag{16}$$

where $C(n,i,j)$ is the (i,j)-th DCT coefficient in the n-th block. $C'(n,i,j)$ is the coefficient with the noise and $S_r(n,i,j)$ is a bipolar random noise of ± 1. The PSNR is used to measure the capability toleration of the JND models. With the same perceptual quality, the lower PSNR is, the more accurate the JND model is. Meanwhile, we adopt the Blind/Referenceless Image Spatial QUality Evaluator (BRISQUE) [13] as our object quality assessment to evaluate the original and distorted images. It is believed that the BRISQUE metric is more conforming to the visual characteristics of human eyes than the PSNR and SSIM [14]. It is worth pointing out that the BRISQUE score is smaller, the perceptual quality of the image is better. As we all know, good performance for a JND profile means that the distorted image by the JND profile shows a low PSNR with the smaller BRISQUE score against other compared JND profiles.

In our performance test, the images chosen from TID 2013 database are shown in Fig. 7. The proposed JND profile is compared to three recently developed JND profiles, Bae (17)'s [7], Wu (13)'s [9], Wu (17)'s [11] in performance. Bae (16)'s JND profile is a DCT-JND profile, Wu (13)'s and Wu (17)'s JND profiles are the pixel-JND models. To prove the necessity of taking into account the disorderly concealment effect, we also test the performance of the original JND profile (Original) which does not incorporate the disorderly concealment effect. Table 2 shows the performance comparison between the four recently developed JND profiles and the proposed in terms of PSNR, and RPISQUE SCORE. The average PSNR result shows that the proposed JND profile has the smallest PSNR values. Specifically, the average PSNR value of the proposed JND profile is1.633 dB, 1.202 dB and 6.637 dB smaller than Original, Bae (17)'s and Wu (17)'s model, respectively. The BRPISQUE SCORE also denotes that the proposed JND profile shows good performance in objective quality assessment. The average BRPISQUE SCORE of Bae (17)'s model and the proposed is approximately equal. Therefore, the proposed JND profile can hide more distortions with same perceptual quality than other JND profile.

Fig. 7. Thumbnail of images $(T_1, T_2, \ldots T_9)$ set for testing.

Table 2. The performance comparisons in TID 2013 database

Image	PSNR (dB)					BRPISQUE SCORE				
	Original	Bae (17)'s	Wu (13)'s	Wu (17)'s	Proposed	Original	Bae (17)'s	Wu (13)'s	Wu (17)'s	Proposed
T_1	29.654	29.193	31.056	34.903	28.111	84.810	84.720	154.944	96.988	85.135
T_2	29.797	29.749	30.436	33.766	28.290	76.545	77.168	147.245	42.102	79.928
T_3	28.401	27.963	30.430	34.612	26.474	92.074	92.445	114.389	52.292	92.246
T_4	27.509	26.732	30.946	33.526	25.940	85.353	86.447	125.143	33.209	85.514
T_5	27.706	27.263	30.498	32.275	26.030	88.806	88.739	90.466	24.723	88.387
T_6	27.453	27.181	30.259	32.841	25.704	81.522	80.932	106.355	23.813	78.970
T_7	28.077	27.396	29.789	34.253	26.916	82.112	83.675	153.689	163.577	83.209
T_8	26.955	26.588	28.484	32.323	24.905	84.637	83.841	85.781	25.144	84.178
T_9	28.266	27.870	31.319	35.233	26.753	81.018	81.149	172.279	63.000	84.640
Average	28.202	27.771	30.357	33.748	26.569	84.097	84.346	127.810	58.316	84.689

Figure 7 shows the subjective performance comparison of four JND profiles for 'T1' image. In generally, the proposed JND model has lowest PSNR value with the better perceptual quality. Bae's JND model shows poor performance in edge region (region A). The contrast intensity calculated by SCI metric is higher relatively in edge area, which the JND threshold in this region is overestimated. Wu (13)'s JND model overestimates the visibility threshold of the smooth region. Compared with our model, Wu (17)'s JND model has highest PSNR value with the similar perceptual quality (Fig. 8).

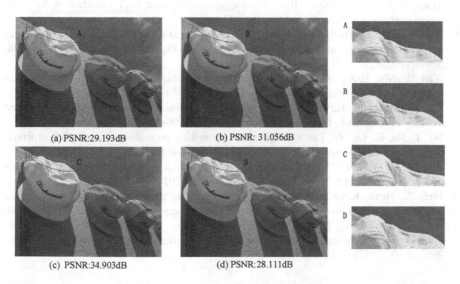

(a) PSNR:29.193dB   (b) PSNR: 31.056dB

(c) PSNR:34.903dB   (d) PSNR:28.111dB

Fig. 8. The performance comparisons of four JND profiles: (a) by Bae (17)'s JND model; (b) by Wu (13)'s JND model; (c) by Wu (17)'s JND model; (d) by the proposed JND model.

6   Conclusion

In this paper, considering the disorderly concealment effect, a novel DCT-JND model was proposed. The proposed disorderly concealment effect factor was used to improve the accuracy of DCT-JND model, as an attempt for overcoming the major shortcoming of the existing DCT-JND models. In particular, the block-level disorder metric was designed to evaluate the disorder intensity of the stimulus. In order to avoid to overestimate the visibility threshold in orderly region and underestimate that of the disorderly region, we have built a disorderly concealment effect factor based on the disorder of the stimulus to adjust the conventional DCT-JND threshold appropriately. The experimental result shows that the proposed model is better in PSNR reduction with the same perceptual quality compared with existing JND profiles.

Acknowledgement. This work was supported in part by National Natural Science Foundation of China (NSFC) (No. 61231010) and National High Technology Research and Development Program (No.2015AA015901).

References

1. Bae, S.-H., Kim, J., Kim, M.: HEVC-based perceptually adaptive video coding using a DCT-based local distortion detection probability model. IEEE Trans. Image Process. **25**(7), 3343–3357 (2016)
2. Wang, H., et al.: MCL-JCV: a JND-based H.264/AVC video quality assessment dataset. In: Proceedings of IEEE, pp. 1509–1513. Phoenix (2016)
3. Ritschel, T., Smith, K., Ihrke, M., Grosch, T., Myszkowski, K., Seidel, H.-P.: 3D unsharp masking for scene coherent enhancement. ACM Trans. Graph. **27**(3), 90:1–90:8 (2008)
4. Ahumada Jr., A.J., Peterson, H.A.: Luminance-model-based DCT quantization for color image compression. Proc. SPIE **1666**, 365–374 (1992)
5. Watson, A.B.: DCTune: a technique for visual optimization of DCT quantization matrices for individual images. In: Sid International Symposium Digest of Technical Papers, vol. 24, p. 946 (1993)
6. Wan, W., Wu, J., Xie, X., Shi, G.: A novel just noticeable difference model via orientation regularity in DCT domain. IEEE Access **5**, 22953–22964 (2017)
7. Bae, S., Munchurl, K.: A DCT-based total JND profile for spatiotemporal and foveated masking effects. IEEE Trans. Image Process. **27**(6), 1196–1207 (2017)
8. Wu, J., Shi, G., Lin, W., Liu, A., Li, F.: Pattern masking estimation in image with structural uncertainty. IEEE Trans. Image Process. **22**(12), 4892–4904 (2013)
9. Wu, J., Lin, W., Shi, G., Wang, X., Qi, F.: Just difference estimation for images with free-energy principle. IEEE Trans. Image Process. **15**(7), 1705–1710 (2013)
10. Wei, Z., Ngan, K.N.: Spatio-temporal just noticeable distortion profile for grey scale image/video in DCT domain. IEEE Trans. Circ. Syst. Video Technol. **19**(3), 337–346 (2009)
11. Wu, J., Li, L., Dong, W., Shi, G., Lin, W., Jay Kuo, C.-C.: Enhanced just noticeable difference model for images with pattern complexity. IEEE Trans. Image Process. **26**(6), 2682–2693 (2017)

12. Ojala, T., Valkealahti, K., Oja, E., et al.: Texture discrimination with multidimensional distributions of signed gray level differences. Pattern Recognit. **34**(3), 727–739 (2001)
13. Mittal, A., Moorthy, A.K., Bovik, A.C.: No-reference image quality assessment in the spatial domain. IEEE Trans. Image Process. **21**(12), 4695–4708 (2012)
14. Wang, Z., Bovik, A.C., Sheikh, H.R., Simoncelli, E.P.: Image quality assessment: from error visibility to structural similarity. IEEE Trans. Image Process. **13**(4), 600–612 (2004)

Breast Ultrasound Image Classification and Segmentation Using Convolutional Neural Networks

Xiaozheng Xie[1], Faqiang Shi[1], Jianwei Niu[1,2(✉)], and Xiaolan Tang[3]

[1] State Key Laboratory of Virtual Reality Technology and Systems, Beihang University, Beijing 100191, China
niujianwei@buaa.edu.cn
[2] Beijing Advanced Innovation Center for Big Data and Brain Computing, Beihang University, Beijing 100191, China
[3] College of Information Engineering, Capital Normal University, Beijing 100048, China

Abstract. Due to the shortage and uneven distribution of medical resources all over the world, breast cancer diagnosis and treatment is a fundamental but vital problem, especially in developing countries. Breast ultrasound image classification and segmentation method by using Convolutional Neural Networks (CNN) can be a new efficient solution in early analysis and diagnosis. What's more, the diagnosing of diversity of cancers is challenge in itself and the training of data-driven based CNN model also highly relay on dataset. In this paper, we first build a breast ultrasound dataset (with 1418 normal and 1182 cancerous samples) labeled by three radiologists from XiangYa Hospital of Hunan Province. And then, we propose a two-stage Computer-Aided Diagnosis (CAD) system to diagnose the breast cancer automatically. Firstly, the system utilize a pre-trained ResNet generated with transfer learning approach to excluded normal candidates, and then use an improved Mask R-CNN model for the accurate tumor segmentation. Experimental results show that the proposed system can achieve 98.72% precision and 98.05% recall for classification, and 85% (1.2% improvement) mAP and 82.7% (3.1% improvement) F1-Measure than the original Mask R-CNN model.

Keywords: Breast ultrasound image · Breast ultrasound dataset
Image classification · Tumor segmentation

1 Introduction

As a leading cause of death for women worldwide, breast cancer has caused a wide public concern, while detection and treatment at an early stage are essencial to effectively overcome this burden. Over the last decades, benefit from the development of Convolutional Neural Networks (CNN) in the field of image classification and segmentation, enhanced the importance of medical imaging

© Springer Nature Switzerland AG 2018
R. Hong et al. (Eds.): PCM 2018, LNCS 11166, pp. 200–211, 2018.
https://doi.org/10.1007/978-3-030-00764-5_19

(such as ultrasound, mammography, X-ray) for the early detection, diagnosis and treatment of this disease.

Ultrasound and mammography imaging technology are often used as the common and effective method in early diagnosis of breast cancer [1]. By contrast, ultrasound is more useful and effective than mammography due to following advantages [6,20,33,35]. (1) It is safer without radiation, cheaper, faster and possible to increase the number of detected nodules. (2) It is more sensitive to detect abnormalities in dense breasts. Traditional breast cancer diagnosis depends on the expertise of radiologists, results in subject diagnosis. Additionally, radiologists are overwhelmed with the data need to be analyzed due to the dramatic increase in the number of patients.

Nowdays, some Computer-Aided Diagnosis (CAD) systems have been developed to assist radiologists to diagnose breast cancer in ultrasound images. Different studies [8,39] have shown that CAD is an important tool to improve the diagnostic sensitivity and specificity. Comparable results are produced with less time, and the inter-/intra- observer variations are reduced to some extent. Considering the different types of complicated breast tumors in ultrasound images, traditional CAD systems [5,9,11] may be not robust enough and do not take heterogenity of different tumors into account. Furthermore, such methods do not balance the accuracy, speed, and level of automation especially when the dataset is complex. Such deviations make it difficult to accurately segment and classify breast images. Several drawbacks of these methods are as follows.

(1) Lack of public dataset in breast ultrasound area limited the development of CAD for breast cancer. Accurate labeling of dataset may also affect the diagnosis performance in this filed.
(2) Classification and segmentation methods in other fields can not be used directly in ultrasound images, especially when complex lesion objects need to be dealt with.

In this paper, to address the aforementioned drawbacks, we first build a breast ultrasound dataset, which consists of 2600 images (1418 and 1182 for normal and cancerous, respectively) obtained from four different devices. Then we conduct the image classification and lesion segmentation based on the state-of-the-art methods in other fields, followed by modifying some structures to adapt to our ultrasound dataset. Specifically, in order to validate the accuracy and sensitivity of classification and segmentation methods, some classification methods: LeNet [18], AlexNet [17], ZFNet [40] and ResNet [14], and segmentation methods: FCN-AlexNet [28], U-Net [26] and Mask-RCNN [13] are used and verified in our dataset.

The contributions of this paper are listed as follows.

(1) A breast ultrasound dataset is constructed, in which the images and tumors are all labeled by three radiologists to reduce the inter-/intra- observer variations.

(2) We investigate a novel CAD system for image classification and segmentation, which modifies ResNet [14] and Mask R-CNN [13] methods to adapt to our dataset.

The rest of this paper is organized as follows. Background and related research of breast ultrasound image classification and segmentation are reviewed in Sect. 2. Our dataset is explained in detail in Sect. 3. Breast ultrasound image classification and segmentation methods are described in Sect. 4. Experiments results of image classification and segmentation are presented in Sect. 5. Finally, this paper is concluded in Sect. 6.

2   Related Work

2.1   Ultrasound Image Classification

Nowdays, some popular classification methods [25,27,31,38] such as SVM, AdaBoost and K-means are employed to learn statical information of tumor regions and background. Besides, many other classification methods and CAD systems are also proposed. Some of them focused on image classification and others are tumor classification. Huynh et al. [3] realize the use of transfer learning approach for ultrasound breast image classification. Shi et al. [29] develop a stacked deep polynomial network (S-DPN) algorithm based representation learning method for tumor classification with small ultrasound imaging. Uniyal et al. [32] classify ultrasound-based breast malignant lesions with considering the ultrasound radio frequency (RF) time series analysis. In addition, Moon et al. [22] propose a CAD classification system based on speckle patterns on automated breast ultrasound imaging and achieves high sensitivity and accuracy. Flores et al. [10] compile distinct morphological and texture features to improve the classification accuracy of breast lesions on ultrasonography. Byra et al. [4] improve classification accuracy in ultrasound breast lesions by using the segmented quantitative ultrasound maps of homodyned K distribution parameters, which showed that the analysis of internal changes in lesion parametric maps lead to a better classification of breast tumors.

Although these methods performance well in tumor classification with high accuracy and sensitivity, the process may be a little cumbersome with considering the features extracted from images and difference between tumor and background. In our work, each image is labeled as cancerous or normal one without considering the tumor category, which is restricted by the label obtained. Some classification network structures, such as LeNet [18], AlexNet [17], ZFNet [40] and ResNet [14] are chosen and adapted to our dataset for automated feature extraction and classification.

2.2   Ultrasound Image Segmentation

Image segmentation methods are categorised into four groups [6]: thresholding-based [15,16], active contour model [24,34], Markov random field [23,37] and neural network [12,21,30]. Despite the good segmentation performance, some limitations still exist in these methods. Threshold based methods are simple but may not perform well for only considering the gray level statistics and the segment performance are vulnerable to threshold chosen. Active contour models can detect edge closed and continuous, but the initialization point is hard to choose and segmentation accuracy may decrease in pseudo-edge and noisy conditions. MRF model may have better segmentation performance, but at the cost of a time-consuming and complex process.

In order to achieve satisfactory segmentation accuracy in complex situations, a tumor segmentation model considering the prior knowledge learning is proposed in [36]. Meanwhile, the computerized image segmentation results often require various case-by-case user interventions to improve the contour correctness, which unfortunately remains quite difficult to address [2].

The neural network methods directly uncover features from the training data, and employ these features in lesion segmentation from ultrasound images. Marcomini et al. [21] realize segmentation and classification of nodules in breast ultrasound digital images using the artificial neural network models. Su et al. [30] apply a fast scanning deep convolutional neural network (fcnn) to pixelwise region segmentation in histopathological breast cancer images. Gomez et al. [12] present an evolutionary segmentation approach for breast lesions on ultrasound based on pulse-coupled neural network (pcnn) model and an adaptive differential evolution algorithm (JADE). Dhungel et al. [7] explore the use of deep learning methods as potential functions in structured prediction models for breast masses segmentation from mammograms. However, some limitations exist in these methods due to the complicated characteristics of breast ultrasound images, especially when the intensity inhomogeneity or coarse texture occurs in the tumor. In addition, the design and tuning of overall performance of the conventional CADx framework tends to be very arduous.

In this study, we investigate three state-of-the-art network structures for breast ultrasound tumor segmentation, Patch-based U-Net [26], transfer learning approach with a pretrained FCN-AlexNet [28] and a modified Mask-RCNN [13]. Then we compare the segmentation performances in our dataset with these algorithms.

3   Dataset Description

Collected from XiangYa Hospital of Hunan Province in 2016 and 2017, our dataset contains a total of 2600 images with a mean image size of $390 * 443$ pixels, 1182 and 1418 with and without tumor area, respectively. From the 1182 cancerous images (collected from 394 patients), 890 were invasive ductal carcinomas, 164 were invasive carcinomas, 73 were non-special type invasive carcinomas, and 55 were other unspecified malignant lesions. It is note that one or more tumors in each cancerous image. Since this dataset was obtained from different systems such as PHILIPS, SIEMENS, HITACHI and ALOKA, each image label and lesion area are annotated and voted by three doctors to reduce

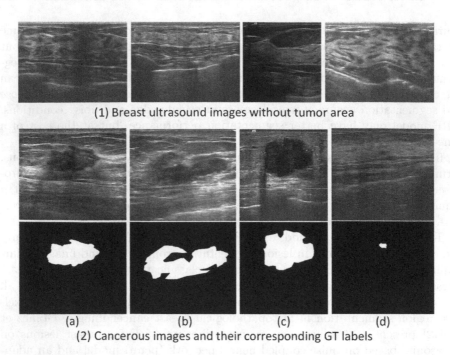

(1) Breast ultrasound images without tumor area

(a)   (b)   (c)   (d)

(2) Cancerous images and their corresponding GT labels

Fig. 1. Breast ultrasound images examples and their ground truth labels. (1) shows the normal images without tumor area in our dataset. (2) exhibits some cancerous images and their ground truth labels of our dataset, where (a), (b), (c), (d) show the example of invasive ductal carcinoma images, non-special type invasive carcinoma images, images with invasive carcinoma and images with other unspecified malignant lesions, respectively.

the difference among them. Some examples of normal and cancerous images in our dataset are illustrated in Fig. 2, where (1) depict the normal images in our dataset, (2a), (2b), (2c), (2d) show the example of invasive ductal carcinoma images, non-special type invasive carcinoma images, images with invasive carcinoma and images with other unspecified malignant lesions, respectively. Our dataset and the respective delineation of the breast lesions will be public for research purposes.

As Fig. 1 shows, tumor areas in our dataset are complicated with irregular shapes and various sizes. Moreover, speckle noise, image quality and image aspect ratio are completely different between different images taken by different devices in different times. Our dataset are used to show the effectiveness of our classification and segmentation method and compare with other state-of-the-art methods in experiments. Due to the limitation in malignant and benign label in our dataset, classification or segmentation of one candidate image mainly depends on whether there is a lesion area.

4   The Proposed CAD System

In this section, we introduce a CAD system based on ResNet and Mask R-CNN to classify and segment ultrasound images in our dataset, where two stages are incorporated: (1) feature extraction and normal candidates exclusion and (2) accurate tumor segmentation using improved Mask R-CNN. The framework of our proposed CAD system is illustrated in Fig. 1.

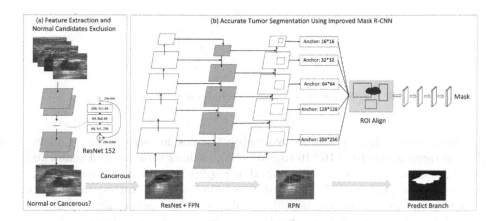

Fig. 2. The framework of our tumor classification and segmentation system.

4.1   Feature Extraction and Normal Candidates Exclusion

As illustrated in most previous works, a good feature extraction network should be deep enough with many convolution layers such that multi-level features can be learned. Further, different filter sizes can output feature maps which accurately represent the spatial arrangement of activations. Inspired by the successful use of ResNet in feature extraction and classification tasks [13], this structure is also used in our image classification tasks (shown in Fig. 2(a)), where a transfer learning approach with the pre-trained parameters from ResNet are used in our dataset. In this way, all images are fed into this network and output their predicted labels, while normal images can be excluded before tumor segmentation.

4.2   Accurate Tumor Segmentation Using Improved Mask R-CNN

After normal candidates exclusion, all images labeled by cancerous as the inputs to the Mask R-CNN for tumor segmentation. Despite the fact that multi-scale (5 anchor sizes for 5 feature maps respectively) and multi-ratio (1:1, 1:2, 2:1) anchors have been taken into account in the original method, there is still a large room for improvement over considering the difference and diversity of different feature maps.

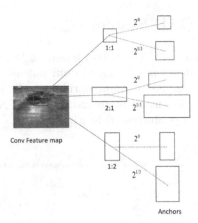

Fig. 3. Illustration of anchors in the improved Mask R-CNN

Owing to the huge difference over object scales in our candidate images (lesions range in size from $16 * 10$ to $956 * 676$), anchors with different scales and ratios are taking into account (illustrated in Fig. 2(b)), where the size of anchors are modified to $(16 * 16, 32 * 32, 64 * 64, 128 * 128, 256 * 256)$ correspondingly. In spired by the anchors setting in [19], we also add two anchors sizes of 2^0 and $2^{1/3}$ of the original set of 3 aspect ratio anchors at each level (shown in Fig. 3). Thus, there are 6 anchors per level and across levels they cover the scale range 16–322 pixels with respect to the resized input.

In this way, the large proportion of big tumor region in small images can also be handled. Thus, each feature map generates 30 scales, and the RoIAlign layer and three branches are followed to generate accurate segmentation results.

5   Experimental Results and Discussion

In this section, a diverse set of experiments are introduced to evaluate the performance of our CAD system on our dataset. The performance of classification are evaluated by using Precision, Recall and F2 (depicts the classification results emphasize on Recall), mAP (mean Average Precision) and F1-meature are selected to measure the segmentation performance. and then 5-flod and 10-fold cross validation are used in classification and segmentation tasks, respectively. The performance metric measures are calculated by following equations:

$$F_2 = \frac{(1 + 2^2) \times P \times R}{2^2 \times P + R} \tag{1}$$

$$mAP = \frac{TPs}{TPs + FPs} \tag{2}$$

$$F1\text{-}measure = \frac{2 \times TPs}{2 \times TPs + FPs + FNs} \tag{3}$$

Where P and R represent the Precision and Recall respectively, TPs refers the number of pixels in tumor area labeled as tumors, FPs is the number of pixels in backgrounds labeled as tumors, FNs is the number of pixels in tumors masked as backgrounds.

5.1 Image Classification Results

All images are resized to 224 * 224 before feeding into different networks, learning rate are set to 0.001 in all methods, while epoches are 20, 20, 20, 65 in LeNet, ZF Net, AlexNet and ResNet networks, respectively. The image classification results of these methods are shown in Table 1. From this table, we observe that the ResNet method achieves the hightest Precision (95.84%), Recall (99.41%) and F_2 measure (99.23%) among these four classification methods, which demonstrate the robust feature extraction and classification ability are performed in this deepest network. By using this structure, most normal images are excluded before tumor segmentation in the next stage.

Table 1. The comparison of different methods in classification

$Method$	$Precision$	$Recall$	F_2
LeNet	0.7532	0.8207	0.8062
ZF net	0.8988	0.9107	0.9082
AlexNet	0.9260	0.9328	0.9314
ResNet	**0.9872**	**0.9805**	**0.9818**

5.2 Image Segmentation Results

To evaluate the segmentation performance of our method, we set hyper-parameters following existing Mask R-CNN [13]. In the training process, a RoI is considered as positive if it has IoU with a ground-truth box of at least 0.5 and negative otherwise. ResNet50 backbone are chosen to adjust to our dataset, all Images are resized as 512 * 512 pixels, each mini-batch has 2 images per GPU and each image has one sampled RoI at least, with a ratio of 1:3 positives to negatives, learning rate and epoches are set as 0.005 and 1500. For the U-net training implementation, learning rate and epoches are 0.0001 and 300, while 0.001 and 60 in the implementation of FCN-AlexNet. In particular, dropout rate is set to 33% in FCN implementation.

The quantitative comparison among our method and the state-of-the-art methods [26,28] is shown in Table 2, where Mask R-CNN achieves competitive results and our method has much higher mAP (85.0% *vs.* 83.8%) and F1-measure (82.7% *vs.* 79.6%) than the original Mask R-CNN method benefit from using more anchor scales. In addition, some special results, such as image with small tumor, image with big tumor, multi-tumor and tumor in complex background,

Table 2. The comparison of different methods in segmentation

Method	mAP	F1-measure
U-net	0.7370	0.6924
FCN-AlexNet	0.8241	0.7743
Mask R-CNN	0.8376	0.7962
Ours	**0.8501**	**0.8270**

are selected to illustrate the segmentation performance in our dataset. As shown in Fig. 4, our method achieves the better segmentation results in most instances, even in complex environment with large background noise.

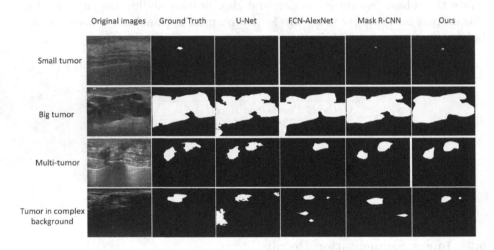

Fig. 4. The segmentation results comparison under four different conditions.

6   Conclusion

In this study, we analyze a CAD system for image classification and segmentation based on our breast ultrasound dataset. We first use a transfer learning approach with the pretrained ResNet structure in image classification, after this stage, all images can be labeled by normal or cancerous one and the normal candidates can be excluded. Then the improved Mask R-CNN method is developed to segment tumor areas in cancerous candidates. Experimental results show that our two-stage method can realize the better performance both in classification and segmentation tasks, where the Precision and Recall in classification tasks can reach to 98.72% and 98.05%, and a slight improvement on mAP of 85.0% vs. 83.8% and F1-Measure of 82.7% vs. 79.6% when compared with using the original Mask R-CNN method.

Acknowledgement. This work has been supported by National Key R&D Program of China (2017YFB1301100), National Natural Science Foundation of China (61572060, 61772060, 61728201) and CERNET Innovation Project (NGII20160316, NGII20170315).

References

1. Akin, O., et al.: Advances in oncologic imaging: update on 5 common cancers. CA: Cancer J. Clin. **62**(6), 364 (2012)
2. Arbelaez, P., Maire, M., Fowlkes, C.C., Malik, J.: Contour detection and hierarchical image segmentation. IEEE Trans. Pattern Anal. Mach. Intell. **33**(5), 898–916 (2011)
3. Huynh, B., Drukker, K., Giger, M.: MO-DE-207B-06: computer-aided diagnosis of breast ultrasound images using transfer learning from deep convolutional neural networks. Med. Phys. **43**, 3705 (2016)
4. Byra, M., Nowicki, A., Wroblewskapiotrzkowska, H., Dobruchsobczak, K.: Classification of breast lesions using segmented quantitative ultrasound maps of homodyned K distribution parameters. Med. Phys. **43**(10), 5561–5569 (2016)
5. Cai, L., Wang, X., Wang, Y., Guo, Y., Yu, J., Wang, Y.: Robust phase-based texture descriptor for classification of breast ultrasound images. Biomed. Eng. Online **14**(1), 26 (2015)
6. Cheng, H.D., Shan, J., Ju, W., Guo, Y., Zhang, L.: Automated breast cancer detection and classification using ultrasound images: a survey. Pattern Recognit. **43**(1), 299–317 (2010)
7. Dhungel, N., Carneiro, G., Bradley, A.P.: Deep learning and structured prediction for the segmentation of mass in mammograms. In: Navab, N., Hornegger, J., Wells, W.M., Frangi, A.F. (eds.) MICCAI 2015. LNCS, vol. 9349, pp. 605–612. Springer, Cham (2015). https://doi.org/10.1007/978-3-319-24553-9_74
8. Drukker, K., Gruszauskas, N.P., Sennett, C.A., Giger, M.L.: Breast us computer-aided diagnosis workstation: performance with a large clinical diagnostic population. Radiology **248**(2), 392–397 (2008)
9. Flores, W.G., Pereira, W.C.A., Infantosi, A.F.C.: Breast ultrasound despeckling using anisotropic diffusion guided by texture descriptors. Ultrasound Med. Biol. **40**(11), 2609–2621 (2014)
10. Flores, W.G., Pereira, W.C.A., Infantosi, A.F.C.: Improving classification performance of breast lesions on ultrasonography. Pattern Recognit. **48**(4), 1125–1136 (2015)
11. Gomez, W., Pereira, W.C.A., Infantosi, A.F.C.: Analysis of co-occurrence texture statistics as a function of gray-level quantization for classifying breast ultrasound. IEEE Trans. Med. Imaging **31**(10), 1889–1899 (2012)
12. Gomez, W., Pereira, W.C.A., Infantosi, A.F.C.: Evolutionary pulse-coupled neural network for segmenting breast lesions on ultrasonography. Neurocomputing **175**, 877–887 (2016)
13. He, K., Gkioxari, G., Dollr, P., Girshick, R.: Mask R-CNN. In: IEEE International Conference on Computer Vision, pp. 2980–2988 (2017)
14. He, K., Zhang, X., Ren, S., Sun, J.: Deep residual learning for image recognition. In: Computer Vision and Pattern Recognition, pp. 770–778 (2016)
15. Horsch, K., Giger, M.L., Venta, L.A., Vyborny, C.J.: Computerized diagnosis of breast lesions on ultrasound. Med. Phys. **29**(2), 157–164 (2002)

16. Joo, S., Yang, Y.S., Moon, W.K., Kim, H.C.: Computer-aided diagnosis of solid breast nodules: use of an artificial neural network based on multiple sonographic features. IEEE Trans. Med. Imaging **23**(10), 1292–1300 (2004)
17. Krizhevsky, A., Sutskever, I., Hinton, G.E.: ImageNet classification with deep convolutional neural networks, pp. 1097–1105 (2012)
18. Lecun, Y., Bottou, L., Bengio, Y., Haffner, P.: Gradient-based learning applied to document recognition. Proc. IEEE **86**(11), 2278–2324 (1998)
19. Lin, T.Y., Goyal, P., Girshick, R., He, K., Dollr, P.: Focal loss for dense object detection, pp. 2999–3007 (2017)
20. Madabhushi, A., Metaxas, D.N.: Combining low-, high-level and empirical domain knowledge for automated segmentation of ultrasonic breast lesions. IEEE Trans. Med. Imaging **22**(2), 155–169 (2003)
21. Marcomini, K.D., Carneiro, A.A.O., Schiabel, H.: Application of artificial neural network models in segmentation and classification of nodules in breast ultrasound digital images. Int. J. Biomed. Imaging **2016**, 7987212 (2016)
22. Moon, W.K., Lo, C.M., Chang, J.M., Huang, C.S., Chen, J.H., Chang, R.F.: Computer-aided classification of breast masses using speckle features of automated breast ultrasound images. Med. Phys. **39**(10), 6465–6473 (2012)
23. Pons, G., Marti, J., Marti, R., Ganau, S., Vilanova, J.C., Noble, J.A.: Evaluating lesion segmentation on breast sonography as related to lesion type. J. Ultrasound Med. **32**(9), 1659–1670 (2013)
24. Rodrigues, R., Braz, R., Pereira, M., Moutinho, J., Pinheiro, A.M.: A two-step segmentation method for breast ultrasound masses based on multi-resolution analysis. Ultrasound Med. Biol. **41**(6), 1737–1748 (2015)
25. Rodrigues, R., Pinheiro, A.M.G., Braz, R., Pereira, M., Moutinho, J.: Towards breast ultrasound image segmentation using multi-resolution pixel descriptors, pp. 2833–2836 (2012)
26. Ronneberger, O., Fischer, P., Brox, T.: U-Net: convolutional networks for biomedical image segmentation. In: Navab, N., Hornegger, J., Wells, W.M., Frangi, A.F. (eds.) MICCAI 2015. LNCS, vol. 9351, pp. 234–241. Springer, Cham (2015). https://doi.org/10.1007/978-3-319-24574-4_28
27. Sadek, I., Elawady, M., Stefanovski, V.: Automated breast lesion segmentation in ultrasound images. Computer Vision and Pattern Recognition. arXiv:1609.08364 (2016)
28. Shelhamer, E., Long, J., Darrell, T.: Fully convolutional networks for semantic segmentation. IEEE Trans. Pattern Anal. Mach. Intell. **39**(4), 640–651 (2017)
29. Shi, J., Zhou, S., Liu, X., Zhang, Q., Lu, M., Wang, T.: Stacked deep polynomial network based representation learning for tumor classification with small ultrasound image dataset. Neurocomputing **194**, 87–94 (2016)
30. Su, H., Liu, F., Xie, Y., Xing, F., Meyyappan, S., Yang, L.: Region segmentation in histopathological breast cancer images using deep convolutional neural network, pp. 55–58 (2015)
31. Takemura, A., Shimizu, A., Hamamoto, K.: Discrimination of breast tumors in ultrasonic images using an ensemble classifier based on the adaboost algorithm with feature selection. IEEE Trans. Med. Imaging **29**(3), 598–609 (2010)
32. Uniyal, N., et al.: Ultrasound RF time series for classification of breast lesions. IEEE Trans. Med. Imaging **34**(2), 652–661 (2015)
33. Wang, D., Shi, L., Heng, P.A.: Automatic detection of breast cancers in mammograms using structured support vector machines. Neurocomputing **72**, 3296–3302 (2009)

34. Wang, W., Zhu, L., Qin, J., Chui, Y.P., Li, B.N., Heng, P.A.: Multiscale geodesic active contours for ultrasound image segmentation using speckle reducing anisotropic diffusion. Opt. Lasers Eng. **54**, 105–116 (2014)
35. Wang, Z., Yu, G., Kang, Y., Zhao, Y., Qu, Q.: Breast tumor detection in digital mammography based on extreme learning machine. Neurocomputing **128**(5), 175–184 (2014)
36. Xi, X., et al.: Breast tumor segmentation with prior knowledge learning. Neurocomputing **237**, 145–157 (2017)
37. Xian, M., Huang, J., Zhang, Y., Tang, X.: Multiple-domain knowledge based MRF model for tumor segmentation in breast ultrasound images, pp. 2021–2024 (2012)
38. Shi, X., Cheng, H.D., Hu, L.: Mass detection and classification in breast ultrasound images using fuzzy SVM. In: Proceedings of Joint Conference on Information Sciences (2006)
39. Yap, M.H., Edirisinghe, E., Bez, H.: Processed images in human perception: a case study in ultrasound breast imaging. Eur. J. Radiol. **73**(3), 682–687 (2010)
40. Zeiler, M.D., Fergus, R.: Visualizing and understanding convolutional networks. In: Fleet, D., Pajdla, T., Schiele, B., Tuytelaars, T. (eds.) ECCV 2014. LNCS, vol. 8689, pp. 818–833. Springer, Cham (2014). https://doi.org/10.1007/978-3-319-10590-1_53

Intra-Image Region Context for Image Captioning

Shihao Wang, Hong Mo, Yue Xu, Wei Wu, and Zhong Zhou$^{(\boxtimes)}$

State Key Laboratory of Virtual Reality Technology and Systems,
Beihang University, Beijing 100191, China
zz@buaa.edu.cn

Abstract. Image captioning is a challenging task involving computer vision and natural language processing. In recent works, visual attention mechanisms have been extensively used. However, they consider little about the correlations among different regions and the attention on regions. This paper is try to make up for the deficiencies in existing approaches and propose a novel captioning model, which extracts the salient region correlations from the image feature, synthesizes intra-image regions' context, and automatically distributes an appropriate attention over regions. The Intra-Image Region Context (IIRC) model proposed in this paper jointly learns regions' semantic correlations in one image. It consists of two main parts. The first is to extract feature vectors of image through convolutional neural work (CNN) and get correlations among regions from feature vectors by recurrent neural network (RNN). The second is to generate the caption according to the synthesis of region contexts from the first network with attention on different region contexts. The model and baseline are evaluated on MSCOCO test server. The experimental results have illustrated that the model is superior over many outstanding models on the metrics of BLEU, METEOR, ROUGE-L and CIDEr. Moreover, the model excels in describing details, especially those related to position and action.

Keywords: Image captioning · Intra-image region · Regions correlations

1 Introduction

Image captioning is a fundamental research issue which aims at automatically generating a natural description of an image. It has received a significant amount of attention in both computer vision and natural language processing research communities [1, 2]. The image captioning's task is to generate semantically and syntactically appropriate target sentence with consecutive words to represent the image content, which can be quite challenging in two ways. First of all, the model needs to learn and capture the semantic information of image with great precision. Secondly, the generation of the target sentence must take into account both the correctness of the syntax and the correlation between the semantics and the image content, which thus requires complex interactions among them.

In recent years, many approaches which achieve impressive results on image captioning [10, 11, 22] have been raised with the availability of larger datasets [3, 4, 9].

© Springer Nature Switzerland AG 2018
R. Hong et al. (Eds.): PCM 2018, LNCS 11166, pp. 212–222, 2018.
https://doi.org/10.1007/978-3-030-00764-5_20

Particularly, a strong and effective approach was proposed to generate captions in high quality [11]. The image features are encoded by the input image with a deep convolutional neural work (CNN), then the encoded feature is used to generate the output caption by the Long Short Term Memory (LSTM) recurrent neural network (RNN) decoder. This encoder-decoder model becomes the baseline of recent research methods.

To improve the quality of the output captions and help the decoder focus on the key image information, the model needs to perform some fine-grained visual processing. Therefore, visual attention mechanisms have been widely applied in image captioning tasks [12, 17, 22]. Most traditional visual attention mechanisms used in image captioning are the top-down variety. These mechanisms are generally trained to selectively attend to the output of one or more layers of a CNN [16, 22]. However, they give little consideration to how the image regions which are subject to attention are selected, and how those different image regions are related with each other.

In this paper we propose a model based on encoder-decoder architecture, which allows the network to generate captions by the correlation of context among image regions. Our mechanism extracts several major regions of the image feature as region features, with each region feature represented by a pooled feature vector. Then we form a sequence including these features in order, and use RNN encoder to read each region feature sequentially. That is encoder maps these image regions sequence into a continuous feature vectors. We call this intra-image region context modelling, which considers the correlations among image regions. After that, the decoder transforms the continuous feature vectors from the encoder to a sequence as the output sentence. Both the encoder and the decoder adopt the LSTM as recurrent neuron. This process with sequence-to-sequence encoding and decoding enables correlation learning of region features in the model. Allowing the decoder units to determine which region features is more helpful and important for each time step, we introduce the non-visual attention mechanism into this framework. In this way, the model can use the context to predict the attention distribution over regions.

In order to evaluate the performance of our method, our model is trained and tested on MSCOCO caption dataset [9] and MSCOCO test server. MSCOCO is a large and popular dataset containing more than 120,000 images. Our results on the test server not only achieve remarkable performance at CIDEr, METEOR, ROUGE-L and BLEU scores, but also outperform current baseline. The scores of evaluation metrics thoroughly reflect the effectiveness of our model.

2 Proposed Model

Our Intra-Image Region Context model consists of two major subnetworks components: intra-image region context (Fig. 1(a)) and language decoder (Fig. 1(b)). The attention module is used between subnetworks. Different from CNN-RNN encoder-decoder architecture like show-and-tell model [11], we use the RNN-RNN encoder-decoder just like sequence-to-sequence model [2]. When our model gets the target image, it will first extract an image feature by the deep CNN model. The center region, top left region, top right region, bottom left region, bottom right region and entire region of the image feature map are serialized, and then they are put into the LSTM encoder to extract the correlations of different region features. After that, the region context feature which is produced at the last time step of LSTM encoder will be put into the first time step to LSTM decoder to generate the caption for the image. To overcome the loss of regional semantic information without fine-grained localisation, we introduce an attention mechanism to our model. Attention mechanism can focus on the most relevant sections of the input region feature vectors sequence and guide our decoder to those sections for feature extraction. An illustration of our complete model for image captioning is provided in Fig. 1.

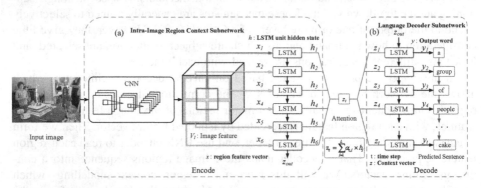

Fig. 1.   An overview of the proposed model for image captioning by intra-image region context.

2.1   Intra-Image Region Context Subnetwork

The Intra-Image Region Context subnetwork includes a deep CNN (e.g. Inception-v4 [7]) and a RNN with LSTM recurrent neuron, as illustrated in Fig. 1(a). The CNN has already been well pre-trained on ImageNet [5] which is a large dataset for image classification mission. The pre-trained network which has a well generalization capability, has already learnt the ability of how to get some useful features. The transfer learning is widely used in lots of computer vision tasks. In our method, we use the well pre-trained CNN to extract the semantic feature of the full image. We get the feature from the last layer before pooling layer and full-connected layer. The feature map V_i will be with the shape like (V_h, V_w, V_c), where V_h, V_w, V_c represents height, width, and channel of the feature separately.

Then we divide the feature map into 5 pieces, which have same shapes of $(V_h/2, V_w/2, V_c)$. We get 5 different important regions of feature map as illustrated in Fig. 2. Their directions are: center, top left, top right, bottom left, bottom left and bottom right. These parts map the semantic information of the corresponding areas of the source image. The attention in the human visual system is able to be focused automatically by top-down and bottom-up signals [18, 19]. When a person wants to observe what the picture is talking about, he usually focuses his attention on the center area of the image first to get the main semantics information of the image. Then he will look at the remaining area of the image to obtain some scene of other semantic information, so that he can provide the caption for the image. Drew on the experience of the method of human observation, we simplify the surrounding area as four corners of image feature map, and center area as the center of image feature map. For keeping the full image semantic information joining the generation of caption, we pool the whole feature map to the size of 5 splits of image feature map. For each feature vector of 6 regions which are mentioned above, we flatten and full-connect them to the m-dimensional feature vector with the length as the number of hidden units. Then, we form the region feature vectors x_n sequence $X = (x_1, \ldots, x_6)$ in order: center, top left, top right, bottom left, bottom right, full.

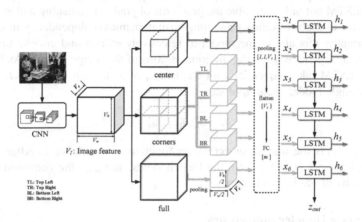

Fig. 2. Intra-Image Region Context Subnetwork. It utilizes the regions of image feature to extract the contexts and connection between them, then puts the region feature into LSTM to get hidden status and the output of our encoder.

LSTM encoder reads region features sequentially, and gives final output z_{out} which represents the context of 6 region semantic features. Recurrent neural network is effective for modeling sequence data. In theory [6], RNN could handle long-term dependencies, but actually it can only remember the limited contents of time steps due to problems of gradient vanishing and explosion. To address this problem, a special RNN neuron called LSTM is proposed and it establishes the state-of-the-art for the sequence task. Therefore, we feed region features sequence X into LSTM. Particularly, at each time step t, the LSTM updates states using the input x_t, previous status h_{t-1} and c_{t-1}, as follows:

$$f_t = \sigma\left(W_{fh}h_{t-1} + W_{fx}x_t + b_f\right) \tag{1}$$

$$i_t = \sigma\left(W_{ih}h_{t-1} + W_{ix}x_t + b_i\right) \tag{2}$$

$$o_t = \sigma\left(W_{oh}h_{t-1} + W_{ox}x_t + b_o\right) \tag{3}$$

$$g_t = tanh\left(W_{gh}h_{t-1} + W_{gx}x_t + b_g\right) \tag{4}$$

$$c_t = f_t \odot c_{t-1} + i_t \odot g_t \tag{5}$$

$$h_t = o_t \odot \tanh(c_t) \tag{6}$$

where all of $\sigma(\cdot)$ refer to the sigmoid function, $tanh(\cdot)$ the hyperbolic tangent function, $\odot$ the operator of element-wise product. The LSTM has five components, four gates and one memory cell: forget gate f, input gate i, output gate o, input modulation gate g, memory cell c_t, with the learned parameters Ws, bs. The cell c_t depends on c_{t-1} which is the previous memory cell, adjusted by forget gate f_t, and g_t adjusted by input gate i_t. Therefore, LSTM not only can solve the problems of gradient vanishing and explosion, but also is able to capture complex and long-term dynamics or dependency in sequence data. Importantly, this allows the model to selectively extract and encode the spatial and semantical dependency among different regions of the image feature. As Fig. 1(a) shown, the LSTM take sequentially an element x_t of X at each time step t. Then, it updates its single hidden state h_t of step t as:

$$h_t = f_\lambda(h_{t-1}, x_t) \tag{7}$$

where f_λ represents the non-linear activation function of parameter λ. After six time steps we will have $h_t(t = 1, \ldots, 6)$, the hidden states, and z_{out}, the comprehensive of region contexts of the image feature.

2.2   Language Decoder Subnetwork

To model the potential high-level region semantic correlation subject to learning a caption sequence generator, we construct a LSTM decoder. Specifically, the LSTM decoder aims at modeling sequential recurrent regions correlation within both intra-image region context z and the comprehensive of region contexts z_{out} and generation dynamic length output as predicted sequence of words y_t over time step t. This is our purpose because of varying co-occurring semantic attributes among regions of the feature. The appropriate caption of the image will be generated from pretreatment list of words. The Language Decoder subnetwork is shown in Fig. 1(b), which consists of one LSTM decoder and attention module. In order to obtain the initial hidden state h_1^2 of decoder, we use the comprehensive of region contexts vector z_{out} to initialize it. This step is for the purpose of incorporating the intra-image region context correlation into the decoding procedure. Different from the encoder, when we infer a caption, the output word and hidden state of time step t, y_{t-1} and h_t^2 rely on the previous h_{t-1}^2 and z_{t-1}, which is initialized by the start token of words (e.g. "<S>"). In fundamental, our model is able to mine the potential high-level region semantic correlation of dynamic sequence precisely because of this recurrent feedback connection in sequence. Different from Eq. (7), h_t^2 is update as follows:

$$h_t^2 = f_\lambda\left(h_{t-1}^2, z_{t-1}\right) \tag{8}$$

Similar to Eq. (1)–Eq. (6), the gates and cells of the decoder LSTM update as following:

$$f_t = \sigma\left(W_{fh}h_{t-1}^2 + W_{fz}z_{t-1} + b_f\right) \tag{9}$$

$$i_t = \sigma\left(W_{ih}h_{t-1}^2 + W_{iz}z_{t-1} + b_i\right) \tag{10}$$

$$o_t = \sigma\left(W_{oh}h_{t-1}^2 + W_{oz}z_{t-1} + b_o\right) \tag{11}$$

$$g_t = tanh\left(W_{gh}h_{t-1}^2 + W_{gz}z_{t-1} + b_g\right) \tag{12}$$

$$c_t = f_t \odot c_{t-1} + i_t \odot g_t \tag{13}$$

$$h_t^2 = o_t \odot tanh(c_t) \tag{14}$$

The decoder LSTM is also updated by previous states and some parameters as the encoder LSTM did before. Notation $y_{1:T}$ refers to a sequence of words $(y_1, \ldots, y_T)$. The conditional distribution over possible result at each time step t, given by:

$$p(y_t|y_{1:t-1}) = \text{softmax}\left(W_p h_t^2 + b_p\right) \tag{15}$$

where W_p and b_p are learned matrixes. The complete sequence is calculated as:

$$p(y_{1:T}) = \prod_{t=1}^{T} p(y_t|y_{1:t-1}) \tag{16}$$

Recurrent Region Attention. Regional correlation patterns in images of real world can have many significant and complex changes. A considerable amount of image semantic information might not be well encoded, because each region context vector could only hold its limited information of semantics. In order to overcome this limitation, we introduce the attention mechanism into our model to improve its performance. So that it will automatically locate at the most relevant sections of the input region feature vector sequence and focus on these sections, when the model is predicting the current words. This is actually a standard sequence-to-sequence alignment mechanism which is different from the attention mechanism in [22]. We implement the mechanism by importing a special structure between the encoder output and reformulated decoder inputs.

Given the output h of the encoder LSTM, at each time step t we generate an attention weight $\alpha_{i,t}$ for each encoder hidden state h_i as:

$$u_{i,t} = h_t^{2^\top} W_u h_i \tag{17}$$

$$\alpha_t = \text{softmax}(u_t) \tag{18}$$

$$z_t = \sum_{i=1}^{L} \alpha_{i,t} h_i \tag{19}$$

where W_u is learned parameters, $u_{i,t}$ is the score at i-th hidden state of time t, $i = 1, \ldots, L(L = 6)$, and n is the splits number of image features we discussed before. Similar to [8], our approach of attention gets the decoder hidden state at time step t. Then we calculate attention scores, and from the calculated scores, we get the context vector z_t which will be concatenated with hidden state h_t^2 of the decoder. After that, we can predict a word of the caption sequence by Eqs. (15) and (16).

At last, the objective of our method is to minimize the cross entropy loss L_{CE} by given target ground truth sequence $y_{1:T}^*$ and captioning model with parameters θ, as follows:

$$L_{CE}(\theta) = - \sum_{t=1}^{T} \log p_\theta \left(y_t^* | y_{1:T}^* \right) \tag{20}$$

We use the stochastic gradient descent (SGD) with gradient decay to optimize the goal function, which is efficient for optimizing our model, and the comprehensive of region contexts just feed at the beginning of the decoder LSTM only once at training time.

3   Experiments

3.1   Dataset

To evaluate our proposed model, a large and high-quality dataset is necessary. In view of this, we use the Microsoft COCO (MSCOCO) 2014 caption dataset [9]. For validation of model parameters and offline evaluation, we use the data splits from the method of 'Karpathy' [10]. These splits have been widely used to demonstrate results of models in the previous woks. The training split contains 113,287 images with five captions each, 5 K images for validation, and 5 K images for testing as well. We also submit our results to MSCOCO test server to get how effective our model is. Following other practicing standard, we slightly filter the model vocabulary. We keep words that appear above five times, convert all captions to lower case and tokenize on space. We end up with a vocabulary of length 10,116. We report results with seven extensively used evaluation metrics: BLEU (1, 2, 3, 4) [23], METEOR [25], ROUGE-L [24], and CIDEr [21].

3.2   Implementation Details

Our proposed IIRC subnetwork consists of two components, CNN and LSTM encoder. Particularly, in this work, we use Inception-v4 [7] CNN model which is well pre-trained on ImageNet [5] to extract the semantic feature of the image for image embedding. We elicit the feature from the layer after Inception-C blocks as our image feature V_l which has the shape of $8 \times 8 \times 1536$ i.e. $V_h \times V_w \times V_w$. Cutting out from V_l, our each region feature vector has the shape of $4 \times 4 \times 1536$. Then region feature vectors are pooled, flattened and full-connected to the 512-dimensional feature vector i.e. m = 512. Determined by experience of others works, both the encoder and the decoder LSTM of our model has 512 hidden state units (neurons). Similarly, word and attention embedding dimension are fixed to 512. Empirically, we set the initial learning rate as 0.5 with learning rate decay factor of 0.5 per 8 epochs for our SGD optimizer, and we find that it is a suitable way for our model optimizing. We initialize our model by the fixed pre-trained parameters of the CNN with given hyperparameters. After the model converges (i.e. we have a nice set of parameters), we unfix parameters of Inception-v4 and fine-tune the model to get the better performance on MSCOCO dataset. The learning rate is fixed to value of 5×10^{-4}.

To quantify the effectiveness of our approach, similar to model in [11], our baseline model uses CNN as encoder and LSTM as decoder in encoder-decoder architecture. The difference is that we upgrade its CNN encoder from Inception-v3 to Inception-v4. The shape of CNN net's last layer output as image feature is $8 \times 8 \times 1536$. This is equivalent to the original net's last layer output in [11] which has the shape of $8 \times 8 \times 2048$. The number of LSTM hidden state units is similarly set to 512. Moreover, we set another model called All Regions Context (ARC) for comparative experiment. The ARC model is similar to our proposed IIRC model. However, it uses 64 regions of size $1 \times 1 \times 1536$ as input region features in Sect. 2.1. To be fair, we trained both the baseline model and the ARC model in the same way as our IIRC model.

Table 1. Results on the online MSCOCO test server.

	BLEU-1		BLEU-2		BLEU-3		BLEU-4		METEOR		ROUGE-L		CIDEr	
	c5	c40	c5	c40	c5	c40	c5	c40	c5	c40	c5	c40	c5	c40
SCA-CNN [16]	71.2	89.4	54.2	80.2	40.4	69.1	30.2	57.9	24.4	33.1	52.4	67.4	91.2	92.1
NIC [11]	71.3	89.5	54.2	80.2	40.7	69.4	30.9	58.7	25.4	34.6	53.0	68.2	94.3	94.6
Review Net [12]	72.0	90.0	55.0	81.2	41.4	70.5	31.3	59.7	25.6	34.7	53.3	68.6	96.5	96.9
ATT_VC [13]	73.1	90.0	56.5	81.5	42.4	70.9	31.6	59.9	25.0	33.5	53.5	68.2	94.3	95.8
MSM [14]	73.9	91.9	57.5	84.2	43.6	74.0	33.0	63.2	25.6	35.0	54.2	70.0	98.4	100.3
PG-BCMR [15]	**75.4**	91.8	59.1	84.1	44.5	73.8	33.2	62.4	25.7	34.0	55.0	69.5	101.3	103.2
Ours: baseline	71.7	88.9	54.5	79.2	40.1	67.5	29.2	55.9	25.3	33.5	52.9	67.1	94.4	96.9
Ours: ARC	71.8	89.2	54.7	79.8	40.3	68.4	29.4	56.7	25.5	33.9	53.1	67.3	95.4	98.4
Ours: IIRC	74.9	**92.0**	**58.5**	**84.4**	**44.8**	**74.3**	**34.2**	**63.5**	**27.0**	**36.3**	**55.5**	**70.8**	**105.7**	**105.5**

3.3 Results and Discussion

To evaluate the effectiveness of our approach, we evaluate our model against prior works as well as our comparative models including baseline and ARC model. The evaluation results of comparison are illustrated in Table 1, where row IIRC is the results of our model. The results in the table show that, our IIRC model has achieved better performance at BLEU (1, 2, 3, 4), METEOR, ROUGE-L as well as CIDEr metrics, and exceeded our baseline and ARC in all metrics. Obviously, the scores of our IIRC model is higher than other models in Table 1 on all metrics only except c5 of BLEU-1 metric. The gap between ARC and IIRC shows the superiority of our approach in choosing salient regions. The approach in [14] utilizes both attributes information and image feature encoding to decode captions, but we only use the image feature from the picture. The approach in [15] uses reinforcement learning in optimizing metrics of its model, and it gives a higher weight on ROUGE metric. The CIDEr metric is different from other evaluation metrics, because it is proposed to aim at image abstract issues and have high matching rate of artificial consensus [21]. Specially, SCST model [17] has an optimizing target of CIDEr score with reinforcement learning, and it has established a state-of-the-art on the caption task. Therefore, the scores of our model on the CIDEr metric are more sufficient to prove the effectiveness of our approach. Simultaneously, the METEOR and ROUGE-L scores can also demonstrate that than BLEUs [20]. The score gap between our model and other models in Table 1 is sufficient to illustrate the validity of our model.

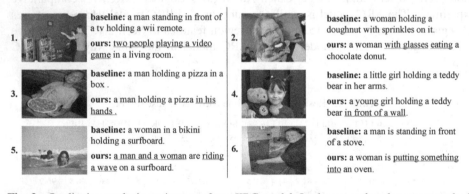

Fig. 3. Qualitative analysis on impact of our IIRC model. In the examples above our method can give more precise details of the picture such as positions, number of objects and the color detail, which baseline fail to do. The red and underlined words are caption details given by our model.

We also conduct a qualitative analysis on the role of intra-image region context in caption generation. We compare our model with baseline, which has similar architecture as [11]. Some samples of the caption generated by our approach method and baseline method are shown in Fig. 3. Our model can get details of position and action from perception of intra-image region context. Moreover, it achieves a well performance relative to our baseline. As examples 1, 2, 5 and 6 in Fig. 3, our model generates captions from the intra-image region context, which are more accurate at action details (e.g. in example 1, baseline just gives 'standing in front a tv holding a wii remote' but ours gives 'playing a video game' which shows intra-image region context information between TV and remote.). As example 3 and 4, our model shows more position details in caption results (e.g. in example 3, from the perception of region context, we get 'in his hands' rather than 'in a box', and the location of pizza is more appropriate). In addition, we can also give a more accurate number of objects in the caption (e.g. example 1 and 5).

4   Conclusion

In this paper, we present an approach that generates captions of images from intra-image region context. Our approach enables the salient region context to be effectively extracted from the image semantic feature, and it is able to automatically perceive the correlation among regions. Applying this approach, we can generate the description of an image based on the fusion of intra-image region contexts. The method is tested on MSCOCO test server. The experiment results demonstrate its superiority on all general caption metrics over other models and its effectiveness of perceiving the intra-image region context. Meanwhile, the IIRC model is able to generate captions with more details on position and action.

Acknowledgements. This work is supported by the Natural Science Foundation of China under Grant No. 61472020, 61572061.

References

1. Russakovsky, O., et al.: Imagenet large scale visual recognition challenge. Int. J. Comput. Vis. **115**(3), 211–252 (2015)
2. Farhadi, A., Endres, I., Hoiem, D., Forsyth, D.: Describing objects by their attributes. In: 2009 Computer Vision and Pattern Recognition, pp. 1778–1785. IEEE (2009)
3. Hodosh, M., Young, P., Hockenmaier, J.: Framing image description as a ranking task: data, models and evaluation metrics. In: International Conference on Artificial Intelligence, pp. 4188–4192 (2015)
4. Young, P., Lai, A., Hodosh, M., Hockenmaier, J.: From image descriptions to visual denotations: new similarity metrics for semantic inference over event descriptions. Nlp.cs. illinois.edu (2014)
5. Deng, J., Dong, W., Socher, R., Li, L.J., Li, K., Fei-Fei, L.: Imagenet: a large-scale hierarchical image database. In: 2009 IEEE Conference on Computer Vision and Pattern Recognition, CVPR 2009, pp. 248–255. IEEE (2009)
6. Bengio, Y., Simard, P., Frasconi, P.: Learning long-term dependencies with gradient descent is difficult. IEEE Trans. Neural Netw. **5**(2), 157–166 (1994)
7. Szegedy, C., Ioffe, S., Vanhoucke, V., Alemi, A.A.: Inception-v4, inception-resnet and the impact of residual connections on learning. In: AAAI, vol. 4, p. 12 (2017)
8. Luong, T., Pham, H., Manning, C.D.: Effective approaches to attention-based neural machine translation. In: Empirical Methods in Natural Language Processing, pp. 1412–1421 (2015)
9. Lin, T.-Y., et al.: Microsoft COCO: common objects in context. In: Fleet, D., Pajdla, T., Schiele, B., Tuytelaars, T. (eds.) ECCV 2014. LNCS, vol. 8693, pp. 740–755. Springer, Cham (2014). https://doi.org/10.1007/978-3-319-10602-1_48

10. Karpathy, A., Fei-Fei, L.: Deep visual-semantic alignments for generating image descriptions. In: Proceedings of the IEEE Conference on Computer Vision and Pattern Recognition, pp. 3128–3137 (2015)
11. Vinyals, O., Toshev, A., Bengio, S., Erhan, D.: Show and tell: a neural image caption generator. In: 2015 IEEE Conference on Computer Vision and Pattern Recognition (CVPR), pp. 3156–3164. IEEE (2015)
12. Yang, Z., Yuan, Y., Wu, Y., Cohen, W.W., Salakhutdinov, R.R.: Review networks for caption generation. In: Advances in Neural Information Processing Systems, pp. 2361–2369 (2016)
13. You, Q., Jin, H., Wang, Z., Fang, C., Luo, J.: Image captioning with semantic attention. In: 2016 IEEE Conference on Computer Vision and Pattern Recognition, pp. 4651–4659 (2016)
14. Yao, T., Pan, Y., Li, Y., Qiu, Z., Mei, T.: Boosting image captioning with attributes. In: Computer Vision and Pattern Recognition, pp. 4894–4902 (2017)
15. Liu, S., Zhu, Z., Ye, N., Guadarrama, S., Murphy, K.: Improved image captioning via policy gradient optimization of spider. In: Proceedings of IEEE Conference on Computer Vision and Pattern, vol. 3 (2017)
16. Chen, L., et al.: SCA-CNN: spatial and channel-wise attention in convolutional networks for image captioning. In: Proceedings of the IEEE Conference on Computer Vision and Pattern Recognition, pp. 5659–5667 (2017)
17. Rennie, S.J., Marcheret, E., Mroueh, Y., Ross, J., Goel, V.: Self-critical sequence training for image captioning. In: CVPR, vol. 1, p. 3 (2017)
18. Corbetta, M., Shulman, G.L.: Control of goal-directed and stimulus-driven attention in the brain. Nat. Rev. Neurosci. 3(3), 201 (2002)
19. Buschman, T.J., Miller, E.K.: Top-down versus bottom-up control of attention in the prefrontal and posterior parietal cortices. Science 318(5847), 1860–1862 (2007)
20. Elliott, D., Keller, F.: Comparing automatic evaluation measures for image description. In: Meeting of the Association for Computational Linguistics, pp. 452–457 (2014)
21. Vedantam, R., Zitnick, C.L., Parikh, D.: Cider: consensus-based image description evaluation. In: Computer Vision and Pattern Recognition, pp. 4566–4575 (2015)
22. Xu, K., et al.: Show, attend and tell: neural image caption generation with visual attention. In: International Conference on Machine Learning, pp. 2048–2057 (2015)
23. Papineni, K., Roukos, S., Ward, T., Zhu, W.J.: Bleu: a method for automatic evaluation of machine translation. In: Proceedings of the 40th Annual Meeting on As-sociation for Computational Linguistics, pp. 311–318. Association for Computational Linguistics (2002)
24. Lin, C.Y.: Rouge: a package for automatic evaluation of summaries. In: Proceedings of Workshop on Text Summarization Branches Out, Post Conference Workshop of ACL 2004 (2004)
25. Denkowski, M., Lavie, A.: Meteor universal: language specific translation evaluation for any target language. In: Proceedings of the Ninth Workshop on Statistical Machine Translation, pp. 376–380 (2014)

Viewpoint Quality Evaluation for Augmented Virtual Environment

Ming Meng[1], Yi Zhou[1,2], Chong Tan[1], and Zhong Zhou[1(✉)]

[1] State Key Laboratory of Virtual Reality Technology and Systems,
Beihang University, Beijing 100191, China
zz@buaa.edu.cn
[2] Beijing BigView Technology Co., Ltd, Beijing, China

Abstract. Augmented Virtual Environment (AVE) fuses real-time video streaming with virtual scenes to provide a new capability of the real-world run-time perception. Although this technique has been developed for many years, it still suffers from the fusion correctness, complexity and the image distortion during flying. The image distortion could be commonly found in an AVE system, which is decided by the viewpoint of the environment. Existing work lacks of the evaluation of the viewpoint quality, and then failed to optimize the fly path for AVE. In this paper, we propose a novel method of viewpoint quality evaluation (VQE), taking texture distortion as evaluation metric. The texture stretch and object fragment are taken as the main factors of distortion. We visually compare our method with viewpoint entropy on campus scene, demonstrating that our method is superior in reflecting distortion degree. Furthermore, we conduct a user study, revealing that our method is suitable for the good quality demonstration with viewpoint control for AVE.

Keywords: Augmented Virtual Environment · Viewpoint quality evaluation
Texture distortion · Depth estimation · Semantic image segmentation

1 Introduction

Augmented Virtual Environment (AVE), known as one part of mixed reality (MR), defined as a dynamic fusion of the real imagery with the 3D models [1]. Broadly speaking, AVE is a virtual-reality environment augmented by fusing real-time, dynamic, multiple information with virtual scenes. The technology was first introduced in 1996 [2], and had made great progress over the last several years. Many kinds of AVE systems have been created, such as Photo Tourism [3] and HouseFly [4], and applied in 3D video surveillance, public security management, city planning and construction [5].

The fusion results directly rely on the texture projection techniques, projecting real-time video onto a 3D model. The 3D model is represented as sample boxes and can't display objects that not belong to this model, resulting unavailable texture distortion, such as the stretch distortion of pedestrians, road facilities, cars and trees. Due to limitation of passive modeling, the modeled depth of each image pixel is not completely accurate. Although the texture distortion, such as texture stretch and object fragment, looks seamless when the user's viewpoint is consistent with the camera's

R. Hong et al. (Eds.): PCM 2018, LNCS 11166, pp. 223–234, 2018.
https://doi.org/10.1007/978-3-030-00764-5_21

viewpoint, it will become obvious as user viewpoint increasingly deviating from camera viewpoint. The illustrations of distortion are shown in Fig. 1.

(a)   (b)

(c)   (d)

Fig. 1. Distortions of images/videos in AVE. (a) (c) Image model from camera's viewpoint. (b) (d) are respectively stretch distortion and object fragment, where the viewpoint deviate from camera's viewpoint.

In this work, we propose a novel viewpoint quality evaluation approach, using texture distortion as the metric of viewpoint quality for AVE. This approach includes stretch distortion and object fragment. We formulate the stretch distortion as accumulated relative error between model depth and "real depth" from depth estimation method, and object fragment as cumulative distance between semantic objects edge to the fragment model boundary. We combine these distortions in a weighted form for VQE. The main contributions of this work include: (1) we propose a new VQE method based on texture distortion. (2) We make a theoretical analysis of distortion phenomenon and the problem is mathematized. (3) We consider the effect of object semantic information on the metric of object fragment.

2   Related Work

Augmented Virtual Environment. AVE system displays still images onto scene models, and observers view them from arbitrary viewpoint. Neumann et al. [1] firstly introduced AVE concept and integrated it into a prototype system, supporting dynamic information extraction and complex scenes analysis through scene models reconstruction, real-time imagery collection and dynamic texture fusion. Sebe et al. [6] made

Neumann's technical extension to AVE, by proposing a novel virtualization system to make observers have an accurate comprehension and perception of dynamic events from arbitrary viewpoints. The Photo Tourism [3] was an end-to-end photo explorer used to interactively browsing 3D photos of popular scenic and historic sites. The HouseFly [4] was developed to project high-resolution videos onto 3D model, and generated the multi-modal virtualization of domestic and retail scene. However, the principle of this system, "directly project", brought insurmountable problems, such as hard to align real with virtual, unexpected video frames distortion. Zhou et al. [7] presented a new AVE video fusion technology based on active image-model technology, extracting video structure to generate image background model of the virtual scene, and projecting the real-time imagery onto model to enable users browse 3D videos from different viewpoint.

Viewpoint quality evaluation. Viewpoint quality is used to describe visual effect from viewpoint, and the higher the score, the better the viewpoint obtaining more detailed visual information in AVE. Generally, viewpoint quality is quantified through the information of 3D scene, such as geometry and texture. Previous methods [8–11] were mostly based on scene geometric information, which are difficult to evaluate high-quality viewpoints in complex scenes with multiple models. Relevant researchers performed the method of VQE based on user's visual perception [12–16], the typical methods include curvature entropy [13] and mesh saliency [14]. The results of these methods are not satisfied for the lack of model geometric information. In order to heighten the user's visual experience to some extent, Christie and Normand [17] investigated VQE method based on semantic information through the basic analysis of geometric and visual information. However, these methods, which are restricted by semantics understanding level of current scene are not suitable for multi-model scenarios.

Single image depth estimation. We compute the degree of stretch distortion through the accumulated relative error between model depth and "real depth", obtained from image depth estimation. Traditional methods of depth estimation were mostly based on geometric priors [18, 19]. Under the rapid development of machine learning, Liu et al. [20, 21] utilized the conditional random fields (CRF) to improve the accuracy of depth estimation for single image. Then Roy and Todorovic [22] adopted neural-random forest for depth estimation of single image, acquiring the same excellent depth estimation result as the above methods. Godard et al. [23] proposed a novel unsupervised depth estimation method, utilizing the unsupervised deep neural network to achieve more accurate results of depth estimation.

Image semantic segmentation. We take the semantic information of object into consideration when measure the degree of object fragment. Previously, the methods of semantic segmentation were mostly classified pixel-wise based on geometric information [24] and statistical methods [25]. The DeepLab [26–28] combined deep-convolutional neural networks (DCNNs) with probability map models without increasing network parameters. The RefineNet [29] aggregated low-level semantic features and high-level semantic classification, to further refine segmentation results with long-range residual links. Zhao et al. [30] proposed PSPNet, extracting multi-scale information through the introduction of pyramid pooling module and achieving more accurate results of semantic segmentation.

Through the analysis of the above work, we extract two main factors that are related to the measurement of VQE, including stretch distortion and object fragment. Taking these distortions into VQE is necessary for improving the roaming experience in AVE. When measuring the stretch distortion, the "real depth" of single image is obtained by Godard's method. And the metric of object fragment is based on the results of semantic image segmentation by Zhao's network structure.

3   Proposed Approach

3.1   Problem Formulation

The key for getting better visual effects lies in how to reduce the visual distortions of AVE. In our scenario, we analyze the following two distortions, stretch distortion and object fragment, to evaluate the viewpoint quality.

Stretch Distortion. The generation schematic diagram of stretch distortion is shown in Fig. 2(a). Suppose we have a source image I for texture projection, captured from a camera viewpoint v_{cam}. When user observes the built image model (or projected image) from a virtual viewpoint v_{usr}, the texture distortion will occur, including stretch distortion $D_{stretch}$ and object fragment $D_{fragment}$. Assuming that the established image model R has a corresponding 3D model C based on true depth, and the spatial point set of C and R is separately denoted as S and S'. The process of projection transformation is defined as

$$\begin{cases} t = M \times S(p_i) \\ t' = M \times S'(p_i) \end{cases}, \tag{1}$$

where t and t' respectively denote the screen position of $S(p_i)$ and $S'(p_i)$ for pixel i. M is perspective transformation matrix, defined as $M = M_w \cdot M_p \cdot M_v \cdot M_m$. The four matrixes respectively indicate viewport matrix, projection matrix, view matrix, and model matrix.

The projection offsets of P pixels cause distortion phenomenon, such as pedestrians and vehicles are stretched. Denoting $L(p_i, v)$ as the distance error of each pixel projected onto screen, and the overall stretch distortion of scene is formulated as

$$D_{stretch}(v, R) = \sum_{i=1}^{P} L(p_i, v), \tag{2}$$

where $v \in v_{usr}$ and $v_{cam} \in v_{usr}$. And $L(p_i, v) = |t - t'|$, if $v \neq v_{cam}$, then $t \neq t'$, indicating that the space coordinates projected on the screen are inconsistent, defined as pixel offset, resulting in stretch distortion. Otherwise, $L(p_i, v) = 0$ represents they are projected to the same screen position, revealing that stretch distortion does not exist.

Object fragment. The generation schematic diagram of object fragment is shown in Fig. 2(b). Each R consists of a group of triangle patches represented as triangle-patch set T_R. The boundary set of R is defined as E_R. The left and right sides of each $e_i \in E_R$ are uniformly sampled, generating pair-wise space coordinates (V_l, V_r), where $V_l \in T_R$ and $V_l \notin E_R$, $V_r \in T_R$ and $V_r \notin E_R$. If $V_l \neq V_r$ and there is no common boundary between them, e_i is defined as fragment boundary, dividing the object into two parts. The projection transformation of each fragment boundary is calculated as

$$\begin{cases} w_1 = M \cdot V_1 \\ w_2 = M \cdot V_2 \end{cases}, \tag{3}$$

where w_1, w_2 respectively represent the screen position of space coordinate $V_1 \in E_R$ and $V_2 \in E_R$.

The projection errors of H pixels of each fragment boundary cause object fragment. Denoting $B(p_j, v)$ as distance error of fragment boundary e_i from v projected onto screen. The overall fragment of image model is formulated as

$$D_{fragment}(v, e_i) = \sum_{j=1}^{H} B(p_j, v), \tag{4}$$

where $B(p_j, v) = |w_1 - w_2|$, if $v \neq v_{cam}$, then $w_1 \neq w_2$ and $B(p_j, v) \neq 0$, indicating that the fragment boundary is projected to different positions on the screen, resulting in object fragment. Otherwise, the fragment boundary is projected to the same screen coordinates, $w_1 = w_2$, and $B(p_j, v) = 0$, symbolizing no object fragment occurs.

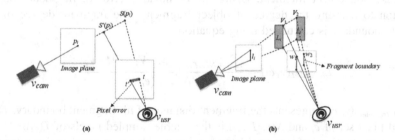

Fig. 2. Generation schematic diagram of distortion. (a) Stretch distortion. (b) Object fragment.

In summary, these two distortions are caused by the inconsistent depth, reflecting in screen when v_{usr} deviates from v_{cam}. The essential reason of stretch distortion is the offset of all pixels in the image, and it is inversely proportional to modeling accuracy. However, object fragment is caused by the offset of fragment boundary in the model, and it is proportional to modeling accuracy.

3.2   VQE Method for Stretch Distortion

Under the analysis of stretch distortion, we utilize the accumulated relative error of pixels projection as the metric of view quality evaluation. Using [23] to calculate the real depth of image, compared with image model depth to obtain projection error. Sampling the image model R to get the sampled pixel set $W(R)$, and the visible sampled pixels from v are denoted as $N(v, W(R))$. The degree of stretch distortion is computed as

$$L_{stretch}(v, R) = \frac{\sum_{p_i \in N(v, W(R))} |M \cdot S(p_i) - M \cdot S'(p_i)|}{|N(v, W(R))|}, \tag{5}$$

where $S(p_i) = l(R) + f(p_i, R) \cdot d$, $l(R)$ represents the location of v_{cam}, $f(p_i, R)$ is the unit vector indicating the orientation looking at p_i of image model, and d is the depth of mapped p_i.

The VQE method based on stretch distortion for single image model is denoted as

$$E_{stretch}(v, R) = \left(1 - \max\left\{\frac{L_{stretch}(v, R)}{L}, 1\right\}\right) \cdot \frac{Vis(v, R)}{r}, \tag{6}$$

where $Vis(v, R)$ denotes projection area of image model, r is screen resolution, and L is a fixed value, representing the acceptable maximum distance of pixel deviation, we take one-fifth of the screen diagonal as L.

3.3   VQE Method for Object Fragment

Analyzing the phenomena of object fragment above, we further propose a method of VQE based on object fragment. Using the cumulative error of fragment boundary projection to measure the degree of object fragment. The fragment degree of each fragment boundary is calculated using equation

$$L_{fragment}(v, e_i) = \frac{\sum_{p_i \in N(v, T(e_i))} |M \cdot V_1 - M \cdot V_2|}{|N(v, T(e_i))|}, \tag{7}$$

where $L_{fragment}(v, e_i)$ represents the fragment distance of ith fragment boundary. $T(e_i)$ is sampled pixel set of e_i and $N(v, T(e_i))$ is the visible sampled pixels of $T(e_i)$.

Different positions of fragment boundary in the object, causing various degree of object fragment. The greatest fragment occurs when the fragment boundary is in the middle of the object. We utilize the distance difference from fragment boundary to the two sides of object to measure the degree of object fragment. This paper obtains the results of semantic image segmentation by PSPNet [30], to get a more accurate distance difference, named as semantic weight λ. Therefore, the above-mentioned calculation function in Eq. (7) is extended into

$$L_{weight}(v, e_i) = \frac{\sum_{p_i \in N(v, T(e_i))} \lambda \cdot |M \cdot V_1 - M \cdot V_2|}{|N(v, T(e_i))|}, \tag{8}$$

where $\lambda = |1 - |d_1 - d_2||$, d_1 and d_2 respectively represent the distance of ith pixel from fragment edge to both segmentation edges of object, normalized to [0–1]. If the fragment boundary is in the middle of object, that is $\lambda \approx 1$, indicating that the fragment degree is most serious. Otherwise, the fragment edge is close to one of the object's segmentation edges, $d_1 \approx 1$ or $d_2 \approx 1$, revealing the fragment degree is not serious and can be ignored.

When determining how distortions affect the viewpoint quality, the score of VQE is in a weighted form, the computational formula is

$$E_{\text{distortion}}(v, R) = \alpha \cdot E_{stretch}(v, R) + \beta \cdot \sum_{i=1}^{N} L_{weight}(v, e_i) \tag{9}$$

where the hyper-parameters α and β are weight factors which control the contributions of the two terms, and we set $\alpha = \beta = 0.5$ empirically. N is the total number of fragment boundary of single image model R.

4 Experiments

4.1 Experimental Setups

We compare our method with viewpoint entropy with four campus scenes. We sample bounding sphere of each scene, getting a viewpoint set with 722 viewpoints. For a better visualization of our results, we utilize 7/8 view sphere with normalized heat map, same as [31], to illustrate viewpoints quality score. The view sphere's center is the source captured location of image, and its radius is the length of the vector from the sphere's center to the built image model's center. The sphere's north-east side is manually removed to make sure the visibility of inside section planes visible.

4.2 Experimental Results and Analysis

The results of VQE based on texture distortion are shown in Fig. 3. We select four image models and the blue cone is field of view (FOV), shown in Fig. 3(b). The red lines of Fig. 3(c) indicate the fragment boundaries.

Each view sphere demonstrates that the optimal viewpoint is the camera viewpoint, locating in the center of view sphere, where the distortion degree is 0. The back view from Fig. 3(d) shows that the more the viewpoint moves towards the rear of camera viewpoint, the larger the spatial range with higher viewpoint quality. The front view from Fig. 3(e) indicates that the viewpoint quality in front of camera viewpoint is deteriorating, that is the distance from viewpoint to image model is inversely proportional to the viewpoint quality.

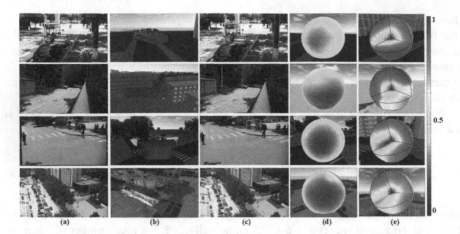

Fig. 3. Results of VQE based on texture distortion. Color values range of each view sphere from blue (good viewpoint) to red (bad viewpoint). (Color figure online)

The comparisons of four VQE methods are shown in Fig. 4.

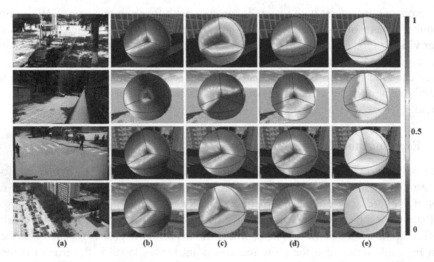

Fig. 4. The results of four VQE methods. The last columns (b) (c) (d) (e) respectively denote our VQE method based on stretch distortion, our VQE method based on object, our VQE method based on texture distortion and the representative viewpoint entropy.

The Fig. 4(b) shows that the distribution of viewpoint quality varies slightly over the section of view sphere, where the quality of viewpoint is poor. This is because the stretch distortion exists in entire image. The Fig. 4(c) indicates that viewpoint quality from section drops rapidly when the viewpoint moves upward, due to the degree of object fragment is more severe from the top. While the viewpoint moves in the left and right direction, the viewpoint quality deteriorates slowly, this is owing to the small fragment area and partial fragment being obscured by the foreground. We weight stretch distortion and object fragment equally, and the results are shown in Fig. 4(d). The last column (e) reveals that the viewpoint quality in the center of viewpoint sphere is lower than the outer edge viewpoint. The above results indicate that the VQE method based on texture distortion can better reflect viewpoint quality than geometric-based method.

In Fig. 5 we list the corresponding scenes from different viewpoints, intuitively displaying the good viewpoint and bad viewpoint by the distortion score of viewpoint.

Fig. 5. The corresponding scenes and distortion degree from different viewpoints. 0 represents the best viewpoint, the closer the score to 1 the worse the viewpoint.

To quantitatively evaluate the effectiveness of our methods, we conduct a user study in AVE. Comparing our method with viewpoint entropy, and each image model is evaluated by 20 participants. Each participant has normal vision and gives a score based on the perceived comfort level. For each method we select five viewpoints to compare the score of user's evaluation and VQE methods (see Fig. 6).

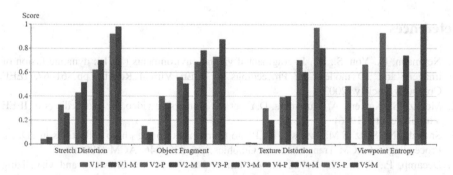

Fig. 6. Results of our user study. The charts report for each method the pair-wise scores of each viewpoint evaluated by user and VQE method.

As shown in Fig. 6, the differences of evaluation score between user and VQE method are not significant in the first three methods. However, the results of the fourth method have significant differences. Our method is better than Viewpoint Entropy for the evaluation of distortion in AVE. We select Bhattacharyya Distance (BC) [32] to measure the similarity of participant evaluation and each method, the similarity value, abbreviated as BC, is shown in Table 1. The closer the user's score is to the score of our method, the greater the BC value is. Obviously, the similarity value of our method is higher than viewpoint entropy, which demonstrates that our method is more suitable for user's virtual perception.

Table 1. Bhattacharyya Distance of the user evaluation and results of VQE methods.

Method	Bhattacharyya Distance
Stretch distortion	0.9912
Object fragment	0.9670
Texture distortion	0.9942
Viewpoint Entropy	0.8952

5   Conclusion and Discussion

With the growth of the size and complexity of AVE, identifying good viewpoints automatically is an important requirement for good visual experience. Our method provides an elegant solution to achieve VQE. Comparing our method with other existing VQE methods, the main contribution of our method is the texture distortion metric for AVE. Experiments illustrated the effectiveness of the quality evaluation of the viewpoints.

Acknowledgement. This work is supported by the Natural Science Foundation of China under Grant No.61572061, 61472020.

References

1. Neumann, U., You, S., Hu, J.: Augmented virtual environments (AVE): dynamic fusion of imagery and 3D models. In: Proceedings of IEEE Virtual Reality, pp. 61–67. IEEE Computer Society (2003)
2. Moezzi, S., Katkere, A., Kuramura, D.Y., et al.: Immersive video. In: Proceedings of IEEE Virtual Reality (1996)
3. Snavely, N., Seitz, S.M., Szeliski, R.: Photo tourism: exploring photo collections in 3D. In: Proceedings of ACM Transactions on Graphics, pp. 835–846. ACM (2006)
4. Decamp, P., Shaw, G., Kubat, R.: An immersive system for browsing and visualizing surveillance video. In: Proceedings of the International Conference on Multimedia, pp. 371–380. ACM (2010)
5. Jian, H., Liao, J., Fan, X.: Augmented virtual environment: fusion of real-time video and 3D models in the digital earth system. Int. J. Digit. Earth **10**(9), 1–20 (2017)

6. Sebe, I.O., Hu, J., You, S.: 3D video surveillance with augmented virtual environments. In: Proceedings of 1st ACM SIGMM International Workshop on Video Surveillance, pp. 107–112. ACM (2003)
7. Zhou, Z., You, J., Yan, J., et al.: Method for 3D Scene Structure Modeling And Camera Registration From Single Image. US 20160249041 A1 (2016)
8. Neumann, L., Sbert, M., Gooch, B.: Viewpoint quality: measures and applications. In: Proceedings of Eurographics Conference on Computational Aesthetics in Graphics, Visualization and Imaging, pp. 185–192. ACM (2005)
9. Polonsky, O., Patané, G., Biasotti, S.: What's in an image? Vis. Comput. **21**(8–10), 840–847 (2005)
10. Zquez, P.P., Feixas, M., Sbert, M.: Viewpoint selection using viewpoint entropy. In: Proceedings of the 6th International Fall Workshop on Vision, Modeling, and Visualization, pp. 273–280 (2001)
11. Freitag, S., Weyers, B., Bönsch, A.: Comparison and evaluation of viewpoint quality estimation algorithms for immersive virtual environments. In: Proceedings of International Conference on Artificial Reality and Telexistence and Eurographics Symposium on Virtual Environments (2015)
12. Yamauchi, H., Saleem, W., Yoshizawa, S.: Towards stable and salient multi-view representation of 3D shapes. In: Proceedings of IEEE International Conference on Shape Modeling and Applications, pp. 40. IEEE Computer Society (2006)
13. Page, D.L., Koschan, A., Sukumar, S.R., Rouiabidi, B., Abidi, M.A.: Shape analysis algorithm based on information theory. In: Proceedings of International Conference on Image Processing, pp. 229–232 (2003)
14. Lee, C.H., Varshney, A., Jacobs, D.W.: Mesh saliency. ACM Trans. Graph. **24**(3), 659–666 (2005)
15. Vázquez, P.-P.: Automatic view selection through depth-based view stability analysis. Vis. Comput. **25**, 5–7 (2009)
16. Miao, Y., Wang, H., Hang, Z.: Best viewpoint selection driven by relief saliency entropy. J. Comput.-Aided Des. Comput. Graph. **23**(12), 2033–2039 (2011)
17. Christie, M., Normand, J.M.: A semantic space partitioning approach to virtual camera composition. Comput. Graph. Forum **24**(3), 247–256 (2005)
18. Tsai, G., Xu, C.H., Liu, J.E.: Real-time indoor scene understanding using Bayesian filtering with motion cues. In: IEEE International Conference on Computer Vision, pp. 121–128. IEEE Computer Society (2011)
19. Zeng, Y., Hu, Y., Liu, S.: GeoCueDepth: exploiting geometric structure cues to estimate depth from a single image. In: IEEE International Conference on Intelligent Robots and Systems (2017)
20. Liu, M.M., Salzmann, M., He, X.M.: Discrete-continuous depth estimation from a single image. In: Proceedings of the IEEE Conference on Computer Vision and Pattern Recognition, pp. 716–723 (2014)
21. Liu, F.Y., Shen, C.H., Lin, G.S.: Deep convolutional neural fields for depth estimation from a single image. In: Proceedings of the IEEE Conference on Computer Vision and Pattern Recognition, pp. 5162–5170 (2015)
22. Roy, A., Todorovic, S.: Monocular depth estimation using neural regression forest. In: Proceedings of the IEEE Conference on Computer Vision and Pattern Recognition, pp. 5506–5514 (2016)
23. Godard, C., Aodha, O.M., Brostow, G.J.: Unsupervised monocular depth estimation with left-right consistency. In: Proceedings of the IEEE Conference on Computer Vision and Pattern Recognition, pp. 270–279 (2017)

24. Lee, D.C., Hebert, M., Kanade, T.: Geometric reasoning for single image structure recovery. In: Proceedings of the IEEE Conference on Computer Vision and Pattern Recognition, pp. 2136–2143 (2009)
25. Zhao, Y.B., Zhu, S.C.: Image parsing via stochastic scene grammar. In: Proceedings of the Conference and Workshop on Neural Information Processing System, pp. 73–81 (2011)
26. Chen, L., Papandreou, G., Kokkinos, I., Murphy, K.: Semantic image segmentation with deep convolutional nets and fully connected crfs. Comput. Sci. **4**, 357–361 (2014)
27. Chen, L.C., Papandreou, G., Kokkinos, I.: Deeplab: semantic image segmentation with deep convolutional nets, atrous convolution, and fully connected crfs. IEEE Trans. Pattern Anal. Mach. Intell. **40**(4), 834–848 (2018)
28. Chen, L.C., Yang, Y., Wang, J.: Attention to scale: scale-aware semantic image segmentation. In: Proceedings of the IEEE Conference on Computer Vision and Pattern Recognition, pp. 3640–3649 (2016)
29. Lin, G.S., Milan, A., Shen, C.H.: RefineNet: multi-path refinement networks with identity mappings for highresolution semantic segmentation. In: Proceedings of the IEEE Conference on Computer Vision and Pattern Recognition, pp. 1925–1934 (2017)
30. Zhao, H.S., Shi, J.P., Qi, X.J.: Pyramid scene parsing network. In: Proceedings of the IEEE Conference on Computer Vision and Pattern Recognition, pp. 2881–2890 (2017)
31. Zhou, Y., Xie, J.Q., Wu, W.: Path planning for virtual-reality integration surveillance system. J. Comput.-Aided Des. Comput. Graph. **30**(3), 514–523 (2018)
32. Guo, R.X., Pei, Q.C., Min, H.W.: Bhattacharyya distance feature selection. In: Proceedings of the 13th International Conference on Pattern Recognition, pp. 195–199 (1996)

A Flower Classification Framework Based on Ensemble of CNNs

Buzhen Huang[1], Youpeng Hu[1], Yaoqi Sun[1], Xinhong Hao[2],
and Chenggang Yan[1(✉)]

[1] Hangzhou Dianzi University, Hangzhou 310018, China
cgyan@hdu.edu.cn
[2] Beijing Institute of Technology, Beijing 100081, China

Abstract. Currently, the classification of flower species has become a hot topic in the field of image classification. Flower classification belongs to the category of fine image classification, and such images are usually represented by multiple visual features. At present, all the flower classification methods based on a single convolutional neural network (CNN) model can hardly extract the features of a flower image as much as possible. In view of the limitation of description methods for flower features and the problem of low accuracy of flower species recognition, this paper proposes a flower classification framework based on ensemble of CNNs. The method consists of the following three parts: (1) The same flower image is processed differently to make the color, texture and gradient of the flower image more prominent; (2) Fine-tune the structure and parameters of the convolutional neural network to adapt it to the extraction of corresponding features. Then use the CNN model with different characteristics to extract the corresponding features; and (3) A framework that can fuse each CNN sub-learner is used to combine various features effectively. We tested the effectiveness of our method on the Oxford Flowers 102 Dataset [2]. The result demonstrates that the proposed approach effectively improves the accuracy of flower classification.

Keywords: Flower classification · Multi-feature · Ensemble learning
Convolutional neural network

1 Introduction

In recent years, the classification and recognition of images has become an important direction in the field of image processing and computer vision. The classification and recognition of flower images as an important part of it has aroused widespread concern. With the continuous development of Internet and information technology, tens of thousands of flower images are collected every day. The classification and information mining of these images has high research and application value [19]. However, flower classification belongs to the category of fine image classification, and such images are usually represented by multiple visual features [11]. Using a single CNN model can only describe a part of the image attributes, and the description of the image content has certain limitations, so the accuracy of image classification is not high. Therefore, we

R. Hong et al. (Eds.): PCM 2018, LNCS 11166, pp. 235–244, 2018.
https://doi.org/10.1007/978-3-030-00764-5_22

propose a flower classification framework based on ensemble of CNNs. Figure 1 gives the illustration of the proposed method. First, the same preprocessing is performed on the training set and the test set. Thus, we got different image sets that contain features such as color, hue, saturation, directional gradient, texture and local features. For datasets containing different features, we fine-tune the structure and parameters of the model, and then use the training set to train the model. Then a novel framework that can fuse each CNN sub-learner is used to combine various features effectively. The experimental results show that our proposed framework has higher accuracy and faster classification speed.

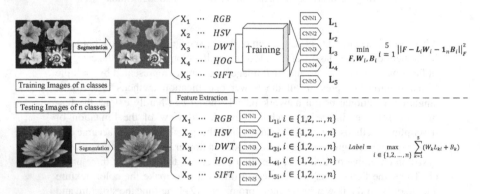

Fig. 1. Overview of the proposed framework. In the above part (the training phase), $X_i (i = 0, 1, \ldots, 5)$ represents the 5 extracted features; $CNN1, CNN2, \ldots, CNN5$ indicate the convolutional neural network model obtained by training each image feature respectively; F denotes the predicted label matrix; L_i denotes the corresponding output vector of the fully connected layer; W_i, B_i represent the weight value and offset value of the i-th feature. In the under part (the testing phase), L_{ki} denotes the i-th dimension of output vector of the k-th feature of the full-connected layer. *Label* indicates the final category of the flower.

2   Related Works

The classification of flower species belongs to the category of fine image classification, and it is one of the research hotspots of current classification recognition. In recent years, people have also done a lot of research.

Nilsback and Zisserman [1] calculated four different characteristics of flowers, including textures, boundary shapes, overall spatial distribution of petals and colors. They used SVM classifiers with a kernel learning framework to combine them. The optimal weight for each feature was based on the specific training set and previous constraint determination [13]. Chai [7] proposed a segmentation algorithm based on the co-distribution of similar flower shapes. By extracting the SIFT features and the Lab features of the entire picture, the corresponding BOW feature vectors were calculated, and image classification was achieved through the SVM classifier. Angelova [4] used a method that eliminates background interference information and had satisfactory results for fine image classification experiments. In addition, Saitoh and Kaneko [14] used clustering methods to extract the characteristics of flowers and leaves from the image, and then identified the flowers using a piecewise linear discriminant function; Mishra [15] proposed an algorithm for identification using multiclass classification based on color, shape volume and cell feature.

The difficulty in identifying the classification of fine images such as flowers lies in the inability to locate the image features. As the method proposed in the above paper, the extraction of key features is particularly important. The common method for the fusion of multiple features is ensemble learning [16] and combining classifiers [17, 18]. Integrating these methods into flower recognition is useful for accuracy.

3   Multi-feature Extraction

In this section, we preprocess images to make their various features more prominent. Then use the deep convolutional neural network to extract feature information. These specific features include: color information, HS (hue, saturation) information, texture information, gradient information, scale-invariant feature information.

3.1   Background Segmentation

For the same flower in different scenes, the background is beyond our control [34]. If we train the image with background, it will inevitably lead to overfitting so that the generalization ability of the generated model is affected [21, 35]. Therefore, the first step in flower classification is to segment the flower image by removing the unwanted background region [24]. In this paper, we use the method proposed by [6].

3.2   Color Features

Color was found to be an indispensable feature for high classification results, especially while preserving spatial correspondence to gray-level features [25]. In a color image, 90% of the edge information is roughly the same as the grayscale image [27]. Consequently, there are still 10% of edge information in a color image, which can be essential in certain computer vision tasks, cannot be found in the corresponding gray image [28]. We use the raw RGB image for feature extraction. The image does not need to be processed. Because the image itself already contains a lot of color, edge, corner and other characteristics information.

3.3   Hue, Saturation (HS) Features

Environmental factors such as changes in lighting cause large variations in the measured color, which in turn leads to confusion between classes. One way to reduce the effect of illumination variations is to use a color space which is less sensitive to it. Hence, we describe the color using the HSV color space [26]. Among them, H and S are more able to reflect the true color characteristics of flowers. Use the original RGB image to convert to HSV image and extract H (hue) and S (saturation) information according to the following formula:

$$H = \begin{cases} 60 \times \frac{G-B}{\max(R,G,B)-\min(R,G,B)}, & R = \max(R,G,B) \\ 120 + 60 \times \frac{B-R}{\max(R,G,B)-\min(R,G,B)}, & G = \max(R,G,B) \\ 240 + 60 \times \frac{R-G}{\max(R,G,B)-\min(R,G,B)}, & B = \max(R,G,B) \end{cases} \tag{1}$$

$$(if \ H < 0 : H = H + 360)$$

$$S = \frac{\max(R,G,B) - \min(R,G,B)}{\max(R,G,B)} \tag{2}$$

3.4 Texture Features

Relatively small vector of Discrete Wavelet Transformation (DWT) [23] features is sufficient for very good texture classification [22]. In [24], Guru et al. proved that the texture features still have a good effect in flower classification. The texture belongs to the high-frequency part of the picture and can represent the details of the picture well. This will be of great help to the improvement of our recognition accuracy. Wavelet transform of image is the basis of wavelet applied to image processing, and its foundation is two-dimensional discrete wavelet transform. The formula for the two-dimensional discrete wavelet transform is:

$$W_{\psi}^i(j,m,n) = \frac{1}{\sqrt{MN}} \sum_{x=0}^{M-1} \sum_{y=0}^{N-1} f(x,y)\psi_{j,m,n}^i(x,y), i = \{H,V,D\} \tag{3}$$

Among them, $f(x,y)$ is the raw image information (here we use the saturation (S) component of the HSV space); $\psi_{j,m,n}^i(x,y)$ is a wavelet function, defined as follows:

$$\psi_{j,m,n}^i(x,y) = 2^{\frac{j}{2}}\psi^i(2^j x - m, 2^j - n), i = \{H,V,D\} \tag{4}$$

When processing an image, the image is transformed into four sub-images by wavelet decomposition: a low-resolution sub-image A, a horizontal-direction sub-image H, a vertical-direction sub-image V, and a diagonal-direction sub-image D. As shown below:

A	W_H
W_V	W_D

The low-frequency resolution sub-image concentrates the main components of the signal, and the other three parts are the high-frequency information of the signal, i.e., the detail information. We use the sum of the three high frequency band outputs of the wavelet transform as texture features:

$$T = W_H + W_V + W_D \tag{5}$$

3.5 Histogram of Oriented Gradient (HOG) Features

Both the detection of flower edges and the characteristics of stems and leaves of flowers can be reflected by gradient information. We use the HOG proposed in [30] to extract the gradient directional features of local regions. In [31], Anna Bosch merged the image pyramid representation proposed by [32] with the HOG descriptor, and proposed an improved Pyramid of Histograms of Orientation Gradients (PHOG) descriptor. Here we use PHOG to embody the local gradient feature of the image, because the appearance and shape of the local object can be well described by the local gradient or the distribution of the edge direction, this was validated in the work of Peter Gehler et al. [29].

The HOG method is based on the calculation of a normalized local direction gradient histogram in a dense grid. Divide the image into small cells. Convolve the cell with gradient operators like Sobel and Laplacian to get the gradient direction and amplitude. The formula is as follows:

$$M(x, y) = \sqrt{I_x^2 + I_y^2} \tag{6}$$

$$\theta(x, y) = \tan^{-1} \frac{I_y}{I_x} \in [0°, 360°) \, or \in [0°, 180°) \tag{7}$$

where I_x and I_y represent the gradient values in the horizontal and vertical directions, $M(x, y)$ represents the gradient values of the gradient, and $\theta(x, y)$ represents the direction of the gradient. For each cell, 36 bin histograms are used to count the gradient information of the pixels in each cell, that is, the gradient direction of the cell is 0 to 360 degrees divided into 36 direction blocks, and the gradient direction of each pixel in the cell is in the histogram. In the weighted projection, the weighted histogram in the gradient direction of this cell can be obtained by mapping it into the corresponding angle range block. Finally, it is normalized to obtain gradient information with 36 dimensions.

3.6 Scale Invariant Feature Transform (SIFT)

Scale-invariant feature transformation is a local feature descriptor that has been proposed by Lowe [12]. This local feature descriptor has scale invariance and can detect key points in the image. The SIFT feature is a point of interest that is independent of the size and rotation of the image based on some local appearance on the object, and the tolerance to changes in light, noise, and microscopic viewing angle is also quite high. They are highly significant and relatively easy to retrieve. In massive feature databases, it is easy to identify objects and there are few misjudgments. SIFT also has an irreplaceable role in [1]. Therefore, we regard it as an important feature of flower recognition.

3.7   Extract Feature from Image

We use convolutional neural networks to perform feature extraction on flower images that have been processed with different methods.

A convolution neural network is a kind of multilayer perceptron, and each layer is composed of two-dimensional planes [10]. The input images convolve with the filter and additive bias by producing sampling layer (N value can be set), then the feature maps get the feature maps of convolutional layer through calculating, weighted value, bias and an activation function. According to the number of convolutional layers, the above work is carried out in cycle. In our work, we use the last layer of full-connection layer as a feature to output.

4   Fusion Framework

Through the previous step, we got multiple individual learners, then we need to combine them by ensemble learning.

In the proposed method, Multi-response Linear Regression (MLR) is used as the secondary learning algorithm.

In this task of MLR, we use the standard error as the loss function. The loss function can be written as:

$$e = ||\mathbf{F} - \mathbf{L}_i\mathbf{W}_i - 1_n\mathbf{B}_i||_F^2, \quad i \in \{1, 2, \ldots, 5\} \tag{8}$$

Where $\mathbf{W}_i, \mathbf{B}_i$ is the weight and offset of the corresponding individual learner, $\mathbf{L}_i$ is output value of each full connected layer and $\mathbf{F}$ is the predicted label matrix.

The task of learning is to figure out the parameters of the model based on the standard error minimization. We can use the least square estimation formula to estimate $\mathbf{W}_i, \mathbf{B}_i$. The formula is as follows:

$$\hat{\mathbf{W}}_i = \left(\mathbf{L}_i^T\mathbf{L}_i\right)^{-1}\mathbf{L}_i^T\mathbf{F} \tag{9}$$

$$\hat{\mathbf{B}}_i = \mathbf{F} - \mathbf{L}_i\hat{\mathbf{W}}_i \tag{10}$$

5   Experiment

In this section, we will test our flower classification framework on the Oxford flowers 102 [2] dataset. Our experiment is mainly divided into three steps. First of all, we do different image processing on the flower segmentation images in Oxford flowers 102 [2] dataset. After preprocessing, the gradient, color, texture, and local shape information in the image are more prominent. And then, the structure and parameters of the CNN model are fine-tuned to make it easier to extract different feature information, and the CNN model is trained separately using the training sets containing different features. The third step is to use the Eq. (8) to find the optimal combination of features. Finally, we test our framework with test sets and get the final label.

5.1 Oxford Flowers 102 Dataset

The Oxford Flowers 102 (Flower-102) [2] is a fine-grained image benchmark, which consists of 8189 images from 102 flower categories. Each category consists of a number of images ranging from 40 to 258. This dataset simultaneously segmented the flower images [6] and provided us with a segmentation dataset. Our experiment is based on segmented image set, and we have adjusted the train/test splits.

5.2 Experimental Method

Image Preprocessing In the image preprocessing section, we do four kinds of processing for the image. (1) Convert the image to HSV color space and remove the V space at the same time. (2) Use DWT to highlight image texture features. (3) Extract the HOG features of the image to obtain a set of 29 * 29 * 36 gradient information. (4) Extracting local features of images using SIFT descriptors. All image processing of the data set is done on MATLAB.

Model Training. The CNN-features are the deep representations of object images generated with a well-trained CNN model [20, 33]. In this paper, AlexNet [3] was chosen to extract the different features of the flower image. AlexNet [3] was presented by Alex in 2012. It has 5 convolutional layers, 3 fully connected layers, and adds LRN, Dropout, ReLu and other functional modules. We resize the flower picture to 224 * 224, and in order to speed up the training speed and the accuracy of the model, we added a batch-normalization layer before the active layer. In training, in order to reduce the training time, we used the parameters of the ImageNet-1000 pre-training model to initialize the parameters and fine-tune the partial structure. The outputs of the 8-th (f8) fully-connected layer of CNN is recognized as the feature in this paper. The dimension of features in f8 is 102.

Multi-feature Fusion. We combine each CNN sub-learner with the established framework and find the optimal linear relationship between them. Finally, we use the test set as input to get the corresponding category label. In addition, we compare the classification effect of a single CNN with our classification framework. The results of our experiment are shown in Fig. 2. The result demonstrates that the accuracy of the proposed method is 10.1% higher than the classification of the raw flower image using a single CNN.

Result. Table 1 shows the performance comparison of the proposed method on the Oxford Flowers 102 dataset [2] with the popular image classification approaches [1, 2, 4, 5, 7–9]. We can have the following conclusions. First, the proposed method has better performance of approximately 12.86% than the method used by Nilsback [1, 2, 7]. This proves that using CNN can extract more features than using SVM. Besides, compared with other CNN method [9], using multiple CNNs to extract features is more advantageous.

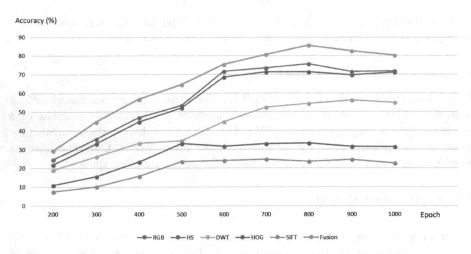

Fig. 2. Comparison of fusion framework and single CNN

Table 1. The comparison of different methods on the Oxford flowers 102 dataset

Method	Accuracy (%)
Nilsback et al. [1]	72.8
M.Seeland et al. [25]	72.8
S. Ito et al. [8]	74.8
Nilsback et al. [2]	76.3
Y. Chai et al. [7]	80.0
Angelova et al. [4]	80.66
N. Murray et al. [5]	81.50
Y. Liu et al. [9]	84.02
Ours	**85.66**

6 Conclusion

This paper presents a flower classification framework based on ensemble of CNNs, which can extract a variety of features from flower images. We established a corresponding CNN model for different flower characteristics to extract the corresponding feature information. After the training of different models is completed, we find the optimal linear combination between them so as to effectively integrate various different features. This method realizes the function of flower classification by using multiple features of the flower image at the same time, and solves the problem of low utilization of image features using a single CNN model. The experimental results show that our method effectively improves the accuracy of flower classification.

References

1. Nilsback, M.E., Zisserman, A.: Automated flower classification over a large number of classes. In: Computer Vision, Graphics & Image Processing, pp. 722–729 (2009)
2. Nilsback, M.E., Zisserman, A.: An automatic visual flora-segmentation and classification of flower images. Ph.D. thesis, University of Oxford (2009)
3. Krizhevsky, A., Sutskever, I., Hinton, G.E.: ImageNet classification with deep convolutional neural networks. In: International Conference on Neural Information Processing System, pp. 1097–1105 (2012)
4. Angelova, A., Zhu, S.: Efficient object detection and segmentation for fine-grained recognition. In: IEEE Conference on Computer Vision and Pattern Recognition, pp. 811–818 (2013)
5. Murray, N., Perronnin, F.: Generalized max pooling. In: IEEE Conference on Computer Vision and Pattern Recognition, pp. 2473–2480 (2014)
6. Nilsback, M.E., Zisserman, A.: Delving into the Whorl of flower segmentation. In: Proceedings of the British Machine Vision Conference, pp. 1049–1062 (2007)
7. Chai, Y., Lempitsky, V., Zisserman, A.: Bicos: a bi-level co-segmentation method for image classification. In: IEEE International Conference on Computer Vision, pp. 2579–2586 (2012)
8. Ito, S., Kubota, S.: Object classification using heterogeneous co-occurrence features. In: European Conference on Computer Vision, pp. 701–714 (2010)
9. Liu, Y., Tang, F., Zhou, D., Meng, Y., Dong, W.: Flower classification via convolutional neural network. In: IEEE International Conference on Functional-Structural Plant Growth Modeling, pp. 110–116(2017)
10. Yu, K., Jia, L., Chen, Y., Xu, W.: Deep learning: yesterday, today, and tomorrow. J. Comput. Res. Dev. **20**(6), 1349 (2013)
11. Wu, X., Gao, L., Yan, M., Zhao, F.: Flower species recognition based on fusion of multiple features. J. Beijing For. Univ. **39**(4), 86–93 (2017)
12. Lowe, D.G.: Distinctive image features from scale-invariant keypoints. Int. J. Comput. Vis. **60**, 91–110 (2004)
13. Varma, M., Ray, D.: Learning the discriminative power-invariance trade-off. In: IEEE International Conference on Computer Vision, pp. 1–8 (2007)
14. Saitoh, T., Kaneko, T.: Automatic recognition of wild flowers. J. Syst. Comput. Jpn. **34**(10), 90–101 (2000)
15. Mishra, P.K., Maurya, S.K., Singh, R.K., Misra, A.K.: A semi automatic plant identification based on digital leaf and flower images. In: International Conference on Advances in Engineering, pp. 68–73 (2012)
16. Liu, Y., Yao, X.: Ensemble learning via negative correlation. J. Neural Netw. **12**(10), 1399–1404 (1999)
17. Kittler, J., Hatef, M., Duin, R.P.W., Matas, J.: On combining classifiers. IEEE Trans. Pattern Anal. Mach. Intell. **20**(3), 226–239 (1998)
18. Kuncheva, L.I.: Combining pattern classifiers: methods and algorithms. IEEE Trans. Neural Netw. **18**, 964 (2007)
19. Qi, G., Hua, X., Zhang, H.: Learning semantic distance from community-tagged media collection. In: International ACM Conference on Multimedia, pp. 243–252 (2009)
20. Qi, G., Aggarwal, C., Huang, T.: Link prediction across networks by biased cross-network sampling. In: IEEE International Conference on Data Engineering, pp. 793–804 (2013)
21. Yan, C., Li, L., Zhang, C., Liu, B., Zhang, Y., Dai, Q.: An effective Uyghur text detector for complex background images. IEEE Trans. Multimed. **99**, 1 (2018)

22. Kociołek, M., Materka, A., Strzelecki, M., Szczypiński, P.: Discrete wavelet transform—derived features for digital image texture analysis. In: International Conference on Signals and Electronic Systems, pp. 514–524 (2001)

23. Furht, B.: Discrete wavelet transform (DWT). In: Furht, B. (ed.) Encyclopedia of Multimedia. Springer, Boston (2006). https://doi.org/10.1007/0-387-30038-4

24. Guru, D.S., Kumar, Y.H.S., Manjunath, S.: Textural features in flower classification. Math. Comput. Model. **54**(3–4), 1030–1036 (2011)

25. Seeland, M., Rzanny, M., Alaqraa, N., Wäldchen, J., Mäder, P.: Plant species classification using flower images—a comparative study of local feature representations. Plos One **12**(2), e0170629 (2017)

26. Nilsback, M.E., Zisserman, A.: A visual vocabulary for flower classification. In: IEEE Conference on Computer Vision and Pattern Recognition, pp. 1447–1454 (2006)

27. Nocak, C.L., Shafer, S.A.: Color edge detection. In: Proceedings DARPA Image Understanding Workshop, pp. 35–37(1987)

28. Dutta, D., Chaudhuri, B.B.: A color edge detection algorithm in RGB color space. In: International Conference on Advances in Recent Technologies in Communication and Computing, pp. 337–340(2009)

29. Gehler, P., Nowozin, S.: On feature combination for multiclass object classification. IEEE International Conference on Computer Vision, pp. 221–228 (2010)

30. He, N., Cao, J., Song, L.: Scale space histogram of oriented gradients for human detection. Int. Symp. Inf. Sci. Eng. **2**, 167–170 (2008)

31. Bosch, A., Zisserman, A., Munoz, X.: ACM International Conference on Image and Video Retrieval, pp. 401–408 (2007)

32. Lazebnik, S., Schmid, C., Ponce, J.: Beyond bags of features: spatial pyramid matching for recognizing natural scene categories. In: IEEE Conference on Computer Vision and Pattern Recognition, pp. 2169–2178 (2006)

33. Wang, P., Li, L., Yan, C.: Image classification by principal component analysis of multi-channel deep feature. In: IEEE Global Conference on Signal and Information Processing, pp. 696–700 (2017)

34. Yan, C., Xie, H., Yang, D., Yin, J., Zhang, Y., Dai, Q.: Supervised hash coding with deep neural network for environment perception of intelligent vehicles. IEEE Trans. Intell. Transp. Syst. **19**, 284–295 (2017)

35. Yan, C., Xie, H., Liu, S., Yin, J., Zhang, Y., Dai, Q.: Effective Uyghur language text detection in complex background images for traffic prompt identification. IEEE Trans. Intell. Transp. Syst. **19**, 220–229 (2017)

Image Translation Between High-Resolution Remote Sensing Optical and SAR Data Using Conditional GAN

Xin Niu[✉], Di Yang, Ke Yang, Hengyue Pan, and Yong Dou

National Laboratory for Parallel and Distributed Processing, College of Computer, National University of Defense Technology, Changsha, China
niuxin@nudt.edu.cn

Abstract. This paper presents a study on a new problem: applying machine learning approaches to translate remote sensing images between high-resolution optical and Synthetic Aperture Radar (SAR) data. To this end, conditional Generative Adversarial Networks (GAN) have been explored. Efficiency of the conditional GAN have been verified with different SAR parameters on three regions from the world: Toronto, Vancouver in Canada and Shanghai in China. The generated SAR images have been evaluated by pixel-based image classification with detailed land cover types including: low and high density residential area, industry area, construction site, golf course, water, forest, pasture and crops. In comparison with an unsupervised GAN translation approach, the proposed conditional GAN could effectively keep many land cover types with compatible classification accuracy to the ground truth SAR data. This is one of first study on multi-source remote sensing data translation by machine learning.

Keywords: Remote sensing · Generative Adversarial Network Deep learning

1 Introduction

Remote sensing images, which are frequently acquired by satellites or airplanes with specific sensors, play an important role in many research and industry fields such as hazard monitoring, geological survey, intelligence collecting and etc.

Within many types of remote sensing data, optical and Synthetic Aperture Radar (SAR) images are two widely used ones. The optical image, which is collected by photosensitive devices, is highly similar to the one by common cameras. However, observation through optical data is often affected by the cloud cover and low illumination conditions. In comparison, SAR image, which is formed by the radar reflection, could be gathered under almost all-weather conditions in

This work was supported by the National Natural Science Foundation of China under Grants U1435219 and 61402507.

both day and night. Nevertheless, high-resolution SAR data is still valuable due to the less share in the sensor deployments.

While the interpretation of optical images is as straight-forward as common photos by camera, SAR images are difficult to understand. The contents of SAR images are more related to the radar scattering mechanisms with the texture, shape and alignment of the observed objects. Interpretation of SAR images often requires expert knowledge.

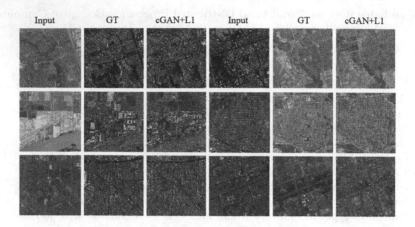

Fig. 1. Optical and SAR remote sensing image translation through cGAN+L1. Optical to SAR: Column 1, 2, 3. SAR to optical: Column 4, 5, 6. Toronto: Row 1. Vancouver: Row 2. Shanghai: Row 3. GT is for ground truth.

Since the two images have complementary functions in target observation, translation between them can be helpful in many ways. First, the simulated image type could contribute to a better understanding of the observed region. For instance, one could explore the generated optical image to interpret the difficult part of the SAR data. Secondly, translation could compensate for the lack of the generated data. For example, optical images under certain conditions or the scarce SAR samples could be generated for machine learning. Nevertheless, remote sensing image translation is still a new research topic. Solutions in the limited studies [3] were mainly focused on low resolution (higher than 30 meters per pixel) data with coarse class replacement approach, by which the region of the source image were classified and replaced by the prepared target textures or customized filters. These early attempts, which rely on few predefined classes and limited handcraft textures, may not suitable for the high-resolution data with land cover details in high variance. Recently, deep learning techniques have shown promising potential for remote sensing applications such as image segmentation, object detection, data fusion and 3D reconstruction [19,21]. However, image translation between multi-source remote sensing data through machine learning has not been effectively explored. With the rise of deep learning approaches, image generation or translation has been greatly developed

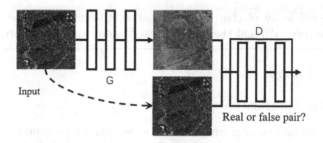

Fig. 2. The conditional GAN translation system.

in fields like face generation [17], style transfer [4,9], colorization [6,12], depth image prediction [11] and etc. Nevertheless, remote sensing image translation through machine learning is still challenging. First, translation between optical and SAR images is an aligned problem with the relation of geo-locations. Object on one image should be transferred to a unique correct form according to its physical property on the other data, and exactly on the same position, as shown in Fig. 1. This problem may not be effectively solved by current style transfer methods [4,9], in which the similar style expression is hard to be evaluated quantitatively. Secondly, the same time SAR and optical observation data are extremely rare. With the time span, mismatches between the two remote sensing images are frequently observed due to the urban development, vegetation seasonal change and etc. This demands tolerance of the translation model for the mismatched sample pairs. Thirdly, optical and SAR images extract diverse object information by different imaging mechanisms. It is required that the model features could represent variant land covers for both two kinds of data types. However, most of the image generation works such as face generation, colorization and etc. are on the same data type: natural image. These methods can explore the pretrained natural image models and the relationship within the same data type to achieve favorable translation results, which is hard for the optical-SAR translation. With the advent of GAN in image generation [8,20], significant progress on multi data type translation has been recognized. Nevertheless, most of the GAN image generation studies focused on the learning scheme or the model structures to avoid collapse or to obtain lively detailed generations. How to improve the translation accuracy between two kinds of observations on the same land covers has been less studied. This paper propose machine learning approaches for optical and SAR remote sensing image translation. To build the learning models, effective techniques from previous GAN based image generations [8,9] have been explored. The major contribution of this paper could be summarized as follow: 1. This is one of the first study to apply machine learning on remote sensing optical and SAR image translation. 2. A Conditional GAN (cGAN) has been explored for optical and SAR translation. Moreover, the supervised cGAN has been compared with a unsupervised GAN for this purpose. 3. Extensive experiments have been conducted on different regions with different SAR parameters to verify the efficiency of the proposed methods in generation accuracy, land cover class preservation and the less discussed sample mismatch problem.

2   Methodology

2.1   Problem Formulation

For an input remote sensing image patch x, we expect to synthesize an image patch y of another data type by the translation generator G. To this end, the cGAN [5] has been employed. In this way, the input image x has been mapped by G with a random noise vector z, to y: $G : \{x, z\} \rightarrow y$. A discriminator D is trained to distinguish the fakes by the generator G from real examples. For the purpose of the aligned data translation, D(x, y) should predict true, if x and y are the real optical and SAR image pair, and report false if y is a generation by x. In the adversarial training, performances of both D and G are trained to be improved, and the final objective is to use the outputs from the well-trained G to fool the well-trained D.

Given the objective of the cGAN

$$\mathcal{L}_{cGAN}(G, D) = \mathbb{E}_{x,y \sim p_{data}(x,y)}[\log D(x, y)] + \\ \mathbb{E}_{x \sim p_{data}(x), z \sim p_z(z)}[\log(1 - D(x, G(x, y)))], \tag{1}$$

such adversarial training process for the G can be formulated as:

$$G^* = arg \min_G \max_D \mathcal{L}_{cGAN}(G, D). \tag{2}$$

As studied by some previous work [8,14], the image generator seems less affected by the noise z, which was also showed in our initial experiments. Therefore, in the final models, we only added the stochasticity by dropout in our generator layers, as the solution in [8].

2.2   Architecture

To obtain the translation generator G, we have designed a training system as shown in Fig. 2.

The cGAN part is similar to that in [8]. We also employed the same convolution-BatchNorm-ReLu [7] modules in our generator and discriminator architectures. In [8], the authors expect to shuttle the low-level information shared between the input and the output of the generator by adding skip connections following a U-Net structure [16]. For the optical and SAR data, although there are deformations on some land cover types, the major spatial structures like the prominent roads, boundaries of flat regions and etc. could be kept. Therefore, the U-Net has also been employed as the network structure of our generator. In experiments, we also found that deformation to a certain extent could also be learned by such U-Net. By denoting a Convolution-BatchNorm-ReLU layer with k filters as Ck, and Ck with a dropout rate of 50% as CDk, The U-Net structure of the generator could be written as: C64-C128-C256-C512-C512-C512-C512-C512-CD512-CD512-CD512-C512-C512-C256-C128-C64. An extra convolution layer is added to the last layer C64 to map the C64 output to the final generation results according to the number of the target channels. In the U-Net, the activations of layer i is concatenated to layer $n - i$, where n is the number of total layers. Similarly, the structure of the discriminator could be written as C64-C128-C256-C512-C512-C512. Convolutions in both generator and discriminator are applied with 4×4 spatial sized filters with stride 2. Particularly, BatchNorm is not used in first layer C64 in the two networks.

It was found in previous studies [8,15], by adding other penalties to the GAN objective, learning of the generator could be guided effectively. In [8], the authors explored L1 distance between the generated image and the target to drive the translation results to be close to the ground truth. In this manner, the L1 distance was defined as:

$$\mathcal{L}_{L1}(G) = \mathbb{E}_{x,y \sim p_{data}(x;y); z \sim p_z(z)}[\|y - G(x;z)\|_1], \qquad (3)$$

and the objective is modified as:

$$G^* = arg \min_G \max_D \mathcal{L}_{cGAN}(G; D) + \lambda L_{L1}(G). \qquad (4)$$

Optimization of the proposed translation models follows the standard approach [5], where gradient descents were alternatively step by step applied between G and D. Minibatch SGD was used with the Adam solver [10]. And the inference process was the same to that in [8].

3   Data and Experiments

To verify the generality of the proposed high-resolution optical and SAR translation models, experiments were conducted on three regions from the world with different SAR parameters. Specifically, the Pauli parameters [2] from RADARSAT-2 C-band data on Toronto in Canada, the amplitudes of three SAR polarization channels [2] HH, HV, VV from RADARSAT-2 C-band on Vancouver in Canada and the amplitude of VV channel from TerraSAR-X X-band data on Shanghai in China. The acquisition times of all the SAR images were in 2008.

The SAR images were ortho-rectified and geo-registered to the optical data. A specific Lee filter [13] has been applied to reduce the speckle on the SAR data.

Corresponding optical images were obtained from the GoogleEarth. Only the Toronto optical data was gathered in 2008. Optical data for Vancouver and Shanghai were gathered in 2016 and 2017 respective. With different time spans between optical and SAR data, tolerance of mismatch sample pairs for the translation models could be evaluated.

In experiments, all the data were spatially resized into 10 m resolution. In each region case, about half of the area was used for training, and another for test. The number of training and test sample pairs were 12000 and 3000 respectively

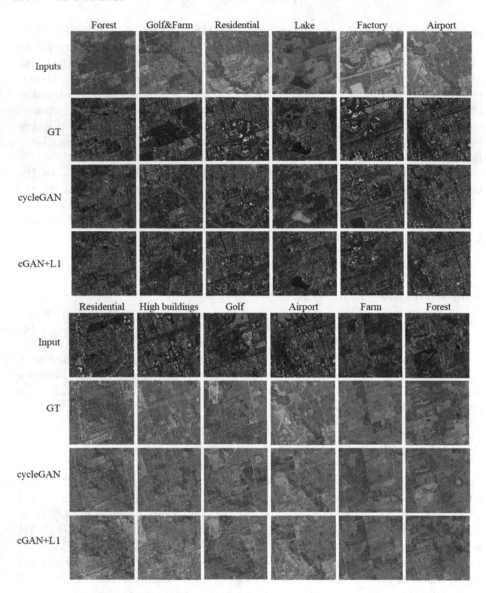

Fig. 3. Comparison on the Toronto area. Optical to SAR: Row 1, 2, 3, 4. Sar to optical: Row 5, 6, 7, 8. GT is for ground truth.

for each region study. These image samples were all with a shape of 256×256 image, and were normalized to input to the GAN.

To explore the performances of the cGAN on SAR and optical remote sensing data translation, a state-of-the-art image translation approach has been compared. For the pair based training model, the traditional cGAN with $\mathcal{L}_{L1}(G)$ objective [8], which is noted as cGAN+L1, was compared with an unsupervised

GAN: cycleGAN [20] to demonstrate the effects of the pair-based training in the multi-type data translation. In each comparison in each region, all the methods were trained with the same samples, the same batch size and the same number of iterations, which is 120 thousands.

To assess the generation results, the commonly used structural similarity index (SSIM) [18] and mean-squared error (MSE) were employed as the evaluation metrics. The window size of SSIM was set to 7. Although SSIM and MSE could be noted as a kind of local measurement, higher SSIM or lower MSE frequently indicates visually better results with correct land cover mapping. Furthermore, the quality of the generated SAR data was also evaluated by land cover classification, where methods with higher SSIM often reported higher classification accuracy.

4   Results and Discussion

4.1   Comparison on Toronto with Short Time Span

Since the Toronto data set has short time span between the optical and SAR data, most of the land covers kept unchanged, and the same object on the two data type could be well matched. Therefore, comparisons on the Toronto region could demonstrate better performances of the proposed method with less influences by the mismatch samples. For each translation direction, we conducted 3 experiments with randomly updated training samples. The average (avg) score and the standard deviation (std) of SSIM and MSE were calculated in Table 1.

Table 1. Translation results on Toronto data using cGAN+L1.

	avg SSIM	std SSIM	avg MSE	std MSE
Optical to SAR	0.2695	0.0034	0.0461	0.0004
SAR to optical	0.2611	0.0020	0.0281	0.0002

Unlike the denoising or compression task, translation between SAR and optical is to represent a image to the other data type with less similar image structure, and there is certain stochasticity in the GAN generator. Therefore, the overall score of the method is relatively lower in comparison with other image applications. However, satisfactory generation can still be observed in both translation directions, as shown in the selected result samples in Fig. 3. It is also found the proposed cGAN+L1 could stably provide better translation in both translation direction.

4.2   Comparison on Different Regions with Large Time Span

In the large time span, vegetation could present great seasonal changes, and the cities may be developed significantly due to the human activities, especially in the faster growing urban area like Shanghai. With lots of mismatched sample pairs, one could expect lower performance of the cGAN+L1. Note that the worse score may also caused by the mismatch in the test pairs, as shown in Fig. 1, where obvious mismatch of the man-made structures between the input and the output ground truth could be observed in Shanghai and Vancouvor. The lower performances could also be observed in Table 2 in the comparison between the tests on Toronto and Vancouvor, which have the same RADARSAT-2 C-band data with almost the same SAR scattering information (the Pauli parameters and the amplitudes are all generated from the three SAR channel raw data, and they all hold most of the information of the source). Especially, with the presence of the mismatched sample pairs, translation from optical to SAR become harder. However, it is also found the SAR to optical translation in Shanghai is even better than that in Toronto which is with short time span. This may also be caused by the less complex problem of only one channel data generation with sharp X-band data. Nevertheless, the cGAN+L1 could also generate visually acceptable results, as shown in Fig. 1.

Table 2. Translation comparison on Toronto, Vancouvor and Shanghai data using cGAN+L1.

	SAR to optical		Optical to SAR	
	SSIM	MSE	SSIM	MSE
Toronto	0.2611	0.0281	0.2695	0.0461
Vancouvor	0.2216	0.0420	0.1300	0.0667
Shanghai	0.3603	0.0247	0.2363	0.0505

4.3   Classification on Generated SAR Images

As mentioned above, short time span data set without lots of the mismatch sample pairs could produce better optical to SAR translation. Therefore, to test the optimal performances of the translation models, we selected Toronto as the test place for the SAR classification.

To this end, a pixel based classification method with K-Nearest Neighbor (KNN) classifier and an adaptive MRF [1] was employed. Totally 19 sub-types were defined for 9 major land cover classes including low density (LD), high density residential area (HD), industry area (IN), golf course (Golf), construction site (Cons.), water, forest, pasture and crops. The final results were formed by merging the classified 19 sub-types into the 9 major classes. Since the GAN only generated results on the test area, training samples for all the compared methods were collected from the same ground truth data in the GAN training area. Approximately 1000 pixels was assigned for each sub-types as the training samples. The validation pixels were randomly selected and evenly distributed over the study area. The number of the validation pixels for each class is assigned accordingly to its proportion and the balance between the other classes. Totally

Table 3. Classification results on the generated SAR images by different methods. OA is the overall accuracy by dividing the number of total samples with the number of all the correct ones

Recall (Percent)									
	Water	Golf	Pasture	Cons.	LD	Crops	Forest	HD	IN	OA
Ground truth	94.94	84.72	48.82	40.23	27.39	47.91	29.62	49.47	50.05	55.3%
cGAN+L1	53.8	36.63	26.07	34.61	21.83	53.44	11.94	49.72	54.35	42.1%
cycGAN	0	3.97	27.41	13.44	18.83	38.32	11.53	59.38	40.86	25.6%
Precision (Percent)									
	Water	Golf	Pasture	Cons.	LD	Crops	Forest	HD	IN	Kappa
Ground truth	95.32	85.35	24.98	36.24	42.26	52.57	27.03	40.42	51.34	0.49
cGAN+L1	99.63	42.54	17.89	13.94	26.68	51.47	20.31	40.5	41.03	0.33
cycGAN	0	18.65	8.8	7.06	18.02	31.4	13.34	37.17	33.9	0.14

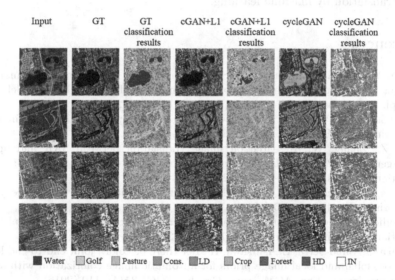

Fig. 4. Classification results on the generated SAR data. GT is for ground truth.

47138 pixels were used for testing the classification results. As shown in Table 3, it is interesting to found cGAN+L1 have also enhance the identification of crops even in comparison with the ground truth data. It might be caused by translating most of the difficult crops into some simple sub types. Nevertheless, low performances on the low scattering types: water, golf, pasture could be found for the conditional GAN. The SAR scatterings of these classes are closely distributed in a low value range (looks dark in the SAR image). Therefore, the differences between these classes are subtle, which increase the difficulty of the GAN translation. In the comparison with and without pair based samples, cycleGAN showed the worst results with a large gap between that by cGAN+L1. Only for the HD, cycleGAN could be comparable with the other methods. For the low scattering types: water, golf and pasture, results by cycleGAN are extremely poor. Deficiency of the cycleGAN could also been observed in Figs. 1, 3 and 4. The above mentioned phenomenon could also be observed from the selected classification samples given in Fig. 4.

5   Conclusion

This paper studied optical and SAR remote sensing image translation, which is a new application problem. Comparing with other advanced translation approaches, the proposed cGAN with the L1 loss could effectively improve the generation results. Such translation model could also tolerate certain time span between the two translation data type and generate satisfactory image for land cover classification. This is one of the first studies on multi type remote sensing image translation by machine learning.

References

1. Ban, Y.: An adaptive contextual SEM algorithm for urban land cover mapping using multitemporal high-resolution polarimetric SAR data. IEEE J. Sel. Top. Appl. Earth Obs. Remote. Sens. **5**(4), 1129–1139 (2012)
2. Cloude, S.R., Pottier, E.: A review of target decomposition theorems in radar polarimetry. IEEE Trans. Geosci. Remote Sens. **34**(2), 498–518 (1996)
3. Fu, Z., Zhang, W.: Research on image translation between SAR and optical imagery. In: ISPRS Annals of Photogrammetry, Remote Sensing and Spatial Information Sciences, pp. 273–278 (2012)
4. Gatys, L.A., Ecker, A.S., Bethge, M.: A neural algorithm of artistic style. arXiv preprint arXiv:1508.06576 (2015)
5. Goodfellow, I., et al.: Generative adversarial nets. In: Advances in Neural Information Processing Systems, pp. 2672–2680 (2014)
6. Iizuka, S., Simo-Serra, E., Ishikawa, H.: Let there be color!: joint end-to-end learning of global and local image priors for automatic image colorization with simultaneous classification. ACM Trans. Graph. (TOG) **35**(4), 110 (2016)
7. Ioffe, S., Szegedy, C.: Batch normalization: accelerating deep network training by reducing internal covariate shift. In: International Conference on Machine Learning, pp. 448–456 (2015)
8. Isola, P., Zhu, J.Y., Zhou, T., Efros, A.A.: Image-to-image translation with conditional adversarial networks. arXiv preprint (2017)
9. Johnson, J., Alahi, A., Fei-Fei, L.: Perceptual losses for real-time style transfer and super-resolution. In: Leibe, B., Matas, J., Sebe, N., Welling, M. (eds.) ECCV 2016. LNCS, vol. 9906, pp. 694–711. Springer, Cham (2016). https://doi.org/10.1007/978-3-319-46475-6_43
10. Kingma, D.P., Ba, J.: Adam: a method for stochastic optimization. arXiv preprint arXiv:1412.6980 (2014)
11. Kuznietsov, Y., Stückler, J., Leibe, B.: Semi-supervised deep learning for monocular depth map prediction. In: Proceedings of the IEEE Conference on Computer Vision and Pattern Recognition, pp. 6647–6655 (2017)

12. Larsson, G., Maire, M., Shakhnarovich, G.: Learning representations for automatic colorization. In: Leibe, B., Matas, J., Sebe, N., Welling, M. (eds.) ECCV 2016. LNCS, vol. 9908, pp. 577–593. Springer, Cham (2016). https://doi.org/10.1007/978-3-319-46493-0_35

13. Lee, J.S.: Refined filtering of image noise using local statistics. Comput. Graph. Image Process. **15**(4), 380–389 (1981)

14. Mathieu, M., Couprie, C., LeCun, Y.: Deep multi-scale video prediction beyond mean square error. arXiv preprint arXiv:1511.05440 (2015)

15. Pathak, D., Krahenbuhl, P., Donahue, J., Darrell, T., Efros, A.A.: Context encoders: feature learning by inpainting. In: Proceedings of the IEEE Conference on Computer Vision and Pattern Recognition, pp. 2536–2544 (2016)

16. Ronneberger, O., Fischer, P., Brox, T.: U-Net: convolutional networks for biomedical image segmentation. In: Navab, N., Hornegger, J., Wells, W.M., Frangi, A.F. (eds.) MICCAI 2015. LNCS, vol. 9351, pp. 234–241. Springer, Cham (2015). https://doi.org/10.1007/978-3-319-24574-4_28

17. Shen, W., Liu, R.: Learning residual images for face attribute manipulation. In: 2017 IEEE Conference on Computer Vision and Pattern Recognition (CVPR), pp. 1225–1233. IEEE (2017)

18. Wang, Z., Bovik, A.C., Sheikh, H.R., Simoncelli, E.P.: Image quality assessment: from error visibility to structural similarity. IEEE Trans. Image Process. **13**(4), 600–612 (2004)

19. Zhang, L., Zhang, L., Du, B.: Deep learning for remote sensing data: a technical tutorial on the state of the art. IEEE Geosci. Remote. Sens. Mag. **4**(2), 22–40 (2016)

20. Zhu, J.Y., Park, T., Isola, P., Efros, A.A.: Unpaired image-to-image translation using cycle-consistent adversarial networks. arXiv preprint arXiv:1703.10593 (2017)

21. Zhu, X.X., et al.: Deep learning in remote sensing: a review. arXiv preprint arXiv:1710.03959 (2017)

A Combined Strategy of Hand Tracking for Desktop VR

Shufang Lu[⊠], Li Cai, Xuefeng Ding, and Fei Gao

College of Computer Science and Technology,
Zhejiang University of Technology, Hangzhou 310023, Zhejiang, China
sflu@zjut.edu.cn

Abstract. Desktop VR has been widely used in data analysis and VR movies. One of the important interactions in VR is to capture and track the 3D motion of hands. Although 3D hand pose estimation has been developed for many years, the trade-off between real-time and accuracy still exists. In this paper, we propose a strategy that combines fast model-based method and Convolutional Neural Network (CNN). Based on the occlusion of the hand depth image captured by Intel RealSense Camera, simple gesture images and complex gesture images are recognized by fast model-based method and CNN, respectively. A large number of experimental results demonstrate that our method achieves real-time performance with high accuracy.

Keywords: Desktop VR · 3D hand tracking · Computer vision
Combined strategy

1 Introduction

A typical representative of an immersive application is desktop VR. It can be widely used in data analysis, watching VR films [1]. Those settings always provide stereoscopy and head-tracking [3], what's more, they also offer gesture recognition for human-computer interaction. Sitting on a chair and placing an elbow on the table make it possible to interact by gestures in a VR environment for a long time. Because head-mounted display (HMD) will block the eyes from observing the real world, the mouse and keyboard are no longer suitable for immersive virtual environments. The handle, the most mature solution with the lowest cost, does not give the user direct control. Currently, gesture sensors on the market, including Leap Motion [17], use the HMD + gesture sensor to implement a self-centered VR interactive forms commonly. But this gesture recognition method is not suitable for the application scenarios of desktop VR because of the following features: (1) **Limited gesture activity range:** In general, the VR experience is viewed from a free perspective, limited by the range of the gesture sensor, it is necessary to attach the gesture sensor to the VR head-mounted display. It ensures that gestures can be tracked in various positions. But the eyes must follow the movement of the hand. (2) **Severe self-occlusion:** Self-occlusions are a common problem in egocentric viewpoint because of 4 degrees-of-freedom (DOF) for a finger. The existing method itself is the cause of further self-occlusion of the gesture interaction in the VR (the gesture sensor is fixed on the VR headset). Although the

© Springer Nature Switzerland AG 2018
R. Hong et al. (Eds.): PCM 2018, LNCS 11166, pp. 256–269, 2018.
https://doi.org/10.1007/978-3-030-00764-5_24

image of the hand detected by the sensor is ensured to be complete, the visual information of the finger location is easily obscured by the back of the hand. In this case. It is still difficult for the current gesture recognition methods to accurately recognize the joint position under a large area of occlusion. (3) **Easy to fatigue:** The desktop VR experience requires users to wear VR devices. A heavy device can easily cause neck fatigue in a short period of time, especially sitting. As manufacturers reduce the weight of HMD equipment, it seems unwise to bundle more sensors on VR headsets.

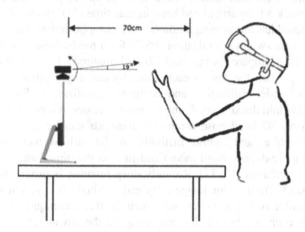

Fig. 1. A desktop VR concept draft.

Based on the deskVR scenario that takes place at an office desk with the user staying seated on the office chair, we propose a bare-hand tracking techniques by using the front-facing camera. Specifically, we employ a computer vision technology and a depth camera for 3D hand tracking which can avoid brightness interference. So the user can focus on the content of the scene without the need for eyes to follow the movement of the hand. And it frees our sight, reduces user restrictions and ensures the naturalness of the interaction.

According to the research of the latest 3D gesture tracking research, the traditional model-based method has faster speeds, but the recognition accuracy of complex gestures is lower. And complex CNN networks [12–14] have high accuracy but poor real-time performance and occupy many GPU computing resources. So the main contribution of this paper is that we propose a method that combines the advantages of both methods. Through a strategy that determines the completeness of the hand image, a large number of simple gesture images are recognized by fast model-based method, complex gesture images hand over to CNN. So many steps of the CNN network and model simulation can be run at interactive speeds. The method gives a good performance in terms of speed and accuracy.

2 Related Work

As the depth sensor is widely used, the field of gesture estimation becomes very active. An excellent summary analysis of existing methods can be found in [8]. In general, gesture estimation methods can be simply divided into model-based methods and data-driven methods. The model-based technique considers an a priori 3D hand model whose pose is determined over time by some tracking processes. But these methods require some kind of accurate initialization. Qian et al. [2] presented a model-based method that can track a fully articulated hand in real-time (25FPS on a desktop without a GPU). This method has a low recognition rate for complicated gestures. Oikonomidis et al. [19] use particle swarm optimization (PSO) for a model-based method, but it has 15 fps with GPU. Compared with model-based methods [2, 19], the data-driven approach directly predicts the pose of each frame by learning depth and image features without complex model calibration and accurate initialization. So a single-frame detection method is initialized every frame to make it easier to recover from estimation errors. Depth-based 3D hand pose estimation methods can be categorized into discriminative, generative, and hybrid methods in data-driven methods. Generative methods assume a pre-defined hand model and fit it to the input image by minimizing hand-crafted cost functions [10, 11]. Recently deep learning is providing new options for estimating hands from depth images. Hybrid methods is a recent trend in hand tracking that combine both generative and discriminative techniques. It can overcome the limitations of each one in isolation and integrate the advantages of both of them. The generative methods is effectively supplemented by the discriminative methods, whether it is initializing or recovering from an error. Moreover, the discriminant component can guide the optimization processing of generating the model to achieve low tracking error and converge to a global minimum in the search space. The detection-based [12, 14] methods perform better than the regression-based [15] methods under normal conditions, but when the self-occlusion is severe under extreme conditions, the regression-based method performs better [8]. While at the extreme viewing angles [70, 120] degrees of viewing angles are conventional viewing angles. When the joint angle is greater than 70°, close to the fist posture, the average error increases to more than 12 mm [8]. This means that the less complete the hand image, the more serious the self-occlusion will be and the more difficult the gesture recognition will be. Recently, several methods have used a 3D voxel grid as input for 3D CNN [12, 13]. Their performance is better than 2D CNN for capturing the spatial structure of depth data. But the *V2V-PoseNet* [12] has only 3.5 fps in a single GPU. The current method has good one-hand pose estimation performance in the training of millions of data sets, but it is difficult to achieve real-time tracking in VR environment for general computer.

Due to the applications in animation, game, human computer interaction and VR/AR, capturing and tracking the 3D motion of hands has been studied for many years [3]. And the need is increasing [4]. Especially with the development of VR application. Different from the general VR game experience, the unique experience of desktop VR makes it more suitable for naked interaction. Several methods use other equipment such as wearable camera [5], Data gloves [6] or Color Glove [7]. These work simplify hand tracking. But this approach is invasive, changing the shape of the hand, and hindering the natural hand movements, reducing the VR operating experience, not only the equipment calibration is cumbersome but also results in overall equipment redundancy. In addition, they need to integrate and synchronize with other systems. Jang et al. [8] use 3D Finger CAPE to solve clicking action and position estimation under self-occlusions in egocentric viewpoint, but the performance reduces much when tracking large divergence of different gesture. The fingertips were identified as the recognition points, and the 3D finger was further analyzed on the basis of the existing gesture recognition method. However, the accuracy of its fingertips depends on the accuracy of joint recognition.

In all of the above methods, the trade-off between real-time and accuracy performance still exists. Techniques that exhibit high accuracy often work at low frame rates and therefore do not apply to interactive systems in spatially-immersive scenarios.

3   Our Proposed Combined Method of 3D Hand Tracking

In the VR experience, a large number of computer computing resources are occupied by image rendering. For real-time tracking of gestures with a high accuracy rate, we require a strategy that many steps of the CNN network and model simulation can be run at an interactive rate. Fortunately, the time-consuming delay of CNN output frames does not affect overall time consistency. The paper systematically combines two hand models, as shown in Fig. 2. It performs fast incremental update for most simple frames and can identify more difficult frames in the CNN for more expensive identification.

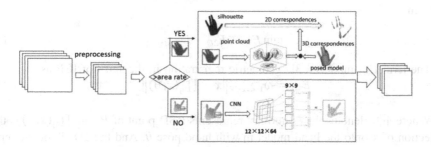

Fig. 2. Diagram of the proposed method. The camera captures continuous depth image frames. After a series of image processing, we extract the hand image. According to the normalized hand image, see Sect. 3.3, the area rate is larger, and the hand image has less self-occlusion area, entering the fast mode-based method (the upper part of the pipeline), see Sect. 3.1. The smaller the rate, the serious the self-occlusion. So the accurate CNN identification method is selected for estimation, see Sect. 3.2. Finally the recognition result of the depth image is the output.

3.1   Fast Model-Based Method

The gesture extraction and real-time recognition method designed in this paper is used for desktop VR. In order to meet the scenario requirements, we assume that the distance between the depth camera and the human hand is within 0.7 m, the position is closest to the camera and the palm of the hand is facing the camera. We use Intel® Real-Sense™ Camera SR300. The depth resolution is 320×240.

Nearest Neighbor. According to the nearest neighbor algorithm (k-NN), the set of points near the point where the classification has been determined also belongs to the same classification. This method is used for hand extraction. That is, the point with the smallest distance from the center is used as the center, and a ball with a radius of 10 cm is made into this sphere. The points are extracted as hand gestures. Analyze the possible situation of the most nearest point. The nearest point may be in the palm, fingertips or wrist. Due to the experience of desktop VR, the person is in a sitting position and the elbow joint is placed on the desktop. The arm is slightly forward so that the possibility of the nearest point can be ignored.

Threshold Segmentation. The hand region is further refined by median filtering and morphological opening, denoted as depth map D. In order to facilitate data processing and reduce subsequent calculations, we convert the actual distance data of the 320×240 pixel depth map into a 3D point cloud, denoted as P. And points are filtered based distance with 0.7 m threshold. It can remove background data with a depth greater than 0.7 m.

Fast Point Cloud Fitting. The two point pairs of the gesture that are pending and candidate gestures to be matched are converted into the same coordinate system to achieve a rigid transformation. Gesture recognition can be understood as a problem of matching between two point pairs, and the algorithm is repeated using ICP. Let $\mathcal{F}$ be the sensor input data consisting of a 3D point cloud P and 2D silhouette S. Give a 3D hand model η with joint parameters $\theta = \{\theta_1, \theta_2, \ldots, \theta_{26}\}$. To solve the optimization problem

$$\min E_{3D} + E_{2D} + E_{wrist}, \tag{1}$$

The term E_{3D} models a 3D geometric registration in the spirit of ICP as

$$E_{3D} = \omega_1 \sum_{x \in P} \|x - \prod_{\eta}(x, \theta)\|_2, \tag{2}$$

Where $\|\cdot\|_2$ denotes the ℓ_2 norm, x represents a 3D point of P, and $\prod_{\eta}(x, \theta)$ is the projection of x onto the hand model η with hand pose θ. And the 2D silhouette term E_{2D} represents the alignment of rendered hand model' 2D silhouette with 2D silhouette extracted from the depth map as

$$E_{2D} = \omega_2 \sum_{p \in S} \left\| p - \prod_S (p, \theta) \right\|_2^2, \tag{3}$$

Where p is the point of the rendered silhouette S, and $\prod_S (p, \theta)$ denotes the projection of p onto the depth frame's silhouette S. With the constraint of the silhouette energy, the occluded fingers can be prevented from moving to the wrong position. Because they are not constrained by any samples in the depth map. At the same time we add the wrist alignment energy E_{wrist} to ensure that the wrist joint can be located along its axis.

$$E_{wrist} = \omega_3 \left\| \prod_{2D} (k_0(\theta)) - \prod_{\ell} (k_0(\theta)) \right\|_2^2, \tag{4}$$

Where k_0 is the 3D position of the wrist joint, and ℓ is the 2D line extracted by PCA of the 3D points corresponding to the wristband. Minimizing the wrist alignment energy helps prevent the palm from rotating/flipping incorrectly during hand tracking.

Minimizing the fitting energies alone easily leads to unrealistic or unlikely hand poses. The common rigid body constraint is simply the match between two bones. For highly complex structures such as hands, additional constraints must be added to prevent the hand model from appearing infeasible. (1) the upper two joints angles on each finger keep match each other (2) make sure base thumb bone collide without anything (3) base bone and mid bone keep move together (4) pinky and mid keep move together (5) The degree of abduction is 0 when the finger is making a fist, otherwise it is the default $45°$ threshold.

3.2 Accurate CNN Identification Method

Modeling: The convolutional network first extract visual features from normalized depth-frame F_t. This gives us a feature map $\emptyset(F_t)$ of the depth-frame F_t as follows,

$$\emptyset(F_t) = \text{CNN}(\text{n}(F_t)) \tag{5}$$

where $\text{n}(F_t)$ denotes the normalized input of depth-frame F_t including hand localization, segmentation and data normalization. The preprocessing steps are described in details in the next. And the dimension of output feature maps are mapped to five feature regions R_1–R_5. Figure 2 graphically illustrate this process. We get results from each separate fully-connected (FC) regressors. Then the results are integrated by another FC layer to input hand joints estimation.

Pre-processing: We segment a fixed-size metric cube from the depth image centered on the hand. And the cube is resized to a 96×96. The depth value in the cube is normalized to $[-1, 1]$. The depth value are clipped to the cube sides front and rear. It provides invariance to different hand-to-camera distances.

Training: The CNN that we constructed to extract visual features consists of six convolutional layers, three max-pooling layers and two fully-connected layers. This is a generic network similar to the one in Ren [16].

Let C denote a convolutional layer, P a pooling layer and F C denote a fully connected layer. Both C and F C layers consist of Rectified Linear Unit (ReLU) [24] activation functions. For C layers, the size is defined as (w $\times$ h $\times$ d), where the first two define the dimension of filters and the last one is the number of filters. For P layers, the size is defined as (w $\times$ h). The output of regression is the 3D world coordinates of hand joints, described as $3 \times J$. The convolutional network can be described concisely as $C(3 \times 3 \times 16) \rightarrow C(3 \times 3 \times 16) \rightarrow P(1 \times 1) \rightarrow C(3 \times 3 \times 32) \rightarrow C(3 \times$

32) → P(1 × 1) → C(3 × 3 × 64) → C(3 × 3 × 64) → P(1 × 1) → F   C(2048) → F C(2048) → F C(3 × J).

Regression for the regional center as a branch feed to two full-connected layers, each with a loss rate of 0.5. We optimize the network parameters by using error back propagation and apply the stochastic gradient descent (SGD) algorithm with a minimum batch size of 128. Set weight attenuation to 0.0005 and momentum to 0.9. The learning rate decays over the epochs at the rate of 0.95 and starts with 0.005. The networks are trained for 80 epochs (Fig. 3).

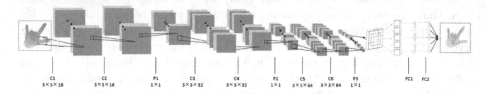

Fig. 3. Structure of the network

3.3   Combined Selection Strategy

When fast model-based hand tracking model is slightly closer to the desired posture, the model will quickly lock in place. However, for complex gestures, it is often easy to identify errors. Once an error occurs, it is difficult to correct it. We hope to establish a strategy that can analyze whether it is an 'easy gesture' based on depth images. Simple gestures can get better recognition accuracy through fast model-based method. As for complex gestures that fingers and palm are severe self-occlusion. Through the trained convolutional neural network model, it can obtain better recognition results than the fast. However, how to measure and determine whether it is an easy gesture is a challenge that needs to be solved urgently from a single depth map. From the current research, the larger the average finger joint angle of the gesture and the greater the number of occluded joints, the hard gesture state. Because we do not need to accurately detect the joint angle or the number of occluded joints, we have found that the smaller area of the hand image (the angle between the hand and the camera direction is 70° to 120°) makes recognition difficulty. Because the joints' average angle assembly increases recognition error. For these image frames, the CNN model can significantly improve the recognition accuracy. When the extreme view results in a smaller hand image, the advantages of this method also exist. Therefore, we choose the size of the hand image as a method selection strategy.

The resolution of the image captured by the depth camera is fixed, but the distance between the hand and camera will also cause the difference of the area of the hand image. Therefore, we need to standardize the image area of the hand. We set the segmented hand image area at a distance of 0.7 m from the sensor to the standard (we set the background depth filter to 0.7 m) when the fingers extend fully and the palm is facing the camera. According to the principle of depth sensor imaging, the image size of each frame captured is standardized as 0.7 m away from the sensor. We define the normalization function as:

$$\partial = \frac{\mu S'}{S} \qquad (6)$$

Where S' is the original image area, S represents the standard area. ∂ is the ratio of normalized image to full gesture image area. Here μ is the conversion factor that transform original image to standard image area (Fig. 4).

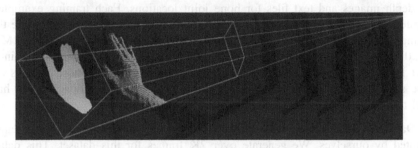

Fig. 4. Standardized hand images size to the same depth

4   Implementation

In order to present the results, we first discuss the considered data and the experimental setup.

4.1   Dataset and Setup

Data Generation: This paper aims to improve the performance of real-time gesture tracking in VR environment through joint recognition strategy. Therefore, whether it is training or testing, the depth map of time series can better reflect the effect of this article. Datasets use synthetic depth images and automatic annotation [22]. For the training method used in this paper, the saved label is the complete pose (position and orientation) 3D transform information for all 17 bones (3 bones per finger, palm and wrist) in the hand. Using this information, CNN is trained to recognize 2D position of the fingertip and some relative angles indicating finger grip and extension. The model was manipulated using a finger bone, each with three bones, distal, middle, and proximal phalanges. Extra bones are used to control the rotation of the palms and wrists. We avoid unreliable gestures by defining each bone rotation constraint. In order to reproduce the gestures in all reasonable gesture interactions, all finger phalanges are set to have only 1-DoF rotation (finger flexion/extension), but the metacarpals have a 2-DoF rotation to allow for adduction/abduction of the fingers. Simulate a 3D joint model of a hand, which can incrementally update its pose to fit the point cloud of each frame. We use a virtual camera to simulate intrinsic parameters of the depth sensor and reproduce the image resolution. And based on this virtual camera we render the ani-mated hand model. In a simulated environment, the virtual camera is always aimed at the hand. It assumes the right hand is the only thing in the near view volume. Save any frame (and current hand pose). For difficult gesture marks, start tracking from a known or reset position and slowly move your finger to other positions. For automatic tag data collection, it is not necessary to recognize the gesture from scratch, as long as it can be moved to this position without losing it. For occlusion gestures, the wrist and palm can still be moved and rotated by keeping the finger in a particular position.

Training Dataset: We have generated a different training dataset. This dataset contains depth images and text files for bone joint locations. Each training example is generated with a fixed view-direction and a random gesture. We generated more than 80 samples for this dataset. They use for training and nearest neighbor extraction. More than 20K frames were generated for this dataset and used to extract clips and train the model. Each test sequence contains 1000 frames that capture various pose changes with severe scale and viewpoint. Then we use the available joint position and our hand model to provide a ground truth partition on this dataset.

Test Dataset: We use two test datasets, including a public dataset and a test dataset generated by ourselves. We generate over 2K frames for this dataset. This dataset consists of some continuously changing gesture frames with a time sequence, and key gestures are randomly selected or selected from a set of predetermined gestures. At the same time, we also set different gesture change speeds and form of movement of the fingers (including opposition-reposition, besides flexion-extension and adduction-abduction). In order to prevent unreliable gestures, we add additional constraints and detections to ensure that action is possible naturally. There are more common datasets for gesture estimation. In the selection of public dataset, we use the ICVL dataset [18]. The dataset shows a variety of hands and viewpoints, and the definition of joint locations is similar to ours.

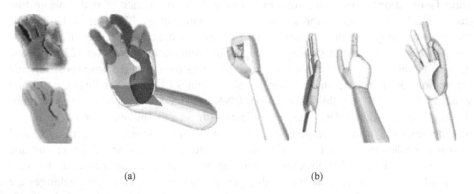

(a)   (b)

Fig. 5. The program captures and saves depth data along with the full hand pose for that frame. (b) is the various gestures we created hand poses

4.2   Results

Area ratio. First, for checking the feasibility of this paper technical contributions as experiments, we tested to find the balance point between accuracy and speed by trying a different ratio of hand area so that we can get an optimal performance. Taking into account the general finger width is 10–20 mm. We set the distance error threshold to 15 mm. We randomly select 2000 frames for experimentation. At first, we have calculated the distribution of hand area ratios in the experimental frames, as shown in Fig. 6. Then we experiment with the effects of different hand area ratios on speed and accuracy. As shown in Fig. 8, we choose the area ratio of 60% as the balance point. At this threshold, both accuracy and speed have a good performance (Fig. 7).

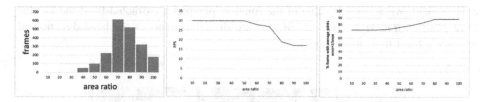

Fig. 6. The experimental results show the accuracy and speed of different area ratio thresholds, and the data distribution in the test set (a).

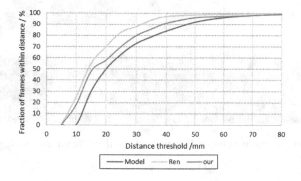

Fig. 7. Comparison of different methods on the ICVL dataset

We choose the success rate, i.e. the percentage of correct frames whose average (or max) joint Euclidean distance (in millimeters) error is lower than a threshold ($\sim$20 mm) [21]. We demonstrate the efficacy of our frameworks by comparing them to the Model [20] and Ren [16]. We execute on the normal test dataset of the real database (i.e. ICVL). This experiment shows the accuracy of each method in estimating joint position in the ICVL test set. The experimental prediction error results are shown in Fig. 6 for ICVL test set. Experimentally, the average processing time of the proposed framework was 23.5 ms (42.5 FPS) in our experimental environment, which has a ten-core Intel i9 7900 CPU, 24 GBs RAM and an NVIDIA GTX 1080Ti.

4.3   Experiment Results by User Test

Experiments were performed to investigate the feasibility of the proposed approach in both quantitative (using our newly gathered hand dataset) and qualitative manners (through user tests in desktop VR environment). Especially for the user test, we migrate the selection mechanism utilizing the combined method into VR environment to show its practical use.

For experiencing a more immersive desktop VR environment, the user need wear a HMD, as shown in Fig. 1. We choose Oculus DK2 as a HMD and Intel® RealSense™ Camera SR300 [23] as a depth camera. In order to adjust to the possible input limitation range of the viewing angle of the camera, we adjust the camera's position and angle as shown in Fig. 1. The VR environment, which is shown in Fig. 8, is implemented by OpenGL. The metric unit in the VR environment is homogenized as a mm. In the environment, we apply the value of the rotation parameter extracted from the Oculus DK2 position tracker so that the VR viewing direction matches the head direction of the user's head movement guide.

Fig. 8. Experimental scenario example. Fingers click on the different cubes triggers a variety of interactive content.

In addition to experimenting with datasets, in order to test the viability of the proposed real-time gesture tracking in a desktop VR environment, we did a user test of gesture operations in a desktop VR environment. The experimental user test environment is the same as our implementation setup. To design such an interaction, when the index finger collides with a cube of a selectable color, it can be rotated 33° and its color will be changed. The user can move the cube with two fingers. We selected 14 participants, ranging from 24 age to 26 age, and average 24 age, all of them graduate students. All participants understand VR-related knowledge, and one-third of them do not experience VR for the first time. Our user study takes approximately 20 min per person, including training, tests, and oral interviews after completing the questionnaire. Before starting the experiment task, we assisted them in wearing the device, calibrated the user's hand size and changed the parameters to standardized hand size and then simply described the experience content. After this, participants were allowed to experience the VR environment for 10 min in an unsupervised environment. The testers were then tasked with requirements, including: (1) index finger clicks on any cube until it spins over 90°. (2) users place four cubes in the same column. (3) users Change the color of the top cube. After the test is completed, complete the questionnaire of the Likert scale (see Table 1) to obtain participants' feedback.

Table 1. Post-questionnaire

Q1	This method is easy to grab the object
Q2	This method is very sensitive to click
Q3	This method is suitable for desktop VR

Finally, by using post-questionnaires applying the 5 Likert scale shown in Table 1, we obtained the tester's preference for the method proposed in this paper and the application device Leap Motion. From Fig. 9, our method is significantly better than Leap Motion. As shown in Fig. 10, based on the answers to the questionnaire, the proposed method is more favored by the testers. Users may consider that gesture sensor such as leap motion is placed on the desktop, affecting the range of their gesture activities, and worry that in the case of wearing HDM, the hand easily moves out of the sensor detection range. And when the front-facing camera is placed in the desktop VR, the elbow can be supported on the desktop. Compared to the suspended arm, this method is less likely to fatigue. At the same time, it naturally reduces the scope of

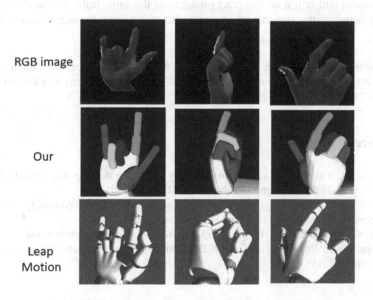

Fig. 9. Comparison of our method with leap motion

gesture activities. Therefore, most participants responded that they preferred the method proposed in this paper.

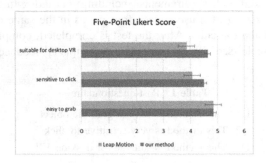

Fig. 10. Mean and standard deviation results of user's preference on each method based on the post-questionnaire.

5   Conclusion

The 3D hand tracking method with fixed-view depth sensor combines the existing technology method [16, 20] with our proposed tracking strategy, resulting in a good performance in speed and accuracy. The method is well suited for gesture tracking in desktop VR. When the main thread is model-based, the model will quickly lock into place when the hand model is slightly close to the desired posture. However, because the model only estimates joints based on depth data, which does not know where a single depth pixel is collectively reflected in the user's hand, the model may fall into a local minimum and return an incorrect posture. At the same time, due to the automatic switching between the two detection methods, slight jitter between frames is easily caused.

Acknowledgements. This work is supported by the Natural Science Foundation of China (No. 61402410) and Zhejiang Provincial Science and Technology Planning Key Project of China (No. 2018C01064).

References

1. Zielasko, D., et al.: Remain seated: towards fully-immersive desktop VR. In: 2017 IEEE 3rd Workshop on Everyday Virtual Reality (WEVR). IEEE (2017)
2. Qian, C., et al.: Realtime and robust hand tracking from depth. In: Proceedings of the IEEE Conference on Computer Vision and Pattern Recognition (2014)
3. Rogez, G., Supancic, J.S., Ramanan, D.: First-person pose recognition using egocentric workspaces. In: Proceedings of the IEEE Conference on Computer Vision and Pattern Recognition (2015)

4. Kim, D., et al.: Digits: freehand 3D interactions anywhere using a wrist-worn gloveless sensor. In: Proceedings of the 25th Annual ACM Symposium on User Interface Software and Technology. ACM (2012)
5. Dipietro, L., Sabatini, A.M., Dario, P.: A survey of glove-based systems and their applications. IEEE Trans. Syst. Man Cybern. Part C **38**(4), 461–482 (2008)
6. Wang, R.Y., Popović, J.: Real-time hand-tracking with a color glove. ACM Trans. Graph. (TOG) **28**(3), 63 (2009)
7. Yuan, S., et al.: Depth-based 3D hand pose estimation: from current achievements to future goals. In: IEEE CVPR (2018)
8. Jang, Y., et al.: 3D finger cape: clicking action and position estimation under self-occlusions in egocentric viewpoint. IEEE Trans. Vis. Comput. Graph. **21**(4), 501–510 (2015)
9. Sharp, T., et al.: Accurate, robust, and flexible real-time hand tracking. In: Proceedings of the 33rd Annual ACM Conference on Human Factors in Computing Systems. ACM (2015)
10. Tang, D., et al.: Opening the black box: hierarchical sampling optimization for estimating human hand pose. In: Proceedings of the IEEE International Conference on Computer Vision (2015)
11. Moon, G., Chang, J.Y., Lee, K.M.: V2V-PoseNet: voxel-to-voxel prediction network for accurate 3D hand and human pose estimation from a single depth map. In: CVPR, vol. 2. no. 3 (2018)
12. Yang, F., Akiyama, K., Wu, Y.: Naist rv's solution for 2017. http://icvl.ee.ic.ac.uk/hands17/author/gui
13. Molchanov, P., Kautz, J., Honari, S.: 2017 hand challenge nvresearch and umontreal team. http://icvl.ee.ic.ac.uk/hands17/challenge
14. Chen, X., et al.: Pose Guided Structured Region Ensemble Network for Cascaded Hand Pose Estimation. arXiv preprint arXiv:1708.03416 (2017)
15. Ge, L., et al.: Hand PointNet: 3D hand pose estimation using point sets. In: Proceedings of the IEEE Conference on Computer Vision and Pattern Recognition (2018)
16. Guo, H., et al.: Region ensemble network: improving convolutional network for hand pose estimation. In: 2017 IEEE International Conference on Image Processing (ICIP). IEEE (2017)
17. Leap motion. http://www.leapmotion.com/. Accessed 10 Sept 2014
18. Tang, D., et al.: Latent regression forest: structured estimation of 3D articulated hand posture. In: Proceedings of the IEEE Conference on Computer Vision and Pattern Recognition (2014)
19. Oikonomidis, I., Kyriazis, N., Argyros, A.A.: Efficient model-based 3D tracking of hand articulations using kinect. In: BmVC, vol. 1, no. 2 (2011)
20. Zhou, X., et al.: Model-based deep hand pose estimation. arXiv preprint arXiv:1606.06854 (2016)
21. Sun, X., et al.: Cascaded hand pose regression. In: Proceedings of the IEEE Conference on Computer Vision and Pattern Recognition (2015)
22. https://github.com/hand_tracking_samples/realtime_annotator
23. Intel's Creative Camera. https://software.intel.com/en-us/realsense/sr300
24. Nair, V., Hinton, G.E.: Rectified linear units improve restricted boltzmann machines. In: Proceedings of the 27th International Conference on Machine Learning (ICML 2010) (2010)

Super-Resolution of Text Image Based on Conditional Generative Adversarial Network

Yuyang Wang, Wenjun Ding, and Feng Su[✉]

State Key Laboratory for Novel Software Technology, Nanjing University,
Nanjing 210023, China
suf@nju.edu.cn

Abstract. To generate high-resolution text images from available low-resolution ones is of great value to many text-related applications, especially text recognition. In this paper, we propose an effective super-resolution method for text images based on Conditional Generative Adversarial Network (cGAN). Specifically, we improve the cGAN model by removing the Batch Normalization layers and introducing the Inception structure to make it more suited to the text image super-resolution task, which contribute to the overall enhanced performances of the proposed method relative to the original cGAN model. Experiment results on public dataset demonstrate the effectiveness of the proposed method.

Keywords: Super-resolution · Text image
Conditional generative adversarial network · Inception
Batch normalization

1 Introduction

Text in images carry rich and direct semantic meaning and play an important role in semantic-based image analysis, indexing, retrieval and many other related applications. Though recent development of deep learning based techniques has led to significant increase on text extraction performance, there still exist large numbers of low-resolution text (especially in natural scene images) that are usually very hard to correctly detect and recognize.

As an effective measure for enhancing image resolutions, the super-resolution (SR) technique aims to generate a high-resolution (HR) image from the given low-resolution (LR) one, which is generally difficult as the LR image usually loses many details needed for the HR image. Accordingly, the super-resolution problem has gained extensive research attention, and many methods have been proposed including those based on learning the mapping between low and high resolution image patches [1,3,10,15,17,21,22] and more recent ones exploiting deep learning techniques [2,8,9,11,19]. Specifically, since the introduction of the Generative Adversarial Network (GAN) model [4], it has been successfully applied

© Springer Nature Switzerland AG 2018
R. Hong et al. (Eds.): PCM 2018, LNCS 11166, pp. 270–281, 2018.
https://doi.org/10.1007/978-3-030-00764-5_25

to the super-resolution problem as well as many other tasks like image-to-image translation, and yielded very promising results. On the other hand, very limited result has been reported on applying GAN architectures to the text image super-resolution task, which can also be considered as a specific image-to-image translation problem.

In this paper, we propose an effective super-resolution method for text images by modifying and improving the Conditional Generative Adversarial Network (cGAN) [6]. The key contributions of the proposed method can be summarized as follows:

- To the best of our knowledge, it is the first work to exploit cGAN for text image super-resolution.
- We remove the Batch Normalization layers from our cGAN generator network, allowing the network to better preserve detailed text cues that are critical to the text super-resolution task.
- We introduce the Inception structure into our cGAN model, which effectively expands the width of the network and allows the generator network to adaptively capture text cues of different sizes, so as to generate more realistic output text images.

The remainder of this paper is organized as follows. Section 2 introduces some related work on image super-resolution. Section 3 describes the proposed cGAN-based method for text image super-resolution in detail. Section 4 presents the results of the experiment.

2   Related Work

Existing work on image super-resolution can generally be divided into two main schemes: SR methods learning LR-HR mapping in the patch space and SR methods exploiting convolutional neural network (CNN) to learn LR-HR mapping directly in the image space.

Most traditional SR approaches try to learn a mapping between LR and HR patches using some supervised learning algorithms. For example, Freeman et al. [3] employed the nearest neighbor (NN) method to learn a dictionary relating LR-HR patches for the SR task. Chang et al. [1] exploited a manifold embedding technique to learn the mapping between LR and HR image patches. Moreover, techniques including kernel ridge regression [10] and sparse representation [22] have also been adopted for learning the LR-HR mapping. On the other hand, as directly modeling the patch space over the entire training set could be very complex, some methods partitioned the training set into several subsets using variant learning algorithms such as K-means [21], sparse dictionary [17] and random forest [15], and learned the LR-HR mapping within respective subsets.

Since the introduction of convolutional neural network, there have appeared a number of SR methods exploiting different CNN models to learn the LR-HR mapping directly in the image space. For example, SRCNN [2] firstly proposed to apply CNN to the super-resolution problem to learn the nonlinear mapping between LR and HR images. Kim *et al.* [8] proposed the VDSR network that predicts the residuals instead of the actual values of image pixels, which allows increasing the depth of convolutional layers to 20 compared to the 3 convolutional layers used in SRCNN and achieves better results with less time cost. Wang *et al.* [19] trained the SCN network that combines sparse coding with a deep CNN to progressively upsample images to the desired size. Kim *et al.* [9] proposed the DRCN model that introduces recursive layers into the CNN model, which reduces the number of parameters and improves the network's performance. Ledig *et al.* [11] firstly applied the Generative Adversarial Network (GAN) to image super-resolution task and proposed the SRGAN model to generate perceptually satisfying SR images with finer texture details.

3 Generative Adversarial Super-Resolution of Text Image

In this work, we propose an enhanced Conditional Generative Adversarial Network [6] model that concentrates on the super-resolution of text images.

3.1 Conditional Generative Adversarial Network

A Generative Adversarial Network (GAN) [4] is a deep neural network model consisting of two adversarial networks: a generator G and a discriminator D. G aims to generate an image as realistic as possible, usually using simple noise signals as input. D classifies an input image to either a real image from the training set or a 'fake' image generated by G. The two networks compete against each other in that, G tries to generate images that fool D to regard them as real ones while D tries to avoid being deceived by G, which altogether effectively learn the data distribution captured by the training set.

Conditional Generative Adversarial Network (cGAN) [6] further extends the GAN model by introducing extra information x (conditioning variable) to both G and D and making them be conditioned on x, which can be any kind of auxiliary information.

In this work, we use an LR image $\mathbf{x}$ as the conditioning variable of the cGAN model, and train the generator network G to generate an SR image $G(\mathbf{z}, \mathbf{x})$ ($\mathbf{z}$ is a simple noise input for cGAN) that realistically approximates the HR image $\mathbf{y}$ of the LR image $\mathbf{x}$. Specifically, G learns to model the distribution $p_{data}(\mathbf{x}, \mathbf{y})$ of the training data which consists of pairs of the input LR image $\mathbf{x}$ and the ground-truth HR image $\mathbf{y}$, and on the other hand, the discriminator network D is learned to distinguish between the generated pair $(\mathbf{x}, G(\mathbf{z}, \mathbf{x}))$ and the pair $(\mathbf{x}, \mathbf{y})$ truly coming from the training data.

3.2 Network Architecture

The proposed cGAN model for text image super-resolution consists of two main parts: the generator network and the discriminator network, which will be described respectively in following sections.

Generator Network. The architecture of the proposed generator network is shown in Fig. 1. Taking an LR image as the input, the proposed generator network is composed of following layers: Conv64LRelu, 16 Residual Blocks [5], Inception [16], Deconv3Tanh, where the number after Conv and Deconv denotes the number of the output channels of the corresponding layer. At the output end, the Deconv layer yields an SR image of twice the size of the input LR image.

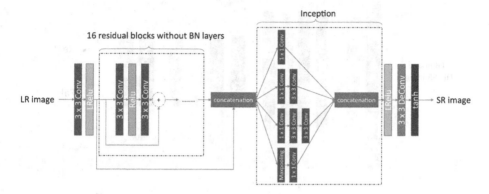

Fig. 1. The structure of the proposed generator network.

Different from the original generator network in the cGAN architecture [6], we make several modifications to the network structure to suit our objective of text super-resolution.

- First, as our task is to reconstruct the HR image from an input LR image whose size is already quite small, we omit the down-sampling operation on the input image, so that as much information of the original LR image as possible can be kept.
- Second, we add residual blocks to the network to better capture the input information, while on the other hand, we remove the Batch Normalization (BN) layer from the residual blocks for the super-resolution task. Specifically, although batch normalization greatly improves the training process of deep networks, it will damage the origin information of the input image like contrast, which could be critical to the super-resolution task. For text images, specifically, BN may impair the original structure of text, resulting in degraded output of the generator network.
- Further, we introduce the Inception structure into the network, which effectively expands the width of the network and allows the generator network to adaptively capture text cues of different sizes, so as to generate more realistic output text images.

Discriminator Network. The discriminator network in the proposed cGAN model is basically a CNN classifier, which, as shown in Fig. 2, consists of following layers: Conv64LRelu, Conv128BNLReluDropout, Conv256BNLReluDropout, Conv1Sigmoid, where the number after Conv denotes the number of filters used by the convolutional layer. The first two convolutional layers use a stride of 2, while the rest use a stride of 1.

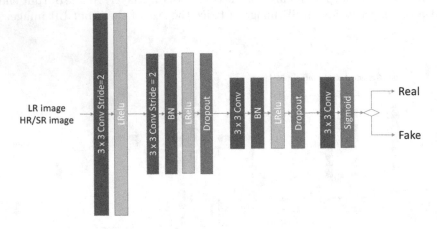

Fig. 2. The structure of the proposed discriminator network.

The input to the discriminator network is a pair of images: (1) Either the ground-truth HR image or the SR image generated by the generator network; (2) The LR image scaled to the same size as the HR or SR image. The discriminator network outputs the probability of the pair of images being matching. For the SR image generated by the generator network, a high probability indicates it's similar to the ground-truth HR image.

Note that, we add dropout layers to the network to avoid overfitting, and more importantly, to keep the discriminator network from becoming too strong, which may cause the training of the generator network to end up with failure because the generator network will find it cannot fool the discriminator any way.

3.3 Loss Function

Given the generator network G and the discriminator network D, the overall loss function of the proposed cGAN model is composed of two main loss components - the *adversarial loss* $\mathcal{L}_{cGAN}$ and the *content loss* $\mathcal{L}_{content}$, as follows:

$$\mathcal{L} = \mathcal{L}_{cGAN}(G, D) + \lambda \mathcal{L}_{content}(G) \tag{1}$$

Adversarial Loss. Adversarial loss measures the generator network's ability to fool the discriminator and the discriminator network's ability to distinguish between the generated and the real images, which is computed as follows:

$$\mathcal{L}_{cGAN}(G, D) = E_{I^{LR}, I^{HR}}[\log D(I^{LR}, I^{HR})] + E_{I^{LR}}[\log(1 - D(I^{LR}, G(I^{LR})))] \tag{2}$$

where, $D(I^{LR}, I^{HR}))$ denotes the probability that the discriminator network correctly identifies the correspondence between the LR image I^{LR} and the ground-truth HR image I^{HR}, $D(I^{LR}, G(I^{LR}))$ denotes the probability that the discriminator network considers the SR image generated by the generator network from the LR image I^{LR} to be a real HR image corresponding to I^{LR}.

Content Loss. Content loss measures the similarity between the generated SR image and the corresponding real HR image in terms of the image content. Two most common choices for the content loss are L_1 (MAE) loss and L_2 (MSE) loss. In this work, we adopt the L_1 loss defined by:

$$\mathcal{L}_{L_1}(G) = \frac{1}{WH} \sum_{x=1}^{W} \sum_{y=1}^{H} |I_{x,y}^{HR} - G(I^{LR})_{x,y}| \tag{3}$$

where, W and H are the width and height of the real HR image I^{HR} and the generated SR image $G(I^{LR})$, respectively.

Furthermore, we incorporate a perceptual loss [7] to the total content loss, which depicts high-level image features to help generate more realistic images and is computed based on the pretrained VGG16 model as follows:

$$\mathcal{L}_{vgg_{i,j}}(G) = \frac{1}{W_{i,j}H_{i,j}} \sum_{x=1}^{W_{i,j}} \sum_{y=1}^{H_{i,j}} |\phi_{i,j}(I_{x,y}^{HR}) - \phi_{i,j}(G(I^{LR})_{x,y})|^2 \tag{4}$$

where, $\phi_{i,j}$ denotes the feature map after the j-th convolution layer and before the i-th maxpooling layer. Specifically, in this work, we exploit $\mathcal{L}_{vgg_{2,2}}$ and $\mathcal{L}_{vgg_{4,3}}$ to compute the perceptual loss, in which, $\phi_{2,2}$ is used to capture low-level text features like strokes, while $\phi_{4,3}$ is used to capture higher level features of characters.

The total content loss of the proposed model is then expressed as:

$$\mathcal{L}_{content}(G) = \lambda_{L_1}\mathcal{L}_{L_1}(G) + \lambda_{vgg_{2,2}}\mathcal{L}_{vgg_{2,2}}(G) + \lambda_{vgg_{4,3}}\mathcal{L}_{vgg_{4,3}}(G) \tag{5}$$

We learn the proposed cGAN model on training LR and HR images by optimizing the following objective function:

$$\min_{G} \max_{D} \mathcal{L}_{cGAN}(G, D) + \lambda\mathcal{L}_{content}(G) \tag{6}$$

4   Experiments

4.1   Dataset

To evaluate the performance of the proposed text image super-resolution method, we adopted the public dataset of ICDAR 2015 Competition on Text Image Super-Resolution [13]. The dataset consists of 708 HD text images extracted from French TV video streams, whose size ranges from 116×36 to

1520×60. To generate the LR and HR pairs of images for the super-resolution task, the dataset downsamples the HD images by factor of 2 to create the ground-truth HR images, and by factor of 4 to create the corresponding LR images as inputs to one super-resolution method. Among the resulting 708 groups of images (each consisting of one LR, HR and original HD image respectively), the dataset uses 567 groups as the training set and the rest 141 groups as the testing set.

4.2   Evaluation Protocol

We adopt the evaluation protocol of the dataset [13], which includes two different categories of measures for the text image super-resolution task: the reconstruction measures and the OCR accuracy score.

The reconstruction measures include three common SR measures: Root Mean Square Error (RMSE), Peak Signal to Noise Ratio (PSNR) and Mean Structural Similarity (MSSIM) [20].

The OCR accuracy score OCR_{ac} is computed based on the OCR results on the super-resolution images by Tesseract OCR engine 3.02:

$$OCR_{ac} = 1 - \frac{1}{K} \sum_{i=1}^{N} (d_i)$$

where, K denotes the total number of characters in the test set, N denotes the total number of the test images, d_i is the Levenshtein (or edit) distance between the recognition result of the text in i-th test image and the ground truth, which is the minimum number of character addition/deletion/substitution required to change one text to the other.

4.3   Training of Network

To construct the training set of LR-HR pairs for the proposed cGAN model, we crop 8×8 patches from the training LR images, 16×16 patches from the training HR images and 32×32 patches from the training HD images at corresponding positions, which are chosen first regularly by scanning in the image with a stride of 8, 16 and 32 respectively, and then randomly in the image to acquire extra patches. Finally, a total of 30333 pairs of LR-HR patches and LR-HD pathes are extracted to train the proposed cGAN models for the super-resolution tasks with scaling factor 2 and 4 respectively.

We employ the ADAM optimizer and a learning rate of 0.0001 in the training of the proposed cGAN model for 20 epochs on a NVIDIA GeForce GTX 1070ti GPU. We set $\lambda_{L_1} = 100$, $\lambda_{vgg_{2,2}} = 0.00005$ and $\lambda_{vgg_{4,3}} = 0.0001$ in Eq. (5).

4.4   Evaluation of the Proposed Network Architecture

We investigate the effectiveness of the proposed modifications to the traditional cGAN model for the text image super-resolution task, specifically, the removal of the Batch Normalization layers and the introduction of the Inception structure.

Table 1 first compares the SR performances with and without the employment of Batch Normalization layers in the generator network, both without the Inception structure. The results show that, the proposed removal of Batch Normalization effectively enhances the performance of the SR model, leading to 2.04 decrease on RMSE, 1.53 increase on PSNR, 0.015 increase on MSSIM and 2.77% increase on OCR accuracy.

Table 1. Comparison of SR performances with and without Batch Normalization (BN) in the proposed model.

Method	RMSE	PSNR	MSSIM	$OCR_{ac}(\%)$
w. BN	13.25	26.21	0.941	72.07
w/o BN	**11.21**	**27.74**	**0.956**	**74.84**

Table 2 further compares the SR performances with and without the Inception structure in the generator network, both with Batch Normalization layers removed as proposed. It can be seen that, the proposed introduction of the Inception structure contributes to 0.37 decrease on RMSE, 0.34 increase on PSNR, 0.002 increase on MSSIM and 0.2% increase on OCR accuracy of the SR model.

Table 2. Comparison of SR performances with and without Inception structure in the proposed model.

Method	RMSE	PSNR	MSSIM	$OCR_{ac}(\%)$
w/o Inception	11.21	27.74	0.956	74.84
w. Inception	**10.84**	**28.08**	**0.958**	**75.04**

We further evaluate the performances of the proposed super-resolution method on the larger scaling factor ($\times 4$) of the LR-HD image pairs in the dataset, compared to the ($\times 2$) scaling factor of the LR-HR image pairs used elsewhere in the experiment. Correspondingly, we introduce an additional deconvolution layer and a leaky ReLU layer before the final deconvolution layer in the generator network shown in Fig. 1 so as to yield the output SR images of 4 times the sizes of the input LR images. Table 3 shows the corresponding SR results. It can be seen that, the larger scaling factor leads to the reduction of the SR image's quality, as it requires more missing details of the high resolution image to be reconstructed from the low resolution image by the model. On the other hand, however, larger scaling factors of the reconstructed images usually result in significant increases of the OCR accuracy, as the larger size of the resulting text, the generally easier for the OCR system to recognize it. Particularly, it seems that, the rough shape of imperfectly reconstructed text with bigger size

Table 3. Comparison of SR performances of the proposed model on scaling factors ×2 and ×4.

Scale	RMSE	PSNR	MSSIM	$OCR_{ac}(\%)$
×2	10.84	28.08	0.958	75.04
×4	12.56	26.85	0.908	78.18

is sometimes more favorable to the OCR system than the clear shape of well reconstructed text but with smaller size.

Figure 3 illustrates the effect of the proposed network architecture on improving the quality of the generated SR image. Specifically, we find that incorporating the BN layers introduces quite a bit blurring to the generated SR image (Fig. 3(b)) that several characters can hardly be figured out. Similarly, excluding the Inception structure also leads to certain degradation of the generated SR text (e.g. character 'E' and 'S' in Fig. 3(c)), compared to the result of the proposed method (Fig. 3(d)).

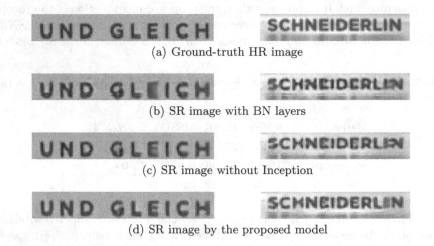

(a) Ground-truth HR image

(b) SR image with BN layers

(c) SR image without Inception

(d) SR image by the proposed model

Fig. 3. Illustration of the effect of Batch Normalization and Inception on text image super-resolution.

4.5   Comparison with State-of-the-Art Methods

We compare the performance of the proposed method with some state-of-the-art image super-resolution methods in Table 4. Note that, besides the result of the *single* proposed model, we also present the result of a *composite* model that selectively combines multiple instances of the proposed model in a similar (but simpler) way to that adopted by the SRCNN-1 and SRCNN-2 methods in

Table 4, which optimally combine 11 models of [2] with different structures or initialization parameters for the text image super-resolution task [13]. Specifically, to construct the composite model, we choose the best 11 proposed models that are trained with different parameter configurations and yield the highest OCR accuracies on the test set. Then, for every test LR image, we average the edit distances between the OCR result by one specific model to those of the rest models, and choose the model yielding the smallest average edit distance as the effective delegate of the composite model for the input test LR image, whose generated SR image is correspondingly taken as the final result of the composite model.

Table 4. Comparison of text image super-resolution performances of variant methods.

Method	RMSE	PSNR	MSSIM	$OCR_{ac}(\%)$
Synchromedia Lab [12]	62.67	12.66	0.623	65.93
SRGAN [11]	14.31	25.43	0.926	68.86
Zeyde [22]	13.05	27.21	0.941	69.72
ASRS [18]	12.86	26.98	0.950	71.25
A+ [17]	10.03	29.50	0.966	73.10
Orange Labs [14]	11.27	28.25	0.953	74.12
Proposed (single model)	10.84	28.08	0.958	75.04
SRCNN-2 [2]	**7.24**	**31.99**	**0.981**	76.10
SRCNN-1 [2]	7.52	31.75	0.980	77.19
Proposed (composite model)	11.27	27.76	0.955	**77.47**

As the results show, our composite model achieves the best OCR accuracy among all methods in comparison and competitive results on the reconstruction measures RMSE, PSNR and MSSIM. Note that, for the text super-resolution task addressed in this work, the OCR accuracy is usually considered a more representative evaluation metric as it focuses on the improvement of text quality by an SR method for text information extraction.

Moreover, the single proposed model achieves close OCR accuracy to those of the second best composite SRCNN models, but has significantly less model complexity and computational cost. For one single SRCNN model, on the other hand, although its network architecture is simpler than ours, the sizes of its convolution kernels (9×9 and 11×11) used in the task [13] are much larger than ours (3×3), resulting in the much higher number of network parameters than our model, as well as the relatively lower computation efficiency of the single SRCNN model (0.12 s per image) than the proposed single model (0.094 s per image) in our experiment.

Table 4 also shows that, the more recent SRGAN model [11] performs not very well on the text image super-resolution task, possibly because it's proposed to generate SR images with finer texture details which are relatively less important and significant for text.

Figure 4 shows some super-resolution results by the proposed method on some text image samples of the dataset, which demonstrate its effectiveness and robustness to text with varied appearances.

(a) LR image   (b) Generated SR image   (c) Ground-truth HR image

Fig. 4. Examples of text image super-resolution results by the proposed method.

5   Conclusion

In this paper, we present a novel cGAN-based super-resolution method for text images. Our method improves the original cGAN network by removing the Batch Normalization layers and introducing the Inception structure, which significantly enhances the network's performance on text image super-resolution. Experiment results on the public dataset demonstrate the effectiveness of the proposed method.

Acknowledgments. Research supported by the Natural Science Foundation of Jiangsu Province of China under Grant No. BK20171345 and the National Natural Science Foundation of China under Grant Nos. 61003113, 61321491, 61672273.

References

1. Chang, H., Yeung, D.Y., Xiong, Y.: Super-resolution through neighbor embedding. In: CVPR 2004, pp. 275–282 (2004)

2. Dong, C., Loy, C.C., He, K., Tang, X.: Learning a deep convolutional network for image super-resolution. In: Fleet, D., Pajdla, T., Schiele, B., Tuytelaars, T. (eds.) ECCV 2014. LNCS, vol. 8692, pp. 184–199. Springer, Cham (2014). https://doi.org/10.1007/978-3-319-10593-2_13
3. Freeman, W.T., Jones, T.R., Pasztor, E.C.: Example-Based Super-Resolution. IEEE Computer Society Press, Washington, D.C. (2002)
4. Goodfellow, I.J., et al.: Generative adversarial nets. In: NIPS. pp. 2672–2680 (2014)
5. He, K., Zhang, X., Ren, S., Sun, J.: Deep residual learning for image recognition. In: CVPR 2016, pp. 770–778 (2016)
6. Isola, P., Zhu, J.Y., Zhou, T., Efros, A.A.: Image-to-image translation with conditional adversarial networks. In: CVPR 2017, pp. 5967–5976 (2017)
7. Johnson, J., Alahi, A., Fei-Fei, L.: Perceptual losses for real-time style transfer and super-resolution. In: Leibe, B., Matas, J., Sebe, N., Welling, M. (eds.) ECCV 2016. LNCS, vol. 9906, pp. 694–711. Springer, Cham (2016). https://doi.org/10.1007/978-3-319-46475-6_43
8. Kim, J., Lee, J.K., Lee, K.M.: Accurate image super-resolution using very deep convolutional networks. In: CVPR 2016, pp. 1646–1654 (2016)
9. Kim, J., Lee, J.K., Lee, K.M.: Deeply-recursive convolutional network for image super-resolution. In: CVPR 2016, pp. 1637–1645 (2016)
10. Kim, K.I., Kwon, Y.: Single-image super-resolution using sparse regression and natural image prior. IEEE Trans. PAMI 32(6), 1127–33 (2010)
11. Ledig, C., et al.: Photo-realistic single image super-resolution using a generative adversarial network. In: CVPR 2017, pp. 105–114 (2017)
12. Moghaddam, R.F., Cheriet, M.: A multi-scale framework for adaptive binarization of degraded document images. Pattern Recognit. 43(6), 2186–2198 (2010)
13. Peyrard, C., Baccouche, M., Mamalet, F., Garcia, C.: ICDAR2015 competition on text image super-resolution. In: ICDAR 2015, pp. 1201–1205 (2015)
14. Peyrard, C., Mamalet, F., Garcia, C.: A comparison between multi-layer perceptrons and convolutional neural networks for text image super-resolution. In: VISAPP, pp. 84–91 (2016)
15. Schulter, S., Leistner, C., Bischof, H.: Fast and accurate image upscaling with super-resolution forests. In: CVPR 2015, pp. 3791–3799 (2015)
16. Szegedy, C., et al.: Going deeper with convolutions. In: CVPR 2015, pp. 1–9 (2015)
17. Timofte, R., De Smet, V., Van Gool, L.: A+: adjusted anchored neighborhood regression for fast super-resolution. In: Cremers, D., Reid, I., Saito, H., Yang, M.-H. (eds.) ACCV 2014. LNCS, vol. 9006, pp. 111–126. Springer, Cham (2015). https://doi.org/10.1007/978-3-319-16817-3_8
18. Walha, R., Drira, F., Lebourgeois, F., Garcia, C., Alimi, A.M.: Resolution enhancement of textual images via multiple coupled dictionaries and adaptive sparse representation selection. Int. J. Doc. Anal. Recognit. 18(1), 87–107 (2015)
19. Wang, Z., Liu, D., Yang, J., Han, W., Huang, T.: Deep networks for image super-resolution with sparse prior. In: ICCV 2015, pp. 370–378 (2015)
20. Wang, Z., Bovik, A.C., Sheikh, H.R., Simoncelli, E.P.: Image quality assessment: from error visibility to structural similarity. IEEE Trans. Image Process. 13(4), 600–612 (2004)
21. Yang, C.Y., Yang, M.H.: Fast direct super-resolution by simple functions. In: ICCV 2013, pp. 561–568 (2013)
22. Zeyde, R., Elad, M., Protter, M.: On single image scale-up using sparse-representations. In: Boissonnat, J.-D., et al. (eds.) Curves and Surfaces 2010. LNCS, vol. 6920, pp. 711–730. Springer, Heidelberg (2012). https://doi.org/10.1007/978-3-642-27413-8_47

Latitude-Based Visual Attention
in 360-Degree Video Display

Huiwen Huang, Yiwen Xu, Jinling Chen, Yanjie Song, and Tiesong Zhao[✉]

College of Physics and Information Engineering, Fuzhou University, Fuzhou, China
t.zhao@fzu.edu.cn

Abstract. With the ever-growing availability of Virtual Reality (VR) products, the 360-degree video has a widespread application prospect, which also brings new challenges to its quality assessment. Considering its unconventional display environment, the users' attention on 360-degree video may be significantly different from its conventional 2D version. Recently, the Weighted-to-Spherically-uniform Peak Signal-to-Noise Ratio (WS-PSNR) was developed to weight PSNR by its covered area in 360-degree video display. In this work, we find that the average attention per area still differs subject to its latitude. Thus we propose to utilize this average attention to formulate a latitude-based quality weight matrix. Through experiments, this new matrix outperforms that of WS-PSNR in pooling of PSNR when latitude-based quality fluctuations exist in 360-degree video display.

Keywords: 360-degree video · Omnidirectional video
Video quality assessment · Visual attention model

1 Introduction

An omnidirectional video, or a 360-degree video allows the consumers to choose their desired viewport by moving heads as they do in the real world [1], in order to have truly immersive feelings. In such a case, the people's perception is still crucial in assessment of video qualities. Correspondingly, there has been a need to develop Video Quality Assessment (VQA) approaches that estimate the average human opinion on 360-video display.

Although there have been numerous researches on VQA of conventional videos [2–4], the subjective and objective VQAs of 360-degree videos have not been well exploited. For subjective VQAs, there are few methods for measuring 360-degree videos quality. Xu *et al.* [5] presented a procedure of subjective test in measuring quality of panoramic videos, including overall and vectorized Differential Mean Opinion Score (DMOS) metrics between different groups of subjects. Experimental results verified that the subjective VQA method was

This research is supported by the National Natural Science Foundation of China (Grant 61671152).

R. Hong et al. (Eds.): PCM 2018, LNCS 11166, pp. 282–290, 2018.
https://doi.org/10.1007/978-3-030-00764-5_26

effective in measuring subjective quality of panoramic videos. Schatz *et al.* [6] developed an approach towards subjective Quality of Experience (QoE) assessment for omnidirectional video streaming on the QoE impact of stalling. However, such subjective methods are time-consuming, uneconomical and hardly to be embedded into real-time system. Therefore, it is desirable to develop objective evaluation methods. For objective VQAs, some works take into account spherical characteristic of 360-degree videos. Sun *et al.* [7] presented a Weighted-to-Spherically-uniform Peak Signal-to-Noise Ratio (WS-PSNR) quality evaluation method which provided different weights for pixel distortions at different locations to realize uniform weights on spherical surface. In [8], a Spherical-PSNR (S-PSNR) quality evaluation metric was proposed by Yu *et al*, which calculated distortions based on a series of uniformly sampled point set on spherical surface. In [9], Zakharchenko *et al.* developed the Craster-Parabolic-Projection-PSNR (CPP-PSNR) quality assessment method, which employed craster parabolic projection to measure resampled pixels' distortions.

To date, WS-PSNR has been utilized as a typical quality metric in 360-degree video coding by Joint Video Exploration Team (JVET) [10]. It is a straightforward approach that weights PSNR by its covered spherical area in 360-degree video display. Nevertheless, it may not apply well in 360-degree VQA due to more degrees of freedom to choose view points. Moreover, it does not seem to take human visual characteristics into account. In order to solve these problems, in this paper, we build a 360-degree video database to study the influence of latitude on the users' average attention. Through it, we derive a latitude-based visual attention map as a quality weight matrix in 360-degree VQA. Primary testing in our database demonstrates the performance of our model in weighting PSNR values, which is superior to the convectional WS-PSNR.

In the following, Sect. 2 presents our subjective database of 360-degree videos. In Sect. 3, we derive the latitude-based visual attention map. The holistic quality metrics are proposed in Sect. 4. Finally, Sect. 5 concludes this paper.

2   360-Degree Video Database

In the experiments we utilize fourteen equirectangular projected 360-degree videos. All of those videos (in YUV 4:2:0 format at resolution 2160 × 1200) are provided from Letin VR Digital Technology Co., Ltd [11] and Audio Video coding Standard (AVS) Workgroup of China. Figure 1 shows one frame of each video in our database. The HM software of High Efficiency Video Coding (HEVC) [12] is used to compress these sequences with three Quantization parameter (Qp) values of 35, 40 and 45. To observe the impacts of latitudes, these Qp values are assigned to 9 different bands under equirectangular projection, respectively. Figure 2 illustrates the variation of the latitudes on the 9 different bands from −90 to +90. We respectively introduce encoding distortions to these bands whilst keeping the overall objective distortion almost intact. The audio tracks are discarded to avoid the impact of acoustic information. Thus, there are in total 14 original sequences and 378 impaired sequences (14 sequences × 3 Qps × 9 bands).

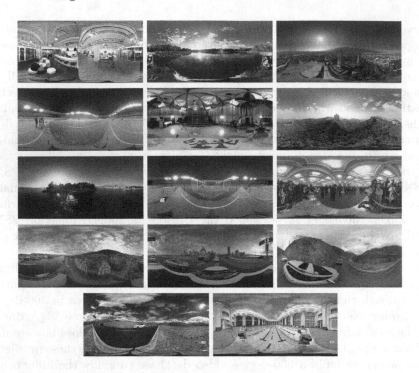

Fig. 1. Dataset samples. From top-left to bottom-right: AVS sequences (*Talk; Lake; The Pagoda; Sport; Performance; The Great Wall; Vessel; Shoot; Auditorium*); JVET sequences (*Aerial City; Driving in City; Driving in Country; HighWay; NataTorium*).

Fig. 2. Nine different bands with equivalent arc length on the sequence.

We use the HTC Vive as the Head Mount Display (HMD) and the Vive Cinema (VC) software as the 360-degree video player. In total, 25 non-expert student subjects (11 males and 14 females) aged from 21 to 25 participate in the experiment. All subjects have a visual acuity above 1.0 and pass the color vision test using the Ishihara eye-chart. To ensure all regions of 360-degree videos are accessible, 25 subjects are allowed to turn around freely. Subjects might have novelty reaction when they are in touch with 360-degree videos in HMD for the first time, thus we expose subjects to a training session with five videos at the beginning of the test in order to establish a stable assessment.

The testing procedure is set up according to the recommendation in ITU-R BT.500-13 [13]. In the formal test, the subjects are asked to orally report the scores of 360-degree sequences. This is to avoid the visual discomfort when they alternately switch between video watching and handwriting. Besides, to avoid visual boredom and fatigue during the test, there is a 15 min break after viewing each session of 30 sequences and is a 5 s mid-grey screen between two sequences. In each session, all 30 sequences are played at a random order, without two successively presented sequences with the same content.

We use the Double Stimulus Impairment Scalend (DSIS) method and each sequence is shown only once. The subjects rate the sequences on a discrete five-level impairment scale ranging from 5=imperceptible to 1=very annoying after watching the sequences [13]. In order to check whether a chosen sample size is sufficient to cause a stable result, the researchers often use "data saturation" [14] as a guiding principle. In our test, the "saturation" occurs in 16 subjects, although a reasonably high degree of consistency has already reached with 12 subjects. It demonstrates that our chosen number of 25 subjects is fairly sufficient to yield stable data.

3   Latitude-Based Visual Attention Model

After collecting all subjective data, we generally follow the rejection criteria in ITU-R BT.1788 [15] to pick out the 22 valid subjects. From the scores of valid subjects, we find that each subject has lower ratings when the same distortion is introduced to lower latitude bands. This implies that the visual attention of subjects become lower from low latitude region to high latitude region. The overall quality tendency of the 360-degree videos with different latitude distortions can be seen in Fig. 3(a), which averaged over all subjects and all video contents. However, we can also observe that the scores of subjects are different for diverse Qp videos. Thus, Qps still have an effect on the curve.

Based on the above analyses, we propose a latitude-based visual attention model that shapes the average quality of 360-degree videos. We formulate this visual attention as the influence of objective loss at different regions on the overall subjective quality (*i.e.*, DMOS):

$$D_S = \frac{\text{DMOS}/\Delta D}{\max\left(\text{DMOS}/\Delta D\right)} = \frac{\text{DMOS}/\Delta Q}{\max\left(\text{DMOS}/\Delta Q\right)}, \tag{1}$$

where D represents the distortion and Q denotes the average quantization step-size within a distorted band. This equation holds because the existing of linear D-Q model, $D = \gamma Q$ in [16]. In the above equation, the visual attention is also normalized by its maximum value at all latitudes. Figure 3(b) shows the visual

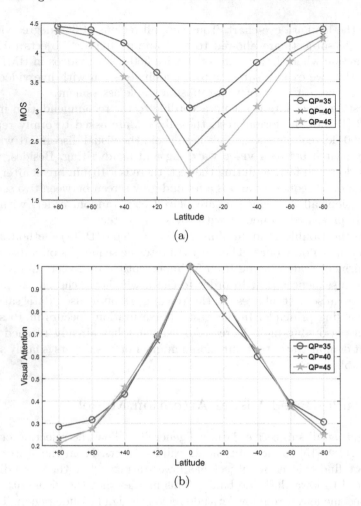

Fig. 3. Latitude-based visual attention. (a) The average MOS at different latitudes. (b) The normalized visual attention.

attention obtained by (1). We observe that the above curves are basically similar for different Qps, thus the influence of Qp can be eliminated. This fully expresses the influence of latitude on perceived distortion and helps to build a quality evaluation model based on latitude.

In order to derive the model for predicting 360-degree video visual attention, we use the hold-out method to randomly select 2/3 samples for training from AVS dataset and other 1/3 for testing. The above processing carry out five times. Eventually, the proposed model is given by:

$$D_S(i,j) = a \times \cos\left[\frac{2\pi}{N} \times \left(j - \frac{N-1}{2}\right)\right] + (1-a), \qquad (2)$$

Table 1. Validation of the proposed visual attention model.

Experiment group	Training	Testing			
	a	PLCC	KRCC	SRCC	RMSE
1	0.4111	0.9413	0.8718	0.9559	0.1247
2	0.4081	0.8915	0.8441	0.9414	0.1597
3	0.4133	0.8488	0.7012	0.8322	0.1757
4	0.4285	0.8193	0.6992	0.8131	0.1773
5	0.4155	0.8849	0.7926	0.9103	0.1482
Average	**0.4153**	**0.8772**	**0.7818**	**0.8906**	**0.1571**

where N denotes the width of 360-degree video and every pixel (i, j) in frame could be calculated as a value. The set of all values corresponds to a weight matrix.

Table 1 shows the values of the parameters obtained from each training group along with their average values. The performance of the proposed measure is evaluated by the Pearson's Linear Correlation Coefficient (PLCC), the Spearman Rank-order Correlation Coefficient (SRCC), the Kendall Rank-order Correlation Coefficient (KRCC) and Root Mean Square Error (RMSE) between predicted scores and MOSs in testing group. We can find that the parameter a is kept almost the same in each training-testing process. The average correlation between our attention model and the users' attentions is very high and our model has a smaller RMSE value. Thus it can be concluded that our attention model achieves a fairly good performance. Meanwhile, to further investigate the validity of the model, we present the Fig. 4 the assessed attention model are highly correlated with the MOSs. In addition, it should be pointed out these correlations are higher than most of correlations between human scores and the MOS. In summary, the proposed attention model is fairly enough to illustrate the characters of human vision.

4   An Application of Our Attention Model

In this section, we utilize the above visual attention model in the weighting of local quality measures and observe its performance. Nowadays, WS-PSNR has also utilized this weighting approach, but with a simple cosine function, as shown in Fig. 5. However, as shown in our above test, the straightforward weighting of WS-PSNR does not fit the visual attention model. This inspires us to change the weighting matrix to our model for an improved VQA performance.

To examine the performance of our method, we use validation group data from JVET dataset to verify the performance of the proposed measure by the PLCC, SRCC, KRCC and RMSE between predicted scores and MOSs. The result is shown in Table 2, where AC, DC, NA, HW and DT represent the

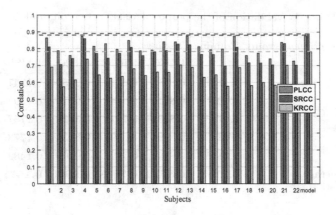

Fig. 4. Comparison between our attention model and the subjects' ratings.

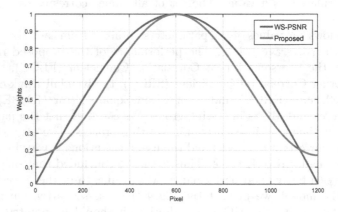

Fig. 5. Quality weighted model of WS-PSNR and the proposed model.

sequences *Aerial City, Driving in City, Driving in Country, HighWay* and *Nata-Torium* in Fig. 1, respectively. For comparison, to further demonstrate the efficiency of the proposed metric, we conduct the comparison experiments by using some existing VQA metrics, including PSNR and WS-PSNR.

From Table 2 we get two conclusions. Firstly, by introducing surface-based weights, WS-PSNR achieves better performance than conventional PSNR, which shows the efficiency of quality weighting to obtain a holistic quality measure. Secondly, by incorporating our latitude-based visual attention, the performance of the proposed measure is better than WS-PSNR in our dataset. Therefore, the efficiency of our latitude-based visual attention model can be justified in 360-degree quality assessment.

Table 2. Comparison between quality assessment metrics.

Validation set		AC	DC	NA	HW	DT	Average
PSNR	PLCC	0.8229	0.7280	0.3739	0.2189	0.7549	**0.5797**
	KRCC	0.6840	0.4234	0.2957	0.2597	0.4719	**0.4270**
	SRCC	0.8332	0.5623	0.3819	0.3694	0.6437	**0.5581**
	RMSE	0.2224	0.1883	0.1664	0.1901	0.2205	**0.1975**
WS-PSNR	PLCC	0.8820	0.7571	0.6294	0.4634	0.8793	**0.7222**
	KRCC	0.7994	0.4526	0.4580	0.4531	0.6043	**0.5535**
	SRCC	0.9225	0.6028	0.6182	0.5846	0.7926	**0.7041**
	RMSE	0.2122	0.1807	0.1557	0.1756	0.2122	**0.1872**
Proposed	PLCC	0.8833	0.7647	0.6801	0.5072	0.8936	**0.7458**
	KRCC	0.7994	0.4526	0.4986	0.4820	0.6158	**0.5697**
	SRCC	0.9280	0.6028	0.6757	0.6020	0.8097	**0.7237**
	RMSE	0.2117	0.1796	0.1543	0.1739	0.2110	**0.1861**

5   Conclusion

This work presents a new dataset of 378 omnidirectional videos for 25 subjects who were asked to explore the videos, as naturally as possible in the subjective experiment. It also proposes a latitude-based visual attention model for 360-degree videos. Simulation results have shown its high correlations to subjective scores. In addition, we apply this model in pooling of PSNR metrics. The obtained holistic measure outperforms WS-PSNR, a widely used 360-degree video quality measurement. A large database will be created in the future that could help to obtain more accurate model parameters and support quality evaluation of all projection formats.

References

1. Chen, Z., Li, Y., Zhang, Y.: Recent advances in omnidirectional video coding for virtual reality: projection and evaluation. Sig. Process. **146**, 66–78 (2018)
2. Wang, Z., Lu, L., Bovik, A.C.: Video quality assessment based on structural distortion measurement. Sig. Process. Image Commun. **19**(2), 121–132 (2004)
3. Chikkerur, S., Sundaram, V., Reisslein, M., Karam, L.J.: Objective video quality assessment methods: a classification, review, and performance comparison. IEEE Trans. Broadcast. **57**(2), 165–182 (2011)
4. Lin, W., Jay Kuo, C.C.: Perceptual visual quality metrics: a survey. J. Vis. Commun. Image Represent. **22**(4), 297–312 (2011)
5. Xu, M., Li, C., Liu, Y., Deng, X., Lu, J.: A subjective visual quality assessment method of panoramic videos. In: Proceedings of the IEEE International Conference on Multimedia and Expo (ICME), pp. 517–522. IEEE (2017)

6. Schatz, R., Sackl, A., Timmerer, C., Gardlo, B.: Towards subjective quality of experience assessment for omnidirectional video streaming. In: Proceedings of the 2017 Ninth International Conference on Quality of Multimedia Experience (QoMEX), pp. 1–6. IEEE (2017)

7. Sun, Y., Lu, A., Yu, L.: Weighted-to-spherically-uniform quality evaluation for omnidirectional video. IEEE Sig. Process. Lett. **24**(9), 1408–1412 (2017)

8. Yu, M., Lakshman, H., Girod, B.: A framework to evaluate omnidirectional video coding schemes. In: IEEE International Symposium on Mixed and Augmented Reality, pp. 31–36. IEEE (2015)

9. Zakharchenko, V., Choi, K. P., Park, J. H.: Quality metric for spherical panoramic video. In: Optics and Photonics for Information Processing X, vol. 9970, pp. 99700–99700-9 (2016)

10. Boyce, J., Alshina, E., Abbas, A.: JVET common test conditions and evaluation procedures for 360 video. In: Joint Video Exploration Team of ITU-T SG 16 WP 3 and ISO/IEC JTC 1/SC 29/WG 11, JVET-E1030, Geneva (2017)

11. Sun, W., Guo, R.: Test sequences for virtual reality video coding from LetinVR. In: Joint Video Exploration Team of ITU-T SG 16 WP 3 and ISO/IEC JTC 1/SC 29/WG 11, JVET-D0179, Chengdu, China (2016)

12. High Efficiency Video Coding (HEVC) reference software HM. Fraunhofer Institute for Telecommunications, Heinrich Hertz Institute (2016). https://hevc.hhi.fraunhofer.de

13. ITU-R BT.500-13: Methodology for the subjective assessment of the quality of television pictures. International Telecommunication Union, Geneva, Switzerland (2012)

14. Zhang, W., Liu, H.: Towards a reliable collection of eye-tracking data for image quality research: challenges, solutions and applications. IEEE Trans. Image Process. **26**(5), 2424–2437 (2017)

15. ITU-R BT.1788: Methodology for the subjective assessment of video quality in multimedia applications, International Telecommunication Union (2007)

16. Wang, H., Kwong, S.: Rate-discortion optimization of rate control for H.264 with adaptive initial quantization parameter determination. IEEE Trans. Circ. Syst. Video Technol. **18**(1), 140–144 (2008)

Branched Convolutional Neural Networks
for Face Alignment

Meilu Zhu[1], Daming Shi[1(✉)], Songkui Chen[1], and Junbin Gao[2]

[1] College of Computer Science and Software Engineering, Shenzhen University,
Shenzhen 518060, People's Republic of China
`dshi@szu.edu.cn`, {`zhumeilu2016,chensongkui2017`}`@email.szu.edu.cn`
[2] The University of Sydney Business School, The University of Sydney, Sydney,
NSW 2006, Australia
`junbin.gao@sydney.edu.au`

Abstract. Face alignment is to localize multiple facial landmarks for a
given facial image. While convolutional neural networks based face alignment methods have achieved superior performance in recent years, the
problem remains unresolved due to the fact that L2 loss function suffers
from imbalance errors of different facial components caused by region-
related changes in pose, illumination and occlusions. In this situation, the
L2 loss function will be dominated by errors from those facial components on which the landmarks are hard predicted. To alleviate this issue,
in this paper, we propose a facial landmarks detection method based
on branched convolutional neural networks, which consists of the shared
layers and component-aware branches. The proposed model first captures more dependencies among facial components through the shared
layers. Then, by virtue of the knowledge of component-aware branches,
different sub-networks can effectively detect the facial landmarks of different components. A series of empirical experiments are carried out on
a benchmark dataset across different facial variations. Compared with
the existing state-of-the-arts, our proposed method not only achieves the
robustness with respect to normal face and occlusion, but also effectively
improves the performance of detecting landmarks on corresponding facial
components.

Keywords: Branched convolutional neural networks
Component-aware mechanism · Face alignment

1 Introduction

Face alignment aims to detect landmarks on different facial components, such
as eyes, mouth and facial contour, which is one of the key techniques for many
face analysis tasks, e.g., face recognition, age prediction and 3D face modeling.
Therefore, it has been a hotspot in computer vision filed over the past decades.

In general, existing methods for the face alignment task mainly rely on three
categories of approaches: template based methods, cascaded regression based

© Springer Nature Switzerland AG 2018
R. Hong et al. (Eds.): PCM 2018, LNCS 11166, pp. 291–302, 2018.
https://doi.org/10.1007/978-3-030-00764-5_27

methods, and deep learning methods. The template based methods, such as Active Contour Model (known as Snakes) [9], Active Shape Model (ASM) [5], and Active Appearance Model (AAM) [4], minimize model parameter errors to learn a parametric model. This way is indirect and sub-optimal because smaller parameter errors are not necessarily equivalent to smaller alignment errors.

In recent years, cascaded regression based methods have achieved better performance than template based methods, learning a cascade of regressors using shape-indexed features to iteratively update the shape estimation stage by stage. For example, Dollar et al. [6] cascaded pose regression in a process of progressively refining a loosely specified initial guess. Cao et al. [3] proposed an explicit shape regression approach which was a two-level boosted regression, shape indexed features and a correlation-based feature selection method. Zhang et al. [20] presented a two-layer shape regression framework that can be integrated with regression-based methods. However, these approaches are not end-to-end learning methods.

As a kind of competitive method, deep learning based methods have also achieved superior performance. Sun et al. [16] first cascaded three-level convolutional neural networks to predict the positions of facial landmarks. However, this model is not appropriate in predicting a large number of landmarks. Zhang et al. [19] also utilized a coarse-to-fine auto-encoder networks to detect facial landmarks, which cascaded a few successive stacked auto-encoder networks. Considering task-constraint can improve prediction performance, Zhang et al. [21] proposed a convolutional neural networks to learn simultaneously multi-tasks.

In this paper, we propose a novel face alignment method via branched convolutional neural networks, which consists of the shared layers part and the components-aware branches part. In our proposed networks architecture, different branches are utilized to predict independently landmarks on different facial components (i.e. eyes, eyebrows, nose, mouth, and contour). To summarize, the main contributions of this work are highlighted as following:

(i) We propose a novel end-to-end CNN framework which is incorporated into the component-aware branches knowledge to extract discriminative features for predicting landmarks of different facial components, respectively;

(ii) We exploit a multi-branches architecture to alleviate the imbalance error problem among facial components during training;

(iii) The model shows the robustness against partial occlusions by means of our proposed component-aware branches;

(iv) Our proposed model achieves state-of-the-art results on 300-W dataset.

The rest of this paper is organized as follows. Section 2 describes the architecture of our proposed face alignment framework and the optimization procedure in detail. In Section 3, we report a series of experiments to validate our contributions of the proposed method on a benchmark dataset. Finally, the paper is closed with the conclusion and future work in Section 4.

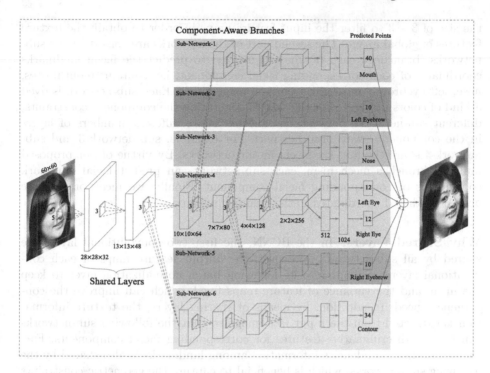

Fig. 1. Branched convolutional neural networks based face landmarks detection. The networks consists of two parts: the shared layers part and the component-aware branches part. For the component-aware branches part, each branch (sub-network) is dyed a kind of colors. (Color figure online)

2   Proposed Method

In this paper, we propose a novel face alignment method via branched convolutional networks. Figure 1 illustrates our scheme of Branched Convolutional Neural Networks (BCNN). BCNN first utilizes the shared layers to capture the component constraints, and then takes advantage of different branches (sub-networks) to predict the landmarks on different facial components. We describe each part of the networks and the learning procedure of the model in detail below.

2.1   Architecture Description

As shown in Fig. 1, the networks consist of two parts: the shared layers part and the component-aware branches part. The networks first extract the texture features from the input facial image based on the first two convolutional layers. In the first convolution layer, 32 convolutional kernels with the size of 5×5 are utilized to filter the input face image, aiming to capture more dependencies among facial components. Following that, we use 48 convolutional kernels with the size of 3×3 to filter the input feature maps in order to obtain the texture features of global image. The second part of the networks are consists of six sub-networks (branches), each of which is utilized to predict the facial landmarks coordinates of each corresponding facial component, i.e. contour, mouth, eyes, nose, left eyebrows, and right eyebrows, respectively. Each sub-network is dyed a kind of colors in Fig. 1. For the sake of generating the component constraints, different sub-networks between one another share different numbers of layers in the component-aware branches part. For example, sub-network-3 and sub-network-4 share the first six layers in our networks. By virtue of our proposed multiple branches mechanism, each sub-network can predict facial landmark coordinates of corresponding facial component. Finally, we incorporate these output predicted coordinates as the final facial shape.

Why Shared Layers. In the BCNN, the first two convolutional layers are shared by all sub-networks. Moreover, the output feature maps of each con-volutional layer are input into its following batch normalization layer to keep the mean and the variance of feature maps fixed, which can improve the con-vergence speed. In addition, by means of sharing layers, the texture informa-tion is extracted from the input face image so that the following sub-networks can extract discriminative features for corresponding facial components. Fur-thermore, the dependencies information among landmarks is also shared by the following sub-networks, which is beneficial to capture the geometric constraints for each component and contributes to reduce the time and space cost.

Why Component-Aware Branches. For the task of facial landmarks detec-tion, the degree of difficulties can change with regard to predict correspond-ing facial landmarks on different facial components. This phenomenon can be attributed to two main reasons, i.e. inherent difference and external impact, respectively. First, the texture information for different facial components is inherently different. For example, it is widely believed that the texture informa-tion around eyes is more discriminative to locate the landmarks easily, while that around facial contour is less. Second, the facial landmarks on some facial com-ponents may be hard detected due to changes in pose, illumination, expression and occlusion. For instance, eyebrows may be occluded when someone wears a pair of glasses, which results in the facial landmarks on eyebrows would be difficulty detected in comparison to those on other facial landmarks without occlusion. These two effects will lead to imbalanced weight for different facial components. In other words, when all components share a set of hyper parame-ters, the L2 loss is dominated by errors from those facial components on which the landmarks are hard predicted. This imbalance phenomenon causes those predicted-easy components to run into overfitting. To alleviate this issue, we incorporate the component-aware branches knowledge into our networks. After performing a number of experiments on a single-pipeline network, we approx-imately rank the degree of prediction difficulties on facial components in the way of observing the convergence curve of each component. Therefore, based on

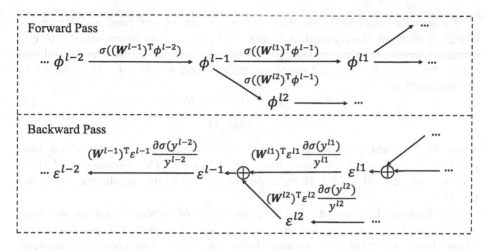

Fig. 2. The forward propagation pass and backward propagation pass of the proposed branched structure.

the rank of difficulties, we design six sub-networks to form different branches for different facial components' landmark detection. Instead of devoting most of capacity of a single-pipeline network to fitting the "predicting-difficult" component, those more predicted-difficult components (e.g. facial contour) share less parameters with others, which enables each branch to extract the discriminative features for corresponding facial landmarks in each sub-network. In return, all sub-networks also promote the shared layers to learn the effective facial features via a component-aware way.

2.2 Optimization Procedure

The dominating representation of a facial shape is the use of facial landmark. Mathematically, a facial shape in an image I can be denoted as a facial landmark vector $S = [x_1, y_1, ..., x_k, y_k, ..., x_K, y_K]$, where x and y are the coordinates of the k-th facial landmark and K denotes the number of the facial key-points. A deep regression network maps an image I into its corresponding face shape S, which can be formulated as:

$$S = \mathcal{G}(I, W),$$ (1)

where $\mathcal{G}(\cdot)$ is a nonlinear function. We can learn the parameters W by applying the gradient-based optimization algorithm to minimize the following objective:

$$E = \sum_{i=1}^{N} \|\hat{S}_i - \mathcal{G}(I_i, W)\|^2 + \lambda \|W\|^2,$$ (2)

where $\hat{S}$ is the ground-truth of the corresponding image. $\|W\|^2$ is the regularization term, controlling the complexity of the model, and λ is the pre-specified parameter.

To demonstrate clearly the learning procedure of our proposed structure, in Fig. 2, we show the forward and backward propagation processes of multiple branches architecture where the $(l-1)$-th layer has one input branch and two output branches. For forward propagation, a response of the $(l-1)$-th layer can be computed as:

$$y^{l-1} = (\boldsymbol{W}^{l-1})^{\mathrm{T}}\phi^{l-2} + \boldsymbol{b}^{l-1}, \tag{3}$$

$$\phi^{l-1} = \sigma(y^{l-1}), \tag{4}$$

where $\boldsymbol{W}^{l-1}$ is the parameters of the $(l-1)$-th convolutional layer whose output feature maps ϕ^{l-1} are input into later branches. The parameters of these two branches are $\boldsymbol{W}^{l1}$ and $\boldsymbol{W}^{l2}$, respectively. $\sigma(\cdot)$ is the non-linear activation function.

For backward propagation, the parameters of unshared layers are only updated by the corresponding sub-network loss. However, the parameters of shared layers are related to multiple facial component branches. For example, for the $(l-1)$-th layers in Fig. 2 , the error of two branches are backward propagated together:

$$\varepsilon^{l-1} = (\boldsymbol{W}^{l1})^{\mathrm{T}}\varepsilon^{l1}\frac{\partial\sigma(y^{l1})}{y^{l1}} + (\boldsymbol{W}^{l2})^{\mathrm{T}}\varepsilon^{l2}\frac{\partial\sigma(y^{l2})}{y^{l2}}, \tag{5}$$

where $\frac{\partial\sigma(y^l)}{y^l}$ denotes the gradient of activation function of l-th layer. The error ε^{l1} also is related to later two branches.

3   Experiments

3.1   Experimental Setting

To verify the efficiency of the proposed BCNN, we evaluate our method on a benchmark dataset 300-W [14] which is a collection of 3,837 faces from existing datasets: LFPW, AFW, HELEN and IBUG. Each face is densely annotated with 68 landmarks. We use 3,148 images as training samples and 689 images as test samples from 300-W dataset. Specifically, these test images are split into three subsets: (i) the challenging subset (135 images from IBUG); (ii) the common subset (554 images, including 224 images from LFPW test set and 330 images from HELEN test set); (iii) the full set (689 images, containing all of test images).

In our experiments, the face bounding box of each image is first detected and cropped from the original image, and then each cropped image is resized to 60×60. Besides, the pixels in the bounding box are normalized to the range of [0,1]. The value of x and y coordinates of the landmarks are mapped into the range of [0,1]. To improve generalization ability, we exploit rotation, translation and flip operators to conduct data augmentation for training set.

For performance evaluation, the normalized root mean squared error (NRMSE) is used to measure the error between the predicted positions and the ground-truth, which is defined as follows:

$$\text{NRMSE} = \frac{1}{N} \sum_{i=1}^{N} \frac{\| \, \boldsymbol{S}_i - \hat{\boldsymbol{S}}_i \, \|^2}{K D_i}, \tag{6}$$

where D denotes the distance between the outer corners of left and right eye. K represents the number of facial key-points and N is the number of test images.

In our experiments, a mini-batch stochastic gradient descent algorithm is used to train the networks. The learning rate is initialized to 0.01 and the momentum of the model is set to 0.9. The size of the mini-batch is set to 64. The network can converge well after about 3×10^5 iterations. All experiments are conducted on Nvidia's Tesla K80 platform using Caffe [8].

3.2   The Effectiveness Evaluation of Component-Aware Branches

In this subsection, we validate the effectiveness of our proposed component-aware branches in a simple case, i.e. test on normal face images. All of testing images are obtained from HELEN, LFPW and AFW datasets whose face images have less changes under pose, illumination and occlusion.

First of all, we test the overall performance of our proposed method on HELEN and LFPW datasets. Table 1 shows the experimental results in comparison with the existing benchmarks, including SDM [18], CFSS [22], RCPR [2], R-DSSD [17], and TCDCN [21].

Table 1. The comparison of different methods on HELEN and LFPW datasets.

Method	LFPW (68-pts)	HELEN (68-pts)
SDM [18]	5.67	5.50
CFSS [22]	4.87	4.63
RCPR [2]	6.56	5.93
R-DSSD [17]	4.52	**4.08**
TCDCN [21]	4.57	4.60
BCNN	**4.22**	4.19

As seen in Table 1, our method can detect the facial key-point effectively on this two datasets. Our proposed BCNN is significantly better than SDM, CFSS, and RCPR that are cascaded regression methods combined with handcrafted features. In other words, handcrafted features from these three methods are less discriminative than features learned from our proposed networks. Compared with CNN-based methods R-DSSD and TCDCN, BCNN outperforms them on the whole, although the NRMSE value of BCNN on testing HELEN dataset is slightly lower than that of R-DSSD. It is noting that these two CNN-based methods do not consider the issue of imbalance error during backward propagation. Therefore, we can conclude that our proposed is able to extract more discriminative features to detect the facial landmarks than above other methods.

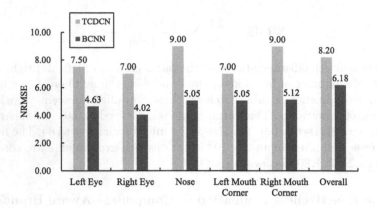

Fig. 3. The comparison of different facial components on AFW dataset.

In order to further verify the effectiveness of our proposed component-aware branches mechanism, we evaluate the performance of different landmarks on corresponding facial components when testing on AFW dataset. We compare our method with TCDCN, which is a state-the-of-art single-pipeline convolutional network based face alignment method. As shown in Fig. 3, our method achieves better performance than TCDCN with regard to all facial inner components, especially the nose and the right mouth corner. In addition, the NRMSE value of overall facial landmarks in our method is also superior to that of TCDCN. These results conclude that our proposed component-aware branches mechanism can effectively improve the performance in comparison with single-pipeline network.

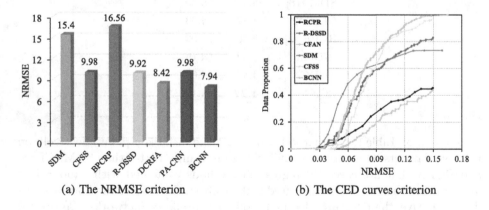

(a) The NRMSE criterion   (b) The CED curves criterion

Fig. 4. Comparison of different models on the IBUG dataset.

3.3   The Robustness Test of Proposed Method Against Occlusion

In this subsection, in order to verify the robustness of our proposed method against occlusion, we evaluate the detection performance on IBUG dataset. IBUG dataset is a extremely challenging dataset, which contains many faces under large variations in pose, expression and partial occlusion. We compare our method with state-of-the-art works, including SDM [18], CFSS [22], BPCRP [15], R-DSSD [11], DCRFA [10] and PA-CNN [7]. Figure 4 shows the comparison results of two kinds of evaluation criteria.

From Fig. 4(a), it can be observed that our proposed method achieves the lowest NRMSE value in all methods. It is noticed that R-DSSD was proposed against occlusion. While R-DSSD can solve the occlusion problem effectively and exclusively, our proposed method has better robustness against occlusion. In addition, PA-CNN utilizes two sub-networks to localize contour landmarks and inner landmarks, respectively, which can improve the performance of inner landmarks. However, it do not consider meticulously the constrained relationship among all facial components. On the contrary, our proposed component-aware branches mechanism is able to relieve the constrains and promote the overall performance under occlusion. We also utilize another evaluation metric to measure the effectiveness of the proposed method, i.e. cumulative error distribution curve. The experimental results are summarized in Fig. 4(b), which illustrates that our method outperforms the other methods. Based on these evaluation metrics, we can conclude that our method is appropriate for the face alignment task under both normal and occlusions.

3.4   Comparison with the State-of-the-Arts on 300-W Dataset

In the previous subsections, we have validated the effectiveness of proposed component-aware branches in both normal and occlusions. In this subsection, we compare our method with three categories of state-of-the-arts. The categories include: template based methods (TSPM [12] and DRMF [1]), cascaded regression based methods (RCPR [2], SDM [18] and LBF [13]), and deep learning based methods (TCDCN [21], DCRFA [10] and R-DSSD [17]). Table 2 provides all of methods testing on different subsets from 300-W dataset.

From Table 2, we can observe that our method outperforms other methods except R-DSSD is slightly better than ours when tested on Common Set. In addition, the results on Challenging Set prove that the branches knowledge incorporated into our networks can effectively under occlusion.

3.5   Facial Landmarks Localization Results on 300-W Dataset

Some examples of the alignment results from our method are shown in Fig. 5. The five columns in Fig. 5 respectively show the alignment results on 300-W dataset under normal conditions, occlusions, expressions, and varying poses, and some failing results. The localization results of the first four columns in Fig. 5 performs very well. However, due to the extremely large variations in facial appearance, the results of the last column in Fig. 5 is inferior.

Table 2. Comparison with the state-of-the-art methods on 300-W dataset.

Method	Common set (68-pts)	Challenging set (68-pts)	Full set (68-pts)
TSPM [12]	8.22	18.33	10.20
DRMF [1]	6.65	19.79	9.22
RCPR [2]	6.18	17.26	8.35
SDM [18]	5.57	15.40	7.50
LBF [13]	4.95	11.98	6.32
TCDCN [21]	4.80	8.60	5.54
DCRFA [10]	4.19	8.42	5.02
R-DSSD [17]	**4.16**	9.20	5.59
BCNN	4.20	**7.94**	**4.93**

Fig. 5. Some examples of the localization results from our proposed method on 300-W dataset.

4 Conclusion and Future Work

In this paper, we proposed a novel face alignment method based on branched convolutional neural networks. In the proposed method, the component-aware branches knowledge is incorporated into our networks, which is able to alleviate the adverse influence of imbalance errors among facial components during training. In the experiments, we test the performance of our method on 300-W dataset when the face images under various conditions, which indicates that our method can detect the facial landmarks in both normal and occlusions. In the future, we plan to move our networks forward the video-based dense landmark detection.

Acknowledgments. This work is supported by Shenzhen Science and Technology Innovation Commission (SZSTI) project (No. JCYJ20170302153752613).

References

1. Asthana, A., Zafeiriou, S., Cheng, S., Pantic, M.: Robust discriminative response map fitting with constrained local models. In: IEEE Conference on Computer Vision and Pattern Recognition, pp. 3444–3451 (2013)
2. Burgos-Artizzu, X.P., Perona, P.: Robust face landmark estimation under occlusion. In: IEEE International Conference on Computer Vision, pp. 1513–1520 (2013)
3. Cao, X., Wei, Y., Wen, F., Sun, J.: Face alignment by explicit shape regression. Int. J. Comput. Vis. **107**(2), 177–190 (2014)
4. Cootes, T.F., Edwards, G.J., Taylor, C.J.: Active appearance models. In: European Conference on Computer Vision, pp. 484–498 (1998)
5. Cootes, T.F., Taylor, C.J., Cooper, D.H., Graham, J.: Active shape models-their training and application. Comput. Vis. Image Underst. **61**(1), 38–59 (1995)
6. Dollar, P., Welinder, P., Perona, P.: Cascaded pose regression. In: IEEE Conference on Computer Vision and Pattern Recognition, pp. 1078–1085 (2010)
7. He, K., Xue, X.: Facial landmark localization by part-aware deep convolutional network. In: Chen, E., Gong, Y., Tie, Y. (eds.) PCM 2016. LNCS, vol. 9916, pp. 22–31. Springer, Cham (2016). https://doi.org/10.1007/978-3-319-48890-5_3
8. Jia, Y., et al:: Caffe: convolutional architecture for fast feature embedding. In: ACM International Conference on Multimedia, pp. 675–678. ACM (2014)
9. Kass, M., Witkin, A.P., Terzopoulos, D.: Snakes: active contour models. Int. J. Comput. Vis. **1**(4), 321–331 (1988)
10. Lai, H., et al.: Deep recurrent regression for facial landmark detection. IEEE Trans. Circuits Syst. Video Technol. (2016)
11. Liu, H., Lu, J., Feng, J., Zhou, J.: Learning deep sharable and structural detectors for face alignment. IEEE Trans. Image Process. **26**(4), 1666–1678 (2017)
12. Ramanan, D.: Face detection, pose estimation, and landmark localization in the wild. In: IEEE Conference on Computer Vision and Pattern Recognition, pp. 2879–2886 (2012)
13. Ren, S., Cao, X., Wei, Y., Sun, J.: Face alignment via regressing local binary features. IEEE Trans. Image Process. **25**(3), 1233 (2016)
14. Sagonas, C., Tzimiropoulos, G., Zafeiriou, S., Pantic, M.: 300 faces in-the-wild challenge: the first facial landmark localization challenge. In: IEEE International Conference on Computer Vision Workshops, pp. 397–403 (2014)

15. Sun, P., Min, J.K., Xiong, G.: Globally tuned cascade pose regression via back propagation with application in 2D face pose estimation and heart segmentation in 3D CT images. In: IEEE International Conference on Computer and Information Technology Workshops, pp. 462–467 (2015)
16. Sun, Y., Wang, X., Tang, X.: Deep convolutional network cascade for facial point detection. In: IEEE Conference on Computer Vision and Pattern Recognition, pp. 3476–3483 (2013)
17. Xing, J., Niu, Z., Huang, J., Hu, W., Xi, Z., Yan, S.: Towards robust and accurate multi-view and partially-occluded face alignment. IEEE Trans. Pattern Anal. Mach. Intell. **PP**(99), 1 (2017)
18. Xiong, X., Torre, F.D.L.: Supervised descent method and its applications to face alignment. In: IEEE Conference on Computer Vision and Pattern Recognition, pp. 532–539 (2013)
19. Zhang, J., Shan, S., Kan, M., Chen, X.: Coarse-to-fine auto-encoder networks (CFAN) for real-time face alignment. In: Fleet, D., Pajdla, T., Schiele, B., Tuytelaars, T. (eds.) ECCV 2014. LNCS, vol. 8690, pp. 1–16. Springer, Cham (2014). https://doi.org/10.1007/978-3-319-10605-2_1
20. Zhang, Q., Zhang, L.: Face alignment with two-layer shape regression. In: Ho, Y.-S., Sang, J., Ro, Y.M., Kim, J., Wu, F. (eds.) PCM 2015. LNCS, vol. 9314, pp. 125–134. Springer, Cham (2015). https://doi.org/10.1007/978-3-319-24075-6_13
21. Zhang, Z., Luo, P., Loy, C.C., Tang, X.: Learning deep representation for face alignment with auxiliary attributes. IEEE Trans. Pattern Anal. Mach. Intell. **38**(5), 918–930 (2016)
22. Zhu, S., Li, C., Chen, C.L., Tang, X.: Face alignment by coarse-to-fine shape searching. In: IEEE Conference on Computer Vision and Pattern Recognition, pp. 4998–5006 (2015)

A Robust Approach for Scene Text Detection and Tracking in Video

Yang Wang, Lan Wang, and Feng Su[✉]

State Key Laboratory for Novel Software Technology, Nanjing University,
Nanjing 210023, China
suf@nju.edu.cn

Abstract. The detection of scene text in videos is of great value in various content-based video applications such as video analysis and retrieval. In this paper, we present a robust scene text detection and tracking method for videos. We first propose an effective deep neural network model for detecting text in individual video frames, which enhances the EAST model by introducing deconvolution layers and inception modules. We then present a correlation filter based tracking algorithm for text in the video and further combine detection and tracking results, which effectively enhances the final video text detection performance. The proposed method outperforms other state-of-the-art methods in experiments on public scene text video datasets.

Keywords: Scene text detection · Tracking · Video · Inception Correlation filter

1 Introduction

Scene text appearing in the video is often an important source of semantic information about the content of the video, and is of great value in many content-based video applications such as video indexing, retrieval and analysis.

Existing methods for detecting scene text in videos can be roughly divided into two major categories - individual frame based and multiple frames based [21]. Individual frame based methods usually exploit the text detectors designed for static images to localize text in every individual video frame, and according to how text candidates are localized, can be further categorized into three groups: connected component based [3], region based [17] and deep neural network (DNN) based [14,18,24].

Multi-frame based text detection methods [7,13,19,23] further exploit the correlations between text appearing in consecutive frames to rectify and complement intra-frame detection results, so as to enhance the final holistic detection performance. Due to the complexity and variety of both scene text and video context, however, the reliable detection of scene text in videos is still a very challenging task.

© Springer Nature Switzerland AG 2018
R. Hong et al. (Eds.): PCM 2018, LNCS 11166, pp. 303–314, 2018.
https://doi.org/10.1007/978-3-030-00764-5_28

In this paper, we propose a robust detection and tracking method for scene text in the video. The key contributions of the proposed method are summarized as follows:

- We propose an effective deep neural network model for text detection, which enhances the original EAST model [24] by introducing deconvolution layers and inception modules for better capturing widely varied text appearances.
- We propose to combine a robust correlation filter based tracking algorithm with the detection model to further improve the holistic video text detection performance by discarding transient false detections and recovering text with temporarily unfavorable detection conditions.
- The proposed method outperforms state-of-the-art methods on public scene text video datasets in the experiment.

The rest of the paper is organized as follows. Section 2 introduces some related work on text detection in videos. Section 3 describes the deep neural network based detection model for text in individual video frames. Section 4 describes the correlation filter based tracking algorithm for text in the video and the combination mechanism of tracking and detection results. Section 5 presents the experimental results on two public scene text video datasets.

2   Related Work

In recent years, a large number of methods have been proposed to detect scene text in videos [21], which can be divided into two major categories: individual frame based and multiple frames based.

Individual-frame text detection methods generally exploit text detectors designed for static images to localize text in every individual frame, which can further be categorized into three groups: connected component based, region based and deep neural network (DNN) based. Connected component based methods exploit connected component analysis such as Stroke Width Transform (SWT) [3] and Maximally Stable Extremal Regions (MSERs) [11] to extract character candidates. Region based methods [17] extract various features by shifting a multi-scale window on the image and employ certain machine learning methods such as SVM and AdaBoost to determine one window region corresponding to text or not.

Text detection methods based on deep neural networks such as convolutional neural network (CNN) and recurrent neural network (RNN) have recently been widely explored [14,18,24] and have yielded very competitive results. For example, Tian et al. [18] proposed a novel Connectionist Text Proposal Network to localize text in the image, which detects a sequence of fixed-width text proposals in the convolutional feature maps and then connectes sequential proposals by a RNN to yield the final detected text line. Zhou et al. [24] proposed a deep Fully Convolutional Network (FCN) based pipeline that abandons unnecessary intermediate components and steps in traditional DNN based detection frameworks and directly targets the final goal of text detection. Shi et al. [14] proposed an oriented text detection method that detects both the segments of word or text line and the links between them simultaneously with a FCN model and combines segments connected by links to produce the final detection result.

On the basis of individual-frame text detection techniques, multi-frame text detection methods further make use of the temporal correlations of text in successive frames by techniques such as tracking [13,19] and spatial-temporal analysis [7,23] to enhance the holistic video text detection performance. For example, Yang et al. [19] utilized MSERs to construct text region candidates that are filtered by a CNN model, and then tracked these regions by a tracking algorithm, finally developed a dynamic programming algorithm for integrating detected and tracked text region to acquire final detection results. Minetto et al. [13] proposed a particle filtering framework based tracking algorithm for text in videos along with an effective text detection algorithm, and employed the Hungarian algorithm to merge tracked regions and new detections. Khare et al. [7] explored temporal information to separate text and background, then analyzed gradient directions of pixels to identify potential text candidates, finally expanded the boundary of potential text candidates to obtain final detection results in videos.

3   Detection of Text Candidates

We propose an effective neural network model for extracting text candidates from one video frame, which modifies and enhances the network architecture proposed in the EAST text detection method [24].

3.1   Network Architecture

The architecture of the proposed text detection network is illustrated in Fig. 1, which consists of three main parts: feature extraction stem, feature merging branch and prediction module.

Feature Extraction Stem. The feature extraction stem of our model aims to obtain the feature representations of the input image. We adopt the ResNet-50 [4] network pretrained on ImageNet [2] for the feature extraction stem, which differs from the PVANet [9] used in the EAST method, for its stronger representability.

As shown in Fig. 1, the feature extraction stem interleaves convolution and pooling layers, yielding four levels of feature maps denoted as $f_{i=1,2,3,4}$, whose sizes are $\frac{1}{32}$, $\frac{1}{16}$, $\frac{1}{8}$, $\frac{1}{4}$ of the input image, respectively.

Feature Merging Branch. The feature merging branch combines location information from lower layers and semantic information in higher layers of the feature extraction stem. Starting from the last feature map generated by the *stem*, the feature merging branch gradually merges the feature maps extracted by the *stem* with the feature maps deconvoluted from previous layer.

There are two main differences between our model and the original EAST model [24]:

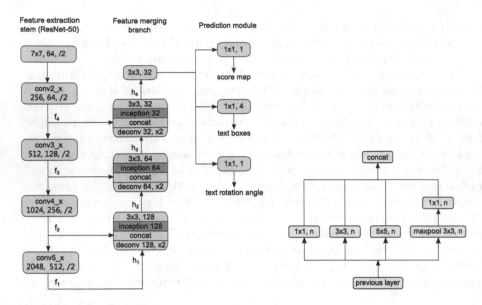

Fig. 1. The architecture of the proposed deep neural network for text detection.

Fig. 2. The inception module. n denotes the number of kernels used in convolution layer and max-pooling layer.

1. We propose to perform *deconvolution* on the feature map instead of the unpool operation in the original EAST network [24], which allows the network to learn more flexible transformations of feature maps.
2. We propose to insert an *inception* module before the last convolution layer of every merging stage as shown in Fig. 1, which increases the width of the network and enables multiple receptive fields of different sizes so as to effectively integrate features of difference scales.

Specifically, as shown in Fig. 1, we take the last feature maps (denoted by f_1) of the feature extraction stem as the input (denoted by h_1) to a deconvolution layer to double their size, which are then concatenated with the current feature maps (denoted by f_2) extracted by the *stem*. We then propagate the merged feature maps through an inception module as shown in Fig. 2, which convolves the feature maps simultaneously with three kernels of 1×1, 3×3, 5×5 sizes respectively. On the other hand, the inception module also applies a max pooling with 3×3 kernel on the input feature maps, followed by a 1×1 convolution. Finally, the inception module concatenates all generated feature maps as the output, which are further convolved with a 3×3 kernel to fuse information from multiple channels. The resulting feature maps (denoted by h_2) are then taken as the input to the next stage. We further propagate similarly the feature maps h_i and f_{i+1} through the feature merging branch for the rest stages $i = 2, 3$. At the end of the last stage of the feature merging branch, we perform a 3×3

convolution on the output feature maps (i.e. h_4) and feed them to the prediction module.

Prediction Module. The prediction module contains 1 channel for the score map F_s, 5 channels for the geometry maps F_g, with 4 channels containing the parameters of the axis-aligned bounding box $\mathbf{R}$ of the potential text, and 1 channel encoding the rotation angle θ. We adopt the same label scheme as the EAST method [24] to convert the quadrangle bounding box of a text to corresponding values in each prediction channel. More details can be found in [24].

Given a video frame fed to the proposed detection model, which outputs a bunch of word or text line candidates on the prediction maps, we first apply thresholding on the score map, discarding candidates with lower scores, and then perform Non-Maximum Suppression (NMS) to obtain the detection results for the frame.

3.2   Loss Functions

The total loss L of the proposed text detection model is formulated as a combination of the loss L_s on the score map and the loss L_g on the geometry maps:

$$L = L_s + \lambda_g L_g \tag{1}$$

where, λ_g balances the importance between two losses and is set to 1.0 in our experiment.

Loss for Score Map. Different to the EAST method, we use the dice coefficient introduced in [12] to calculate the classification loss as follows:

$$L_s = 1 - \frac{2\sum_i^N \hat{\mathbf{Y}}_i \mathbf{Y}_i^*}{\sum_i^N \hat{\mathbf{Y}}_i + \sum_i^N \mathbf{Y}_i^*} \tag{2}$$

where, $\hat{\mathbf{Y}}$ and $\mathbf{Y}^*$ denote the prediction values and the ground truth on the score map respectively, N denotes the total number of the score map's elements.

Loss for Geometry Map. We adopt the geometry loss L_g presented in [24], which is formulated as the weighted sum of a boundary loss and a rotation angle loss as follows:

$$L_g = -log\text{IoU}(\hat{\mathbf{R}}, \mathbf{R}^*) + \lambda_\theta(1 - cos(\hat{\theta} - \theta^*)) \tag{3}$$

where, $\text{IoU}(\hat{\mathbf{R}}, \mathbf{R}^*)$ denotes the IoU loss in [22], which is the ratio of intersection area to union area between the predicted boundary $\hat{\mathbf{R}}$ and the ground truth boundary $\mathbf{R}^*$. $\hat{\theta}$ and θ^* denote the predicted and the ground truth rotation angle, respectively. λ_θ is the weight balancing the boundary loss and the rotation angle loss, which is set to 20 in our experiment.

3.3 Training of Network

To augment the training data, we randomly sample 512×512 corps from the training images to form minibatches of size 14. We then train the proposed network with the ADAM optimizer, whose learning rate starts from 10^{-4} and stops at 10^{-5} with a decay of 0.94 every 10000 steps. When the network performance no longer improves, the training is terminated.

4 Text Tracking in Video

Due to motion blur, varied lighting conditions and complex scene context, the text in a video is sometimes difficult to be detected precisely exploiting information in one video frame solely, resulting in a number of false alarms and missed text candidates. Therefore, we adopt a robust tracking algorithm for text in the video to exploit temporal correlation of text cues across frames to improve text detection performance on the basis of the intra-frame detection.

4.1 Tracking by Correlation Filter

The basic idea behind correlation filter based tracking is to dynamically learn a robust filter capturing the characteristic of the target object in the previous frame, which is then convolved over possible regions in the current frame where the target object may appear. The location with highest response will then be regarded as the tracking result of the current frame.

In this work, we exploit a robust correlation filter based tracking method - the Staple [1] tracker for tracking text in the video. It combines HOG and color histogram descriptions of image patches for learning the correlation filter, which is robust to color changes and deformations while keeping the real-time processing speed.

To initialize the Staple tracker, which is essentially a single-object tracking method, we inspect each text candidate d detected in the current frame. If it overlaps with an existing tracked text candidate t by Staple, we merge d with t, otherwise we feed d to Staple so as to start a new tracking trajectory for d.

4.2 Combination of Tracking and Detection

We propose an effective combination algorithm for detected and tracked text candidates, which improves both the accuracy and completeness of the detection results.

First, we eliminate the transient false text candidates in the detection results based on matching between the tracked and the detected candidates. Specifically, for a tracked text candidate t, we look for its matched detected peer d ($IoU(t, d) > 0.8$ in this work) in every frame in t's trajectory maintained by the tracker. For a frame that such d can be found, it's considered as a *matched* frame, otherwise it's denoted as a *unmatched* frame. If we encounter 3 successive unmatched frames (starting from frame f^*) during the tracking of a text candidate t, we terminate the tracking and end t's trajectory right before f^*. We then compute the proportion of the matched frames to all frames in the trajectory. If this proportion does not exceed 0.7, the trajectory is regarded as invalid and we discard all tracked text candidates and matched detected candidates in the trajectory.

We then merge tracked text candidates with detected ones in every frame. For each detected text candidate d, we compute the overlapping ratio between d and every tracked text candidate t in the same frame, which is defined as the proportion of the overlapped area to that of the smaller candidate. Denoting the tracked candidate with the maximum overlapping ratio r_o as t^*, if r_o is greater than a threshold $\mathcal{T}_o$ (0.8 in this work), we take t^* as the merged candidate. Otherwise, if r_o is greater than 0, d is taken as the merged candidate. Finally, we take the merged candidates in every video frame as the final detection results.

5 Experiments

5.1 Dataset

We adopt the public scene text video dataset of ICDAR 2013 Robust Reading Competition Challenge 3: Text in Videos [6] and the dataset presented in [13] for evaluating the video text detection performance of the proposed method and comparing it with several recent state-of-the-art methods.

The ICDAR 2013 dataset consists of 28 video sequences - 13 videos for training and the rest for testing, obtained from real-life situations including outdoor and indoor scenes. One video sequence typically lasts from 10 seconds to 1 minute with the image size ranging from 307×93 to 3888×2592 pixels.

The dataset [13] consists of 5 video sequences of different outdoor scenes, lasting from around 6 seconds to 41 seconds with an uniform image size of 640×480 pixels. As no specific training set is provided in dataset [13], we use a total of 1229 training images from ICDAR 2013 [6] and ICDAR 2015 [5] Robust Reading Competition as the training set for this dataset.

5.2 Evaluation Protocol

We adopt the standard evaluation protocol for text detection, which measures the detection performance by precision p, recall r, and f measure. The precision rate p is evaluated as the percentage of text regions correctly detected within all text regions detected, and the recall rate r is defined as the ratio of text regions correctly detected to the ground truth text regions. The f measure is the harmonic mean value of p and r (i.e. $\frac{2*p*r}{p+r}$).

5.3 Evaluation of Text Detection

To verify the effectiveness of the proposed enhancements to the EAST model [24], which is trained in our experiment on the same datasets in Sect. 5.1 as used by the proposed model, we compare the individual-frame text detection performances with and without introducing the deconvolution layers and inception modules into the detection network architecture in Table 1.

The result shows, the proposed introduction of the deconvolution layers and inception modules into the detection model effectively enhances its detection performance on individual video frames. For example, the deconvolution layers and the inception modules respectively increase the f-measure by 1.71% and 1.09% on dataset [13], and 1.65% and 1.82% on ICDAR 2013 dataset, on the basis of the EAST model. By combining both components, the proposed model achieves overall increases on all performance metrics (2.21%/4.05% on f-measure, 2.10%/4.51% on precision, 2.40%/3.72% on recall) on dataset [13] and ICDAR 2013 dataset.

Table 1. Comparison of individual-frame detection performances with and without introduction of the deconvolution layers and inception modules into the detection network on dataset [13] and ICDAR 2013 dataset.

Method	Dataset [13]			ICDAR 2013 dataset		
	Precision	Recall	F-measure	Precision	Recall	F-measure
EAST	74.76	75.89	75.02	64.13	53.22	56.44
EAST + deconvolution	76.82	77.36	76.73	65.57	55.09	58.09
EAST + inception	75.21	78.02	76.11	66.86	54.62	58.26
EAST + both (**Proposed**)	**76.86**	**78.29**	**77.23**	**68.64**	**56.94**	**60.49**

5.4 Evaluation of Text Tracking

To inspect the effect of the proposed text tracking measures on improving the text detection result, we compare the final detection performances with and without exploiting the tracking measures in Table 2. Specifically, the latter result is obtained by performing individual-frame detection on video frames as presented in the previous section.

Table 2. Comparison of detection performances with and without text tracking on dataset [13] and ICDAR 2013 dataset.

Method	Dataset [13]			ICDAR 2013 dataset		
	Precision	Recall	F-measure	Precision	Recall	F-measure
w/o tracking	76.86	78.29	77.23	68.64	56.94	60.49
w. tracking (**Proposed**)	**83.03**	**84.22**	**83.30**	**71.90**	**58.67**	**62.65**

The result shows, compared to the individual-frame detection scheme, the proposed approach combining tracking with detection achieves significant

Fig. 3. Illustration of the effect of text tracking on video text detection.

Table 3. Comparison of video text detection performances on dataset [13] and ICDAR 2013 dataset (%).

Method	Precision	Recall	F-measure
Dataset [13]			
Proposed	83.03	**84.22**	**83.30**
Yang *et al.* [19]	**85**	77	81
Zuo *et al.* [25]	84	68	75
Minetto *et al.* [13]	61	60	63
ICDAR 2013 dataset			
Proposed	**71.90**	**58.67**	**62.65**
Khare *et al.* [7]	57.91	55.9	51.7
Yin *et al.* [20]	48.62	54.73	51.56
Shivakumara *et al.* [16]	51.15	53.71	50.67
Shivakumara *et al.* [15]	51.10	50.10	50.59
Zhao *et al.* [23]	47.02	46.30	46.65
Khare *et al.* [8]	41.4	47.6	44.3
Liu *et al.* [10]	44.60	38.91	41.62
Epshtein *et al.* [3]	39.80	32.53	35.94

increases on all performance metrics on both datasets, by discarding transient false text candidates and helping recover missed text candidates with temporarily unfavorable detection conditions via propagating proper detection results across consecutive frames.

Fig. 4. Examples of video text detection results by the proposed method.

Figure 3 further illustrates how the proposed text tracking measures help improve the individual-frame detection results. In example (a), the temporary false text candidate detected in the upper-right part of the image can be effectively eliminated based on its over-short tracking trajectory. In example (b), on the other hand, the incomplete bounding box of the text acquired by the detector as a result of the temporary local highlight on the text can be recovered by combining tracking results from previous frames without such interference.

5.5   Comparison with State-of-the-Art Methods

We compare the performance of the proposed method with some state-of-the-art video text detection methods in Table 3.

Our method achieves the highest f-measure and recall among all methods being compared on both datasets. Specifically, on the more complicated ICDAR 2013 dataset with largely varied text qualities, the proposed method significantly precedes the second best method on all performance metrics - 13.99%

on precision, 2.77% on recall and 10.95% on f-measure, which demonstrate the effectiveness of the proposed detection and tracking method for scene text in videos.

Figure 4 shows some scene text detection results in videos by the proposed method, which exhibits sufficient robustness to largely varied appearances and qualities of video text.

6 Conclusion

In this paper, we propose a novel deep neural network model for detecting scene text in video frames, which enhances the EAST model by introducing effective deconvolution layers and inception modules. We then propose to combine a robust correlation filter based tracking method with the detection model, which effectively further enhances the holistic text detection performance on videos.

Acknowledgments. Research supported by the Natural Science Foundation of Jiangsu Province of China under Grant No. BK20171345 and the National Natural Science Foundation of China under Grant Nos. 61003113, 61321491, 61672273.

References

1. Bertinetto, L., Valmadre, J., Golodetz, S., Miksik, O., Torr, P.H.S.: Staple: complementary learners for real-time tracking. In: 2016 IEEE Conference on Computer Vision and Pattern Recognition (CVPR), pp. 1401–1409 (2016)
2. Deng, J., Dong, W., Socher, R., Li, L., Li, K., Li, F.: Imagenet: a large-scale hierarchical image database. In: 2009 IEEE Computer Society Conference on Computer Vision and Pattern Recognition (CVPR), pp. 248–255 (2009)
3. Epshtein, B., Ofek, E., Wexler, Y.: Detecting text in natural scenes with stroke width transform. In: 2010 IEEE Computer Society Conference on Computer Vision and Pattern Recognition (CVPR), pp. 2963–2970 (2010)
4. He, K., Zhang, X., Ren, S., Sun, J.: Deep residual learning for image recognition. In: 2016 IEEE Conference on Computer Vision and Pattern Recognition (CVPR), pp. 770–778 (2016)
5. ICDAR 2015 robust reading competition. http://rrc.cvc.uab.es/
6. Karatzas, D., et al.: ICDAR 2013 robust reading competition. In: 2013 12th International Conference on Document Analysis and Recognition, pp. 1484–1493 (2013)
7. Khare, V., Shivakumara, P., Paramesran, R., Blumenstein, M.: Arbitrarily-oriented multi-lingual text detection in video. Multimedia Tools Appl. **76**(15), 16625–16655 (2017)
8. Khare, V., Shivakumara, P., Raveendran, P.: A new histogram oriented moments descriptor for multi-oriented moving text detection in video. Expert Syst. Appl. **42**(21), 7627–7640 (2015)
9. Kim, K., Cheon, Y., Hong, S., Roh, B., Park, M.: PVANET: deep but lightweight neural networks for real-time object detection. CoRR **abs/1608.08021** (2016)
10. Liu, C., Wang, C., Dai, R.: Text detection in images based on unsupervised classification of edge-based features. In: Eighth International Conference on Document Analysis and Recognition (ICDAR), vol. 2, pp. 610–614 (2005)

11. Matas, J., Chum, O., Urban, M., Pajdla, T.: Robust wide-baseline stereo from maximally stable extremal regions. Image Vis. Comput. **22**(10), 761–767 (2004)

12. Milletari, F., Navab, N., Ahmadi, S.A.: V-net: fully convolutional neural networks for volumetric medical image segmentation. In: 2016 Fourth International Conference on 3D Vision (3DV), pp. 565–571 (2016)

13. Minetto, R., Thome, N., Cord, M., Leite, N.J., Stolfi, J.: Snoopertrack: text detection and tracking for outdoor videos. In: 2011 18th IEEE International Conference on Image Processing, pp. 505–508 (2011)

14. Shi, B., Bai, X., Belongie, S.: Detecting oriented text in natural images by linking segments. In: 2017 IEEE Conference on Computer Vision and Pattern Recognition (CVPR), pp. 3482–3490 (2017)

15. Shivakumara, P., Phan, T.Q., Tan, C.L.: New fourier-statistical features in rgb space for video text detection. IEEE Trans. Circ. Syst. Video Technol. **20**(11), 1520–1532 (2010)

16. Shivakumara, P., Sreedhar, R.P., Phan, T.Q., Lu, S., Tan, C.L.: Multioriented video scene text detection through bayesian classification and boundary growing. IEEE Trans. Circ. Syst. Video Technol. **22**(8), 1227–1235 (2012)

17. Tian, S., Pan, Y., Huang, C., Lu, S., Yu, K., Tan, C.L.: Text flow: a unified text detection system in natural scene images. In: 2015 IEEE International Conference on Computer Vision (ICCV), pp. 4651–4659 (2015)

18. Tian, Z., Huang, W., He, T., He, P., Qiao, Y.: Detecting text in natural image with connectionist text proposal network. In: Leibe, B., Matas, J., Sebe, N., Welling, M. (eds.) ECCV 2016. LNCS, vol. 9912, pp. 56–72. Springer, Cham (2016). https://doi.org/10.1007/978-3-319-46484-8_4

19. Yang, C., et al.: Tracking based multi-orientation scene text detection: a unified framework with dynamic programming. IEEE Trans. Image Process. **26**(7), 3235–3248 (2017)

20. Yin, X.C., Yin, X., Huang, K., Hao, H.W.: Robust text detection in natural scene images. IEEE Trans. Pattern Anal. Mach. Intell. **36**(5), 970–983 (2014)

21. Yin, X.C., Zuo, Z.Y., Tian, S., Liu, C.L.: Text detection, tracking and recognition in video: a comprehensive survey. IEEE Trans. Image Process. **25**(6), 2752–2773 (2016)

22. Yu, J., Jiang, Y., Wang, Z., Cao, Z., Huang, T.: Unitbox: an advanced object detection network. In: 2016 ACM Conference on Multimedia, MM 2016, pp. 516–520. ACM, New York (2016)

23. Zhao, X., Lin, K.H., Fu, Y., Hu, Y., Liu, Y., Huang, T.S.: Text from corners: a novel approach to detect text and caption in videos. IEEE Trans. Image Process. **20**(3), 790–799 (2011)

24. Zhou, X., et al.: EAST: an efficient and accurate scene text detector. In: 2017 IEEE Conference on Computer Vision and Pattern Recognition (CVPR), pp. 2642–2651 (2017)

25. Zuo, Z.Y., Tian, S., Pei, W., Yin, X.C.: Multi-strategy tracking based text detection in scene videos. In: 2015 13th International Conference on Document Analysis and Recognition (ICDAR), pp. 66–70 (2015)

Improving Intra Block Copy with Low-Rank Based Rectification for Urban Building Scenes

Qijun Wang[✉]

Anhui University, Hefei, China
wanqijun308@163.com

Abstract. In this paper, we propose an enhanced intra block copy method for intra prediction in HEVC through low-rank based rectification for urban building scenes. Since the viewpoint is not always perpendicular to the facade of buildings (object plane), repetitive patterns on object plane in 3D space appear with scale shift in the image, and the corresponding redundancy cannot be removed through conventional intra block copy based on translational motion model. To solve this problem, we start with theoretic analysis to perspective transformation, and get the lemma that the two-parametered pure perspective transformation is the cause of scale shift, and can be determined by low-rank based rectification. Therefore, intra block copy can be enhanced via low-rank based rectification for repetitive patterns on object plane. To achieve the best performance, rate-distortion optimization is utilized to determine whether prediction from rectified domain would be used in intra block copy for coding unit (CU) or not. Experimental results show that our proposed method can achieve as the highest as 9.54% and averagely 2.13% bit-rate saving on Urban-100 dataset for urban building scenes compared to conventional intra block copy on HEVC platform.

Keywords: Video coding · Intra block copy · Perspective transformation Low rank

1 Introduction

Intra prediction plays an important role for the whole coding performance in video coding, and is designed elaborately in the most recent and advanced video coding standard known as HEVC [1, 2] which includes up to 35 intra modes. As shown in Fig. 1(a), the prediction of each HEVC intra mode for current block is generated through an extrapolation from the upper and left boundary pixels, and this way performs well in predicting local, continuous, and directional image content with a low computational complexity.

Motion estimation and compensation is the basic technique for inter-picture prediction, and can find its potential in intra distant prediction, as depicted in Fig. 2(b). Recently, this technique was adopted to the screen content coding (SCC) extension of HEVC [3], referred to as intra block copy (IBC). IBC is good at predicting non-local, repetitive image content, for example text characters in screen content for desktop sharing between different devices, but cannot do well in natural video. The following

© Springer Nature Switzerland AG 2018
R. Hong et al. (Eds.): PCM 2018, LNCS 11166, pp. 315–325, 2018.
https://doi.org/10.1007/978-3-030-00764-5_29

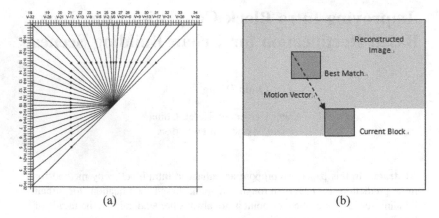

(a)   (b)

Fig. 1. (a) The directional modes of intra prediction in HEVC. (b) Intra block copy is designed for repeated features.

variants of IBC are also investigated, such as using flip or rotation of the copied blocks for prediction [4, 5], but the performance is still limited, due to the high overhead for transmitting flip or rotation parameters.

For urban building scenes, there are many man-made regular structures, taking the test images as shown in Fig. 2 for example, and it is intuitive that the windows appearing in the image are of the same size in 3D space, but actually they have different sizes in the image. Following the orientation of building façade (object plane), there is an effect of scale-shift, and the under-lying reason is that the imaging plane of the camera is not always fronto-parallel to object plane. In these situations, IBC and its variants with the philosophy of translational motion model cannot work well. This challenging problem can be conquered when object plane's orientation in 3D space is known, and motion estimation and compensation of IBC could be done in rectified domain from which the actual structure in 3D space can be restored.

Fig. 2. From a viewpoint, scale shift appears following the orientation of object plane.

From theoretic analysis into perspective transformation of different points on the same object plane in 3D space, a pure homography matrix with only two parameters is a significant factor to scale-shift. Based on low-rank based optimization, the repetitive patterns can be rectified to eliminate scale shift, and hence the two-parametered pure perspective transformation can be determined. The prediction of IBC is generated in rectified domain via the pure perspective transformation. The image area within the same plane share the same transformation parameters (the two parameters for construction of pure perspective transformation), bringing an acceptable overhead comparing to [4, 5]. To achieve the best performance, the best motion vector must be approached. For each motion vector candidate, the prediction of IBC with pure perspective transformation is generated, and the best motion vector is determined under the criteria of Sum of Square Difference (SSD) or Sum of Absolute Difference (SAD). Meanwhile, rate-distortion optimization (RDO) is applied to determine whether prediction from rectified domain is used or not for current coding unit. In this way, we enable much more flexible prediction for both non-local translational and non-local perspective repeated image content.

The remainder of this paper is organized as follows: Sect. 2 gives a detailed description of low-rank based rectification; Sect. 3 discusses the implementation of intra block copy with low-rank based rectification. Experimental results and analysis are presented in Sect. 4, followed by conclusions in Sect. 5.

2 Low-Rank Based Rectification

2.1 Perspective Transformation

Perspective transformation is a mapping from a projective plane to another, and is commonly represented by a homography matrix H:

$$H = \begin{pmatrix} h_1 & h_2 & h_3 \\ h_4 & h_5 & h_6 \\ h_7 & h_8 & 1 \end{pmatrix} = \begin{pmatrix} h_1 - h_3 h_7 & h_2 - h_3 h_8 & h_3 \\ h_4 - h_6 h_7 & h_5 - h_6 h_8 & h_6 \\ 0 & 0 & 1 \end{pmatrix} \begin{pmatrix} 1 & 0 & 0 \\ 0 & 1 & 0 \\ h_7 & h_8 & 1 \end{pmatrix} = A * \widehat{H}$$

$$(1)$$

which can be further factored into an affine transformation matrix A and a pure perspective transformation matrix $\widehat{H}$. The pure perspective transformation will change repeated patterns on object plane with different scales in the image. For a point with coordinate of (x, y) under the assumption that all positions on object plane in 3D space have the same z coordinate of one, the mapped position in the image through left multiplexing $\widehat{H}$:

$$\widehat{H} * \begin{pmatrix} x \\ y \\ 1 \end{pmatrix} = \begin{pmatrix} 1 & 0 & 0 \\ 0 & 1 & 0 \\ h_7 & h_8 & 1 \end{pmatrix} \begin{pmatrix} x \\ y \\ 1 \end{pmatrix} \tag{2}$$

and the normalization form is:

$$\begin{cases} x' = \frac{x}{h_7 x + h_8 y + 1} \\ y' = \frac{y}{h_7 x + h_8 y + 1} \end{cases} \tag{3}$$

If (x, y) have a minor shift to $(x + \delta x, y + \delta y)$, the mapped one becomes:

$$\begin{pmatrix} x_{new} \\ y_{new} \end{pmatrix} \approx \begin{pmatrix} x \\ y \end{pmatrix} + J_{\widehat{H}} \begin{pmatrix} \delta x \\ \delta y \end{pmatrix}, \text{ with } J_{\widehat{H}} = \begin{pmatrix} \frac{\partial x'}{\partial x} & \frac{\partial x'}{\partial y} \\ \frac{\partial y'}{\partial x} & \frac{\partial y'}{\partial y} \end{pmatrix} \tag{4}$$

in which $J_{\widehat{H}}$ represents the Jacobian matrix of $\widehat{H}$ in the form of (x', y') differentiating x and y respectively. In 3D space, the coordinate change of $(\delta x, \delta y)$ is stretched by $J_{\widehat{H}}$, and the distance of $\|(\delta x, \delta y)\|_2$ is scaled by the determinant of $J_{\widehat{H}}$. Combining Eq. (3) with (4), the scale change in the image can be measured by

$$\det(J_{\widehat{H}}) = \det\left((h_7 x + h_8 y + 1)^{-2} \begin{pmatrix} h_8 y + 1 & -h_8 x \\ -h_7 y & h_7 x + 1 \end{pmatrix} \right) = (h_7 x + h_8 y + 1)^{-3} \tag{5}$$

From Eq. (5), we can see that scale change depends on detailed position, and then a rectangle in object plane, which is not fronto-parallel to the imaging plane of camera in 3D space, may be projected to a non-parallelogram. The parallel property losing would hinder the performance of intra block copy based on translational motion model under this circumstance.

The same theoretic analysis can also be applied to matrix A, yielding that:

$$\det(J_A) = \begin{vmatrix} h_1 - h_3 h_7 & h_2 - h_3 h_8 \\ h_4 - h_6 h_7 & h_5 - h_6 h_8 \end{vmatrix}$$
$$= h_1 h_5 - h_2 h_4 + h_3 (h_4 h_8 - h_5 h_7) + h_6 (h_2 h_7 - h_1 h_8) \tag{6}$$

which is a constant, meaning affine transformation does not change the parallel property, and can preserve the similarity among repeated patterns (with the same sizes in 3D space).

From the equations of (5) and (6), we can see that scale shift brought by perspective transformation is due to the two-parametered pure transformation form composed of h_7 and h_8, which are enough for rectifying scale-shift.

2.2   Low-Rank Based Rectification

Many surfaces or structures in 3D exhibit low-rank textures. It is assumed that such a texture $I^0(x, y)$ lies approximately on a planar surface in 3D space, the image $I(x, y)$ that we observe from a certain viewpoint is a transformed version of the original low-rank texture function $I^0(x, y)$:

$$I(x,y) = H * I^0(x,y) + E(x,y) \tag{7}$$

E is the error matrix, and it is assumed that E is a sparse matrix. Hence, the problem is formulated as:

$$\min_{I^0,E,H} rank(I^0) + \lambda \|E\|_0 \ \ subject\ to\ I = H * I^0 + E \tag{8}$$

$rank(I^0)$ is the rank of I^0, and $\|E\|_0$ denotes the number of non-zero entries in E. That is, we aim to find the texture I^0 of the lowest rank and the error E of the fewest nonzero entries that agrees with the observation I up to a domain transformation. λ is a weighting parameter that trades off the rank of the texture versus the sparsity of the error.

The rank function and the ℓ^0-norm are extremely difficult (in general NP-hard), and can be relaxed by the matrix nuclear norm $\|I^0\|_*$ (sum of all singular values) for $rank(I^0)$ and the ℓ^1-norm $\|E\|_1$ (the sum of absolute values of all entries) for $\|E\|_0$ respectively, and the problem is evolved into:

$$\min_{I^0,E,H} \|I^0\|_* + \lambda \|E\|_1 \ \ subject\ to\ I = H * I^0 + E \tag{9}$$

Inspired by the transform invariant low-rank textures [6], the Augmented Lagrangian Multipliter (ALM) method can be used to solve this type of convex optimization problem effectively. But this method need an initialization of a rectangular region indicating the position of regular and near-regular structures. The method [9] based on vanishing point (VP) is applied. Firstly, line segments are detected, and then are clustered using a RANSAC-based voting approach. Assuming there are only up to three vanishing points corresponding plane orientations in the 3D scene. We estimate the spatial support of each VP by diffusing its corresponding line segments using a wide Gaussian kernel. Then, we estimate the spatial support for the planes by performing element-wise multiplication of its VP's support line density maps. These

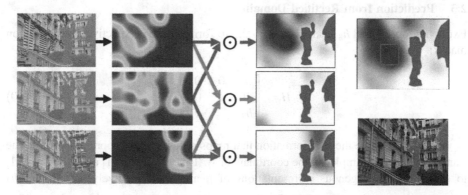

Fig. 3. The rectangular region with the maximum average response value on the most dominant plane (the top row) is taken as the initialization of regular structures.

product maps have a high response where the two sets of the line segments overlapped with each other. We select a rectangular region with the maximum average response value on the most dominant plane, as shown in Fig. 3.

Taking the image of the first frame of sequence "BasketballDrill" shown in Fig. 4. for example, the H, I^0 and E can be derived through optimizing the objective function (9) via ALM with the regular region with red rectangle.

(a)

(b)   (c)

Fig. 4. The perspective transformation H is used to map the rectangular region (indicated by red rectangle) to low-rank form (indicated by green lines) (a). The rectified form I^0 (b), and the error E (c). (Color figure online)

2.3   Prediction from Rectified Domain

Extracting out h_7 and h_8 from H in Sect. 2.2 to form a pure perspective transformation matrix:

$$\widehat{H} = \begin{bmatrix} 1 & 0 & 0 \\ 0 & 1 & 0 \\ h_7 & h_8 & 1 \end{bmatrix} \tag{10}$$

We first compute the transformation that maps a block at the coordinate of t_i to the transformed block sampled at the coordinate of s_i. Take $\tilde{t}_i = [t_i^x, t_i^y, 1]$ and $\tilde{s}_i = [s_i^x, s_i^y, 1]$ to denote the homogenous representations of t_i and s_i, respectively. Let h_1, h_2, h_3 represent the row vectors of $\widehat{H}$. The source and target block positions in rectified domain are computed by:

$$\tilde{t}'_i = \widehat{H} * \tilde{t}_i = [h_1\tilde{t}_i, h_2\tilde{t}_i, h_3\tilde{t}_i]^T \tag{11}$$

$$\tilde{s}'_i = \widehat{H} * \tilde{s}_i = [h_1\tilde{s}_i, h_2\tilde{s}_i, h_3\tilde{s}_i]^T \tag{12}$$

We define (d_x, d_y) as the displacement vector from target to source block positions in rectified domain. For homogenous coordinate $\tilde{t}'_i$, the scale is $h_3\tilde{t}_i$. Therefore, combining with Eq. (11), homogenous coordinate $\tilde{s}_i$ can be represented as:

$$\tilde{s}'_i = \tilde{t}'_i + \begin{bmatrix} h_3\tilde{t}_i d_x \\ h_3\tilde{t}_i d_y \\ 0 \end{bmatrix} = \begin{bmatrix} h_1 + h_3 d_x \\ h_2 + h_3 d_y \\ h_3 \end{bmatrix} \tilde{t}_i \tag{13}$$

Combining Eqs. (12) and (13) by applying the inverse of the pure homography matrix $\widehat{H}^{-1}$, we can get:

$$\tilde{s}_i = \widehat{H}^{-1}\tilde{s}'_i = \widehat{H}^{-1} \begin{bmatrix} h_1 + h_3 d_x \\ h_2 + h_3 d_y \\ h_3 \end{bmatrix} \tilde{t}_i \tag{14}$$

For each pixel in block $\tilde{t}_i$, the corresponding prediction pixel coordinates are achieved through Eq. (14). If the coordinates are not integral, bilinear/bicubic interpolation is applied to get the pixel value with fractional coordinates. The current square block can be mapped to an irregular area by aggregating all mapped pixel positions.

3 Intra Block Copy with Low-Rank Based Rectification

For the image content with repetitive textures, IBC exploits this non-local repetitive property by finding a matching block with the same size to be the predictor for current block. IBC extends the prediction source from one-dimensional boundary pixels to two-dimensional non-local pixels. For easy implementation, IBC is realized by inter-picture prediction through substituting reference frame with already reconstructed image region. The motion vector indicating the best matching block should be recorded in bit-stream and transmitted to the decoder. The best matching block is determined by greedy search in the allowed area accounting for computational complexity and image availability in current encoder. Different positions in the allowed area correspond different motion vectors, each of which is composed of horizontal component mvx and vertical component mvy, and the best motion vector is determined under the criteria of SSD or SAD or their derived forms.

Similarly, intra block copy with low-rank based rectification is based on perspective motion model instead of translational motion model, and generates all the predictions of current block at every candidate position in the allowed area through transformation defined by Eq. (13). Nevertheless, motion vector in original image cannot be directly taken as (d_x, d_y) in Eq. (13), since (d_x, d_y) is a normalized form defined in rectified domain. During the search process, every motion vector candidate is transformed to the one in rectified domain by $\hat{H}$ matrix:

$$\begin{cases} d_x = mvx/(h_7 * mvx + h_8 * mvy + 1) \\ d_y = mvy/(h_7 * mvx + h_8 * mvy + 1) \end{cases} \tag{15}$$

As same as that in IBC, the best motion vector is determined under the criteria of SSD or SAD and is written to bit-stream. As depicted in Fig. 5, IBC with low-rank based rectification is taken as an additional mode to be integrated into hybrid video coding framework, and RDO is utilized to decide which mode in candidate set including IBC and conventional intra prediction modes is the best coding mode.

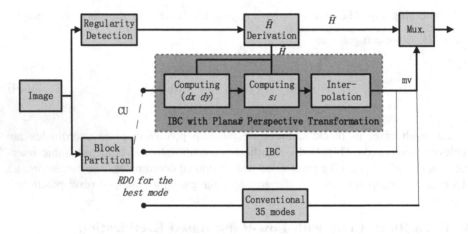

Fig. 5. The diagram for integrating IBC with low-rank based rectification into hybrid video coding framework.

Besides, in syntax level, a new flag lcu_affine_flag is added in the largest coding unit (LCU). When lcu_affine_flag is set to 0, no coding unit in this LCU will adopt low-rank based rectification to generate prediction, and when lcu_affine_flag is set to 1, no less than one coding unit will adopt low-rank based rectification to generate prediction. Another flag cu_affine_flag is used to indicate whether all the prediction units in current coding unit to generate prediction from rectified domain.

4   Experimental Results

The proposed intra block copy with low-rank based rectification is implemented on the latest HM-SCC-extensions software [7], and compared with HM-SCC-extensions software with intra block copy (IBC) configuration on. Considering the fact that small quantization parameter (QP) will cause motion estimation/compensation in the unit of small block and the derived parameters of pure perspective transformation by low-rank based optimization is not accurate enough, we made a little adjustment by setting QP to {27, 32, 37, 42} instead of {22, 27, 32, 37} in common test conditions. Meanwhile, we select 10 typical images of urban building scenes from Urban-100 dataset (a common dataset containing urban views), as depicted in Fig. 6. The content of these images covers a variety of urban building types, such as common street view (008, 012 and 020), building surface with single facet (003, 030 and 080)/with multiple facets (001, 005 and 060), and other man-made structures (002). The search range of intra motion estimation is restricted to |mvx| + |mvy| < 382, only integer motion vector is used.

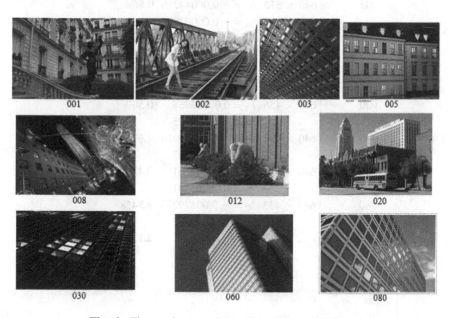

Fig. 6. The test images chosen from Urban-100 dataset

The BD-rate [8] reduction of all the test sequences is shown in Table 1. From Table 1, our proposed intra block copy with low-rank based rectification can achieve as high as 9.54% (on image 080) and averagely 2.13% BDrate reduction, compared with conventional intra block copy as the anchor, and outperforms intra block copy on all test images. The rate reduction on complex scenes such as 001, 002 and 020 is minor, because the complex texture hinders the convergence of objective function. On the contrary, clean and regular structures are robust enough to get an acceptable perspective parameters, and the coding gain is obvious, reaching the maximum on image 080. Not all clean and regular structure will bring obvious gain, for example image 005, the leaning degree of camera view is not very big, in this situation, conventional IBC can also do well, and then the coding gain of intra block copy with low-rank based rectification is not as large as that of image 003, 030 and 080.

The rate-distortion (R-D) curves on image 80 are illustrated in Fig. 7. From the R-D curves, we find the coding gain is higher in low bit-rate than that in high bit-rate, and the reason can be attributed to the rate-distortion optimization (RDO) utilized in the

Table 1. Coding result of IBC with low-rank based rectification comparing with IBC

Test image	Size	Parameters	Rate reduction
001	640 × 424	$h_7 = -0.00319206$ $h_8 = 0.00053392$	0.07%
002	640 × 424	$h_7 = -0.00008306$ $h_8 = -0.00033536$	0.23%
003	640 × 640	$h_7 = 0.00075896$ $h_8 = -0.00130970$	3.39%
005	640 × 512	$h_7 = -0.00044256$ $h_8 = -0.00002827$	0.86%
008	640 × 424	$h_7 = -0.00385141$ $h_8 = 0.00356244$	0.89%
012	640 × 480	$h_7 = 0.00193544$ $h_8 = -0.00014878$	0.76%
020	640 × 456	$h_7 = 0.00193968$ $h_8 = 0.00013125$	0.50%
030	640 × 424	$h_7 = -0.00036910$ $h_8 = 0.00375863$	4.67%
060	640 × 424	$h_7 = 0.00048061$ $h_8 = 0.00005431$	0.43%
080	640 × 432	$h_7 = -0.00179273$ $h_8 = 0.00015862$	**9.54%**
Avg			**2.13%**

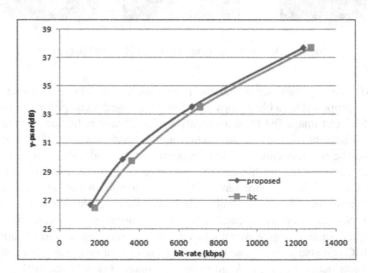

Fig. 7. The R-D curves on image 080.

codec. When the bit-rate is low, the QP would be large, and RDO tend to select the mode that will bring low bit-rate overhead and relax the requirement of prediction accuracy. In this situation, intra block copy with low-rank based rectification with large block partition will outperform conventional intra block copy.

5   Conclusions

In this paper, we start with theoretic analysis into general perspective transformation, and get that the two-parametered perspective transformation is the cause of scale shift through low-rank based rectification. We also propose an enhanced intra block copy with low-rank based rectification for intra prediction in HEVC, the new method can achieve as high as 9.54% and averagely 2.13% bit-rate saving on Urban-100 dataset representing urban building scenes, compared to HEVC with IBC anchor. Especially on the image with clean, regular, and large leaning degree of camera view, the coding gain of our method is significant. For complex urban scene, the building planes should be segmented intensively, and handle each image region with different parameters of perspective transformation.

Acknowledgements. This work is funded by NSFC of China under Grant No. 61302111 and Co-Innovation Center for Information Supply & Assurance Technology, Anhui University under Grant No. ADXXBZ201611

References

1. Sullivan, G.J., Ohm, J., Han, W.-J., Wiegand, T.: Overview of the high efficiency video coding (HEVC) standard. IEEE Trans. Circuits Syst. Video Technol. **22**(12), 1649–1668 (2012)
2. Lainema, J., Bossen, F., Han, W.-J., Min, J., Ugur, K.: Intra coding of the HEVC standard. IEEE Trans. Circuits Syst. Video Technol. **22**(12), 1792–1801 (2012)
3. Xu, J., Joshi, R., Cohen, R.A.: Overview of the emerging HEVC screen content coding extension. IEEE Trans. Circuits Syst. Video Technol. **26**(1), 50–62 (2015)
4. Xu, X., Liu, S., Ye, J., Lei, S.: Pu level intra block copying with flipping mode. In: Asia-Pacific Signal and Information Processing Association, 2014 Annual Summit and Conference (APSIPA), pp. 1–7 (2014)
5. Zhang, Z., Sze , V.: Rotate intra block copy for still image coding. In: IEEE International Conference on Image Processing (ICIP), pp. 4102–4106 (2015)
6. Zhang, Z., Ganesh, A., Liang, X., Ma, Y.: TILT: transform invariant low-rank textures (2010). arXiv:1012.3216
7. HM, HEVC test Model. https://hevc.hhi.fraunhofer.de/svn/
8. Jontegaard, G.: Calculation of average PSNR differences between RD-curves. Document VCEG-M33, Austin, Texas, USA (2001)
9. Huang, J.-B., Singh, A., Ahuja, N.: Single image super-resolution from transformed self-exemplars. In: IEEE Conference on Computer Vision and Pattern Recognition (CVPR), pp. 5197–5206 (2015)

Assembly-Based 3D Modeling Using Graph Convolutional Neural Networks

Xufeng Lang, Zhengxing Sun$^{(\boxtimes)}$, Qian Li, and Jinlong Shi

State Key Laboratory for Novel Software Technology, Nanjing University,
Nanjing 210046, China
szx@nju.edu.cn

Abstract. Assembly-based methods make 3D shape modelling convenient and effective even for non-expert users. However, it is still difficult to choose a reasonable component from an unlabeled shape dataset. In this work, the spectral graph convolutional neural networks (graph CNNs) are used to label a subcomponent in the given shape by their context information and geometry features using convolution operation. Then an appropriate component to replace the above labeled component is found by the same network according to their labels from shapes database. After replacing the component, reasonable results can be obtained in most experiments, which prove the reliability of our method. In addition, we found that the use of dropout and residual could greatly improve the training and performance. The context information, compared with the geometry features, is more effective in creating new shapes.

Keywords: 3D shape assembly · Graph convolutional neural networks
Shape modeling

1 Introduction

Many efforts have been made to create new shapes [1–8], among which, assembling existing parts from a given object set is easy and effective [7], i.e., selecting some components from a given collection of shapes, and then placing them with a reasonable arrangement [8]. In reality, it is a challenge to select relevant and plausible components from a given collection of shapes, and then place them with a reasonable arrangement.

Many approaches have been proposed to address this challenge, e.g., interfaces for part-wise shape assembly was reported [1], which can greatly reduce user's interaction in component selection and placement. However, their suggested models rely on a supervised dataset, whose build is a time-demanding process, and their results are similar to the original existing shapes. Another method was proposed by defining some fixed spatial substructures artificially and then interchanging among them, such as support substructure [2] and symmetric functional arrangements (SFARR-s) substructure [3]. Modeling through substructure methods create non-trivial and functionally plausible shapes, which motivate us to use the structure rather than the geometry features of the components to create new shapes. Besides, the granularity of component in the above two kinds of methods have a great influence on the effectiveness. In this

© Springer Nature Switzerland AG 2018
R. Hong et al. (Eds.): PCM 2018, LNCS 11166, pp. 326–337, 2018.
https://doi.org/10.1007/978-3-030-00764-5_30

work, the selection of component was performed by comparing label after labeled each component in the shape.

In this article, we automatically over-segment both shapes from the dataset and given shape to a uniform geometric granularity (e.g., cylinder, cuboid and so on), rather than semantic level. The shape in our work is represented by a graph structure, in which, each node represents a subcomponent that is smaller than or equal to semantic components, and one edge denotes the connection between two subcomponents. The spectral graph CNNs (convolutional neural networks) was used to fuse the local spatial structure and geometric features of the subcomponent, then max pooling layer was used to construct mappings of intra-subcomponent features. After several layers of convolution and max pooling, the feature map of each subcomponent could be obtained, and then each subcomponent was labeled using the softmax layer. To train the network, Laplacian matrix was set by each mini-batch, and then the convolution was performed on the graph in the spectral domain. Then the entire network is trained using softmax layer and cross-entropy loss. For a particular label, we merge the subcomponents with this label in the entire shape into a component, and then swap it with subcomponents that have the same label in the given shape.

Key contributions of our approach are as follows:

- The graph CNNs are extended into shape structure analysis, by fusing the geometric features and contextual relationships of the components, we can get new feature of them and label them.
- We propose to replace components with the same label (contains context information), therefore, the overall structure of the shape is preserved after replacement according to their labels. Meanwhile the plausibility and diversity can be guaranteed. Besides, components could be replaced between different shape families.
- Since we automatically over-segment all models to the same level, our method does not need to consider the component granularity.

2   Related Work

3D Modeling by Part Assembly. Funkhouser et al. presented example-based modeling techniques firstly [4]. Then probabilistic model was proposed, by which new shapes are assembled automatically using reasonable labeled components [1]. Shen et al. proposed to convert consumer-level scanning data to high-quality 3D shapes with labeled semantic parts [5]. However, these part-based methods rely on a database with consistently segmented shapes. Jaiswal et al. [6] used factor graphs to create new components by considering the relationship between the component pairs. A symmetric functional arrangements substructure was proposed by Zheng et al. to create functionally plausible shape variations [3]. Su et al. created 3D shapes using a support substructure via re-combination of cross-category object parts from an existing database in different shape families [2]. These above methods all achieve impressive results in shape assembly with in-class shapes, or shapes across the categories. However, there are several issues exist. Firstly, hand-crafted substructure they used might have to be adjusted when switching to a new dataset. Secondly, the shape still need to be segmented into semantic levels, though the label information is not necessary. While in our method, both geometry feature and context information can be combined with the convolutional neural networks to get appropriate feature maps. Therefore, we can obtain similar components with structural and geometric features for replacement. The subcomponent context information is considered, rather than the shape global structure, therefore, it is possible to exchange subcomponents between different types of models and maintain the original structure of the generated results.

Convolutional on Graph. CNNs on graph was first proposed by Bruna [11], however, for a specific component, this spatial convolution should consider all the neighborhood feature in the graph, which is difficult to realize. It was reported that the Chebyshev polynomials and approximate evaluation scheme could be used to achieve localized filtering in spectral domain, which reduces the computational cost [12]. In addition, a first-order approximation to the Chebyshev polynomials as the graph filter could further reduce the training parameters [14]. In this work, we use the Spectral Graph Convolution to combine the features of a component with its neighbor features.

Deep Models of 3D Shapes. Deep learning has achieved impressive performances in the computer vision, speech recognition, and natural language processing. Recently, deep models have also be applied to 3D shape analysis. For example, in PointNet, uniformly 1024 points were sampled on mesh faces, then they were put into their network for further calculation [9]; a generative model on voxelized 3D shapes were trained by 3D ShapeNets, then it was applied to shape assembly and shape completion [10]. To share weight and achieve optimization in convolutional architectures, 3D voxel grids or projections were used to transform unstructured and unordered point clouds or meshes into regular representations before feeding them to the network [8]. In this work, we directly feed shape structure graph into graph CNNs and minimize information loss.

3 Overview

As show in Fig. 1, given a shape, we first over-segment it automatically, and then process these segments to sub-components with reasonable sizes. After deciding the structure graph of each shape, we can label subcomponents using Graph CNNs (which will be discussed later); therefore, the subcomponent can be selected manually or automatically according to their labels. Given a shape by user, to create new shapes, we need to select the shapes from the dataset and put them into the same network to find the subcomponent with the same label with that from the given shape. The subcomponents selected in the shape dataset are interchanged with the components in the given shape to create new shapes. After that, we put new shapes back to the network to see whether the original labels are changed or not.

The process of labeling component is realized using the Graph CNNs: the geometric feature of a subcomponent and its context structure information can be combined through the spectral domain convolution operation, then the feature maps are obtained through convolution and max pooling layers, finally, the entire network is trained using softmax layer and cross-entropy loss.

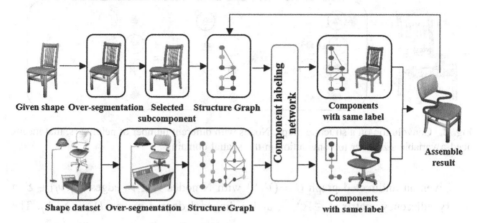

Fig. 1. An overview of our component assembling flow. Firstly, the graph CNNs model is trained. For a given shape, we perform over-segmentation on it. Once the user selects a subcomponent, we can label it with the trained network, and then select the components with the same label in dataset. Then the new shape can be created by replacing the subcomponents between given shape and dataset shapes. Finally, the overall reasonability is evaluated by relabeling the subcomponent in the generated result.

4 Method

In this section, we introduce how to use the convolutions do shape structure analysis and subcomponent labeling. Because the numbers of nodes and neighbors are uncertain, we use spectral convolutions to generate nodes feature maps.

4.1 Spectral Graph Convolutions

In this work, a structural graph is built for each 3D shape. The graph nodes represent sub-components, which are over segmentation of semantic components, and the graph edges define connectivity between these sub-components. As show in Fig. 2, in our shape structural graph, the number of nodes varies, and the number and order of neighbors per node is uncertain. Therefore, in order to convolve directly on these nodes, a variable-size filter is needed.

In order to apply convolutional on structural graph, two strategies can be used to define convolutional filters, i.e., spatial domain or spectral domain [12]. Spatial approach provides filter localization via the finite size of the kernel [12]; however, the adjacent nodes number of a specific node in our graph is variable. While the spectral approach provides a well-defined localization operator on graphs via convolutions with a Kronecker delta [13], by which we can enter the graph directly into the network. Therefore, a spectral like spatial method for the graph convolution operation is used in our work.

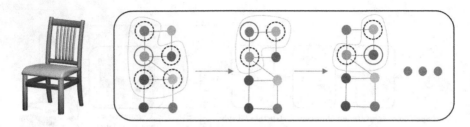

Fig. 2. Convolution on a structure graph. Nodes with different number of neighbors, that means need a variable-size filter to convolution in the spatial domain.

Given an undirected graph $\mathcal{G} = (\mathcal{V}, \mathcal{E})$ with N nodes $v_i \in \mathcal{V}$, edges $(v_i, v_j) \in \mathcal{E}$, a binary adjacency matrix $A \in \mathbb{R}^{N \times N}$, and a diagonal degree matrix $D_{ii} = \sum_j A_{ij}$. The graph Laplacian matrix L have been normalized as follow:

$$L = I - D^{-1/2} A D^{-1/2} \tag{1}$$

where I is the identity matrix. The node feature and its contextual information can be combined using L. According to the convolution theorem, a convolution is a linear operator of the Fourier basis, which is represented by the eigenvectors of the Laplacian operator. Because L is a symmetric positive definite matrix, it has a complete set of orthonormal eigenvectors $\{u_s\}_{s=0}^{N-1} \in \mathbb{R}^{N \times N}$. Eigenvectors are the graph Fourier basis, and depends on the set of eigenvalues $\{\lambda_l\}_{l=0}^{N-1}$, which are frequencies of the graph. Then graph Laplacian is diagonalized as $L = U \Lambda U^T$, where the Fourier basis $U = [u_0, u_1 \ldots u_n]$ and $\Lambda = diag([\lambda_0 \ldots \lambda_n])$.

In a shape, the functionality of a component always depends on context. Therefore, we only need to consider the relationship between it and the nearest K localized neighbor nodes. Given a node with C-dimensional feature $x \in \mathbb{R}^C$, its graph Fourier transform is defined as $\hat{x} = U^T x$ [13], and its inverse is $x = U\hat{x}$. Then we can obtain a spectral filter $g_\theta(\Lambda) = diag(\theta)$, where the parameter θ is a vector of Fourier coefficients, however, this filter is not localized in space and its learning complexity is in $O(n)$. To obtain convolved matrix of a graph, we use the approximate Chebyshev polynomial [14].

$$C = \tilde{D}^{-\frac{1}{2}} \tilde{A} \tilde{D}^{-\frac{1}{2}} X \Theta \tag{2}$$

where $C \in \mathbb{R}^{N \times F}$ is the convolved signal matrix, $X \in \mathbb{R}^{N \times C}$ denotes the features of N nodes in a structural graph, $\Theta \in \mathbb{R}^{C \times F}$ is a matrix of filter parameters, $\tilde{A} = A + I_N$ and $\tilde{D}_{ii} = \sum_j \tilde{A}_{ij}$.

Finally, for a layer l in multi-layer graph convolutional networks, its input is $H^{(l)}$ ($H^{(1)} = X$) and output is:

$$H^{(l+1)} = \sigma\left(\tilde{D}^{-\frac{1}{2}}\tilde{A}\tilde{D}^{-\frac{1}{2}}H^{(l)}\Theta^{(l)} + b^{(l)}\right) \tag{3}$$

where the $\sigma(\cdot)$ is the activation function of layer l, and $b^{(l)}$ is bias of layer l. Given an adjacency matrix A, we first calculate $\tilde{D}^{-\frac{1}{2}}\tilde{A}\tilde{D}^{-\frac{1}{2}}H^{(l)}$ in a pre-processing step.

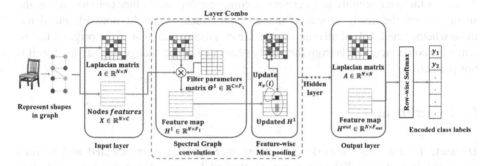

Fig. 3. Schematic depiction of a graph CNNs for shape subcomponents reorganization. The number of nodes and graph structure are remained throughout the process.

4.2 Network Configuration

As show in Fig. 3, the spectral graph convolution layer is the core layer in our network. Through this layer, the component feature and its contextual information can be fused. In addition, we have the batch normalization layer and max-pooling layer. The max pooling layer is performed on each node by feature [20]. For node v and its neighbors $N(v)$, x_v is the feature of v, which can be updated by replacing the i-th feature with the maximum one among the i-th feature of $N(v)$ and v. Then the new feature of v is represented as $\hat{x}_v(i) = \max(\{x_v(i), x_u(i), \forall i \in N(v)\})$. For the last output layer, we use a row-wise softmax activation function, which is defined as $softmax(x_i) = \frac{1}{Z}\exp(x_i)$ with $Z = \sum_i \exp(x_i)$.

Then the cross-entropy error is used as our loss function:

$$\mathcal{L} = -\sum_{m \in Y_m}\sum_{f=1}^{F} Y_{mf} ln Z_{mf} \tag{4}$$

where Y_m is the set of node that have been labeled. In this work, the number of convolution layers is set to five.

4.3   Batch Training of Graphs

In this work, the weights of network layers are trained by gradient descent. Training is done with initial learning rate of 10^{-3}, and dropout probability of 0.5 to prevent overfitting in the training process [21].

In our experiment, the influence of model depth (number of layers) on classification performance is investigated. A cross-validation experiment is performed. In addition, the model variant is considered using residual connections [22] between hidden layers, which can facilitate training of deeper models by enabling the model to carry over information from the previous layer's input.

4.4   Components Compatibility Assessment

To assess the compatibility of the selected component(s) with other components in the shape, we put the newly generated shape graph structure into the network and determine whether the labels of all components are consistent with that of the original labels. Once the component labels from the newly generated shape change, we say the result is not perfect.

5   Experiments

Dataset. To train our network, 200 chairs in ShapeNet were selected and marked manually, in addition, 200 high topological models were considered [16], including chairs, tables, lamps, beds, carts, robots, airplanes, bikes, boats. Our method was tested using the same categories from ModelNet repository [15]. For quantitative evaluation and comparison of placement algorithms, we randomly split every category into two parts, 80% for training and 20% for test sets.

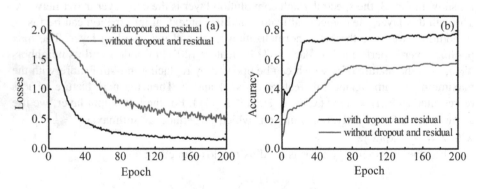

Fig. 4. (a) Training losses and (b) verification accuracy of our Graph CNNs.

Data Preprocessing. Consistently segmenting shape sets into semantic level is time consuming, especially when the amount of data is large [8]. In our method, instead of segmenting to semantic level (e.g., seat, back in a chair), we segment all shapes in the dataset into minimal meaningful segmentation granularity (e.g., cylinder, cuboid and so on). We choose rich topological dataset shapes [16] with consideration of structural variations from ModelNet [15]. These shapes were aligned roughly using the PCA (Principal Components Analysis) algorithm firstly, followed by pre-segment using a convexity-based over-segmentation method [17]. Then, the subcomponents of shapes in the dataset were generated by merging over-segmented patches [8]. Finally, we check each segmented result, and adjust failed results manually.

To train our network, we obtained the training samples by labeling some meshes. We use shape correspondence method to perform an one-to-one correspondence between shapes in database [16], then select two closest shapes, perform a depth first search and spread the labels between them until all subcomponents have a labeled.

We calculate the Light Field Descriptor [18], SDF histograms [19], and Oriented Bounding Box of each subcomponent as their geometry features; and then get the connection slot for each subcomponent to determine the location of subsequent component connections. We also get the adjacency matrix of subcomponents in each shape.

Performance Boosted of Network. The hidden layer consists of a convolution layer, a batch normalization layer and a max pooling layer. To accelerate the training, the dropouts and residuals also be used, whose effect on the network performance is given in Fig. 4. It is seen that, the training losses drop quickly during the first 50 epoch when using residual and dropout, and the verification accuracy is higher with the residual and dropout used. Therefore, the convergence speed and verification accuracy of training can be significantly improved with dropout and residual.

Functional Compatibility. Assembly-based modeling methods are difficult to evaluate, and the rationality of the generated results is often related to design goals and personal preferences, which has a strong subjectivity. In this article, using a rational segmentation benchmark [8], the functionality of a component (e.g., a chair includes back, seat, leg, and armrest) is denoted by a component label. For a particular component with a certain label from the given shape, we need to find out the components with the same label from dataset using our network.

Two experiments to search proper component in dataset are given in Fig. 5. In Fig. 5(a), the aircraft is as a given shape, and the aircraft fin is chosen as a component, which would be labeled through our trained network. Using this label, we search similar components from unlabeled dataset, which consists of shapes from ShapeNet and our per-segment shapes. The result is given in the box of Fig. 5(a), it is seen that most of the components we get are the same semantics. Among them, components with large differences in context structure rank in the lower priority. For example, the fin with the elevator at the top ranks in the lower priority. In Fig. 5(b), the chair back is selected as the component from a given shape (chair). It is seen from the results in the box, we obtain components that exhibit large geometric differences but similar context structures. Meanwhile, some of the results consist of several subcomponents. Therefore, the obtain of functionally similar results rely more on component context rather than geometric features.

The assembly results of four experiments are given in Fig. 6. For each experiment, the part in the left is selected as a component from the given shape; the parts at the top four shapes in the right are the obtained components from dataset using our training

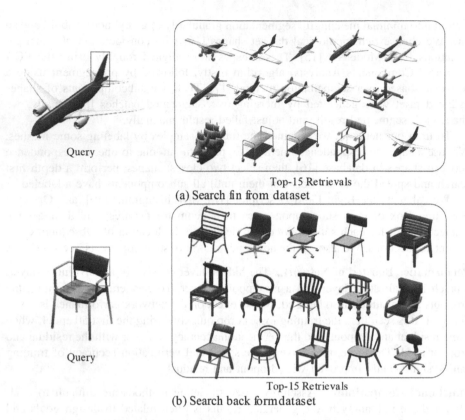

Query

Top-15 Retrievals
(a) Search fin from dataset

Query

Top-15 Retrievals
(b) Search back form dataset

Fig. 5. Query results: in each test case, component in box (left) indicates selected component from a given shape and components (right) are selected components from dataset. Please note that many of the components selected composed of subcomponents. In the box, the closer to the left and the upper, the more similar to component in the given shape.

network for replacement; and the new shapes created are given in the bottom four shapes in the right. As given in Fig. 6, when one leg of a chair or table was selected, due to symmetry information, we can search and replace all legs in it. When the hull of boat is selected, we will replace it with other hull-like components. It is seen that, all results of the boats, chairs and tables are good. The components we have chosen from the dataset not only have similar structures in their original models, but also make the new shape created maintain the structure of the given shape. Unfortunately, failed result also exists (see the lamp experiment in Fig. 6(b)). For example, the strut of the lamp is chosen, originating from the selected component's context and geometric features are similar to those of the given component.

Comparisons. Our method is compared with the support structure method by Su et al. [2] with respect to the back reshuffling test. To generate new results, their approach relies on good-quality pre-segmentation. While in our method, for a given shape (chair), its back is selected as the component to be replaced. Then the new component

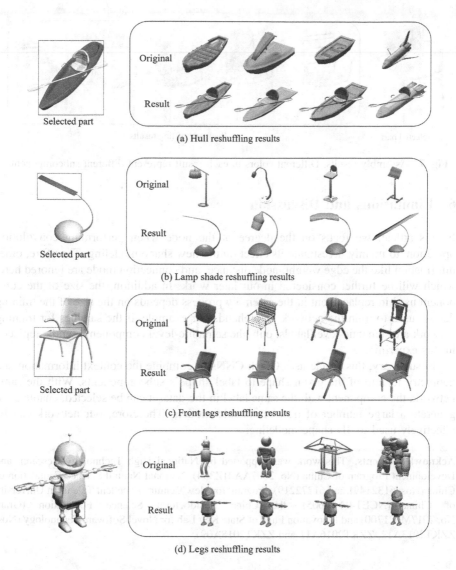

(a) Hull reshuffling results

(b) Lamp shade reshuffling results

(c) Front legs reshuffling results

(d) Legs reshuffling results

Fig. 6. For each experiment, with a given shape in the left, we select a component (component in the left), then select appropriate components (component at the top in the right) from a multi-classes model set for replacement, then assemble them with remained components to generate new shapes (bottom four shapes in the right).

used in replacement can be obtained by identifying subcomponents labels, which just requires a simple segmentation of the shapes in dataset. Our results are given in Fig. 7, different colors in each result represent different subcomponents. Our results are more diverse and reliable than Su et al.'s, since only number 2 and 3 in Fig. 7 could be generated in their work [2].

Selected part   Back reshuffling results

Fig. 7. Assembly results. Different colors in each result represent different subcomponents.

6   Limitations and Discussion

In this article, we focus on the degree of the node when performing convolution operation, to mainly investigate its effect on the new shape modeling. Therefore, other information like the edge weight, node distance, and connection mode are ignored here, which will be further considered in our later work. In addition, the size of the component used in replacement in the assembly process depends on the size of the training dataset used to train the network beforehand. For example, if the samples for training network are semantic-level labels, only the semantic-level components can be replaced in this experiment.

In summary, this article uses Graph CNNs to combine the context information and geometry feature of the given shape to label shape's subcomponents. With the same network, the component with the same label in the dataset can be selected, enable us to generate a large number of new reasonable models. Therefore, our network can be effectively used to 3D shape modeling.

Acknowledgements. This work was supported by National High Technology Research and Development Program of China (No. 2007AA01Z334), National Natural Science Foundation of China (Nos. 61321491 and 61272219), Program for New Century Excellent Talents in University of China (NCET-04-04605), the China Postdoctoral Science Foundation (Grant No. 2017M621700) and Innovation Fund of State Key Lab for Novel Software Technology (Nos. ZZKT2013A12, ZZKT2016A11 and ZZKT2018A09).

References

1. Kalogerakis, E., Chaudhuri, S., Koller, D., et al.: A probabilistic model for component-based shape synthesis. ACM Trans. Graph. (TOG) **31**(4), 55 (2012)
2. Su, X., Chen, X., Fu, Q., et al.: Cross-class 3D object synthesis guided by reference examples. Comput. Graph. **54**, 145–153 (2016)
3. Zheng, Y., Cohen-Or, D., Mitra, N.J.: Smart variations: functional substructures for part compatibility. Comput. Graph. Forum **32**(2pt2), 195–204 (2013)
4. Funkhouser, T., Kazhdan, M., Shilane, P., et al.: Modeling by example. ACM Trans. Graph. (TOG) **23**(3), 652–663 (2004)
5. Shen, C.H., Fu, H., Chen, K., et al.: Structure recovery by part assembly. ACM Trans. Graph. (TOG) **31**(6), 180 (2012)

6. Jaiswal, P., Huang, J., Rai, R.: Assembly-based conceptual 3D modeling with unlabeled components using probabilistic factor graph. Comput. Aided Des. **74**, 45–54 (2016)
7. Xu, K., Kim, V.G., Huang, Q. et al.: Data-driven shape analysis and processing. SIGGRAPH ASIA 2016 Courses, p. 4. ACM (2016)
8. Sung, M., Su, H., Kim, V.G., et al.: Complementme: weakly-supervised component suggestions for 3D modeling. ACM Trans. Graph. (TOG) **36**(6), 226 (2017)
9. Qi, C.R., Su, H., Mo, K., et al.: Pointnet: deep learning on point sets for 3D classification and segmentation. Proc. CVPR IEEE **1**(2), 4 (2017)
10. Wu, Z., Song, S., Khosla, A. et al.: 3D shapenets: a deep representation for volumetric shapes. In: Proceedings of the IEEE Conference on Computer Vision and Pattern Recognition, pp. 1912–1920 (2015)
11. Bruna, J., Zaremba, W., Szlam, A., LeCun, Y.: Spectral networks and locally connected networks on graphs (2013). arXiv preprint arXiv:1312.6203
12. Defferrard, M., Bresson, X., Vandergheynst, P.: Convolutional neural networks on graphs with fast localized spectral filtering. Adv. Neural Inf. Process. Syst. **29**, 3844–3852 (2016)
13. Shuman, D.I., Narang, S.K., Frossard, P., et al.: The emerging field of signal processing on graphs: extending high-dimensional data analysis to networks and other irregular domains. IEEE Signal Process. Mag. **30**(3), 83–98 (2013)
14. Kipf, T.N., Welling, M.: Semi-supervised classification with graph convolutional networks (2016). arXiv preprint arXiv:1609.02907
15. Princeton ModelNet. http://modelnet.cs.princeton.edu/. Accessed 21 May 2018
16. Alhashim, I., Xu, K., Zhuang, Y., et al.: Deformation-driven topology-varying 3D shape correspondence. ACM Trans. Graph. (TOG) **34**(6), 236 (2015)
17. Kaick, O.V., Fish, N., Kleiman, Y., et al.: Shape segmentation by approximate convexity analysis. ACM Trans. Graph. (TOG) **34**(1), 4 (2014)
18. Chen, D.Y., Tian, X.P., Shen, Y.T., et al.: On visual similarity based 3D model retrieval. Comput. Graph. Forum. **22**(3), 223–232 (2003)
19. Shapira, L., Shamir, A., Cohen-Or, D.: Consistent mesh partitioning and skeletonisation using the shape diameter function. Vis. Comput. **24**(4), 249 (2008)
20. Li, R., Wang, S., Zhu, F. et al.: Adaptive graph convolutional neural networks (2018). arXiv preprint arXiv:1801.03226
21. Srivastava, N., Hinton, G., Krizhevsky, A., et al.: Dropout: a simple way to prevent neural networks from overfitting. J. Mach. Learn. Res. **15**(1), 1929–1958 (2014)
22. He, K., Zhang, X., Ren S, et al.: Deep residual learning for image recognition. In: Proceedings of the IEEE Conference on Computer Vision and Pattern Recognition, pp. 770–778 (2016)

Blur Measurement for Partially Blurred Images with Saliency Constrained Global Refinement

Xianyong Fang[1], Qingqing Guo[1], Cheng Ding[1], Linbo Wang[1(✉)],
and Zhigang Deng[2]

[1] Institute of Media Computing, Anhui University, Hefei 230601, China
{fangxianyong,wanglb}@ahu.edu.cn
[2] Department of Computer Science, University of Houston,
Houston, TX 77204-3010, USA
zdeng4@uh.edu

Abstract. Blur measurement of partially blurred image is still far from being resolved. This calls for more distinctive blur features and, even more importantly, a global refinement strategy that has not been considered by existing studies. In this paper we propose a new spatial and frequencial coupled blur descriptor by composing the number of extreme points, the vector of all singular values and the entropy-weighted pooling of the high frequency DCT coefficients. We also introduce a global refinement scheme to explore the merits of saliency for further refining the initial measurements. Consequently, we propose a novel saliency constrained blur measurement method by integrating a neural network based blur metric and a superpixel-scale blur refinement together. Experimental results show the efficiency of our method qualitatively and quantitatively, especially for the images with flat textures.

Keywords: Partial blurred image · Blur measurement · Saliency

1 Introduction

Partially blurred images widely exist in our daily lives due to the intentionally or unintentionally motion or defocusing capture. Detecting the blur degree for each pixel in these images is important for many tasks, including image segmentation, depth estimation, image deblurring, and refocusing. There still lacks of an efficient method to this inverse problem although there have been many studies on it. In this work, we first introduce a new blur descriptor so that an initial measure for each pixel can be efficiently obtained. It is easy to see that the estimated metric usually appears inconsistent between neighborhoods if only local features are used and the global information is ignored. To address this problem, we further present a global refinement to improve the local metric by exploiting global information, which can effectively remove wrong measurements especially

© Springer Nature Switzerland AG 2018
R. Hong et al. (Eds.): PCM 2018, LNCS 11166, pp. 338–349, 2018.
https://doi.org/10.1007/978-3-030-00764-5_31

for regions with flat textures. To the best of our knowledge, the global refinement has not been reported in the previous literature.

Our method does not assume particular blurring models [1,2] and thus is general and applicable for real scenes. Our proposed blur descriptor includes both spatial and frequencial features so that spatially visual appearance and anti-noisy frequency characteristics can be complementary to each other. It is effective from three aspects in comparison with existing multi-feature oriented methods [3–5].

First, spatially, we introduce the extreme point [6] as a blur feature, which was previously applied to texture analysis due to its embedded structure information. Extreme point is more sparse and thus distinguishable by counting its numbers than the widely used gradient [7,8] among spatial features [9,10].

Secondly, we take the complete set of singular values as a spatial complementary to the counting based extreme point because all singular values may vary significantly after blurred. A larger singular value often captures larger scale information, and the blurred image keeps the shape structures at large scale and discards the image details at small scale. The idea is also simple to apply without the burden of manually selecting the high-scale singular values presented in exiting methods [11,12] which may obtain biased results for different images.

Thirdly, we adopt the state-of-art frequencial feature [13] which is the entropy-weighted pooling of a novel DCT based High-frequency multiscale Fusion and Sort Transform (HiFST) of gradient magnitudes. This metric is robust to noise distortion in comparison with existing frequencial features [14,15].

However, wrong measurements seem unavoidable because each blur metric is local and can be easily miscalculated when there are locally clear flat textures. Some existing refinement studies [1,9,12,13] are still local because they only refer to the neighboring pixels. The low-frequency properties are shared by both the blurred region and the flat texture and thus makes the locally spatial or frequencial features fail. Therefore, a global refinement strategy is highly desirable so that the overall scene distribution can be exploited to constrain the measurement, considering that flat texture appears very often in real scenes.

We can clearly see the object without difficulty if observing a blurred scene even with many flat textures. This insight inspires us to consider the importance of visual saliency [16] for a seemingly messy scene. Salient object [17] is the only one standing out in the scene considering the fact that humans have difficulty in paying attention to more than one simultaneously. The partially blurred image is also of no exception, where both the blurred and the unblurred regions in a nature scene can be either salient or unsalient. Therefore, we propose a saliency constrained measurement refinement method so that saliency is used as a global representation of an image and thus helps refining the blur measurement globally and eliminating wrong measures.

In the method, we adopt a neural network (NN) based learning approach to obtain the blur measure which is good at fuzzy inference by simulating the work of human brain and thus fits well with the visual perception of blur. We also take a superpixel based strategy to accelerate the saliency constrained blur refinement process. A bilateral filter is used to obtain a structure-preserved pixel-wise consistent blur measurement finally.

2 Our Blur Metric

2.1 Spatial and Frequencial Coupled Blur Descriptor

The proposed descriptor consists of three components: the number of extreme points, the vector of all singular values, and the entropy-weighted pooling of the high frequency DCT coefficients. Therefore, the blur descriptor, $B_{i,j}$, for each pixel at (i,j), $p_{i,j}$, in the blurred $M \times N$ image I can be defined as

$$B_{i,j} = (L_{i,j}, V_{i,j}, D_{i,j}) \tag{1}$$

with $L_{i,j}$, $V_{i,j}$ and $D_{i,j}$ denoting the number of extreme points, the vector composed of all singular values, and the entropy-weighted pooling of the high frequency DCT coefficients.

Extreme Points. Local structures represent the local directions and intrinsic signal dimensions [6] and image extrema can be divided into five structure classes by intensity distribution: '|', '–', '/', '\' and '*'. The first four classes are one-dimensional except the last one being two-dimensional. The local minimum is weak for a blurred image and therefore the local maximum is considered in our work as extreme point. The extreme points show the high frequency information spatially and thus can depict the blur degree of each pixel.

Figure 1b shows the total number of each ideal extreme class in the four blurred or clear 40×40 regions specified in Fig. 1a. The first four structure classes are few for all the regions while the blurred regions (Regions #1 and #2) are null and thus more sparse. However, for the fifth class, the clear regions (Regions #3 and #4) contain apparently more extreme points than the blurred regions (Regions #1 and #2). The map of the number of extreme points of the fifth structure class (Fig. 1c) also shows this observation. Therefore, the fifth class can be more efficient and is adopted by us as a part of the blur descriptor.

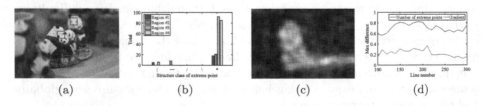

(a)   (b)   (c)   (d)

Fig. 1. An example of the extreme point feature. (a): The example image with four specified clear or blurred regions; (b) the numbers of extreme points for the four regions in (a); (c): the map of the number of the extreme points belonging to the fifth structure class; and (d): the blur measurement abilities of extreme points and gradient.

Let us denote the intensity of $p_{i,j}$ being $E_{i,j}$. The noise-free definition of extreme point belonging to the fifth class, '*', for the center of a 3×3 area is

$$\bigcap_{\substack{i=-1 \\ |i|+|j| \neq 0}}^{+1} \bigcap_{j=-1}^{+1} E_{i,j} > E_{i+k,j+q} \tag{2}$$

where k and q are the numbers constraining the position of $E_{i+k,j+q}$ inside the area, and $|k| + |q| \neq 0$ so that the maximum condition is kept.

Figure 1d shows the performance of the extreme points in comparison with the widely adopted gradient. The number of extreme points belonging to the fifth structure class for a 8×8 region surrounding each pixel is counted. The max differences of either gradient or number of extreme points for each line between the top and the bottom lines of the clear object, No. 100 and 300, are computed, considering that the maximum and minimum of either gradient or number of extreme points in each line represent the clear pixel and the blurred one respectively. It can be seen that the extreme points are more discriminative than gradient and can be an effective feature for blur measurement.

Singular Values. An image I can be decomposed into the weighted sum of n eigen-images by Singular Value Decomposition (SVD) as follows,

$$I = \sum_{i=1}^{n} \lambda_i E_i \tag{3}$$

where $\lambda_i (1 \leq i \leq n)$ are the eigen values in a decreasing order and $E_i (1 \leq i \leq n)$ are rank-1 matrices called eigen-images. The eigen-images capture different detailed information: The more significant eigen-images, the larger scale information captured. A blurred image keeps the shape structure at large scale while discarding the image detail at small scales. Therefore, singular values can depict the subtle differences between blurred and non-blurred pixels.

We take the vector of all the singular values. Figure 2a shows this principle. Each region can have relatively high λ_1 for the largest scale, but its remaining eigen values for smaller scales vary differently: they are more significant in the clear region (Regions #4-#6) than the blurred one (Regions #1-#3) (most of them are nearly null). In addition, λ_1 for the blurred regions varies significantly with some regions (*e.g.*, Regions #5 or #6) even close to the clear ones. Consequently, manually selecting the most significant singular values [11,12] can be invalid. Therefore, the complete set of singular values is adopted by us as an efficient part of the blur descriptor, *i.e.*, $V_{i,j} = (\lambda_1, \ldots, \lambda_n)$.

DCT Coefficients. DCT aims to transform a signal from spatial to frequencial space with cosine functions. High frequencies are sparse with mostly null in comparison with the other ones for a blurred image. Therefore the high frequency DCT coefficients can be used to measure the blur degree. We adopt the efficient

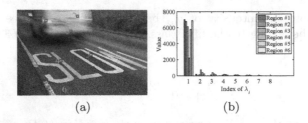

Fig. 2. Demonstration of the singular value feature. (a): The example image with six specified rectangular regions (three blurred regions and three clear ones); (b) comparison of the singular values among the six specified regions in (a). *Note*: The singular values are ordered in a decreasing order.

DCT feature, HiFST [13] which extracts and combines the high-frequency DCT coefficients of the local area in the gradient image multiresolutionally into a vector with their absolute values sorted in an increasing order. Assume the patch size as $W \times W$ and the vector consisting of the absolute high-frequency DCT coefficients of $p_{i,j}$ as $H_{i,j}^W$. The multi-scale HiFST decomposition is

$$F_{i,j} = sort(\bigcup_{r=1}^{m} H_{i,j}^{M_r}) \qquad (4)$$

where $sort(\cdot)$ is a function for incremental ranking, m is the lay total, $M_r = 2^{2+r} - 1$ controls the patch size in different layers.

$F_{i,j}$ is normalized to effectively differentiate clear and blurred regions and the final blur measure is computed through a entropy-weighted max pooling,

$$D_{i,j} = T_{i,j}.\omega_{i,j} \qquad (5)$$

where $T_{i,j}$ is obtained by the max pooling of $F_{i,j}$ among $\frac{W^2+W}{2}$ frequency layers and $\omega_{i,j}$ is the entropy of T computed by $k \times k$ neighborhood surrounding the corresponding pixel in T. Please refer to [13] for more details on HiFST.

2.2 Learning the Blur Measure with Neural Network

We take the NN approach to train a classifier as the metric to estimate the initial blur degree with the blur descriptor. Especially, backpropagation (BP) neural network [18] containing only three layers (input, output, and hidden layers) is used. It is simple to deploy, fast to train, and also experimentally robust.

3   Saliency Constrained Refinement of Blur Measurement

3.1   Relationship Between Saliency and Blur Measure

The blur metric based on the above NN classifier may fail due to flat texture which constitutes a common difficulty for all existing blur measurement methods.

Interestingly, we can see clearly the whole blurred object(s) when observing a partially blurred image even with clear flattened texture. This observation intrigues us to look into the role of of visual saliency in separating the blurred and clear regions. A salient object is the only one viewed by a human due to the difficulty of viewing more than one simultaneously. Therefore, the corresponding saliency map can be exploited to refine the measurement globally.

Figure 3 illustrates the principle of saliency for blur refinement. The accurate blur measurement (Fig. 3b) is obtained with the state-of-the-arts method [13] for fair demonstration. This measurement and the saliency map (Fig. 3c) show the same intensity distributions which can be justified by comparing them directly (Fig. 3d): The difference of blur measures increases when the difference of the saliencies increases. The bigger difference of the saliencies means the more possibility of the pixels belonging to different objects and having different blur degrees. Therefore, saliency can be used to refine the blur measures so that the pixels having the similar saliencies can be refined to have similar measures.

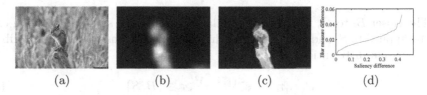

(a) (b) (c) (d)

Fig. 3. Illustration of the relationship between blur measure and saliency. (a): The example image; (b) the accurate measurement of (a); (c): the saliency map; (d): the statistical relation between saliency and blur measure.

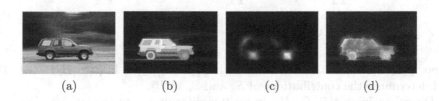

(a) (b) (c) (d)

Fig. 4. An example of the saliency constrained blur measurement refinement. (a): The example image; (b): the saliency map; (c): the inaccurate measurement by Golestaneh and Karam [13]; (d): the refined measurement.

The observation from Fig. 3 is important for images with flat texture which show low frequency distributions as the blurred regions. Saliency consistencies can then be adopted to refine the measures. Figure 4 shows the example car has apparent flat texture (Fig. 4a). The state-of-the-art method (Fig. 4c) cannot correctly measure it with many false estimations in the car, while our saliency constrained method (Fig. 4d) obtains more visually correct results. The details of the saliency constrained refinement is described as follows.

3.2   The Saliency Constrained Refinement Strategy

Assuming the blur measure and saliency of the current pixel p_c being B_c and S_c and its N neighbors being $\mathbb{B}_c = \{B_1, B_2, \ldots, B_N\}$ and $\mathbb{S}_c = \{S_1, S_2, \ldots, S_N\}$ respectively. The refined B_c, $\hat{B}_c$, relies on not only its own initial blur measurement (prior) $P(B_c) = B_c$, but also S_c and $\mathbb{S}_c$, considering that the neighbors sharing the similar saliency should be similarily blurred. Therefore, the refinement process can be formulated as a maximum a posteriori (MAP) estimation,

$$
\begin{aligned}
P(B_c | \mathbb{B}_c, S_c, \mathbb{S}_c) &\propto P(\mathbb{B}_c; S_c; \mathbb{S}_c | B_c) P(B_c) \\
&\propto P(\mathbb{B}_c | B_c) P(S_c, \mathbb{S}_c | B_c) P(B_c)
\end{aligned}
\tag{6}
$$

by assuming the conditional independency between the blur measure and the saliency. It can be further rewritten as:

$$
P(B_c | \mathbb{B}_c, S_c, \mathbb{S}_c) \propto \prod_i P(B_i | B_c) (\prod_i P(S_i | B_c)) P(S_c | B_c) P(B_c).
\tag{7}
$$

The closer B_i to B_c is, the more similar they are. The same observation can also be applied to S_i and S_c. Therefore, this type of relationship can be described as a Gaussian distribution,

$$
P(\tau_j | \tau_c) \propto e^{-(\tau_j - \tau_c)^2}, \tau \in \{B, S\}
\tag{8}
$$

Now taking the posterior distributions, $P(B_j | B_c)$ and $P(S_j | S_c)$ (Eq. 8) into Eq. 7 and deriving its right side w.r.t B_c, we can then deduce the solution of $\hat{B}_c$ by Eq. 7 via MAP estimation as

$$
\begin{aligned}
\hat{B}_c &\propto \underset{B_c}{\arg\max} \ln \prod_i P(B_i | B_c) + \ln \prod_i P(S_i | B_c) + \ln P(S_c | B_c) + \ln P(B_c) \\
&\cong \frac{\sum_i B_i}{N} + g(S_c, \mathbb{S}_c, B_c)
\end{aligned}
\tag{9}
$$

where $g(S_c, \mathbb{S}_c, B_c)$ is the derivative of $\ln \prod_i P(S_i | B_c) + \ln P(S_c | B_c)$ w.r.t. B_c and determines the contributions of S_c and $\mathbb{S}_c$ to B_c.

Directly solving $g(S_c, \mathbb{S}_c, B_c)$ in Eq. 9 is difficult due to the lacking of proper formulations of the embedded distributions. Intuitively, the smaller the saliency difference between S_i and S_c is, the smaller the difference between B_i and B_c is and thus the more $\hat{B}_c$ relies on B_i. In addition, a neighboring pixel is invalid as the reference when its saliency is significantly different from S_c, i. e., they lie on different objects. Therefore, $g(\cdot)$ can be formulated as

$$
g(S_c, \mathbb{S}_c, B_c) = \sum_i max((e^{-\alpha(\|S_i - S_c\|)} - \beta), 0) B_c
\tag{10}
$$

where α translates the weight to be in a more distinugishable range and β controls the threshold for a valid reference.

3.3 Our Measurement Algorithm

Directly applying the above pixel oriented idea can be slow and, therefore, we take a superpixel strategy to accelerate the measurement process, considering the regional constancy in the image. Accordingly, the average measure and saliency of each superpixel from the initial blur measurement and saliency map are adopted for a superpixel-scale refinement. Then a bilateral filtering is used so that pixel-consistent but also structure-preserving results can be obtained.

Figure 5 gives the pipeline. Both Fig. 5b and c are obtained in the pixel scale by the pre-trained NN classifier and the saliency detection algorithm, respectively. Then the superpixel-scale refinement (Fig. 5e) is iteratively applied according to the superpixel segmentation (Fig. 5d) until the saliency consistency is satisfied. Finally, bilateral filtering is applied to obtain the measurement output (Fig. 5f).

Fig. 5. The pipeline of the proposed measurement algorithm. (a): The input partially blurred image; (b): the blur measurement; (c): the saliency map; (d): the superpixel segmentation; (e): the superpixel-scale refinement; and (f): the final output by the bilateral filter.

The supperpixel segmentation is obtained by the simple linear iterative clustering (SLIC) [19] and the saliency is detected by the discriminative regional feature integration (DRFI) [20]. A surrounding superpixel can be from a different object in a high probability if its saliency is larger than that of the center suprpixel. In this case, the surrounding superpixel is invalid as the reference for Eq. 10. Therefore, only the saliency of the superpixel with the max weight among neighbors is considered in Eq. 10.

4 Experimental Results

Our method is formatted into two versions to show the performance of the saliency constrained refinement: *Our method without saliency* and *our method*. The former does not include the saliency constrained refinement while the latter is the complete implementation of the proposed method. Previous algorithms to be compared include Liu *et al.* [3], Chakrabarti *et al.* [21], Su *et al.* [11], Shi *et al.* [4], Shi *et al.* [9], Tang *et al.* [15], Yi and Eramian [10] and Golestaneh and Karam [13]. The experiments were performed on the dataset of Shi *et al.* [4] which includes 294 motion blurred images and 704 defocus ones. In the implementation of our method, there are 20 nodes in the middle layer of BP with the max iteration, error threshold and learning ratio are set to 5000, 0.00006 and 0.004, respectively; the approximate size of each supperpixel, and α and β in Eq. 10 are set to 10, 10 and 0.7, respectively; and the max iterations of the refinement is set to 3.

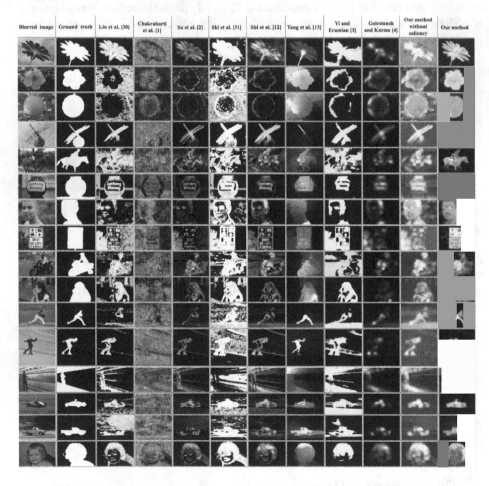

Fig. 6. Qualitative comparison of the blur measurement performance.

Figure 6 shows the qualitative results of different methods. The top-listed four images are richly textured while the remaining 12 images contain some flat textures (Those nearer to the bottom generally contain larger flat textures). The results of *our method without saliency* are comparable with the best ones of the previous methods and *our method* produced consistently better results without significant wrong measure than all other methods, partially due to the benefits of the saliency constrained global refinement.

Statistical comparison of different methods was performed using the precision-recall curve. Two types of comparison were taken: One is the comparison with all images in the dataset of Shi *et al.* [4] and the other is with only the images containing apparently flat textures (Fig. 7) in the dataset (There are 82 images in total). Figure 8 shows *our method* achieved better results than all other methods for both comparisons and even *our method without saliency* performed better for the flatten textures images than Golestaneh and Karam [13] (Fig. 8b). The robust extreme points introduced in our approach are significantly sparse in the blurred area and thus are extraordinarily effective for the low textured images.

Fig. 7. Examples of the image with flat textures for the statistical comparison.

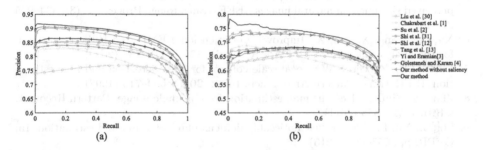

Fig. 8. Statistical comparison of the blur measurement performance. (a): Comparison with all images in the dataset; and (b): comparison with only the images containing flat textures (Fig. 7).

5   Conclusion

This paper proposes three contributions: (1) A novel blur descriptor consisting of the number of extreme points, the vector of all singular values, and the entropy-weighted pooling of high frequency DCT coefficients; (2) a global refinement strategy using the saliency map to update the initial blur measurement; and (3) a novel blur measurement method which integrates the NN classifier as a blur metric for the initial blur measurement, the superpixel-scale refinement with the saliency constraint and the bilateral filtering for final structure-preserving oriented output. Experiments show the proposed method is effective, especially for the images with flat textures.

Our method can be very slow for a large image and thus we will study faster algorithm in the future. Current method may not work well when saliency detection fails. Therefore, more effective global refinement strategy is also one of our future directions for a better blur measurement.

Acknowledgement. This work is co-supported by Key Science & Technology Program of Anhui Province (1604d0802004), Nature Science Foundation of China (61502005) and Nature Science Foundation of Anhui Province (1608085QF129, 1708085MF151).

References

1. Tai, Y.W., Brown, M.S.: Single image defocus map estimation using local contrast prior. In: ICIP, pp. 1797–1800 (2009)
2. Park, J., Tai, Y.W., Cho, D., Kweon, I.S.: A unified approach of multi-scale deep and hand-crafted features for defocus estimation. In: CVPR, pp. 1063–6919 (2017)
3. Liu, R., Li, Z., Jia, J.: Image partial blur detection and classification. In: CVPR, pp. 1–8 (2008)
4. Shi, J., Xu, L., Jia, J.: Discriminative blur detection features. In: CVPR, pp. 2965–2972 (2014)
5. Vu, C.T., Phan, T.D., Chandler, D.M.: S_3: a spectral and spatial measure of local perceived sharpness in natural images. IEEE Trans. Image Process. **21**(3), 934–945 (2012)
6. Xu, G., Wang, X., Xu, X.: Improved bi-dimensional EMD and Hilbert spectrum for the analysis of textures. Pattern Recognit. **42**(5), 718–734 (2009)
7. Elder, J.H., Zucker, S.W.: Local scale control for edge detection and blur estimation. IEEE Trans. Pattern Anal. Mach. Intell. **20**(7), 699–716 (1998)
8. Zhuo, S., Sim, T.: Defocus map estimation from a single image. Pattern Recognit. **44**(9), 1852–1858 (2011)
9. Shi, J., Xu, L., Jia, J.: Just noticeable defocus blur detection and estimation. In: CVPR, pp. 657–665 (2015)
10. Yi, X., Eramian, M.: LBP-based segmentation of defocus blur. IEEE Trans. Image Process. **25**(4), 1626–1638 (2016)
11. Su, B., Lu, S., Tan, C.L.: Blurred image region detection and classification. In: ACM Multimedia, pp. 1397–1400 (2011)
12. Fang, X., Shen, F., Guo, Y., Jacquemin, C., Zhou, J., Huang, S.: A consistent pixel-wise blur measure for partially blurred images. In: ICIP, pp. 496–500 (2014)
13. Golestaneh, S.A., Karam, L.J.: Spatially-varying blur detection based on multiscale fused and sorted transform coefficients of gradient magnitudes. In: CVPR, pp. 5800–5809 (2017)
14. Javaran, T.A., Hassanpour, H., Abolghasemi, V.: Automatic estimation and segmentation of partial blur in natural images. Vis. Comput. **33**(2), 151–161 (2017)
15. Tang, C., Wu, J., Hou, Y., Wang, P., Li, W.: A spectral and spatial approach of coarse-to-fine blurred image region detection. IEEE Signal Process. Lett. **23**(11), 1652–1656 (2016)

16. Bhattacharya, S., Venkatesh, K., Gupta, S.: Visual saliency detection using spatiotemporal decomposition. IEEE Trans. Image Process. **27**(4), 1665–1675 (2018)

17. Borji, A.: What is a salient object? A dataset and a baseline model for salient object detection. IEEE Trans. Image Process. **24**(2), 742–756 (2015)

18. Hecht-Nielsen, R.: Theory of the backpropagation neural network. Neural Netw. **1**(1), 445–445 (1988)

19. Achanta, R., Shaji, A., Smith, K., Lucchi, A., Fua, P., Ssstrunk, S.: SLIC superpixels compared to state-of-the-art superpixel methods. IEEE Trans. Pattern Anal. Mach. Intell. **34**(11), 2274–2282 (2012)

20. Jiang, H., Wang, J., Yuan, Z., Wu, Y.: Salient object detection: a discriminative regional feature integration approach. In: CVPR, pp. 2083–2090 (2013)

21. Chakrabarti, A., Zickler, T., Freeman, W.T.: Analyzing spatially-varying blur. In: CVPR, pp. 2512–2519 (2010)

SCAN: Spatial and Channel Attention Network for Vehicle Re-Identification

Shangzhi Teng[1(✉)], Xiaobin Liu[2], Shiliang Zhang[2], and Qingming Huang[1]

[1] University of Chinese Academy of Sciences, Beijing, China
shangzhi.teng@vipl.ict.ac.cn, qmhuang@ucas.ac.cn
[2] Peking University, Beijing, China
{xbliu.vmc,slzhang.jdl}@pku.edu.cn

Abstract. Most existing methods on vehicle Re-Identification (ReID) extract global features on vehicles. However, as some vehicles have the same model and color, it is hard to distinguish them only depend on global appearance. Compared with global appearance, some local regions could be more discriminative. Moreover, it is not reasonable to use feature maps with equal channels weights for methods based on Deep Convolutional Neural Network (DCNN), as different channels have different discrimination ability. To automatically discover discriminative regions on vehicles and discriminative channels in networks, we propose a Spatial and Channel Attention Network (SCAN) based on DCNN. Specifically, the attention model contains two branches, *i.e.*, spatial attention branch and channel attention branch, which are embedded after convolutional layers to refine the feature maps. Spatial and channel attention branches adjust the weights of outputs in different positions and different channels to highlight the outputs in discriminative regions and channels, respectively. Then feature maps are refined by our attention model and more discriminative features can be extracted automatically. We jointly train the attention branches and convolutional layers by triplet loss and cross-entropy loss. We evaluate our methods on two large-scale vehicle ReID datasets, *i.e.*, *VehicleID* and *VeRi-776*. Extensive evaluations on two datasets show that our methods achieve promising results and outperform the state-of-the-art approaches on *VeRi-776*.

Keywords: Vehicle Re-Identification
Deep Convolutional Neural Network · Attention

1 Introduction

Vehicle Re-Identification (ReID) aims to match a query vehicle image against a large-scale vehicle image gallery set [3,11–13,27]. The ability to quickly find and track suspect vehicles makes vehicle ReID important for traffic surveillance and applications on smart city. Vehicle ReID is related with several extensively studied tasks on vehicle identification, such as vehicle attribute prediction [29] and fine-grained vehicle classification [8,11,29]. Different from these tasks that

© Springer Nature Switzerland AG 2018
R. Hong et al. (Eds.): PCM 2018, LNCS 11166, pp. 350–361, 2018.
https://doi.org/10.1007/978-3-030-00764-5_32

Fig. 1. Illustration of challenging issues on vehicle ReID. The first row shows four images that capturing the same vehicle. The second row shows different vehicles with the same model and color.

Fig. 2. Examples of attention regions on different vehicles. The first row shows input vehicle images. The second row shows salient image regions which are learned by our attention model.

mainly focus on identifying the fine-grained categories of vehicles, vehicle ReID focuses on the instance-level identification. As different instances of the same maker and model may be similar with each other, vehicle ReID is more challenging and far from being solved.

Vehicle ReID task requires highly discriminative features to precisely identify different vehicles. However, in surveillance scenario, the quality of vehicle images could be easily affected by many factors, such as illumination, view point, and occlusion. This makes hand-crafted features unstable and prevents them from working. Recently, Deep Convolutional Neural Networks (DCNNs) have made breakthrough in many tasks including person ReID [9,10,19,21,24–26] and fine-grained categorization [6,22,30]. Existing works have designed many DCNN based model for vehicle feature learning and deep model have dominated the methods of vehicle ReID. More details of related works will be summarized in Sect. 2.1.

Although previous works have achieved significant success, there still remain several open issues that make vehicle ReID a challenging task. Firstly, different views of the same vehicle may capture little common region, as shown in the first row of Fig. 1. This results in large intra-class distance and false negative samples in ReID. Secondly, lots of vehicles with the same model (maker, product year and type) and color are quite similar. For example, Fig. 1 shows some different vehicles with similar appearance in the second row. It can be observed that only from some small regions, *e.g.*, the annual inspection marks on the front window, can we tell the difference. Lastly, the misalignment problem is serious in vehicle ReID. Vehicle image is difficult to align each part with a fixed order. As most existing DCNN based approaches [3,11,13] extract features from the whole vehicle image, they fail to conquer aforementioned issues very well.

Inspired to conquer these issues, we propose a Spatial and Channel Attention Network (SCAN) based on DCNN. SCAN contains two branches, *i.e.*, spatial attention branch and channel attention branch, to explore discriminative regions on vehicles and discriminative channels in networks, respectively. SCAN produces saliency weight maps to highlight discriminative areas and channels. Some examples of generated spatial saliency maps are shown in Fig. 2. Compared with forcing the model to extract regional features from rigid local regions and some certain channels, SCAN uses a automatical soft attention strategy and thus can adaptively explores discriminative regions and channels for different input vehicles. The convolutional layers and attention branches are jointly trained in an end-to-end manner to simultaneously learn feature maps and spatial-channel attention, respectively.

Our methods are evaluated on two large-scale vehicle ReID datasets. Experimental results show that our methods achieve promising performance compared with recent works. The contributions of our method is two-fold. (1) A deep learning attention network is proposed for vehicle ReID. The attention module is designed to refine the feature maps in CNN. This reinforces the useful details and weakens the useless information. Thus, our model can make the feature representation more discriminating. As far as we know, this is the first attempt of learning attention network for solving the vehicle ReID problem without any extra annotation. (2) We test the effectiveness of our attention network on two vehicle ReID datasets. Experiments show that our attention model outperforms the state-of-the-art methods on *VeRi*.

2   Related Work

2.1   Vehicle ReID

With the development of smart city and public security, vehicle ReID has gained more and more attentions. Liu *et al.* [13,14] propose a vehicle ReID dataset *VeRi-776* which contains over 50,000 images of 776 vehicles captured by 20 cameras covering an elliptical area. Number plate, vehicle appearances and spatio-temporal relation are separately used in [14] to learn the similarity scores between pairs of images. Shen *et al.* [18] also use spatio-temporal information for improving the ReID results. A chain MRF model is used to generate a visual-spatio-temporal path which gives a similarity score by LSTM. Wang *et al.* [23] pre-train a region proposal module in order to produce the response maps of 20 vehicle key points. They then extract regional features based on key points prediction. And the spatial-temporal constraints is also adopted to refine the retrieval results. This work needs extra notation of key points on a large-scale dataset to pre-train the model, and the model is complex. Moreover, in most cases we do not have spatial and temporal information. So the above methods may not work out. Liu *et al.* [11] propose a Coupled Clusters Loss (CCL) which modifies traditional triplet loss. And a vehicle ReID dataset is proposed. Yan *et al.* [28] use multi-grain relations to improve vehicle search performance. These two methods only extract features from global appearance, resulting in features lack discrimination power.

2.2 Fine-Grained Vehicle Classification

Fine-grained classification is relevant to ReID task. They all focus on learning discriminative feature representations. Wang *et al.* [22] propose a patch-based framework. It acquires triplets of patches with geometric constraints and automatically mines discriminative geometrically-constrained triplets for classification. However, this method is not an end-to-end framework and it can not handle large view changes. Huang *et al.* [6] propose another part based models which pre-trains a region proposal network to capture the discriminative object regions. Different from these part-based methods, Zhang *et al.* [31] and Qian *et al.* [17] use distance metric learning to reduce the intra-class distance and increase inter-class distance. Metric learning is also widely used in ReID task. Yang *et al.* [29] propose a fine-grained vehicle model classification dataset (CompCars) which is the largest vehicle model dataset.

2.3 Attention Modelling

Recently, deep attention learning methods have been proposed in many tasks to handle the matching misalignment challenge. Many tasks have demonstrate that the attention model is valid. Such as semantic segmentation [2], visual question answering [15], tracking [33] and person ReID [9,10,19]. Most of these attention models are designed to select several parts from the whole image. Local and global characteristics are concatenated together to get a more comprehensive representation. In our work, SCAN produces saliency weight maps to explore discriminative regions on vehicles and discriminative channels in networks.

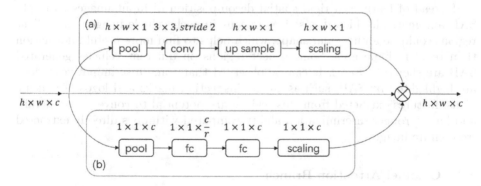

Fig. 3. The structure of SCAN consists of (a) spatial attention branch and (b) channel attention branch. The ReLU and Reshape operation are not shown for brevity.

3 Spatial and Channel Attention Network

This section presents the end-to-end trainable framework of Spatial and Channel Attention Network (SCAN). SCAN consists of two branches: one for spatial attention learning and one for channel attention learning, respectively. The structure of proposed attention module is shown in Fig. 3. These two attention branches learn weights separately and then fuse weights with the original feature maps. The SCAN module can be integrated into any modern DCNN architectures.

3.1 Spatial Attention Branch

We employ a Spatial Attention Branch (SAB) to automatically discover discriminative regions on vehicles. The structure of SAB is illustrated in Fig. 3(a). The input of SAB is a set of feature maps denoted as $f \in R^{h \times w \times c}$, where h, w, and c denote the size of height, width and channels of feature maps, respectively. SAB first uses a channel-wise global average pooling layer without involving more parameter. The channel-wise global average pooling at the spatial location (i, j) is defined as follows:

$$s_{i,j} = \frac{1}{c} \sum_{k=1}^{c} f_{i,j,k},$$

(1)

where $f_{i,j,k}$ is the value of feature maps at the location (i, j) of kth channel. SAB then uses a convolutional layer of 3×3 filter with stride 2, and an up sampling layer is then added after the convolutional layer. A convolutional layer of 1×1 filter is added to automatically learn an adaptive attention scale. The ReLU function is applied as active function to each convolutional layer. A deconvolutional layer is finally used to generated spatial attention feature maps. We use sigmoid function to normalise the value of each output of the generated attention feature maps into the range between 0.5 and 1.

Instead of learning a rigid spatial decomposition of input images as in [4], SAB automatically identifies salient regions in each vehicle image. The salient regions could be an irregular shape, which helps to find more useful information than regular shape area. Some salient regions on different vehicles generated SAB are shown in Fig. 2. It can be observed that some discriminative regions are highlighted by SAB, such as annual inspection marks and logos of makers. Thus, features extracted from this regions are potential to convey more details and have stronger discrimination ability compared with ones directly extracted from entire images.

3.2 Channel Attention Branch

The purpose of designing Channel Attention Branch (CAB) is to improve the discrimination power of the network by explicitly modelling the interdependencies between the channels of its convolutional features [5]. The structure of CAB is illustrated in Fig. 3(b). The input to CAB is the same as SAB. CAB first uses a global average pooling layer to integrate spatial information in each feature map. The global average pooling is defined as:

$$c_k = \frac{1}{h \times w} \sum_{i=1}^{h} \sum_{j=1}^{w} f_{i,j,k}$$

(2)

Then two fully connected layers with $\frac{c}{r}$ and c outputs are used. Parameter r is designed for reducing the model parameter number from c^2 to ($\frac{c^2}{r} + \frac{c^2}{r}$), e.g. only $\frac{c}{8}$ parameters are involved when r = 16. A 1×1 convolutional layer is used to scale the output, and a deconvolutional layer is finally used to produce the attention weight map of the same size as the input feature map.

3.3 Vehicle ReID by SCAN

We add proposed two attention branch, *i.e.*, SAB and CAB, after the conv5 layer in VGG_CNN_M_1024 and the conv5_3 layer in VGG16. After the SCAN we add a global average pooling layer, a 512-D fully connected layer and two loss layers as illustrated in Fig. 4. Given a trained SCAN model, we use the 512-D fully connected layer as the vehicle feature. For vehicle ReID , we calculate L_2 distance between query images and each gallery image using this 512-D deep feature. We then rank all gallery images in ascendant order by their L_2 distances to the probe image. Based on ranking results, we could find and track vehicles in surveillance video analysis.

4 Experiments

4.1 Datasets and Base Model

We use two existing vehicle datasets to validate SCAN. *VeRi-776* [12] is a benchmark dateset for vehicle ReID that is collected from real-world surveillance scenarios, with over 50,000 images of 776 vehicles in total. Each vehicle is captured by 2 to 18 cameras in an urban area of 1 km^2 during a 24-hour time period. *VehicleID* [1] is a surveillance dataset, which consists 26,267 vehicles and 22,1763 images in total. The numbers of identities and images for training and testing are listed in Table 1.

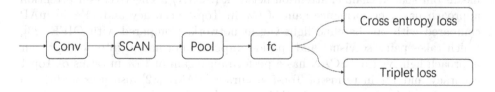

Fig. 4. We use cross entropy loss and triplet loss to train our spatial channel attention network.

Experiments are conducted based on the widely-used deep learning framework Caffe [7]. We use VGG_CNN_M_1024 [1] network and VGG16 [20] network as the basic network. Different from original VGG_CNN_M_1024 and VGG16 network, our baseline network replace all fully connected layers of original network with a 512-D layer. This change greatly reduces the number of parameters in the network. For VGG_CNN_M_1024, our model size is 12 times smaller than the original one. For VGG16, our model size is 16 times smaller than the original one. Model size are illustrated in Table 4. And then we call our baseline network light_vgg_m and light_vgg16. The mean Average Precision (mAP), Top-1 accuracy and Top-10 accuracy are chosen as the evaluation metric.

Table 1. Statistics of the two datasets used in our experiment.

Dataset	Train ID/image	Probe ID/image	Gallery ID/image
VeRi-776	576/37778	200/1678	200/11579
VehicleID	13164/100182	2400/17638	2400/2400

The batch size used to train the network in our experiments is set to 64. The initial learning rate is set to 0.001. The learning rate decay factor is 0.8 for every 2,0000 iterations. The weight decay factor is set to 0.0004. The momentum is set to 0.9. The loss weight for cross entropy loss is set to 1 and the loss weight for triplet loss is also set to 1.0.

4.2   Results on VeRi-776

The experimental results on *VeRi* are summarised in Table 2. It can be observed that our proposed attention approach achieves the best performance on *VeRi-776* Dataset. light_vgg_m+SCAN denotes the light_vgg_m network with proposed SCAN. And light_vgg16+SCAN denotes the light_vgg16 network with proposed SCAN. Note that the Top-1 accuracy of our baseline model light_vgg_m is higher than most existing methods by large margins. But the mAP of our baseline model is lower than OIFE [23] and VAMI [32]. Significant performance gain can be observed from light_vgg_m+SCAN compared with light_vgg_m, which means our spatial-channel-attention network (SCAN) is effective. Our attention approach has a performance gain of 6% in Top-1 accuracy and 11% in mAP compared with our baseline light_vgg_m network. Compared with OIFE [23], which takes pairwise visual and spatio-temporal information into account, our approach light_vgg_m+SCAN has a performance gain of 15% in terms of Top-1 accuracy and 1% in terms of Top-5 accuracy. VAMI [32] also uses additional perspective information to train their network and the Top-1 accuracy is lower than our attention model. The same improvements can also be observed in performance by light_vgg16 and light_vgg16+SCAN. The performance of our model light_vgg_m+SCAN is a little higher than light_vgg16+SCAN. One possible reason is that the image resolution in *VeRi-776* dataset is small and the small size network can learn its parameters well with low resolution images.

Table 2. Experiment results of the proposed method and other compared methods on *VeRi-776* dataset.

VeRi-776	Top-1 (%)	Top-5 (%)	mAP (%)
KEPLER [16]	48.2	64.3	33.53
FACT [14]	50.95	73.48	18.49
FACT+Plate-SNN+STR [14]	61.44	78.78	27.77
OIFE [23]	68.3	89.7	51.42
VAMI [32]	77.03	90.82	50.13
light_vgg_m	76.02	86.05	38.94
light_vgg_m+SCAN	**82.24**	**90.76**	49.87
light_vgg16	76.82	86.71	39.91
light_vgg16+SCAN	79.92	88.32	50.15

4.3 Results on VehicleID

The experimental results on *VehicleID* are summarised in Table 3. On this dataset we also trained two baseline networks: light_vgg_m and light_vgg16. Because our baseline model has fewer parameters than normal VGG_M and VGG16 networks, the performance of our baseline on *VehicleID* dataset is not very good. However, after adding our attention module the performance has been significant improved. Our attention approach light_vgg_m+SCAN has a improvements gain of 11% in terms of Top-1 accuracy and 6% in terms of Top-5 accuracy compared with light_vgg_m network. light_vgg16+SCAN has a gain of 3% in terms of Top-1 accuracy and 5% in terms of Top-5 accuracy compared

Table 3. Experiment results of the proposed method and other compared methods on *VehicleID* dataset. light_vgg16+SCAN* indicates we use the SCAN feature vector and light_vgg16 feature vector together.

VehicleID	Top-1 (%)	Top-5 (%)
KEPLER [16]	45.4	68.9
VGG + Triplet Loss [11]	31.9	50.3
VGG + CCL [11]	32.9	53.3
Mixed Diff + CCL [11]	38.2	61.6
OIFE [23]	67.0	82.9
VAMI [32]	47.34	70.29
light_vgg_m	44.14	65.21
light_vgg_m+SCAN	55.73	71.73
light_vgg16	60.63	72.67
light_vgg16+SCAN	63.52	77.53
light_vgg16+SCAN*	**65.44**	**78.47**

Fig. 5. Visualisation of our spatial attention in vehicle ReID. The odd rows show the original vehicle. The even rows show the attention weight maps learned from our spatial attention module. Best viewed in color.

with light_vgg16 network. If we concatenate features extracted by light_vgg16 and light_vgg16+SCAN, the performance may have further improved. The final Top-1 accuracy of light_vgg16+SCAN* is 65.44%, which is a little lower than OIFE [23]. However, [23] needs extra annotation information to pre-train a key point regressor network. And four vehicle datasets are used in [23]. We only use *VehicleID* without extra annotation to train our model. So our method has been proved to be effective.

Table 4. Comparisons of model size. NP denotes the number of parameters in each model.

Model	NP (million)
VGG_M_1024	86.2
VGG16	127.2
light_vgg_m	6.8
light_vgg16	7.9
SCAN	0.033

4.4 Visualisation of Spatial Attention and Model Size

We visualise our learned spatial attention of SCAN in Fig. 5. It can be observed that spatial attention branch locates some spatial regions of vehicles, which approximately corresponds to headlights, taillights, vehicle signs, and vehicle marks. This compellingly shows the effectiveness of our spatial attention learning. We compare model size of original models and our light models in Table 4. It is clear that our proposed light models use less parameters than original models, while achieve better performance.

5 Conclusions

In this work, we focus on the problem of vehicle re-identification, which aims at finding out the images belonging to exactly the same vehicle with the query image. To address this problem, we proposed an end-to-end trainable framework, namely Spatial Channel Attention Network (SCAN), for joint learning attention weights and feature representation. SCAN consists of two branches, $i.e.$, spatial attention branch and channel attention branch, to adjust the weight of outputs in different positions and channels. With our SCAN model we could explore discriminative regions and channels for powerful feature extraction. The proposed SCAN does not need bounding box or part annotations for training. We evaluated our proposed approach on two vehicle ReID datasets and a series of experiments show the validity of our model. Our two baseline network are all lightweight CNN architectures. So it's easy to embed our model in mobile devices.

Acknowledgments. This work was supported in part by National Natural Science Foundation of China under Grant: No. 61572050, 91538111, 61429201, 61620106009, 61332016, U1636214, 61650202, and the National 1000 Youth Talents Plan, in part by National Basic Research Program of China (973 Program): 2015CB351800, in part by Key Research Program of Frontier Sciences, CAS: QYZDJ-SSW-SYS013.

References

1. Chatfield, K., Simonyan, K., Vedaldi, A., Zisserman, A.: Return of the devil in the details: delving deep into convolutional nets. arXiv preprint arXiv:1405.3531 (2014)
2. Chen, L.C., Yang, Y., Wang, J., Xu, W., Yuille, A.L.: Attention to scale: scale-aware semantic image segmentation. In: Proceedings of the IEEE Conference on Computer Vision and Pattern Recognition, pp. 3640–3649 (2016)
3. Feris, R.S., et al.: Large-scale vehicle detection, indexing, and search in urban surveillance videos. IEEE Trans. Multimed. **14**(1), 28–42 (2012)
4. Fu, J., Zheng, H., Mei, T.: Look closer to see better: recurrent attention convolutional neural network for fine-grained image recognition. In: CVPR, vol. 2, p. 3 (2017)
5. Hu, J., Shen, L., Sun, G.: Squeeze-and-excitation networks, vol. 7. arXiv preprint arXiv:1709.01507 (2017)
6. Huang, S., Xu, Z., Tao, D., Zhang, Y.: Part-stacked CNN for fine-grained visual categorization. In: Proceedings of the IEEE Conference on Computer Vision and Pattern Recognition, pp. 1173–1182 (2016)
7. Jia, Y., et al.: Caffe: convolutional architecture for fast feature embedding. In: Proceedings of the 22nd ACM International Conference on Multimedia, pp. 675–678. ACM (2014)
8. Krause, J., Stark, M., Deng, J., Fei-Fei, L.: 3D object representations for fine-grained categorization. In: Proceedings of the IEEE International Conference on Computer Vision Workshops, pp. 554–561 (2013)
9. Li, S., Bak, S., Carr, P., Wang, X.: Diversity regularized spatiotemporal attention for video-based person re-identification. In: Proceedings of the IEEE Conference on Computer Vision and Pattern Recognition, pp. 369–378 (2018)
10. Li, W., Zhu, X., Gong, S.: Harmonious attention network for person re-identification. In: CVPR, vol. 1, p. 2 (2018)
11. Liu, H., Tian, Y., Yang, Y., Pang, L., Huang, T.: Deep relative distance learning: tell the difference between similar vehicles. In: Proceedings of the IEEE Conference on Computer Vision and Pattern Recognition, pp. 2167–2175 (2016)
12. Liu, X., Zhang, S., Huang, Q., Gao, W.: Ram: a region-aware deep model for vehicle re-identification. In: ICME (2018)
13. Liu, X., Liu, W., Ma, H., Fu, H.: Large-scale vehicle re-identification in urban surveillance videos. In: 2016 IEEE International Conference on Multimedia and Expo (ICME), pp. 1–6. IEEE (2016)
14. Liu, X., Liu, W., Mei, T., Ma, H.: A deep learning-based approach to progressive vehicle re-identification for urban surveillance. In: Leibe, B., Matas, J., Sebe, N., Welling, M. (eds.) ECCV 2016. LNCS, vol. 9906, pp. 869–884. Springer, Cham (2016). https://doi.org/10.1007/978-3-319-46475-6_53
15. Lu, J., Yang, J., Batra, D., Parikh, D.: Hierarchical question-image co-attention for visual question answering. In: Advances In Neural Information Processing Systems, pp. 289–297 (2016)
16. Martinel, N., Micheloni, C., Foresti, G.L.: Kernelized saliency-based person re-identification through multiple metric learning. IEEE Trans. Image Process. **24**(12), 5645–5658 (2015)
17. Qian, Q., Jin, R., Zhu, S., Lin, Y.: Fine-grained visual categorization via multi-stage metric learning. In: Proceedings of the IEEE Conference on Computer Vision and Pattern Recognition, pp. 3716–3724 (2015)

18. Shen, Y., Xiao, T., Li, H., Yi, S., Wang, X.: Learning deep neural networks for vehicle re-ID with visual-spatio-temporal path proposals. In: 2017 IEEE International Conference on Computer Vision (ICCV), pp. 1918–1927. IEEE (2017)
19. Si, J., et al.: Dual attention matching network for context-aware feature sequence based person re-identification. arXiv preprint arXiv:1803.09937 (2018)
20. Simonyan, K., Zisserman, A.: Very deep convolutional networks for large-scale image recognition. arXiv preprint arXiv:1409.1556 (2014)
21. Su, C., Li, J., Zhang, S., Xing, J., Gao, W., Tian, Q.: Pose-driven deep convolutional model for person re-identification. In: ICCV (2017)
22. Wang, Y., Choi, J., Morariu, V., Davis, L.S.: Mining discriminative triplets of patches for fine-grained classification. In: Proceedings of the IEEE Conference on Computer Vision and Pattern Recognition, pp. 1163–1172 (2016)
23. Wang, Z., et al.: Orientation invariant feature embedding and spatial temporal regularization for vehicle re-identification. In: Proceedings of the IEEE Conference on Computer Vision and Pattern Recognition, pp. 379–387 (2017)
24. Wei, L., Liu, X., Li, J., Zhang, S.: VP-ReID: vehicle and person re-identification system. In: ICMR (2018)
25. Wei, L., Zhang, S., Gao, W., Tian, Q.: Person transfer GAN to bridge domain gap for person re-identification. In: CVPR (2018)
26. Wei, L., Zhang, S., Yao, H., Gao, W., Tian, Q.: Glad: global-local-alignment descriptor for pedestrian retrieval. In: ACM MM (2017)
27. Xu, Q., Yan, K., Tian, Y.: Learning a repression network for precise vehicle search. arXiv preprint arXiv:1708.02386 (2017)
28. Yan, K., Tian, Y., Wang, Y., Zeng, W., Huang, T.: Exploiting multi-grain ranking constraints for precisely searching visually-similar vehicles. In: The IEEE International Conference on Computer Vision (ICCV) (2017)
29. Yang, L., Luo, P., Change Loy, C., Tang, X.: A large-scale car dataset for fine-grained categorization and verification. In: Proceedings of the IEEE Conference on Computer Vision and Pattern Recognition, pp. 3973–3981 (2015)
30. Yao, H., Zhang, S., Zhang, Y., Li, J., Tian, Q.: Coarse-to-fine description for fine-grained visual categorization. IEEE Trans. Image Process. **25**(10), 4858–4872 (2016)
31. Zhang, X., Zhou, F., Lin, Y., Zhang, S.: Embedding label structures for fine-grained feature representation. In: Proceedings of the IEEE Conference on Computer Vision and Pattern Recognition, pp. 1114–1123 (2016)
32. Zhou, Y., Shao, L.: Aware attentive multi-view inference for vehicle re-identification. In: Proceedings of the IEEE Conference on Computer Vision and Pattern Recognition, pp. 6489–6498 (2018)
33. Zhu, Z., Wu, W., Zou, W., Yan, J.: End-to-end flow correlation tracking with spatial-temporal attention. arXiv preprint arXiv:1711.01124 (2017)

Speech Data Enhancement Based on Hybrid Neural Network

Xinyue Cao[1], Xiao Sun[1(✉)], and Fuji Ren[1,2]

[1] School of Computer and Information, HeFei University of Technology,
Hefei, Anhui, China
cxy35366@163.com, sunx@hfut.edu.cn
[2] School of Computer and Information, Tokushima University, Tokushima, Japan

Abstract. With the rapid development of artificial intelligence, the recognition of speech, text, physiological signals and facial expressions has drawn more and more attention from scholars at home and abroad. Therefore, we cannot just study the problems of one area, but more we look for the similarities across fields. In this paper, the method of image enhancement is adapted according to the speech characteristics, and several feasible methods of speech data enhancement are proposed to avoid the problems of data collection and corpus limitation in speech emotion recognition. Based on the Hybrid neural network (Convolution Neural Network, CNN and Recurrent neural network, RNN) model, the feasibility and performance of the method are verified and showed through several sets of comparative experiments in different methods while a high recognition accuracy is obtained.

Keywords: Speech emotion recognition · Data enhancement
Speech time-frequency map · Hybrid neutral network

1 Introduction

In the field of speech emotion recognition, the traditional method is based on speech characterization information and standard sound features, such as mfcc, short-term energy, spectral entropy, formant and so on [3]. Zhou et al. [14] proposed a speech emotion recognition system using both spectral and prosodic features different from the traditional methods. These are based on the inherent properties of the speech data, but this method will be limited by the amount of speech data. It is really difficult to collect and process speech fragments.

This article is primarily inspired by converting the speech segment to the time-frequency map through Fourier transform, replacing the speech with a picture and then use the Hybrid neural network to extract features. Attempt to combine traditional image enhancement methods with the features of speech segments to construct several different speech data enhancement methods. Aiming at new image sets generated by several different data enhancement methods, the model is trained by the CNN and RNN hybrid model and finally, the results

© Springer Nature Switzerland AG 2018
R. Hong et al. (Eds.): PCM 2018, LNCS 11166, pp. 362–372, 2018.
https://doi.org/10.1007/978-3-030-00764-5_33

are obtained by several groups of contrast experiments. The method described in this paper can solve the problem of data limitation and improve the accuracy of speech emotion recognition effectively.

At the same time, in this paper, we use the data that is completely different from all train sets as the verification set, and the robustness of the model is analyzed through the classification effect at different times. On the other hand, we use a variety of evaluation indicators such as pricision, recall rate and F1-score for comprehensive data analysis. This article starts from the three points above and makes the following contributions:

1. Feature extraction using speech time-frequency map instead of traditional acoustic features. 2. Considering the feature-based fusion image enhancement method for data enhancement and solving the problem of difficult collection of corpus data. 3. A hybrid network model is proposed, taking into account both the speech timing characteristics and image convolution features.

2   Related Work

2.1   Traditional Speech Emotion Recognition

Traditional speech emotion recognition is mainly composed of two parts: first is to extract features, and the second is to design a classifier based on the extracted features. At present, the most used classifiers are support vector machines and hidden Markov models. On the other hand, due to the wide variety of speech features, how to extract discriminative features have gradually become the focus of research.

Traditional emotional features of speech can be divided into two parts: acoustic-based emotional features and semantic-based emotional features. Among them, the most commonly used prosodic features are pitch, energy, fundamental frequency, and duration, which have been widely accepted by researchers because of their strong emotional discrimination ability. Jin used both acoustic and lexical features achieve four-class emotion recognition accuracy of 69.2% on the USC-IEMOCAP database [6]. Kandali presents new features based on GMMs and MFCC in speech emotion recognition [1].

2.2   Speech Emotion Recognition Based on Neural Network

With the advancement and development of deep learning, more and more researchers begin to use deep neural networks to extract features. Chao et al. [2] used unsupervised pre-training to denoise the auto-encoder to reduce the speaker's influence on emotional features. Mao et al. [5] proposed a semi-supervised CNN model to extract emotion-related features. Experiments prove that it has strong robustness to speaker changes, environmental interference, and language changes. Mariooryad et al. [10] constructed a phoneme-level trajectory model for the feature and decomposed the speaker's characteristics from the acoustic features to make up for the speaker's influence on speech emotion recognition. George Trigeorgis [12] and others proposed a method of identifying the audio signal by using the raw audio waveform as an input and using the model of a deep convolutional recurrent network. The invariability of convolution can also be used to overcome the diversity of speech itself. Lim [7] proposed to convert the audio fragments into time-frequency spectra, then construct the Time Distributed CNN model with CNN and LSTM networks and compare with the results of CNN and LSTM models. Finally, the average of the seven categories trained by the Time Distributed model is the best, which can reach 88.01%. Mao et al. [9] proposed an approach to learning affect-salient features in Speech Emotion Recognition. Local invariant features from unlabeled samples are used as the input to a feature extractor of salient discriminative feature analysis (SDFA). Mu [11] came up with a distributed Convolutional-Recurrent Neural Networks with attention model to learn features from raw spectral information automatically. The weighted accuracy up to 64.08% for four-emotion classification in IEMOCAP dataset.

3   The Data Sets

There are three datasets used in this paper, the first one is CASIA Mandarin emotional corpus provided by the Institute of Automation, Chinese Academy of Sciences, which includes four professional speakers. Each speaker is divided into six emotions: angry, frightened, happy, neutral, sad and surprised. Each category has 300 identical texts with a total of 7200 sentences. The length of each sentence is about 1–2 s with good consistency. Figure 1 shows an example of a time-frequency spectrum of an audio fragment. As the second one is validation set, there are totally 697 speech fragments in the TV play and the talk show that are totally different from the length, content, form of the training set. They are also divided into the above six categories. The number of each classification is shown in Fig. 2.

In addition, the German speech emotion database (DMO-DB) is used to test the performance of this model method. The dataset was obtained by simulating 7 emotions (neutral, anger, fear, happiness, sadness, disgust, boredom) on 5 sentences by 5 men and 5 women. The sentence is a daily colloquial style without excessive written language modification. In this paper, a total of 535 sentences are used, and the distribution of the specific language of each emotion is shown in Fig. 3.

4   Methodology

4.1   Fourier Transform

All of the waveforms can be viewed as a result of the overlapping of many sine waves. Because the sine wave is periodic, we can mark the position of the sine wave by setting the peak closest to the frequency axis and projecting the position

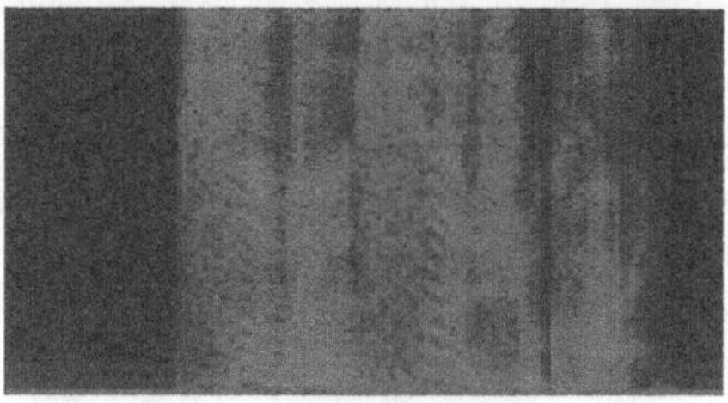

Fig. 1. An example of a time-frequency spectrum of an audio fragment.

Fig. 2. The number of each classification in the validation dataset.

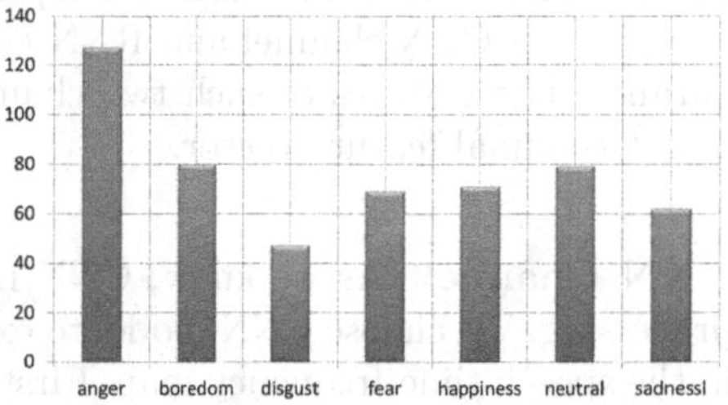

Fig. 3. The number of each classification in DMO-DB.

to the lower plane, that is, the distance from the frequency axis to the peak. So through the Fourier transform can be the spectrum, time and phase information for any waveform.

The fourier transform is the process of converting a non-periodic, time-domain signal into a non-periodic, continuous signal in the frequency domain. Although we can clearly see from Fourier transform how much each frequency in a message contains, it is difficult to see information such as the start time and duration of different signals. Therefore, in order to further understand the digital signal, Short Time Fourier Transform (STFT) have been introduced as follow:

The basic idea of short-time Fourier transform is to consider the non-stationary process as a series of short-term stationary signal superposition, short-term by windowing in time. In Fourier integrals, the time window function $g(t - \mu)$ is multiplied with the signal $f(t) \in L^2(R)$ to achieve fenestration and translation near μ, and then to Fourier transform. In the linear space, there is a measurable, squarely integrable function $f(t) \in L^2(R)$ that performs a windowed Fourier transform on it:

$$G_f(\xi, \mu) = \int f(t)g(t - \mu)e^{-j\xi t}dt \tag{1}$$

The inverse transformation of window Fourier transform formula is:

$$f(t) = \frac{1}{2A} \iint G_f(\xi, \mu)e^{-j\xi t}g(t - \mu)\, d\xi\, d\mu \tag{2}$$

It is equivalent to using a "magnifying glass" with the same shape, size and magnification to move on the time-frequency phase plane to observe the frequency characteristics within a fixed length of time. Short-Time Fourier is the windowing operation in a section of the speech, in the local window function among the discrete Fourier transform operation. This can be ac certain degree of effective attention to the local data information.

4.2 Model Presentation

Figure 4 shows the framework of the proposed model. The model is composed of two parts of CNN channel and RNN channel. We trained these two channels in parallel. The features of such two channels are integrated together and trained to get the final feature vector.

CNN Channel. As we know, CNN model is widely used in natural language processing. We choose CNN model to extract the deep emotional features hidden in the speech time-frequency map. First, the size of the input picture is set to 208 * 208, through the convolutional layer, then we can obtain a feature map M_{ij} and we use ReLu as an activation function. Afterward, the maxpooling layer will extract the most important feature by taking the maximum value of the results:

$$pooling(M_{ij}) = max(M_{ij}) \tag{3}$$

RNN Channel. The RNN model is very suitable for feature extraction of sequence data, and the speech data has temporal characteristics. Therefore, in this work, we choose select the RNN to perform feature extraction on the sequence pictures generated by speech. We divide each time-frequency graph along the time axis into 16 sequence pictures and place them into the RNN for training. We use the normal distribution with a mean of 0 and a standard deviation of 0.05 as the initialization weight of the model and initialize the cycle kernel with the identity matrix. In the process, we also select ReLu as the activation function.

Feature Fusion. CNN can effectively extract the local effective features of time-frequency maps, while RNN can extract temporal characteristics based on sequence fragments. Therefore, combining the output of the two can generate a representative feature vector.

$$output = output(RNN) + output(CNN) \tag{4}$$

Later, we use the Dense layer to compress the output vectors into a specific length and finally, use the softmax function to convert it into a 6-dimensional one-hot vector for classification. The result is a 6-dimensional vector, which the final class will be 1, the other classes will be 0.

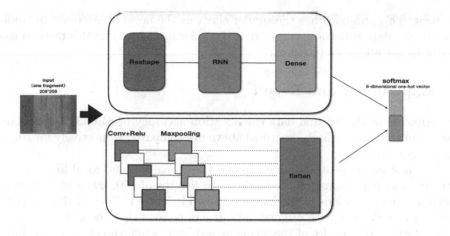

Fig. 4. The proposed model in this paper.

4.3   Data Enhancement Methods

After converting speech segment into images, various image enhancement methods can be effectively introduced. However, not all image enhancement methods are suitable owing to the nature of the speech segments. Currently, the tried and true method of image enhancement is based on preserving the maximum of the original audio characteristic, including image flipping, stitching and the like.

Global Flip. The global flip of the time-frequency map can be essentially understood as the inversion of the audio segment, that is, the normal audio is played back in reverse order. After manual testing, we found that this operation can still determine the emotional tendency of the audio segment to some extent based on the tone.

Local Flip. Cut the middle part of the existing time-frequency map and flip it back to the original place. Compared with the former enhancement method, this method has less modification to the audio segment and can be said very similar to the original audio. In this experiment, there are seven different sizes of clips for this comparison.

Random Splicing. Each picture is divided into about half, in the same classification of the picture sequence for random splicing. This approach is equivalent to splice audio clips of the same emotional orientation, and the emotional sentiment can be smoothly distinguished even if the semantic reserved semantics are excluded.

Gaussian Noise. Adding Gaussian noise of different standard deviations in each time-frequency map can be regarded as adding noise to the original audio segment. This approach has the similar effect as the noise interference in traditional voice data collection, except that the noise introduced by this method has less influence and is more regular.

5   Experiment and Result

Experiments to the original data classification accuracy of 81.00% as the standard, and the four methods described above are compared respectively for single and combination.

The first experiment mainly compared the global flip and local flip. A comparison of global flipping and local flip with lengths of 100, 200 and 300 pixels taken at the central axis of the image is shown in Table 1. Each of these experiments was performed with original data. It can be seen that the global flip have little effect on the results of the whole experiment, while the effect of the local flip is significantly greater than the global flip.

Table 1. A comparison between global flip and local flip.

Method	Parameter	Accuracy
Original data		81.00%
Global flip		81.00%
Local flip	100 pixels	**91.00%**
	200 pixels	**89.50%**
	300 pixels	**91.00%**

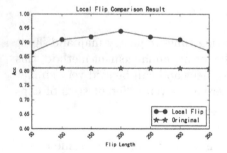

Fig. 5. Local flip comparison result.

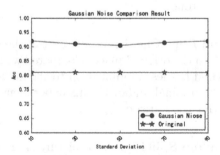

Fig. 6. Influence of different levels of noise on emotion recognition.

The second experiment is to refine the central pixel size into seven cases from 50 to 350 respectively. The local flip comparison result is shown in Fig. 5.

In Fig. 5, the same order of magnitude of the local flip real results are always higher than the baseline value. 200 of them peaked, however, as can be shown from Fig. 1, the key information of the time-frequency map is generally concentrated in the middle of the horizontal axis. Therefore, the size of the interception should not be more than one-third of the length of the key information is the best. (The length of the image used in this paper is 900).

In the third experiment, the standard deviation in Gaussian noise was modified to 10, 30, 50, 70 and 90 respectively to analyze the influence of different degrees of noise on emotion recognition. The experimental results are recorded in Fig. 6. We find that different levels of Gaussian noise classification accuracy are maintained at 90% or more, and the overall accuracy is much higher than the previous.

The fourth experiment mainly focuses on the comparative experiments of three different orders of magnitude for the data of local flipping, random splicing, and Gaussian noise. The final results are shown in Fig. 7. Because random splicing is uncontrollable, taking the average of multiple groups of experiments among different orders of magnitude as its real accuracy.

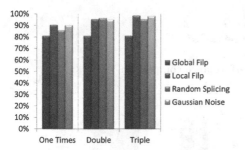

Fig. 7. Comparative experiments of three different orders of magnitude for the data of local flipping, random splicing and gaussian noise on CASIA.

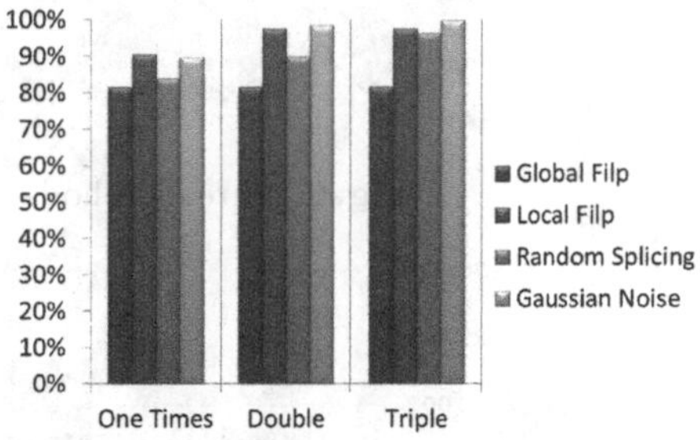

Fig. 8. Comparative experiments of three different orders of magnitude for the data of local flipping, random splicing and gaussian noise on DMO-DB.

Three different orders of magnitude experimental results show that: in the low order of magnitude, the effect of local flip and Gaussian noise much better than random splicing. When the data reach twice, the effect of them is not much different. Then when the data reaches three times, random splicing effect is not improved, the accuracy of the remaining two categories has been significantly improved. In the same way, we also conducted verification experiments in three different orders of magnitude in the above three methods in DMO-DB. The results are shown in Fig. 8. As can be seen in Fig. 8, the distribution of the accuracy of sentiment classification in DMO-DB is basically the same as CASIA. We can conclude that in the large-scale data enhancement, the method of preserving the original data information is most effective when the intrinsic properties of the audio fragments are effectively taken into consideration. In addition, when dealing with an audio fragment, it is effective to operate in the part where the key information is distributed.

Table 2. Accuracy of emotional classification for different models on CASIA datasets

Model	Accuracy
Spectral+prosodic features	88.35%
Extreme learning machine decision tree	89.60%
Deep auto-encoder	83.50%
Weighted linear discriminate analysis	88.78%
CNN+RNN	**91.00%**

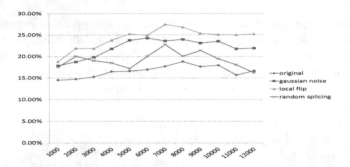

Fig. 9. Three methods performance comparison.

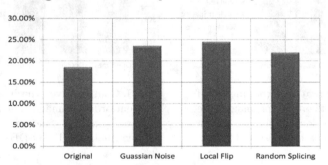

Fig. 10. Model performance comparison.

In Table 2, the model proposed in this paper is compared with many traditional speech recognition models on CASIA [4,8,13]. Most of these models are based on the improvement and fusion of traditional speech features. However, this paper proposes a new idea in the field of speech emotion recognition. It has achieved good results by mapping the speech, and we will also consider the fusion of these features with traditional features in the future work.

6   Conclusion

In the end, we use one person's speech data in the original data set as a validation set, and a completely different set as the validation set, for model performance evaluation. The performance trends of the several enhancement methods proposed in this paper on the verification set are shown in Fig. 9. The average performance is shown in Fig. 10. We found that in the second case, enhanced experiment results are generally higher than the unenhanced one, and the local flip effect is the best, reaching 24.475%. Similarly, the same phenomenon occurs in the verification of one's speech data, the unenhanced result is 30.79%, and the accuracy of the highest local flip can reach 44.24%. This shows that during the process of using the neural network model, due to the influence of data volume and other factors, it will cause overfitting in the training process, resulting in poor robustness. Therefore, our next work will also study this issue. Under the premise of depth feature information and the advantages of traditional speech recognition methods in overfitting problems, and significantly improves the model's robustness.

Acknowledgment. The work is supported by the State Key Program of National Natural Science of China (61432004, 71571058, 61461045). This work was partially supported by the China Postdoctoral Science Foundation funded project (2017T100447). This research has been partially supported by National Natural Science Foundation of China under Grant No. 61472117. This work is also supported by the foundational application research of Qinghai Province Science and Technology Fund (No. 2016-ZJ-743).

References

1. Kandali, A.B., Routray, A., Basu, T.K.: Emotion recognition from Assamese speeches using MFCC features and GMM classifier. https://doi.org/10.1109/TENCON.2008.4766487
2. Chao, L., Tao, J., Yang, M., Li, Y.: Improving generation performance of speech emotion recognition by denoising autoencoders. In: International Symposium on Chinese Spoken Language Processing, pp. 341–344 (2014)
3. Chen, L., Mao, X., Xue, Y., Cheng, L.L.: Speech emotion recognition: features and classification models. Digit. Signal Process. **22**(6), 1154–1160 (2012). https://doi.org/10.1016/j.dsp.2012.05.007
4. Fei, W., Ye, X., Sun, Z., Huang, Y., Zhang, X., Shang, S.: Research on speech emotion recognition based on deep auto-encoder. In: 2016 IEEE International Conference on Cyber Technology in Automation, Control, and Intelligent Systems (CYBER), pp. 308–312. IEEE (2016)
5. Huang, Z., Dong, M., Mao, Q., Zhan, Y.: Speech emotion recognition using CNN, pp. 801–804 (2014)
6. Jin, Q., Li, C., Chen, S., Wu, H.: Speech emotion recognition with acoustic and lexical features. https://doi.org/10.1109/ICASSP.2015.7178872
7. Lim, W., Jang, D., Lee, T.: Speech emotion recognition using convolutional and recurrent neural networks. In: Signal and Information Processing Association Summit and Conference, pp. 1–4 (2017)
8. Liu, Z.T., Wu, M., Cao, W.H., Mao, J.W., Xu, J.P., Tan, G.Z.: Speech emotion recognition based on feature selection and extreme learning machine decision tree. Neurocomputing **273**, 271–280 (2018)

9. Mao, Q., Dong, M., Huang, Z., Zhan, Y.: Learning salient features for speech emotion recognition using convolutional neural networks. IEEE Trans. Multimed. **16**(8), 2203–2213 (2014)
10. Mariooryad, S., Busso, C.: Compensating for speaker or lexical variabilities in speech for emotion recognition. Speech Commun. **57**(1), 1–12 (2014)
11. Mu, Y., Gómez, L.A.H., Montes, A.C., Martínez, C.A., Wang, X., Gao, H.: Speech emotion recognition using convolutional-recurrent neural networks with attention model. DEStech Transactions on Computer Science and Engineering (CII) (2017)
12. Trigeorgis, G., et al.: Adieu features? End-to-end speech emotion recognition using a deep convolutional recurrent network. In: IEEE International Conference on Acoustics, Speech and Signal Processing (2016)
13. Zhou, Y., Sun, Y., Zhang, J., Yan, Y.: Speech emotion recognition using both spectral and prosodic features. In: International Conference on Information Engineering and Computer Science, pp. 1–4 (2009)
14. Zhou, Y., Sun, Y., Zhang, J., Yan, Y.: Speech emotion recognition using both spectral and prosodic features. https://doi.org/10.1109/ICIECS.2009.5362730

Unified Data Hiding and Scrambling Method for JPEG Images

Zhaoxia Yin[1], Youzhi Xiang[1], Zhenxing Qian[2],
and Xinpeng Zhang[2(✉)]

[1] Anhui University, Hefei 230601, China
yinzhaoxia@ahu.edu.cn
[2] Shanghai University, Shanghai 200072, China
{zxqian,xzhang}@shu.edu.cn

Abstract. This paper proposes a novel method of unified data hiding and JPEG scrambling. This differs from the traditional image encryption algorithms in that, we propose to construct a privacy-preserved JPEG image during the data hiding procedure. That is while the scrambling operation masks the JPEG image content by the secret key, the additional messages is also represented by the sequence of the shuffled RLE (Run Length Encoding) pairs. Therefore, images uploaded to the cloud server unable to reveal the original content. Meanwhile, the owner can use the embedded messages for fragile watermarking, JPEG labeling and image management. The proposed method is reversible. After downloading a copy from the cloud, an authorized user can extract the hidden message and recover the original JPEG losslessly and simultaneously. Experimental results show that the proposed method provides a large embedding payload and has a good capability for preserving privacy.

Keywords: JPEG · RLE · Reversible · Preserving privacy

1 Introduction

Privacy-preservation in the cloud environment has attracted much attention in recent years. In cloud computing, users who are unwilling to reveal the contents of their multimedia may send an encrypted copy to a remote server for privacy protection. The server accomplishes different applications in the encrypted domain, including compressing encrypted images [1], signal transformation in cipher-texts [2], pattern recognition in the encrypted domain [3], watermarking in encrypted multimedia [4], data searching in encrypted dataset [5], and RDH in encrypted images [6–8].

To solve the privacy issues of image content, some works have focused on designing access control protocols so that the shared images can only be accessed by a group of selected users [9–11]. Another category of privacy protection schemes aims to encrypt the image data before uploading them to the public storage servers or social network. Along these line, Ra et al. proposed P3 [12], a privacy preserving image sharing system divides images into public and private parts. The public part is shared over the social network as plaintext, and the secret part is encrypted and stored in a cloud storage server. However, the public part of P3 leaks significant visual information of the image content.

© Springer Nature Switzerland AG 2018
R. Hong et al. (Eds.): PCM 2018, LNCS 11166, pp. 373–383, 2018.
https://doi.org/10.1007/978-3-030-00764-5_34

Alternatively, Tierney et al. designed a system called Cryptagram [13], enabling users to encrypt photos with traditional block ciphers and then embed the encrypted bit-streams into JPEG files. However, as pointed out by [14], the desirable properties of Cryptagram are realized at the cost of a significantly increased storage burden, as the use of a cover image results in significant file expansion.

Sun et al. identified the issue of uploading images to social media, in particular Facebook. When uploading images, social networks like Facebook modify the images. To ensure that encrypted images are robust to these lossy activities, in ACM MM 2016, Sun et al. proposed a good privacy protection method specifically designed for online photo sharing over Facebook [14]. Based on an in-depth study on the manipulations that Facebook performs on uploaded images, they suggested a domain (DCT-based) image encryption scheme that is robust to lossy Facebook operations by setting the JPEG Quality Factor (QF) to 71. They achieved very good performance in terms of visual security, reconstructed image quality, and storage costs. However, this method is tailor-made for Facebook and does not necessarily apply to other social networks.

In this work, we suggest a privacy-preserving, error-free reversible and storage-efficient solution by proposing a unified reversible data embedding and scrambling method for JPEG images, which aims to achieve image scrambling and additional data insertion simultaneously. The inserted data can be an image label for retrieval, fragile watermarking for content integrity authentication, or error-correcting codes for robust transmission. Therefore, in addition to privacy-preservation of image content, this proposed solution has significant implications for big image data storage and management. As validated by our experimental results, our method achieves strong performance in terms of visual security, payload, and storage cost.

The remainder of the paper is organized as follows. Section 2 describes the proposed framework. Section 3 introduces the unified data hiding and scrambling method. Experimental results are given in Sect. 4. Section 5 concludes this paper.

2 Proposed Framework

Here we introduce the new framework proposed in this paper at first. As shown in Fig. 1, given an initial JPEG image I, which has already been compressed, and therefore has DC and AC encoding information available, we can scramble this image and embed additional data in a unified way. Firstly, we apply entropy decoding to the JPEG image to restore the DC coefficients, encoded by using DPCM (differential pulse code modulation), and AC coefficients, which are subject to the zigzag ordering and encoded in the form of RLE (Run Length Encoding) pair sequence (This will be described in more depth in Sect. 3.1). We can then recode RLE sequence within each block to embed data, while also taking the block as the unit to scramble the DC and AC at the same time, as will be discussed in Sect. 3.2. After entropy encoding, the privacy-preserving version I' of the JPEG image I with an additional message embedded is obtained, which is then suitable for cloud storage. When the image I' is downloaded from the cloud storage, as long as the viewer has the secret key used for unified data embedding and scrambling, they can extract the additional message while recovering the original JPEG file.

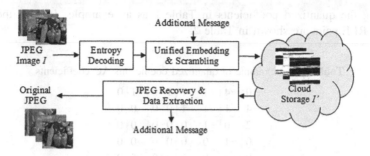

Fig. 1. Proposed framework

The design goals are summarized as follows. First, the algorithm must have a reasonably high level of security. Second, the high embedding payload is one of the focuses of the mothod. Then, the reconstructed image should be of high quality. Finally, we also would like to minimize the storage cost on the cloud, which is high efficiency of storage.

3 Unified Data Hiding and Scrambling Method

In this section, we propose a unified data embedding and scrambling method. The propose algorithm adopts a dynamic mapping method to embed data. At the same time, the algorithm uses the distortion generated during the rearrangement of the quantized run-length codes to encrypt the image and protect the original content.

3.1 JPEG Parsing

JPEG image compression [18] is a very widely used approach for compressing images for transmission, storage, or sharing.

Assuming an 8 bit image, the pixel values in the input image are first shifted by-128, and the processed image is partitioned into non-overlapping blocks, each of 8 × 8 pixels. Each block is transformed into the frequency domain by using the DCT (discrete cosine transformation) and each DCT coefficient is then subject to quantization w. r.t. (with respect to) the QT (quantization table) scaled by the specified QF. DC coefficients are encoded by using DPCM where only the difference between the current and the previously coded value is stored.

On the other hand, the AC coefficients are subject to entropy encoding, which means that zigzag ordering is applied to the matrix of values, and these are encoded in the form of RLE pair sequences:

$$\{p_j = (r_j, v_j)\}_{j=1}^{L}$$

Where L is the number of the nonzero AC coefficients in a block, v_j is the value of the j-th nonzero AC coefficient and r_j is the zero run before v_j.

Taking the quantized coefficients in Table 1 as an example, $L = 13$, the corresponding RLE pairs are shown in Table 2.

Table 1. An example of quantized coefficients AC coefficients

−10	−4	1	−1	1	0	−1	0
−4	1	1	−1	0	0	0	0
2	0	−1	0	0	0	0	0
0	−1	0	0	0	0	0	0
0	1	0	0	0	0	0	0
0	0	0	0	0	0	0	0
0	0	0	0	0	0	0	0
0	0	0	0	0	0	0	0

Table 2. Corresponding RLE pairs of Table 1

p1	p2	p3	p4	p5	p6	p7	p8	p9	p10	p11	p12	p13
(0, −4)	(0, −4)	(0, 2)	(0, 1)	(0, 1)	(0, −1)	(0, 1)	(3, −1)	(0, −1)	(0, −1)	(0, 1)	(4, 1)	(7, −1)

3.2 Data Embedding and JPEG Re-coding

For the proposed unified data embedding and JPEG Re-coding approach, RLE pairs within each block are shuffled with a way determined by K_b (which is a component of the secret key K_{em}) and the data to be embedded. While the shuffling operation masks the image content, the embedded data are also represented by the order of shuffled RLE pairs. Here, the embedded data are made up of two parts: the additional data, and the compressed side information for image recovery. The detailed side information generation algorithm is addressed in Algorithm 1.

Algorithm 1: Side Information Generation

Input: L RLE pairs sequence $P: p_1, p_2, ..., p_L$

Output: Side Information n_i, $i = 1, 2, ..., L$

Step 1: Calculate $e: e_i = |r_i| - |v_i|$, $i = 1, 2, ..., L$

Step 2: i=1

Step 3: Sort $P: p_1, p_2, ..., p_L$ by e_i, r_i, v_i ascending and denote as $Q: q_1, q_2, ..., q_L$.

       Find the RLE pair in Q corresponding to p_1, denote as q_x.

Step 4: $n_i = x$, remove p_1 from P and $i = i + 1$, $L = L - 1$.

Step 5: Repeat step 3-4 until $L = 1$.

For each RLE sequence within each block, it can be seen there are two properties for these RLE pairs: large magnitude and short zero run for low frequency RLE pairs, small magnitude and long zero run for high frequency RLE pairs. This is the key point behind Algorithm 1 and also for unified block RLE sequence re-coding with data embedding in our method.

Based on the above introduction, RLE Pairs re-coding within each block with data embedding can be processed as below.

Suppose there is a block containing L-consecutive RLE pairs: $P : p_1, p_2, \ldots, p_L$

Step 1: RLE group generation.

Assign the given RLE pairs to K groups: $g_1, g_2, \ldots, g_K$ according to different r and v. That means for $i \neq j$, if $r_i = r_j$ and $v_i = v_j$, then we assign p_i, p_j to the same group. Continuing with the previous example in Table 2, there are 7 groups, as shown in Table 3, in which g_i represents each group indexed by i and $i = 1, 2, \ldots, K$, f_i is the pair frequency of group g_i.

Table 3. Group assignment of RLE pairs in Table 2

g_1	g_2	g_3	g_4	g_5	g_6	g_7
(0, −4)	(0, 2)	(0, 1)	(0, −1)	(3, −1)	(4, 1)	(7, −1)
f_1	f_2	f_3	f_4	f_5	f_6	f_7
2	1	4	3	1	1	1

Step 2: RLE group coding.

For all K ($K \geq 2$) groups, assign each group g_i a binary code according to K. Here we continue with the previous example in Table 3 and the corresponding code assignment of each group is shown in Table 4, in which c_i represents the embedded data of RLE pairs in each group and $i = 1, 2, \ldots, K$. The detailed algorithm is shown in Algorithm 2: Code Assignment.

Table 4. Code assignment of each group

g_1	g_2	g_3	g_4	g_5	g_6	g_7
(0, −4)	(0, 2)	(0, 1)	(0, −1)	(3, −1)	(4, 1)	(7, −1)
c_1	c_2	c_3	c_4	c_5	c_6	c_7
00	010	011	100	101	110	111

Algorithm 2: Code Assignment

Input: The number of groups K ($K \geq 2$)

Output: c_i of each g_i, $i = 1, 2, ..., K$

Step 1: Calculate the bit-number of each code $n = \lceil \log_2 K \rceil$;

Step 2: Enumerate all the possibilities of n bits: $a_1, a_2, ..., a_{2^n}$

Step 3: Enumerate all the possibilities of $n-1$ bits: $b_1, b_2, ..., b_{2^{n-1}}$

Step 4: Enumerate the K possibilities of c_i, $i = 1, 2, ..., K$

  1) If $(\log_2 K)$ is not an integer: Choose $2^{\lfloor \log_2 K \rfloor + 1} - K$ front possibilities from $b_1, b_2, ..., b_{2^{n-1}}$ and $2(K - 2^{\lfloor \log_2 K \rfloor})$ possibilities from $a_1, a_2, ..., a_{2^n}$ starting from the end, then the K possibilities are saved as c_i, $i = 1, 2, ..., K$.

  2) If $(\log_2 K)$ is an integer: Choose all possibilities from $a_1, a_2, ..., a_{2^n}$, then the K possibilities are saved as c_i, $i = 1, 2, ..., K$.

Example for Algorithm 2 with input $K = 5$:

Step 1: $n = \lceil \log_2 5 \rceil = 3$;

Step 2: All the possibilities of 3 bits: 000, 001, 010, 011, 100, 101, 110, 111.

Step 3: All the possibilities of 2 bits: 00, 01, 10, 11.

Step 4: Choose 3 front possibilities from {00, 01, 10, 11}, that is {00, 01, 10}, and choose 2 end possibilities from {000, 001, 010, 011, 100, 101, 110, 111}, that is {110, 111}. The 5 codes are saved as {c1=00, c2=01, c3=10, c4=110, c5=111}.

Step 3:   RLE sequence re-coding.

If the embedded data corresponds to binary data c_i, then select an RLE pair from the group g_i to embed and $f_i = f_i - 1$. Please note that if $f_i = 0$ then $K = K - 1$ and new code c_i should be reassigned for each group g_i where $i = 1, 2, ..., K - 1$ according to Algorithm 2.

To further protect the image content, DC coefficients can be encrypted by shuffling the DPCM prediction error. In addition, AC coefficients can be encrypted by shuffling the RLE sequences by taking block as the unit. In other words, both the RLE sequences of AC and DPCM prediction error of DC are shuffled by taking the block as the unit with a method determined by K_b.

3.3   JPEG Recovery with Data Extraction

Given a privacy-preserving JPEG image I' as shown in Fig. 1, inverse scrambling for the DPCM prediction error of DC and the RLE sequences of AC by taking block as the unit. Then, for each RLE sequence within each block, data extraction and block RLE sequence re-coding can be conducted according to Algorithm 3.

Algorithm 3: Data extraction and block RLE sequence re-coding

---

Input: L recoded RLE sequence: $P': p_1', p_2', ..., p_L'$.

　　　Side information n_i , $i = 1, 2, ..., L$.

Output: The embedded data c_i of each p_i' .

　　　Recovered RLE sequence $P: p_1, p_2, ..., p_L$.

Step 1: Calculate $e_i' = |r_i'| - |v_i'|$, $i = 1, 2, ..., L$.

Step 2: i=1

Step 3: Calculate the bits carried by P' according to RLE group generation and RLE group coding in Subsection 3.2, then extract the embedded data c_i carried by p_i' .

Step 4: Sort $P': p_1', p_2', ..., p_L'$. by e_i' ascending and denote as $Q': q_1', q_2', ..., q_L'$.

　　　Find the n_i -th RLE pair in Q' , $p_i = q_{mi}'$.

Step 5: Remove q_{mi}' from Q' and $i = i+1$, $L = L-1$.

Step 6: Repeat step 3-5 until $i = L$. Then all of the embedded data are extracted and the original RLE sequence $P: p_1, p_2, ..., p_L$ is recovered.

---

4 Experimental Results

In this section, we experimentally evaluate the performance of our proposed scheme by using a set of commonly used grayscale images and 100 images randomly selected from UCID-v2 [15]. We first compare our method with other JPEG encryption joint data hiding methods [7–9] in terms of payload. Additionally, we also compare our method with other image privacy-preserving methods such as P3 [12], and Sun et al.'s method [14] in terms of security and storage cost respectively.

　　To verify the proposed method, we use a set of grayscale images sized 512×512, and compress them to JPEG format with different QF (quality factors). An example is given in Fig. 2, in which (a) shows the original images Pepper and Boat compressed with a quality factor 80. After unified data embedding and scrambling with an encryption key K_b, format-compliant bit-streams are generated. The encrypted bit-streams can be decoded by a JPEG decoder, and are shown in Fig. 2(b). The encrypted bit-streams have the same lengths as the original and contain additional 80229 and 111831 bits, respectively. On the recipient side, additional messages can be extracted error-free if the key is available, and the original bit-streams can be recovered losslessly as shown in Fig. 2(c).

　　In Tables 5 and 6, we show the embedding payloads measured in bits, bpp (bit per pixel) and bpAC (bit per nonzero-AC) with QF = 80 and QF = 90, respectively. We can see that, either Gross Capacity or Net Payload are different according to different images and different QF. The experimental results also shows the proposed method can achieve larger net payload compared with the methods [7–9], as shown in Table 7. The Average is obtained by the 100 images from UCID [15].

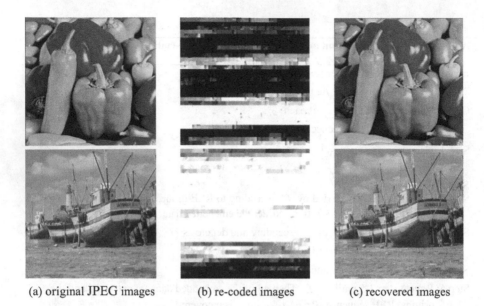

(a) original JPEG images   (b) re-coded images   (c) recovered images

Fig. 2. Image visual quality generated by our method

Table 5. Embedding payloads with QF = 80

Test images	Gross capacity (bits)	Net payload (bits)	Net payload (bpp)	Net payload (bpAC)
Airplane	52274	43183	0.16	1.35
Baboon	273116	229832	0.88	2.14
Boat	132056	111831	0.43	1.80
Lena	90697	76515	0.29	1.62
Pepper	94744	80229	0.31	1.61

Table 6. Embedding payloads with QF = 90

Test images	Gross capacity (bits)	Net payload (bits)	Net payload (bpp)	Net payload (bpAC)
Airplane	112985	93796	0.36	1.63
Baboon	422534	356733	1.36	2.38
Boat	234177	199573	0.76	2.11
Lena	167574	142264	0.54	1.91
Pepper	189092	161218	0.61	1.96

Table 7. Comparison of net payloads (bits) with QF = 80

Image set	[7]	[8]	[9]	Proposed method
UCID average	10238	1160	1364	111393

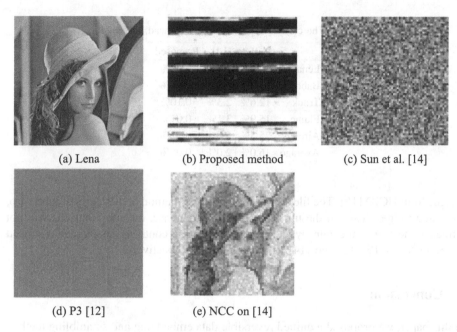

(a) Lena (b) Proposed method (c) Sun et al. [14]

(d) P3 [12] (e) NCC on [14]

Fig. 3. The encrypted version and sketch attack for the test image Lena

In addition, we compare our proposed privacy-preserving scheme with P3 [12] and Sun et al.'s method from MM2016 [14] in terms of visual security and robustness against sketch attacks. The test image Lena (a) and the different encrypted versions (b), (c), (d) are shows in Fig. 3. As can be seen, it is difficult to get any meaningful information from the encrypted version of our proposed scheme (b) and Sun et al.'s method (c). In contrast, for P3 [12], the public part stored in plaintext reveals significant visual information of the original image (d).

Furthermore, to examine the robustness of the proposed and Sun et al.'s method [14] further, four sketch attack techniques, namely, DCM (DC category mapping) [16], NCC (non-zero AC coefficient count) [17], EAC (energy of AC coefficients block) [17], and PLZ (position of last non-zero AC coefficients) [17], are considered. Table 8 summarizes the outcome of sketch attacks. As we can see, our proposed method survives all sketch attacks and shows stronger security.

Finally, storage overhead, also called file-size extension, is an important evaluation criterion. Another advantage of this work is that the JPEG file-size can be preserved, after recoded. In Table 9, we show the overhead incurred by different privacy protection methods respect to the JPEG with QF = 71. The Average is obtained by the 100

Table 8. Comparative results for all sketch attacks

Method	DCM	NCC	EAC	PLZ
Sun et al.'s method [14]	Pass	Fail	Pass	Pass
Proposed method	Pass	Pass	Pass	Pass

Table 9. The comparison of storage overhead, QF = 71

	[12]	[14]	Proposed
Lena	20.2%	3.2%	0.08%
Baboon	13.8%	1.4%	0.04%
Truck	12.6%	2.3%	−0.10%
Elaine	15.9%	3.1%	0.01%
Airplane	25.5%	6.0%	0.02%
Average	16.0%	2.0%	0.05%

images from UCID [15]. The file size of the proposed method consistently around zero, with an average across all the images of a 0.05% increase, meaning that this will not influence the size of the entropy encoding bitstream. In contrast, the average overhead of the methods, P3 [12] and [14], is 16% and 2%, respectively.

5   Conclusion

In this paper, we proposed a unified reversible data embedding and scrambling method for JPEG images, which the contribution is achieve image scrambling and additional data insertion simultaneously. In addition to privacy-preservation of image content, it also has significant implications for big image data storage and management. Experimental results shows that the proposed method have strong performance in image content security, storage cost and embedding capacity. In particular, the proposed method also can achieve robustness over Facebook if we set the JPEG QF to 71 and handle the pixel overflow as suggested by [14].

References

1. Liu, W., Zeng, W., Dong, L., Yao, Q.: Efficient compression of encrypted grayscale images. IEEE Trans. Image Process. **19**(4), 1097–1102 (2010)
2. Zheng, P., Huang, J.: Discrete wavelet transform and data expansion reduction in homomorphic encrypted domain. IEEE Trans. Image Process. **22**(6), 2455–2468 (2013)
3. Rahulamathavan, Y., Phan, R., Chambers, J., Parish, D.: Facial expression recognition in the encrypted domain based on local fisher discriminant analysis. IEEE Trans. Affect. Comput. **4**(1), 83–92 (2013)
4. Bianchi, T., Piva, A.: TTP-free asymmetric fingerprinting based on client side embedding. IEEE Trans. Inf. Forensics Secur. **9**(10), 1557–1568 (2014)

5. Fu, Z., Sun, X., Liu, Q., et al.: Achieving efficient cloud search services: multi-keyword ranked search over encrypted cloud data supporting parallel computing. IEICE Trans. Commun. **98**(1), 190–200 (2015)
6. Ong, S.Y., Wong, K.S., Qi, X., et al.: Beyond format-compliant encryption for JPEG image. Sig. Process. Image Commun. **31**, 47–60 (2015)
7. Qian, Z., Zhou, H., Zhang, X., et al.: Separable reversible data hiding in encrypted JPEG bitstreams. IEEE Trans. Dependable Secure Comput. (2016). https://doi.org/10.1109/TDSC.2016.2634161
8. Chang, J.C., Lu, Y.Z., Wu, H.L.: A separable reversible data hiding scheme for encrypted JPEG bitstreams. Sig. Process. **133**, 135–143 (2017)
9. Cutillo, L.A., Molva, R., Onen, M.: Privacy preserving picture sharing: enforcing usage control in distributed on-line social networks. In: Proceedings SNS 2012, p. 6. ACM (2012)
10. Klemperer, P., et al.: Tag, you can see it! Using tags for access control in photo sharing. In: Proceedings CHI 2012, pp. 377–386. ACM (2012)
11. Mazurek, M.L., et al.: Toward strong, usable access control for shared distributed data. In: Proceedings FAST 2014, pp. 89–103. USENIX Association (2014)
12. Ra, M.R., Govindan, R., Ortega, A.: P3: toward privacy-preserving photo sharing. In: Proceedings NSDI 2013, pp. 515–528. USENIX Association (2013)
13. Tierney, M., Spiro, I., Bregler, C., Subramanian, L.: Cryptagram: photo privacy for online social media. In: Proceedings COSN 2013, pp. 75–88. ACM (2013)
14. Sun, W., Zhou, J., Lyu, R., et al.: Processing-aware privacy-preserving photo sharing over online social networks. In: ACM on Multimedia Conference, pp. 581–585. ACM (2016)
15. Schaefer, G., Stich, M.: UCID—an uncompressed colour image database. In: Proceedings SPIE: Storage and Retrieval Methods and Applications for Multimedia, vol. 5307, pp. 472–480 (2004)
16. Ong, S.Y., Minemura, K., Wong, K.S.: Progressive quality degradation in JPEG compressed image using DC block orientation with rewritable data embedding functionality. In: IEEE International Conference on Image Processing, pp. 4574–4578. IEEE (2013)
17. Minemura, K., Moayed, Z., Wong, K., Qi, X., Tanaka, K.: JPEG image scrambling without expansion in bitstream size. In: IEEE International Conference on Image Processing, pp. 261–264. IEEE (2012)
18. International Telecommunication Union: CCITT Recommendation T.81, - Information technology-digital compression and coding of continuous-tone still—requirements and guidelines (1992)

Cross-Modal Retrieval with Discriminative Dual-Path CNN

Haoran Wang, Zhong Ji$^{(\boxtimes)}$, and Yanwei Pang

School of Electrical and Information Engineering, Tianjin University,
Tianjin 300072, China
jizhong@tju.edu.cn

Abstract. Cross-modal retrieval aims at searching semantically similar examples in one modality by using a query from another modality. Its typical applications including image-based text retrieval (IBTR) and text-based image retrieval (TBIR). Due to the rapid growth of multimodal data and the success of deep learning, cross-modal retrieval has received increasing attention and achieved significant progress in recent years. Dual-path CNN is a novel framework in this domain, which yields competitive performance by utilizing instance loss and inter-modal loss. However, it is still less discriminative in modeling the intra-modal relationship, which is also important in bridging a more discriminative cross-modal embedding network. To this end, we propose to incorporate an additional intra-modal loss into the framework to remedy this problem by preserving the intra-modal structure. Further, we develop a novel batch flexible sampling approach to train the entire network effectively and efficiently. Our approach, named Discriminative Dual-Path CNN (DDPC), achieves the state-of-the-art results on the MS-COCO dataset, improving IBTR by 4.9% and TBIR by 5.9% based on Recall@1 on the 5K test set.

Keywords: Cross-modal retrieval · Convolutional neural network
Intra-modal · Inter-modal · Embedding space

1 Introduction

In the era of big data, the amount of data from different modalities such as image and text, are increasing at an unprecedented rate. This makes cross-modal retrieval, which performs retrieval across multi-modal data, an urgent demand in people's daily life. For example, in some scenarios, people always want to search some relevant images given a description of words, such as "A little girl is playing tennis with her father", "A couple walks near the seashore at sunset".

Cross-modal retrieval is a relatively new paradigm [1–8], and is more challenging than traditional single-modal retrieval [9, 10] (e.g., text-based text retrieval, image-based image retrieval). Its main challenge lies in that there is a "heterogeneity gap" between different modalities since their representations and distributions are inconsistent [5, 11, 12]. Take image and text for example. Generally, images are represented with Convolutional Neural Networks (CNN) features, while texts are with Word2vec or

Bow features. They are distributed in different representation spaces. To match them, we have to resort to a common embedding space to bridge this "heterogeneity gap".

Some promising methods have been put forward, rising from linear embedding approaches, such as CCA [13, 14] to deep frameworks, such as [15–17]. For example, as one of the pioneering studies, Mao *et al.* [15] present to learn a common embedding space between the two modalities via the combination of CNN and recurrent neural networks (RNNs), trained with Softmax loss regularized by the L2 norm. Kiros *et al.* [16] propose a similar framework, however, they employ Gated Recurrent Unit (GRU) as the text encoder instead of RNNs and adopt ranking loss to train the model. Niu *et al.* [17] propose to replace the chain-structured RNNs with a tree-structured LSTM as text encoder to learn the hierarchical relations between images and sentences. Generally, these deep frameworks leverage CNN for image feature extraction and RNNs for text feature learning.

Recently, Zheng *et al.* [18] presents a novel Dual-Path CNN (DPC) approach, where two individual CNN are used for modeling the image and text representations, respectively, in an end-to-end framework. In specific, DPC has two losses in its objective function, one is instance loss and the other is inter-modal loss. The former is actually a Softmax loss by considering image/text pairs as categories, with the purpose of enhancing the discriminative property for different instances in each modality. The latter is a ranking loss to reflect the relationship between image and text. Although great success has been achieved, it does not lay enough emphasis on intra-modal relationship, which helps in bridging a more discriminative cross-modal embedding network. As pointed out in [19], making the distances among instances in intra-class more compact usually may lead to notable improvement. To this end, we propose to incorporate an additional intra-modal loss into the whole network, strengthening the intra-modal discriminative capability by preserving the structure of each modality.

The Contributions of this Paper are Twofold. First, we propose to utilize an intra-modal loss to preserve the structure of each modality, aiming at strengthen the intra-modal discriminative capability. The proposed intra-modal loss, together with original inter-modal loss and instance loss, ensure the well discriminative capability of the dual-path CNN approach. Thus, we call it Discriminative Dual-Path CNN (DDPC), as shown in Fig. 1. Second, to solve the contradiction of sampling strategies between instance loss and ranking loss, we develop a flexible batch sampling approach for training the network with all three loss terms simultaneously and effectively. The extensive experiments on MSCOCO dataset demonstrate that DDPC achieves competitive cross-modal retrieval accuracies compared with the state-of-the-art methods.

The rest of the paper is organized as follows: Sect. 2 describes the dual-path CNN framework briefly, and the proposed intra-modal loss and batch sampling method is described in details in Sect. 3. Extensive experiments are presented in Sect. 4, and conclusions are drawn in Sect. 5.

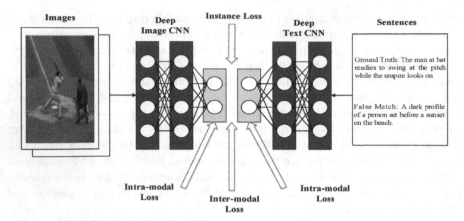

Fig. 1. A brief framework of our DDPC method. It consists of two CNN branches for image and text, respectively. The overall loss function is composed of an instance loss, an inter-modal loss and an intra-modal loss

2 Two-Branch CNN Architecture

In this paper, we use a two-branch CNN framework [18] to learn visual and textual representations simultaneously in an end-to-end way. Specifically, one branch is a deep Image CNN for processing visual input and the other branch is a deep Text CNN for processing textual input. As pointed in [18], it is able to extract more fine-grained information of images and texts by an end-to-end fine-tuning.

2.1 Image CNN

We use ResNet-50 [20] pre-trained on ImageNet as a basic model before implementing fine-tuning operation for learning more fine-grained visual feature. Given an input image of size 224 × 224, we get a 2,048-dimensional feature vector. We denote the 2,048-dim vector f_v as the visual feature of the input X.

2.2 Text CNN

Given a sentence, we first represent it as a code C of size $m \times k$, where m denotes the length of the sentence, and k denotes the size of the dictionary. C is used as the input of the sentence for the Text CNN. Word2vec [21] is utilized as a general dictionary to filter out some rare words, indicating that a word that doesn't exist in the word2vec dictionary (3,000,000 words) will be discarded. For instance, we eventually have 29,141 words as the dictionary on MSCOCO dataset. Every word in MSCOCO corresponds to an index $n \in [1, k]$ in the dictionary. For instance, a sentence of m words can be converted to $m \times k$ matrix. The text input C can thus be formulated as:

$$C(i, j) = \begin{cases} 1 & if \, j = n_i \\ 0 & otherwise \end{cases} \tag{1}$$

where $i \in [1, m], j \in [1, k]$. Because the text CNN requires a fixed-length input, we set a fixed length in this paper since most sentences in MSCOCO contain less than 32 words. Similar to [18], we pad with zeros to the columns of if the length of the sentence is shorter than 32. We thus obtain the $32 \times k$ sentence code C. Then, we further reshape C into the format, whose three dimension could be considered as height, width and depth known in the image CNN. Finally, the textual features are represented via a deep CNN similar to the structure of Image CNN. The details are referred to [18].

3   Loss Functions

3.1   Instance Loss and Inter-modal Loss

We first utilize the losses in [18] as our initial losses. One is the novel instance loss, which takes each image and its descriptions as a class. In this way, all the multi-modal data pairs are treated as classes. Besides, the instance loss is actually a Softmax loss to classify these classes. The other one is the inter-modal loss, which explicitly models the relationship between two modalities with a ranking loss. As demonstrated in [18], both losses are quite effective in reflecting the complicated relationships among images and texts. For brevity, we denote the instance loss and the inter-modal ranking loss as L_{ins} and $L_{inter\text{-}modal}$, respectively.

3.2   The Proposed Intra-modal Loss

When training the network with instance loss alone (this stage will be described in Subsect. 3.4), we encounter an unsatisfactory phenomenon that both convergence rate and classification accuracy of the Image CNN are much better than those of Text CNN. In their experiments, Zheng et al. [18] validate the two CNN could converge after a long time (more than 70 epochs for MSCOCO). However, we observe the performance of Text CNN is poor compared with that of Image CNN. This motivates us to consider whether we can improve the performance of Text CNN by utilizing intra-modal structural information.

Inspired by existing work about preserving intra-modal structure [19, 26, 27], we impose an intra-modal constraint by preserving neighborhood structure within each modal. Specifically, in the cross-modal retrieval task, we want sentences (resp. images) with similar meaning to be close to each other. Let y_p denote the set of sentences describing the same image, and $M(y_p)$ denote the neighborhood of y_p composed of sentences corresponding to the same image. Then, we want to ensure a relatively large margin between $M(y_p)$ and other textual instances which are out range of its neighborhood by coordinating parameter α:

$$D(y_p, y_q) + \alpha < D(y_p, y_s) \quad \forall y_q \in M(y_p), \forall y_s \notin M(y_p), \tag{2}$$

Similar to (2), we define the constraints for the visual modality as

$$D(x_{p'}, x_{q'}) + \alpha < D(x_{p'}, x_{s'}) \quad \forall x_{q'} \in M(x_{p'}), \forall x_{s'} \notin M(x_{p'}), \tag{3}$$

where $M(x_{p'})$ is composed of images described by the same sentences.

Figure 2 provides an intuitive illustration of how intra-modal structure preservation help to cross-modal matching. The common embedding space in the Fig. 2(a) satisfies the constraint condition of inter-modal matching. That is, each circle (representing an image) is closer to all rectangles with the same color (representing its corresponding sentences) than to any rectangles with different color. Analogously, for any rectangles (sentences), the closest circle (image) has the same color. However, without preserving the intra-modal structure, it is possible to lead to a situation that the two rectangles (sentences) are very close to each other although they match with right circle (image). Then if given a new image query (orange circle), it can't guarantee to be matched to its corresponding sentences since blue and red rectangles are both very close to it. This problem can be tackled in the embedding space shown in the Fig. 2(b), in which intra-modal structure preservation are implemented, enforcing semantically similar sentences (rectangles with same color) closer to each other. This leads to the result that the new image query can easily find its corresponding sentences by measuring distance in the common embedding space.

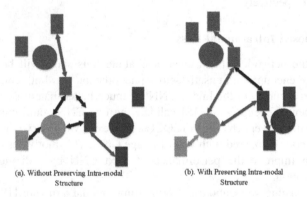

(a). Without Preserving Intra-modal Structure   (b). With Preserving Intra-modal Structure

Fig. 2. Illustration of how intra-modal structure preservation affect the procedure of learning the joint embedding space. Rectangles represent textual instances and circles denote visual instances. Same color indicates images and sentences are from the same image/text group. (Color figure online)

3.3 Final Loss

We first formulate the aforementioned inter-modal loss and intra-modal loss with ranking hinge loss. The loss function is:

$$L_{inter-modal} = \lambda_1 \sum_{p,q,s} [\alpha - D(x_p, y_q) + D(x_p, y_s)]_+ +$$

$$\lambda_2 \sum_{p',q',s'} [\alpha - D(x_{p'}, y_{q'}) + D(x_{p'}, y_{s'})]_+ \tag{4}$$

$$L_{intra-modal} = \lambda_3 \sum_{p,q,s} [\alpha - D(x_p, x_q) + D(x_p, x_s)]_+ +$$

$$\lambda_4 \sum_{p',q',s'} [\alpha - D(y_{p'}, y_{q'}) + D(y_{p'}, y_{s'})]_+ \tag{5}$$

where $L_{inter-modal}$ and $L_{intra-modal}$ denote inter-modal loss and intra-modal loss, respectively. The margin α could be varied according to different types of distance or even different instances. To make it easy to optimize, we fix $\alpha = 0.5$ for all terms across all training instances in our experiments.

The weights λ_1 and λ_2 balance the strengths of both ranking terms of inter-modal constraint. Following other work with a bi-directional ranking loss [16, 28, 29] we set both to be 1. The weights λ_3 and λ_4 control the importance of the structure-preservation terms, which serve as regularizers for the bi-directional retrieval tasks [29]. Note that there are a number of sentences for an image for most existing cross-modal retrieval datasets. And is rare for two different images to be described by an identical sentence. Thus, the visual-modality constraints Eq. (5) is trivial and we set $\lambda_3 = 0$ in our experiment. In addition, we set $\lambda_4 = 0.3$, since it usually produces the best result.

Then, we combine the instance loss and the above two ranking losses to form the final loss function, which is defined as:

$$L = w_1 L_{ins} + w_2 L_{inter-modal} + w_3 L_{intra-modal} \tag{6}$$

where w_1, w_2, w_3 are weights for different losses. As described in Sect. 3.4, in the first training stage, the ranking loss is not used ($w_1 = 1$, $w_2 = w_3 = 0$); in the second training stage, all three losses are used.

3.4   The Proposed Flexible Batch Sampling Strategy

It encounters a problem when training the final loss, that is "how could we optimize the instance loss and ranking losses simultaneously?" If we optimize the ranking loss, the usual solution is selecting a certain number of triplets to form a triplet mini-batch since optimizing over all triplets is computationally infeasible. It means only low proportion of instances can participate in the training stage when we use ranking loss. Nonetheless, in contrast, we want to incorporate a maximum number of the training instances into the training stage to avoid overfitting.

To resolve the contradiction, we design a special batch sampling method to select suitable instances for model training, which is called flexible batch sampling method. Figure 3 provides an intuitive illustration of how to select the instances to form the flexible batch. The circle and rectangle denote image input and text input, respectively. The elements enclosed in curly brackets above denote an image batch. The image batch is divided into four equal parts (we call them sub-batch), denoted by $\{X_a, X_b, X_c, X_d\}$. Analogously, the text batch below is also divided into four equal parts, denoted by $\{Y_a, Y_b, Y_c, Y_d\}$.

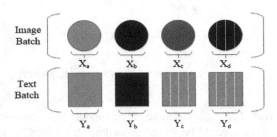

Fig. 3. Illustration of the proposed flexible batch sampling method. Rectangles represent sub-batches and circles denote sub-batches of images. Note that the same color and pattern indicates two sub-batches are from the same image/text group. Same color but different pattern indicates two sub-batches are very close to each other in the embedding space but not in the same image/text group. Different color shapes indicate images and texts are not matched.

Now, we describe how to build the flexible batch. First, we select three sub-batches $\{X_a, X_b, X_c\}$ by random sampling. Then, we select two sub-batches $\{Y_a, Y_b\}$, according to the prerequisite that $\{X_a, Y_a\}$ and $\{X_b, Y_b\}$ are from the same image/text group respectively. Next, referring to the method of selecting hard negative instances proposed in [16], we select Y_c by searching the top N (N denotes the size of sub-batches) hardest negative instances for X_a. Finally, we select X_d by searching the top N hardest instances for Y_b.

When we get the flexible batch, we can utilize it to train the final loss simultaneously. As shown in Fig. 3, we can observe that most sub-batches have no class intersection in their own modality except for $\{Y_c, Y_d\}$. It means the amount of instances belonging to various classes are approximately balanced. Meanwhile, we can reduce the impact of slight imbalance problem between textual instances by setting w_2 to be a small value. Besides, we can utilize $\{X_a, Y_a, Y_c\}$, $\{X_a, Y_a, Y_d\}$ and $\{X_b, Y_b, X_d\}$ to form triplets, then compute the inter-modal ranking loss $L_{inter\text{-}modal}$ by Eq. (4) over these triplets. Meanwhile, we take $\{Y_c, Y_d, Y_a\}$ as triplets and compute the intra-modal loss $L_{intra\text{-}modal}$ by Eq. (5) over them.

In addition, the training procedure is split into two stages like that in [18]. In state I, we fix the pre-trained weights in the Image CNN and use the instance loss to tune the remaining part. In stage II, we end-to-end fine-tune the entire network based on the proposed flexible batch sampling strategy.

4 Experiments

4.1 Dataset and Implementation Details

We evaluate our approach on the MSCOCO dataset [22], totally containing 123,287 images and 616,767 description sentences. For cross-modal retrieval, we follow the setting in [26], which consists of 113,287 training images with about five captions each, 5,000 images for validation and 5,000 images for testing. The performance evaluation is reported on 1K test set (5 fold) and 5K test set. We use the widely-used Recall@K as evaluation metric. Recall@K means the possibility of at least one true match appears among the top K ranked matches for one query.

The entire network is trained by stochastic gradient descent (SGD) with a mini-batch size of 32 in all our experiments. We implement our approach based on the Matconvnet package [23]. Most implementation details of training procedure can be referred to [18]. Note that, in the second training stage, all three terms in the final loss are considered, we set $w_1 = 1$, $w_2 = 2$ and $w_3 = 0.6$ in Eq. (6). To fine-tune the entire network, we use a learning rate starting with 0.001 and the learning rate is dropped to 0.0001 after 10 epochs. According to our observation, the final loss converges after about 20 epochs.

4.2   Result Analysis

We compare our DDPC approach with the state-of-the-art methods on MSCOCO dataset. The compared methods include recent models for the cross-modal retrieval. Besides, we also summarize the visual and textual embeddings used in these works in both Tables 1 and 2.

Table 1. Experiment on MSCOCO 1 K test set. R@K (%) is Recall@K (higher is better).

Method	Visual	Textual	IBTR		TBIR	
			R@1	R@5	R@1	R@5
DVSA [26]	ft RCNN	w2v + ft RNN	38.4	69.9	27.4	60.2
GMM-FV [14]	fixed VGG-16	w2v + HGLMM	39.4	67.9	25.1	59.8
m-RNN [15]	fixed VGG-16	ft RNN	41.0	73.0	29.0	42.2
RNN-FV [24]	fixed VGG-19	feature from [26]	41.5	72.0	29.2	64.7
sm-LSTM [25]	fixed VGG-19	ft RNN	53.2	83.1	40.7	75.8
DPC [18]	ft ResNet-50	ft ResNet-50	65.6	89.8	47.1	79.9
DDPC (Stage I)	fixed ResNet-50	ft ResNet-50	52.0	80.9	37.3	70.1
DDPC (Stage II)	ft ResNet-50	ft ResNet-50	**67.2**	**91.3**	**48.5**	**83.4**

We can see that DDPC achieves the best performance in all metrics, which outperform sub-optimal results of DPC approach by 4.9% in IBTR and by 5.9% in TBIR based on Recall@1 on the 5K test set, and by 2.4% in IBTR and by 3.0% based on Recall@1 on the 1K test set, respectively. These clearly demonstrate that DDPC is competitive in cross-modal retrieval.

4.3   Ablation Study

We implement ablation study to further validate the effectiveness of intra-loss on MSCOCO 5k validation set. As shown in Table 3, when adding inter-modal loss term

Table 2. Experiment on MSCOCO 5 K test set

Method	Visual	Textual	IBTR		TBIR	
			R@1	R@5	R@1	R@5
DVSA [26]	ft RCNN	w2v + ft RNN	16.5	39.2	10.7	29.6
VQA-A [30]	fixed VGG-19	ft RNN	23.5	50.7	16.7	40.5
DPC [18]	ft ResNet-50	ft ResNet-50	41.2	70.5	25.3	53.4
DDPC (Stage I)	FixedResNet-50	ft ResNet-50	28.3	55.3	18.2	41.7
DDPC (Stage II)	ft ResNet-50	ft ResNet-50	**43.2**	**70.6**	**26.8**	**53.6**

Table 3. Ablation study on MSCOCO 5K validation set.

Method	Stage	Image-to-text		Text-to-image	
		R@1	R@5	R@1	R@5
Only instance loss	II	38.5	67.8	22.8	49.8
Instance loss + inter-modal ranking loss	II	41.7	70.6	25.9	53.7
Full model	II	43.6	71.2	27.1	54.8

and intra-modal loss term to the final loss function successively, the performance improves steadily. It further proves that the intra-modal loss is effective.

5 Conclusion

In this paper, we present an intra-modal loss into the framework of dual-path CNN to learn the discriminative cross-modal embeddings from training image/text pairs end-to-end. Also, we develop a flexible batch sampling method for training. The proposed DDPC approach achieves competitive performance on MSCOCO dataset.

Acknowledgements. This work was supported by the National Basic Research Program of China under Grant 2014CB340400, and the National Natural Science Foundation of China under Grants 61771329, 61472273, and 61632018.

References

1. Ge, J., Liu, X., Hong, R., et al.: Deep graph Laplacian hashing for image retrieval. In: Pacific Rim Conference on Multimedia, pp. 3–13 (2017)
2. Hong, R., Li, L., Cai, J., et al.: Coherent semantic-visual indexing for large-scale image retrieval in the cloud. IEEE Trans. Image Process. **26**(9), 4128–4138 (2017)
3. Wang, B., Xu, Y., Han, Y., et al.: Movie question answering: remembering the textual cues for layered visual contents. arXiv preprint arXiv:1804.09412 (2018)
4. Li, C., Deng, C., Li, N., et al.: Self-supervised adversarial hashing networks for cross-modal retrieval. arXiv preprint arXiv:1804.01223 (2018)

5. Peng, Y., Qi, J., Huang, X., et al.: CCL: cross-modal correlation learning with multigrained fusion by hierarchical network. IEEE Trans. Multimed. **20**(2), 405–420 (2018)
6. Hong, R., Zhang, L., Zhang, C., et al.: Flickr circles: aesthetic tendency discovery by multi-view regularized topic modeling. IEEE Trans. Multimed. **18**(8), 1555–1567 (2016)
7. Hong, R., Zhang, L., Tao, D.: Unified photo enhancement by discovering aesthetic communities from flickr. IEEE Trans. Image Process. **25**(3), 1124–1135 (2016)
8. Hong, R., Hu, Z., Wang, R., et al.: Multi-view object retrieval via multi-scale topic models. IEEE Trans. Image Process. **25**(12), 5814–5827 (2016)
9. Liu, J., Zha, Z., Tian, Q., et al.: Multi-scale triplet CNN for person re-identification. In: ACM International Conference on Multimedia, pp. 192–196 (2016)
10. Liu, J., Zha, Z., Chen, X., et al.: Dense 3D-convolutional neural network for person re-identification in videos. ACM Trans. Multimed. Comput. Commun. Appl. **14**(4) (2018)
11. Ji, Z., Zhen, W., Pang, Y.: Deep pedestrian attribute recognition based on LSTM. In: IEEE International Conference on Image Processing (2017)
12. Liu, J., Zha, Z., Zhang, H., et al.: Context-aware visual policy network for sequence-level image captioning. In: ACM International Conference on Multimedia (2018)
13. Gong, Y., Ke, Q., Isard, M., et al.: A multi-view embedding space for modeling internet images, tags, and their semantics. Int. J. Comput. Vis. **106**(2), 210–233 (2014)
14. Klein, B., Lev, G., Sadeh, G., et al.: Fisher vectors derived from hybrid Gaussian–Laplacian mixture models for image annotation. In: IEEE Computer Vision and Pattern Recognition, pp. 4437–4446 (2015)
15. Mao, J., Xu, W., Yang, Y., et al.: Deep captioning with multimodal recurrent neural networks (m-RNN). In: International Conference on Learning Representations (2015)
16. Kiros, R., Salakhutdinov, R., Zemel, R.: Unifying visual-semantic embeddings with multimodal neural language models. arXiv preprint arXiv:1411.2539 (2014)
17. Niu, Z., Zhou, M., Wang, L., et al.: Hierarchical multimodal LSTM for dense visual-semantic embedding. In: IEEE International Conference on Computer Vision, pp. 1899–1907 (2017)
18. Zheng, Z., Zheng, L., Garrett, M., et al.: Dual-path convolutional image-text embedding. In: IEEE International Conference on Computer Vision (2018)
19. Weinberger, K., Saul, L.: Distance metric learning for large margin nearest neighbor classification. J. Mach. Learn. Res. **10**, 207–244 (2009)
20. He, K., Zhang, X., Ren, S., et al.: Deep residual learning for image recognition. In: IEEE Conference on Computer Vision and Pattern Recognition, pp. 770–778 (2016)
21. Mikolov, T., Chen, K., Corrado, G., et al.: Efficient estimation of word representations in vector space. arXiv preprint arXiv:1301.3781 (2013)
22. Lin, T.-Y., et al.: Microsoft COCO: common objects in context. In: Fleet, D., Pajdla, T., Schiele, B., Tuytelaars, T. (eds.) ECCV 2014. LNCS, vol. 8693, pp. 740–755. Springer, Cham (2014). https://doi.org/10.1007/978-3-319-10602-1_48
23. Vedaldi, A., Lenc, K.: MatConvNet: convolutional neural networks for MATLAB. In: ACM International Conference on Multimedia, pp. 689–692 (2015)
24. Lev, G., Sadeh, G., Klein, B., Wolf, L.: RNN Fisher vectors for action recognition and image annotation. In: Leibe, B., Matas, J., Sebe, N., Welling, M. (eds.) ECCV 2016. LNCS, vol. 9910, pp. 833–850. Springer, Cham (2016). https://doi.org/10.1007/978-3-319-46466-4_50
25. Huang, Y., Wang, W., Wang, L.: Instance-aware image and sentence matching with selective multimodal LSTM. In: IEEE Computer Vision and Pattern Recognition, pp. 2310–2318 (2017)
26. Karpathy, A., Li, F.F.: Deep visual-semantic alignments for generating image descriptions. In: IEEE Computer Vision and Pattern Recognition, pp. 3128–3137 (2015)

27. Karpathy, A., Joulin, A., Li, F.F.: Deep fragment embeddings for bidirectional image sentence mapping. In: Advances in Neural Information Processing Systems, pp. 1889–1897 (2014)
28. Shaw, B., Jebara, T.: Structure preserving embedding. In: ACM International Conference on Machine Learning, pp. 937–944 (2009)
29. Wang, L., Li, Y., Lazebnik, S.: Learning deep structure-preserving image-text embeddings. In: IEEE Computer Vision and Pattern Recognition, pp. 5005–5013 (2016)
30. Lin, X., Parikh, D.: Leveraging visual question answering for image-caption ranking. In: Leibe, B., Matas, J., Sebe, N., Welling, M. (eds.) ECCV 2016. LNCS, vol. 9906, pp. 261–277. Springer, Cham (2016). https://doi.org/10.1007/978-3-319-46475-6_17

Deep Learning for Ovarian Tumor Classification with Ultrasound Images

Chengzhu Wu[1], Yamei Wang[2], and Feng Wang[1(✉)]

[1] Shanghai Key Laboratory of Multidimensional Information Processing,
Department of Computer Science and Technology, East China Normal University,
Shanghai, China
fwang@cs.ecnu.edu.cn

[2] The International Peace Maternity and Child Health Hospital of China Welfare
Institute affiliated with School of Medicine, Shanghai Jiaotong University,
Shanghai, China

Abstract. Deep learning has shown great potentials for medical image analysis and computer-aided diagnosis of some diseases such as MRI brain tumor segmentation, mammogram classification, and diabetic macular edema classification. In this paper, we explore deep learning approaches for ovarian tumor classification based on ultrasound images. First, considering the lack of public ultrasound images, we annotate an ultrasound image dataset consisting of 988 image samples of three types of ovarian tumors. Second, we evaluate the generalization ability of different convolutional neural network (CNN) models on ultrasound images. Our experiments show that deep learning approaches achieve considerably high accuracies on the classification of ovarian tumors which are competitive with professional medical staffs.

Keywords: Convolution neural network · Ovarian tumor
Ultrasound diagnosis

1 Introduction

At present, ovarian malignancies suffer from the highest mortality rate among all gynecologic malignancies. In Europe, the patient's five-year survival rate for ovarian cancer is 35% [2]. The early diagnosis and detection of malignant tumors is important to reduce the mortality rate. However, most patients with ovarian cancers have no obvious early symptoms and lack relevant clinical manifestations. Screening for ovarian malignancies in high-risk groups is still the most effective way of detecting ovarian cancers, while transvaginal ultrasound is the most commonly used as a first-line screening tool since other tools such as computed tomography (CT), magnetic resonance imaging (MRI), and angiography are more complicated and have certain impact on human body. However, due to the complexity of the ultrasound images of ovarian tumors, and the overlaps between features of benign and malignant tumors as shown in Fig. 1, the

© Springer Nature Switzerland AG 2018
R. Hong et al. (Eds.): PCM 2018, LNCS 11166, pp. 395–406, 2018.
https://doi.org/10.1007/978-3-030-00764-5_36

misdiagnosis rate of the experienced medical staffs based on the conventional two-dimensional ultrasound images can reach 20% [6].

The diagnosis of ovarian tumors based on two-dimensional ultrasound images can be treated as a classification problem. Recently, tremendous success has been made in image classification with deep learning. Different networks are proposed [7,8,11,20,22]. Due to the ability of automatically extracting more comprehensive features, deep learning significantly improves the recognition accuracies compared with the traditional hand-crafted features. This motivates us to explore deep learning approaches for computer-aided diagnosis of ovarian tumors. Compared with human, computer has some advantages. First, computer can find the small difference in ultrasound images which is hard to capture by human's perception system. Second, well-trained neural networks can extract more comprehensive features automatically which have proved effective in the classification of images. However, large datasets with labeled samples are necessary for network training. For instance, the commonly used ImageNet dataset has 1 million training images and thousands of categories. For medical images, especially for ultrasound images of ovarian tumors, there lacks public datasets for training and testing.

In this paper, we explore deep learning for automatic diagnosis of ovarian tumors based on ultrasound images. First, considering the lack of ultrasound image datasets, we annotate a dataset of ovarian tumors, which contains 988 labelled samples including three categories of ovarian tumors (Benign, Borderline, and Malignant). Second, we employ deep learning approaches to automatically classify the ovarian tumors. The performances of four networks are evaluated and compared including VGGNet [20], DenseNet [8], ResNet [7], and GoogleNet(V3) [22]. Furthermore, we study the effects of different learning strategies including data augmentation and transfer learning approaches on the generalization ability of the models.

2   Related Works

In recent years, there have been quite a number of works on automatic diagnosis based on different kinds of medical images. In [9], an algorithm is proposed to segment brain tumors from MRI images by extracting the patches of different sizes. In [1], a new hybrid system is proposed to take eye screening for diabetic macular edema patients. In [24], deep learning is employed to identify breast metastases. In [19], a U-Net model is proposed for segmentation of neuronal structures in electron microscopic stacks. In [5], GoogleNet(V3) architecture is used to detect skin cancers with the dermoscopy images. In [25], an image-to-image RNN architecture is proposed to locate the vertebra centroids in 3D CT volumes. The performances of the automatic diagnosis are basically competitive with professionals for some diseases such as brain tumors, skin cancers, and breast cancers.

For the classification of ovarian tumors, Lotfi et al. [14] proposed a new prognostic method for predicting time to tumor recurrence in ovarian cancer and used genes as the independent variables. In [17], multi-category classification models are built to identity ovarian cancer subclass based on DNA microarray technology. In [3], a classifier is designed to distinguish four ovarian tumor varieties based on Bayesian least squares support vector machines and kernel logistical regression. The data set comes from Phase I of the International Ovarian Tumor Analysis (IOTA) group, contains cancer history, demographic information, and treatment records. In [13], a tumor sensitive matching flow (TSMF) technique is proposed to detect and segment the ovarian cancer metastases, and then locate the lesions. This technique is validated on contrast-enhanced CT data from 30 patients. In [21], different genomic characteristics information is used to analyze the interaction of gene expression based on a sparse regression. In [23], multi-layer perception architecture is used to identify the malignant ovarian tumors. Although different features are employed to classify the ovarian tumors, the diagnosis accuracy remains unsatisfactory. Ultrasound is the most commonly used for the diagnosis of ovarian tumors in medical practice. However, the echoes in the ultrasound images are unevenly distributed with various internal structures, irregular shapes, and smooth or rough echoes at the edges. This brings much difficulty to medical staffs. In this paper, we employ deep learning to classify three types of ovarian tumors, and evaluate the most studied networks for this task. To the best of our knowledge, this is the first work to address the classification of ovarian tumors with ultrasound images using deep learning approaches.

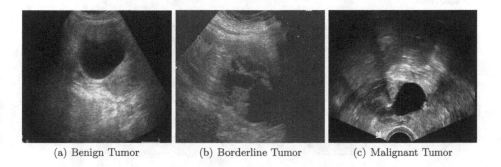

(a) Benign Tumor   (b) Borderline Tumor   (c) Malignant Tumor

Fig. 1. Three examples of ultrasound images with different types ovarian tumors.

3   Ultrasound Image Dataset of Ovarian Tumors

In this paper, we focus on the classification of ovarian tumors with 2D-ultrasound images. Our dataset consists of 988 images of three types of ovarian tumors, i.e. benign tumors, borderline tumors, and malignant tumors. All the samples were collected at the International Peace Maternity & Child Health Hospital of China Welfare Institute from May 2008 to May 2018. The patients were 7 to 84 years old. All cases were not treated with radiotherapy, chemotherapy, hormone therapy, or immunotherapy before the examination. For the screening of ovarian

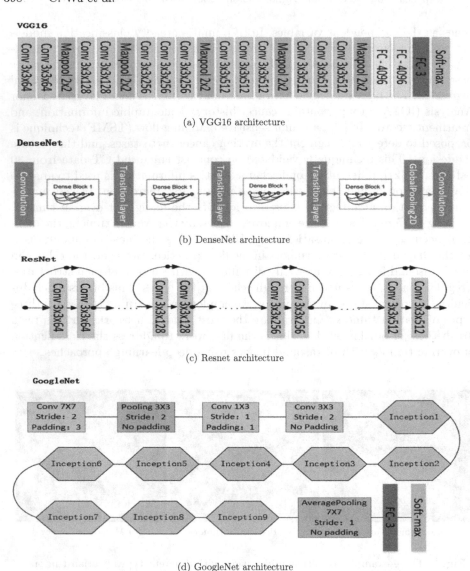

(a) VGG16 architecture

(b) DenseNet architecture

(c) Resnet architecture

(d) GoogleNet architecture

Fig. 2. A simplified illustration of the four evaluated CNN structures.

tumors, the ultrasonic instruments were operated by the professional medical staffs. The models of the ultrasonic diagnostic instruments include Voluson E6, Voluson E8 produced by GE company, and HD11, iU 22 produced by PHILPS. The examination was carried out within 1–2 weeks before the resection operation, and the patients in childbearing age were examined within 5–10 days of the menstrual cycle. The type of each ovarian tumor was then annotated according to the result of pathological diagnosis on the slices of the tumor after the resection operation. This can guarantee the accuracy of the annotations.

Figure 1 illustrates three examples of different types ovarian tumors in our dataset. With the ultrasound images, the medical staffs diagnose the ovarian tumors mainly according to a few features of the tumors, including the morphology, the boundary, the thickness of the tumor wall, the wall structure, the posterior sound shadow, the solid component of the tumor, the internal echo, the single occurrence, the peripheral relationship, and the metastasis observed by the two-dimensional ultrasound instrument. However, these features are usually qualitative and there lacks quantitative measure to accurately differentiate between the three types of tumors. The diagnose is much heavily dependent on the experience of the medical staffs. Since the features usually overlap between different types of tumors, the misdiagnosis rate is quite high which can reach 20% in medical practice. This brings much higher cost to the patients. For instance, misclassifying a benign tumor as a malignant tumor may cause unnecessary surgical operations, while misclassifying a malignant tumor as a benign tumor may put the patients in danger. Our aim is to explore deep learning for automatically extracting comprehensive features from the ultrasound images for the classification of ovarian tumors, which can aid the diagnosis in medical practice so as to reduce the misdiagnosis rate.

4   Network Architectures and Learning Strategies

In this section, we evaluate the most studied convolutional neural network architectures for the classification of ovarian tumors. Four networks are employed. We study the performances of different networks and parameter settings. Meanwhile, we evaluate the transfer learning approaches which use the pre-trained models from ImageNet dataset [4] to the ultrasound images.

4.1   Convolutional Neural Network Architectures

In this paper, the following networks are evaluated: VGGNet [20], DenseNet [8,18], ResNet [7], and GoogleNet(V3) [22]. These models have achieved great success in a number of image classification problems. Here, we briefly recall the architectures of these networks.

- **VGGNet:** The VGGNet proposed in [20] is a most studied model, which is much deeper than other models such as AlexNet [10]. It consists of a linear stack of convolution layers and pooling layers as illustrated in Fig. 2(a). The VGGNet model achieved the first and the second places in the localization and classification tracks respectively among all submissions in the ImageNet Challenge 2014.
- **DenseNet:** The DenseNet model proposed in [8] introduces direct connections from each layer to all the subsequent layers. This model could improve the information flow between the layers and implement feature reuse to make

the model training easier. The connections between all layers lead to an implicit deep supervision [12]. Figure 2(b) illustrates the layout of DenseNet structure.

- **ResNet:** The ResNet model proposed in [7] (see Fig. 2(c)) adds a skip-connection that uses the non-linear transformations, and the gradient can flow directly from later layers to earlier layers. This model achieved the first place in the classification, detection, and localization tasks of ImageNet Challenge 2015.

- **GoogleNet(V3):** The GoogleNet(V3) model proposed in [22] is of significantly more complicated and deeper than all previous convolutional neural network models as shown in Fig. 2(d). It proposes a new module called "Inception". This module consists of connected filters of different sizes. In our implementation, the Inception (V3) architecture [22] is employed.

4.2 Learning Strategies and Settings

Data Augmentation. The training of deep neural networks usually requires a large number of labeled samples. However, the cost of collecting medical images is quite high, and professionals are needed to label them. Thus, most medical image datasets are relatively small. To deal with the lack of training images, we implement several data augmentation operations. We rotate each input image by 15°, flip the photo horizontally and vertically, and crop it randomly. Through these operations, data distribution can be diversified. Meanwhile, this increases the numbers of training samples so as to avoid over-fitting. These operations are carried out in real-time during the whole process of training.

Learning from Scratch vs. Transfer Learning. In many tasks with convolutional neural networks, transfer learning is widely used to cope with the lack of training samples. A common way is to pre-train a model on a larger dataset of different or similar domains, and then fine-tune the model with the target dataset. For instance, in [26], the convolution neural network is employed to identify the scene images and transfer learning is used via pre-trained models learned from ImageNet, and then fine-tuned in the Places Database. In some works on medical images, transfer learning is also used. In [27], a new algorithm named Active Incremental Fine-Tuning is proposed by employing the continuous fine-tuning and active learning to select the appropriate labeled samples. In [16], it is argued that transfer learning can sometimes have a negative impact since the spatial distributions of data vary a lot among different domains. The nature of medical images is different from other optical photos and images, and thus pre-trained models from other image domains may not be appropriate for medical image domains. In this section, we evaluate the two learning strategies: learning from scratch and transfer learning for the classification of ovarian tumors.

Our experiment settings are listed in Table 1. For learning from scratch, all the parameters of convolutional neural networks are initialized with random distributions. For transfer learning, we employ the transfer learning approach in [15]. The models are pre-trained on ImageNet dataset. Next, we remove the fully connected layers of the pre-trained models and connect them to the softmax layer in order to match our classification model. Thus we can fine-tune the layers of the convolution neural network so as to save much time during the fine-tuning of the model. The learning rates are empirically set when the models achieve the highest accuracies. As can be seen in Table 1, for transfer learning, the learning rates are basically much smaller since only fine-tuning is required based on the pre-trained models. As a result, the training with transfer learning is much faster with less epochs and time for the models to converge (Note that for transfer learning, we directly use the models pre-trained from ImageNet and the pre-training time is not taken into account in Table 1).

Table 1. Experiment settings for different networks and learning strategies. **-Scratch**: Learning from Scratch; **-TF**: Transfer Learning.

CNN models	Epoch	Learning rate	Training time(hours)
VGG-16-Scratch	1000	0.1	15.6
ResNet-50-Scratch	1000	0.0001	19.9
DenseNet-Scratch	3000	0.01	21.0
GoogleNet(v3)-Scratch	3000	0.01	32.6
VGG-16-TF	200	0.0001	0.66
ResNet-50-TF	500	0.005	1.00
DenseNet-TF	100	0.001	0.64
GoogleNet(v3)-TF	150	0.005	0.62

5   Experimental Results and Discussions

In this section, we present and discuss the experimental results using the four networks and different settings presented in Sect. 4.

Learning from Scratch vs. Transfer Learning. Table 2 lists the classification accuracies of four networks with two learning strategies. By comparing the two learning strategies, only for VGG-16 model, transfer learning significantly outperforms learning from scratch. One reason might be that the accuracy of this network is much lower than the other three models. For all the other networks, the accuracies are significantly reduced. Our experimental results show that transfer learning using a pre-trained model on ImageNet does not improve the training of networks. The reason mainly lies in two aspects. First, our work focus on the ultrasound images which are generated according to the echoes of human organs to the ultrasound signals, while ImageNet classes are optical images of animals and objects. These are two completely different domains with different imaging process and content. Thus, the feature distributions of these two image domains are different from each other. Second, the relatively small dataset limits the ability of transferring from the pre-trained models to the optimal ones. Once most parameters are learned based on ImageNet data, it gets even harder to find the optimal parameters with fine-tuning. To better employ transfer learning for ultrasound image analysis, it might be better to use images of similar domains such as other kinds of medical images (e.g. MRI and CT) for the pre-training of models.

Table 2. Classification accuracies using four models with learning from scratch (Scratch) and with transfer learning (TF).

CNN models	Accuracy
VGG-16-Scratch	65.00%
VGG-16-TF	75.56%
ResNet-50-Scratch	83.50%
ResNet-50-TF	69.44%
DenseNet-Scratch	**87.50%**
DenseNet-TF	48.68%
GoogleNet(v3)-Scratch	**92.50%**
GoogleNet(v3)-TF	64.50%

Comparison Between Different Networks. As presented in Table 2, among the four evaluated models, DenseNet and GoogleNet(V3) generally yield pretty high accuracies of 87.50% and 92.50% respectively. DenseNet [8] introduces direct connections from each layer to all subsequent layers, and reuses the feature maps from all layers. GoogleNet(V3) [22] can handle richer spatial features and increase feature diversities. Our experiments show the effectiveness of DenseNet and GoogleNet(V3) for ultrasound image analysis.

The best CNN model (GoogleNet(V3)) yields an accuracy of 92.50% which is quite competitive compared with professionals. The training process of GoogleNet(V3) model is illustrated in Fig. 3 (a) and (b). The trend of change in classification accuracy and loss values gradually converge. To better present the performance of this model, we further evaluate it with some more metrics which are commonly used in medical research. The results are presented in Fig. 3(c). The ROC curve and Precision-Recall curve are illustrated in Fig. 3(c) and (e) respectively. The area-under-the-ROC-curve (AUC), true positive rate (TPR), precision and recall are used as the performance metrics. The confusion matrix (Fig. 3(d)) is also presented. Among all three types of ovarian tumors, it is important to accurately recognize the malignant tumors since they are the most mortal. The GoogleNet(V3) model achieves a high accuracy of 97% for malignant tumor classification with a very low miss rate. This is of great value in medical practice. For borderline tumor, there lacks a precise definition and it shares the characteristic of both malignant tumor and benign tumor. Thus, the accuracy

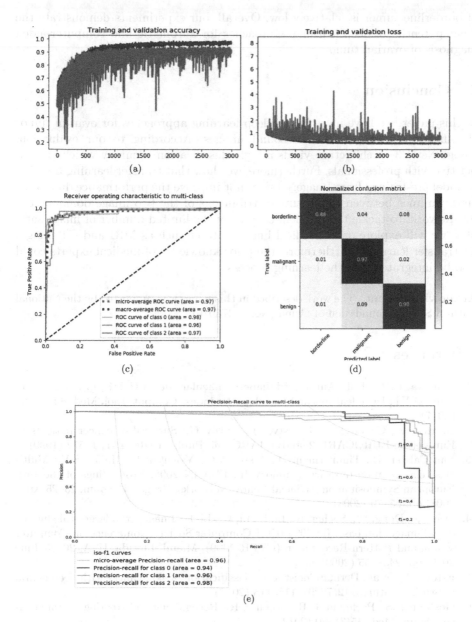

Fig. 3. The learning process and performance of GoogleNet(V3) model for ovarian tumor classification using ultrasound images. (a) Training accuracy and validation accuracy (blue and red lines) are plotted against the training process of 3000 epochs; (b) Training and validation cross-entropy loss (blue and red lines) are plotted against the training process; (c) Receiver operating characteristic (ROC) curve on the three classes; (d) The confusion matrix on the classification of three ovarian tumor classes; (e) Precision-Recall curve on the three classes. (Color figure online)

of borderline tumor is relatively low. Overall, our experiments demonstrate the great potential of deep learning approaches for automatic and computer-aided diagnosis of ovarian tumors.

6   Conclusion

In this paper, we have investigated deep learning approaches for ovarian tumor classification based on 2D-ultrasound images. According to our evaluation, GoogleNet(V3) architecture yields a high classification accuracy which is competitive with professionals. Furthermore, we show that transfer learning by using a model pre-trained from ImageNet does not improve the performance due to the large variance between the feature distributions of the two image domains. For future work, to tackle the problem caused by the limited number of image samples, we will explore other medical image dataset such as MRI and CT images for transfer learning. Furthermore, the prior knowledge of medical experts could also be integrated into the learning process.

Acknowledgement. The work described in this paper was supported by the National Natural Science Foundation of China (No. 61375016).

References

1. Acharya, U.R., et al.: Automated diabetic macular edema (DME) grading system using DWT, DCT features and maculopathy index. Comput. Biol. Med. **84**, 59–68 (2017)
2. Berrino, F., Capocaccia, R., Estve, J., Gatta, G.: Survival of cancer patients in Europe(the EUROCARE-2 study). IARC Sci. Publ. - IARC **151**, 1–572 (1999)
3. Van Calster, B., Timmerman, D., Testa, A.C., Valentin, L., Huffel, S.V.: Multiclass classification of ovarian tumors. In: ESANN 2008, Proceedings of the 16th European Symposium on Artificial Neural Networks, Bruges, Belgium, 23–25 April 2008, pp. 65–70 (2008)
4. Deng, J., Dong, W., Socher, R., Li, L., Li, K., Li, F.: Imagenet: a large-scale hierarchical image database. In: 2009 IEEE Computer Society Conference on Computer Vision and Pattern Recognition (CVPR 2009), Miami, Florida, USA, 20–25 June 2009, pp. 248–255 (2009)
5. Esteva, A., et al.: Dermatologist-level classification of skin cancer with deep neural networks. Nature **542**(7639), 115–118 (2017)
6. Gostout, B.S., Pachman, D.R., Lechner, R.: Recognizing and treating ovarian cancer. Minn. Med. **95**(3), 40 (2012)
7. He, K., Zhang, X., Ren, S., Sun, J.: Deep residual learning for image recognition. In: 2016 IEEE Conference on Computer Vision and Pattern Recognition, CVPR 2016, Las Vegas, NV, USA, 27–30 June 2016, pp. 770–778 (2016)
8. Huang, G., Liu, Z., van der Maaten, L., Weinberger, K.Q.: Densely connected convolutional networks. In: 2017 IEEE Conference on Computer Vision and Pattern Recognition, CVPR 2017, Honolulu, HI, USA, 21–26 July 2017, pp. 2261–2269 (2017)

9. Hussain, S., Anwar, S.M., Majid, M.: Segmentation of glioma tumors in brain using deep convolutional neural network. Neurocomputing **282**, 248–261 (2018)
10. Krizhevsky, A., Sutskever, I., Hinton, G.E.: Imagenet classification with deep convolutional neural networks. In: Advances in Neural Information Processing Systems 25: 26th Annual Conference on Neural Information Processing Systems 2012, Proceedings of a meeting held 3–6 December 2012, Lake Tahoe, Nevada, United States, pp. 1106–1114 (2012)
11. Krizhevsky, A., Sutskever, I., Hinton, G.E.: Imagenet classification with deep convolutional neural networks. Commun. ACM **60**(6), 84–90 (2017)
12. Lee, C., Xie, S., Gallagher, P.W., Zhang, Z., Tu, Z.: Deeply-supervised nets. In: Proceedings of the Eighteenth International Conference on Artificial Intelligence and Statistics, AISTATS 2015, San Diego, California, USA, 9–12 May 2015 (2015)
13. Liu, J., Wang, S., Linguraru, M.G., Yao, J., Summers, R.M.: Augmenting tumor sensitive matching flow to improve detection and segmentation of ovarian cancer metastases within a PDE framework. In: Proceedings of the 10th IEEE International Symposium on Biomedical Imaging: From Nano to Macro, ISBI 2013, San Francisco, CA, USA, 7–11 April 2013, pp. 652–655 (2013)
14. Lotfi, M., Misganaw, B., Vidyasagar, M.: Prediction of time to tumor recurrence in ovarian cancer: comparison of three sparse regression methods. In: Cai, Z., Daescu, O., Li, M. (eds.) ISBRA 2017. LNCS, vol. 10330, pp. 1–11. Springer, Cham (2017). https://doi.org/10.1007/978-3-319-59575-7_1
15. Ntalampiras, S.: Bird species identification via transfer learning from music genres. Ecol. Inform. **44**, 76–81 (2018)
16. Pan, S.J., Yang, Q.: A survey on transfer learning. IEEE Trans. Knowl. Data Eng. **22**(10), 1345–1359 (2010)
17. Park, J.S., Choi, S.B., Chung, J.W., Kim, S.W., Kim, D.W.: Classification of serous ovarian tumors based on microarray data using multicategory support vector machines. In: 36th Annual International Conference of the IEEE Engineering in Medicine and Biology Society, EMBC 2014, Chicago, IL, USA, 26–30 August 2014, pp. 3430–3433 (2014)
18. Pleiss, G., Chen, D., Huang, G., Li, T., van der Maaten, L., Weinberger, K.Q.: Memory-efficient implementation of densenets. CoRR abs/1707.06990 (2017)
19. Ronneberger, O., Fischer, P., Brox, T.: U-Net: convolutional networks for biomedical image segmentation. In: Navab, N., Hornegger, J., Wells, W.M., Frangi, A.F. (eds.) MICCAI 2015. LNCS, vol. 9351, pp. 234–241. Springer, Cham (2015). https://doi.org/10.1007/978-3-319-24574-4_28
20. Simonyan, K., Zisserman, A.: Very deep convolutional networks for large-scale image recognition. CoRR abs/1409.1556 (2014)
21. Sohn, K., Kim, D., Lim, J., Kim, J.H.: Relative impact of multi-layered genomic data on gene expression phenotypes in serous ovarian tumors. BMC Syst. Biol. **7**(S–6), S9 (2013)
22. Szegedy, C., Vanhoucke, V., Ioffe, S., Shlens, J., Wojna, Z.: Rethinking the inception architecture for computer vision. In: 2016 IEEE Conference on Computer Vision and Pattern Recognition, CVPR 2016, Las Vegas, NV, USA, 27–30 June 2016, pp. 2818–2826 (2016)
23. Verrelst, H., Moreau, Y., Vandewalle, J., Timmerman, D.: Use of a multi-layer perceptron to predict malignancy in ovarian tumors. In: Advances in Neural Information Processing Systems 10: NIPS Conference, Denver, Colorado, USA, pp. 978–984 (1997)
24. Wang, D., Khosla, A., Gargeya, R., Irshad, H., Beck, A.H.: Deep learning for identifying metastatic breast cancer. CoRR abs/1606.05718 (2016)

25. Yang, D., et al.: Deep image-to-image recurrent network with shape basis learning for automatic vertebra labeling in large-scale 3D CT volumes. In: Proceedings of the 20th International Conference on Medical Image Computing and Computer Assisted Intervention - MICCAI 2017, Part III, Quebec City, QC, Canada, 11–13 September 2017, pp. 498–506 (2017)

26. Zhou, B., Lapedriza, À., Xiao, J., Torralba, A., Oliva, A.: Learning deep features for scene recognition using places database. In: Advances in Neural Information Processing Systems 27: Annual Conference on Neural Information Processing Systems 2014, Montreal, Quebec, Canada, 8–13 December 2014, pp. 487–495 (2014)

27. Zhou, Z., Shin, J.Y., Zhang, L., Gurudu, S.R., Gotway, M.B., Liang, J.: Fine-tuning convolutional neural networks for biomedical image analysis: actively and incrementally. In: 2017 IEEE Conference on Computer Vision and Pattern Recognition, CVPR 2017, Honolulu, HI, USA, 21–26 July 2017, pp. 4761–4772 (2017)

Arbitrary Image Emotionalizing with Style Transfer

Xing Luo(iD), Jiajia Zhang(iD), Guang Wu(iD), and Yanxiang Chen(✉)(iD)

School of Computer Science and Information Engineering,
Hefei University of Technology, Hefei 230601, China
chenyx@hfut.edu.cn

Abstract. Endowing user-specified emotions on arbitrary images will be an interesting application. The emotion transfer of landscape photograph has shown greater appeal than the emotion transfer of facial image in the field of art design or painting. In this work, we present a novel method to synthesize arbitrarily specified emotions into the inputted landscape images. With style transfer aiming to generate the style of the source image while preserving the content of the target image, an Image Emotionalizing Network is developed for emotional rendering based on style transfer by using convolutional neural networks. Other than the feed-forward transfer networks which may be limited to single emotion transfer, an extension version of instance normalization is adopted to make our method suitable for the transfer of multiple emotions. As the datasets that we conduct are sufficient to train a well-performing network, our experimental results demonstrate that the proposed method can synthesize impressive emotional images and the experimental speed is equivalent to the previous effective methods.

Keywords: Image emotionalizing · Feature alignment · Style transfer

1 Introduction

In recent decades, with the popularity of image and video sharing website, people are susceptible to various emotions by watching the images or videos. At the same time, the emotion analysis mainly focuses on the study of facial emotional image. Recently, the emotion analysis of landscape photographs has shown great appeal in research community. In addition, the landscape photographs with different colors or different textures may affect audiences with different emotions, including contentment, excitement, awe, warmth, fear, sadness, etc. With the rapid increment of audiences' social needs, it is necessary to explore solutions to transfer user-specified emotions to input target images. Moreover, when we want to share our photos in social networks, it is funny to paint these photos with certain emotions before sharing. Similarly, these solutions can be applied in the field of art design or painting.

Yuan et al. [12] proposed a novel emotional prediction framework based on mid-level attributes of an image to predict its emotion. Meanwhile, they proposed an eigenface-based emotion detection to detect facial emotion. Jindal et al. [26] also built an image prediction framework using deep neural network. Machajdik et al. [25]

© Springer Nature Switzerland AG 2018
R. Hong et al. (Eds.): PCM 2018, LNCS 11166, pp. 407–416, 2018.
https://doi.org/10.1007/978-3-030-00764-5_37

developed a method that extract and combine low-level features of an emotion image to apply for image emotion classification. Gao et al. [24] designed a deep neural framework to predict visual emotions by utilizing the ambiguity and relationship between emotional categories.

Although there are many works of emotion image analysis, there are few research of image emotionalizing. Xu et al. [22] developed a system for image emotionalizing. In this system, Xu et al. designed the emotion models that are constructed from color feature of image pixels, and then defined a liner transformation to match the features of the target image to the emotion models. In [23], Li et al. presented a system to endow the specified-emotions to arbitrary input images, in which a framework based on data-driven is proposed to extract the emotion information of the labeled emotion images.

With the sharp increase for data on social network, the traditional image emotionalizing methods may no longer be appropriate. At present, convolutional neural networks have led to satisfactory results in processing big data. In this paper, we introduce a novel method based on style transfer for arbitrary image emotionalizing by using convolutional neural networks method. Arbitrary image emotionalizing is a generalized multiple emotions method based on a trained convolutional neural networks. Style transfer is a novel technique that aims to generate the style of the source image while preserving the content of the target image. Our method adopts an extension version of Instance Normalization and the fast feed-forward network [3, 14]. The former has been proven to have a pleasant effect on image generation, while the latter optimizes the loss function between the generated images and the source images. Note that, we mainly select landscape photographs (e.g., architecture, grasslands, wilderness, rivers, etc.) as our emotional training data. The experimental results show the effectiveness for synthesizing emotional images. And the test speed of our method achieves several orders of magnitude improvement than previous work and maintains the flexibility for image emotionalizing. The emotional images are shown in Fig. 2 and the statistics of the emotional images are shown in Table 1.

Table 1. Images statistic of emotional dataset.

Emotions	Contentment	Warmth	Excitement	Awe	Fear	Sadness
Statistics	7,617	828	4,872	263	3,211	3344

2 Related Works

2.1 Image Analysis and Image Emotionalizing

Image emotion analysis has always been concerned and challenged. Yuan et al. [12] propose an emotion prediction approach that uses the middle level attributes to predict the emotion of the given images. Compared to the previous emotion analysis, [12] proves that the results of the emotion analysis algorithm are easier to understand and show adaptation to higher levels. Borth et al. [11] extract 1200 adjective noun pairs that correspond to different emotions.

Before the works of image emotionalizing, color transfer studies [18–21] have been proposed for images color transformation. In [18, 21], Chang, et al. presented an example-based color stylization to stylize the color of images and video frames. Reinhard et al. [20] propose a transformation to generate a synthetic image that borrows the color of a target image to render the source image. However, Reinhard et al. focus on the color transfer while our method not only focus on image color transfer, but also focus on image semantic transfer. In the field of Image Emotionalizing, Xu et al. [22] present an algorithm to learn emotion-related knowledge (e.g. image local appearance distribution), then propose a system that synthesizes emotions into any input image. Li et al. [23] build a novel framework to render specific emotions to arbitrary images or videos. Our method has the same goal as Xu et al. [22] and Li et al. [23] method. For image emotionalizing, our network generate the arbitrary synthesis images though aligning the mean and standard variance of input images. More importantly, our method is different from that of [18–23]: first, we adopt deep learning method while other methods employ traditional modeling methods; second, color transfer studies [18–21] focus on color-channel transformation without any semantic information transformation, our algorithm focuses not only on color transformation but also on semantic transformation. Compared with [22], our method still has a certain degree of artistic style transfer, which is a feature that the second method does not have.

2.2 Style Image Transfer

Gatys et al. [7] proposed an interesting task that renders the semantic content of an image in different styles. They use the image features derived from a pre-trained neural network for target recognition to build the loss function. The loss function contains the content loss calculated on high-level activation layer and the style loss calculated on all layers of the network. Using SGD optimize the loss function to generate a stylized image from a white noise image. However, the [7] algorithm consumes a lot of time cost and memory cost. In order to solve this problem, Ulyanov et al. [3] propose a feed-forward transfer network that trained with a large dataset for a single style. This method can achieve considerable experimental quality and appealing speed. Li and Wand [15] propose a Markovian Generative Adversarial Network to train texture synthesis, style transfer, and video stylization. The experiments of Li and Wand [15] prove their runtime exceeds a few orders of magnitude than previous style transfer. However, the efficient feed-forward networks are unable to apply to various arbitrary style transfer. Chen et al. [10] propose a style-swap method that carry out transfer arbitrary style by combining the content structure with closest-match style texture in style-swap layers.

3 The Approach of Image Emotionalizing

In this section, we first introduce Image Emotionalizing Network that generates the synthesis images with specific emotion. This network contains three parts, including Feature Extraction Layer, Alignment Layer and Reconstruct Layer. Then we propose our optimization objective that aims at training the Reconstruct Layer to find the optimal solutions, leading to the generation of emotionalizing images.

3.1 Image Emotionalizing Network

As shown in Fig. 1, we introduce a feed-forward network of image emotionalizing. Our Image Emotionalizing Network takes an emotion image and a content image as input, and combines the emotion with the content image as output. At the end of the emotionalizing network, we perform an optimization procedure to optimize the loss function to train the reconstruct network.

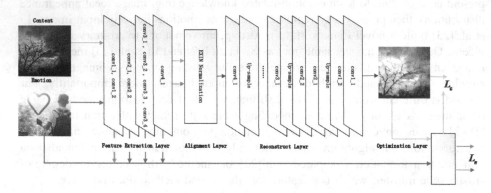

Fig. 1. The overall architecture of our Image Emotionalizing Network. For the Feature Extractive Layer, the first few layers of VGG19 network are adopted to obtain the feature statistics of input images. For the Alignment Layer, the function is to implement image emotionalizing by aligning the feature statistics of the content image and emotion images. For the Reconstruct Layer, this layer is used to recreate the emotionalizing images in the pixel space using the feature statistics generated by the Alignment Layer. The L_c and L_e represent the content loss and emotion loss respectively (e.g. Eqs. 6 and 7).

Feature Extraction Layer. The network that we adopt is a part of VGG-19 network that pre-trained on ImageNet [1], for simplicity, we called this kind of network Feature Extraction Layer. In the original image style transfer [7], in lower layers of CNN, we found that there is almost no difference between reconstruction content image and the original content image. In higher layers of the network, the details of the image are lost while the high-level as known as the abstract information is preserved. We also found that most appealing images are usually generate with all layers included. In order to guarantee the quality of the reconstruction, in our image emotionalizing approach, we merely adopt the first few network layers up to the 4-th convolution layer (conv4_1), just like [7].

Alignment Layer. The input of the alignment layer is the feature statistics extracted from the above Feature Extraction Layer, which are the mean and variance of the content and the emotional images. Then we deliver the feature statistics to Alignment Layer, generating target feature maps that are used to reconstruct synthesis images. Our Alignment Layer follows the method below:

Ioffe and Szegedy [2] introduced batch normalization that significantly accelerate the training of deep network by performing the normalization for each training mini-batch. In the past decades, BN has become a very common method in many generation tasks, such as texture synthesis [3], image generation [4]. This shows that BN brought a lot of inspiration to synthesis and image generation. However, in the improved texture synthesis of feed-forward transfer network [5], Ulyanov et al. found that Instance Normalization works much better than Batch Normalization because Instance Normalization helps normalize the feature information from a pair of content and style image. In the later work [6] it can be proved that the IN layers are able to train high-performance feed-forward network.

We propose to replace the affine parameters γ, β of BN with standard deviation and mean of input emotion image; we call *EXIN*:

$$EXIN(x, y) = \sigma(y)\left(\frac{x - \mu(x)}{\sqrt{\sigma^2(x) + \varepsilon}}\right) + \mu(y) \tag{1}$$

$$\mu(x) = \frac{1}{HW}\sum_{h=1}^{H}\sum_{w=1}^{W} x_{hw} \tag{2}$$

$$\sigma^2(x) = \frac{1}{HW}\sum_{h=1}^{H}\sum_{w=1}^{W} (x_{hw} - \mu(x))^2 \tag{3}$$

In Eq. 1, $\mu(y)$, $\sigma(y)$ indicate the mean and standard variance of input emotion images. In the optimization layer, the x and y represent the feature map of images.

Reconstruct Layer. Straight after the Alignment Layer, we build a Reconstruct Layer that is the mirror image of the Feature Extraction Layer, in which up-sampling layers replace the pooling layers of the Feature Extraction Layer. This layer is mainly used to recreate images in the pixel space using target features generated by Alignment Layer. In the experiments, the Reconstruction Layer is the training object that optimizes the following loss function.

3.2   Optimization Function

In order to optimize the loss of the generated image and the original image, we added an Optimization Layer (shown in Fig. 1). Here, we denote $f(c)$, $f(s)$ as feature maps of content image and emotion images respectively. Specifically, we denote f as Feature Extraction Layer, and T denotes the target feature produced by Alignment Layer. As for the generated image, combining content image with emotion, denoted as $G(c,s)$. Our emotion loss only computes the mean and standard deviation of the emotion style feature. In Eq. 7, l_i denotes the layers in Feature Extraction Layer. Similar to [7], in our experiments, we use the layers L = {relu1_1, relu2_1, relu3_1, relu4_1}.

$$T = EXIN(f(c), f(s)) \tag{4}$$

$$G(c, s) = Recon(e) \tag{5}$$

$$L_c = \|f(G(c,s)) - T\|_2 \tag{6}$$

$$L_e = \sum_i^L \|\mu(l_i(G(c,s))) - \mu(l_i(s))\|_2$$
$$+ \sum_i^L \|\sigma(l_i(G(c,s))) - \sigma(l_i(s))\|_2 \tag{7}$$

4   Experiments

4.1   Datasets Construction

As mentioned, we mainly apply landscape images as emotional image for training. We adopt the MS-COCO 2014 dataset [8] as content images; also, we construct a new dataset as emotional dataset. The emotional dataset contains six emotions, including Contentment, Warmth, Excitement, Awe, Fear, and Sadness. Figure 2 shows the example images of our emotional dataset, Table 1 shows the statistics of the six emotions. Because the awe and warmth emotional images are too few to train our network, so we only choose two positive emotions (Contentment, Excitement) and two negative emotions (Fear, Sadness).

Fig. 2. The example images of our emotional dataset. Top Two Rows: two positive emotions, Bottom Two Rows: two negative emotions.

4.2   Implementation Details

The full Feature Extraction layers are conv1, conv2, conv3, conv4. One Relu layer follows behind each convolution layer in Feature Extraction Layer. We resize each of images to 512×512 and train for roughly 150,000 iterations with a batch size of 4 content-emotion image pairs. Training used Adam optimizer [9] with learning rate of 1×10^{-3} and learning rate decay 5×10^{-5}. We compute target content loss on relu4_1 layer and compute emotion loss on relu1_1 layer, relu2_1 layer, relu3_1 layer, and relu4_1 layer. In our experiments, we compute the loss function (Eq. 8) to train the

Reconstruct Layers, in which α represents the weight of the content image and the emotion image:

$$L = L_c + \alpha L_e \tag{8}$$

4.3   Experiments Results

As mentioned, we conduct our experiments for four emotions, including Contentment, Excitement, Fear, and Sadness. The details of the results will be described in the following sections.

In Fig. 3, we show qualitative evaluations and quantitative evaluations by comparing our experiments with other representative methods: (1) the original transfer network [7]; (2) the fast feed-forward for single style transfer [3]; (3) the fast feedforward transfer for arbitrary style [10]. We generate our experiments results with a single GPU GeForce GTX 1080; training costs roughly 14 h.

Fig. 3. Qualitative examples results of image emotionalizing on different methods. All tested and generated images are 512 × 512 pixels.

Experimental Evaluation. In Fig. 3, we randomly select several qualitative examples from experiment results about sadness image emotionalizing by comparing our results with other representative methods. Overall, the quality of our results are quite comparable to other methods. Obviously, we can conclude that the quality of our method achieves more satisfying results than [7] (e.g., row 1, 2, 4). For example, the lenna of our results has more details while Gatys et al. appears unusually blurry in the mouth and hair (row 4, column 4). However, in some situation (e.g., row 5), the style of Gatys et al. [7] is more in line with the style of emotion images, our method is not as effective as the method of [7]. Compared with Ulyanov et al. [3], our results are more vivid in emotion expression, while the results of Ulyanov et al. are more realistic in terms of

Table 2. Comparison of the average computation times of our methods. We conduct two kind of resolution experiments, which are 512 and 256 pixels. The timing unit is seconds

Method	Time cost (512 × 512)	Time cost (256 × 256)
Gatys et al.	145.776	52.513
Ulyanov et al.	2.105	2.052
Chen and Schmidt	11.114	5.768
Ours	2.304	2.17

emotion expression. This obviously does not meet the requirements of our image emotionalizing. From these experimental results, we can conclude that our experiments yielded attractive results that are comparable to many methods.

Speed Analysis. To test the real-time nature of our method, we perform a speed analysis by comparing the time spent on image emotionalizing on various methods, including our method, Gatys et al. [7] method, Ulyanov et al. [3] method and Chen and Schmidt method [10].

In Table 2, we show the speed comparison of our method with mentioned methods [3, 7, 10]. Comparing with the original algorithm [7], our method is nearly 1–2 orders of magnitude faster than Gatys. In addition, comparing with the fastest feed-forward network method, our method speed is a bit slower than [3]. It is worth considering that the feed-forward networks [3] just conduct single image stylization and need less calculation than ours. Similarly, when compared to arbitrary emotion image, we found our method performed better in real-time image emotionalizing. In general, our approach has appealing speed in real-time and flexibility in arbitrary emotion image emotionalizing (Fig. 4).

Fig. 4. More examples of the four emotions. The horizontal rows are emotion images. The first column of each emotion is the content images. The first row of each emotion is the emotional images

5 Discussion and Conclusion

In this work, we developed an emotion synthesis algorithm for emotion image emotionalizing. As mentioned, our method are adapted to transfer emotion statistical information for arbitrary emotion images. At the same time, our experimental results are helpful for understanding image features expression of convolution neural network.

Our method adopt an extension Instance Normalization that enables align the mean and variance of the input images. Our method is difference with previous neural optimization process [7] and feed-forward approach [3, 10, 14] that match statistic feature to approximate original inputs in pixel-level. To be specific, we align the statistics of the input content image and the input emotion image in feature space.

Although our experimental results and response speed are very promising at the application level, our method also has a few shortcomings and space for improvement. One of the shortcomings is we do not propose more ways to measure emotion in emotion image emotionalizing. In future work, we will develop more evaluation measures for the quality of image emotionalizing and study more emotions (E.g. amusement, anger, awe, disgust).

Acknowledgement. This work was partially supported by National Nature Science Foundation of China (61772201), and Key Projects of Anhui Province Science and Technology Plan (15czz02074).

References

1. Simonyan, K., Zisserman, A.: Very deep convolutional networks for large-scale image recognition. arXiv preprint arXiv:1409.1556 (2014)
2. Ioffe, S., Szegedy, C.: Batch normalization: accelerating deep network training by reducing internal covariate shift. In: Proceedings of the 32nd International Conference on Machine Learning, pp. 448–456 (2015)
3. Ulyanov, D., Vedaldi, A., Lempitsky, V.: Texture networks: feed-forward synthesis of texture and stylized images. In: Proceedings of the 33rd International Conference on Machine Learning, pp. 1349–1357. (2016)
4. Radford, A., Metz, L., Chintala, S.: Unsupervised representation learning with deep convolutional generative adversarial networks. arXiv preprint arXiv:1511.06434 (2015)
5. Ulyanov, D., Vedaldi, A., Lempitsky, V.: Improved texture networks: maximizing quality and diversity in feed-forward stylization and texture synthesis. In: CVPR, pp. 4105–4113 (2017)
6. Ulyanov, D., Vedaldi, A., Lempitsky, V.: Instance normalization: the missing ingredient for fast stylization. CsCV. arXiv:1607.08022 (2017)
7. Gatys, L.A., Ecker, A.S., Bethge, M.: Image style transfer using convolutional neural networks. In: Conference on Computer Vision and Pattern Recognition (2016)
8. Lin, T.-Y., et al.: Microsoft COCO: common objects in context. In: Fleet, D., Pajdla, T., Schiele, B., Tuytelaars, T. (eds.) ECCV 2014. LNCS, vol. 8693, pp. 740–755. Springer, Cham (2014). https://doi.org/10.1007/978-3-319-10602-1_48
9. Kingma, D.P., Ba, J.: Adam: a method for stochastic optimization. arXiv preprint arXiv: 1412.6980 (2014)

10. Chen, T. Q., Schmidt, M.: Fast patch-based style transfer of arbitrary style. arXiv preprint arXiv:1612.04337 (2016)
11. Borth, D., Chen, T., Ji, R., Chang, S.F.: Sentibank: large-scale ontology and classifiers for detecting sentiment and emotions in visual content. In: ACM on Multimedia, pp. 459–460 (2013)
12. Yuan, J., Mcdonough, S., You, Q., Luo, J.: Sentribute: image sentiment analysis from a mid-level perspective. In: Proceedings of the Second International Workshop on Issues of Sentiment Discovery and Opinion Mining, pp. 1–8. ACM (2013)
13. Gatys, L.A., Ecker, A.S., Bethge, M.: A neural algorithm of artistic style. arXiv preprint arXiv:1508.06576 (2015)
14. Johnson, J., Alahi, A., Li, F.F.: Perceptual losses for real-time style transfer and super-resolution. In: European Conference on Computer Vision, pp. 694–711. (2016)
15. Li, C., Wand, M.: Precomputed real-time texture synthesis with markovian generative adversarial networks. In: Leibe, B., Matas, J., Sebe, N., Welling, M. (eds.) ECCV 2016. LNCS, vol. 9907, pp. 702–716. Springer, Cham (2016). https://doi.org/10.1007/978-3-319-46487-9_43
16. Cao, D., Ji, R., Lin, D., Li, S.: Visual sentiment topic based microblog image sentiment analysis. Multimed. Tools Appl. 75(15), 8955–8968 (2016)
17. Dumoulin, V., Shlens, J., Kudlur, M.: A learned representation for artistic style. CoRR, abs/1610.07629 2.4, 5 (2016)
18. Chang, Y., Saito, S., Nakajima, M.: Example-based color transformation of image and video using basic color categories. IEEE Trans. Image Process. 16(2), 329–336 (2007)
19. Gupta, M.R., Upton, S., Bowen, J.: Simulating the effect of illumination using color transformation. In: Computational Imaging III. International Society for Optics and Photonics, pp. 248–259 (2005)
20. Reinhard, E., Ashikhmin, M., Gooch, B., Shirley, P.: Color transfer between images. IEEE Comput. Graph. Appl. 21(5), 34–41 (2001)
21. Chang, Y., Saito, S., Uchikawa, K., Nakajima, M.: Example-based color stylization of images. ACM Trans. Appl. Percept. 2(3), 322–345 (2005)
22. Xu, M., Ni, B., Tang, J., Yan, S.: Image re-emotionalizing. In: PCM, pp. 3–14 (2011)
23. Li, T., Ni, B., Xu, M., Wang, M., Gao, Q., Yan, S.: Data-driven affective filtering for images and videos. IEEE Trans. Cybern. 45(10), 2336–2349 (2017)
24. Gao, W., Li, S., Lee, S.Y.M., Zhou, G., Huang, C.R.: Joint learning on sentiment and emotion classification. In ACM, pp. 1505–1508 (2013)
25. Machajdik, J., Hanbury, A.: Affective image classification using features inspired by psychology and art theory. In: Proceedings of the 18th ACM international Conference on Multimedia, pp. 83–92 (2010)
26. Mathews, A., Xie, L., He, X.: SentiCap: generating image descriptions with sentiments. In: 13th AAAI Conference on Artificial Intelligence, pp. 3574–3580 (2016)
27. Hong, R., Zhang, L., Zhang, C., Zimmermann, R.: Flickr circles: aesthetic tendency discovery by multi-view regularized topic modeling. IEEE Trans. Multimed. 18(8), 1555–1567 (2016)
28. Hong, R., Zhang, L., Tao, D.: Unified photo enhancement by discovering aesthetic communities from Flickr. IEEE Trans. Image Process. 25(3), 1124–1135 (2016)
29. Hong, R., Hu, Z., Wang, R., Wang, M., Tao, D.: Multi-view object retrieval via multi-scale topic models. IEEE Trans. Image Process. 25(12), 5814–5827 (2016)

Text-to-Image Synthesis via Visual-Memory Creative Adversarial Network

Shengyu Zhang[1], Hao Dong[2], Wei Hu[3], Yike Guo[2], Chao Wu[1], Di Xie[4], and Fei Wu[1(✉)]

[1] Zhejiang University, Hangzhou, China
{wufei,chao.wu}@zju.edu.cn, light.e.gal@gmail.com
[2] Imperial College London, London, UK
{hao.dong11,y.guo}@imperial.ac.uk
[3] Baidu Research, Beijing, China
elweihu@gmail.com
[4] Hikvision Research Institute, Hangzhou, China
xiedi@hikvision.com

Abstract. Despite recent advances, text-to-image generation on complex datasets like MSCOCO, where each image contains varied objects, is still a challenging task. In this paper, we propose a method named visual-memory Creative Adversarial Network (vmCAN) to generate images depending on their corresponding narrative sentences. vmCAN appropriately leverages an external visual knowledge memory in both multimodal fusion and image synthesis. By conditioning synthesis on both internally textual description and externally triggered "visual proposals", our method boosts the inception score of the baseline method by 17.6% on the challenging COCO dataset.

Keywords: Text-to-image · Visual knowledge · Adversarial model

1 Introduction

Realistic image generation from natural language descriptions is an active research task. The technique is applicable to many practical applications such as image editing and sketch or game designing. Models based on Generative Adversarial Networks (GAN) [6] have achieved promising results on datasets merely consisting of single category objects in images like CUB [35] and Oxford Flower [20]. However existing methods are far from promising on complex dataset like MSCOCO [16], in which generally one image contains varied objects and objects are rarely centered in the image [24,38]. In order to generate complex scenes, existing approaches attempt to utilize word level attention to fine-grain image [37], establish hierarchical text-to-image mapping [8] and enhance the text description in a manner of dialog [5]. However, little work has been carried out using auxiliary visual knowledge. According to the human painting process,

© Springer Nature Switzerland AG 2018
R. Hong et al. (Eds.): PCM 2018, LNCS 11166, pp. 417–427, 2018.
https://doi.org/10.1007/978-3-030-00764-5_38

real-world scenes or some references may help a painter learn quickly during training and improve the generation quality during inference. That is to say, one sophisticate painter in general triggered many of the relevant visual cues during his/her painting.

Based on these intuitions, we suggest using sub-images as the visual cues to enhance text-to-image generation. More specifically, we use proposals extracted by Region Proposal Network [26] as visual cues which are stored in the external visual-knowledge memory. A proposal feature vector can be viewed as a visual summary of a meaningful sub-image, especially the one with the highest probability containing a real-world object.

The extracted proposals (i.e., visual cues) bears many of visual details such as texture, shape, color, size, etc., and they can potentially be inspired together to synthesize images after they are triggered by corresponding textual descriptions.

In this paper, given one textual sentence and the external visual-knowledge memory, we first utilize the multi-modal Encoder to encode the textual sentence into a multi-modal hidden vector. The multi-modal encoder is similar to the Memory Network model proposed by [31]. However, this paper uses semantic embeddings instead of bag-of-words representation.

The key contributions of our work are listed as following: we propose a model named visual-memory Creative Adversarial Network (vmCAN) for generating complex real-world images in a synthesis manner via the appropriate integration of an external visual-knowledge memory (i.e., visual cues). We employ a multi-modal encoder to encode visual cues and textual description into a multi-modal hidden vector to trigger the relevant visual counterparts of sentence descriptions. Knowledge retrieval process is stacked along with the stacked image generation process. We conduct experiment and evaluations on MSCOCO and our proposed approach boosts the inception score by 17.6% than the baseline.

2   Related Work

2.1   Memory Network

First proposed by [31,32], Memory Network has been utilized to augment neural networks for different tasks, such as algorithm inference [7], conversational systems [5,29,34] and question answering [31,36]. Memory helps extend the capability to capture long-term dependencies and provides a way to model relevant information inside their surroundings. As for unconditioned image generation area, [14] presents a deep generative model (DGMs) with memory and attention to capture the local detail information and [12] successfully applied a life-long memory network [11] to adversarial models.

Compared with these memory augmented networks, we propose to employ an external visual-knowledge and memory network to model and leverage the correlations between visual images and textual sentences. The uniqueness of our model will be specified in the next section.

2.2   Generative Adversarial Networks

Recently, Generative Adversarial Networks have shown the ability to generate appealing images with conditions. Generated images are required not only realistic but also well aligned with the condition constraints. The condition variable can be simple discrete class labels [3,19,21,22] and language sequence [4,24,38] which is complex in structure and plentiful in expression. Constrained on visual domain, GAN model has been applied to domain transfer [9,17], image editing [2,39], super-resolution [10,13] and style transfer [10]. [25] managed to draw pictures conditioned on object location. [4] proposed a method to edit a given image with specific textual description. Compared with these Conditional GAN models, our proposed vmCAN attempts to synthesize images conditioned on the textual description and multiple relevant sub-images, which can be seen as an appropriate extension to CGAN framework.

3   Knowledge Grounded Synthesis

We formulate the *sentence-to-image* problem as following: given an image description t, which may remark objects, properties of objects and relations between objects, we aim to learn a series of stacked multi-modal encoders $ME_0, ..., ME_{gn}$ and stacked generators $G_0, ...G_{gn}$. The final output is one corresponding image $s = G_{gn}(ME_{gn}(...(G_0(ME_0(t, P_0))...), P_{gn}))$, where P_i is one group of m proposals sampled from Kownledge Proposal Memory and gn is the number of stacked processes. We set $gn = 2$ in our experiment.

Compared with textual-knowledge based system [5,28], one primary challenge in leveraging visual knowledge is that images relevant to the target synthesized image still contain much irrelevant information. In our opinion, objects are typically the most important part of an image and they can be easily extracted using Region Proposal Network.

Another problem is that relevant sub-images cannot be directly applied to the target image. For example, virtual viewpoint synthesis [30] requires large viewpoint inputs like video sequence which can provide important structure and texture information. As a result, we use semantic vectors to represent these sub-images and employ attention mechanism to leverage them.

Our model will be demonstrated in three parts: 1. Proposal Extraction for knowledge preparation. 2. Multi-modal Encoder to encode text and relevant proposals (i.e. visual cues) into a multi-modal hidden vector. 3. Stacked Adversarial Generation to generate the realistic image in a stacked manner conditioned on the multi-modal hidden vector.

3.1   Proposal Extraction

Region Proposal Network proposed by [26] ranks and refines region boxes called anchors to generate high-quality region proposals which most likely contain an object. After RPN, a Region of Interest Pooling layer is used to normalize different sized CNN feature map into the same size. The output of ROI pooling is used as visual cues in our visual-memory knowledge.

We extract about 320000 proposals from MSCOCO **training** dataset images to build our Proposal Knowledge Memory. Each proposal is a semantic vector of dimension 1024. When there are more than 5 proposals in one image, we just keep top 5 proposals with the highest predicted objectness score. Extraction can be finished in an offline fashion (Fig. 1).

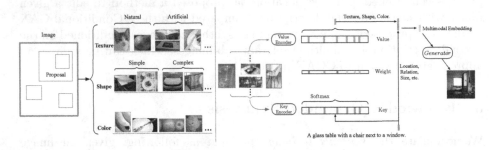

Fig. 1. We extract proposals which may provide texture, shape and color cues to build our visual-knowledge memory. After given a sentence "a glass table with a chair next to a window", the multi-modal encoder triggers some of useful visual cues (i.e. proposals in visual-knowledge memory) to generate a multi-modal hidden vector w.r.t. the given sentence.

3.2   Multi-modal Encoder

Based on memory networks, our Multi-modal Encoder ME uses two encoders and attention mechanism to model proposals (i.e., visual cues) and textual descriptions.

We first encode text description t into a continuous representation $\varphi(t)$ using a pre-trained text encoder φ [24]. We further augment the text embedding using a method proposed by [38]. This augmentation helps generate a large number of additional text embeddings for adversarial training [4]. More formally, a fully connected layer is applied over the input text embedding to generate μ and σ. The augmented text embedding is computed as $\varphi(t)' = \mu + \sigma \odot \varepsilon$ where ς is sampled from $\mathcal{N}(0, I)$ and $\odot$ is the element-wise multiplication operator. Augmented text vector is of dimension d. We randomly sample m proposals $P = \{p_1, p_2, p_3, ..., p_m\}$ from Proposal Knowledge Memory. Based on [31], these proposals are encoded into key representations and value representations respectively:

$$k_i = \kappa_0(p_i)$$
$$v_i = \nu_0(p_i) \tag{1}$$

Where p_i is a proposal feature vector of dimension 1024, k_i and v_i are of dimension d. Key representation is used to attend and weight retrieved knowledge. Value representation contains useful guidance necessary for generation.

Both key and value encoders are neural networks which are simple fully connected layers with ReLU activation function in our model. The multi-modal hidden vector is produced as follows:

$$a_i = Softmax(\varphi(t)'^T k_i)$$
$$o = \sum_i v_i a_i$$
$$c = (o, \varphi(t)')$$
$$\hat{c} = ReLU(Wc + b)$$

$\hat{c} = ME(t, P)$ is the final multi-modal hidden vector of dimension d.

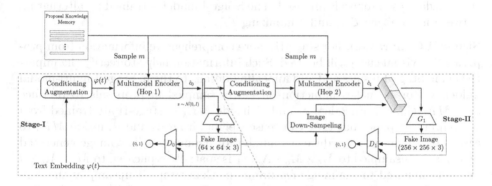

Fig. 2. Pipeline of the proposed vmCAN. At the first stage of generation, augmented text embedding $\varphi(t)'$ and sampled m proposals will be encoded to a multi-modal representation $\hat{c}_0$. This feature vector will guide the first sketch-like image generation. For stage-II, multimodal tensor $\hat{c}_0$ from the first stage will make induction on newly sampled m proposals. The output $\hat{c}_1$ and the downsampled first-stage generated image will be concatenated and used for real-world image generation.

3.3   Stacked Adversarial Generation

Building upon StackGAN-v1 [38], the whole pipeline model is defined in Fig. 2.

Stage-I Generation. For stage-I Generation, more sketch-like information like shapes from proposals (i.e. visual cues) will be used. The multimodel condition vector $\hat{c}_0$ helps generator produce a low-resolution sketch-like image. A noise vector is sampled from a normal distribution p_z. Concatenated by this noise vector, the multi-modal hidden vector goes through several upsampling blocks to generate a $W_0 \times H_0$ color image. Then Discriminator D_0 downsamples this image to $M_d \times M_d \times N_d$ feature map. Meanwhile, the augmented text embedding $\varphi(t)'$ is spatially replicated to $M_d \times M_d \times d$ and then concatenated with above image feature map. The concatenated tensor is further downsampled to a fake-real score whose range is between 0 and 1. The loss functions of G_0 and D_0 are defined as follows:

$$\mathcal{L}_{D_0} = \mathop{\mathbb{E}}_{(I,t)\sim p_{data}} [log D_0(I, \varphi(t)')] +$$

$$\mathop{\mathbb{E}}_{z\sim p_z, t\sim p_{data}} [log(1 - D_0(G_0(z, \hat{c}_0), \varphi(t)'))]$$

$$\mathcal{L}_{G_0} = \mathop{\mathbb{E}}_{z\sim p_z, t\sim p_{data}} [log(1 - D_0(G_0(z, \hat{c}_0), \varphi(t)'))] \tag{3}$$

$$+ \lambda D_{KL}(\mathcal{N}(\mu_0(\varphi(t)')), \Sigma_0(\varphi(t)') \| \mathcal{N}(0, I))$$

where I is the real image, t is the pre-trained text embedding, z is a noise vector sampled from a given distribution such as Gaussian distribution in our experiment and $\hat{c}_0 = ME_0(t, P_0)$ is the multi-modal hidden vector. In our model, the discriminator is not conditioned on proposals. That is to say, the generated image doesn't need to be well aligned with proposals. Only some useful visual cues inside these proposals are used. The Stage-I model is trained by alternating between maximizing $\mathcal{L}_{D_0}$ and minimizing $\mathcal{L}_{G_0}$.

Stage-II Generation. For stage-II, more comprehensive information from proposals (i.e. visual cues) will be used. Such information helps to rectify the imperfection in Stage-I results and add appealing details to them. Stage-I multi-modal hidden vector $\hat{c}_0$ makes induction on newly sampled m proposals to produce $\hat{c}_1 = ME_1(G_0, P_1)$. Encoders κ_1 and ν_1 inside ME_1 in stage-II are trained from scratch in this paper but they can reuse weights from κ_0 and ν_0 inside ME_0 to ease training since they reduce the number of parameters. The image generated by G_0 is downsampled to $M_g \times M_g \times N_g$. $\hat{c}_1$ is spatially replicated to $M_g \times M_g \times d$ and concatenated to the image feature map. Generator G_1 upsamples the concatenated feature map to a $W_1 \times H_1$ image. Discriminator D_1 downsampling process is the same as Stage-I except that the input image is larger and downsampling networks are more complex. Similar to stage-II, the loss functions of G_1 and D_1 are:

$$\mathcal{L}_{D_1} = \mathop{\mathbb{E}}_{(I,t)\sim p_{data}} [log D_1(I, \varphi(t)')] +$$

$$\mathop{\mathbb{E}}_{s_0\sim p_{G_0}, t\sim p_{data}} [log(1 - D_1(G_1(s_0, \hat{c}_1), \varphi(t)'))]$$

$$\mathcal{L}_{G_1} = \mathop{\mathbb{E}}_{s_0\sim p_{G_0}, t\sim p_{data}} [log(1 - D_1(G_1(s_0, \hat{c}_1), \varphi(t)'))] \tag{4}$$

$$+ \lambda D_{KL}(\mathcal{N}(\mu_0(\varphi(t)')), \Sigma_0(\varphi(t)') \| \mathcal{N}(0, I))$$

where s_0 is the generated image by Stage-I G_0. To make it more directly comparable to baseline StackGAN model, we set the model parameters as $N_z = 100$, $W_0 = 64$, $H_0 = 64$, $M_d = 4$, $N_d = 512$, $N_d = 128$, $M_g = 16$, $N_g = 512$, $d = 128$, $W_1 = 256$, $H_1 = 256$, and $\lambda = 1$.

a yellow pickup truck stopped at an intersection

a small laptop is sitting on a desk

a living room filled with living room furniture and decor

a person cutting a pizza with a pair of scissors

a couple of plates that have some food on them

this hazy picture depicts traffic on a busy street

Fig. 3. Some samples of descriptions, top 2 relevant proposals and generated images. Generated images are highlighted using red boxes and relevant proposals are underlined in blue. (Color figure online)

4 Experiments

4.1 Experimental Setup

Datasets. Our model is evaluated on COCO captioning 2015 dataset. By default, it contains 80k images for training and 40k images for validation with 5 captions per image. There are over 80 semantic object categories in total and each image contains varied objects which are rarely centered in the image.

Evaluation. In order to measure images generation recognizability and generation diversity, we use inception score. Moreover, we generate captions to quantitatively measure how well the generated images are conditioned on the textual descriptions.

Inception Score - Inception score is first proposed by [27] and has been acknowledged to be well correlated with human evaluation on the quality and diversity of generated images.

Caption Quality [8] - Since the inception score cannot reflect whether the generated images are well conditioned on the given text descriptions, we generate captions using a pre-trained caption model [18] trained on MS-COCO. Then we measure how similar these generated captions are to textual input using four standard language similarity metrics: BLEU [23], METEOR [1], ROGUE_L [15] and CIDEr [33].

In addition to quantitative evaluations above, we also conduct qualitative evaluations in terms of visualization.

4.2 Quantitative Results

Ablative Analysis. In order to better understand the impact of visual knowledge, we conduct ablative analysis by using Ground Truth knowledge for each textual description. Given a text-image training pair, proposals extracted from the paired image are considered as Ground Truth knowledge in terms of the

Table 1. Inception Score by different models on MSCOCO test sets. Higher is better.

	Inception score
StackGAN [38]	8.35 ± 0.03
vmCAN-R	9.94 ± 0.12
vmCAN-GT	$\mathbf{10.36 \pm 0.17}$
chatPainter [28]	9.74 ± 0.02
Hong et al. [8]	11.46 ± 0.09
AttnGAN [37]	$\mathbf{25.89 \pm 0.47}$

corresponding textual description. Instead of training a multi-modal encoder to encode proposals and text to a meaningful multi-modal vector, we simply average this group of proposal vectors, linearly transform the averaged vector to a tensor of dimension d, concatenate this tensor with augmented text embedding $\varphi(t)'$ and finally non-linearly encode them into a multi-modal hidden vector of dimension d. This vector will be used for further generation. This is, in fact, a weak upper-bound because we simply average these proposal feature vectors. Even though, we get substantial improvement both on the Inception Score and Caption Generation BLEU. This weak upper-bound model is noted as vmCAN-GT and the previous model is named vmCAN-R.

Table 1 shows the image recognisability analysis. In detail, we compare the test set image generation Inception Score between our method, the baseline method, and some other approaches based on conditional GANs. Our model boosts the baseline method by 17.6%. This quantitatively shows that proposals (i.e. visual cues) can help enhance the generation quality and potentially increase the generation variety. We will point out that the proposed vmCAN does not achieve a better performance compared to attnGAN [37]. However, attnGAN, which also has a stacked generation process, can be enhanced by incorporating the visual-knowledge memory and the multi-modal encoders.

Table 2. Evaluation metrics based on caption generation to measure whether the generated images are well conditioned on the given text descriptions. Higher is better in all columns.

	Caption generation						
	BLEU-1	BLEU-2	BLEU-3	BLEU-4	METEOR	ROUGE_L	CIDEr
StackGAN [38]	0.400	0.188	0.078	0.037	0.092	0.267	0.039
vmCAN-R	0.399	0.187	0.079	0.038	0.093	0.266	0.039
vmCAN-GT	**0.467**	**0.261**	**0.137**	**0.075**	**0.124**	**0.307**	**0.145**
Real Image	0.743	0.577	0.427	0.313	0.273	0.488	0.946

Table 2 shows the caption generation result. By conditioning both on text description and additional visual knowledge, our method yields little loss on text-image (generated) relevance with randomly sampled knowledge and an improvement with Ground Truth knowledge. This result further shows the effectiveness of the utilization of proposals (i.e. visual cues) and the necessity of building an efficient and accurate text-proposal retrieval system.

4.3 Qualitative Results

Figure 3 shows some examples of generated image and Top 2 relevant proposals. In detail, we compute the cosine similarity between the text embedding $\varphi(t)'$ and key encodings of K sampled proposals and visualize the top 2 relevant proposals represented using bounding box. This result shows that our multi-modal encoder is able to activate relevant visual knowledge although some of them are irrelevant from the human perspective.

Figure 4 shows some creative generation examples generated by our proposed method. These results illustrate that our method is able to generate novel images which do not exist in the source dataset.

Fig. 4. Some novel images generated by our model.

5   Conclusions

In this paper, we propose a visual-memory augmented approach named vmCAN for *sentence-to-image* synthesis. Our model obtains substantial improvement over the baseline method on the challenging MSCOCO dataset.

Acknowledgement. This work is partially supported by the National Natural Science Foundation of China (Nos. 61751209, U1611461), the Key Program of Zhejiang Province, China (No. 2015C01027), and Artificial Intelligence Research Foundation of Baidu Inc.

References

1. Banerjee, S., Lavie, A.: METEOR: an automatic metric for MT evaluation with improved correlation with human judgments. In: Proceedings of the ACL Workshop on Intrinsic and Extrinsic Evaluation Measures for Machine Translation and/or Summarization, pp. 65–72 (2005)
2. Brock, A., Lim, T., Ritchie, J.M., Weston, N.: Neural photo editing with introspective adversarial networks. In: ICLR (2017)

3. Chen, X., Duan, Y., Houthooft, R., Schulman, J., Sutskever, I., Abbeel, P.: Info-GAN: interpretable representation learning by information maximizing generative adversarial nets. In: NIPS (2016)
4. Dong, H., Yu, S., Wu, C., Guo, Y.: Semantic image synthesis via adversarial learning. In: ICCV (2017)
5. Ghazvininejad, M., et al.: A knowledge-grounded neural conversation model. In: AAAI (2017)
6. Goodfellow, I., et al.: Generative adversarial nets. In: NIPS (2014)
7. Graves, A., Wayne, G., Danihelka, I.: Neural turing machines. arXiv preprint arXiv:1410.5401 (2014)
8. Hong, S., Yang, D., Choi, J., Lee, H.: Inferring semantic layout for hierarchical text-to-image synthesis. arXiv preprint arXiv:1801.05091 (2018)
9. Isola, P., Zhu, J.Y., Zhou, T., Efros, A.A.: Image-to-image translation with conditional adversarial networks. In: CVPR (2017)
10. Johnson, J., Alahi, A., Fei-Fei, L.: Perceptual losses for real-time style transfer and super-resolution. In: Leibe, B., Matas, J., Sebe, N., Welling, M. (eds.) ECCV 2016. LNCS, vol. 9906, pp. 694–711. Springer, Cham (2016). https://doi.org/10.1007/978-3-319-46475-6_43
11. Kaiser, L., Nachum, O., Roy, A., Bengio, S.: Learning to remember rare events. In: ICLR (2017)
12. Kim, Y., Kim, M., Kim, G.: Memorization precedes generation: learning unsupervised GANs with memory networks. arXiv preprint arXiv:1803.01500 (2018)
13. Ledig, C., et al.: Photo-realistic single image super-resolution using a generative adversarial network. In: CVPR (2017)
14. Li, C., Zhu, J., Zhang, B.: Learning to generate with memory. In: ICML (2016)
15. Lin, C.Y.: ROUGE: a package for automatic evaluation of summaries. In: Marie-Francine Moens, S.S. (ed.) Text Summarization Branches Out: Proceedings of the ACL 2004 Workshop, pp. 74–81. Association for Computational Linguistics, Barcelona, July 2004
16. Lin, T.-Y., et al.: Microsoft COCO: common objects in context. In: Fleet, D., Pajdla, T., Schiele, B., Tuytelaars, T. (eds.) ECCV 2014. LNCS, vol. 8693, pp. 740–755. Springer, Cham (2014). https://doi.org/10.1007/978-3-319-10602-1_48
17. Liu, M.Y., Breuel, T., Kautz, J.: Unsupervised image-to-image translation networks. In: NIPS (2017)
18. Luo, R.: An image captioning codebase in pytorch (2017). https://github.com/ruotianluo/ImageCaptioning.pytorch
19. Mirza, M., Osindero, S.: Conditional generative adversarial nets. arXiv preprint arXiv:1411.1784 (2014)
20. Nilsback, M.E., Zisserman, A.: Automated flower classification over a large number of classes. In: Sixth Indian Conference on Computer Vision, Graphics and Image Processing, ICVGIP 2008, pp. 722–729. IEEE (2008)
21. Odena, A., Olah, C., Shlens, J.: Conditional image synthesis with auxiliary classifier GANs. In: ICML (2017)
22. van den Oord, A., Kalchbrenner, N., Espeholt, L., Vinyals, O., Graves, A., et al.: Conditional image generation with pixelCNN decoders. In: NIPS (2016)
23. Papineni, K., Roukos, S., Ward, T., Zhu, W.J.: BLEU: a method for automatic evaluation of machine translation. In: ACL. Association for Computational Linguistics (2002)
24. Reed, S., Akata, Z., Yan, X., Logeswaran, L., Schiele, B., Lee, H.: Generative adversarial text to image synthesis. In: ICML (2016)

25. Reed, S.E., Akata, Z., Mohan, S., Tenka, S., Schiele, B., Lee, H.: Learning what and where to draw. In: NIPS (2016)
26. Ren, S., He, K., Girshick, R., Sun, J.: Faster R-CNN: towards real-time object detection with region proposal networks. In: NIPS (2015)
27. Salimans, T., Goodfellow, I., Zaremba, W., Cheung, V., Radford, A., Chen, X.: Improved techniques for training GANs. In: NIPS (2016)
28. Sharma, S., Suhubdy, D., Michalski, V., Kahou, S.E., Bengio, Y.: Chat-Painter: improving text to image generation using dialogue. arXiv preprint arXiv:1802.08216 (2018)
29. Sordoni, A., et al.: A neural network approach to context-sensitive generation of conversational responses. In: NAACL-HLT (2015)
30. Starck, J., Hilton, A.: Virtual view synthesis of people from multiple view video sequences. Graph. Model. **67**(6), 600–620 (2005)
31. Sukhbaatar, S., Weston, J., Fergus, R., et al.: End-to-end memory networks. In: NIPS (2015)
32. Sukhbaatar, S., Weston, J., Fergus, R., et al.: Memory networks. In: ICLR (2015)
33. Vedantam, R., Lawrence Zitnick, C., Parikh, D.: CIDEr: consensus-based image description evaluation. In: CVPR (2015)
34. Vinyals, O., Le, Q.: A neural conversational model. In: ICML (2015)
35. Wah, C., Branson, S., Welinder, P., Perona, P., Belongie, S.: The caltech-UCSD birds-200-2011 dataset (2011)
36. Weston, J., et al.: Towards AI-complete question answering: a set of prerequisite toy tasks. In: ICLR (2016)
37. Xu, T., et al.: AttnGAN: fine-grained text to image generation with attentional generative adversarial networks. arXiv preprint arXiv:1711.10485 (2017)
38. Zhang, H., et al.: StackGAN: text to photo-realistic image synthesis with stacked generative adversarial networks. In: ICCV (2017)
39. Zhu, J.-Y., Krähenbühl, P., Shechtman, E., Efros, A.A.: Generative visual manipulation on the natural image manifold. In: Leibe, B., Matas, J., Sebe, N., Welling, M. (eds.) ECCV 2016. LNCS, vol. 9909, pp. 597–613. Springer, Cham (2016). https://doi.org/10.1007/978-3-319-46454-1_36

Sequence-Based Recommendation
with Bidirectional LSTM Network

Hailin Fu, Jianguo Li$^{(\boxtimes)}$, Jiemin Chen, Yong Tang, and Jia Zhu

School of Computer Science, South China Normal University,
Guangzhou 510000, China
{hailin,jianguoli,chenjiemin,ytang,jzhu}@m.scnu.edu.cn

Abstract. In modern recommendation systems, most methods often neglect the sequential relationship between items. So we propose a novel Sequence-based Recommendation model with Bidirectional Long Short-Term Memory neural network (BiLSTM4Rec) which can capture the sequential feature of items to predict what a user will choose next. By collecting consumed items of a user in a sequence with time ascending order, fitting the model with the last item as the label, the rest items as the features, we regard this recommendation assignment as a super multiple classification task. Once trained well, the output layer of our model will export the probabilities of the next items with given sequence. In the experiments, we compare our approach with several commonly used recommendation methods on a real-world dataset. Experimental results indicate that our sequence-based recommender can perform well for short-term interest prediction on a sparse, large dataset.

Keywords: Recommendation system · Social media data mining
Sequential prediction · Bidirectional recurrent neural network
Deep learning

1 Introduction

The primary assignments of recommendation systems faced with consist of two parts: ratings predicting and products recommendation. So to predict what a user will choose next given his consumed history is one of the crucial mission [17] in recommendation area. In many websites and applications, such as online electronic business, news/videos website, music/radio station, they need an excellent service for users to recommend what they will like in future. Existing recommenders mainly concentrate on finding the neighbor sets for users or items, or leveraging other explicit/implicit information (such as tags, reviews, item contents and user profiles) for neighborhood-aware. However, to the best of our knowledge, few works use the sequential feature of data to build recommender. We find that the sequence of data implicates much exciting and relevant information, for example in a video website, the user who watched "Winter is coming" (S01, E02 of Game of Thrones) will be more likely to watch "The Kingsroad"

© Springer Nature Switzerland AG 2018
R. Hong et al. (Eds.): PCM 2018, LNCS 11166, pp. 428–438, 2018.
https://doi.org/10.1007/978-3-030-00764-5_39

(S01, E02 of Game of Thrones). Even at the 2011 Recsys conference, Pandora[1]'s researchers gave a speech about music recommendation and said they found many users consumed music in sequences.

Our work was inspired by the previous study of Lai et al. [13], where a neural network is proposed to capture the sequence of words in a sentence. We took a similar approach by considering one item as a word, the catalog of items as a vocabulary, and the historical consumed items of one user as a sentence, to capture the sequence of the user consumed items. The main contributions of our work are as follows:

- We propose a novel Sequence-based Recommendation model with Bidirectional Long Short-Term Memory neural network, or BiLSTM4Rec for short, which can capture the sequential features of data, as well scales linearly with the number of objectives (both of users and items).
- We regard item sequence as a sentence and use an $M \times d$ embedding matrix E to represent M items which reduces memory cost evidently when faced big data.

2 Related Work

2.1 Traditional Methods in Recommendation

There are three main classes of traditional recommendation systems. Those are collaborative filtering systems, content-based filtering systems, and hybrid recommendation systems [1]. Collaborative filtering (CF) can be generally classified into Memory-based [12,14] CF and Model-based [5,10,16] CF. Memory-based CF [5,12,14,16] systems generate recommendations for users by finding similar user or item groups. Model-based CF systems, e.g., Bayesian networks [5], clustering models [16], Probabilistic Matrix Factorization [10], mainly use the rating information to train corresponding model then use this model to predict unknown data. Content-based systems [3] generate recommendations for users based on a description of the item and a profile of the user's preference. Hybrid recommendation systems [4] combine both collaborative and content-based approaches. However, above traditional methods are not suitable for sequential features capturing. Association rules, e.g., Apriori [2] and FP-Growth [9], can be used to mine sequential features, but they suffer from sensitivity to threshold settings, high time and space complexity, missing of minority rules.

2.2 Deep Learning in Recommendation

Deep learning can efficiently capture unstructured data, and extract more complex abstractions into higher-level data representation [18]. Okura et al. [11] presented a Recurrent Neural Network (RNN) based news recommender system for Yahoo News. Covington et al. [6] used historical query, demographic and

---

[1] www.pandora.com.

other contextual information as features, presented a deep neural network based recommendation algorithm for video recommendation on YouTube. Hidasi et al. [7] presented a Session-based recommendation with an RNN variant, i.e., GRU. Wan et al. [17] also used RNN to build a next basket recommendation. Zhu et al. [19] used a LSTM variant, i.e., Time-LSTM, to model users' sequential actions. Tang et al. [15] propose Convolutional Sequence Embedding Recommendation Model which incorporates the Convolutional Neural Network (CNN) to learn sequential features, and Latent Factor Model (LFM) to learn user specific features to personalized top-N sequential recommendation. Those works show deep neural networks suit recommendation and RNN architectures are good at modeling the sequential order of objects.

3   Proposed Approach

We propose a novel deep neural model to capture the sequence feature of the user's consumed data. Our model mainly consists of five layers: embedding, recurrent structure, fully-connected layer, pooling layer and output layers. Figure 1 shows the structure of our sequence-based recommender.

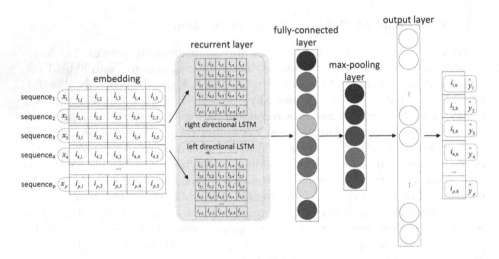

Fig. 1. Framework of sequence-based recommendation

3.1   Notations

Let us assume that $\mathbb{U} = \{u_1, u_2, \ldots, u_N\}$ is the user set and $\mathbb{I} = \{i_1, i_2, \ldots, i_M\}$ is the item set. For each user u, there is an observed consumed items sequence $\mathbb{S}_u = \{s_u^1, s_u^2, \ldots, s_u^{t-1}, s_u^t\}$ in ascending order of time, where s_u^t is the item consumed by user u at time t. The sequential prediction problem is to predict s_u^{t+1} for each user u.

We consider any item can appear only once in the history of a user. Therefore model could recommend items that the user had not yet selected. The output of the network is a softmax layer with a neuron mapping an item in the catalog. We use $p(i_k | S_u, \theta)$ to represent the probability that user u who had a historical consumed items sequence S_u would select item i_k at next time, where θ is the parameters in the network.

3.2 Embedding Layer

We use the latest sequences of user consumed items as the features, and the last item as the label, to build a super multiple classification supervised learning model. So in the period of the feature engineering, we need to convert the features into vectors and map them with labels. One-hot vector representation is the most common method to discrete every item. However, One-hot encoded vectors are high-dimensional and sparse. If we use One-hot encoding to deal with 1000 items, each item will be represented by a vector containing 1000 integers, 999 of which are zeros. In a big dataset, this approach is unacceptable considering computational efficiency. Word Embedding shines in the field of Natural Language Processing, instead of ending up with huge One-hot encoded vectors we also can use an embedding matrix to keep the size of each vector much smaller:

$$e(I_i) = E I_i \tag{1}$$

where $E \in \mathbb{R}^{|e| \times |M|}$, $|e|$ is the size of the embedding layer, $|M|$ is the number of items in the training set. So $e(I_i)$ is the embedding of consumed item I_i, which is a dense vector with $|e|$ real value elements. Compared with the One-hot encoder (size of $|M| \times |M|$), our item sequence embedding matrix (size of $|e| \times |M|$) reduces memory cost rapidly when dealt with big data.

3.3 User Short-Term Interest Learning

We combine a user consumed item I_u^t and other items previous and subsequent I_u^t to present the current interest of user u at time t. The behavior sequences help us to indicate a more precise short-term interest of user. In this recommender, we use a recurrent structure, which is a bidirectional Long Short-Term Memory neural network, to capture the short-term interest of the user.

We define $h_b(I_i)$ as the user's interest before consuming an item I_i and $h_a(I_i)$ as the user's interest after consuming an item I_i. Both $h_b(I_i)$ and $h_a(I_i)$ are dense vectors with $|h|$ real value elements. The interest $h_b(I_i)$ before item I_i is calculated using Eq. (1). $W^{(b)}$ is a matrix that transforms the hidden layer (interest) into the next hidden layer. $W^{(cb)}$ is a matrix that is used to combine the interest of the current item with the next item's previous interest. σ is a non-linear activation function. The interest $h_a(I_i)$ after consuming item I_i is calculated in a similar equation. Any user's initial interest uses the same shared parameters $h_b(I_1)$. The subsequent interest of the last item in a user's history share the parameters $h_a(I_n)$.

$$h_b(I_i) = \sigma(W^{(b)} h_b(I_{i-1}) + W^{(cb)} e(I_{i-1})) \tag{2}$$

$$h_a(I_i) = \sigma(W^{(a)} h_a(I_{i+1}) + W^{(ca)} e(I_{i+1})) \tag{3}$$

where the initial interest $h_b(I_1), h_a(I_n) \in \mathbb{R}^{|h|}$, $W^{(b)}, W^{(a)} \in \mathbb{R}^{|h| \times |h|}$, $W^{(cb)}, W^{(ca)} \in \mathbb{R}^{|e| \times |h|}$.

As shown in Eqs. (2) and (3), the interest vector captures the interest in user's previous and subsequent behavior. We define the temporary status of the interest when a user takes a behavior I_i as the Eq. (4) shown. This manner concatenates the previous temporary status of interest $h_b(I_i)$ before user consuming item I_i, the embedding of behavior I_i consumed item $e(I_i)$, and the subsequent temporary status of interest $h_a(I_i)$ after user consuming item I_i.

$$x_i = [h_b(I_i); e(I_i); h_a(I_i)] \tag{4}$$

So using the consumed behavior sequences $\{i_1, i_2, \ldots, i_{n-1}, i_n\}$, if our model learned the temporary interest status x_{n-1}, users who consumed item i_{n-1} would have a bigger probability to get a recommended item i_n. The recurrent structure can obtain all h_b in a forward scan of the consumed items sequences and h_a in a backward scan of the consumed items sequences. After we obtain the representation x_i of the temporary status of interest when user taking an item I_i, we apply a linear translation together with the $tanh$ activation function to x_i and send the result to the next layer.

$$y_i^{(2)} = tanh(W^{(2)} x_i + b^{(2)}) \tag{5}$$

where $W^{(2)} \in \mathbb{R}^{H \times (|e| + 2|h|)}$, $b^{(2)} \in \mathbb{R}^H$ are parameters to be learned, H is the recurrent layer size, $y_i^{(2)}$ is a latent interest vector, in which each interest factor will be analyzed to determine the most useful factor for representing the users consumed items sequences.

3.4   Popularity Trend Learning

When all of the sequences of user's consumed items calculated, we apply a max-pooling layer.

$$y^{(3)} = \max_{i=1}^{n} y_i^{(2)} \tag{6}$$

Max pooling is done by applying a max filter to non-overlapping subregions of the upper representation. With the pooling layer, the number of parameters or weights within the model reduced rapidly, which could reduce the spatial dimension of the upper input volume drastically and lessen the computation cost. We could capture the attribute throughout the entire sequence and find the most popular sequences combination in the whole users' history using the max-pooling layer. The last part of our model is an output layer as follows:

$$y^{(4)} = W^{(4)} y^{(3)} + b^{(4)} \tag{7}$$

where $W^{(4)} \in \mathbb{R}^{O \times H}$, $b^{(4)} \in \mathbb{R}^O$ are parameters to be learned, O is the fully-connected layer size.

Finally, a softmax activation function applied to $y^{(4)}$, which can convert the output values to the probabilities of next items.

$$p_i = \frac{e^{y_i^{(4)}}}{\sum_{k=1}^{n} e^{y_k^{(4)}}} \tag{8}$$

3.5 Training

We define all of the parameters to be trained as θ.

$$\theta = \left\{ E, b^{(2)}, b^{(4)}, h_b(B_1), h_a(B_n), W^{(2)}, W^{(4)}, \; W^{(b)}, W^{(a)}, W^{(b)}, W^{(cb)}, W^{(ca)} \right\} \tag{9}$$

The training target of the network is to minimize the categorical cross entropy loss:

$$\mathcal{L}(y, S, \theta) = -\sum_{u \in \mathbb{U}} [y_u \log p(y_u | S_u, \theta) + (1 - y_u) \log(1 - p(y_u | S_u, \theta))] \tag{10}$$

3.6 Time Complexity Analysis

In the embedding layer, it has a matrix multiplication operation with time complexity of $O(n)$. In the recurrent structure, for every sequence, the Bi-directional LSTM structure will apply a forward and a backward scan, and based on the citation [13], the time complexity of the BiLSTM we can know is $O(n)$. The time complexity of the pooling layer and the fully-connected layer is also $O(n)$. The overall model is a cascade of those layers, therefore, our sequence-based recommendation model appears a time complexity of $O(n)$, which is linearly correlated with the number of sequences. The overall time complexity of the model is more acceptable than collaborative filtering ($O(n^2)$) [12,14], so that big data can be effectively processed.

4 Experiments

4.1 Evaluation and Metrics

As a recommendation model, we will recommend top N item(s) for each user, denoted as $\hat{I}_u^{t+1}$. We adopt *Precision@N*, *Recall@N* scores to evaluate our model and baseline models. We can define the measures as following equations:

$$Precision@N = \frac{\sum_u |\hat{I}_u^{t+1} \cap I_u^{t+1}|}{|\mathbb{U}| * N} \tag{11}$$

$$Recall@N = \frac{\sum_u |\hat{I}_u^{t+1} \cap I_u^{t+1}|}{|\sum_u |I_u^{t+1}||} \tag{12}$$

In order to get a harmonic average of the precise and recall, the $F1@N$ score was measured here, where a higher F1 score reaches, a more effective result gets, which is shown as Eq. (13):

$$F1@N = \frac{2 \times Precision@N \times Recall@N}{Precision@N + Recall@N} \tag{13}$$

4.2 Dataset and Experimental Settings

Since fewer existing recommendation datasets reserve the continuity of user behavior information, to verify our approach is feasible, we perform experiments on a real-world dataset and make it public: LiveStreaming[2] dataset, which collected users' behavior data from a live streaming website in China. Each line in LiveStreaming records a sequence of browsed items of a user in ascending order of time. The initial collected LiveStreaming dataset contains 1806204 lines, which means that contacts 1806204 unique users. The length of each line ranges from 1 to 1060. Total 541772 different items contained in that dataset. We remove those users who are annotated by less than 15 items then randomly select 10 thousand users as the experimental part denoted as LiveStreaming-10M. Finally, 10000 users and 12292 items contain in LiveStreaming-10M. We randomly split 80% of this part into training set, and keep the remaining 20% as the validation set.

Our model is implemented on Keras with TensorFlow-gpu backend, trained on a single GeForce GTX 1050 with 4 GB memory. For hyper-parameter settings, we set: embedding layer size $|e| = 100$, recurrent layer size $H = 200$, fully-connected layer size $O = 100$, batch size as 32 and initial number of epochs as 100, learning rate α as 0.01, momentum β as 0.9, we adopt classification top 20 accuracy as the evaluation metric. Our code is publicly available[3].

We also explore how many data the BiLSTM4Rec model needs for every user to learn the global sequences tendency and make a good recommendation, so we design an experiment using LiveStreaming-10M with different Time Windows ranges from 1 to 10. When Time Windows equals to 1, means we only use the last item in the sequence as label, the latest one item as feature to fit our model. Figure 3 shows that when Time Windows is less than 6, the results tend to be stable. When Time Windows is too small, much information will lose; too big, it will be hard to train the model. So we choose Time Windows as five in the next experiment.

4.3 Compared Algorithms

In this paper, We compare our model with several existing baselines, and the state-of-the-art approaches in the area of recommendation system:

---

[2] Data is available in https://www.kaggle.com/hailinfu/livestreaming.
[3] Code is available in https://github.com/fuhailin/BiLSTM4Rec.

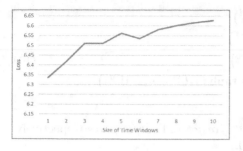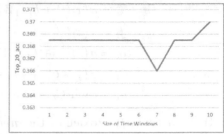

Fig. 2. The training results of BiLSTM4Rec model on 80% LiveStreaming-10M with different Time Windows

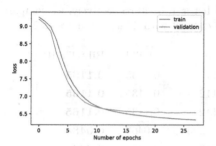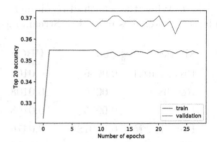

Fig. 3. BiLSTM4Rec training on 80% LiveStreaming-10M with Time Windows as 5, and recommends items with the top 20 probability values

- **POP:** Popularity predictor that always recommends the most popular items of the training set, it feedbacks the global popularity. Despite its simplicity, it is often a strong baseline in certain domains.
- **IBCF:** As the most classical methods of recommendation, Item-based Collaborative Filtering (IBCF) [14] and User-based Collaborative Filtering (UBCF) [12], both yet still strong baselines for top-N recommendation. In this task, we set the rating 1 if user consumed the corresponding item, or 0 if not. Similarity between u_i and u_j was measured using cosine angle:

$$sim(u_i, u_j) = \cos(\boldsymbol{u_i}, \boldsymbol{u_j}) = \frac{\boldsymbol{u_i} \cdot \boldsymbol{u_j}}{\|\boldsymbol{u_i}\|_2 * \|\boldsymbol{u_j}\|_2} \tag{14}$$

- **UBCF:** User-based Collaborative Filtering which evaluates the similarity between items by different users' ratings on the item, recommending items similar to those items consumed already for user. In UBCF, we also use cosine angle to measure similarity between items.
- **LSTM:** We also compared with the basic Long Short Term Memory (LSTM) [8] model. In the embedding stage, LSTM and BiLSTM4Rec share the same size of the embedding layer. The commonly-used update equations of LSTM are as follows:

$$i_t = \sigma\left(x_t U^i + h_{t-1} W^i\right)$$
$$f_t = \sigma\left(x_t U^f + h_{t-1} W^f\right)$$
$$o_t = \sigma\left(x_t U^o + h_{t-1} W^o\right) \tag{15}$$
$$C_t = \sigma\left(f_t * C_{t-1} + i_t * \tanh\left(x_t U^g + h_{t-1} W^g\right)\right)$$
$$h_t = \tanh(C_t) * o_t$$

Here, we use i, f, o to denote the *input*, *forget* and *output* gates respectively. We also set the units of LSTM as 200.

Table 1. When the length of recommendation list is 1, 5, 20 respectively, we compare different approaches using LiveStreaming-10M dataset with Time Windows as 5

Metrics	POP	UBCF	IBCF	LSTM	BiLSTM4Rec
Precision@1	0.0056	0.0108	0.0112	0.1135	**0.1165**
Recall@1	0.0056	0.0108	0.0112	0.1135	**0.1165**
F1@1	0.0056	0.0108	0.0112	0.1135	**0.1165**
Precision@5	0.00018	0.0476	0.006223	0.0463	**0.0488**
Recall@5	0.0009	0.0476	0.0313	0.2315	**0.244**
F1@5	0.0003	0.0476	0.010433333	0.077167	**0.081333**
Precision@20	0.00015	**0.1154**	0.004705	0.01825	0.01895
Recall@20	0.003	0.1154	0.0941	0.365	**0.379**
F1@20	0.000285714	**0.1154**	0.008961905	0.034762	0.036095

Our experiments firstly compare BiLSTM4Rec model with itself in different length of sequence. From the Fig. 2 we can conclude that our sequence-based recommendation also can make recommendations for someone even though with a little information about him. Furthermore, from the results (Table 1), we can see that the performances of our approach are better than other traditional recommendation baselines on this living broadcast dataset, especially when faced with short-term predictions. Our experiments have also demonstrated that LSTM is effective for sequence modeling, and our BiLSTM4Rec model does have a certain improvement over the basic LSTM model in the sequence based recommendation.

5   Results and Conclusion

Overall, BiLSTM4Rec is a novel recommendation by modeling recent consumed items as a "sentence" to predict what users will choose next. We introduced the Bidirectional Long Short-Term Memory neural network to a new application domain: recommendation system. We dealt with consumed items sequences by embedding matrix to save memory cost, and the final model can learn short-term interest of the user. Experimental results show that our approach outperforms existing methods to a great extent, and shows it is suitable to do a short-term prediction. Moreover, it doesn't have an unacceptable time complexity.

In future work, we want to use neural networks to capture other information not only the sequences of data, and to generate a more accurate and longer term prediction.

Acknowledgement. This work was supported by the National Natural Science Foundation of China (No. 61772211, No. 61750110516), and Science and Technology Program of Guangzhou, China (No. 201508010067).

References

1. Adomavicius, G., Tuzhilin, A.: Toward the next generation of recommender systems: a survey of the state-of-the-art and possible extensions. Springer (2013)
2. Agrawal, R., Imieliński, T., Swami, A.: Mining association rules between sets of items in large databases. In: ACM SIGMOD Record, vol. 22, pp. 207–216. ACM (1993)
3. Balabanović, M., Shoham, Y.: Fab: content-based, collaborative recommendation. Commun. ACM **40**(3), 66–72 (1997)
4. Burke, R.: Hybrid recommender systems: survey and experiments. User Model. User-Adapt. Interact. **12**(4), 331–370 (2002)
5. Chien, Y.H., George, E.I.: A Bayesian model for collaborative filtering. In: AIS-TATS (1999)
6. Covington, P., Adams, J., Sargin, E.: Deep neural networks for YouTube recommendations. In: Proceedings of the 10th ACM Conference on Recommender Systems, pp. 191–198. ACM (2016)
7. Hidasi, B., Karatzoglou, A., Baltrunas, L., Tikk, D.: Session-based recommendations with recurrent neural networks. Comput. Sci. (2015)
8. Hochreiter, S., Schmidhuber, J.: Long short-term memory. Neural Comput. **9**(8), 1735–1780 (1997). https://doi.org/10.1162/neco.1997.9.8.1735
9. Li, H., Wang, Y., Zhang, D., Zhang, M., Chang, E.Y.: PFP: parallel FP-growth for query recommendation. In: Proceedings of the 2008 ACM Conference on Recommender Systems, pp. 107–114. ACM (2008)
10. Mnih, A., Salakhutdinov, R.R.: Probabilistic matrix factorization. In: Advances in Neural Information Processing Systems, pp. 1257–1264 (2008)
11. Okura, S., Tagami, Y., Ono, S., Tajima, A.: Embedding-based news recommendation for millions of users. In: Proceedings of the 23rd ACM SIGKDD International Conference on Knowledge Discovery and Data Mining, pp. 1933–1942. ACM (2017)
12. Resnick, P., Iacovou, N., Suchak, M., Bergstrom, P., Riedl, J.: GroupLens: an open architecture for collaborative filtering of netnews. In: Proceedings of the 1994 ACM Conference on Computer Supported Cooperative Work, CSCW 1994, pp. 175–186. ACM, New York (1994). http://doi.acm.org/10.1145/192844.192905
13. Sak, H., Senior, A., Beaufays, F.: Long short-term memory recurrent neural network architectures for large scale acoustic modeling. Computer Science, pp. 338–342 (2014)

14. Sarwar, B., Karypis, G., Konstan, J., Riedl, J.: Item-based collaborative filtering recommendation algorithms. In: Proceedings of the 10th International Conference on World Wide Web, pp. 285–295. ACM (2001)
15. Tang, J., Wang, K.: Personalized top-n sequential recommendation via convolutional sequence embedding. In: Proceedings of the Eleventh ACM International Conference on Web Search and Data Mining, WSDM 2018, pp. 565–573. ACM, New York (2018). http://doi.acm.org/10.1145/3159652.3159656
16. Ungar, L.H., Foster, D.P.: Clustering methods for collaborative filtering. In: AAAI Workshop on Recommendation Systems, vol. 1, pp. 114–129 (1998)
17. Wan, S., Lan, Y., Wang, P., Guo, J., Xu, J., Cheng, X.: Next basket recommendation with neural networks. In: RecSys Posters (2015)
18. Zhang, S., Yao, L., Sun, A.: Deep learning based recommender system: a survey and new perspectives. CoRR abs/1707.07435 (2017)
19. Zhu, Y., et al.: What to do next: modeling user behaviors by time-LSTM. In: Proceedings of the Twenty-Sixth International Joint Conference on Artificial Intelligence, IJCAI 2017, pp. 3602–3608 (2017). https://doi.org/10.24963/ijcai.2017/504

Natural Scene Text Detection Based on Deep Supervised Fully Convolutional Network

Nan Zhang, Xiaoning Jin[(⊠)], and Xiaowei Li

Beijing Advanced Innovation Center for Future Internet Technology,
Beijing University of Technology, Beijing, China
jinxn@bjut.edu.cn

Abstract. In the past few years, text detection in natural scenes has attracted increasing attention due to many real-world applications. Most existing methods only detect horizontal or nearly horizontal texts and have complicated processes. When using the neural network to detect text in the image, some ambiguity and small words are easy to be ignored because of many pooling operations. Therefore, this paper proposes an end-to-end trainable neural network for detecting multi-oriented text lines or words in natural scene images. The network fuses multi-level features and is guided by deep supervision during training. In this way, richer hierarchical representations can be learned automatically. The network makes two kinds of predictions: text/no text classification and location regression, thus we can directly locate multi-oriented words or text lines without other unnecessary intermediate steps. Experimental results on the ICDAR 2015 datasets and MSRA-TD500 datasets have proven that the proposed method outperforms the state-of-the-art methods by a noticeable margin on F-score.

Keywords: Scene image · Multi-oriented text · Deep supervision

1 Introduction

Text detection in natural scenes is an important part of computer vision. As a prerequisite for text recognition, text detection directly impacts the accuracy of text recognition. In general, because of the complex environment and variety of the natural scene, the captured image may be tilted, distorted, blurred, unevenly illuminated, etc., which pose a great challenge to text detection. Most previous studies focused on the detection of horizontal or nearly horizontal texts [1–5]. However, text in the real-world may be in any orientation, which largely limits the practicability and adaptability of these methods.

This paper proposes a novel deep supervised detection framework for multi-oriented scene text. The network fuses multi-level features and is guided by deep supervision during training. Our model's output has two branches. The first branch predicts whether each pixel in the image is a text pixel. The second branch predicts text regions, which have 5 channels represent the distance of each text pixel from corresponding location to the top, right, bottom and left boundary of text geometry and the rotation angle of the text regions in the image. Then we use the routine Non-Maximum Suppression mechanism (NMS) to remove overlapped text lines. Through experiments, the proposed

© Springer Nature Switzerland AG 2018
R. Hong et al. (Eds.): PCM 2018, LNCS 11166, pp. 439–448, 2018.
https://doi.org/10.1007/978-3-030-00764-5_40

method performs better than other relative methods, such as CTPN [1], siglink [6], EAST [7]. The proposed method achieves an F-Score of 80.81% on ICDAR2015 Incidental Scene Text Dataset [8] and 78.36% on MSRA-TD500 datasets [9].

Taken the research contributions altogether, the main points of innovation lie in two aspects below:

(1) Propose a novel multi-oriented text detection framework for in-depth supervision. The multi-level features are utilized to achieve a good detection result by supervising the features extracted from multiple stages of the fully convolutional network

(2) The framework is an end-to-end trainable neural network. We can directly get words or text lines of multi-orientation after text predictions sent to Non-Maximum Suppression.

The rest of the article is structured as follows: We introduce the related work on text detection in Sect. 2. Then, in Sect. 3, the proposed algorithm is explained in detail. The experiments and comparisons are presented in Sect. 4. At last, the conclusion is drawn in Sect. 5.

2   Related Work

Traditional methods for text detection, such as Maximally Stable Extremal Regions (MSER) [6] and Stroke Width Transform (SWT) [10], must come with many complicated hand-crafted features. They first detected characters or components candidates and then grouped them into a word or text line according to the height, angle and other information. These methods have many false positives in blurred or low-resolution characters conditions, while the process of group characters is complicated. Another traditional method is based on slide window [11, 12], which first selected box regions where potentially contained text, then used a classifier to determine whether the regions included the text. Although some improvements in terms of time and efficiency have been made, they still have limited performance in multi-oriented text detection.

Recently, many works tend to use deep convolutional neural networks and have been seen human performance levels reached or surpassed in classification of image in ImageNet [13] and games such as Go [14]. Convolutional neural networks have also advanced general object detection substantially. For example, using Faster R-CNN [15] or YOLO [16] could directly predict the coordinates of each object. Unfortunately, these methods still did not perform well for small text or multi-oriented text prediction. Another approach based on fully convolutional network (FCN) [17] could predict the classification of each pixel and perform well on object detection. However, it is difficult to identify each individual text line by using the method when multiple text lines flock together.

3   Method

The architecture of the proposed method is depicted in Fig. 1. It has two stages that text region prediction based on deep convolutional neural network model and final results production by sending to Non-Maximum Suppression. The neural network model consists of four parts: feature extractor, feature upsampling, feature merging and pre-diction. The feature extractor is based on the Resnet-50 backbone [14]. We choose three stages of Resnet-50 and upsamples their features. Each of upsampling feature predicts text regions. Different stage can extracts different granular features which is important to detect multi-oriented and multi-scale text lines. Thus the final output merges the upsampling features of three stages and produces final text predictions. The model calculate the loss of each prediction from side-outputs and output when training the network. It achieves deep supervision to the network. Last, we eliminate duplicate prediction regions with Non-Maximum Suppression.

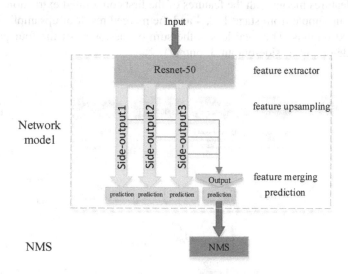

Fig. 1. Architecture of the proposed method.

3.1   Network Architecture

When using convolutional neural network to extract features, features extracted from high-level convolutional layer have more abstract and richer information and the receptive fields are also wider than low-level convolutional layer. But the spatial location information of these features is lost seriously. Correspondingly, the features information obtained from the low-level convolutional layer is more specific and retains richer spatial location information. In natural scene image, there are many kinds of text, the size is not fixed, and there are still cases of rotation, distortion and so on. Although large text regions can be detected more accurately, it is easy to over-look the small text if using convolutional neural network to detect text directly. In order to detect small-sized words, we are inspired by the idea of FCN, texts in natural scenes may be detected by fusing the features obtained from the low-level and high-level layers of convolutional neural networks. Fully convolutional network can output a text/no text map of the same size as the input. Our goal is to detect words or lines of text in the image. When many words or lines of text in an image are close together, the predicted text regions may stick together. It is also the important issue of FCN detecting text in natural scene that we want to solve.

The method is mainly inspired by the Holistically-Nested Edge Detection (HED) [18] and an Efficient and Accurate Scene Text Detector (EAST) [7]. HED is an edge detection algorithm based on fully convolutional network and deeply-supervised Nets that extract features of multiple scales and levels. The EAST algorithm outputs the text score map and geometry of each pixel in text region for each input image. As shown in Fig. 2, we use resnet-50 to extract features from the image. The features extracted from stage 1, stage 2 and stage 4 are upsampled to the same size of the features which extract from the first convolutional operation. The results of the upsampled features merge with the features of the first convoluted extraction. All of the features that upsample from stage 1, 2, 4 and the merged result of upsampling are used to predict text regions. The total loss is the sum of the losses of the four parts (side-output 1, side-output 2, side-output 3, output).

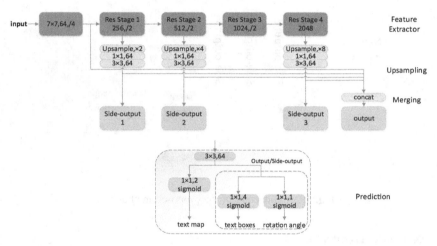

Fig. 2. Architecture of the proposed text detection network. Top: the design of text detection network. Bottom: detailed components of side-outs and output.

The prediction has two parts: the text map and the geometry of the corresponding text region. In order to reduce the computational complexity, we upsample quarter of the size of the input image $S \times S$, then we can obtain a $\frac{S}{4} \times \frac{S}{4}$ text map and five $\frac{S}{4} \times \frac{S}{4}$ graphs. The graphs contains geometry information of the input image and is represented by the distance between output pixel location with the boundary of target bounding box and a rotation angle of the text box. The label generation follows the design of EAST.

We made improvements in prediction of the text map. For each cell of text map, the method predicts the probability of text and the probability of non-text. By comparing the results, we can know whether pixels are text. It is more convenient than setting the threshold manually in EAST [16], and the prediction is relatively more accurate.

3.2   Loss Function

The model calculates the loss of each prediction which contains side-output1, side-output, side-output3 and output.

$$Loss = \sum_{i=1}^{M} \alpha_i loss^{(i)} \tag{1}$$

where $M = 4$, *loss* means the loss of prediction and consists of two parts: loss of text map $loss_{text}$ and loss of geometry $loss_{geo}$.

$$loss = loss_{text} + loss_{geo} \tag{2}$$

Text images in natural scene have a very important feature: the proportion of textual pixels to non-textual pixels is seriously unbalanced, which poses a great challenge to the training of neural networks. A cost-sensitive loss function is proposed in [19] for highly unbalanced segmentation. Practical experiments show that Generalized Dice Loss (GDL) performs well on unbalanced semantic segmentation. Thus we use GDL to calculate the loss of text region probability $loss_{GDL}(Pr_{y=1})$ and the loss of non-text region probability $loss_{GDL}(Pr_{y=0})$

$$loss_{GDL} = 1 - 2\sum_{i=1,j=1}^{\frac{s}{4},\frac{s}{4}} \frac{Y_{ij}^* \hat{Y}_{ij}}{\sum_{i=1,j=1}^{\frac{s}{4},\frac{s}{4}} Y_{ij}^* + \sum_{i=1,j=1}^{\frac{s}{4},\frac{s}{4}} \hat{Y}_{ij}} \tag{3}$$

$$loss_{text} = loss_{GDL}(Pr_{y=1}) + loss_{GDL}(Pr_{y=0}) \tag{4}$$

where $\hat{Y}$ is the prediction of the text score map and Y^* is the ground truth.

In order to calculate the predicted loss of quadrangle of the text, we use the IOU loss function [20]. Its convergence rate is faster and it has more accurate text localization than others.

$$loss_{geo} = loss_{rec} + loss_{angle} \tag{5}$$

$$loss_{rec} = -log\frac{I}{U} = -log\frac{\hat{R} \cap R^*}{\hat{R} \cup R^*} \tag{6}$$

where $\hat{R}$ is the geometry map that can be calculated by the distance from text pixel location to the top, right, bottom and left boundary of geometry, and R^* is the ground truth.

The loss function of rotation angle is as follows:

$$loss_{angle} = 1 - cos\left(\hat{\theta} - \theta^*\right) \tag{7}$$

where $\hat{\theta}$ represents the rotation angle of the predicted text geometry and θ^* represents its corresponding ground truth.

4 Experiments

4.1 Datasets and Data Augmentation

ICDAR2015 Incidental Scene Text (IC15). ICDAR2015 contains 1,500 natural scene images randomly shot with Google Glass, which includes 1000 training sets and 500 test sets. Unlike other datasets, the ICDAR2015 includes a large number of natural scene texts in multi-size, multi-direction, multi-type. Some text in the dataset is ambiguous, and human eye is hard to identify, so it is indicated by "###" in the label. For this part of the labeled data, we do not consider both positive and negative samples and do not calculate this label loss when calculating the predicted loss in training.

The MSRA Text Detection 500 Database (MSRA-TD500) [9]. MSRA-TD500 contains 500 nature scene images taken with a pocket camera, which includes 300 training sets and 200 test sets. Different from ICDAR 2015, it includes both English and Chinese text, and the label is text line rather than word. MSRA-TD500 are taken from indoor (office and mall) and outdoor (street) scenes, the text may be in different fonts, sizes, colors and orientations, which presents a significant challenge in detecting text lines. Because the number of training data is too small, we added 400 pictures from HUST-TR400 dataset [21] during training.

4.2 Implementation Details

The number of real data is limited, so data augmentation is important for training our model. We randomly resize the images to 0.5, 1.0, 2.0, and 3.0 scales and randomly crop from the images. Then resize the cropped image into 512×512 before inputting to the model. In the data set, most of images contain text. If the input image contains text, we consider the image as a positive sample, otherwise as a negative sample. We set the ratio of positive to negative samples to 5:3.

We use resnet-50 checkpoints provided by Tensorflow slim to initialize the network and train end-to-end using ADAM optimizer [22]. The learning rate is 1e-4 and decay every 10000 steps with a base of 0.94.

Through testing at multiple scales using the same network, we found that inputting images with different sizes can detect text regions of different sizes, especially on the MSRA-TD500 dataset. So during testing, we resize the original image to 0.2x, 0.3x, 0.5x, and 1.0x. When the results of text region in 1.0x image is large, we remove it. For 0.2x, 0.3x, 0.5x, only the results of the larger text box is retained.

After Non-Maximum Suppression processing, we can get some text bounding boxes. Comparing the ratio of the area of the text bounding box to the area of the corresponding text-mapped region, we can eliminate some of the unreasonable predictions.

Our methods are implemented with Tensorflow [23] on Python, and are carried out on GTX1080

4.3 Result

Comparing with other output results, side-output1 is the first output which through fewer convolutional layers and fewer pooling layers. From Fig. 3(a), we can see that side-output1 can only detect part of the text regions and have a significant deviation compare with the actual text box. Side-output1 also has the worst result in Table 1. Side-output2 can detect all the text regions in Fig. 3(b), but significant offset still exists. Side-output3 improves the precision of detection, but it still ignores a small portion of text regions. The recall of side-output3 to side-output2 is reduced by 2.94%. Output who fuse other side-output's result achieves the best recall and accuracy. From Table 1, we can know that higher side-output have a higher the precision of prediction, of which side-output3 layer has the highest accuracy. Output layer fuses multiple levels of features and its recall and accuracy are the highest, which has confirmed the scientific and reasonable detection framework we have proposed.

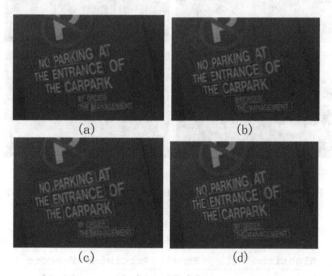

(a)   (b)

(c)   (d)

Fig. 3. Example detection results of each output. (a) side-output1 (b) side-output2 (c) side-output (d) output.

Table 1. Side-result on ICDAR2015. MS means multi-scale testing

Algorithm	Recall	Precision	F-score
MS	**77.23**	**84.73**	**80.81**
Output	77.23	83.94	80.44
Side-output3	72.94	82.52	77.43
Side-output2	75.88	81.91	78.78
Side-output1	42.27	78.60	54.98

Figure 4 shows some detecting examples of our method. It's easy to see that our detecting method can detect multi-oriented, multi-style, multi-size text from the test images. We also compare our method's result with others in testing single scale images on ICDAR2015. From Table 2, our method significantly outperforms state-of-the art methods. At the meantime, the precision and F-score are also better than others. As shown in Table 3, under the scheme of MSRA-TD500, our proposed algorithm achieves 73.71% in recall, 83.63% in precision, and 78.36% in F-score. The performance of our proposed method achieves the state-of-the-art in terms of F-score.

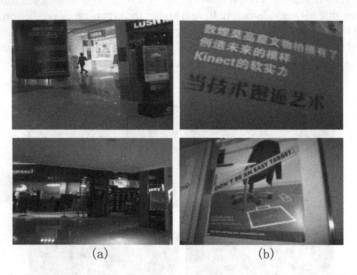

(a)   (b)

Fig. 4. Detection example of our model. (a): ICDAR2015. (b): MSRA-TD500.

Table 2. Result on ICDAR2015 incidental scene text

Algorithm	Recall	Precision	F-score
Proposed	**77.23**	**83.94**	**80.44**
EAST [7]	73.47	83.57	78.20
He et al. [24]	73.00	80.00	77.00
SegLink [6]	76.80	73.10	75.00
Yao et al. [25]	58.69	72.26	64.77
Tian et al. [1]	51.56	74.22	60.85
Zhang et al. [26]	43.09	70.81	53.58

Table 3. Result son MSRA-TD500

Algorithm	Recall	Precision	F-score
Proposed MS	73.71	83.63	**78.36**
SegLink [6]	70.00	86.00	77.00
EAST [7]	67.43	**87.28**	76.08
Yao et al. [25]	**75.31**	76.51	75.91
Yin et al. [27]	63.00	81.00	71.00
Kang et al. [28]	63.00	71.00	66.00

5 Conclusion

This paper propose a new text detection algorithm for detecting multi-oriented texts in natural scenes. The algorithm based on the HED model can predict quadrangles and rotate angle of text regions. After eliminating duplicate prediction regions by using Non-maximum suppression, words or lines predictions can be directly produced from full images without other unnecessary intermediate steps. Fewer processing steps reduce the possibility of mistakes. The results of experiments on standard benchmarks confirm the correctness of our method.

References

1. Tian, Z., Huang, W., He, T., He, P., Qiao, Yu.: Detecting text in natural image with connectionist text proposal network. In: Leibe, B., Matas, J., Sebe, N., Welling, M. (eds.) ECCV 2016. LNCS, vol. 9912, pp. 56–72. Springer, Cham (2016). https://doi.org/10.1007/978-3-319-46484-8_4
2. Gupta, A., Vedaldi, A., Zisserman, A.: Synthetic data for text localisation in natural images. In: 2016 IEEE Conference on Computer Vision and Pattern Recognition (CVPR), pp. 2315–2324 (2016)
3. Zhong, Z., Jin, L., Zhang, S., Feng, Z.: DeepText: a unified framework for text proposal generation and text detection in natural images. In: IEEE International Conference on Acoustics, Speech and Signal Processing, pp. 1–18 (2017)
4. Neumann, L., Matas, J.: A method for text localization and recognition in real-world images. In: Kimmel, R., Klette, R., Sugimoto, A. (eds.) ACCV 2010. LNCS, vol. 6494, pp. 770–783. Springer, Heidelberg (2011). https://doi.org/10.1007/978-3-642-19318-7_60
5. Tian, S., Pan, Y., Huang, C., Lu, S., Yu, K., Tan, C.L.: Text flow: a unified text detection system in natural scene images. In: 2015 IEEE International Conference on Computer Vision (ICCV), pp. 4651–4659 (2015)
6. Shi, B., Bai, X., Belongie, S.: Detecting oriented text in natural images by linking segments. In: IEEE Conference on Computer Vision and Pattern Recognition (2017)
7. Zhou, X., et al.: EAST: an efficient and accurate scene text detector (2017)
8. Karatzas, D., et al.: ICDAR 2015 competition on robust reading. In: Proceedings of the International Conference on Document Analysis and Recognition, ICDAR, pp. 1156–1160 (2015)

9. Yao, C., Bai, X., Liu, W., Ma, Y., Tu, Z.: Detecting texts of arbitrary orientations in natural images. In: Proceedings of the IEEE Computer Society Conference on Computer Vision and Pattern Recognition, pp. 1083–1090 (2012)
10. Epshtein, B., Ofek, E., Wexler, Y.: Detecting text in natural scenes with stroke width transform (2010)
11. Jaderberg, M., Vedaldi, A., Zisserman, A.: Deep features for text spotting. In: Fleet, D., Pajdla, T., Schiele, B., Tuytelaars, T. (eds.) ECCV 2014. LNCS, vol. 8692, pp. 512–528. Springer, Cham (2014). https://doi.org/10.1007/978-3-319-10593-2_34
12. Wang, K., Belongie, S.: Word spotting in the wild. In: Daniilidis, K., Maragos, P., Paragios, N. (eds.) ECCV 2010. LNCS, vol. 6311, pp. 591–604. Springer, Heidelberg (2010). https://doi.org/10.1007/978-3-642-15549-9_43
13. He, K., Zhang, X., Ren, S., Sun, J.: Deep residual learning for image recognition. In: 2016 IEEE Conference on Computer Vision and Pattern Recognition (CVPR), pp. 770–778 (2016)
14. Silver, D., et al.: Mastering the game of Go without human knowledge. Nature 550, 354–359 (2017)
15. Girshick, R.: Fast R-CNN. In: Proceedings of the IEEE International Conference on Computer Vision, pp. 1440–1448 (2015)
16. Impiombato, D., Giarrusso, S., et al.: You only look once: unified, real-time object detection Joseph. Nucl. Instrum. Methods Phys. Res. Sect. A 794, 185–192 (2015)
17. Long, J., Shelhamer, E., Darrell, T.: Fully convolutional networks for semantic segmentation PPT. In: Proceedings of the IEEE Conference on Computer Vision and Pattern Recognition, vol. 8828, pp. 3431–3440 (2015)
18. Xie, S., Tu, Z.: Holistically-nested edge detection. Int. J. Comput. Vis. 125, 3–18 (2017)
19. Sudre, C.H., Li, W., Vercauteren, T., Ourselin, S., Jorge Cardoso, M.: Generalised dice overlap as a deep learning loss function for highly unbalanced segmentations. In: Cardoso, M.J., et al. (eds.) DLMIA/ML-CDS -2017. LNCS, vol. 10553, pp. 240–248. Springer, Cham (2017). https://doi.org/10.1007/978-3-319-67558-9_28
20. Jiang, Y., Cao, Z., et al.: UnitBox: an advanced object detection network. In: ACM on Multimedia Conference, pp. 516–520 (2016)
21. Yao, C., Bai, X., Liu, W.: A unified framework for multioriented text detection and recognition. IEEE Trans. Image Process. Publ. IEEE Signal Process. Soci. 23, 4737–4749 (2014)
22. Kingma, D.P., Ba, J.L.: Adam: a method for stochastic optimization. Int. Conf. Learn. Represent. 2015, 1–15 (2015)
23. Abadi, M., et al: TensorFlow: a system for large-scale machine learning. In: 12th USENIX Symposium on Operating Systems Design and Implementation (OSDI 2016), pp. 265–284 (2016)
24. He, P., Huang, W., He, T., Zhu, Q., Qiao, Y., Li, X.: Single shot text detector with regional attention. In: IEEE International Conference on Computer Vision (2017)
25. Yao, C., Bai, X., Sang, N., Zhou, X., Zhou, S., Cao, Z.: Scene text detection via holistic, multi-channel prediction, pp. 1–10 (2016)
26. Zhang, Z., Zhang, C., Shen, W., Yao, C., Liu, W., Bai, X.: Multi-oriented text detection with fully convolutional networks. CVPR 363, 4159–4167 (2016)
27. Yin, X.-C., Pei, W.-Y., Zhang, J., Hao, H.-W.: Multi-orientation scene text detection with adaptive clustering. IEEE Trans. Pattern Anal. Mach. Intell. 37, 1930–1937 (2015)
28. Kang, L., Li, Y., Doermann, D.: Orientation robust text line detection in natural images. In: 2014 IEEE Conference on Computer Vision and Pattern Recognition, pp. 4034–4041 (2014)

MFM: A Multi-level Fused Sequence Matching Model for Candidates Filtering in Multi-paragraphs Question-Answering

Yang Liu[1], Zhen Huang[1](✉), Minghao Hu[1], Shuyang Du[2], Yuxing Peng[1], Dongsheng Li[1], and Xu Wang[1]

[1] Science and Technology on Parallel and Distributed Laboratory, National University of Defense Technology, Changsha, China
{liuyang12a,huangzhen,huminghao09,pengyuhang,dsli,wangxu16}@nudt.edu.cn
[2] Tongji University, Shanghai, China
dushuyang@126.com

Abstract. Text-based question-answering (QA in short) is a popular application on multimedia environments. In this paper, we mainly focus on the multi-paragraphs QA systems, which can retrieve many candidate paragraphs to feed into the extraction module to locate the answers in the paragraphs. However, according to our observations, there are no real answer in many candidate paragraphs. To filter these paragraphs, we propose a multi-level fused sequence matching (MFM in short) model through deep network methods. Then we construct a distant supervision dataset based on Wikipedia and carry out several experiments on that. Also we use another popular sequence matching dataset to test the performance of our model. Experiments show that our MFM model can outperform recent models not only on the filtering candidates in multi-paragraphs QA task but also on the sequence matching task.

Keywords: Question-answering · Sequence matching · Fused encoding

1 Introduction

Question-answering (QA in short) on text is a longtime hot topic in Natural Language Processing (NLP in short). We can simply categorize it into *single-paragraph* QA and *multi-paragraphs* QA. On the *single-paragraph* datasets such as Stanford Question Answering Dataset (SQuAD), top models can even exceed human performance using read comprehension (*RC* in short) models [4,9,13], in which every answer to the questions must be located in a given paragraph. So recently *multi-paragraphs* QA turns to be a popular research topic, where the datasets can be very large-scale and complex such as Wikipedia[1]. Such QA systems usually attempt to combine the search engine with *RC* models, such as

---

[1] https://www.wikipedia.org.

© Springer Nature Switzerland AG 2018
R. Hong et al. (Eds.): PCM 2018, LNCS 11166, pp. 449–458, 2018.
https://doi.org/10.1007/978-3-030-00764-5_41

DrQA [1], in which question-related documents can be retrieved back as candidates by the search engine and the candidates answers can be extracted by the *RC* models. Since the documents are too long for the *RC* models to extract answers, document usually will be split into *paragraphs* one by one.

However, according to our observations, feeding the retrieval candidate paragraphs directly to the *RC* model will lead to a sharp reduction on precision of answer. To validate that, we have carried out series experiments on the codes of *Drqa-master*[2], for which the result is depicted in the Fig. 1. In doing so, we carefully analyze the cause of precision reduction. Actually the problem is that few parts of paragraphs may cover the answer and a smaller part of paragraphs can make sure *RC* model extracts the answer correctly. In other words, lots of paragraphs returned by *search* are noise for *RC* model, while all of them are considered to contain the correct answer. Consequently, they disturb the selection of correct answer and cause the precision reduction. So we aim to filter these question-unrelated candidate paragraphs for this QA task.

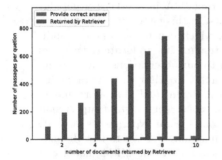 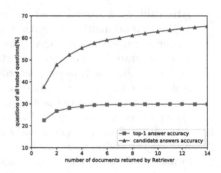

Fig. 1. The number of candidate paragraphs is denoted as N_{cp}, while the ones that can provide correct answer by the *RC* model is denoted as $N_{cp\sim\{c\}}$. We can observe that $N_{cp} \gg N_{cp\sim\{c\}}$. The percentage of candidate paragraphs with correct answers after *retrieving* is denoted as $P_{cp\sim\{r\}}$ and the percentage of candidate paragraphs with correct answers after *extracting* is denoted as $P_{cp\sim\{e\}}$. Also we can observe that $P_{cp\sim\{r\}} \gg P_{cp\sim\{e\}}$

Actually, using the traditional retrieval methods (ex. BM25 variants.) [5,8,11] of search engine, question-related documents can be ranked so as to filter those with lower scores. However, these methods mainly rely on statistical features of words rather than the semantic features of sentences or paragraphs. In contrast, neural network models are suitable for such task by deep learning method to capture the complex semantic features. And the main task is to compute the similarity between question and paragraphs, which is a classical task of text matching in NLP. Unfortunately, existing neural network models [2,6,10,14–16] for text matching tasks (Ex. textual entailment) can't be directly adapted to this candidates filtration task. Because these tasks such as textual entailment

---

[2] https://github.com/facebookresearch/DrQA.

and answer selection usually process short sequences and the two sequences to match are of similar lengths. However, the candidates filtration task is to match long sequences of paragraphs with the short sequences of questions, which is difficult to fuse their semantic information.

To address the problem, we proposes a multi-level fused sequence matching model for candidates filtering, which is referred to as MFM model. Compared with previous text matching model, the structure of MFM model is more complex which is more proper to fully capture semantic feature of paragraphs. This novel model is mainly consist of three parts: (i) *multi-level fused encoding* step for fusing word-embedding features, manual features and contextual features, (ii) *alignment and comparison* step which obtain the comparison features between paragraph and question through an attention mechanism, and (iii) *aggregation and prediction* step which aggregate all the features and make a final classification on the candidate paragraph. To validate our claim, we construct a distant supervised dataset for *multi-paragraphs* QA. Then we carry out several experiments on this dataset and public dataset on SNIL, which is a popular dataset for textual entailment task. Results show that our MFM model can outperform recent methods in precision scores on the two datasets.

The rest of this paper is organized as follows. Section 2 presents the overview and design details of our MFM model. Section 3 illustrates the experimental results. Section 4 discusses the related works. And finally Sect. 5 concludes the paper.

2   Model

In the multi-paragraphs QA system, *search* will return many paragraphs depend on question (recorded as $Q = \{q_1, q_2, q_3, \ldots, q_j \ldots, q_m\}$), while most candidate paragraphs are noise that will be feed into RC model to extract answers. So our MFM model aims to discriminate each paragraph (recorded as $P = \{p_1, p_2, p_3, \ldots, p_i, \ldots, p_n\}$) of them that is noise. Here, p_i and q_j are token(word) of sequence.

2.1   Multi-level Fused Encoding

Initial Encoding. Before precessing sequences with neural network, it is necessary to initially encode each token. Thus, we use the 300-dimensional GloVe [3] word embeddings as the representation of each token on word-level. Word embeddings can reflect the semantic relevance among words. Moreover, we append manual features to the embeddings for the tokens of sequence P, which have been proved to be effective on RC task [1]. Manual features of each token we used include part-of-speech (POS), name entity recognition (NER) tags, term frequency (TF) and a flag, which indicates whether the token appeared in the sequence Q (Fig. 2).

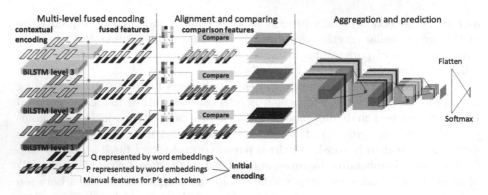

Fig. 2. In our MFM model, features contain fused features and comparison features. All the features will be aggregated by CNN to make final prediction.

Multi-level Contextual Encoding. In order to obtain higher-level semantic information of each word in the context, we employ Bidirectional Long Short-Term Memory [7] (BiLSTM for short) to encoding each token from word-level embeddings. Considering that only one BiLSTM can not fully capture the contextual information of long sequence [4], we use multi-layer BiLSTM to obtain the contextual features for each token of sequence on multiple levels, which is formulated as Eqs. (1) and (2).

$$\boldsymbol{H}_0\left(P\right) = [\begin{bmatrix} \boldsymbol{emb}(p_1) \\ \boldsymbol{f}(p_1) \end{bmatrix}, \begin{bmatrix} \boldsymbol{emb}(p_2) \\ \boldsymbol{f}(p_2) \end{bmatrix}, ..., \begin{bmatrix} \boldsymbol{emb}(p_n) \\ \boldsymbol{f}(p_n) \end{bmatrix}]$$

$$\boldsymbol{H}_0\left(Q\right) = [\boldsymbol{emb}(q_1), \boldsymbol{emb}(q_2), ..., \boldsymbol{emb}(q_m)] \tag{1}$$

$$\boldsymbol{H}_{k+1}\left(P\right) = BiLSTM_P\left(\boldsymbol{H}_k\left(P\right)\right)$$

$$\boldsymbol{H}_{k+1}\left(Q\right) = BiLSTM_Q\left(\boldsymbol{H}_k\left(Q\right)\right) \tag{2}$$

Here $\boldsymbol{emb}(.)$ is corresponding word embeddings while $\boldsymbol{f}(.)$ is the manual features. Thus $\boldsymbol{H}_0(.)$ is the initial encoding. $\boldsymbol{H}_k(.)$ is the kth level contextual encoding for each token and $\boldsymbol{H}_{k+1}(.)$ depends on $\boldsymbol{H}_k(.)$ shown as Eq. (2). Compared with $\boldsymbol{H}_k(.)$, $\boldsymbol{H}_{k+1}(.)$ gathers more contextual information.

Fused Encoding. To obtain abundant features, we concatenate the initial word embeddings with each level contextual encoding as Eq.(3),

$$\tilde{\boldsymbol{H}}_k\left(P\right) = \begin{bmatrix} \boldsymbol{H}_k\left(P\right) \\ \boldsymbol{emb}\left(P\right) \end{bmatrix}, \quad \tilde{\boldsymbol{H}}_k\left(Q\right) = \begin{bmatrix} \boldsymbol{H}_k\left(Q\right) \\ \boldsymbol{emb}\left(Q\right) \end{bmatrix}, \quad k = 1, 2, ... \tag{3}$$

2.2   Attention Model Based Alignment and Comparison

Alignment. To compare the sequences of question and paragraph, we use the attention mechanism to align them [14,16]. Taking into account the difference in length between P and Q, we employ a unidirectional attention to align question

(Q) to paragraph (P). Before alignment, we measure similarity between token $p_i \in P$ and $q_j \in Q$ as Eqs. (4, 5) shown.

$$P_k = ReLU(W\tilde{H}_k(P) + b)$$
$$Q_k = ReLU(W\tilde{H}_k(Q) + b) \tag{4}$$

$$SimMatrix_k := [s_{ij}^{(k)}] = P_k^T Q_k \tag{5}$$

W and b are the learnable parameters, while $SimMatrix$ is a $m \times n$ sized matrix whose element s_{ij} means the similarity between p_i and q_j.

$$A_k := [\alpha_{ij}^{(k)}], \quad \alpha_{ij}^{(k)} = \frac{exp(s_{ij}^{(k)})}{\sum_j exp(s_{ij}^{(k)})} \tag{6}$$

$$P_k' = A_k Q_k^T \tag{7}$$

P_k' is the kth level attentional representation of sequence Q, which is an alignment representation of sequence P.

Comparison. After alignment, we compare the features of the two sequences on corresponding level as follows:

$$sub_k = (P_k - P_k') \odot (P_k - P_k') \tag{8}$$

$$mul_k = P_k \odot P_k' \tag{9}$$

For each level k, we use sub_k and mul_k as its comparison features.

2.3   Aggregation and Prediction

In the previous section, we obtained abundant information at different levels. In this section we aggregate the captured features and make a final prediction.

Multi-level Input. The features of kth level can be treated as a 4-channel 2D feature map, whose channels contain P_k, P_k', sub_k and mul_k. We stack up all the layers of feature maps and make up a 4K-channel 2D feature map which contains features (fusion features and comparison features) of all levels for the two sequences, which is formulated as Eqs. (10, 11). Here, K is the number of BiLSTM. $Concat(.,.)$ means stacked together on channel.

$$FM_k = Concat(P_k, P_k', sub_k, mul_k) \tag{10}$$

$$FeatureMap = Concat(FM_1, FM_2, ..., FM_K) \tag{11}$$

Convolution and Prediction. We aggregate all features by multi-layer 2D-CNN, which can effectively reduce the size of the feature map and capture the fine-grained features.

$$v = \max_{row} CNN(FeatureMap) \tag{12}$$

$CNN(.)$ is a multi-layer CNN. After CNN processing, the $FeatureMap$ is converted into a small sized map with x channels, and x is the number of convolutional kernels in CNN's last layer. In each channel, the map is compressed into a vector by max pooling along rows which corresponds to the length of sequence. So, v is a x channels 1 dimensional image used to make prediction.

$$v' = Flatten(v) \tag{13}$$

$$p = Softmax(W_p v' + b_p) \tag{14}$$

Through a full connection layer, we employ Softmax to predict the relation between two sequences. $Flatten(.)$ means concatenate each row of the matrix into vector and (W_p, b_p) is parameters to be trained.

3 Experiments and Evaluations

In this section, we evaluate our model on two different datasets: $SNLI$ and a paragraphs filter dataset (PF) constructed from $DrQA$ by distant supervision. $SNLI$ is a traditional dataset on the task of textual entailment while the other dataset is for paragraphs filter of multi-paragraphs QA. Our work is mainly focused on the latter dataset, for which we add our work as a $Filter$ to the $DrQA$.

3.1 Distant Supervised Dataset

In the $DrQA$, $Retriever$ will search many paragraphs for each question, and $Reader$ (the RC model) will extract the answer for each paragraph. For a given question, only few paragraphs can get the right answer, we give these paragraphs a $right$ label, and the other paragraphs a $wrong$ label shown as Table 1. We use questions in dev-set and train-set of $SQuAD$ to construct dev-set and train-set of PF respectively.

Considering that $wrong$ candidates are more than the $right$ one, we replicate the $right$ samples and the number of them will be equal to the $wrong$ samples.

3.2 Baseline

Here, we will introduce two baselines for our work: **DrQA-Reader** [1] and **CA Model** [2]. The former is a RC model employed by DrQA and we make a slight change based on its network structure. The latter is a typical model for text matching task such as textual entailment. For $SNLI$, there are many other work, we just simply take the reported performance for the purpose of comparison.

DrQA-Reader. This model contains multi-level BiLSTM encoding and manual features. We only change the final prediction part, which merge the token features to obtain a single sequence feature by a self-attention mechanism for classification.

$$\beta = Softmax(w_{merge}^T P + b_{merge}) \tag{15}$$

Table 1. An example of *PF* dataset.

Question	Paragraphs	Answer	Label
Which NFL team repre- sented the AFC at Super Bowl 50? (The true answer is "**Denver Bron- cos**")	In Super Bowl 50, the **Denver Broncos**, led by the league's top-ranked defense, defeated the Carolina Panthers, who had the league's top-ranked offense, in what became the final game of quarterback Peyton Manning's career	Denver Broncos	Right
	The Baltimore Ravens snapped the NFC's three-game winning streak by winning Super Bowl XLVII in a 34-31 nail-biter over the **San Francisco** 49ers	San Francisco	Wrong
	The 1979 season was the last season of the dynasty. Bradshaw threw for over 3,700 yards and 26 touchdowns and John Stallworth had 1183 yards receiving. The Steelers finished 12–4, once again tops in the AFC Central. In the playoffs they defeated the Dolphins 34–14 and the Oilers 27–13, to meet the **Los Angeles** Rams in their fourth Super Bowl	Los Angeles	Wrong

$$v' = \sum_i \beta_i p_i, \quad i \in 1, 2, ..., m. \tag{16}$$

Here, $w_{merge} \in \Re^h$, $b_{merge} \in \Re^m$ are learnable parameters and $P \in \Re^{h \times m}$ is the feature representation of paragraph (P in our work) while $p_i \in \Re^h$ is the feature vector for i^{th} token of paragraph, where h is the size of feature vector and m is the length of paragraph. Finally, v' will be used to make prediction as Eq. (14).

CA-Model. Wang et al. in [2] have proposed a generic Compare-Aggregate framework for text matching task such as textual entailment and answer selection. They apply a single BiLSTM to encode P and Q and then use unidirectional attention to compare them. Finally, only comparison features are aggregated by 1D CNN for prediction. We use the CA-model framework as benchmark on datasets *SNLI* and *PF*.

3.3   Result and Anlysis

We use accuracy as the evaluation metric for the text matching task, since there is only one correct label for each instance (shown as Table 2). But for the evaluation of multi-paragraphs QA pipeline, it will return many answers for each question. So we use recall for all questions which check whether the correct answer appears in returned top-n answers (shown as Table 3). To demonstrate each component's property of our model, we performed ablation experiment (shown as Table 4).

We observe following from the results. Firstly, our model achieves the best performance on dataset *PF* and very competitive on the popular dataset *SNLI*.

Table 2. Performance for textual entailment task on *SNLI* and paragraphs filter task on distant supervised dataset *PF*.

	SNLI			PF	
	Train	Dev	Test	Train	Dev
LSTMN [19]	0.885	-	0.863	-	-
Match-LSTM [20]	0.920	-	0.861	-	-
Decomp Attention [25]	0.905	-	0.868	-	-
EBIM+TreeLSTM [24]	0.930	-	**0.877**	-	-
DrQA-Reader [1]	-	-	-	0.791	0.729
CA-model [2]	0.894	-	0.868	0.865	0.771
Our work	0.935	0.862	0.865	0.933	**0.832**

Table 3. Performance for the pipeline of Open domain QA on *SQuAD-dev*'s questions.

	DrQA-pipeline (Retriever+ Reader) [1]	DrQA-new-pipeline (Retriever+ **Filter**+Reader)
top1	0.299	0.325
top2	0.360	0.383
top3	0.393	0.411
top4	0.417	0.430
top5	0.431	0.438

Table 4. Ablation studies on the distant supervised dataset *PF*'s dev set

Model	PF-dev
Without manual features	0.754
Without fusion	0.740
Only use top level contextual encoding	0.763
Use 1D-CNN aggregation	0.827
Full model	0.832

Secondly, our model as a *Filter* can improve the recall of pipeline of multi-paragraphs QA system, which indicates that the property of the pipeline of multi-paragraphs QA can be improved by removing amount of noise candidates. So our model is a generally for many text-matching cases and our work is effective for the multi-paragraphs QA systems.

4   Related Works

We review related work in two types of tasks for semantic information analysis of neural network.

4.1   Sequence Matching

Sequence matching task is very similar to our work on objective. The difference is the data situation. Mainstream sequence matching task such as textual entailment and answer selection usually faced with short sequence, which can be analyzed with involuted comparison features. [15] used fine-grained pair-wise comparison, and use CNN to capture the features of similarity matrix of two sequences. [6] tried to use soft-attention to compare two sequences and use 1D-CNN to capture features of sequence. Comparison and CNN works alternately. Soft-attention can mix different sequences' information well, so it is widely used for TE [21] and AS [14,22]. Based on previous works, [2] summed up a generic Compare-Aggregate framework for AS, TE and so on. And [10] used bidirectional attention to compare two sequences under Compare-Aggregate framework.

4.2 Machine Reading Comprehension

Reading comprehension (*RC* in short) task is similar to our work on data situation but the objective of task is different. On the task of *RC*, two sequences (question and paragraph) have different statuses because the prediction of answer is mainly implement on paragraph. Thus, [12] tackle cloze-type *RC* task (dataset CNN and DailyMail [23]) by encoding question into one feature vector and select answer in paragraph based on it. [13] tried to use bidirectional attention flow, however, in order to fuse all features into paragraph's representation, it merge attentional features of question into one feature vector which will be affixed to paragraph's each token vector. On account of information loss by merging question's attentional features, [9] used two rounds of attentional alignment so as to avoid summarizing question.

5   Conclusion

In this work, we propose a multi-level fused sequence matching model (MFM model) to filter the noise candidates. Our model encode two sequences of question and paragraph on multi-levels and make fusion with initial encodings. Thus, it can fully capture contextual features for long sequences. Experimental results show that our model can get an outstanding performance compared with baseline on *PF* and be competitive on benchmark of *SNLI*.

References

1. Chen, D., Fisch, A., Weston, J., et al.: Reading wikipedia to answer open-domain questions, 1870–1879 (2017)
2. Wang, S., Jiang, J.: A compare-aggregate model for matching text sequences. In: Conference on ICLR 2017 (2017)
3. Pennington, J., Socher, R., Manning, C.: Glove: global vectors for word representation. In: Conference on Empirical Methods in Natural Language Processing, pp. 1532–1543 (2014)
4. Huang, H.Y., Zhu, C., Shen, Y., et al.: FusionNet: fusing via fully-aware attention with application to machine comprehension (2017)
5. Robertson, S., Zaragoza, H.: The probabilistic relevance framework: BM25 and beyond. Found. Trends® Inf. Retr. **3**(4), 333–389 (2009)

6. Yin, W., Schütze, H., Xiang, B., et al.: ABCNN: attention-based convolutional neural network for modeling sentence Pairs. Comput. Sci. (2015)
7. Hochreiter, S., Schmidhuber, J.: Long short-term memory. In: Supervised Sequence Labelling with Recurrent Neural Networks, pp. 1735–1780. Springer, Heidelberg (1997)
8. Zaragoza, H., Craswell, N., Taylor, M.J., et al.: Microsoft Cambridge at TREC 13: web and hard tracks. In: TREC 2004 (2004)
9. Xiong, C., Zhong, V., Socher, R.: Dynamic coattention networks for question answering (2016)
10. Wang, Z., Hamza, W., Florian, R.: Bilateral multi-perspective matching for natural language sentences (2017)
11. Ponte, J.M., Croft, W.B.: A language modeling approach to information retrieval. In: Research and Development in Information Retrieval, pp. 275–281 (1998)
12. Kadlec, R., Schmid, M., Bajgar, O., et al.: Text understanding with the attention sum reader network, 908–918 (2016)
13. Seo, M., Kembhavi, A., Farhadi, A., et al.: Bidirectional attention flow for machine comprehension (2016)
14. Tan, M., Xiang, B., Zhou, B.: LSTM-based deep learning models for non-factoid answer selection. Comput. Sci. (2015)
15. He, H., Lin, J.: Pairwise word interaction modeling with deep neural networks for semantic similarity measurement. In: Conference of the North American Chapter of the Association for Computational Linguistics: Human Language Technologies, pp. 937–948 (2016)
16. Yu, L., Hermann, K.M., Blunsom, P., et al.: Deep learning for answer sentence selection. Comput. Sci. (2014)
17. Bowman, S.R., Angeli, G., Potts, C., et al.: A large annotated corpus for learning natural language inference. Comput. Sci. (2015)
18. Feng, M., Xiang, B., Glass, M.R., et al.: Applying deep learning to answer selection: a study and an open task, 813–820 (2015)
19. Cheng, J., Dong, L., Lapata, M.: Long short-term memory-networks for machine reading (2016)
20. Wang, S., Jiang, J.: Learning natural language inference with LSTM (2015)
21. Rocktaschel, T., Grefenstette, E., Hermann, K.M., et al.: Reasoning about entailment with neural attention (2015)
22. Tan, M., Santos, C.D., Xiang, B., et al.: Improved representation learning for question answer matching. In: Meeting of the Association for Computational Linguistics, pp. 464–473 (2016)
23. Hermann, K.M., Kociský, T., Grefenstette, E., et al.: Teaching machines to read and comprehend, 1693–1701 (2015)
24. Chen, Q., Zhu, X., Ling, Z., et al.: Enhancing and combining sequential and tree LSTM for natural language inference, 1657–1668 (2016)
25. Parikh, A.P., Täckström, O., Das, D., et al.: A decomposable attention model for natural language inference, 2249–2255 (2016)

Multiview CNN Model for Sensor Fusion Based Vehicle Detection

Zhenchao Ouyang[1,3], Chunyuan Wang[1], Yu Liu[1(✉)], and Jianwei Niu[2,3]

[1] State Key Laboratory of Software Development Environment,
School of Computer Science and Engineering, Beihang University,
Beijing 100191, China
buaa_liuyu@buaa.edu.cn

[2] Beijing Advanced Innovation Center for Big Data and Brain Computing (BDBC),
School of Computer Science and Engineering, Beihang University,
Beijing 100191, China

[3] State Key Laboratory of Virtual Reality Technology and Systems,
School of Computer Science and Engineering,
Beihang University, Beijing 100191, China

Abstract. The self-diving vehicle is a critical revolution in the automobile industry, transportation and people's daily life. Generally, self-driving vehicles utilize the LIDAR, IMU, radar along with traditional optical cameras for sensing surrounding environment. Therefore, on road object detection can be improved by utilizing multiview of LIDAR and camera instead of single sensor. In this paper, we propose a sensor data fusion-based multiview CNN model for on road vehicle detection. We first up-sample the sparse LIDAR point cloud as gray images, and then combine the up-sampled gray image (F channel) with the camera image (RGB channels) as four-channel fusion data to train a CNN model. To reduce the redundant data, we tested the performance of CNN models by different fusion schemes. In this way, we can reduce the consumption of the CNN model in both training and on board testing for vehicular environment. Our final fusion model (RBF) can reach an average accuracy of 82% on the real trace dataset of KITTI.

Keywords: Data fusion · LIDAR · Camera · Deep learning
Object detection · Self-driving

1 Introduction

With the advent of self-driving or autonomous vehicle, we may have the potential to reduce human factor-based traffic accidents, and improve the efficiency of traveling, transportation and other related areas [1]. Business magnates, technique companies, scientific research institutions and the government are working together to advance their respective means to improve the development of self-driving vehicle. Moreover, a large amount of newly established companies on self-driving have appeared in the past three years. The self-driving vehicle relies mainly

© Springer Nature Switzerland AG 2018
R. Hong et al. (Eds.): PCM 2018, LNCS 11166, pp. 459–470, 2018.
https://doi.org/10.1007/978-3-030-00764-5_42

on the perception data of surrounding environment gained from different sensors, and makes control decisions according to these data and current vehicle dynamics [2]. This means that the more data we have, the better the vehicle can understand the environment and drive better. Without sufficient data, it is often too hard to guarantee the safety and reliability of a self-driving vehicle. We also have to handle so many factors and faces so many challenges of software, hardware, automobile industry, sensors, controllers and even the 'outdated' traffic rules [3].

The whole mechanism of self-driving can be treated as a wheeled robot with the ability of perception, cognition, decision and execution. The first and foremost task is to perceive the surrounding environment (such as lane lines, traffic signals, other vehicles and pedestrians, obstacles, etc.) with different kinds of sensors, and select the critical information for cognition. Due to the dynamic motion pattern and high speed of a vehicle, the real time on road detection and tracking of objects constitutes an important part of the control loop in self-driving.

Optical camera-based (including monocular, binocular, and array of camera) object detection has long been studied in related areas, such as Advanced Driving Assistance System (ADAS) [4] and robot vision [5,29,30]. However, the camera may be seriously affected by environment lighting and weather conditions. In the night or heavy rain, the poor light condition may lead to a lower detection accuracy or failure of algorithms [6]. For new sensor technologies used in self-driving, several different sensing methods, such as the laser, MilliMeter Waves (MMW) and ultrasonic wave, are available for self-driving in no light environment, due to active sensors. Sensor data fusion-based perception can be a feasible way to improve traditional camera-based detection by utilizing the features observed from different sensor views, and several different multiview based sensor fusion methods are proposed and tested in self-driving cars [1,3,6].

In this paper, we propose a novel multiview based Convolutional Neural Network (CNN) model by fusing LIDAR point cloud and traditional camera image for on road vehicle detection. In our system, the up-sampled LIDAR point cloud is used as an additional channel along with the RGB image are evaluated. We also optimize the fusion scheme by testing different combinations of the four channels of fused data. The simulations are tested based on a modified CNN model from Yolo [24] to solve the problem of on road dynamic object detection for self-driving vehicles.

The contribution of this paper can be summarized as follows:

– We propose sensor data fusion scheme based on LIDAR point cloud and camera image to meet the requirement of efficient data processing for self-driving.
– We deploy our fusion scheme on a real trace dataset of KITTI [8], and train the fused data with CNN models to test its performance of vehicle detection.

This paper is organized as follows. Section 2 describes related research on the multiview based sensor fusion for self-driving and objection detection. Details of our multiview based sensor fusion detection model is proposed in Sect. 3. Section 4 briefs the structure of the deep learning CNN model we used and how we deal with the real trace data set for sensor fusion. Section 5 shows the experimental results of our detection system by different fusion strategies. We make a short conclusion in the last section.

2   Related Work

In this section, we review some of the existing work related to multiview based applications in terms of object detection methods, deep learning models and data sets.

2.1   Object Detection Methods

Single Sensor-Based Methods. MPF [2] is a multiview-based crowd motion detection framework for capturing the pedestrian group in crowded spaces. This method can characterize the structural property of different scales of pedestrians using different camera images, and automatically determine the group number. Song et al. [7] present a monocular-based 3D localization for fast and highly accurate 3D object detection. They try to learn the 3D features of objects from structure from motion (SFM), and reach a high localization accuracy in both near- and far-distance views on the dataset of KITTI. The Multi-spectral Aggregated Channel Features (MACF) is introduced in [9] for on road pedestrian detection. In testing of different pedestrian detection tasks, the MACF achieves 15% lower average miss rate than the single ACF-based methods.

Multi-sensor Fusion-Based Methods. Xue et al. [10] propose a multi-sensor fusion approach that considers three kinds of sensor data: the camera image, LIDAR point cloud, and the GIS. With the fusion data, their robotic car can make self-localization and detect surrounding obstacles. However, they fuse the camera and IMU for visual location, and use the approach in [11] to fuse the image and LIDAR for pedestrian detection. In [11], the authors try to extract image features from different resolutions of an image to form a multiview feature pyramid for object detection. With a multi-resolution feature pyramid, they can get richer image HOG features. Therefore, their algorithm can reach higher precision for object detection. A co-interested person detection method is proposed in [12], which can capture the co-interest person from several camera channels. This kind of multiview camera image fusion can achieve stable target tracking with a Conditional Random Field (CRF) model for motion object. In [13], a multiview learning based colored object recognition method is presented. With the RGB-Depth data collected with the calibrated binocular camera, they can build the domain-adaptive dictionary pair for learning the features of objects. A novel discriminative cross-domain dictionary learning-based object recognition framework is also proposed in their work. A CNN-based lane classification model is presented in [14]. Considering the drawbacks of 1-D or 2-D filters that extract features of lane markings from either camera images or LIDAR, they fuse the color images and LIDAR data together, and train a convolutional neural networks (CNNs) to handle this kind of multiview data for classification of lane markings and non-markings. This approach also helps to automatically learn data features and is more robust for pixel-level classification. For the reason of efficient feature extraction, Zhang et al. [15] also try to fuse the camera and LIDAR data for vehicle detection in an ADAS. They first use the stereo camera for detection of the vehicle object, and then verify the data according to the LIDAR point cloud. Different from directly fusing raw Lidar data [26], DepthCN [28] also proposed a local point cloud upsampling algorithm based on Delaunay Triangulation [27] to generate coarse 66×112 mesh channel for fusion.

However, all the fusion-based detections simply add different sensor data together and miss to evaluate the contribution of different data sources or channel. Additionally, a large amount of redundant data may exist in different sensors, and may lead to additional computation during data processing. We thus prefer to reduce this kind of inefficiency by removing some of the sensor data before fusion.

2.2 Deep Learning Models and Datasets

Due to the powerful learning ability, a series of deep learning models (such as Bimodal Deep Boltzmann Machines [18], CNNs [19], and GAN [17]) are applied in the area of self-driving. Extended from traditional CNN model, the R-CNN [20] tries to apply high-capacity CNNs to bottom-up region proposals so as to localize and segment objects. With this improvement, the network structure can deal with large multi-object detection tasks at hundred-level. In Fast R-CNN [21] and Faster R-CNN [22], the authors propose a region proposal network to share image features. With this process, the computational bottleneck can be easily overcome in the detection network. Their detection system consumes less running time, and can also predict object bounds and 'objectness scores' at each position simultaneously. To overcome the disadvantage of preprocess resizing in the deep CNN, the SPP-net [23] uses spatial pyramid pooling to eliminate the problem of resizing. This network is also robust to object deformations and can improve all CNN-based visual recognition. The YOLO [24] treats object detection as a regression problem of spatially separating bounding boxes and associated class probabilities. In this way, the YOLO can handle multi-object detection in one evaluation and can achieve a much higher processing rate than most of the previous work. The YOLO9000 [25] is an extension of the YOLO with a series of improvements (such as multi-scale training, batch normalization and convolutional with anchor boxes). Due to the fast processing rate and high precision, we try to use the YOLO to handle on road real time vehicle detection for self-driving on the benchmark of KITTI.

3 Multiview Based Sensor Fusion Model

We present a multiview based sensor fusion CNN model that can fuse the LIDAR point cloud and camera image for on road vehicle detection. With the fused data from different sensors, the CNN model can extract more features of other vehicle from views of the camera and LIDAR, leading to better detection accuracy than single sensor based models.

3.1 Up-Sampling and Sensor Data Fusion

Both [16,17] utilize the Bilateral filter to up-sample raw 3D LIDAR point and get the dense gray image. We also up-sample LIDAR data into a gray channel image (right bottom of Fig. 1) with the Bilateral filter. The gray image as another channel is combined with the three channels of RGB image (as shown in Fig. 1) to obtain the four-channel fusion data.

Fig. 1. Separated RGB channels and up-sampled LIDAR gray channel. The bottom left one is an up-sampled gray image from raw LIDAR point cloud using the Bilateral filter. (Color figure online)

The most common way to fuse the camera and LIDAR is simply by extracting the features from the two sources and putting them together. In this process, two different data patterns have to be dealt with: (1) 2D colored and dense camera image, and (2) 3D sparse LIDAR point cloud with intensity and distance information. With a good calibration, we can up-sample the LIDAR point cloud and match it with image as another data channel. Therefore, we can get the four-channel fusion data (as shown in Fig. 2) at each time slot, and feed the data into the CNN for model training.

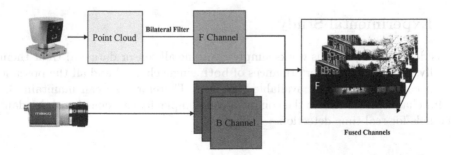

Fig. 2. Fusion process of LIDAR point cloud and RGB image.

4   Yolo Net and KITTI Dataset

In this paper, we use the YOLO v2 as the basic network structure for real time on road vehicle detection. The network structure is shown in Fig. 3. The whole net includes 23 **convolution** layers, 5 **pool** layers, 2 **route** layers, and 1 **reorg** layer. The additional **route** layer and the **reorg** layer can combine different features of front levels and constitute a multi-level feature map.

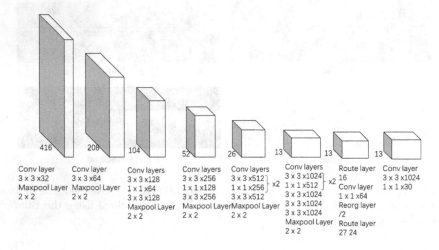

Fig. 3. Model structure of YOLO v2, including 23 convolution layers, 5 pool layers, 2 route layers, and 1 reorg layer

We test our system by using the image and LIDAR data from the KITTI object dataset collected in real road conditions. In addition, considering that our detector only detects on road vehicles, we merge the four types (cars, trucks, vans, and trams) of vehicles into one class, and label them as 'car'. The other objects are ignored.

5   Experimental Study

Considering that previous works simply combine all sensor data and treat them equally, we analyze the performances of both single channel and all the possible fusion combinations of all available channels. Therefore, we can maintain the useful channels and reduce the computation complexity by removing redundant channels for real time detection.

5.1   Fusion Strategy

As mentioned above, we have 4-channel fusion data with up-sampling and data fusion. According to the number of fused channels, we carry out the following experiments. Details are shown in Fig. 4. Four models with different combinations of fused data channels are used for training the YOLO v2 models.

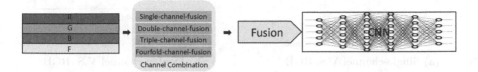

Fig. 4. All possible fusion combinations: single, using only one of the channels for detection; double, using two of the four channels for detection; triple, using three of the four channels for detection; and fourfold, using all channels for detection.

The original KITTI object dataset is divided into a training set and a verification set by a ratio of 1:1. The positive sample requirement threshold is set as $IOU = 0.5$ (the ground-truth-over-union). Values lower than this threshold will be treated as negative. The intersection of the detector result (R) and groundtruth (G) divided by the union of the detector result (R) and groundtruth (G) is the value of IOU, as Eq. 1.

$$IOU = (R \cap G)/(R \cup G) \tag{1}$$

After the training, we test each model to evaluate the detection performance of the verification set mentioned above. The model for each generated detection box will give a probability of whether each box contains the car. We only keep the detection box when the detection probability is greater than the predefined threshold (0.5). Otherwise, the result will be discarded.

5.2 Evaluation of Single Channel Evaluation

We first evaluate the single channel to see how each channel contributes to the final vehicle detection.

Figure 5(a) shows the Precision-Recall curve of single channel-based model and a control group of full RGB image. It can be seen that the detection contribution of each characteristic channel is different when the model is trained to use only one channel. The up-sampled LIDAR point cloud (F channel) achieves the best effect, even better than the image (RGB channel). Another strange result is that except for the R channel, single channels of G and B outperform the full RGB image.

5.3 Evaluation of Channel Fusion

We then train and test the YOLO model with data of double-, triple- and four-channel fusion. Results are as shown in Fig. 5.

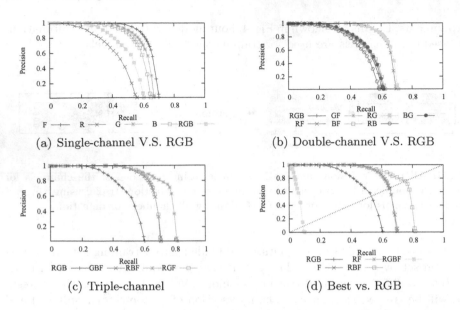

Fig. 5. PR curves of models for different channel combinations.

As we have four different single channels, we have six combinations of double-channel fusion: RF, GF, BF, RG, RB, and BG. Figure 5(b) shows the Precision-Recall curve of the model using two channels. It can be seen that the detection results of RG-channel and RF-channel are similar to each other performance, and are better than the other combinations. Moreover, the RG-channel is slightly better than RF-channel at low precision values.

For triple-channel fusion, except the original RGB, we have three other fusion combinations. Figure 5(c) shows the Precision-Recall curves of triple-channel fusion. It is easy to see that when one of the RGB-channel is replaced, the model achieves better detection accuracy. The combination of RBF performs the best than the others.

Table 1. Detection performance.

Model	F	RF	RBF	RGBF
Time(s)	43.9241	49.7094	52.2190	56.4959
fps	11.35	10.1	8.59	8.1

In Fig. 5(d), we compare the best schemes of different fusion channels (single, double and triple) in the previous experiments with four-channel fusion. It can be seen that the PR curve of RBF covers all the other curves, and the point of intersection between the PR curves and the $x = y$ (the balance point when $Precision = Recall$) is also larger than the rest of the others. This means that

Fig. 6. Several examples of the detection results.

when all the channels are used, there may be data redundancy and data collision. Therefore, the RGBF-channel does not perform as well as expected. Besides, the RBF-channel shows the best detection accuracy.

Table 1 shows the performance of the detectors trained with different fusion combinations. We evaluate each model with subset by randomly selecting 1000 frames from the KITTI test set, and evaluate the total process time and frame per second (fps). It can be seen that the detectors with less channels consume less time. The triple channel model in previous tests with the best performance can achieve 8.59 fps on Nvidia Tx1, which is close to the requirement of 10 fps for commonly used sample frequency in the self-driving area.

5.4   Improvement of Fusion Channel

According to the KITTI dataset standard, we divide the bounding box into three categories according to the size and degree of occlusion as follows: (1) Easy, less than 15%, completely visible, with the smallest bounding box being 40Px (pixel); (2) Moderate, for partial occlusion, with the block less than 30% and the smallest bounding box height of 25Px; (3) Hard, for most of the block, with the minimum block less than 50% minimum and bounding box height of 25Px. Figure 7 shows the maximum recall rate of the three different bounding boxes using the RBF model (RBF-original, RBF-o) and a modified version (RBF-modified, RBF-m).

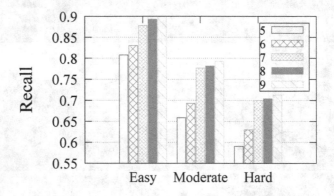

Fig. 7. Recall of Easy, Moderate and Hard modes with 5–9 candidate boxes.

It can be seen that the performance of the models drops when smaller bounding boxes and higher occluded blocks are used. Therefore, we adjust the model to fit 512×512 pixels image. Keep adding the candidate boxes can slightly improve the performance, however, we should also considering the time consumption on mobile platform. Figure 7 shows the maximum recall rate of the three different bounding boxes after adjustment of the model. By adding two candidate boxes (RBF-m), the recall rates of models trained with Easy, Moderate and Hard labeled bounding boxes increase by 0.0701, 0.1195 and 0.10934 when using 7-candidate-boxes, respectively. This can also guarantee near 10 FPS on Nvidia Tx2 for autonomous vehicle.

6   Conclusion

In this paper, we analyze the multiview based sensor fusion with a deep learning CNN model for on road vehicle detection. We first introduce the fusion strategies of combining the up-sampled LIDAR point cloud and RGB camera image as four-channel fusion data. Then, we evaluate all possible combinations of channel-based fusion on Yolo with a dataset collected from real on-road environment. Then we choose the minimum channel fusion scheme with the highest detection accuracy as our final fusion strategy (RBF). Simulation results show that our fusion based MultiView model can reach an average accuracy of 82% under the 0.5 threshold on real trace images and LIDAR point cloud from KITTI. By removing the useless data channels, we also reduce the training time and detection time by 28% of the full-four-channel model, and the final model can reach an average process rate of 8.59 fps for the self-driving car on Nvidia Tx1.

Acknowledgement. This work has been supported by National Key R&D Program of China (2017YFB1301100), National Natural Science Foundation of China (61572060, 61772060, 61728201), State Key Laboratory of Software Development Environment(No. 2016YFC1300205), and CERNET Innovation Project (NGII20160316, NGII20170315).

References

1. Borenstein, J., Herkert, J., Miller, K.: Self-driving cars: ethical responsibilities of design engineers. IEEE Technol. Soc. Mag. **36**(2), 67–75 (2017)
2. Li, X., Chen, M., Nie, F., et al.: A multiview-based parameter free framework for group detection. In: AAAI Conference on Artificial Intelligence (2017)
3. Chen, Z., Huang, X.: End-to-end learning for lane keeping of self-driving cars. In: 2017 IEEE Intelligent Vehicles Symposium (IV), pp. 1856–1860 (2017)
4. Ouyang, Z., Niu, J., Liu, Y., et al.: Multiwave: a novel vehicle steering pattern detection method based on smartphones. In: IEEE International Conference on Communications, pp. 1–7. IEEE (2016)
5. Jafari, O.H., Yang, M.Y.: Real-time RGB-D based template matching pedestrian detection. In: IEEE International Conference on Robotics and Automation. IEEE (2016)
6. Tung, F., Chen, J., Meng, L., et al.: The raincouver scene parsing benchmark for self-driving in adverse weather and at night. IEEE Robot. Autom. Lett. **2**(4), 2188–2193 (2017)
7. Song, S., Chandraker, M.: Joint SFM and detection cues for monocular 3D localization in road scenes. In: Computer Vision and Pattern Recognition, pp. 3734–3742. IEEE (2015)
8. Geiger, A.: Are we ready for autonomous driving? The KITTI vision benchmark suite. In: IEEE Conference on Computer Vision and Pattern Recognition, pp. 3354–3361. IEEE Computer Society (2012)
9. Hwang, S., Park, J., Kim, N., et al.: Multispectral pedestrian detection: benchmark dataset and baseline. In: Computer Vision and Pattern Recognition, pp. 1037–1045. IEEE (2015)
10. Xue, J.R., Wang, D., Shao-Yi, D.U., et al.: A vision-centered multi-sensor fusing approach to self-localization and obstacle perception for robotic cars. Front. Inf. Technol. Electron. Eng. **18**(1), 122–138 (2018)
11. Dollar, P., Appel, R., et al.: Fast feature pyramids for object detection. IEEE Trans. Pattern Anal. Mach. Intell. **36**, 1532–1545 (2014)
12. Lin, Y., Abdelfatah, K., Zhou, Y., et al.: Co-interest person detection from multiple wearable camera videos, pp. 4426–4434 (2015)
13. Huang, Y., Zhu, F., Shao, L., et al.: Color object recognition via cross-domain learning on RGB-D images. In: IEEE International Conference on Robotics and Automation, pp. 1672–1677. IEEE (2016)

14. Gu, X., Zang, A., Huang, X., et al.: Fusion of color images and LIDAR data for lane classification. In: Proceedings of the 23rd SIGSPATIAL International Conference on Advances in Geographic Information Systems, pp. 1–4 (2015)
15. Zhang, F., Clarke, D., Knoll, A.: Vehicle detection based on LIDAR and camera fusion. In: IEEE International Conference on Intelligent Transportation Systems, pp. 1620–1625. IEEE (2014)
16. Premebida, C., Carreira, J., Batista, J., et al.: Pedestrian detection combining RGB and dense LIDAR data. In: IEEE/RSJ International Conference on Intelligent Robots and Systems, pp. 4112–4117. IEEE (2014)
17. Ouyang, Z., Niu, J., Liu, Y., et al.: A cGANs-based scene reconstruction model using LIDAR point cloud. In: Proceedings of the 15th IEEE International Symposium on Parallel and Distributed Processing with Applications (2017)
18. Liu, W., Ji, R., Li, S.: Towards 3D object detection with bimodal deep Boltzmann machines over RGBD imagery. In: Computer Vision and Pattern Recognition, pp. 3013–3021. IEEE (2015)
19. Qi, C.R., Su, H., Nießner, M., et al.: Volumetric and multi-view CNNs for object classification on 3D data, pp. 5648–5656 (2016)
20. Girshick, R., Donahue, J., Darrell, T., Malik J.: Rich feature hierarchies for accurate object detection and semantic segmentation. In: CVPR (2014)
21. Girshick R.: Fast R-CNN. Computer Science (2015)
22. Ren, S., Girshick, R., Girshick, R., et al.: Faster R-CNN: towards real-time object detection with region proposal networks. IEEE Trans. Pattern Anal. Mach. Intell. **39**(6), 1137–1149 (2015)
23. He, K., Zhang, X., Ren, S., et al.: Spatial pyramid pooling in deep convolutional networks for visual recognition. IEEE Trans. Pattern Anal. Mach. Intell. **37**(9), 1904–1916 (2015)
24. Redmon, J., Divvala, S., Girshick, R., et al.: You only look once: unified, real-time object detection. In: IEEE Conference on Computer Vision and Pattern Recognition, pp. 779–788. IEEE Computer Society (2016)
25. Redmon, J., Farhadi, A.: YOLO9000: better, faster, stronger, pp. 6517–6525 (2016)
26. Li, B., Zhang, T., Xia, T.: Vehicle detection from 3D lidar using fully convolutional network. In: Robotics: Science and Systems (2016)
27. Asvadi, A., Garrote, L., Premebida, C., Peixoto, P., Nunes, U.J.: Real-time deep convnet-based vehicle detection using 3D-LIDAR reflection intensity data. In: Ollero, A., Sanfeliu, A., Montano, L., Lau, N., Cardeira, C. (eds.) ROBOT 2017. AISC, vol. 694, pp. 475–486. Springer, Cham (2018). https://doi.org/10.1007/978-3-319-70836-2_39
28. Asvadi, A., Garrote, L., Premebida, C., et al.: DepthCN: vehicle detection using 3D-LIDAR and ConvNet. In: IEEE International Conference on Intelligent Transportation Systems. IEEE (2017)
29. Wang, X., Nie, L., Song, X., et al.: Unifying virtual and physical worlds: learning toward local and global consistency. ACM Trans. Inf. Syst. **36**(1), 1–26 (2017)
30. Chen, J., Song, X., Nie, L., et al.: Micro tells macro: predicting the popularity of micro-videos via a transductive model. In: ACM on Multimedia Conference, pp. 898–907. ACM (2016)

Color Image Super Resolution by Using Cross-Channel Correlation

Kan Chang[1,2](✉), Caiwang Mo[1], Minghong Li[1], Tianyi Li[1], and Tuanfa Qin[1,2]

[1] School of Computer and Electronic Information, Guangxi University,
Nanning 530004, China
changkan0@gmail.com
[2] Guangxi Key Laboratory of Multimedia Communications and Network
Technologies, Guangxi University, Nanning 530004, China

Abstract. Single image super resolution (SR) aims to reconstruct a high resolution (HR) image from a low resolution (LR) image. However, many SR methods are only designed for the grayscale images. As a result, when dealing with the color images, the cross-channel information is often ignored by those approaches. In this paper, we propose a color image SR method by taking the cross-channel correlation of color images into consideration. In our method, the gradients of the differences between color channels are required to be sparse. In addition, to make our SR framework more robust, the average signal of the three color channels is also enforced to be sparse in the gradient domain. Finally, to solve the optimization problem which includes the cross-channel-correlation-based regularization terms, an efficient algorithm is presented. Experimental results demonstrate the effectiveness of the proposed method quantitatively and visually.

Keywords: Color image · Single image super resolution
Cross-channel correlation · Optimization · Bregman iteration

1 Introduction

Super resolution (SR) is a post-processing technique required in many multimedia applications, such as intelligent surveillance, medical imaging, remote sensing, etc. Given a low resolution (LR) measurement, the task of single image SR is to reconstruct the corresponding high resolution (HR) version of image.

One commonly studied type of single image SR methods is the learning-based methods, whose basic idea is to learn the mapping function from the LR space to the HR space. The representative approaches include the sparsity-based approaches [27,30], the ridge-regression-based methods [22,23,26], the self-similarity-based methods [14,28], etc. With the development of the parallel

---

This work was supported in part by Natural Science Foundation (NSF) of China under Grants 61761005 and 61761007, and in part by the NSF of Guangxi under Grants 2016GXNSFAA380154 and 2016GXNSFAA380216.

R. Hong et al. (Eds.): PCM 2018, LNCS 11166, pp. 471–481, 2018.
https://doi.org/10.1007/978-3-030-00764-5_43

technologies, several deep-learning-based methods [8,9,15] have been proposed recently. However, with a single GPU, the training phase of the deep-learning-based methods usually lasts for days.

Another important kind of methods considers the SR problem as an inverse problem, and solve the problem by applying prior information to regularize the solution spaces. The typical types of prior information include the local priors [20, 21], non-local priors [3,4], sparsity-based priors [10–12], etc. To achieve superior quality of results, some approaches jointly make use of complementary priors, such as [5,10,11,19,31]. Nevertheless, incorporating sophisticated priors usually leads to a high computational burden.

Though multiple types of methods have been proposed, most of them are designed to enhance the resolution of a single channel. When handling the color images, one typical approach is to apply the SR methods to each color channel independently. Another solution is to transform the LR image to the YCbCr space. Due to the fact that human eyes are more sensitive to the luminance component, the sophisticated SR methods are only utilized to the luminance channel. For the chromatic channels, the simple bicubic interpolation is used. In the above two solutions, the cross-channel correlation of color images is not fully exploited. Consequently, false colors and artifacts may exist in the results [16].

A few methods have been proposed to better describe the cross-channel correlation [6,7,16–18,25,29]. For example, the grayscale dictionary learning method K-SVD [1] is extended to learn correlation among different channels in [16]. In [25,29], the quaternion-based sparse representation model is proposed and employed to the color image SR task. Recently, Mousavi *et al.* [17,18] extend the sparsity-based SR method [27] to multiple color channels by enforcing edge similarities among different channels. However, in [17] and [18], because of the newly assembled cross-channel information, the sparse coding problem becomes much harder to solve than the conventional one. Thus the method MCcSR [17,18] has a relatively high computational burden.

In this paper, we propose a color image SR method, and the main contributions are twofold: First of all, a cross-channel-correlation-based (3C) regularization approach is presented, which is able to well characterize the cross-channel information within color images. Secondly, an efficient optimization algorithm named cross-channel-correlation-based color image SR (3C-CSR) is developed so as to solve the minimization problem incorporating the 3 C regularization terms.

The remainder of the paper is organized as follows. The proposed 3C-CSR algorithm is introduced in Sect. 2. The experimental results are presented in Sect. 3. Finally, we conclude this paper in Sect. 4.

2   The Proposed Method

2.1   Cross-Channel-Correlation-Based Color Image SR Framework

In the case of color image SR, usually, we have the following degradation model:

$$\mathbf{Y} = \mathbf{HX} + \mathbf{n} \tag{1}$$

where $\mathbf{X} = [\mathbf{x}_R^T, \mathbf{x}_G^T, \mathbf{x}_B^T]^T$, $\mathbf{Y} = [\mathbf{y}_R^T, \mathbf{y}_G^T, \mathbf{y}_B^T]^T$; $\mathbf{x}_R$, $\mathbf{x}_G$, and $\mathbf{x}_B$ are three column vectors representing the HR red (R), green (G) and blue (B) channels, respectively, while $\mathbf{y}_R$, $\mathbf{y}_G$, and $\mathbf{y}_B$ are their corresponding LR measurements; $\mathbf{H} = \text{diag}(\mathbf{h}_R, \mathbf{h}_G, \mathbf{h}_B)$, with $\mathbf{h}_R$, $\mathbf{h}_G$, and $\mathbf{h}_B$ denoting the degrading matrices on R, G, and B channels, respectively. Generally, the degradation matrices involve the operations of blurring and down-sampling. $\mathbf{n}$ stands for the additive Gaussian noise.

If the correlations among different color channels can be properly captured and incorporated into the SR framework, the quality of the SR results will be further improved. To address this problem, one common way is to assume that the high-frequency components on different color channels are similar to each other [6,17,18]. To better understand this phenomenon, the gradients of different signals are compared in Fig. 1. It can be seen from Fig. 1 that, the differences between color channels, i.e., $(\mathbf{x}_G - \mathbf{x}_B)$, and $(\mathbf{x}_B - \mathbf{x}_R)$, are sparse and small in gradient domain, which indicates that the above assumption holds. Based on this observation, we propose the following SR reconstruction framework:

$$\hat{\mathbf{X}} = \underset{\mathbf{X}}{\arg\min} \frac{1}{2}\|\tilde{\mathbf{X}} - \mathbf{X}\|_2^2 + \beta\|\mathbf{D}(\mathbf{x}_R + \mathbf{x}_G + \mathbf{x}_B)\|_1$$
$$+ \alpha\big[\|\mathbf{D}(\mathbf{x}_G - \mathbf{x}_B)\|_1 + \|\mathbf{D}(\mathbf{x}_B - \mathbf{x}_R)\|_1\big] \tag{2}$$

where $\mathbf{D} = [\mathbf{D}_h^T, \mathbf{D}_v^T]^T$, with $\mathbf{D}_h \in \mathbb{R}^{N \times N}$ and $\mathbf{D}_v \in \mathbb{R}^{N \times N}$ denoting the matrices to produce the horizontal and vertical first-order derivatives, respectively. N stands for the number of pixels in an image. $\tilde{\mathbf{X}}$ represents the initial estimation of the HR image, α and β are trade-off parameters to balance among different regularization terms.

In problem (2), the first regularization term requires the reconstructed $\mathbf{X}$ to be close to an initial guess $\tilde{\mathbf{X}}$. Any other SR methods, e.g., ANR [22] and ScSR [27], can be applied to generate $\tilde{\mathbf{X}}$ from the given LR image $\mathbf{Y}$. Therefore, our SR reconstruction framework can also be regarded as a quality enhancement method for the other SR approaches.

The rest terms in problem (2) exploit the cross-channel correlation of color images. There are two main differences between our model and the existing MCcSR model [17,18]:

(1) We use the L_1 norm, rather than the L_2 norm to measure the differences between color channels in the gradient domain. As shown in Fig. 1, both $(\mathbf{x}_G - \mathbf{x}_B)$ and $(\mathbf{x}_B - \mathbf{x}_R)$ are very sparse in the gradient domain. Due to this reason, the L_1 norm is more suitable for characterizing these signals than the L_2 norm.

(2) Besides the differences between color channels, we also consider the average signal of three color channels. As observed from Fig. 1, the gradient of the average signal is more sensitive to the edges in images than the differences between color channels. Consequently, including this term is able to make the SR reconstruction framework more robust.

Fig. 1. The horizontal gradient comparison for different images. From top to bottom: images *butterfly*, *bike*, and *flowers*. From left to right: original image, the horizontal gradients of $(x_G - x_B)$, $(x_B - x_R)$, and $(x_R + x_G + x_B)/3$. Note that the dark areas suggest that the gradients at these areas are large.

2.2   Optimization Algorithm for Color Image SR

The objective function of problem (2) has three L_1 norms, which makes the problem difficult to solve. However, by merging all the L_1 portions together, we are able to get the following re-written problem

$$\hat{X} = \underset{X}{\operatorname{argmin}} \frac{1}{2}\|\tilde{X} - X\|_2^2 + \alpha\|\tilde{D}SX\|_1 \qquad (3)$$

where $\tilde{D} = \operatorname{diag}(D, D, D)$, and the matrix S is defined as

$$S = \begin{bmatrix} \beta I/\alpha & \beta I/\alpha & \beta I/\alpha \\ 0 & I & -I \\ -I & 0 & I \end{bmatrix} \qquad (4)$$

where I stands for the identity matrix with a size of $N \times N$.

To decouple X from the L_1 portion in problem (3), a new variable d is introduced. By letting d to be equal to $\tilde{D}SX$, problem (3) is further modified as

$$\hat{X} = \underset{X}{\operatorname{argmin}} \frac{1}{2}\|\tilde{X} - X\|_2^2 + \alpha\|d\|_1 \qquad s.t. \quad d = \tilde{D}SX \qquad (5)$$

Algorithm 1. 3C-CSR

Input: $\tilde{\mathbf{X}}$, $\mathbf{Y}$, $\mathbf{H}$, α, β, μ
Set $\mathbf{d}^0 = \tilde{\mathbf{D}}\mathbf{S}\tilde{\mathbf{X}}$, and $\mathbf{b}^0 = 0$
For $k = 0, 1, ..., K$
    Use the CG method to compute (8), so as to update $\mathbf{X}^{k+1}$
    Update $\mathbf{d}^{k+1}$ via (9)
    $\mathbf{b}^{k+1} = \mathbf{b}^k + \tilde{\mathbf{D}}\mathbf{S}\mathbf{X}^{k+1} - \mathbf{d}^{k+1}$
End for
Use the BP algorithm to get a refined result $\hat{\mathbf{X}}$ from $\mathbf{X}^{K+1}$.
return $\hat{\mathbf{X}}$

After that, we change the problem (3) to a new unconstrained optimization problem by using a quadratic penalty function:

$$\{\hat{\mathbf{X}}, \hat{\mathbf{d}}\} = \operatorname*{argmin}_{\{\mathbf{X},\mathbf{d}\}} \frac{1}{2}\|\tilde{\mathbf{X}} - \mathbf{X}\|_2^2 + \alpha\|\mathbf{d}\|_1 + \frac{\mu}{2}\|\mathbf{d} - \tilde{\mathbf{D}}\mathbf{S}\mathbf{X} - \mathbf{b}\|_2^2 \qquad (6)$$

where μ denotes a trade-off parameter. Note that, to ensure a quick convergence of the final algorithm, an intermediate variable $\mathbf{b}$ is introduced to the problem (6) and will be updated according to Bregman iteration [13].

The minimization of problem (6) can be efficiently performed by iteratively minimizing the sub-problems with respect to $\mathbf{X}$ and $\mathbf{d}$. As a result, at the $(k+1)$th iteration, we have

$$\begin{cases} \mathbf{X}^{k+1} = \operatorname{argmin}_{\mathbf{X}} \|\tilde{\mathbf{X}} - \mathbf{X}\|_2^2 + \mu\|\mathbf{d}^k - \tilde{\mathbf{D}}\mathbf{S}\mathbf{X} - \mathbf{b}^k\|_2^2 \\ \\ \mathbf{d}^{k+1} = \operatorname{argmin}_{\mathbf{d}} \alpha\|\mathbf{d}\|_1 + \frac{\mu}{2}\|\mathbf{d} - \tilde{\mathbf{D}}\mathbf{S}\mathbf{X}^{k+1} - \mathbf{b}^k\|_2^2 \qquad (7) \\ \\ \mathbf{b}^{k+1} = \mathbf{b}^k + \tilde{\mathbf{D}}\mathbf{S}\mathbf{X}^{k+1} - \mathbf{d}^{k+1} \end{cases}$$

As the "$\mathbf{X}$" sub-problem only contains L_2 norms, by setting the derivation of the objective function to be zero, the closed-form solution of this sub-problem can be obtained by

$$\mathbf{X}^{k+1} = \left[\mathbf{I} + \mu(\tilde{\mathbf{D}}\mathbf{S})^T(\tilde{\mathbf{D}}\mathbf{S})\right]^{-1}\left[\tilde{\mathbf{X}} + \mu(\tilde{\mathbf{D}}\mathbf{S})^T(\mathbf{d}^k - \mathbf{b}^k)\right] \qquad (8)$$

Nevertheless, the computational complexity of directly computing (8) is high. Therefore, we use the conjugate gradient (CG) method to calculate $\mathbf{X}^{k+1}$.

For the "$\mathbf{d}$" sub-problem, it also has the closed-form solution:

$$\mathbf{d}^{k+1} = \operatorname{shrink}(\tilde{\mathbf{D}}\mathbf{S}\mathbf{X}^{k+1} + \mathbf{b}^k, \alpha/\mu) \qquad (9)$$

where $\operatorname{shrink}(x, \tau)$ stands for the shrinkage formula, which is computed as

$$\operatorname{shrink}(x, \tau) = \frac{x}{|x|}\max(|x| - \tau, 0) \qquad (10)$$

Since the two sub-problems derived from (6) have been addressed, we can summarize the proposed 3C-CSR algorithm as Algorithm 1. To guarantee that

Table 1. PSNR(dB) and SSIM results comparison for images in *Set5* (×3)

Images	Bicubic	Zeyde [30]	ScSR [27]	ANR [22]	MCcSR [18]	3C-CSR
baby	38.39	39.29	39.27	39.41	39.36	**39.65**
	0.8839	0.8981	0.9028	0.9029	0.9040	**0.9078**
bird	36.35	37.80	37.71	38.09	38.01	**38.28**
	0.9101	0.9276	0.9286	0.9315	0.9313	**0.9328**
butterfly	28.73	29.80	29.97	30.06	30.58	**31.39**
	0.7932	0.8252	0.8265	0.8294	0.8478	**0.8697**
head	36.00	36.36	36.41	**36.50**	36.46	**36.50**
	0.7259	0.7408	0.7518	0.7495	0.7531	**0.7532**
woman	33.29	34.48	34.39	34.69	34.82	**35.25**
	0.8783	0.8985	0.8985	0.9017	0.9041	**0.9109**
Average	34.55	35.55	35.55	35.75	35.85	**36.21**
	0.8383	0.8580	0.8616	0.8630	0.8681	**0.8749**

the result of the $(K+1)$th iteration, i.e., $\mathbf{X}^{K+1}$, conforms with the requirement $\mathbf{Y} = \mathbf{HX}$, the back-projection (BP) algorithm [27] is applied in 3C-CSR as a post-processing stage.

3   Experimental Results

In this section, we conduct experiments to verify the effectiveness of the 3C-CSR. The experiments are carried out under the MATLAB 2014b environment and on a PC with Intel(R) Core(TM) i5-4460 CPU and 8G RAM.

Four representative methods are compared, including Zeyde's method [30], ScSR [27], ANR [22] and MCcSR [18]. Note that except MCcSR [18], the other three benchmark methods only work on the luminance channel, while on the chromatic channels, the basic bicubic interpolation is applied. Thus for a fair comparison, in our experiments, their implementations are modified accordingly, so as to work on all the R, G and B channels. For the above methods, the patch size in the LR space is set as 5×5, and the overlap between adjacent patches is 4 pixels; the size of the dictionary is 512, which is the same as in [18]; in the training phase, the training set of images proposed by Yang *et al.* [27] is used.

The basic settings for 3C-CSR are: $\alpha = 16.0$, $\beta = 40.0$ and $\mu = 1.0$; the number of iterations for updating $\mathbf{X}$ is $K = 15$, and the number of iterations in the BP algorithm is 15. The SR results of ScSR [27] is selected as the initial estimation, i.e., the input $\tilde{\mathbf{X}}$ of 3C-CSR.

The widely used datasets *Set5* [2] and *Set14* [30] are chosen for testing[1]. The peak signal to noise ratio (PSNR) and structure similarity index (SSIM) [24]

---

[1] As *bridge* and *man* in *Set14* are gray images, they are excluded from our experiments.

Table 2. PSNR(dB) and SSIM results comparison for images in *Set14* (×3)

Images	Bicubic	Zeyde [30]	ScSR [27]	ANR [22]	MCcSR [18]	3C-CSR
baboon	26.71	26.91	26.99	**27.01**	27.00	**27.01**
	0.4812	0.5100	0.5383	0.5280	**0.5396**	0.5383
barbara	30.75	31.16	31.24	31.19	31.27	**31.38**
	0.7221	0.7442	0.7542	0.7489	0.7559	**0.7591**
coastguard	31.31	31.71	31.78	31.75	31.79	**31.92**
	0.5801	0.6109	0.6333	0.6212	0.6333	**0.6339**
comic	27.50	28.07	28.24	28.28	28.40	**28.56**
	0.6856	0.7209	0.7411	0.7349	0.7483	**0.7557**
face	35.95	36.33	36.35	**36.47**	36.41	**36.47**
	0.7238	0.7393	0.7498	0.7479	0.7513	**0.7518**
flowers	30.92	31.68	31.85	31.91	32.07	**32.40**
	0.7797	0.8034	0.8143	0.8122	0.8193	**0.8237**
foreman	35.75	36.93	36.87	37.22	37.54	**38.09**
	0.8906	0.9067	0.9062	0.9106	0.9134	**0.9198**
lenna	35.25	36.02	36.05	36.23	36.26	**36.42**
	0.7874	0.8018	0.8079	0.8079	0.8104	**0.8124**
monarch	34.06	35.04	35.18	35.33	35.66	**36.34**
	0.9064	0.9181	0.9189	0.9206	0.9236	**0.9303**
pepper	35.23	36.13	35.93	36.10	36.30	**36.48**
	0.7899	0.8014	0.8031	0.8031	0.8060	**0.8080**
ppt3	28.02	29.04	29.16	29.09	29.32	**29.77**
	0.8623	0.8836	0.8844	0.8784	0.8899	**0.9016**
zebra	31.33	32.65	32.81	32.89	33.16	**33.66**
	0.7787	0.8131	0.8295	0.8241	0.8340	**0.8431**
Average	31.90	32.64	32.70	32.79	32.93	**33.21**
	0.7490	0.7711	0.7818	0.7781	0.7854	**0.7898**

results obtained by different methods for images in *Set5* and *Set14* at a magnification factor of 3 are listed in Tables 1 and 2[2], respectively. As can be seen, 3C-CSR achieves the best performance among all the tested methods. For *Set5*, the average PSNR and SSIM gains of 3C-CSR over the second best method, i.e., MCcSR [18] are 0.36dB and 0.0068, respectively. For *Set14*, 3C-CSR outperforms MCcSR [18] by 0.28dB and 0.0044 on average.

The subjective results of *butterfly* and *zebra* are respectively provided in Figs. 2 and 3. It can be observed that the results generated by bicubic

---

[2] The PSNR is measured on all the color channels; the SSIM is calculated on each channel, and the average value of three channels is reported.

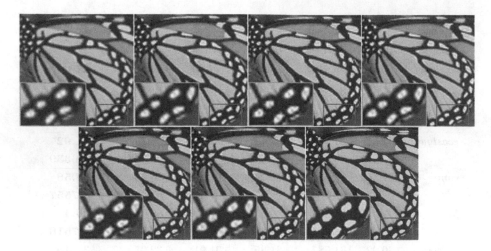

Fig. 2. The SR results of *butterfly* obtained by different methods. From left to right and from top to bottom: bicubic, Zeyde [30], ScSR [27], ANR [22], MCcSR [18], 3C-CSR and original image. Please zoom in for better viewing.

Fig. 3. The SR results of *zebra* obtained by different methods. From left to right and from top to bottom: bicubic, Zeyde [30], ScSR [27], ANR [22], MCcSR [18], 3C-CSR and original image. Please zoom in for better viewing.

interpolation and Zeyde's method [30] are relatively blurry. Although ScSR [27], ANR [22] and MCcSR [18] are able to produce sharper edges, obvious artifacts can be found. Compared with the other methods, 3C-CSR is capable of well preserving the edges and details in images. Moreover, the artifacts generated by ScSR [27] have been largely reduced by 3C-CSR (as mentioned before, ScSR [27] is applied to provide the initial estimation for 3C-CSR in our experiments).

The average running times of different methods on *Set5* are shown in Table 3. It can be found that Zeyde's method [30] and ANR [22] are much faster than the other methods. Because a complex optimization problem is considered for each image patch, it costs a long time to compute MCcSR [18]. Compared with ScSR [27] and MCcSR [18], 3C-CSR requires much less running time. The main complexity of 3C-CSR comes from using the CG method to compute (8), which needs $O(N^2 N_g)$, where N_g is the number of iterations of the CG method.

Table 3. Average running time (s) on test images in *Set5*.

Methods	Zeyde [30]	ScSR [27]	ANR [22]	MCcSR [18]	3C-CSR
Time(s)	3.5	233.5	1.9	10474.1	17.9

4 Conclusion

In this paper, we propose an algorithm called 3C-CSR for single image SR problem. To achieve a better performance, the cross-channel correlation of color images is exploited in 3C-CSR. The experimental results show that 3C-CSR outperforms the other benchmark methods on the widely used datasets *Set5* and *Set14*. Meanwhile, the computational burden of 3C-CSR is acceptable.

References

1. Aharon, M., Elad, M., Bruckstein, A.: K-SVD: an algorithm for designing overcomplete dictionaries for sparse representation. IEEE Trans. Signal Process. **54**(11), 4311–4322 (2006)
2. Bevilacqua, M., Roumy, A., Guillemot, C., Morel, A.: Low-complexity single-image super-resolution based on nonnegative neighbor embedding. In: British Machine Vision Conference (BMVC), pp. 1–12 (2012)
3. Buades, A., Coll, B., Morel, J.M.: A review of image denoising algorithms, with a new one. SIAM Multiscale Model. Simul. **4**(2), 490–530 (2005)
4. Chang, K., Ding, P.L.K., Li, B.: Single image super-resolution using collaborative representation and non-local self-similarity. Signal Process. **149**, 49–61 (2018)
5. Chang, K., Ding, P.L.K., Li, B.: Single image super resolution using joint regularization. IEEE Signal Process. Lett. **25**(4), 596–600 (2018)
6. Chang, K., Ding, P.K., Li, B.: Color image demosaicking using inter-channel correlation and nonlocal self-similarity. Signal Process.: Image Commun. **39**, 264–279 (2015)
7. Chang, K., Li, B.: Joint modeling and reconstruction of a compressively-sensed set of correlated images. J. Vis. Commun. Image Represent. **33**, 286–300 (2015)
8. Dong, C., Loy, C.C., He, K., Tang, X.: Image super-resolution using deep convolutional networks. IEEE Trans. Pattern Anal. Mach. Intell. **38**(2), 295–307 (2016)
9. Dong, C., Loy, C.C., Tang, X.: Accelerating the super-resolution convolutional neural network. In: Leibe, B., Matas, J., Sebe, N., Welling, M. (eds.) ECCV 2016. LNCS, vol. 9906, pp. 391–407. Springer, Cham (2016). https://doi.org/10.1007/978-3-319-46475-6_25
10. Dong, W., Zhang, L., Shi, G., Li, X.: Nonlocally centralized sparse representation for image restoration. IEEE Trans. Image Process. **22**(4), 1620–1630 (2013)
11. Dong, W., Zhang, L., Shi, G., Wu, X.: Image deblurring and super-resolution by adaptive sparse domain selection and adaptive regularization. IEEE Trans. Image Process. **20**(7), 1838–1857 (2011)

12. Elad, M., Aharon, M.: Image denoising via sparse and redundant representations over learned dictionaries. IEEE Trans. Image Process. **15**(12), 3736–3745 (2006)

13. Goldstein, T., Osher, S.: The split Bregman method for $l1$ regularized problems. SIAM J. Imaging Sci. **2**(2), 323–343 (2009)

14. Huang, J., Ahuja, A.S.N.: Single image super-resolution from transformed self-exemplars. In: IEEE Conference on Computer Vision and Pattern Recognition (CVPR), pp. 5197–5206. IEEE, Boston (2015)

15. Kim, J., Lee, J., Lee, K.M.: Accurate image super-resolution using very deep convolutional networks. In: IEEE Conference on Computer Vision and Pattern Recognition (CVPR), pp. 1646–1654. IEEE, Las Vegas (2016)

16. Mairal, J., Elad, M., Sapiro, G.: Sparse representation for color image restoration. IEEE Trans. Image Process. **17**(1), 53–69 (2008)

17. Mousavi, H., Monga, V.: Sparsity based super resolution using color channel constraints. In: IEEE International Conference on Image Processing (ICIP), pp. 579–583. IEEE, Phoneix (2016)

18. Mousavi, H., Monga, V.: Sparsity-based color image super resolution via exploiting cross channel constraints. IEEE Trans. Image Process. **26**(11), 5094–5106 (2017)

19. Ren, C., He, X., Teng, Q., Wu, Y., Nguyen, T.Q.: Single image super-resolution using local geometric duality and non-local similarity. IEEE Trans. Image Process. **25**(5), 2168–2183 (2016)

20. Rudin, L., Osher, S., Fatemi, E.: Nonlinear total variation based noise removal algorithms. Phys. D: Nonlinear Phenom. **60**(1), 259–268 (1992)

21. Takeda, H., Farsiu, S., Milanfar, P.: Kernel regression for image processing and reconstruction. IEEE Trans. Image Process. **16**(2), 349–366 (2007)

22. Timofte, R., Smet, V.D., Gool, L.V.: Anchored neighborhood regression for fast example-based super resolution. In: IEEE International Conference on Computer Vision (ICCV), pp. 1920–1927. IEEE, Sydney (2013)

23. Timofte, R., De Smet, V., Van Gool, L.: A+: adjusted anchored neighborhood regression for fast super-resolution. In: Cremers, D., Reid, I., Saito, H., Yang, M.-H. (eds.) ACCV 2014. LNCS, vol. 9006, pp. 111–126. Springer, Cham (2015). https://doi.org/10.1007/978-3-319-16817-3_8

24. Wang, Z., Bovik, A.C., Sheikh, H.R., Simoncelli, E.P.: Image quality assessment: from error visibility to structural similarity. IEEE Trans. Image Process. **13**(4), 600–612 (2004)

25. Xu, Y., Yu, L., Xu, H., Zhang, H., Nguyen, T.: Vector sparse representation of color image using quaternion matrix analysis. IEEE Trans. Image Process. **24**(4), 1315–1329 (2015)

26. Yang, C.Y., Yang, M.H.: Fast direct super-resolution by simple functions. In: IEEE International Conference on Computer Vision (ICCV), pp. 561–568. IEEE, Sydney (2013)

27. Yang, J., Wright, J., Huang, T.S., Ma, Y.: Image super-resolution via sparse representation. IEEE Trans. Image Process. **19**(11), 2861–2873 (2010)

28. Yang, M., Wang, Y.: A self-learning approach to single image super-resolution. IEEE Trans. Multimed. **15**(3), 498–508 (2013)

29. Yu, M., Xu, Y., Sun, P.: Single color image super-resolution using quaternion-based sparse representation. In: IEEE International Conference on Acoustics. Speech and Signal Processing (ICASSP), pp. 5804–5808. IEEE, Florence (2014)

30. Zeyde, R., Elad, M., Protter, M.: On single image scale-up using sparse-representations. In: Boissonnat, J.-D., Chenin, P., Cohen, A., Gout, C., Lyche, T., Mazure, M.-L., Schumaker, L. (eds.) Curves and Surfaces 2010. LNCS, vol. 6920, pp. 711–730. Springer, Heidelberg (2012). https://doi.org/10.1007/978-3-642-27413-8_47
31. Zhang, K., Gao, X., Tao, D., Li, X.: Single image super-resolution with non-local means and steering kernel regression. IEEE Trans. Image Process. **21**(11), 4544–4556 (2012)

Stereoscopic Video Quality Prediction Based on End-to-End Dual Stream Deep Neural Networks

Wei Zhou, Zhibo Chen$^{(\boxtimes)}$, and Weiping Li

CAS Key Laboratory of Technology in Geo-spatial
Information Processing and Application System,
Department of Electronic Engineering and Information Science,
University of Science and Technology of China, Hefei 230027, China
weichou@mail.ustc.edu.cn, {chenzhibo,wpli}@ustc.edu.cn

Abstract. In this paper, we propose a no-reference stereoscopic video quality assessment (NR-SVQA) method based on an end-to-end dual stream deep neural network (DNN), which incorporates left and right view sub-networks. The end-to-end dual stream network takes image patch pairs from left and right view pivotal frames as inputs and evaluates the perceptual quality of each image patch pair. By combining multiple convolution, max-pooling and fully-connected layers with regression in the framework, distortion related features are learned end-to-end and purely data driven. Then, a spatiotemporal pooling strategy is employed on these image patch pairs to estimate the entire stereoscopic video quality. The proposed network architecture, which we name End-to-end Dual stream deep Neural network (EDN), is trained and tested on the well-known stereoscopic video dataset divided by reference videos. Experimental results demonstrate that our proposed method outperforms state-of-the-art algorithms.

Keywords: Convolutional neural network · Stereoscopic video
No-reference video quality assessment · Spatiotemporal pooling

1 Introduction

Stereoscopic video quality assessment (SVQA) is challenging because the left and right views of 3D/stereoscopic videos can synthetically generate depth perception, which leads to an additional perceptual dimension to be considered. Through the whole 3D media processing chain from acquisition, compression, to transmission, reconstruction, and display, etc., original stereoscopic videos undergo a variety of quality degradations. Consequently, it is important to effectively predict and optimize the quality of experience (QoE) throughout the processing chain.

According to the existence of non-distorted reference videos, SVQA algorithms can be divided into three categories: full-reference SVQA (FR-SVQA),

© Springer Nature Switzerland AG 2018
R. Hong et al. (Eds.): PCM 2018, LNCS 11166, pp. 482–492, 2018.
https://doi.org/10.1007/978-3-030-00764-5_44

reduced-reference SVQA (RR-SVQA), and no-reference SVQA (NR-SVQA). The main purpose of SVQA is to design an objective criterion to accurately predict the perceptual subjective quality of 3D videos. During recent years, some 3D video quality metrics have been proposed. Early FR-SVQA models requiring pristine stereoscopic videos were studied. For example, the perceptual quality metric (PQM) used conventional 2D objective metrics to assess 3D video quality [9]. The modified PSNR, called PHVS-3D, exploited 3D discrete cosine transform (3D-DCT) to evaluate the perceptual quality of stereoscopic videos [8]. In [18], the spatial frequency dominance (SFD) model considered the phenomenon that spatial frequency determines view domination based on the human visual system (HVS). The 3D spatial-temporal structural (3D-STS) metric utilized the inter-view correlation of spatial and temporal structural information [4]. In [21], an objective metric named SJND-SVA was designed by integrating the stereoscopic visual attention (SVA) with the stereoscopic just-noticeable difference (SJND).

The drawback of these FR models is that original 3D videos are not always available in most practical situations. Thus, NR-SVQA models should be developed to assess stereoscopic video quality without needing reference 3D videos. In [26], an NR optical flow-based method was developed to predict 3D video quality. Recently, the motion feature based no reference stereo video quality metric (MNSVQM) [7] and the blind stereoscopic video quality evaluator (BSVQE) [2] have been proposed. However, most of SVQA algorithms still extract hand-crafted features from stereoscopic videos, and then yield visual quality evaluation. In other words, one of the advantages of applying deep neural network (DNN) to SVQA is that it can directly input raw image/video and combine feature learning with quality regression in the training stage. Moreover, the DNN-based SVQA metrics are robust and can be trained end-to-end with little prior domain knowledge. Therefore, this paper focuses on studying the NR-SVQA method for 3D videos using DNN-based end-to-end learning.

Deep learning has achieved remarkable results in object detection, image classification, and recognition [3,11,14]. At the same time, the application of DNN in image/video quality assessment has also been started to be explored in several studies. However, how to effectively predict the quality of image/video using DNN is quite different from traditional computer vision tasks. Specifically, DNN based image/video quality prediction is a challenging problem, mainly due to most of the existing data augmentation as well as patch preprocessing techniques are not suitable for image/video quality assessment [13].

From the perspective of training strategies, the relevant works of applying convolutional neural network (CNN) to no-reference image quality assessment (NR-IQA) can be broadly divided into two kinds of categories, including patch-wise training and image-wise training. The patch-wise training strategy [10,16, 17,27] partitions the image into patches, and then independently predicts the quality of each patch through regression. In contrast, the image-wise training strategy [1,6,12,19] obtains the image quality by aggregating and pooling patch features or predicted scores.

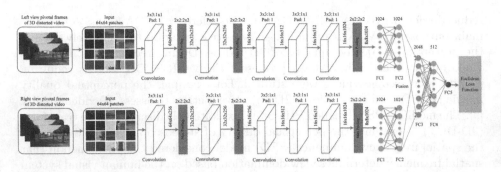

Fig. 1. Architecture of the End-to-end Dual stream deep convolutional Neural network (EDN) for quality regression of image patches from both left and right view videos. There are two sub-networks that all output 1024 − *dim* vectors as representations of image patches. These two sub-networks have identical configuration. The predicted score is fed into the final Euclidean loss function for comparison with ground truth quality label. The numbers shown above each arrow give the size of the corresponding output. The numbers shown above each box (*pad* = 1) indicate the size of kernel as well as the size of stride for the corresponding layer.

The datasets of 2D/3D video quality assessment (VQA) are smaller than that of IQA. Thus, existing 2D VQA models exploit extracted video features, and then feed these features into the proposed deep learning framework. In [15], the 3D shearlet transform was applied to extract spatial-temporal feature from 2D videos and input the feature for 1D CNN to predict the score. In [24], several features were extracted from video stream. The unsupervised restricted Boltzmann machine (RBM) [5] was then employed to predict 2D video quality. These 2D VQA algorithms take hand-crafted features as inputs to deep learning models. However, how to directly input raw image/video to develop an end-to-end DNN architecture for 3D VQA, which integrates both feature extraction and quality regression, has not been proposed.

In this paper, to our best knowledge, this is the first study for the end-to-end learning of stereoscopic video quality. Specifically, we present a DNN-based NR-SVQA approach named End-to-end Dual stream deep Neural network (EDN) to predict the perceptual quality of stereoscopic videos. Our basic idea is that each stereoscopic video has left and right views, which motivates our proposed dual stream network. Moreover, compression distortions introduced in the existing 3D video quality dataset are usually homogeneous which is also consistent in most real application scenarios. Therefore, we can divide the stereoscopic videos into patches and then assign the score of whole stereoscopic video to cropped patches, which solves insufficient training data effectively. Note that both data augmentation and patch preprocessing are not used according to the characteristic of perceptual quality assessment. Extensive experimental results demonstrate that the proposed EDN can effectively predict stereoscopic video quality, in addition achieve highly competitive correlation with human subjective scores compared with many state-of-the-art quality prediction algorithms.

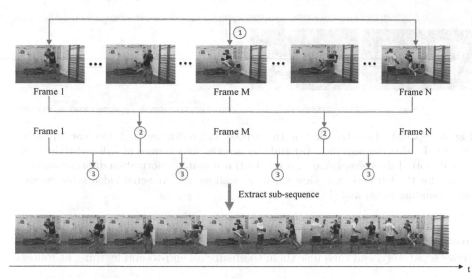

Fig. 2. Demonstration of three-step approach to extract intermediate frame as pivotal frames in temporal domain. The numbers in blue circles represent each step. Frame 1, Frame M and Frame N are the first frame, the intermediate frame and the last frame, respectively. (Color figure online)

The rest of this paper is organized as follows. In Sect. 2, we present the proposed deep CNN-based NR-SVQA method step by step. Section 3 shows the experimental results and analysis. The conclusion and future work are given in the final section.

2   Proposed Method

The proposed architecture of the End-to-end Dual stream deep convolutional Neural network (EDN) for quality regression is shown in Fig. 1. Due to each stereoscopic video has left and right views, this network contains two sub-networks which are two CNNs with identical configuration and shared parameters. They are designed to automatically learn image visual information through the lower level to higher level feature learning. The image patch pairs for left and right views are taken as inputs to the two sub-networks respectively. The architecture and settings of these two sub-networks are inspired by AlexNet [14] which achieves promising effects in solving image visual related tasks.

For each sub-network, the input to this sub-network is the pixel data of RGB channels of an image patch. Five convolutional layers are then applied to the input image patch. The convolutional layers use convolution kernels (with $stride = 1, pad = 1$), and reduce the size of feature maps only through max pooling. The output of each sub-network is a $1024 - dim$ feature vector (i.e. $FC2$). Then, we utilize a fusion layer for the outputs of these two sub-networks

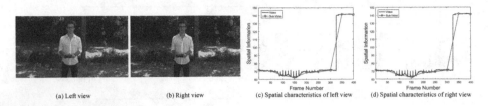

(a) Left view   (b) Right view   (c) Spatial characteristics of left view   (d) Spatial characteristics of right view

Fig. 3. Spatial characteristics of the whole video frames and the corresponding extracted video sub-sequence. (a) and (b) are the first frames of left and right views for a distorted stereoscopic video; (c) and (d) are spatial information distribution of all frames for the left and right view videos as well as the extracted video sub-sequences corresponding to (a) and (b), respectively.

to obtain a $2048 - dim$ feature vector. In other words, we train the left and right views separately, and then fuse them to ensure an end-to-end learning as follows:

$$V = [V_l, V_r], \tag{1}$$

where V_l and V_r denote the output feature vectors of left view and right sub-networks, respectively. The fusion layer, which synthesizes the left and right views, may reflect depth perception as the human brain works [20]. Afterwards, two full-connection layers are applied to perform regression onto a single quality score for each input image patch pair. Finally, the predicted score is fed into the Euclidean loss function to be verified by the ground truth quality score.

2.1 Stereoscopic Video Preprocessing

In order to generate more training data to solve the problem of small stereoscopic video datasets and ensure the content diversity and category balance of training data, we preprocess stereoscopic videos as follows.

First, we conduct three-step approach to extract intermediate frames as pivotal frames in temporal domain, as illustrated in Fig. 2. From Fig. 2, we can see that the first frame and the last frame of an input video are preferential, which represent probable video content generally. Suppose for the moment that we extract m frames from the input video, we then compute the remaining frame indices of pivotal frames by:

$$index = \left\lfloor \frac{n}{m-1} \times N \right\rfloor, \tag{2}$$

where $n = 1, 2, \ldots, m - 2$, N is the number of frame for the input video. Specifically, the first step is that the intermediate frame between the first frame and the last frame is picked. Likewise, the second step is that the intermediate frame between the first frame and previous intermediate frame is extracted. The intermediate frame between the previous intermediate frame and the last frame is also picked. Additionally, the third step is also to extract the intermediate frames, which is similar to the second step. In other words, the frames are extracted equidistantly.

In addition, the spatial characteristics of the whole video frames and the corresponding extracted video sub-sequence can be seen in Fig. 3. In Fig. 3(a) and (b) show the first frames of left and right views for a distorted stereoscopic videos. Meanwhile, (c) and (d) are the spatial information (SI) [22] distribution of all frames for the left and right view videos as well as the extracted video sub-sequences corresponding to (a) and (b), respectively. In general, we can find that the left and right views of a specific stereoscopic video have similar spatial information distribution. For each view, the image content of each extracted frame in the sub-sequence differs from each other. Moreover, these extracted frames constitute a video sub-sequence, which can represent the input video in a sense.

Second, the extracted pivotal frames are divided into 64×64 non-overlapped patches spatially. Moreover, the image patches are not preprocessed, such as resizing, to maintain the originally perceptual quality. Finally, we input the generated image patch pairs to our proposed EDN.

2.2   Patch Pair-Wise Learning

Let a stereoscopic video be represented by left and right view videos V_l and V_r. Each video has N frames. We extract m pivotal frames from the left and right views respectively. For each extracted frame, we then divide it into 64×64 non-overlapped image patches, i.e. Pl_i and Pr_i patches, $i = 1, 2, \ldots, p$. The predicted quality score for each patch pair of the left and right views is given by the output of EDN with network weights w. Here, the ground truth quality label for each patch pair is assigned the same as the global subjective score of the corresponding stereoscopic video. Our learning objective is defined by mean square error (MSE) as:

$$\min_{w} \|q_i - y_i\|_F^2, \tag{3}$$

where q_i, $i = 1, 2, \ldots, p$ denote the outputs of our proposed EDN, which are the predicted quality scores for p patch pairs in the same image frame. Moreover, y_i, $i = 1, 2, \ldots, p$ are the corresponding ground truth quality scores for the input image patches. Then, we apply a spatiotemporal pooling strategy to these image patch pairs to estimate the entire stereoscopic video quality. Specifically, the predicted quality score for each frame-pair of the left and right views is computed as follows:

$$Q_j = \frac{1}{p} \sum_{i=1}^{p} q_i, \tag{4}$$

where $j = 1, 2, \ldots, m$ are the temporal indices of pivotal frames. Finally, the quality of the stereoscopic video is averaged by:

$$Q = \frac{1}{m} \sum_{j=1}^{m} Q_j, \tag{5}$$

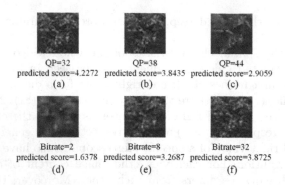

QP=32   QP=38   QP=44
predicted score=4.2272   predicted score=3.8435 predicted score=2.9059
(a)   (b)   (c)

Bitrate=2   Bitrate=8   Bitrate=32
predicted score=1.6378   predicted score=3.2687 predicted score=3.8725
(d)   (e)   (f)

Fig. 4. Illustration of some patch examples for left view and the predicted quality scores for the corresponding image patch pairs. The unit for bitrate is Mb/s.

where Q_j represents the predicted quality for the jth frame, and $j = 1, 2, \ldots, m$. In our experiment, there are totally nine frames for a given video (i.e. $m = 9$). Then, the patches of extracted frames are exploited to train the EDN model.

3   Experimental Results and Analysis

3.1   Dataset Description and Evaluation Methodology

The NAMA3DS1-COSPAD1 dataset [23] is used in our experiments. This dataset has 10 reference stereoscopic video sequences with a variety of texture, structure, temporal, and depth information. The video sequences have a resolution of 1920 × 1080 pixels. The frame rate of these videos is 25 fps. The duration of 9 reference videos is 16 seconds, and the remaining one reference video has 13 s. In other words, each stereoscopic video has either 400 or 325 frames. Additionally, distortions in the dataset contain H.264 video compression artifacts and JPEG2000 still image compression artifacts. These artifacts are introduced symmetrically to left and right view videos in the NAMA3DS1-COSPAD1 dataset. The H.264 video compression artifacts are produced using JM reference software by varying the quantization parameter (QP) setting as 32, 38, and 44. Moreover, the JPEG2000 still image compression artifacts, which use 2, 8, 16, and 32 Mb/s, are applied to video frames. We can predict the stereoscopic video quality from its pivotal frames and corresponding patch pairs. The ground truth is the mean opinion score (MOS) obtained by human subjective ratings. Two commonly used criteria including Spearman rank-order correlation coefficient (SROCC) and Pearson linear correlation coefficient (PLCC) are adopted for quantitative performance comparison. SROCC is evaluated according to the rank of scores which measures the prediction monotonicity. PLCC is used to evaluate the prediction accuracy. Higher correlation coefficients means better correlation with human quality judgement.

Table 1. Comparison of 2D IQA and 3D VQA state-of-the-art metrics on NAMA3DS1-COSPAD1 stereoscopic video dataset.

Metrics	SROCC	PLCC
PSNR	0.6470	0.6699
SSIM	0.7492	0.7664
PQM	0.6006	0.6340
PHVS-3D	0.5146	0.5480
SFD	0.5896	0.5965
3D-STS	0.6214	0.6417
SJND-SVA	0.6229	0.6503
Optical flow-based method	0.8552	0.8949
MNSVQM	0.8394	0.8611
BSVQE	0.9086	0.9239
NR CNN method	0.8570	0.8926
Proposed EDN	**0.9334**	**0.9301**

3.2   Performance Comparison

The performance of the proposed metric is conducted on NAMA3DS1-COSPAD1 dataset. Due to the lack of publicly available distorted stereoscopic video datasets, we limit our evaluation on the NAMA3DS1-COSPAD1 dataset. We report the SROCC and PLCC performance values between the obtained quality scores through different metrics and the ground truth MOS for stereoscopic videos. Meanwhile, higher SROCC and PLCC performance values indicate the better agreement with the perceptual quality scores rated by viewers.

Our proposed EDN is compared with two classic 2D IQA metrics which are peak signal to noise ratio (PSNR) and structural similarity (SSIM) [25]. In addition, several 3D VQA algorithms are performed including PQM [9], PHVS-3D [8], SFD [18], 3D-STS [4], SJND-SVA [21], an optical flow-based method [26], MNSVQM [7], and BSVQE [2]. We also employ another proposed CNN architecture for no-reference image quality assessment (i.e. NR CNN method) [10] and feed left and right view image patches to train the network separately. Then, the predicted quality score can be computed by averaging left and right view scores.

In the experiment, we randomly choose 80% of the reference videos for training and the other 20% for testing. We then obtain 331040 training patch pairs and 77760 testing patch pairs by preprocessing stereoscopic videos. We conduct 100000 iterations in patch pair-wise training stage, and the batch size is 64. The results are shown in Table 1. From Table 1, we can see that our EDN framework outperforms both 2D IQA and 3D VQA state-of-the-art metrics. Furthermore, some patch examples for left view and the predicted quality scores for the corresponding image patch pairs using our proposed EDN can be seen in Fig. 4. From

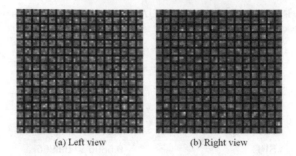

(a) Left view   (b) Right view

Fig. 5. Visualized learned kernels in the first convolutional layer. (a) Left view; (b) Right view.

Fig. 4, we can find that the patches with different quality scores, i.e. figures (a-f), are distinguished well through the proposed EDN framework, both for H.264 video compression artifacts and JPEG2000 still image compression artifacts.

The quality scores of patch pairs for each stereoscopic video can be predicted through our trained model, which reaches around 30 fps and is suitable for real-time applications. Therefore, our NR-SVQA trained model is accurate, real-time, and adaptable to new video content.

3.3   Visualize Learned Kernel

In order to analyze the characteristics of the trained model regarding to extracting features from the deep learning structure, which is generally taken as an unknown black-box, we visualize the learned kernels in the first convolutional layer.

Figure 5 depicts the 256 pairs of each 3×3 kernel at the first convolutional layer both for the left view and right view. We can observe that the left and right views of training patches have different spatial texture characteristics since the 3D depth perception exists between these two views and the stereoscopic pairs are fused by the human eye based on spatial correlation in 3D perception [20].

4   Conclusion and Future Work

In this paper, we propose the first study of an end-to-end deep CNN based NR-SVQA framework. The lack of training data for stereoscopic videos is effectively resolved by the stereoscopic video preprocessing method and the spatiotemporal pooling strategy. In addition, the content diversity and category balance of training data are also ensured through the proposed three-step approach to extract pivotal frames from input left and right view videos. Experimental results demonstrate that the proposed method outperforms state-of-the-art approaches. In the future research, the local ground truth target generated for each training patch needs to tackle the non-stationary characteristic of perceptually spatiotemporal quality for stereoscopic videos and more HVS characteristics such as attention mechanism should be taken into consideration.

Acknowledgement. This work was supported in part by the National Key Research and Development Program of China under Grant No. 2016YFC0801001, the National Program on Key Basic Research Projects (973 Program) under Grant 2015CB351803, NSFC under Grant 61571413, 61632001, 61390514, and Intel ICRI MNC.

References

1. Bosse, S., Maniry, D., Wiegand, T., Samek, W.: A deep neural network for image quality assessment. In: 2016 IEEE International Conference on Image Processing (ICIP), pp. 3773–3777. IEEE (2016)
2. Chen, Z., Zhou, W., Li, W.: Blind stereoscopic video quality assessment: from depth perception to overall experience. IEEE Trans. Image Process. **27**(2), 721–734 (2018)
3. Girshick, R., Donahue, J., Darrell, T., Malik, J.: Rich feature hierarchies for accurate object detection and semantic segmentation. In: Proceedings of the IEEE Conference on Computer Vision and Pattern Recognition, pp. 580–587 (2014)
4. Han, J., Jiang, T., Ma, S.: Stereoscopic video quality assessment model based on spatial-temporal structural information. In: 2012 IEEE Visual Communications and Image Processing (VCIP), pp. 1–6. IEEE (2012)
5. Hinton, G.E., Osindero, S., Teh, Y.W.: A fast learning algorithm for deep belief nets. Neural Comput. **18**(7), 1527–1554 (2006)
6. Hou, W., Gao, X., Tao, D., Li, X.: Blind image quality assessment via deep learning. IEEE Trans. Neural Netw. Learn. Syst. **26**(6), 1275–1286 (2015)
7. Jiang, G., Liu, S., Yu, M., Shao, F., Peng, Z., Chen, F.: No reference stereo video quality assessment based on motion feature in tensor decomposition domain. J. Vis. Commun. Image Represent. **50**, 247–262 (2018)
8. Jin, L., Boev, A., Gotchev, A., Egiazarian, K.: 3D-DCT based perceptual quality assessment of stereo video. In: 2011 18th IEEE International Conference on Image Processing, pp. 2521–2524. IEEE (2011)
9. Joveluro, P., Malekmohamadi, H., Fernando, W.C., Kondoz, A.: Perceptual video quality metric for 3D video quality assessment. In: 2010 3DTV-Conference: The True Vision-Capture, Transmission and Display of 3D Video, pp. 1–4. IEEE (2010)
10. Kang, L., Ye, P., Li, Y., Doermann, D.: Convolutional neural networks for no-reference image quality assessment. In: Proceedings of the IEEE Conference on Computer Vision and Pattern Recognition, pp. 1733–1740 (2014)
11. Kavukcuoglu, K., Sermanet, P., Boureau, Y.L., Gregor, K., Mathieu, M., Cun, Y.L.: Learning convolutional feature hierarchies for visual recognition. In: Advances in Neural Information Processing Systems, pp. 1090–1098 (2010)
12. Kim, J., Lee, S.: Fully deep blind image quality predictor. IEEE J. Sel. Top. Signal Process. **11**(1), 206–220 (2017)
13. Kim, J., Zeng, H., Ghadiyaram, D., Lee, S., Zhang, L., Bovik, A.C.: Deep convolutional neural models for picture quality prediction. IEEE Signal Process. Mag. **34**, 130–141 (2017)
14. Krizhevsky, A., Sutskever, I., Hinton, G.E.: Imagenet classification with deep convolutional neural networks. In: Advances in Neural Information Processing Systems, pp. 1097–1105 (2012)

15. Li, Y., et al.: No-reference video quality assessment with 3D shearlet transform and convolutional neural networks. IEEE Trans. Circuits Syst. Video Technol. **26**(6), 1044–1057 (2016)

16. Li, Y., Po, L.M., Feng, L., Yuan, F.: No-reference image quality assessment with deep convolutional neural networks. In: 2016 IEEE International Conference on Digital Signal Processing (DSP), pp. 685–689. IEEE (2016)

17. Li, Y., et al.: No-reference image quality assessment with shearlet transform and deep neural networks. Neurocomputing **154**, 94–109 (2015)

18. Lu, F., Wang, H., Ji, X., Er, G.: Quality assessment of 3D asymmetric view coding using spatial frequency dominance model. In: 2009 3DTV Conference: The True Vision-Capture, Transmission and Display of 3D Video, pp. 1–4. IEEE (2009)

19. Lv, Y., Yu, M., Jiang, G., Shao, F., Peng, Z., Chen, F.: No-reference stereoscopic image quality assessment using binocular self-similarity and deep neural network. Signal Process.: Image Commun. **47**, 346–357 (2016)

20. Parker, A.J.: Binocular depth perception and the cerebral cortex. Nat. Rev. Neurosci. **8**(5), 379 (2007)

21. Qi, F., Zhao, D., Fan, X., Jiang, T.: Stereoscopic video quality assessment based on visual attention and just-noticeable difference models. Signal, Image Video Process. **10**(4), 737–744 (2016)

22. Rec, I.: P. 910: Subjective video quality assessment methods for multimedia applications. International Telecommunication Union, Geneva (2008)

23. Urvoy, M., et al.: NAMA3DS1-COSPAD1: subjective video quality assessment database on coding conditions introducing freely available high quality 3D stereoscopic sequences. In: 2012 Fourth International Workshop on Quality of Multimedia Experience (QoMEX), pp. 109–114. IEEE (2012)

24. Vega, M.T., Mocanu, D.C., Famaey, J., Stavrou, S., Liotta, A.: Deep learning for quality assessment in live video streaming. IEEE Signal Process. Lett. **24**(6), 736–740 (2017)

25. Wang, Z., Bovik, A.C., Sheikh, H.R., Simoncelli, E.P.: Image quality assessment: from error visibility to structural similarity. IEEE Trans. Image Process. **13**(4), 600–612 (2004)

26. Yang, J., Wang, H., Lu, W., Li, B., Badiid, A., Meng, Q.: A no-reference optical flow-based quality evaluator for stereoscopic videos in curvelet domain. Inf. Sci. **414**, 133–146 (2017)

27. Zhang, W., Qu, C., Ma, L., Guan, J., Huang, R.: Learning structure of stereoscopic image for no-reference quality assessment with convolutional neural network. Pattern Recognit. **59**, 176–187 (2016)

Fast and Robust 3D Numerical Method for Coronary Artery Vesselness Diffusion from CTA Images

Hengfei Cui[1,2(✉)]

[1] Shaanxi Key Lab of Speech and Image Information Processing (SAIIP),
School of Computer Science, Northwestern Polytechnical University,
Xi'an 710072, China
hfcui@nwpu.edu.cn
[2] Centre for Multidisciplinary Convergence Computing (CMCC),
School of Computer Science, Northwestern Polytechnical University,
Xi'an 710072, China

Abstract. Optimized anisotropic diffusion is commonly used in medical imaging for the purpose of reducing background noise and tissues and enhancing the vessel structures of interest. In this work, a hybrid diffusion tensor is developed, which integrates Frangi's vesselness measure with a continuous switch, suitable for filtering both tubular and planar image structures. Besides, a new 3D diffusion discretization scheme is proposed, in which we apply Gaussian kernel decomposition for computing image derivatives. This scheme is rotational invariant and shows good isotropic filtering properties on both synthetic and real Computed Tomography Angiography (CTA) data. In addition, segmentation approach is performed over filtered images obtained by using different schemes. Our method is proved to give better segmentation result and more thin branches can be detected. In conclusion, the proposed method should garner wider clinical applicability in Computed Tomography Coronary Angiography (CTCA) images preprocessing.

Keywords: Frangi's filter · Vesselness measure · Diffusion tensor Anisotropic diffusion

1 Introduction

Computed Tomography (CT) imaging is becoming an increasingly utilized imaging modality in quantitative evaluation of coronary artery stenosis [3,4]. Numerous automatic vessel segmentation methods from literature have been tested. These methods are found to fail on CT scans because of higher noise, missing ridges and less contrast between coronary arteries and surrounding structures [9]. Therefore, as a common prerequisite of segmentation and analysing of coronary arteries, strong and durable vessel enhancement is essential in biomedical

© Springer Nature Switzerland AG 2018
R. Hong et al. (Eds.): PCM 2018, LNCS 11166, pp. 493–502, 2018.
https://doi.org/10.1007/978-3-030-00764-5_45

imaging. Currently, a large number of medical imaging techniques has been propounded for 3D vessel structures enhancement and extraction [1,2]. One way of discriminating vascular structures from other structures that have similar CT values is to enhance tubular structures by use of a multi-scale Hessian filter [7,8]. However, due to neighboring bony structures and plaques, direct application of line filters, such as the Hessian filter, will cause severe artifacts, i.e. false positive responses, at step edges (object boundaries). Besides, there is one principle limitation of tensors, they cannot model complex image structures only symmetric spherical shapes. Using the structure tensor to find the orientations is more robust against noise than the Hessian used by Frangi, but it fails to give the best cylindrical structure detection.

In the present context, there are mainly three advantages of vessel enhancement diffusion (VED). Enhanced continuity of vessels is important for vessel and centerline tracking or vessel segmentation algorithms based on region growing. VED is able to overcome significant intensity drops due to the strong anisotropic diffusion along the vessel. Moreover, VED is able to improve vessel separation in MIPs, since the intensity distribution of each filtered vessel becomes more homogeneous while the gap between vessels becomes smaller. In clinical case, VED can be used to suppress non-vessel structures before vessel segmentation.

Despite such a wide range of approaches, the robust enhancement of 3D curvilinear structures remains challenging due to the intensity variations along the 3D curve. A medical image is often assumed to have piecewise smooth regions with oscillatory noise, separated by sharp edges. There are many methods available in the literature to denoise such an image [3], in this work we focus on vesselness enhancing diffusion filtering. The described method to find the image structure orientations is comparable to Frangi's vesselness filter [6]. Using the structure tensor to find the orientations is more robust against noise than the Hessian used by Frangi, but a combination of both methods gives the best cylindrical structure detection [7].

We focus on improving the CT image quality by filtering the data to remove noise and enhance the edges, with smoothing which adapts to the underlying image structure to preserve edges. We introduce 3D nonlinear anisotropic diffusion filtering which is based on the 2D coherence enhancing diffusion introduced by Weickert. The diffusion tensor in this method is oriented using an image structure tensor, with a kernel which is elongated in the direction of the underlying image edges. The focus of this paper are the diffusion tensor and discretization schemes of the anisotropic diffusion. We will develop a new diffusion tensor in which the Frangi's vesselness measure is incorporated. Furthermore, we evaluate the performance of the standard discretization scheme and the optimized scheme of Kroon [7], and introduce a new scheme in which the image derivatives are computed by Gaussian kernel decomposition. The remaining part of this paper is organized as follows. In Sect. 2 we introduce the new vesselness enhancing diffusion tensor and discretization scheme. The results on synthetic and real CT images are presented in Sect. 3. Finally, conclusions are given in Sect. 4.

2   Methods

2.1   Diffusion Filtering

The diffusion equation is commonly given by

$$\frac{\partial u}{\partial t} = \nabla \cdot (\mathbf{D}\nabla u) \tag{1}$$

where $u(u = u(t, x, y, z))$ is the image, x, y, z the voxel coordinates and t the diffusion time. The eigenvectors of the diffusion tensor $\mathbf{D}$ are set equal to the eigenvectors $\mathbf{v_1}, \mathbf{v_1}, \mathbf{v_1}$ with $\mathbf{v_1} = [v_{11}, v_{12}, v_{13}]$ of the structure tensor:

$$\mathbf{D} = \begin{bmatrix} D_{11} & D_{12} & D_{13} \\ D_{12} & D_{22} & D_{23} \\ D_{13} & D_{23} & D_{33} \end{bmatrix} \quad with \quad D_{ij} = \sum_{n=1...3} \lambda_n v_{ni} v_{nj} \tag{2}$$

The eigenvalues are calculated with a from 2D to 3D extended equation of Weickert [1]. Extension from 2D to 3D gives two possibilities. The first one is edge enhancing diffusion (EED):

$$\lambda_1 = 1 \tag{3}$$
$$\lambda_2 = 1 \tag{4}$$
$$\lambda_3 = 1 - \exp\left(\frac{-3.31488}{(GMS/\lambda_e^2)^4}\right) \tag{5}$$

where GMS is the gradient magnitude square $GMS = ||\nabla_\theta L||^2$ and λ_e planar structure contrast parameter. In this way, diffusion perpendicular to an edge is inhibited, but along the edge diffusion still takes place.

Another one is coherence enhancing diffusion (CED):

$$\lambda_1 = \alpha - (1 - \alpha) \exp\left(-\frac{C}{k}\right) \tag{6}$$
$$\lambda_2 = \alpha \tag{7}$$
$$\lambda_3 = \alpha \tag{8}$$

with $\alpha \in (0, 1)$ a global smoothing constant, and C the edge enhancing smoothing constant. The described edge enhancing diffusion filtering is repeated in an iterative way. The number of iterations is set by the user, and will determine the amount of smoothing.

Our goal is to develop a vessel enhancement diffusion tensor D that satisfies these three requirements and enables multiscale approaches to vessel analysis. In the literature, smoothed vesselness function $\mathcal{V}$ has been used to construct several scale spaces based on the eigensystem of the Hessian. In this work, a novel anisotropic diffusion filter is developed, which uses the vesselness measure to determine the diffusion strength to be applied for each pixel. More specifically, the diffusion tensor D is defined for the purpose of steering main diffusion along the vessel and inhibiting diffusion perpendicular to the vessel. The diffusion tensor is therefore given by:

$$D \triangleq Q\Lambda'Q^T \tag{9}$$

where Q represents the eigenvectors of the Hessian $\mathcal{H}$, and Λ' is a diagonal matrix whose diagonal entries are given by:

$$\lambda_1' \triangleq 1 + (\omega - 1) \cdot \mathcal{V}^{\frac{1}{s}} \tag{10}$$

$$\lambda_2' \triangleq 1 + (\epsilon - 1) \cdot \mathcal{V}^{\frac{1}{s}} \tag{11}$$

$$\lambda_3' \triangleq 1 + (\epsilon - 1) \cdot \mathcal{V}^{\frac{1}{s}} \tag{12}$$

with $\mathcal{V} \in [0, 1]$ and parameters $\omega > \epsilon$, $\epsilon > 0$ and $s \in \mathbb{R}^+$. Normally, ω is used to control the diffusion strength, which should be a large value (larger than one). To ensure the tensor's positive definiteness, ϵ should be a small positive value. In addition, small values of time discretization step should be chosen to prevent instabilities.

2.2 Numerical Discretization

The original CTA image I is a discrete function thus the equations must be discretized. First we describe derivate discretization, secondly different diffusion schemes. The common way is to replace spatial differences with central differences [5] and use $\frac{\partial u}{\partial t}$ for a forward difference approximation. For the standard discretization of the divergence operator central differences are used:

$$\partial_y(D_{12}(\partial_x u)) = \frac{1}{2}\left(D_{12(i,j+1,k)}\frac{u_{(i+1,j+1,k)} - u_{(i-1,j+1,k)}}{2} \right. \\ \left. -D_{12(i,j-1,k)}\frac{u_{(i+1,j-1,k)} - u_{(i-1,j-1,k)}}{2}\right) \tag{13}$$

Thus convolution with a spatially and temporally varying mask (stencil) gives the diffusion update.

Kroon has proposed a new rotational invariant numerical discretization scheme. First, divergence operator can be re-written using the product rule:

$$\nabla \cdot (\mathbf{D}\nabla u) = div(\mathbf{D})\nabla u + trace(\mathbf{D}(\nabla\nabla^T u)) \tag{14}$$

Divergence part of the equation:

$$div(\mathbf{D})\nabla u = (\partial_x u)(\partial_x D_{11} + \partial_y D_{12} + \partial_z D_{13}) \\ + (\partial_y u)(\partial_x D_{12} + \partial_y D_{22} + \partial_z D_{23}) \\ + (\partial_z u)(\partial_x D_{13} + \partial_y D_{23} + \partial_z D_{33}) \tag{15}$$

Hessian part of the equation:

$$trace(\mathbf{D}(\nabla\nabla^T u)) = (\partial_{xx} u)D_{11} + (\partial_{yy} u)D_{22} + (\partial_{zz} u)D_{33} \\ + 2(\partial_{xy} u)D_{12} + 2(\partial_{xz} u)D_{13} + 2(\partial_{yz} u)D_{23} \tag{16}$$

However, Kroon's optimized scheme is based on numerical optimization, which has several shortcomings. First, the target minimization cost function $e_f(\mathbf{p})$ must be carefully selected. For example, two filtering performance terms e_f and e_g are defined, which represent the differences between the filtered images and an image with circles of varying spatial frequencies and a least squares fitted Gaussian kernel, respectively. Second, the final kernel values vary depending on different optimization methods and parameters, which makes the scheme non-reproducible to all the datasets.

A 26-nearest neighbours discretization of the Laplacian operator can be used in this work:

$$u_{i,j,k}^{t+1} = u_{i,j,k}^t + \lambda[c_1 \cdot T_1 \cdot \delta_1 u + c_2 \cdot T_2 \cdot \delta_2 u + ... + c_{26} \cdot T_{26} \cdot \delta_{26} u] \qquad (17)$$

where the subscripts 1–26 are used for the 26-nearest neighbors. The symbol δ indicates the nearest-neighbor difference, and T_i indicates the diffusion tensor element. The conduction coefficients are defined as $c_i = 1/d_{0,i}^2$, where $d_{0,i}$ represents the Euclidean distance between central voxel and its nearest neighbor d_i. Only a small time step is allowed for the scheme to be stable, i.e., $0 \le \delta t \le 1/4$.

3   Results

The results in this section are divided into two parts. The first is to evaluate the proposed diffusion tensor, the discretization scheme and the influence of the involved parameters. The second is filtering the CTA images from real patients, to evaluate the performance on coronary arteries.

3.1   Evaluation

In this section, quantitative analysis of the accuracy of the proposed algorithm was performed. First, we evaluate the performance of the proposed scheme regarding to noise removal, compared to the standard scheme and Kroon's optimized scheme. A 3D spherical image $I_{perfect}$ ($65 \times 65 \times 65$) with varying spatial frequencies was generated using $u = sin(x^2 + y^2 + z^2)$, to test rotational invariant performance. Besides, uniform noise is added to $I_{perfect}$ to compare filtering performance. We use a diffusion time of 5 s and small time step of 0.1 s, the diffusion parameters are chosen: $c_1 = 0.001$, $c_2 = 10^{-10}$, $\rho = 1$, $\sigma = 1$. The eigenvalues are defined as CED type.

Figure 1 shows the image results and difference between a least squares fitted Gaussian 2D function and the diffusion result. Ideal uniform diffusion is equal to Gaussian filtering, thus the standard diffusion and the optimized scheme perform well. In addition to visual inspection, quantitative analysis is also conducted to compare the performance of the 3D diffusion schemes. We may use the generated circle image as the ground truth image for evaluating the accuracy of our algorithms. We start with u_{noise} as initial image ($t = 0$) and perform one iteration of our algorithms. Mean square root of deviations per pixel between the numerical

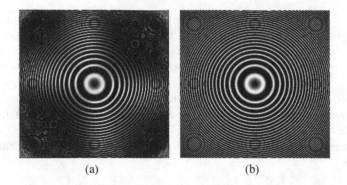

Fig. 1. Filtered results. (a) Standard scheme. (b) Novel scheme.

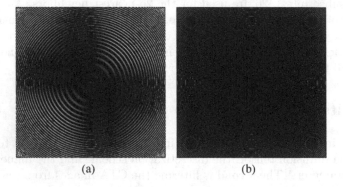

Fig. 2. Error images. (a) Standard scheme. (b) Novel scheme.

solution and the analytical solution is plotted in Fig. 2 for varying wave numbers and different explicit schemes.

The summed squared pixel distance between the ground truth spherical image and original diffusion results is used as a performance value. A steep edge contains high frequencies which will be removed by the low pass filter, resulting in a large pixel distance. In uniform regions high frequency noise will also be removed, thus a large pixel distance is a sign of noise which is not removed by the diffusion filtering. We calculate the smoothing pixel distance values for the edge pixels and for the uniform regions. The results are shown in Table 1.

In order to explore the maximum time step for which the proposed stencil is stable, we compute the image variance for each value of step size. A constant decreasing image variance is one of the main principles of iterative noise filtering. By looking at the image variance while filtering with a number of diffusion step sizes we will find the maximum time step for which our stencil is stable, Fig. 3. The maximum stable step size is found to be 1s, with a higher value small regions with very high and very low values start to occur.

Table 1. A comparison of errors between ground truth image and filtered image for different schemes.

	L_1	L_2	L_∞	Time (s)
Raw	0.461	0.623	2.087	0.293
Standard	0.3584	0.5551	1.8860	0.1240
Optimized	0.2419	0.3048	1.6159	0.1681
New	0.1269	0.1583	0.7057	0.1041

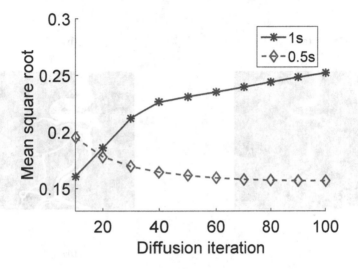

Fig. 3. Stability test of the proposed scheme.

Table 2. Size, resolution and processing time of each dataset.

	Size	Resolution (mm^3)	Time (min)
Dataset 22	$512 \times 512 \times 211$	$0.283 \times 0.283 \times 0.700$	25
Dataset 23	$512 \times 512 \times 206$	$0.328 \times 0.328 \times 0.750$	23
Dataset 27	$512 \times 512 \times 54$	$0.391 \times 0.391 \times 2.000$	19
Dataset 29	$512 \times 512 \times 266$	$0.305 \times 0.305 \times 0.450$	27

3.2   CT Images Diffusion

In this study, the novel numerical vesselness diffusion scheme was applied into 4 cardiac CTA datasets acquired by a MDCT scanner with ECG gating in DICOM format. All the experiments were performed on a HP Z800 Windows Server. Table 2 summarizes the size, resolution and computation time of each DICOM dataset. The parameters of all the schemes are selected as: time step 0.5 s total diffusion time 15 s, $c_1 = 10^{-3}$, $c_2 = 10^{-5}$, $\rho = 2$, $\sigma = 1$. The eigenvalues are defined as VED type. The average computational time per dataset is

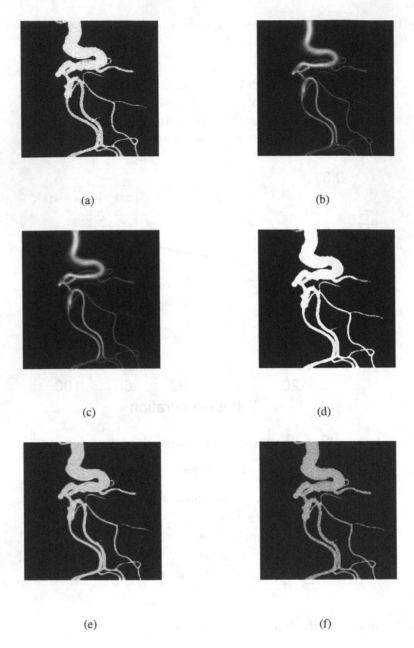

(a)   (b)

(c)   (d)

(e)   (f)

Fig. 4. MIP displays of different filtered images.

approximately 23.5 min, which demonstrates our method's clinical potential as a real-time cardiac assessment tool.

Figure 4 shows the comparison of different diffusion schemes on 3D vessel structures. Gaussian noise is added to the original vessel segmentation result

(Fig. 4(a)). Figure 4(b) presents the maximal intensity projection (MIP) by Frangi's vesselness filter. By applying the multi-scale vesselness diffusion, the vessel structures can be detected. However, some thin vessel branches are not clearly visualised. This is because thin vessels with width less than or equal to one voxel may be smoothed out during the diffusion process. The enhanced Frangi's vesselness image is shown in Fig. 4(c). More thin vessels are smoothed out since Frangi's filter is not capable of preserving thin vessels. The filtered result by using the standard scheme is presented in Fig. 4(e). The Gaussian noise is smoothed out in vessel regions. However, the standard scheme may induce some shrinkage effect, which smoothes out some thin vessels (see arrow). Figure 4(f) gives the result filtered by Kroon's optimized scheme. The scheme is even not able to smooth out the noise in vessel regions. Besides, the scheme takes two times longer computation time than the standard scheme. The MIP of the 3D vessel by using the proposed scheme is given in Fig. 4(d). It is observed that the new diffusion method is capable of enhancing the vessel structures and removing background noises.

4 Conclusion

We developed an efficient and robust vesselness diffusion scheme for vessel enhancement and background noise reduction. A new diffusion filter incorporated with Frangi's vesselness measure is proposed and the technique of Gaussian gradient separation is applied when computing image derivatives. The proposed 3D anisotropic diffusion scheme shows better vessel enhancement and noise reduction in our synthetic and CTA data, compared to the standard and Kroon's optimized scheme. Filtering is Gaussian in uniform image regions without checkerboard artifacts. The results show that the better edge preservation also causes high noise structures to be preserved. Compare to traditional methods like vesselness and neuriteness, our proposed method resists to noisy background and also strong to enhance curve-like features in the CT images. Based on the filtered CT images, the proposed method is able to give better coronary artery segmentation results, which include more thin vessel branches and exclude less noisy artefacts. Given the speed and accuracy, the proposed diffusion scheme may garner wider clinical applicability in CTCA images pre-processing.

Acknowledgment. The study is supported by the National Natural Science Foundation of China under Grants 61471297, the China Postdoctoral Science Foundation under Grant 2017M623245 and the Fundamental Research Funds for the Central Universities under Grants 3102018zy031. We are very grateful to the National Heart Centre Singapore for the DICOM datasets.

Conflict of Interest
The authors declare that they have no conflict of interest.

References

1. Arbab-Zadeh, A., Hoe, J.: Quantification of coronary arterial stenoses by multidetector CT angiography in comparison with conventional angiography: methods, caveats, and implications. JACC. Cardiovasc. Imaging **4**, 191–202 (2011)
2. Butcher, J.C.: The Numerical Analysis of Ordinary Differential Equations: Runge-Kutta and General Linear Methods. Wiley-Interscience, Hoboken (1987)
3. Cai, W., Harris, G.J., Yoshida, H.: Computation of vesselness in CTA images for fast and interactive vessel segmentation. Int. J. Image Graph. **7**, 159–176 (2007)
4. Campeau, L., Corbara, F., Crochet, D., Petitclerc, R.: Left main coronary artery stenosis: the influence of aortocoronary bypass surgery on survival. Circulation **57**, 1111–1115 (1978)
5. Cornea, N.D., Silver, D., Min, P.: Curve-skeleton properties, applications, and algorithms. IEEE Trans. Vis. Comput. Graph. **3**, 530–548 (2007)
6. Cui, H., et al.: Coronary artery segmentation via hessian filter and curve-skeleton extraction. In: 2014 IEEE Conference on Biomedical Engineering and Sciences (IECBES), pp. 93–98. IEEE (2014)
7. Dodge, J., Brown, B.G., Bolson, E.L., Dodge, H.T.: Lumen diameter of normal human coronary arteries. Influence of age, sex, anatomic variation, and left ventricular hypertrophy or dilation. Circulation **86**, 232–246 (1992)
8. Hassouna, M.S., Farag, A., et al.: Variational curve skeletons using gradient vector flow. IEEE Trans. Pattern Anal. Mach. Intell. **31**, 2257–2274 (2009)
9. Kitslaar, P.H., Frenay, M., Oost, E., Dijkstra, J., Stoel, B., Reiber, J.: Connected component and morpholgy based extraction of arterial centerlines of the heart (cocomobeach). In: The MIDAS Journal: MICCAI Workshop-Grand Challenge Coronary Artery Tracking (2008)

Multi-view Viewpoint Assessment
for Architectural Photos

Jingwu He[1], Linbo Wang[2], Wanqing Zhao[1], Yang Zhang[1], Xu Han[1],
Chenghao Guo[3], and Yanwen Guo[1,3(✉)]

[1] The National Key Lab for Novel Software Technology, Nanjing University,
Nanjing 210023, China
ywguo@nju.edu.cn
[2] The MOE Key Laboratory of Intelligent Computing and Signal Processing,
Anhui University, HeFei 230039, China
[3] Science and Technology on Information Systems Engineering Laboratory,
Nanjing 210007, China

Abstract. This paper proposes a robust method of viewpoint assessment for taking a good photograph of architecture. Unlike the conventional works devoted to assessing the aesthetic of a photograph mainly relying on the image features, both the image and geometric features are extracted from the architecture photos in our method. Furthermore, we explore the mutual knowledge between these two aspects of features with multi-view learning. With the learner trained by multi-view learning, the viewpoint goodness of architecture photograph can be assessed by either aspect of the features. Experiments suggest that the multi-view learning with kernel canonical correlation analysis achieves superior performance over using solely traditional image features. With the help of multi-view learning, we can harness the geometric cues with image features effectively.

Keywords: Viewpoint assessment · Multi-view learning
Image features

1 Introduction

During last several years, cell phones have a significant improvement of the camera sensor quality, which enables ordinary people to take photos more and more easily. When traveling to a resort, people usually want to take photos with architecture as the background for memory. However, how to take a visually-pleasing architecture photo is still a common puzzle for novice photographers. In past decades, many visual image features have been explored to formulate the photo aesthetic problem ranging from image content [1–3], simplicity [4], and basic photographic rules [5] to high level photographic compositional rules [6,7] such as the rule of thirds. Image quality measures that incorporate aesthetics as well as other structure based and statistical models are also developed in [8]. Furthermore, deep learning approaches are also employed to learn image aesthetic

© Springer Nature Switzerland AG 2018
R. Hong et al. (Eds.): PCM 2018, LNCS 11166, pp. 503–513, 2018.
https://doi.org/10.1007/978-3-030-00764-5_46

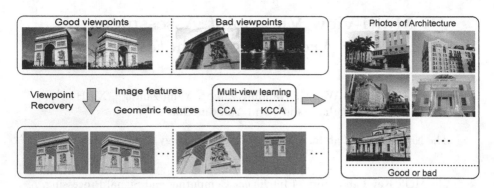

Fig. 1. The systematic overview of our multi-view learning framework for viewpoint assessment.

representations [9,10]. Geometric features, such as the gaussian curvature [11], viewpoint entropy [12], mesh saliency [13], and surface regions of interest [14] are widely used for the viewpoint assessment and selection of a 3D model. However, these features cannot be extracted accurately and efficiently with only the architecture photos available directly, and it is not suitable to assess the viewpoint quality of these photos. We observe that photographers always underline selecting the shooting place to achieve a good geometric layout when taking a photograph. This inspires us to assess the viewpoint of architecture by considering both image and geometric features. Moreover, the image features can be extracted more accurately and efficiently than the geometric features, and we intend to propose a robust method to assess the viewpoint with image features solely. To achieve this, we explore the mutual knowledge embedded in the image features and the geometric features. With the mutual knowledge we learned, viewpoint quality of the photo can be assessed when only image features are available.

For the problem of viewpoint assessment of an architecture photo, image and geometric features form a two-view feature space. It is suitable to employ multi-view learning such as canonical correlation analysis (CCA) [15] and kernel canonical correlation analysis (KCCA) [16]. The experimental results suggest that considering the geometric features is helpful to take an architecture photograph with good viewpoint.

A schematic overview of our system is given in Fig. 1. Fifteen world famous architectures and their 3D models are collected from the Internet. Each architecture contains hundreds of photographs. To get the geometric features, the viewpoint of each image with respect to its 3D model is estimated by a semi-automatic method, which is effective and efficient. By this way, both the 2D image and the 3D geometric features can be extracted to form the two-view feature space (Sect. 2.1). Multi-view learning is employed for the assessment task with the features extracted from both of the aspects (Sect. 2.2). Comparison experiments with different features and different methods on two datasets are conducted in Sect. 3. Section 4 concludes the whole paper.

In summary, the main contributions of this paper are three-fold. First, we implement multi-view learning to explore the mutual knowledge between the image and geometric features for viewpoint assessment. Second, with our multi-view learner, we can assess the viewpoint goodness with image features solely. At last, the experimental results suggest that our multi-view learner achieves superior performance over using solely the 2D image or 3D geometric features.

2   Learning Good Viewpoint Quality

Given an image of architecture I_q, the purpose of our method is to predict its viewpoint goodness y_q. More specifically, for each image I, the image features V and geometric features G are extracted for the multi-view learning.

2.1   Two-View Features Setup

The sense of viewpoint can be depicted more efficiently in geometric aspect than image aspect. To this end, we explore the relationship between 3D geometric and 2D image features and set up a robust viewpoint assessment system. Before stepping into the feature extraction and two-view features correlation analysis, we estimate the viewpoint of each photo for extracting the geometric features.

Viewpoint Recovery. We collect 15 world famous architecture models with their corresponding photographs. For each of architectures, the collected photos are fed to the algorithm of Structure from Motion (SfM) to reconstruct the point cloud model, and to get the viewpoints corresponding to the point cloud. However, the point cloud is not suitable for extracting the geometric features, because it is partial and coarse. To relieve this puzzle, the model registration is processed to get viewpoints with respect to the mesh model. After that, the point cloud model and the mesh model can be coincided. Meanwhile, we can get the viewpoint corresponding to the mesh model for each photo. The registration process is demonstrated in Fig. 2.

Image Features. Considering image appearance, local structure, composition, etc., various image features are extracted including color, brightness, contrast, vanishing lines, and photo composition. More specifically, it includes:

- $\mathbf{v}_{vl}$: *Vanishing lines.* Vanishing lines are important visual features and can be used to describe the geometry property about the viewpoint. $\mathbf{v}_{vl}$ is defined as the three angles between every two of them.
- $\mathbf{v}_{ds}$: *Directions of lines.* The histogram of lines directions are various for different viewpoints. Such that the directions of lines are mostly vertical and horizontal when photographing the architecture from the front side.
- $\mathbf{v}_{HOG}$: *HOG (Histogram of oriented gradient).* HOG describes the local histograms of gradient directions. It is various for different viewpoints.

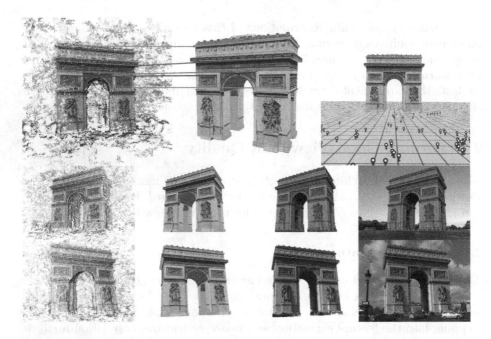

Fig. 2. Viewpoint recovery with model registration. The first column and the second column are the point cloud model and the mesh model, respectively. We also show the camera locations with respect to the mesh model on the top row of the third column. In addition, the projection results of point cloud model and the mesh model, the foreground of the architecture in the corresponding photo, and the original photo are shown on last two rows.

– $\mathbf{v}_{color}$: *Color entropy and distribution.* Each of the RGB channels is quantized into 8 values to minimize computation and reduce noise. Then we can get a histogram with $512 = 8^3$ bins. These two features are computed as follows:

$$c_e = \sum P_i \log P_i, \quad \text{where } P_i = \frac{\mathbf{H}(i)}{\sum \mathbf{H}(i)},$$
$$c_d = 1 - \sum P_i^2, \quad \text{where } P_i = \frac{\mathbf{H}(i)}{\sum \mathbf{H}(i)}, \tag{1}$$

where H is the histogram normalized to unit length. c_e and c_d represent color entropy and distribution, respectively.

– $\mathbf{v}_{hue}$: *Hue count.* The hue count measures the simplicity of a photo [4].
– $\mathbf{v}_{rule}$: *Rule of Thirds.* Rule of Thirds is used for optimizing the composition of a photo. It is always used as a guideline for the salient regions. The details are in [7].
– $\mathbf{v}_{bright}, \mathbf{v}_{contrast}$: *Brightness and Contrast.* Both of them are low level features [4]. In addition, contrast and brightness can be changed with different viewpoints.

Geometric Features. Given a 3D model, a lot of features are proposed to depict the viewpoints [12,13,17,18]. We extract these features including area, silhouette, depth, surface curvature, angles and some empirical rules.

- $g_{area}, g_{surface}$: *Project area and Surface visibility.* Project area is defined as the ratio of projected area of the model to the overall image area. Surface visibility determines the ratio of hidden surface of a shape [17]. Both of these features are related to the area of the shape for a particular viewpoint.
- g_{sl}, g_{sc}, and g_{sce}: *Silhouette length, Silhouette curvature and Silhouette curvature extrema.* Silhouette length defines the overall length of the silhouette for the shape in the image plane. Meanwhile, silhouette curvature is used as a visual attribute. It generates significant information to the viewer [18,19]. Moreover, silhouette attributes are regarded as the first index into the human memory of shapes.
- g_{dm}, g_{dd}: *Depth max and Depth distribution.* g_{dm} is defined as the maximum depth value of any visible points of the shape with respect to the viewpoint. Depth distribution is introduced to encourage a broad, even distribution of depths in the scene [17]. Besides, depth features are introduced to avoid degenerate viewpoints.
- g_{mc} *and* g_{gc}: *Mean curvature and Gaussian curvature.* Surface curvatures are common feature in geometric processing. They are assumed related to the semantic of the shapes [17].
- g_{ap}: *Above preference.* People tend to prefer views from slightly above the horizon [20], and this feature is computed to value the preference of viewpoint. It is defined as $g_{ap} = \mathcal{G}(\phi; \frac{3\pi}{8}, \frac{\pi}{4})$, $\mathcal{G}(x, \mu, \sigma) = e^{\frac{-(x-\mu)^2}{\sigma^2}}$ where ϕ is the latitude with 0 at the north pole and $\pi/2$ at the equator, and $\mathcal{G}$ is the non-normalized Gaussian function $exp(-(x-\mu)^2/\sigma^2)$.
- g_{angles}: *Axes angles.* Given a model, it has its model axes x, y and z. For a viewpoint, after applying Model-View transformation on these three axes, we can get x_m, y_m and z_m in camera coordinate system. g_{angles} is a 9-dimension vector defined as $g_{angles} = \angle(\mathbf{a}, \mathbf{b})$, where $\mathbf{a} \in \{x_m, y_m, z_m\}$, and $\mathbf{b} \in \{x_c, y_c, z_c\}$. $\{x_c, y_c, z_c\}$ are camera's axes with respect to camera coordinate system.
- g_{sphere}: *Sphere coordinate.* Given a 3D scene, the location of camera will also affect the preference of users. We describe the camera location with sphere coordinate with respect to the architecture model. It is described as $[r, \theta, \phi]$, and we take $g_{sphere} = [\theta, \phi]$ as features for a given viewpoint.
- g_{ar}: *Area ratio.* To avoid the degenerated viewpoints, this feature is computed as the ratio of the projection area of the 3D model in the photo g_{area} to the whole projected area of the model g_{aw}. It is further demonstrated in Fig. 3.

2.2 Viewpoint Assessment with Multi-view Learning

The CCA and the KCCA are usually employed to explore the mutual knowledge of these two views for viewpoint assessment.

Fig. 3. The first image suggests the architecture of the photo, and the second image shows the whole projected area of the mesh model. The feature of $\mathbf{g}_{ar}$ is shown in the third image as the area in the first image to the area in the second image.

Linear Canonical Correlation Analysis (CCA). CCA can be seen as a technique to seek a pair of linear transformations, one for each view of features. When the view of features is transformed, the correlation of these two views is mutually maximized.

Let $\mathbf{v} \in V$ and $\mathbf{g} \in G$ as the feature vectors of the input image $\mathbf{I}_i$ ($i = 1, 2, \ldots n$) with $\mathbf{v}$ and $\mathbf{g}$ formed by all the image and geometric features extracted in Sec. 2.1, where V and G are the feature spaces of the image and the geometric views. Two projections u and v are defined as $u = \mathbf{w}_v^T \mathbf{v}$ and $v = \mathbf{w}_g^T \mathbf{g}$, where $\mathbf{w}_v$ and $\mathbf{w}_g$ are two projecting vectors. CCA aims to choose $\mathbf{w}_v$ and $\mathbf{w}_g$ to maximize the correlation between the two univariates u and v. It also means that the correlation to be maximized by

$$\rho = \max_{\mathbf{w}_v, \mathbf{w}_g} \frac{\mathbb{E}[uv]}{\sqrt{\mathbb{E}[u^2]\mathbb{E}[v^2]}}. \tag{2}$$

Since ρ is invariant to the scale of $\mathbf{w}_v$ and $\mathbf{w}_g$, $\mathbb{E}[u^2]$ and $\mathbb{E}[v^2]$ are constrained to 1. The corresponding Lagrangian is

$$L = \mathbb{E}[uv] - \frac{1}{2}\lambda_v(\mathbb{E}[u^2] - 1) - \frac{1}{2}\lambda_g(\mathbb{E}[v^2] - 1). \tag{3}$$

Taking derivations with respect to $\mathbf{w}_v$ and $\mathbf{w}_g$,

$$\begin{aligned} \frac{\partial L}{\partial \mathbf{w}_v} &= \sum_i \mathbf{v}_i \mathbf{g}_i^T \mathbf{w}_g - \lambda_v \sum_i \mathbf{v}_i \mathbf{v}_i^T \mathbf{w}_v = 0, \\ \frac{\partial L}{\partial \mathbf{w}_g} &= \sum_i \mathbf{g}_i \mathbf{v}_i^T \mathbf{w}_v - \lambda_g \sum_i \mathbf{g}_i \mathbf{g}_i^T \mathbf{w}_g = 0. \end{aligned} \tag{4}$$

Simplifying the equation of Eq. 4, $\lambda_v = \lambda_g = \lambda$. Denoting $S_{\mathbf{vv}} = \sum_i \mathbf{v}_i \mathbf{v}_i^T$, $S_{\mathbf{vg}} = \sum_i \mathbf{v}_i \mathbf{g}_i^T$, and $S_{\mathbf{gg}} = \sum_i \mathbf{g}_i \mathbf{g}_i^T$, Eq. 4 can be rewrited as

$$B^{-1}A\mathbf{w} = \lambda\mathbf{w}, \tag{5}$$

where

$$B = \begin{bmatrix} S_{\mathbf{vv}} & 0 \\ 0 & S_{\mathbf{gg}} \end{bmatrix}, A = \begin{bmatrix} 0 & S_{\mathbf{vg}} \\ S_{\mathbf{vg}} & 0 \end{bmatrix}, \mathbf{w} = \begin{bmatrix} \mathbf{w}_v \\ \mathbf{w}_g \end{bmatrix}. \tag{6}$$

It is a generalized eigenproblem with the form of $A\mathbf{x} = \lambda\mathbf{x}$, and we can find the projecting vectors of $\mathbf{w}_g$ and $\mathbf{w}_v$, respectively.

Kernel Canonical Correlation Analysis (KCCA). CCA finds a linear transformation for the features of two views such that the correlation coefficient is maximized. However, it cannot always extract useful features because of its linearity. KCCA is proposed to solve this dilemma. KCCA projects feature vectors into a higher-dimension feature space with the kernel functions. $\Phi(\mathbf{v})$ and $\Psi(\mathbf{g})$ are two kernelized feature vectors of the image features in v and geometric features in g, respectively. The projecting vectors are defined as $\mathbf{w}_v = \sum_i \alpha_i \Phi(\mathbf{v}_i), \mathbf{w}_g = \sum_i \beta_i \Psi(\mathbf{g}_i)$. Similar to CCA, the Lagrangian function is defined as

$$L = \alpha^T K_v K_g \beta - \frac{1}{2}\lambda(\alpha^T K_v^2 \alpha - N) - \frac{1}{2}\lambda(\beta^T K_g^2 \beta - N), \tag{7}$$

where $K_v = \sum_i \Phi(\mathbf{v}_i)^T \Psi(\mathbf{v}_i)$, and K_g is defined in the similar form. Taking derivations with respect to α and β,

$$\frac{\partial L}{\partial \alpha} = K_v K_g \beta - \lambda K_v^2 \alpha = 0,$$
$$\frac{\partial L}{\partial \beta} = K_g K_v \alpha - \lambda K_g^2 \beta = 0. \tag{8}$$

Similar to Eqs. 4 and 8 is written in the matrix form of Eq. 5, where

$$B = \begin{bmatrix} K_v K_v & 0 \\ 0 & K_g K_g \end{bmatrix}, A = \begin{bmatrix} 0 & K_v K_g \\ K_g K_v & 0 \end{bmatrix}, \mathbf{w} = \begin{bmatrix} \alpha \\ \beta \end{bmatrix}. \tag{9}$$

Therefore, KCCA is solved as a generalized eigenproblem.

3   Experiments

In this section, we demonstrate the performance of multi-view learning with image and geometric features for viewpoint assessment of photographing architectures. Besides, we also evaluate the performance with the common features in both image and geometric aspects. To start the experiments, we collect 15 world famous architectures with 5894 photos and their corresponding 3D models from Internet. In addition, viewpoint goodness value of each photo is collected by a user study conducted on the Amazon Mechanical Turk (AMT). Each photo is scored from 1 to 5 (1 for the worst and 5 for the best viewpoint). To avoid the bias, each photo is scored 20 times by different subjects, and the average score defines the viewpoint goodness value. To train the multi-view learner more effectively, we further rule out some of the photos having conflicting scores, and split the remaining images into training set with 1303 images and testing set with 695 images. The photos with high and low score are labelled as good and bad, respectively.

Table 1. Accuracy comparison of different feature projection methods.

	Image view	Geometric view
Original	75.68%	78.42%
CCA	75.83%	78.27%
KCCA	**76.12%**	**78.85%**

Table 2. Accuracy comparison of different features.

Geometry feature	MC [17]	GC [17]	VE [12]	KCCA
Accracy	70.79%	70.36%	67.63%	**78.85%**
Image feature	Blur [4]	Rule of third [7]	Hue count [4]	KCCA
Accracy	69.93%	62.73%	69.21%	**76.12%**

SVM classifier with RBF kernel are chose to train on our dataset and compare the performance with solely 2D image or 3D geometric features. Besides, we also compare the performance of the CCA and KCCA, and the accuracy comparison results are shown in Table. 1. The results suggest that both view of features projected with KCCA achieve superior accuracy than the features projected with linear CCA and the original features. In addition, we also compare the performance of several common features in both image and geometric aspects. They are Mean Curvature (MC) [17], Gaussian Curvature (GC) [17], Viewpiont Entropy (VE) [12], Blur [4], Rule of Third [7], and the Hue Count [4]. The result is shown in Table 2. It demonstrates that both aspects of our features with KCCA perform much better than traditional features.

Furthermore, to set up a robust method for photographing architectures, we aim to assess the viewpoint quality with the image features solely. To this end, we also collect 100 pairs of photos $P = \{(I_i, I'_i) | 1 \leq i \leq 100\}$ with the context of architectures, and each pair has the same building with different viewpoints. Moreover, 20 preference choices for each pair of the photos are also collected, and the goodness score for a pair of them is defined as $S = \{(s_i, s'_i) | 1 \leq i \leq 100, s_i + s'_i = 1\}$.

The proposed method is extended by defining the viewpoint score as

$$g(x) = \frac{1}{1 + e^{-f(x)}}, \tag{10}$$

where $f(x)$ is the distance to the super plane of the SVM classifier. Given a pair of photos with the same architecture, our method can be used to select the better one with the higher score. The consistency value is defined as

$$\ell = \frac{1}{N} \sum_{i=1}^{N} (h((g(I_i) - g(I'_i)))s_i + h((g(I'_i) - g(I_i)))s'_i), \tag{11}$$

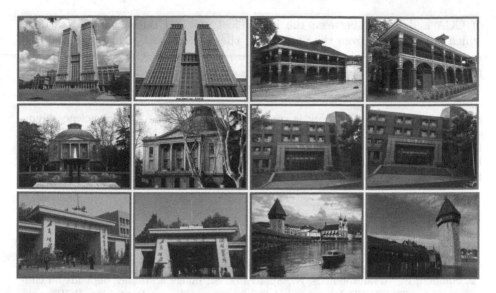

Fig. 4. Some paired photos with good and bad viewpoint quality evaluated by our KCCA learner with only image features. The photos framed with red and green for each of the photos are with good and bad viewpoints. Specifically, for each pair, the first and second photos are with the good and bad viewpoints, respectively.

Table 3. Performance comparison on paired photos with image features.

	Accuracy	ℓ
Original	76.0%	65.1%
CCA	72.0%	63.8%
KCCA	**78.0%**	**66.5%**
Human	**79.6%**	**68.9%**

where $N = 100$, and $h(x) = \begin{cases} 1 & x > 0 \\ 0 & x \leq 0. \end{cases}$

The comparison results of the accuracy and consistency values of the paired photos are shown in Table 3. Besides, we further collect 5 preference choices for each pair of the photos. The result suggests that human selection is with the highest accuracy and the value of ℓ. The KCCA performs better than the other two classifiers, while linear CCA is with less comparative performance. Several examplar photos are shown in Fig. 4.

4   Conclusions

A robust scheme is proposed for assessing the viewpoint of architecture based on the image features. We explore the mutual knowledge with multi-view learning of CCA and KCCA between the image features and the geometric features, which are extracted from the photos and the collected 3D architecture models, respectively. When only image features are available, KCCA achieves superior performance than other learning methods. With this knowledge, the viewpoint quality of the given photo can be assessed with the image features efficiently.

Acknowledgments. This work is supported in part by the Natural Science Foundation of Jiangsu Province under Grants BK20150016, the National Natural Science Foundation of China under Grants 61772257, 61502005, 61672279, the Fundamental Research Funds for the Central Universities 020214380042, and the Anhui Science Foundation under Grant 1608085QF129.

References

1. Kao, Y., He, R., Huang, K.: Deep aesthetic quality assessment with semantic information. IEEE Trans. Image Process. **26**(3), 1482–1495 (2017)
2. Simond, F., Arvanitopoulos, N., Süsstrunk, S.: Image aesthetics depends on context. In: 2015 IEEE International Conference on Image Processing (ICIP), pp. 3788–3792, September 2015
3. Tang, X., Luo, W., Wang, X.: Content-based photo quality assessment. IEEE Trans. Multimed. **15**(8), 1930–1943 (2013)
4. Ke, Y., Tang, X., Jing, F.: The design of high-level features for photo quality assessment. In: 2006 IEEE Computer Society Conference on Computer Vision and Pattern Recognition (CVPR 2006), vol. 1, pp. 419–426, June 2006
5. Mavridaki, E., Mezaris, V.: A comprehensive aesthetic quality assessment method for natural images using basic rules of photography. In: 2015 IEEE International Conference on Image Processing (ICIP), pp. 887–891, September 2015
6. Guo, Y., Liu, M., Gu, T., Wang, W.: Improving photo composition elegantly: considering image similarity during composition optimization. Comput. Graph. Forum **31**(7), 2193–2202 (2012)
7. Luo, Y., Tang, X.: Photo and video quality evaluation: focusing on the subject. In: Forsyth, D., Torr, P., Zisserman, A. (eds.) ECCV 2008. LNCS, vol. 5304, pp. 386–399. Springer, Heidelberg (2008). https://doi.org/10.1007/978-3-540-88690-7_29
8. Temel, D., AlRegib, G.: A comparative study of computational aesthetics. In: 2014 IEEE International Conference on Image Processing (ICIP), pp. 590–594, October 2014
9. Tian, X., Dong, Z., Yang, K., Mei, T.: Query-dependent aesthetic model with deep learning for photo quality assessment. IEEE Trans. Multimed. **17**(11), 2035–2048 (2015)
10. Chang, H., Yu, F., Wang, J., Ashley, D., Finkelstein, A.: Automatic triage for a photo series. ACM Trans. Graph. **35**(4), 148:1–148:10 (2016)
11. Polonsky, O., Patané, G., Biasotti, S., Gotsman, C., Spagnuolo, M.: What's in an image? Vis. Comput. **21**(8), 840–847 (2005). https://doi.org/10.1007/s00371-005-0326-y
12. Vázquez, P., Feixas, M., Sbert, M., Heidrich, W.: Automatic view selection using viewpoint entropy and its application to image-based modelling. Comput. Graph. Forum **22**(4), 689–700 (2003)

13. Lee, C.H., Varshney, A., Jacobs, D.W.: Mesh saliency. ACM Trans. Graph. **24**(3), 659–666 (2005)
14. Leifman, G., Shtrom, E., Tal, A.: Surface regions of interest for viewpoint selection. IEEE Trans. Pattern Anal. Mach. Intell. **38**(12), 2544–2556 (2016)
15. Härdle, W.K., Simar, L.: Canonical correlation analysis. In: Härdle, W.K., Simar, L. (eds.) Applied Multivariate Statistical Analysis, pp. 443–454. Springer, Heidelberg (2015). https://doi.org/10.1007/978-3-662-45171-7_16
16. Cai, J., Huang, X.: Robust kernel canonical correlation analysis with applications to information retrieval. Eng. Appl. Artif. Intell. **64**, 33–42 (2017)
17. Secord, A., Lu, J., Finkelstein, A., Singh, M., Nealen, A.: Perceptual models of viewpoint preference. ACM Trans. Graph. **30**(5), 109:1–109:12 (2011)
18. Feldman, J., Singh, M.: Information along contours and object boundaries. Psychol. Rev. **112**(1), 243–252 (2005)
19. Vieira, T., et al.: Learning good views through intelligent galleries. In: Computer Graphics Forum, vol. 28, pp. 717–726. Wiley Online Library (2009)
20. Blanz, V., Tarr, M.J., Bülthoff, H.H.: What object attributes determine canonical views? Perception **28**(5), 575–599 (1999)

Underwater Image Enhancement Using Stacked Generative Adversarial Networks

Xinchen Ye[1,2(✉)], Hongcan Xu[1,2], Xiang Ji[1,2], and Rui Xu[1,2]

[1] DUT-RU International School of Information Science and Engineering,
Dalian University of Technology, Dalian, China
`yexch@dlut.edu.cn`
[2] Key Laboratory for Ubiquitous Network and Service Software of Liaoning Province,
Dalian, China

Abstract. This paper addresses the problem of jointly haze detection and color correction from a single underwater image. We present a framework based on stacked conditional Generative adversarial networks (GAN) to learn the mapping between the underwater images and the air images in an end-to-end fashion. The proposed architecture can be divided into two components, i.e., haze detection sub-network and color correction sub-network, each with a generator and a discriminator. Specifically, a underwater image is fed into the first generator to produce a hazing detection mask. Then, the underwater image along with the predicted mask go through the second generator to correct the color of the underwater image. Experimental results show the advantages of our proposed method over several state-of-the-art methods on publicly available synthetic and real underwater datasets.

Keywords: Underwater · Image enhancement · GAN
Haze detection · Color correction

1 Introduction

Nowadays, underwater activities, e.g., underwater culture, monitoring and autonomous underwater detection, become more and more important for the purpose of the development of the ocean science. With more distinct vision, underwater robot can do better detection; the observation and research of underwater organisms will be more convenient; water environment monitoring will be improved. Apparently, all of the underwater visual tasks can benefit from the enhanced underwater images. However, most of underwater experiments rely on the expensive sensors to provide clear images with little color deviation compared to the in-air images. Obtaining satisfactory visibility of underwater images has

This work was supported by National Natural Science Foundation of China (NSFC) under Grant 61702078, 61772106, and by the Fundamental Research Funds for the Central Universities.

been historically difficult due to the absorptive and scattering properties of sea-water. Hence, the research to enhance the underwater images is necessary and becomes a hotspot.

The scattering of light from water particles, along with the attenuation in the color of different wavelengths of ambient light, cause a hazing effect in captured underwater images, which need to be removed from the images to provide a clear visualized picture of the underwater scene. Traditional underwater image enhancement methods can be divided into three categories: image based algorithms, model based algorithms, and learning based methods. Image based algorithms [1–4] estimate the transmission map directly from captured underwater image and then use it for haze removal and color correction. Model-based algorithms [5–8] are to model the imaging process under the underwater optical properties. The common limitations of these two categories are that they both use various assumptions and handcrafted constrains, which leads to a unsatisfactory recovery results. Due to the powerful learning ability, CNN presents promising solutions to in-air dehazing [9], which also can be transfer to the applications of underwater image enhancement [10–12].

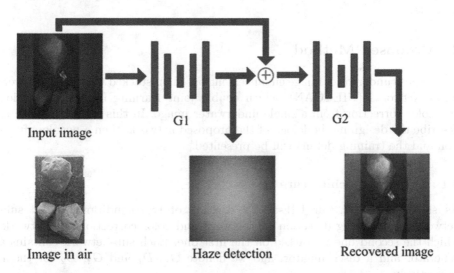

Fig. 1. Our joint learning architecture for haze detection and color correction of underwater enhancement.

Recently, Generative Adversarial Networks (GAN) [13] embody the huge advantage on image restoration tasks. The GAN is a minimax two-player game between the generator G and the discriminator D via an adversarial training process, in which the discriminator D learns to distinguish the real image and the synthesized one, and the generator learns to fool the discriminator. Several methods [14,15] restore the underwater scenes under the architecture of GANs. These methods use a single GAN to model the mapping between the underwater image and the in-air one, which maintain a global view to reason the scene semantic structure and light conditions.

As we observe, haze mask detection and color correction in underwater image enhancement share a fundamental characteristic essentially. As shown in Fig. 1, the haze mask is represented by a combination of transmission map and ambient light that can be used as an auxiliary information for color correction, which indicates the strong correlations on possible mutual benefits between these two tasks. Since no existing approaches have explored joint learning of hazing detection and color correction from a single underwater image based on GAN architecture, in this work, we propose to learn both tasks aiming at enjoying the mutually benefits from each other via stacked conditional GAN. The proposed stacked architecture can be divided into two components, i.e., hazing detection sub-network and color correction sub-network, each with a generator and a discriminator. As shown in Fig. 1, a underwater image is fed into the first generator (G_1) to produce a hazing detection mask. Then, the underwater image along with the predicted mask go through the second generator (G_2) to correct the color of the underwater image. Experimental results show the advantages of our proposed method over several state-of-the-art methods on publicly available datasets.

2   Proposed Method

We propose underwater image enhancement based on stacked Generative Adversarial Networks (UIE-sGAN), which enable joint learning for haze detection and color correction from a single underwater image. In this section, we'll first describe the design methodology of the proposed network, then a new loss function and the training details will be presented.

2.1   Network Architecture

As shown in Fig. 2, our UIE-sGAN consists of two conditional GAN sub-networks, i.e., hazing detection sub-network and color correction sub-network, which the second one is stacked on the first one. Each sub-network contains a generator and a discriminator, and we denote G_1, D_1 and G_2, D_2 for them, respectively.

Hazing Detection Sub-network. In this sub-network, both the generator G_1 and discriminator D_1 are conditioned on the input RGB underwater image. G_1 is trained to output the corresponding haze detection mask. Then, both the underwater image and the generated haze mask are sent to D_1 to distinguish from the real corresponding ground-truth pairs. Specifically, the generator G_1 adopts the standard U-Net architecture [16], which is made up of 7 convolution (Conv) layers and 7 deconvolution (DeConv) layers. Batch normalization (BN) and Leaky ReLU (LReLU) are used after each Conv or Deconv. Outside the U-Net structure, the first layer is a combination of Conv and LReLU, while the last layer contains Dconv and another nonlinear function 'tanh'. For the

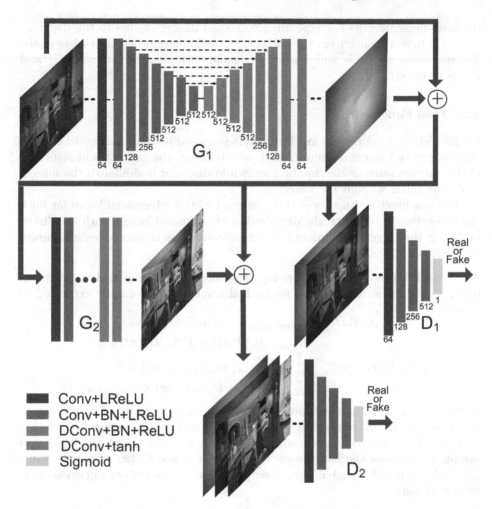

Fig. 2. Our network architecture. It consists of two stacked conditional GANs for haze detection and color correction, respectively. Both the sub-networks have the same generator and discriminator structures. G_2 is sketched out briefly for saving place. Layers are marked as different colors for easy recognition.

discriminator D_1, one Conv+LReLU block is placed at first, followed by three Conv+BN+LReLU blocks and a sigmoid layer to output the real or fake label. The number of feature maps in each layer are shown in Fig. 2.

Color Correction Sub-network. This sub-network takes the pair of underwater image and the generated haze mask from G_1 as the generator G_2's input. The haze mask from G_1 has extracted the message that includes light scattering and light attenuation preliminarily, and while we unite the water images and mask, it will get better consequence absolutely. Then, the G2 outputs the final RGB image with no color deviation. A triplets containing the underwater image, the haze mask, and its corresponding corrected image, are fed to the discriminator D2 to distinguish from the real corresponding ground-truth triplets in the datasets. Note that, G_2 and D_2 has the same network structure with G_1 and D_1, respectively.

2.2 Loss Function

In this section, a multi-term loss function is proposed to allow our model capturing context and semantic information, which makes the content and structure of the outputs same as the inputs, meanwhile the color is similar to the images that were taken without the water.

The loss function consists of three terms, i.e., two adversarial losses for both the sub-networks to match the distribution of generated images with the distribution in the target domain, and one consistency loss to encourage a accurate regression of the target image.

Adversarial Loss. For the mapping functions G_1, G_2 and their discriminators D_1, D_2, the adversarial loss for both the sub-networks can be expressed as follows:

$$
\begin{aligned}
\mathcal{L}_1(G_1, D_1) = E_{x,m \sim p_{data}(x,m)}[\log D_1(x, m)] + \\
E_{x \sim p_{data}(x)}[\log(1 - D_1(x, G_1(x)))]
\end{aligned}
\tag{1}
$$

$$
\begin{aligned}
\mathcal{L}_2(G_2, D_2|G_1) = E_{x,y,m \sim p_{data}(x,y,m)}[\log D_2(x, y, m)] + \\
E_{x \sim p_{data}(x)}[\log(1 - D_2(x, G_1(x), G_2(x, G_1(x))))]
\end{aligned}
\tag{2}
$$

where x, y, m denotes the input underwater image, the output in-air image, and the haze mask, respectively. G_1 tries to generate haze mask $G_1(x)$ that looks similar to the real haze mask, while D_1 aims to distinguish the translated sample $G_1(x)$ from the real sample m. G_2 takes x and $G_1(x)$ as inputs, while D_2 distinguishes the combination of outputs from G_1 and G_2 conditioned on x, from real pairs.

Consistency Loss. We add a consistency loss to encourage a accurate regression of the generated images, including haze detection mask and the final recovered image. The consistency loss can be expressed as follows:

$$
\begin{aligned}
\mathcal{L}_c(G_1, G_2|G_1) = E_{x,m \sim p_{data}(x,m)}||m - G_1(x)||_2 + \\
\lambda E_{x,y \sim p_{data}(x,y)}||y - G_2(x, G_1(x))||_2
\end{aligned}
\tag{3}
$$

where λ is a weighting parameter. Finally, the total loss is the linear combination of the above mentioned three losses with weights as follows:

$$
\mathcal{L} = \lambda_1 \mathcal{L}_1(G_1, D_1) + \lambda_2 \mathcal{L}_2(G_2, D_2|G_1) + \lambda_3 \mathcal{L}_c(G_1, G_2|G_1)
\tag{4}
$$

where the weights λ, λ_1, λ_2, λ_3 are all set at 1 based on heuristic experiments on our training data, which makes the order of magnitude of these losses equal to contribute to the final loss function.

Fig. 3. Synthetic training dataset.

Note that, the design of the proposed stacked sub-networks offers a novel perspective for multi-task learning. It can not only focus on one task once a time in different stages, but also share mutual improvements through forward and backward information prorogation.

2.3 Training Details

As for the training dataset, we use the synthetic underwater dataset established by [12], a novel method that uses a model based GAN to generate underwater images based on indoor Kinect dataset NYU depth [17], discriminated against the real captured underwater images on a artificial testbed (a man-made rock platform 3 ft submerged in a pure water test tank, shown in Fig. 1). The groundtruth haze masks are also obtained from [12]. The synthetic datasets for training our network are shown in Fig. 3. There are totally 1449 synthetic images. We use 1200 and 249 images pairs for training and testing separately. We resize the training images to 286×286, and randomly crop them to 256×256. We augment the training data in both datasets with rotation and flipping operations.

In the training procedure, an alternating update scheme is used for training the whole network. Specifically, we first update D_1, D_2 with G_1, G_2 fixed, then G_1, G_2 are updated with D_1, D_2 fixed. We use the xavier initializer to initialize the network. The model is trained using ADAM with a learning rate to 0.00005. The batch size is set to 1. We implement our network with the TensorFlow framework and train it using NVIDIA 1080Ti GPU.

3   Experimental Results

We evaluate our results by comparing with five state-of-the-art methods: (1) the combination of Gray World algorithm [2] and Non-local Image Dehazing [8]

(GW-N for short)[1], Fast Image Processing algorithm (FIP) [18], CycleGAN [19], pix2pix [20] and WaterGAN [12]. For the learning based methods, i.e., Cycle-GAN, pix2pix and WaterGAN, the same training dataset are used to train their network for fair comparison.

Table 1. Objective comparison using the PSNR and SSIM measurements on the average of the tested images from synthetic dataset.

	GW-N	FIP	CycleGAN	pix2pix	Ours
PSNR	11.80	14.80	20.49	24.65	25.40
SSIM	0.50	0.77	0.78	0.89	0.94

3.1 Experiment on Synthetic Datasets

We choose Peak Signal to Noise Ratio (PSNR) and structural similarity index (SSIM) to assess the recovered results for all of the compared methods. PSNR (dB) is the objective criterion to measure the image distortion, while SSIM predicts the perceived quality of images (ranging from -1 to 1, the higher the better). The numerical comparison is demonstrated in Table 1. Our method is far better than other methods both in PSNR and SSIM, which demonstrate our superior performance.

The visual comparison is demonstrated in Fig. 4, verifying the qualitative superiority of UIE-sGAN. Compared against other methods, our approach achieves the clear vision, balanced color, stretched contrast, and the most similar results to the groundtruth images. GW-N fails to recover the right color of the scenes. CycleGAN successfully removes the green haze mask, but cannot maintain the semantic content for lack of supervision, resulting in a very low PSNR and SSIM. Both the FIP and pix2pix method obtain the relatively better haze-free results, however, they could not preserve the overall color, tone, and brightness for a recovered image in some cases. For example, the bottom examples of FIP and pix2pix looks excessively reddish than the groundtruth, Besides, FIP are also subject to severe blocking artifacts, which leads to the worse objective measures than pix2pix and ours. All the compared baseline methods produce unsatisfactory results.

3.2 Experiment on Real Datasets

To test the generalization of our framework, we choose a real underwater dataset captured by [12], i.e., MHL dataset captured by the Marine Hydrodynamics Laboratory of University of Michigan, which include over 15000 underwater images, to validate stronger generalization of our UIE-sGAN. The visual comparison is

---

[1] Note that, Gray World algorithm aims at correcting the color, while Non-local Image Dehazing can be recast as a post processing to deblur the corrected image. The combination of both algorithms can achieve a relatively high performance.

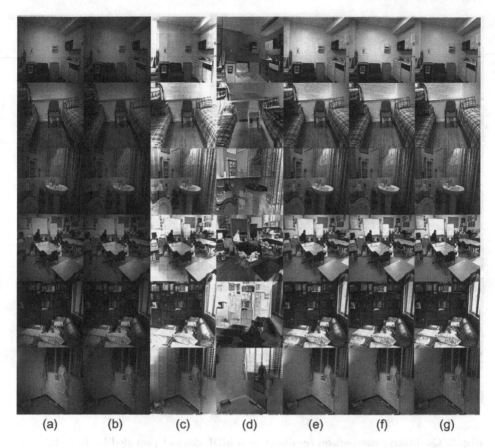

(a)   (b)   (c)   (d)   (e)   (f)   (g)

Fig. 4. Qualitative comparison on synthetic underwater datasets: (a) original underwater images; results recovered by (b) GW-N, (c) FIP, (d) CycleGAN, (e) pix2pix, (f) our UIE-sGAN, and (g) in-air groundtruth images. (Color figure online)

demonstrated in Fig. 5[2]. The similar conclusion can be obtained from the results of real dataset. Both the FIP and WaterGAN method achieve a haze-free performance, but they could not preserve the overall color, presenting excessively reddish. The pix2pix method generates comparable results to ours, but are subject to a little yellowish on the rock. The original scene captured in-air can be seen in Fig. 1 as a reference.

---

[2] There is no in-air ground truth for comparison, so we just show the visual comparison, and give some explanations.

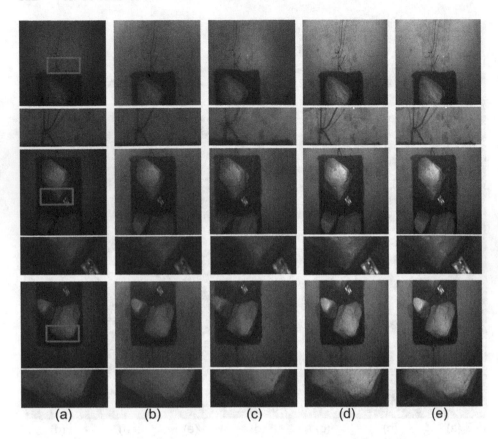

Fig. 5. Qualitative comparison results on real MHL dataset provided by Ref. [12]: (a) three underwater images chosen from MHL dataset; Results recovered by (b) FIP, (c) WaterGAN, (d) pix2pix, (e) our UIE-sGAN.

4   Conclusion

We present a framework to jointly learn the haze detection and color correction in an end-to-end fashion, based on stacked conditional GAN. The proposed architecture can be divided into two components, i.e., hazing detection sub-network and color correction sub-network, each with a generator and a discriminator. Specifically, a underwater image is fed into the first generator to produce a hazing detection mask. Then, the underwater image along with the predicted mask go through the second generator to correct the color of the underwater image. Experimental results show the advantages of our proposed method over several state-of-the-art methods on publicly available datasets.

References

1. Iqbal, K., Salam, R.A., Osman, M., Talib, A.Z., et al.: Underwater image enhancement using an integrated colour model. IAENG Int. J. Comput. Sci. **32**(2), 239–244 (2007)
2. Provenzi, E., Fierro, M., Rizzi, A.: A spatially variant white-patch and gray-world method for color image enhancement driven by local contrast. IEEE Trans. Pattern Anal. Mach. Intell. **30**(10), 1757–1770 (2008)
3. Ancuti, C., Ancuti, C.O., Haber, T., Bekaert, P.: Enhancing underwater images and videos by fusion. In: 2012 IEEE Conference on Computer Vision and Pattern Recognition (CVPR), pp. 81–88. IEEE (2012)
4. Li, C.-Y., Guo, J.-C., Cong, R.-M., Pang, Y.-W., Wang, B.: Underwater image enhancement by dehazing with minimum information loss and histogram distribution prior. IEEE Trans. Image Process. **25**(12), 56645677 (2016)
5. Schechner, Y.Y., Averbuch, Y.: Regularized image recovery in scattering media. IEEE Trans. Pattern Anal. Mach. Intell. **29**(9) (2007)
6. Chiang, J.Y., Chen, Y.-C.: Underwater image enhancement by wavelength compensation and dehazing. IEEE Trans. Image Process. **21**(4), 1756–1769 (2012)
7. Zhang, S., Zhang, J., Fang, S., Cao, Y.: Underwater stereo image enhancement using a new physical model. In: 2014 IEEE International Conference on Image Processing (ICIP), pp. 5422–5426. IEEE (2014)
8. Berman, D., Treibitz, T., Avidan, S.: Non-local image dehazing. In: 2016 IEEE Conference on Computer Vision and Pattern Recognition (CVPR), 27–30 June 2016 (2016)
9. Zhang, H., Patel, V.M.: Densely connected pyramid dehazing network. In: 2018 IEEE Conference on Computer Vision and Pattern Recognition (CVPR), 18–22 July 2018 (2018)
10. Shin, Y.S., Cho, Y., Pandey, G., et al.: Estimation of ambient light and transmission map with common convolutional architecture. In: Oceans, pp. 1–7. IEEE (2016)
11. Wang, Y., Zhang, J., Cao, Y., et al.: A deep CNN method for underwater image enhancement. In: IEEE International Conference on Image Processing, pp. 1382–1386. IEEE (2017)
12. Li, J., Skinner, K.A., Eustice, R.M., Johnson-Roberson, M.: WaterGAN: unsupervised generative network to enable real-time color correction of monocular underwater images. IEEE Robot. Autom. Lett. **3**, 387–394 (2018)
13. Goodfellow, I.J., et al.: Generative adversarial networks (2014)
14. Li, C., Guo, J., Guo, C.: Emerging from water: underwater image color correction based on weakly supervised color transfer. IEEE Signal Process. Lett. **PP**(99), 1 (2017)
15. Chen, X., Yu, J., Kong, S., et al.: Towards quality advancement of underwater machine vision with generative adversarial networks (2018)
16. Ronneberger, O., Fischer, P., Brox, T.: U-Net: convolutional networks for biomedical image segmentation. In: Navab, N., Hornegger, J., Wells, W.M., Frangi, A.F. (eds.) MICCAI 2015. LNCS, vol. 9351, pp. 234–241. Springer, Cham (2015). https://doi.org/10.1007/978-3-319-24574-4_28
17. Silberman, N., Hoiem, D., Kohli, P., Fergus, R.: Indoor segmentation and support inference from RGBD images. In: Fitzgibbon, A., Lazebnik, S., Perona, P., Sato, Y., Schmid, C. (eds.) ECCV 2012. LNCS, vol. 7576, pp. 746–760. Springer, Heidelberg (2012). https://doi.org/10.1007/978-3-642-33715-4_54

18. Chen, Q., Xu, J., Koltun, V.: Fast image processing with fully-convolutional networks. In: ICCV (2017)
19. Zhu, J.-Y., Park, T., Isola, P., Efros, A.A.: Unpaired image-to-image translation using cycle-consistent adversarial networks. In: ICCV (2017)
20. Isola, P., Zhu, J.-Y., Zhou, T., Efros, A.A.: Image-to-image translation with conditional adversarial networks. arXiv:1611.07004

An Efficient Complexity Reduction Scheme for CU Partitioning in Quality Scalable HEVC

Bo Liu$^{(\boxtimes)}$, Qiang Li, and Jianlin Song

Chongqing Key Laboratory of Signal and Information Processing,
Chongqing University of Posts and Telecommunication,
Chongqing, People's Republic of China
liubo068@hotmail.com

Abstract. The scalable extension of HEVC (known as SHVC), uses Inter-layer predictions with multiple HEVC layers in addition to the advanced coding tools of HEVC, which causes huge computational complexity. One of the main reasons that result in the SHVC encoder complexity is selecting the best coding unit (CU) depth level. This paper aims to develop a complexity reduction scheme for CU depth prediction and CU partitioning termination of Quality SHVC. In this regard, first, the CU depth correlation degree is used to predict the most probable depths. Then, a hypothesis testing for the residuals distribution of current CU is introduced to terminate the depth selection early. Experimental results demonstrate that the proposed scheme significantly reduces the enhancement layer (EL) execution time of SHVC encoder by 58.19% on average compared with unmodified SHVC encoder while maintaining the overall coding efficiency.

Keywords: SHVC · Depth decision · Complexity
Scalable video compression

1 Introduction

High Efficiency Video Coding (HEVC), also known as H.265, is a latest video compression standard, which is developed by Joint Collaborative Team on Video Coding (JCT-VC) [1]. HEVC encoder provides twice the compression efficiency of the previous standard approximately, H.264/AVC, while maintaining the same overall video quality. However, the computational complexity of HEVC is two to four times that of H.264/AVC [2]. As a scalable extension of HEVC, SHVC supports different kind of scalability types, including temporal scalability, spatial scalability, and quality scalability [3]. SHVC encoder needs to encode multiple layers and each layer must perform HEVC coding process. A whole SHVC bit

This work is supported by the National Natural Science Foundation of China (No. 61571071) and Nature Science Foundation Project of Chongqing (No. cstc2016jcyjA0543 and No. cstc2017jcyjXB0037).

© Springer Nature Switzerland AG 2018
R. Hong et al. (Eds.): PCM 2018, LNCS 11166, pp. 525–532, 2018.
https://doi.org/10.1007/978-3-030-00764-5_48

stream consists of a base layer (BL) and one or more enhancement layers (ELs). Since the coding complexity of HEVC is already extremely complex, the complexity of SHVC is mainly caused by the additional Inter-layer prediction during HEVC coding of multiple layers.

Many previous works have been proposed to reduce the computational complexity of Quality SHVC. Tohidypour et al. [4] use the already encoded CTUs in BL and EL to predict the coding unit size of to-be-encoded CTU in EL, which establishes the model based on Bayes rule and trains it using a small train dataset results in the coding speed improvement. Wang et al. [5] propose a multi-strategy Intra prediction for Quality SHVC, which skips depth levels with low probabilities and unnecessary Intra prediction. This method also allows checking a subset of Intra modes instead of all 35 Intra modes and terminate the CU depth decision early. The above methods focus on the Intra prediction and significantly reduce the computational complexity. However, Inter prediction algorithms could be an effective research direction to speed up the coding procedure. The method proposed in [6] select the optimal motion search range for current CU based on the correlation between the BL and EL. This algorithm terminates the mode search early and avoids the redundant computations. Bailleul et al. [7] determine the best depth level in current CU by copying the BL depth information directly and disallow the Intra prediction and orthogonal block modes. The coding speed is improved but the coding efficiency is decreased severely simultaneously. Tohidypour et al. [8] present a complexity reduction scheme which combines a mode prediction method and a reference layer mode-information based mode prediction method for two enhancement layers.

In this study, our objective is to develop a complexity reduction scheme for Quality SHVC. Considering that the depth correlation exists in the spatial and Inter-layer CUs, it is efficient to predict the to-be-encoded CU depth level by measuring the CU depth correlation degree. Therefore, some low likelihood depth levels can be skipped. First, we design a model for correlation degree to predict the most probable depths. The depth prediction method utilizes the depth information of CUs neighboring to the current CU in BL and EL. Then, residual coefficients distribution of current depth level is tested to terminate the CU partitioning early. To evaluate the performance of proposed scheme, we implement our method to an SHVC encoder with quality scalability. Experimental results show that the proposed scheme can significantly reduce the coding execution time of SHVC encoder, while maintaining almost the original coding efficiency.

The rest of this paper is organized as follows. Section 2 presents the proposed complexity reduction scheme. Experimental results and conclusions are shown in Sects. 3 and 4, respectively.

2   Proposed Algorithm

2.1   Correlation Degree Based Depth Prediction

SHVC encoder employs the hybrid coding framework and the coding process would generally be executed as follows. Each to-be-encoded picture is split into the small coding unit (CU). In addition, SHVC allows a partitioning of CUs into sub-CUs based on the quadtree structure and rate-distortion (RD) cost calculations. More specifically, SHVC encoder supports the four sizes of CU, which are 64×64, 32×32, 16×16, 8×8, and the corresponding depth level is 0, 1, 2, and 3, respectively. Since CUs in BL and the collocated CUs in EL correspond with the same video contents, strong depth correlations exist between the BL and the EL. Thus, the depth level of current CU can be predicted by its available neighboring CUs. The correlation degree between the current CU and its available neighboring CUs should be defined. Figure 1 shows the available CUs of current CU for depth prediction. C is the current CU in EL, L is the left CU, U is the upper CU, UL is the upper-left CU, and UR is the upper-right CU. And, BC, BL, BU, BUL, and BUR is the collocated CU of C, L, U, UL, and UR in BL, respectively.

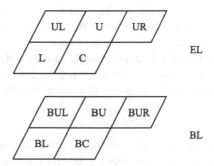

Fig. 1. Available CUs for depth prediction of current CU

The more similar depth values of available neighboring CUs in EL and its collocated CUs in BL are, the more similar depth value of current CU in EL and its collocated CU in BL will be. And it should be very high correlation degree of depth values between BL and EL. In addition to current CU, the depth values of neighboring CUs in BL can be used as an estimator of the depth values in EL. The errors between estimator from BL and depth values in EL is measured by the Mean Squared Error (MSE). Obviously, the Inter-layer correlation degree is inversely proportional to the MSE value. Note that the maximum depth level of CU is 3. The correlation degree is indicated by the Inter-layer depth weight. In order to distinguish different correlation degree, a power of two is used. Furthermore, other powers can be used as long as they can distinguish the different Inter-layer correlation degrees. Accordingly, the Inter-layer depth weight is expressed as follows:

$$\omega_d^l = 2^{3-round\left[\sqrt{\frac{1}{4}\sum_{i=1}^{4}(d_{bi}-d_{ei})^2}\right]} \tag{1}$$

where Inter-layer correlation is record as superscript l, subscript d denotes the d-th depth ($d = 0, 1, 2, 3$), ω_d^l represents the Inter-layer weight for neighboring CU of d-th depth level, and d_{bi} and d_{ei} refer to the depth values for available neighboring CUs of current CU in BL and EL as shown in Fig. 1, respectively. If the neighboring CUs of current CU are not available (for the edge CUs), the Inter-layer depth weight is set to 0. Otherwise, the Inter-layer depth weight can be obtained from Eq. 1.

As for the to-be-encoded CU, the content of CU is same as that of collocated CU in BL. Therefore, the CU in EL always uses the same or similar depth of BL. When a CU in EL checks a certain depth, the correlation degree of current CU in BL and EL can also be indicated by the CU depth weight. The current CU depth weight is expressed as:

$$\omega_d^c = 2^{3-|d_{ec}-d_{bc}|} \tag{2}$$

where the correlation of current CU in BL and EL is recorded as superscript c, d_{ec} and d_{bc} refer to the depth values of current CU in EL and the collocated CU in BL. ω_d^c represents the current CU depth weight.

The Inter-layer and current CU depth weight can be obtained by Eqs. 1 and 2. Since the depth weight is related to the correlation degree, and the correlation degree is correlated with depth decision, the total weight ω_d of d-th depth is derived as:

$$\omega_d = \omega_d^l + \omega_d^c \tag{3}$$

The weight of all depths can be calculated by applying the above method. The all possible depths then are sorted according to the corresponding weight values in a descending order. In order to reduce the coding complexity and maintain the coding efficiency, the depth selection method can be summarized based on the calculated weight values and sorted depth levels as follows:

(a) The depth 3 is set as the default depth.
(b) If the first three depths are equal to 0, the depth 1, 2, and 3 are skipped and will not be checked.
(c) If the first two depths are less than or equal to 1, the depth 2 and 3 are skipped.
(d) If the first three depths are less than or equal to 2, the depth 3 is skipped.
(e) If the first two depths are equal to 2 or 3, the depth 0 will be skipped.

2.2   Residuals Distribution Based Depth Early Termination

As mentioned above, SHVC supports a partitioning of CUs into sub-CUs based on the quadtree structure and RD-cost calculations. If CU is well partitioned according to the main features of video content, the residual coefficients will follow a certain distribution. The residual coefficients are usually modeled using a Gaussian distribution due to its superior performance [9,10]. Moreover, if the residual coefficients obey a Gaussian distribution, the current depth level may be the best depth, it is no need to further check the next depth level. Otherwise, the next depth level should be checked when residual coefficients for current CU do not obey a Gaussian distribution. To this end, a hypothesis testing is introduced to determine whether the residual coefficients from current CU follow a Gaussian distribution. For this purpose, Anderson-Darling (A-D) test is introduced, which tests whether a Gaussian distribution describes the residual coefficients adequately.

As it was stated, the residual coefficients obey a Gaussian distribution. Suppose that the residual coefficients of current CU can be modeled as:

$$X \sim N(\mu, \sigma^2) \tag{4}$$

where μ and σ^2 are respectively the mean value and the variance. X represents residual sample of current CU, the n observations values X_i, for $i = 1, \ldots, n$, of X are standardized to generate new values Y_i. The test statistic based on A-D test is calculated using [11]:

$$A^2 = -n - \frac{1}{n} \sum_{i=1}^{n} (2i - 1)(\ln \Phi(Y_i) + \ln(1 - \Phi(Y_{n+1-i}))) \tag{5}$$

where $\Phi(Y_i)$ is the standard normal Cumulative Distribution Function (CDF), n is the number of residual coefficients of current CU. The test can be adopt for size of 64×64, 32×32 and 16×16, the corresponding value n are 64^2, 32^2 and 16^2.

According to statistical hypothesis testing, the significance level α is the probability of rejecting the null hypothesis which the residual coefficients follow a Gaussian distribution. Each significance level α corresponds to the only critical value z_α. More precisely, if A^2 exceeds a given critical value β, then the hypothesis of normality is rejected with the significant level α. Thus, the decision of Gaussian distribution can be express as:

$$A^2 <= z_\alpha \tag{6}$$

If the condition defined in Eq. 6 are satisfied, residual coefficients can be assumed to follow a Gaussian distribution. Then, the current depth level can be assumed to be the best depth, and the next depth level does not need to be checked. The key to use Eq. 6 is specify the critical value z_α. To this end, some significance level values are selected for testing, i.e., 1%, 2.5%, 5%, 10%, 15%. As a result, the value 2.5%, 5%, and 5% are selected as the significance level for current CU size of 64×64, 32×32, and 16×16, respectively. The corresponding critical values z_α can be obtained from Percentage Points Table for A^2 of the literature [11].

$$z_\alpha = \begin{cases} 0.918, 64 \times 64 \\ 0.787, 32 \times 32 \\ 0.787, 16 \times 16 \end{cases} \tag{7}$$

3 Experimental Results

In this experiment, in order to evaluate the performance of the proposed complexity reduction scheme, the proposed methods have been implemented in the SHVC reference software (SHM-11.0 [13]). Random-Access (RA) and Low-Delay (LD) GOP structure is used for the experiment. The maximum CU size is 64×64 pixels, the minimum CU size is 8×8 pixels, and the corresponding depth level lies between 0 and 3. According to common SHM test conditions [14], the quantization parameter (QP) values for the BL are $QP_{BL} \in \{26, 30, 34, 38\}$, and the corresponding QP values for the EL are $QP_{EL} \in \{22, 26, 30, 34\}$. The performance the proposed complexity reduction scheme is compared with the unmodified SHVC encoder. Coding efficiency is measured with PSNR and bitrate, and computational complexity is measured with execution time saving. BDPSNR (dB) and BDBR (%) are used to represent the average PSNR and bitrate differences [15]. The total coding time saving and EL coding time saving for percentage are represented by DT-Total (%) and DT-EL (%).

Table 1. Coding performance and complexity reduction compared with SHM-11.0

Class	Sequence	RA structure				LD structure			
		BDBR (%)	BDPSNR (dB)	DT-Total (%)	DT-EL (%)	BDBR (%)	BDPSNR (dB)	DT-Total (%)	DT-EL (%)
A	PeopleOnStreet	0.73	−0.042	22.47	43.86	0.66	−0.028	24.58	43.66
	Traffic	0.36	−0.026	28.53	53.58	0.29	−0.031	27.35	52.49
B	BasketballDrive	0.22	−0.012	31.58	60.44	0.31	−0.019	32.37	62.55
	BQTerrace	0.25	−0.019	35.66	64.40	0.33	−0.033	36.29	71.64
	Cactus	0.63	−0.038	34.73	64.68	0.81	−0.042	35.87	66.34
	Kimono1	0.38	−0.029	31.72	60.13	0.36	−0.021	31.28	58.67
	ParkScene	0.26	−0.011	32.46	60.24	0.24	−0.023	32.61	59.37
Average		0.40	−0.025	31.02	58.19	0.43	−0.028	31.48	59.25

Table 1 demonstrates the performance of the proposed scheme compared with SHM-11.0 in RA and LD structure respectively. With RA structure configure, it can be observed that the proposed scheme achieves the average total time reduction percentage of 31.02%. The method reduces the coding execution time by 58.19% for the EL at a cost of 0.40% bitrate increase and 0.025 dB PSNR decrease on average. With LD structure configure, the proposed scheme achieves the average total coding time reduction and EL coding time reduction by 31.48% and 59.25% on average respectively at a cost of 0.43% bitrate increase and 0.028 dB PSNR decrease on average. The minimum loss of bitrate and PSNR is 0.22% and 0.011 dB for "BasketballDrive" (1920×1080, RA) and "ParkScene" (1920×1080, RA), respectively. For coding time gain of the proposed method, the maximum EL coding time reduction is 71.64% for "BQTerrace" (1920×1080, LD), and the minimum EL coding time reduction is 43.66% for "PeopleOnStreet" (2560×1600, LD).

Table 2. Coding performance and complexity reduction compared with the state-of-the-art algorithms

Class	Sequence	EMD			CACRS			PAPS			Proposed		
		BDBR (%)	BDPSNR (dB)	DT-EL (%)	BDBR (%)	BDPSNR (dB)	DT-EL (%)	BDBR (%)	BDPSNR (dB)	DT-EL (%)	BDBR (%)	BDPSNR (dB)	DT-EL (%)
A	PeopleOnStreet	0.46	-0.020	42.71	0.35	-0.012	9.62	0.97	-0.060	56.89	0.73	-0.042	43.86
	Traffic	1.32	-0.038	64.64	0.44	-0.021	20.08	1.30	-0.038	69.69	0.36	-0.026	53.58
B	BasketballDrive	2.23	-0.042	54.20	0.15	-0.014	11.21	1.35	-0.026	66.46	0.22	-0.012	60.44
	BQTerrace	2.38	-0.042	53.96	0.23	-0.019	18.12	0.68	-0.010	61.11	0.25	-0.019	64.40
	Cactus	1.06	-0.017	55.35	0.31	-0.016	14.11	0.96	-0.019	68.51	0.63	-0.038	64.68
	Kimono1	0.95	-0.026	51.78	0.21	-0.035	21.07	0.77	-0.021	62.75	0.38	-0.029	60.13
	ParkScene	2.39	-0.075	62.75	0.41	-0.045	19.84	1.17	-0.038	66.98	0.26	-0.011	60.24
	Average	1.54	-0.037	55.06	0.30	-0.023	16.29	1.03	-0.030	64.63	0.40	-0.025	58.19

Table 2 illustrates the performance of proposed algorithm compared to the state-of-the-art coding complexity reduction schemes for the EL: (1) Early MERGE Mode Decision (EMD) [12] (just for the EL), (2) Content Adaptive Complexity Reduction Scheme (CACRS) [6], (3) Probabilistic Approach for Predicting the Size of Coding Units (PAPS) [4] (just for the EL1). As it is observed from Table 2, the EMD method reduces coding execution time by 55.06% for the EL with 1.54% BDBR increase and 0.037 dB PSNR decrease. The coding execution time of the proposed scheme is improved by 3.13% compare with EMD method, while achieving a gain of 0.012 dB PSNR and 1.14% bitrate saving. And the proposed scheme is more efficient than EMD method. Compared with the CACRS method, the proposed method achieves more EL execution time reduction of 41.9% on average with negligible coding efficiency losses. Although the coding execution time reduction of PAPS method is slightly higher than the proposed scheme, it achieves a better coding efficiency with 0.005 dB PSNR increase and 0.63% bitrate saving. In summary, the proposed complexity reduction scheme significantly reduces the coding execution time of Quality SHVC encoder while maintaining the overall coding efficiency.

4 Conclusion

In this paper, an efficient complexity reduction scheme for Quality SHVC, which can speed up the coding process, has been proposed. The proposed scheme uses the depth correlation degree to predict the possible CU depths of to-be-encoded CU in the current EL. At every depth decision process, a hypothesis testing for the residuals distribution of current CU is introduced to terminate the depth selection early. The performance evaluation of the proposed scheme was tested over a set of video sequences and was compared with three state-of-the-art methods as well as the unmodified SHVC encoder. Simulation results show that the proposed scheme significantly reduces the EL execution time by 58.19% on average at a cost of negligible coding efficiency losses (with 0.40% bitrate increase and 0.025 dB PSNR decrease). Note that the proposed method may be implemented to the spatial SHVC encoder, this will be our ongoing work.

References

1. Sullivan, G.J., Ohm, J., Han, W.-J., Wiegand, T.: Overview of the high efficiency video coding (HEVC) standard. IEEE Trans. Circuits Syst. Video Technol. **22**(12), 1649–1668 (2012)
2. Yan, S., Hong, L., He, W., Wang, Q.: Group-based fast mode decision algorithm for intra prediction in HEVC. In: 2012 Eighth International Conference on Signal Image Technology and Internet Based Systems (SITIS), pp. 225–229. IEEE (2012)
3. Boyce, J.M., Ye, Y., Chen, J., Ramasubramonian, A.K.: Overview of SHVC: scalable extensions of the high efficiency video coding standard. IEEE Trans. Circuits Syst. Video Technol. **26**(1), 20–34 (2016)
4. Tohidypour, H.R., Pourazad, M.T., Nasiopoulos, P.: Probabilistic approach for predicting the size of coding units in the quad-tree structure of the quality and spatial scalable HEVC. IEEE Trans. Multimed. **18**(2), 182–195 (2016)
5. Wang, D., Zhu, C., Sun, Y., Dufaux, F., Huang, Y.: Efficient multi-strategy intra prediction for quality scalable high efficiency video coding. IEEE Trans. Image Process. (2017)
6. Tohidypour, H.R., Pourazad, M.T., Nasiopoulos, P.: Content adaptive complexity reduction scheme for quality/fidelity scalable HEVC. In: 2013 IEEE International Conference on Acoustics, Speech and Signal Processing (ICASSP), pp. 1744–1748. IEEE (2013)
7. Bailleul, R., De Cock, J., Van De Walle, R.: Fast mode decision for SNR scalability in SHVC digest of technical papers. In: 2014 IEEE International Conference on Consumer Electronics (ICCE), pp. 193–194. IEEE (2014)
8. Tohidypour, H.R., Pourazad, M.T., Nasiopoulos, P.: An encoder complexity reduction scheme for quality/fidelity scalable HEVC. IEEE Trans. Broadcast. **62**(3), 664–674 (2016)
9. Cho, S., Kim, M.: Fast CU splitting and pruning for suboptimal CU partitioning in HEVC intra coding. IEEE Trans. Circuits Syst. Video Technol. **23**(9), 1555–1564 (2013)
10. Nan, H., Yang, E.-H.: Fast motion estimation based on confidence interval. IEEE Trans. Circuits Syst. Video Technol. **24**(8), 1310–1322 (2014)
11. Stephens, M.A.: EDF statistics for goodness of fit and some comparisons. J. Am. Stati. Assoc. **69**(347), 730–737 (1974)
12. Pan, Z., Kwong, S., Sun, M.-T., Lei, J.: Early merge mode decision based on motion estimation and hierarchical depth correlation for HEVC. IEEE Trans. Broadcast. **60**(2), 405–412 (2014)
13. SHVC Test Model:SHM-11.0 [Online]. https://hevc.hhi.fraunhofer.de/svn/svn_SHVCSoftware/tags/SHM-11.0. Accessed 10 May 2018
14. Seregin, V., He, Y.: Common SHM test conditions and software reference configurations. Document JCTVCQ1009, pp. 1–4 (2014)
15. Bjøntegaard, G.: Calculation of average PSNR differences between RD-curves. In: ITU-T Q.6/SG16 VCEG, 15th Meeting, Austin, Texas, USA, April 2001

Extended Multi-column Convolutional Neural Network for Crowd Counting

Zhiyuan Xue[✉] [ID], Jie Shen[ID], Xin Xiong[ID], Chong Yuan[ID],
and Yinlong Bian[ID]

University of Electronic Science and Technology of China, Chengdu, China
uestcxzy@gmail.com, sjie@uestc.edu.cn, cumul.x@gmail.com,
201621060738@std.uestc.edu.cn, uestcbyl@gmail.com

Abstract. With the rapid growth of population all over the world, crowd analysis has become a vital way to maintain public safety in crowded scenes like outdoor sports events. In this paper, we propose a method that can accurately estimate number of people and their distribution in a crowded scene. Inspired by current state-of-the-art methods, we use multi-column CNN with different reception fields as basic regressor. Also, VGG-16 pre-trained on ImageNet is used to generate deep features. These two parts were then merged together for the final 1×1 convolutional layer and thus density maps are generated. Since all layers in our model is convolutional layers, the input image can be of any size, and the model is easy to implement and train. After experimenting on several major crowd counting datasets, our method turns out to have higher accuracy comparing to other existing methods.

Keywords: Crowd counting · Density estimation · Crowd analysis

1 Introduction

Due to exponential growth in the world population, crowded scene like Fig. 1 are becoming more and more common. Crowd counting has always been a popular field of computer vision. However, many challenges make this problem extremely difficult to solve, such as variation in scale and perspective, and non-uniform distribution and illumination [14]. Numerous solutions have been proposed. Earlier methods are mostly based on detection. [3] used histogram oriented gradients as features and trained a classifier to extract full bodies. Also, other machine learning methods like random forest [5] and boosting [17] are used to improve accuracy of body detection. [4,8] used part-based detection to detect specific body parts and estimate crowd counts. Apart from detection-based methods, regression is also used to map from features like edges and texture to crowd counts [16]. Combining detection and regression methods, Idrees et al. [6] proposed to extract multiple features with different methods and employed Fourier analysis and Markov Random Field to generate final results.

© Springer Nature Switzerland AG 2018
R. Hong et al. (Eds.): PCM 2018, LNCS 11166, pp. 533–540, 2018.
https://doi.org/10.1007/978-3-030-00764-5_49

Fig. 1. Sample images of crowded scenes from ShanghaiTech dataset.

Since the success of Alexnet, Convolutional Neural Network (CNN) has become a vital method in computer vision research. It has also been widely used in crowd counting problems. Wang et al. [19] first implemented Alexnet. They replaced the last fully connected layer of 4096 neurons with a single neuron to predict the final count. Shang et al. [13] proposed an end-to-end model which takes the whole image as input and output the count directly. Pre-trained GoogleNet and lone-short time memory (LSTM) are used in their network. Zhang et al. [20] proposed a CNN based method that can be used in multiple scenes. Their model is trained on a certain scene, and can be easily fine-tuned with perspective information given. However, perspective information cannot be easily acquired in most applications. Walach and Wolf [18] used layered boosting and selective sampling. CNN layers is added iteratively so that every layer is trained to estimate the residual error of the earlier prediction. Zhang et al. [21] proposed a multi-column CNN structure to overcome the problem with perspective information. Inspired by [2], three columns with different filter size is used as regressor. However, simply merge the three column together is not proper, since regressors with the inappropriate scales can greatly influence those with appropriate scales. The solution by Switching-CNN [12] used a classifier based on VGG-16 to perform a 3-way classification, and only the result of the chosen regressor is used to estimate the final density map. The problem of this approach is that different parts of the network are trained separately in multiple stages, making it difficult to implement.

2   The Proposed Method

Our approach first converts labeled training images to density maps, then use a specific CNN structure to train the map from original images to density maps. After training, the network is able to generate density maps from any input images and head counts can be achieved by integration of the density map.

2.1   Density Map

The first stage of our method is to generate density maps for training images. Instead of using a regressor to directly estimate head counts from the image, density map based counting has some advantages. Apart from total counts, density maps also contain information like crowd distribution. In order to generate

Fig. 2. An example of density map generation.

density maps from labeled images from the datasets, we use a method proposed in [21]. In this method, a delta function convolved with a Guassian kernel is used to represent every labeled head. So density map can be generated with function

$$H(x) = \sum_{i=1}^{N} \delta(x - x_i) * G_{\sigma_i}(x) \tag{1}$$

in which

$$\sigma_i = \beta \bar{d}_i \tag{2}$$

Here $\bar{d}_i$ is the average distance from head i to k nearest neighbors of head i. According to [21], the result is the best with $\beta = 0.3$. Figure 2 shows an example of the original image and density map generated with this method.

2.2 Extended Multi-column CNN

Multi-column Regressor. Since lack of perspective information, the head size and distances between heads in the image can be different. So we use a modified version of multi-column CNN proposed in [21] with different reception fields as basic regressors. The size of convolutional filters are from 3 to 7 (5 to 9 for the first layer), so that they can adapt to different head scales. We added one 2×2 max pooling layer to the original structure to match the size of the output of the VGG-based feature extractor, which has three 2×2 max pooling layers. The numbers of output channels of three columns are 8, 10 and 12.

VGG-Based Feature Extractor. The purpose of this part is to extract features representing head sizes and perspective information of the given image. In order to achieve the goal, we use part of VGG-16, which has a good reputation for extracting deep features from images. Other structures like ResNet and GoogLeNet can also do the work but VGG is faster. The first 4 parts of VGG-16 is implemented, with 10 convolutional layers. We initialize their parameters with those pre-trained on ImageNet. After the 10th convolutional layer with 512 channels, a list of convolutional layers with 256 to 30 channels is followed to gradually reduce the number of channels.

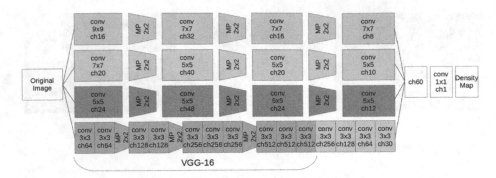

Fig. 3. The structure of the proposed extended multi-column convolutional neural network. "conv n × n" means convolutional layer with filter size n × n, "ch" means number of output channels. "MP 2 × 2" means 2 × 2 max-pooling layer.

Final Stage. After the regressors and the feature extractor, all output channels are concatenated into a layer with 60 channels. Half of these channels represents head counts of 3 different scales and the other half contains the information of head size and other important features. So we use one 1 × 1 convolutional layer, so that these two kinds of information can be combined and thus get the final density map. The structure of the whole network is shown in Fig. 3. Rectified linear unit (ReLu) is used as activation function because of its outstanding performance in computer vision tasks.

2.3 Training Details

As is mentioned above, our proposed network is easy to train for its simple all-convolutional structure. Training images can be directly used as input even if they have different sizes. However, in order to achieve more accurate results, we first train the multi-column regressor alone and use the trained parameters to initialize parameters of the original network. The VGG part is initialized with values trained on ImageNet. Learning rate is set to 1e−6 and adam optimization algorithm is used for parameter tuning. The loss function is

$$L(\theta) = \frac{1}{2N} \sum_{i=1}^{N} ||D(X_i; \theta) - D_i||_2^2 \tag{3}$$

in which $D(X_i; \theta)$ is the estimated density map generated with input X_i and parameters θ, D_i is the original density map, N is the number of images and L is the loss.

3   Experiments

Our model is evaluated in three major crowd counting datasets: ShanghaiTech [21], UCF_CC_50 [6] and Mall [1]. In this section we train and test our model on these datasets and compare the results with existing state-of-the-art methods.

3.1 Evaluation

The purpose of all crowd counting algorithms is to estimate number of people in a certain scene. After generating the density map, crowd count is achieved by integration, and compared with ground truth value. The two metrics normally used to evaluate crowd counting methods are

$$MAE = \frac{1}{n} \sum_{1}^{N} |z_i - \hat{z}_i| \tag{4}$$

$$MSE = \sqrt{\frac{1}{n} \sum_{1}^{N} (z_i - \hat{z}_i)^2} \tag{5}$$

in which z_i represents the number of labeled heads in the image, $\hat{z}_i$ represents the estimated number, and N is the number of images.

3.2 ShanghaiTech Dataset

ShanghaiTech dataset [21] consists of 1198 annotated images from both online and real streets. This dataset is divided into Part A with extreme dense crowds and Part B with less people. Both parts are already split into training and testing images. To train our network for ShanghaiTech, we first crop 9 samples of every training image, each of 1/9 the size of the original image. Then density maps of the same region are resized to 1/8 height and width, in order to match the output of the network. When testing the model, whole test images were sent directly into the network and density maps of 1/8 height and width are generated. The examples of results generated by our trained network is shown in Fig. 4, and the comparison with other methods are shown in Table 1.

Table 1. Comparison of extended MCNN with other methods on ShanghaiTech dataset

Method	Part A		Part B	
	MAE	MSE	MAE	MSE
Zhang et al. [20]	181.8	277.7	32.0	49.8
MCNN [21]	110.2	173.2	26.4	41.3
Lu et al. [9]	93.4	144.5	20.1	32.3
Switching-CNN [12]	90.4	135.0	21.6	33.4
exMCNN	**88.2**	**127.9**	**15.9**	**27.1**

Fig. 4. The example of ground truth (left) and density map generated by our network (right) from an image in ShanghaiTech Part A. The actual count is 467 and the predicted count is 498.

3.3 UCF_CC_50

UCF_CC_50 [6] is the first truly challenging crowd counting dataset. It is created from the Internet and consists 50 images with 63075 labeled heads in total. The number of labeled heads varies from 94 to 4543. Considering the number of images available, we perform a cross-validation to train and test our method. The images are randomly divided into 5 folds with 10 in each fold. 4 folds are used for training while 1 used for testing each stage. The result is the average of 5 stages. Same as above, we crop 9 samples from each image for training. Table 2 shows the average MAE and MSE are better than existing methods.

3.4 Mall Dataset

Mall dataset [1] is collected from a surveillance camera in a shopping mall. It has 2000 labeled frames from a video sequence. 800 of the images are treated as training samples while the remaining 1200 as testing samples. The comparison of our method and existing ones are shown in Table 3.

Table 2. Comparison of extended MCNN with other methods on UCF_CC_50 dataset

Method	MAE	MSE
Zhang et al. [20]	467.0	498.5
MCNN [21]	377.6	509.1
Walach and Wolf [18]	364.4	-
MoCNN [7]	361.7	493.3
Lu et al. [9]	355.0	532.1
Hydra-CNN [11]	333.7	425.2
Marsden et al. [10]	338.6	424.5
Switching-CNN [12]	318.1	439.2
exMCNN	**276.3**	**394.7**

Table 3. Comparison of extended MCNN with other methods on Mall dataset

Method	MAE	MSE
Walach and Wolf [18]	**2.01**	-
MoCNN [7]	2.75	13.4
Sheng et al. [15]	2.41	9.12
exMCNN	2.45	**3.08**

4 Conclusion

In this paper, we proposed a novel extended multi-column convolutional neural network structure specifically for crowd counting problems. The network consists of two parts: multi-column regressor and VGG-based feature extractor. We described the detailed structure of the network and training details. While testing on three challenging crowd counting datasets, our model achieves state-of-the-art results, with lower MAEs and MSEs than existing methods.

References

1. Chen, K., Loy, C.C., Gong, S., Xiang, T.: Feature mining for localised crowd counting. In: BMVC, vol. 1, p. 3 (2012)
2. Ciregan, D., Meier, U., Schmidhuber, J.: Multi-column deep neural networks for image classification. In: 2012 IEEE Conference on Computer vision and pattern recognition (CVPR), pp. 3642–3649. IEEE (2012)
3. Dalal, N., Triggs, B.: Histograms of oriented gradients for human detection. In: 2005 IEEE Computer Society Conference on Computer Vision and Pattern Recognition, CVPR 2005, vol. 1, pp. 886–893. IEEE (2005)
4. Felzenszwalb, P.F., Girshick, R.B., McAllester, D., Ramanan, D.: Object detection with discriminatively trained part-based models. IEEE Trans. Pattern Anal. Mach. Intell. **32**(9), 1627–1645 (2010)

5. Gall, J., Yao, A., Razavi, N., Van Gool, L., Lempitsky, V.: Hough forests for object detection, tracking, and action recognition. IEEE Trans. Pattern Anal. Mach. Intell. **33**(11), 2188–2202 (2011)
6. Idrees, H., Saleemi, I., Seibert, C., Shah, M.: Multi-source multi-scale counting in extremely dense crowd images. In: 2013 IEEE Conference on Computer Vision and Pattern Recognition (CVPR), pp. 2547–2554. IEEE (2013)
7. Kumagai, S., Hotta, K., Kurita, T.: Mixture of counting CNNs: adaptive integration of CNNs specialized to specific appearance for crowd counting. arXiv preprint arXiv:1703.09393 (2017)
8. Lin, S.F., Chen, J.Y., Chao, H.X.: Estimation of number of people in crowded scenes using perspective transformation. IEEE Trans. Syst. Man Cybern.-Part A: Syst. Hum. **31**(6), 645–654 (2001)
9. Lu, M., Yan, B.: Deep residual convolution neural network for single-image robust crowd counting. In: Zeng, B., Huang, Q., El Saddik, A., Li, H., Jiang, S., Fan, X. (eds.) PCM 2017. LNCS, vol. 10736, pp. 654–662. Springer, Cham (2018). https://doi.org/10.1007/978-3-319-77383-4_64
10. Marsden, M., McGuiness, K., Little, S., O'Connor, N.E.: Fully convolutional crowd counting on highly congested scenes. arXiv preprint arXiv:1612.00220 (2016)
11. Oñoro-Rubio, D., López-Sastre, R.J.: Towards perspective-free object counting with deep learning. In: Leibe, B., Matas, J., Sebe, N., Welling, M. (eds.) ECCV 2016. LNCS, vol. 9911, pp. 615–629. Springer, Cham (2016). https://doi.org/10.1007/978-3-319-46478-7_38
12. Sam, D.B., Surya, S., Babu, R.V.: Switching convolutional neural network for crowd counting. In: Proceedings of the IEEE Conference on Computer Vision and Pattern Recognition, vol. 1, p. 6 (2017)
13. Shang, C., Ai, H., Bai, B.: End-to-end crowd counting via joint learning local and global count. In: 2016 IEEE International Conference on Image Processing (ICIP), pp. 1215–1219. IEEE (2016)
14. Shen, J., Cai, Y.J., Luo, L.: A context-aware mobile web middleware for service of surveillance video with privacy. Multimed. Tools Appl. **74**(18), 8025–8051 (2015)
15. Sheng, B., Shen, C., Lin, G., Li, J., Yang, W., Sun, C.: Crowd counting via weighted VLAD on dense attribute feature maps. IEEE Trans. Circuits Syst. Video Technol. (2016)
16. Sindagi, V.A., Patel, V.M.: A survey of recent advances in CNN-based single image crowd counting and density estimation. Pattern Recognit. Lett. **107**, 3–16 (2017)
17. Viola, P., Jones, M.J., Snow, D.: Detecting pedestrians using patterns of motion and appearance. In: Null, p. 734. IEEE (2003)
18. Walach, E., Wolf, L.: Learning to count with CNN boosting. In: Leibe, B., Matas, J., Sebe, N., Welling, M. (eds.) ECCV 2016. LNCS, vol. 9906, pp. 660–676. Springer, Cham (2016). https://doi.org/10.1007/978-3-319-46475-6_41
19. Wang, C., Zhang, H., Yang, L., Liu, S., Cao, X.: Deep people counting in extremely dense crowds. In: Proceedings of the 23rd ACM International Conference on Multimedia, pp. 1299–1302. ACM (2015)
20. Zhang, C., Li, H., Wang, X., Yang, X.: Cross-scene crowd counting via deep convolutional neural networks. In: 2015 IEEE Conference on Computer Vision and Pattern Recognition (CVPR), pp. 833–841. IEEE (2015)
21. Zhang, Y., Zhou, D., Chen, S., Gao, S., Ma, Y.: Single-image crowd counting via multi-column convolutional neural network. In: Proceedings of the IEEE Conference on Computer Vision and Pattern Recognition, pp. 589–597 (2016)

ECG Classification Algorithm Using Shape Context

Xin Liu[✉] and Zhiqiang Wei

College of Information Science and Engineering, Ocean University of China,
Qingdao, China
liuxinouc@126.com

Abstract. ECG classification algorithm based on machine learning is often required to obtain a classification model by analyzing and studying a large number of sample data. Whether the distribution of sample data is uniform or whether the coverage of the sample is comprehensive determines the accuracy of the classification model finally obtained. The method proposed in this paper to classify ECG waveforms using shape context features combined with labeling information, zoom matrix mechanism, sliding comparison mechanism, standard deviation distance and Hamming distance, requires only a few typical samples. Experiments have shown that this method can obtain fairly accurate classification results for typical common heart diseases. Compared with the traditional diagnosis method based on the discriminant tree, this method has less dependence on prior knowledge, it does not need to measure the width and amplitude of each wave, and it has good robustness, and will not cause jumps in the classification results due to wave group boundary measurement errors. Compared with the machine learning method, this method does not need a large number of training sets and does not need to train the model. It is also relatively easy to extend the new disease classification type.

Keywords: ECG classification · Shape context

1 Introduction

ECG (Electrocardiograph) is a visual image that reflects the heart's electrical activity. Its use is simple and non-invasive, and it has been widely used in clinical practice. There are many kinds of morbid ECGs and there are great differences. There are also significant differences in the ECG of different patients with the same pathology. The existing ECG classification methods based on pattern recognition are not very effective in clinical application and are mainly characterized by poor robustness. The existing classification algorithm is implemented by waveform measurement and decision tree. That is, the ECG signal is first decomposed into a P-QRS-T wave group, and then the characteristics such as the shape, amplitude, and time limit of each wave group are obtained, and then the classification of the disease symptoms is distinguished by the method of the discriminant tree. When the P-QRS-T limit is blurred, the boundary error can cause more serious errors in subsequent calculation steps. The problems that result are: The judgment of normal electrocardiogram is very accurate, but the judgment of

R. Hong et al. (Eds.): PCM 2018, LNCS 11166, pp. 541–553, 2018.
https://doi.org/10.1007/978-3-030-00764-5_50

abnormal ECG is very inaccurate. Users urgently need to make accurate judgments about abnormal ECG.

There are many ideas for solving this problem. The first idea is to try to eliminate ECG noise. Xiong et al. [1] proposed a shrinkage denoising technique and proposed a superimposed shrinkage denoising automatic coder (CDAE) to construct a deep neural network (DNN) for denoising. The ECG noise reduction method has better performance, in particular, the signal to noise ratio (SNR) is improved by 2.40 dB and the root means squared error (RMSE) is improved by approximately 0.075 to 0.350. The second idea is to adopt new noise-insensitive classification methods, such as neural networks. Muhammad et al. [2] proposed an ECG beat classification system using convolutional neural networks (CNN). The model integrates the two main parts of the ECG pattern recognition system, feature extraction, and classification. The test results demonstrate that the proposed method is more accurate than most prior art classifications. Kiranyaz et al. [3] proposed a one-dimensional convolutional neural network (CNN) with an adaptive function that can use a relatively small amount of training data to train a dedicated CNN for each patient in detecting ventricular ectopic beats. (VEB) and supraventricular ectopic beat (SVEB) achieve excellent classification performance. The third idea is to use new features and feature combinations to improve the classification accuracy. Ge [4] proposed a method based on the improved Fisher standard (IFC), which can obtain more effective features. The classification accuracy based on IFC is higher than the traditional LDA-based method. Das and Ari [5] proposed two different feature extraction methods for ECG heartbeat classification: (1) S-transform based features and temporal features and (2) ST and WT based features and temporal features. Extracted feature sets are individually classified using Multilayer Perceptron Neural Networks (MLPNN). The experimental results show that this method has better performance than the existing methods.

In the above methods, choosing features or features combinations is a key point to solve the problem. Their methods can improve the classification accuracy of specific cases, but does not generally improve the classification accuracy of all cases, due to one feature can only show significant differences between certain specific cases. In theory, it is also feasible to combine several weak classifiers into one strong classifier. However, it is better to find a more general feature among various cases to avoid dimensional explosion caused by too many features. The method proposed in this paper attempts to describe ECG signals using two-dimensional features. In the field of 2D graphics processing, the shape context feature is a feature that can stand the test of time and is widely used in handwriting recognition, symbol recognition, shape recognition and other fields. The method proposed in this paper is based on the extraction of shape context features, combined with related electrocardiogram medical knowledge, and the classification result is obtained based on the position and band position of the lead in the specific electrocardiogram waveform. Experiments have shown that this method can obtain fairly accurate classification results for typical common heart conditions.

2 Algorithm Design

In the doctor's eyes, the electrocardiogram is a two-dimensional image. The doctor usually first makes a preliminary classification based on the shape of the electrocardiogram and then measures the specific wave segment to obtain a more accurate diagnosis. In electrocardiogram textbooks, a large number of knowledge points are not described by magnitude and time limit but are described by visualized vocabulary. For example, the coronal T wave, ST segment elevation, ST-T hook-like changes. These morphological features are highly specific in the diagnosis and can often be directly diagnosed based on morphology.

There are many ways to process two-dimensional images. Considering that ECG is a two-dimensional data with a line structure, the best-known feature for such data is the graphics context feature.

2.1 ECG Shape Context Characteristics

The graphics context feature [6] describes the distribution of other pixels in the neighborhood of the pixel. The data format of the sample is the set of log-polar coordinates of the points. First, the shape information of a point is represented by a set of relative vectors formed by all the other points and divided into a number of fan-shaped regions, and then the number of other points falling in these fan-shaped regions is counted to form a shape context feature vector. Such a feature can have a high sensitivity to the distribution of nearby points and reduce the impact of distant points.

Consider the degree of similarity between two shapes. The distance between the point p_i in the first shape and the point q_i in the second shape is defined as C_{ij}, and the chi-square test is used to test the degree of similarity between the two shapes. Its formula is shown as Eq. 1. The result of the calculation is an N × N matrix called the Cost Matrix.

$$C_{ij} = C(p_i, q_i) = 0.5 \sum_{k=1}^{N} \frac{[h_i(k) - h_j(k)]^2}{h_i(k) + h_j(k)} \tag{1}$$

C_{ij} changes with the correspondence between p_i and q_i. Using the Hungarian algorithm it is possible to find an optimal match, minimizing the overall cost C_{ij} even if the value of Eq. 2 is minimized.

$$H(\pi) = \sum_i C(p_i, q_{\pi(i)}) \tag{2}$$

The result of the minimization is the cost between the two shapes (i.e., the Shape Context Cost value, referred to as the cost value). The more similar the two shapes, the smaller the cost value. In order to obey the expression that the smaller the distance, the higher the degree of similarity, the 1-cost value is used as the graphics context similarity S_{sc}, as shown in Eq. 3.

$$S_{sc} = 1 - H(\pi) \tag{3}$$

2.2   Zoom Matrix Mechanism

The shape context is usually used in the fields of target object recognition, handwriting recognition, symbol sign recognition, and the like. ECG and handwriting are similar to single line graphics. However, ECG differs from handwriting. First, the size of the handwriting is relatively arbitrary, while the ECG signal has a strict space-time scale. On the electrocardiogram drawing, the longitudinal direction represents 0.1 mV per mm; the lateral direction represents 0.04 S time per mm. Some types of ECG abnormalities, such as chest leads 0.5 mV (0.5 mm), can be diagnosed as a disease. Heart rate also affects the ECG. For example, a healthy adult man whose heart rate is in the range of 60 to 100 is normal. When the heart rate is 100, the RR interval, PR interval, and QT interval will be shorter than the heart rate of 60. Therefore, in the process of ECG contrast, certain methods need to be adopted to eliminate the deviation caused by non-pathological factors such as heart rate, bodily form, gender, and age.

Changes in the amplitude and time frame of the sample point have an impact on the cost value. As shown in Fig. 1.

$$cost = 0.021 \qquad cost = 0.013 \qquad cost = 0.097$$

Fig. 1. Effect of sample amplitude on cost.

If you simply compare the shapes, you can compare the two comparison samples according to the maximum and minimum values. However, the ECG characteristics of some conditions are just too high or too low, such as ventricular hypertrophy. In order to search for the most similar comparison samples within a reasonable range, the method controls the amplitude range of the voltage scaled by the sample within the $\pm 35\%$ range, and the time range is controlled within the $\pm 25\%$ range to form a scaling matrix. The sample ECG is compared with each element in the scaling matrix and takes its minimum value as the final cost result.

2.3   Sliding Comparison Mechanism

The phase difference of the sample has an effect on the cost value. In order to eliminate the influence of the phase difference of the sample, the method uses a sliding method to compare two waveform samples. During the sliding process, the cost value goes down firstly, then up. Take the minimum cost as the final result. Due to the slower computational speed of the graphics context, the Euclidean distance of the down-sampling can be compared in the actual project to find the nearest position of the Euclidean distance, and then use the graphics context distance to find the extreme value in a small range. If the two samples are closer, the minimum cost is smaller, as shown in Fig. 2. If there is a gap between the two samples, the final cost value will not be too small.

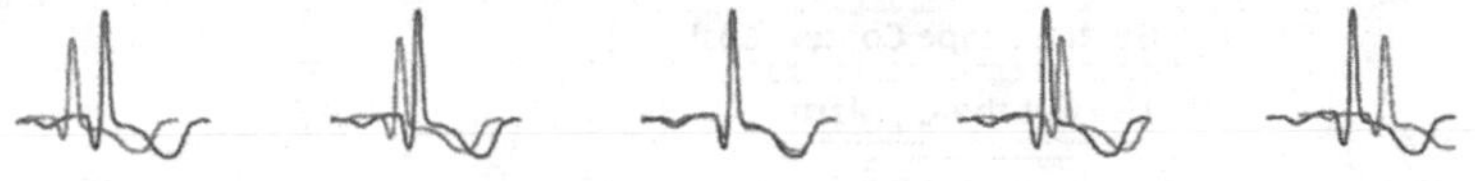

cost = 0.171   cost = 0.143   cost = 0.021   cost = 0.118   cost = 0.163

Fig. 2. Comparison of two similar samples.

2.4 The Weighting of Tagging Information

The position of the waveform with a special shape is also related to the classification result of the electrocardiogram. There are two types of annotation information in medical literature. The first type of annotation information indicates which leads are closely related to the type of disease. For example, the ECG appears "the ST segment can be elevated in the II, III, and aVF leads" (II, III, aVF are lead names. aVF is a lead with its positive pole at the left foot) suggests obstruction in the inferior wall of the heart. If this morphological change occurs on other leads, it has nothing to do with the inferior wall lesions. In this case, a calculation rule needs to be added in the algorithm, and the S_{sc} value is only calculated for the relevant lead, and then the mean value or the maximum value is selected as the similarity value S of the entire sample. The choice of whether to use the average value or the maximum value depends on the content of the literature. If the bibliographic word such as "The best lead to use for this assessment is V1 ..." then the maximum value is chosen; if it is "It can be seen in every lead ..." then choose the mean value. The similarity S is selected as shown in Eq. 4.

$$S = \begin{cases} \max(S_{sc} \text{ of denoted lead })......(on \ any) \\ \text{mean}(S_{sc} \text{ of denoted lead })......(on \ every) \end{cases} \tag{4}$$

The second type of annotation information indicates which segments of the waveform have the key features to identify the disease. This method uses Eq. 5 to add a weight of $\beta \times S_{key}$ to the S_{sc} value. S_{key} is the Euclidean distance between the key segments of the two comparison samples (the key segments can be resampled so that they are equal in length). The weight β can be measured by a large number of experiments. S_{key} is very important in some types of checkouts. For example, in the diagnosis of atrial abnormalities, the shape and amplitude of p-waves are critical diagnostic indicators. Since the amplitude of the p-wave is only one-tenth of the qrs complex, if the labeling information is not considered, then the qrs value will be approximately equal to the similarity of this ECG sample, thus completely masking the p-wave morphological difference. The complete process for calculating similarity S is shown in Fig. 3.

$$S_{key} = \sqrt{\sum_{k=1}^{N} [h_i(k) - h_j(k)]^2}$$
$$S = S_{sc} + \beta \cdot S_{key} \tag{5}$$

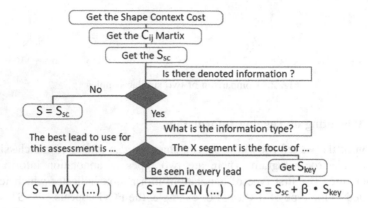

Fig. 3. Complete processes for calculating similarity S.

2.5   Prediction Method Based on Standard Deviation Distance

According to the above method, the distance $[S_1, S_2, ..., S_n]$ between the new sample Q and the standard group sample $[K_1, K_2, ..., K_n]$ is calculated, and the maximum value S_m is found out, indicating that between Q and K_m The highest degree of similarity, the introduction of Q belongs to m category. But in fact, the problem is not that simple. The differences between some illnesses are very subtle and even almost mixed. In ECG reports, doctors are allowed to report more than one suspected cases at the same time.

If the sample to be tested does not belong to any one of the standard groups, it needs to be excluded. Due to the complexity of ECG data and the limited nature of standard group samples, it is difficult to determine the standard group sample distribution. Assume that the test sample Q belongs to a certain type in the standard group. The sample in the standard group is $[x_1, x_2, ..., x_n]$, and its distribution radius is $[r_1, r_2, ..., r_n]$. According to the law of triangles, the constraint condition of formula 6 is obtained.

$$\begin{cases} 1. & d(Q, x_i) + d(Q, x_j) \geq d(x_i, x_j) \\ 2. & r_i + d(x_i, x_j) \geq d(Q, x_j) \end{cases} \tag{6}$$

Equation 6(1) states that the distance from Q to x_i to x_j cannot be too close; Eq. 6 (2) specifies that the distance from Q to x_i and x_j cannot be too far. The distance between Q and each sample will shift. Close to one, away from the other. The range of the overall distance from Q to all samples cannot exceed the range covered by the distribution radius of all samples. The standard deviation of the distance between the sample to be tested and all standard group samples should be within a certain range. If it is beyond this range, whether too close or too far, it means that the sample to be tested should not belong to any one of the standard groups. Based on this, the exclusion formula based on the standard deviation is shown as Eq. 7. The minimum standard deviation σ_{min} and the maximum standard deviation σ_{max} need to be measured by tests.

$$\sigma_{min} \leq \sigma(d(Q, x_i)) \leq \sigma_{max} \qquad (7)$$

There is still one such exception: There may be holes in the center of the sample set. If the sample to be tested is in the center cavity, it cannot be excluded by the standard deviation. However, this situation can be eliminated by using the Hamming distance described below.

2.6   Forecasting Method Based on Hamming Distance

To append an English alphabetic alias L to each category tag in the standard group, the ordered arrangement of elements in the category tag set $Y^{(K)}$ can be considered as a character sequence $L^{(K)}$.

Take the sample Q to be tested and compare it with the standard group sample $[K_1, K_2, ..., K_n]$ to obtain a sequence consisting of similarity and labels $[[S_1, Y_1], [S_2, Y_2], ..., [S_K, Y_K]]$. According to the value of similarity S, it is arranged in descending order. Its label Y_n can form an ordered sequence. Replace Y_i with L_i to obtain the character sequence $[L^{(K)}]$. E.g:

$$\text{Let} \quad Y_1 := A, \quad Y_2 := B, \quad Y_3 := C$$
$$\text{Then} \quad Y_1 Y_1 Y_2 Y_3 Y_2 Y_2 := AABCBB$$

Examining the distance between Q and the standard group sample, we can find a law that moves Q within a small range, and there is a jump in the character sequence relationship between samples with high similarity (nearby) and the samples with lower similarity (remote) has a relatively stable character sequence relationship. This phenomenon can be illustrated by Fig. 4.

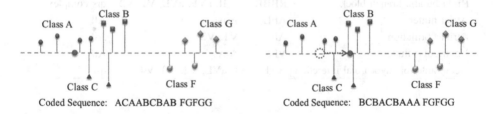

Fig. 4. Relationship between sample distance and alias sequence.

Calculate the ordered label alias sequence $[L^{(K)}]$ between the test sample and the standard group sample, and compare the labels with the standard group to verify the obtained label alias sequence $H[L^{(K)}]^{(K)}$ Distance, the type of sample with the smallest Hamming distance is taken as the classification result. However, the classification result is lower than the classification result based on the standard deviation. Therefore, the report should be marked as "suspected".

Set a threshold for the Hamming distance. If the Hamming distances of all sequences in $[L^{(K)}]$ and $H[L^{(K)}]^{(K)}$ are greater than the threshold, then the location of the data space where the sample is to be tested is indicated. There is a large difference from all standard group samples, most likely in the center hollow area of the standard group sample. This sample to be tested should not belong to any one of the standard groups.

3　Experiments

Standard group ECG data was produced according to the authoritative ECG teaching materials and related literature. This article uses digitization technology to convert printed images into waveform data. The standard group dataset contains 72 groups of electrocardiographic patterns including normal ECG and is divided into 14 categories, as shown in Table 1. Each disease contains 4 to 7 cases. Each case has 12-lead data and its length varies from 1 to 3 cycles.

Table 1. Disease classification and annotated information table.

Classification name	Abbr.	Key leads	Key segment
Normal ECG	NORM	All	All
Clockwise rotation	CR	V1, V2, V3, V4, V5, V6	qrs complex
Right atrial abnormalities	RAA	II, III, aVF, V1, V2	p-q
Left ventricular hypertrophy	LVH	I, III, aVF, V1,V5,V6	q-t
Right ventricular hypertrophy	RVH	aVR, V1, V2, V3, V5	q-t
Double ventricle hypertrophy	DVH	aVR, V1, V5, V6	q-t
Ventricular septal defect	VSD	All	q-t
Atrial septal defect	ASD	All	All
Hyperkalemia	HPEKA	All	t-u
Right bundle branch block	RBBB	I, II, aVF, aVL, V1, V2	qrs complex
Atrial flutter	AFL	Any	All
Atrial fibrillation	AF	V1, V2	All
Ventricular flutter	VFL	Any	All
Acute anterior myocardial infarction	AAI	I, aVL, V2, V3, V4	q-t

3.1　Verification Results with Tagging Information

Conduct cross-validation of the ECG samples in the standard group to verify the suitability of the graphics context method on this data set. If we do not consider the labeling information in the literature, we can only rely on the shape context features for classification. In the Confusion Matrix results, only the NORM type shows a clear square dark border, while other types of square borders are not obvious and do not show the theoretically expected distribution.

　　The reason is that in most cases, there is no strong similarity between multiple samples belonging to the same disease. If it is not clearly stated in the literature to determine the key waveform characteristics of the disease, it is impossible to rely on the visual effects to classify them into the same category. Therefore, this algorithm designs an "annotated information table" to indicate the Key Leads and Key Segment information in each sample, as shown in Table 1. The specific method is to add a tag to the key leads and mark the start and end points of the key segment on the data of the standard group sample. The computer can then automatically process the labeling information according to the flow of Fig. 3.

It can be seen from the experimental results shown in Fig. 5 that the labeling information has significantly improved the classification effect, and an ideal square pattern is presented on the diagonal of the Confusion Matrix.

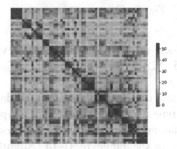

Fig. 5. Confusion Matrix after adding denoted information.

Among them, the shape context similarity between the normal class, right bundle branch block type, hyperkalemia class, ventricular flutter class, and acute anterior wall myocardial infarction class was very high. However, the classification results in the clockwise, right atrial abnormalities, and various ventricular hypertrophy categories were not good. One of the main reasons is that because the shape context features are sensitive to the shape of the waveform, the discrimination rate will increase if there is a certain shape of a certain type of disease. If the shape characteristics of certain waveforms are not obvious, the sensitivity is not high enough.

3.2   Prediction Process Verification

A series of similarity values can be obtained by comparing the samples to be tested and the standard library samples one by one. The similarity is normalized to the 0–1 range and arranged in a descending order to obtain a columnar graph as Fig. 6.

In general, the type with the highest similarity is usually used as a result of the prediction system. However, in this system, we must also consider such a situation: This sample does not belong to any type of standard group. Therefore, first of all, it should be identified whether a sample to be tested belongs to a certain category in the standard group.

Due to the complexity of ECG data and the limitations of standard group samples, it is impossible to measure the distribution of its graphics context distance. The method adopted in this paper is: cross-testing $N = 72$ cases in the standard group, each case can

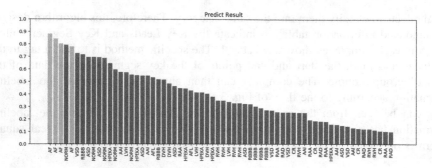

Fig. 6. Similarity histogram.

obtain N − 1 similarity values, calculate the Standard Deviation of the N − 1 simi-larities, and the result is 0.45–0.29 range. We using $STD_{min} = 0.26$ and $STD_{max} = 0.48$ as the discrimination threshold.

Then, the N − 1 similarities obtained by the cross-validation of the standard group samples are sorted, and the type name is replaced with a type tag to obtain an ordered character sequence. The Hamming distance between these character sequences is calculated and the result is in the range of 0–14. Slightly relax the limit we set $H_{max} = 20$. The Standard Deviation Distance and the Hamming Distance are integrated to formulate the judgment flow shown in Fig. 7.

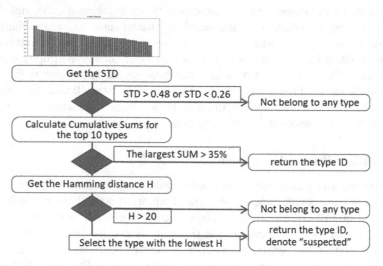

Fig. 7. Classification result judgment process.

3.3   Model Prediction Experiment

Test the classification accuracy of the entire model. Through the prediction of the 133 groups of data to be measured from other documents, the accuracy of classification of ECG classifications of various types is shown in Table 2. From the test results, the classification accuracy of the normal type and VFL (ventricular flutter) type is the highest. From the morphology of the ECG waveform, the normal type is regular and the VFL waveform is similar to a sine wave, belonging to two extreme forms. The classification accuracy of LVH, RVH, HPEKA, RBBB, and AAI is also high. Each of these conditions has a typical characteristic waveform. In addition, the elimination accuracy rate for the types not belonging to the standard group also reached 97%.

This method does not have high accuracy in the classification of DVH (double ventricle hypertrophy), AFL (atrial flutter), and AF (atrial fibrillation). The diagnostic criteria for AF are "Heart rate is absolutely uneven", and the graphical context method and Sliding Window mechanism are not sensitive to periodic changes.

In theory, if you add more typical samples to the standard group, you can further increase the number of classifications for this method. If you break down the existing categories into several subtypes, you can further improve the classification accuracy.

Table 2. Test results.

Abbr.	Count	Correct	Error	Precision rate
NORM	24	24	0	100%
CR	8	5	3	63%
RAA	6	4	2	66%
LVH	8	7	1	88%
RVH	4	3	1	75%
DVH	2	1	1	50%
VSD	6	4	2	66%
ASD	3	1	2	33%
HPEKA	11	8	3	73%
RBBB	7	6	1	86%
AFL	6	3	3	50%
AF	7	3	4	43%
VFL	4	4	0	100%
AAI	5	4	1	80%
Others	32	31	1	97%
Figure out	133	108	25	81%

4 Conclusion

This article proposes a new ECG classification algorithm. The algorithm uses the shape context feature as a key feature to measure the difference in electrocardiogram morphology, and combines the prior knowledge to specify key leads and key bands. Through the combination of these features, the ability to classify samples is improved. In addition, the scaling matrix mechanism and the sliding comparison mechanism are used to improve the generalization ability of the model. The standard deviation and Hamming distance are used to predict the sample. Through experiments, the accuracy of the ECG classification algorithm proposed in this paper can reach 81%. This method can be sensitive to identify the abnormal shape of the ECG.

Acknowledgements. This work is supported by The Aoshan Innovation Project in Science and Technology of Qingdao National Laboratory for Marine Science and Technology (No. 2016ASKJ07).

References

1. Xiong, P., et al.: A stacsked contractive denoising auto-encoder for ECG signal denoising. Physiol. Meas. **37**(12), 2214 (2016)
2. Muhammad, Z., Kim, J., Yoon C.: An automated ECG beat classification system using convolutional neural networks. In: International Conference on It Convergence and Security IEEE, pp. 1–5 (2016)
3. Kiranyaz, S., et al.: Convolutional neural networks for patient-specific ecg classification. In: Engineering in Medicine and Biology Society IEEE, p. 2608 (2015)
4. Ge, D.: Application of an improved fisher criteria in feature extraction of similar ECG patterns. In: Huang, D.-S., Ma, J., Jo, K.-H., Gromiha, M.M. (eds.) ICIC 2012. LNCS (LNAI), vol. 7390, pp. 358–365. Springer, Heidelberg (2012). https://doi.org/10.1007/978-3-642-31576-3_46
5. Das, M.K., Ari, S.: ECG beats classification using mixture of features. Int. Sch. Res. Not. **2014**(3) (2014)
6. Belongie, S., Malik, J., Puzicha, J.: Shape matching and object recognition using shape contexts. Lect. Notes Comput. Sci. **2**(4), 509–522 (2002)
7. Han, M., Zheng, D.C.: Shape recognition based on fuzzy shape context. Acta Autom. Sin. **38**(1), 68–75 (2012)
8. Schwartz, W.R., Menotti, D.: ECG-based heartbeat classification for arrhythmia detection. Comput. Methods Prog. Biomed. **127C**, 144 (2016)
9. Goh, K.W., et al.: Issues in implementing a knowledge-based ECG analyzer for personal mobile health monitoring. In: International Conference of the IEEE Engineering in Medicine and Biology Society. p. 6265 (2006)
10. Kiranyaz, S., Ince, T., Gabbouj, M.: Real-time patient-specific ECG classification by 1D convolutional neural networks. IEEE Trans. Biomed. Eng. **63**(3), 664 (2016)
11. Romero, D., et al.: Ensemble classifier based on linear discriminant analysis for distinguishing Brugada syndrome patients according to symptomatology. In: Computing in Cardiology Conference IEEE (2017)
12. Shaji, S., Vinodini Ramesh, M., Menon, V.N.: Real-time processing and analysis for activity classification to enhance wearable wireless ECG. Syst. Comput. **45**, 55 (2015). https://doi.org/10.1007/978-81-322-2523-2_3
13. Rajpurkar, P., et al.: Cardiologist-level arrhythmia detection with convolutional neural networks. arXiv:1707.01836v1 (2017)
14. Chen, S.-L., Tuan, M.-C., Chi, T.-K., Lin, T.-L.: VLSI architecture of lossless ECG compression design based on fuzzy decision and optimisation method for wearable devices. Electron. Lett. **51**(18), 1409–1411 (2015)
15. Chu, C.-T., Chiang, H.-K., Hung, J.-J.: Dynamic heart rate monitors algorithm for reflection green light wearable device. IEEE (2015). 978-1-4799-8562-3/15

16. Muduli, P.R., Gunukula, R.R., Mukherjee, A.: A deep learning approach to fetal-ECG signal reconstruction. In: Communication IEEE, pp. 1–6 (2016)
17. Srivastava, V.K.: DWT-based feature extraction from ECG signal. Am. J. Eng. Res. **2**, 3 (2013)
18. Tang, X., Shu, L.: Classification of electrocardiogram signals with RS and quantum neural networks. Int. J. Multimed. Ubiquit. Eng. **9**(2), 363–372 (2014)

Small Object Detection Using Deep Feature Pyramid Networks

Zhenwen Liang, Jie Shao$^{(\boxtimes)}$, Dongyang Zhang, and Lianli Gao

Center for Future Media, School of Computer Science and Engineering,
University of Electronic Science and Technology of China, Chengdu 611731, China
{zhenwenliang,dyzhang}@std.uestc.edu.cn,
{shaojie,lianli.gao}@uestc.edu.cn

Abstract. Recent studies have achieved great progress on the object detection in terms of accuracy and speed using convolutional neural networks (CNNs). However, no matter the one-stage detector or the two-stage detector, usually it is still a challenging task for them to detect small objects because of the low resolution and fuzzy feature representation. When the training set only contains small objects, the performance degrades drastically. To improve the performance of small object detection, we develop a new two-stage detector similar to Faster-RCNN. At region proposal stage, we adopt the feature pyramid architecture with lateral connections, which makes the semantic feature of small objects more sensitive. Meanwhile, we design specialized anchors to detect the small objects from large resolution image and train the network with focal loss. At classification stage, dense convolutional network is used to strengthen the feature transmission and multiplexing, which leads to more accurate classification with fewer parameters. We experiment with the challenging Tsinghua-Tencent 100K benchmark, and the evaluations demonstrate a significant performance improvement compared with other state-of-art methods.

Keywords: Object detection · Convolutional neural network
Feature pyramid

1   Introduction

Recognizing objects in images is an important but challenging computer vision task, which requires to correctly give the object locations and their accurate categories. Because of the rapid development of convolutional neural networks (CNNs), solving computer vision tasks through deep learning has become a hot spot. No matter in terms of accuracy or speed, the performance of object detection has been significantly improved by the means of CNN architecture.

Current state-of-the-art object detectors can be divided into two categories, **one-stage detector** and **two-stage detector**. The one-stage detector, such as Retinanet [14], SSD [15] and YOLO9000 [16], after one feed-forward run, can

© Springer Nature Switzerland AG 2018
R. Hong et al. (Eds.): PCM 2018, LNCS 11166, pp. 554–564, 2018.
https://doi.org/10.1007/978-3-030-00764-5_51

give the object location and the result of its classification simultaneously through a single network. As to the two-stage detector, such as Faster R-CNN [17], R-FCN [2] and FPN [13], all of them consist of region proposal network (RPN) and classification network. A sparse set of candidate bounding boxes produced by the first stage are fed into the second stage network to classify the proposals into specific categories. Thanks to the high accuracy, the two-stage framework becomes the dominant paradigm in modern object detectors. However, in terms of speed, the one-stage framework surpasses most two-stage based methods.

Despite the great success of all detectors above, actually they usually fail to detect very small objects, especially in a high-resolution image. For small objects, the rich feature representations are difficult to learn because of the poor-quality appearance and less pixel information. Compared with large objects, the feature representations of small objects are less discriminative for detector, which is the key reason to impair the detection accuracy [11]. However, in real-life, small objects are very common. For example, how to detect traffic signs from distance is a typical application of small object detection. Therefore, it is of great importance to enhance the performance of small object detection.

There are a number of studies [1,10,19,20] devoted to solve this problem. One of the most common methods [1] is to increase the scale of input images, so the pixels of a small object are super-resolved accordingly. Image pyramid [20] is also introduced to deal with the problem. The large-scale input image is resized to multiple scales to form an image pyramid which is used for inference, and then non-maximum suppression can be used to combine the prediction results. However, image pyramid brings about heavy time consumption during training and test phase. Some other researchers [10,19] design a novel network architecture to integrate the multiple high-level and low-level feature layers. Some specific illustrations are shown in Fig. 1. Note that, all the methods mentioned here are just the trick of simply increasing the feature or data dimensionality. However, more semantically stronger features need to be utilized for the small object detection task.

In this work, we propose a novel two-stage based approach to address the challenging problem of small object detection, which can accurately detect very small object from an image with large resolution. In region proposal stage, to compute multiple convolutional feature maps over an entire input image, ResNet-50 [5] is used as the backbone. Besides, we adopt feature pyramid network (FPN) [13] to get semantically strong features at all levels. For accurate small object detection, features from the layers near the input (referred to as shallow layers) combined with the deep feature layers are used for feature pyramid. We also modify the anchors to fit in with the size of small object. Extensive evaluations are conducted on the challenging Tsinghua-Tencent 100K dataset [21], which is a challenging traffic sign benchmark (some typical traffic signs might just occupy 0.2% of an image with resolution of over 2048 × 2048, and several examples of the detection results for small objects are illustrated in Fig. 2). Compared with the existing detectors, experimental results show that the proposed approach achieves substantial performance improvement.

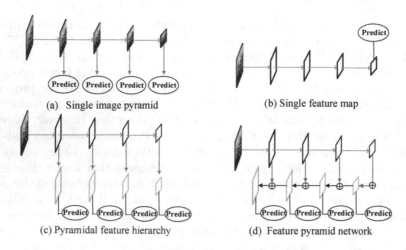

(a) Single image pyramid   (b) Single feature map

(c) Pyramidal feature hierarchy   (d) Feature pyramid network

Fig. 1. Different ways of feature reconstruction for object detection. (a) Image is resized to different scales, which generate different features to make predictions independently. (b) Just like Fast R-CNN [3] and Faster R-CNN [17], only the last layer of the network is used. (c) Extracting features of different scales from hierarchical layers to make predictions, only one layer is used to make each prediction. (d) Feature pyramid network adopts the vertical connection and the pyramidal pathway, merged by addition, and two layers are used to make each prediction.

2   Related Work

2.1   CNN-based Object Detectors

Object detection is a hot research topic in computer vision. The central problem is to judge "where" and "what" for each object instance when given an image [12]. Since the emergence of the ConvNets [8], the CNN-based detectors such as Faster R-CNN [17], R-CNN [4] and FPN [13] show a strong ability to improve accuracy. In R-CNN [4], the classical region proposal based method, each proposal adopts the scale normalization strategy and then uses a ConvNet to give the category. Following the diagram established by R-CNN, Fast R-CNN [3] and Faster R-CNN [17] are other famous two-stage networks, which only compute feature maps from a single scale. They achieve a good trade-off between accuracy and speed. Some other one-stage (also known as proposal-free) methods, such as Retinanet [14], SSD [15] and YOLO9000 [16], generally can produce comparable accuracy compared with two-stage methods. Besides, they have faster detection speed by avoiding an additional object proposal generation stage.

2.2   Small Object Detection

Traffic sign detection is the most representative application of small object detection. Color and shape based features were widely used to address this problem

Fig. 2. Detection results of the proposed approach on Tsinghua-Tencent 100K. We can find that our approach is sensitive to small objects. "p26","po" and others represent the names of different traffic signs.

before the prevalent of CNN [9], but now, CNN-based methods play an important role in traffic sign detection and classification due to their outstanding performance. Zhu et al. [21] propose a single network which can determine an object's bounding box together with its class label simultaneously. Hinge loss is proposed by Jin et al. [7], which helps to improve the accuracy and makes the network convergence faster. In particular, Sermanet et al. [18] use the multi-stage features from the CNN to boost the discrimination. Generally speaking, multi-scale detector [14–16], which uses multiple layers of the feature hierarchy performs much better, especially for small objects.

3   Our Approach

To detect small objects from a high-resolution image, we design a two-stage architecture as shown in Fig. 3. For region proposal stage, two top-down feature pyramids are built to obtain the high-level semantic feature maps at all scales, which boost the discriminative power of feature representation for small object.

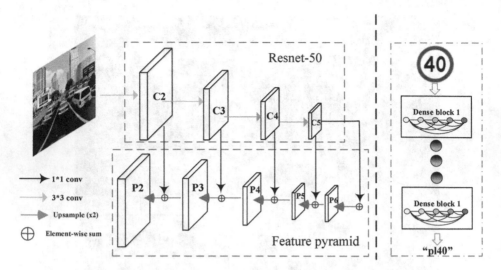

Fig. 3. Overview of our two-stage network. The region proposal stage is a feature pyramid network based on Resnet-50, and different from Fig. 1(d), we use the last layer C_5 twice to make full use of high-level features. For classification network we construct our network by borrowing the idea from dense block [6].

For classification stage, we adopt the idea of Densenet [6]. To improve accuracy, focal loss [14] is used to supervise the proposal network.

3.1 Feature Pyramid Network Architecture

For our two-stage network, the feature pyramid architecture is mainly designed for the region proposal network. We adopt ResNet-50 [5] as the backbone. Given an image as the input, we can get a set of feature maps from each stage's last residual block, denoted as $\{C_2, C_3, C_4, C_5\}$, corresponding to conv2, conv3, conv4 and conv5 [13]. The feature map of conv1 is not included by the consideration of large GPU memory cost. Intuitively, the bottom features have stronger semantic information and larger receptive field than the top, so it is wise to use the bottom features for large object detection. The feature maps near the input are high resolution and low-level semantic feature, but they have less receptive field which is helpful for small object detection. Different from (a), (b) and (c) shown in Fig. 1, we leverage a set of features with pyramidal shape, from each stage of ResNet-50, to create a feature pyramid that has strong semantics at all scales.

As show in Fig. 3, FPN augments another top-down pathway besides the backbone through combining the low-resolution, semantically strong features from the top with the high-resolution, semantically weak features from the bottom. Thus, the multi-scale feature pyramid is simple and efficient built, with rich semantic representation at all levels. For the low-resolution feature map on the top, nearest neighbor upsampling method is used to super-resolved the spatial resolution by a factor of 2, and then they are merged with the front high-resolution feature map through element-wise addition. This process is iterated until the lowest level of features merge to the highest level. From Fig. 3, we can learn that the most bottom feature map denoted as C_5 is used twice to enhance the semantic representation of small object. Finally, we can get a set of feature maps denoted as $\{P_2, P_3, P_4, P_5, P_6\}$, corresponding to $\{C2, C3, C4, C5\}$. Generally, feature pyramid network architecture is easy to built, which leads to high-level semantic feature maps at all scales, but with marginal extra cost.

3.2 Anchors

In RPN [17], the anchors are a set of reference boxes with pre-defined scales and aspect ratios. Because they cover objects of different shapes, if some target objects exist, the anchors can be used for bounding box regression. Traffic signs occupy only a small fraction of an image, so we carefully design the anchors for small object detection. Formally, we define the anchors with the areas of $\{8^2, 16^2, 32^2, 64^2, 128^2\}$ pixels on $\{P2, P3, P4, P5, P6\}$ respectively. At each level, we design the multiple aspect ratios with $\{1 : 2, 1 : 1, 2 : 1\}$ and anchors of sizes $\{2^0, 2^{1/3}, 2^{2/3}\}$. Therefore, there are $3 \times 3 \times 5 = 45$ anchors over the whole pyramidal feature hierarchy. In order to adapt to the lower features of the dimension, the anchor stride is changed to $\{4, 8, 16, 32, 64\}$. As to detection, a tiny fully convolutional sub-network is designed for the purpose of regressing the offset from each anchor box to a nearby ground-truth object, which operates on each level of the pyramid feature at multi-scale.

3.3 Loss Function

Focal loss [14] achieves great success on solving the problem of class imbalance during training. Before focal loss, cross entropy (CE) loss is a common choice to classify the foreground and background classes, which is define as:

$$CE(p, y) = -(y \log(p) + (1 - y)(\log(1 - p))) \tag{1}$$

where $y \in \{1, -1\}$ specifies the ground-truth class and $p \in [0, 1]$ stands for the estimated probability of the class with label $y = 1$. For convenience, we defined p_t as:

$$p_t = \begin{cases} p & if \ y = 1 \\ 1 - p & otherwise, \end{cases} \tag{2}$$

so the equation can be rewrite as $CE(p_t) = -\log(p_t)$. Following [14], the focal loss can be defined as:

$$\mathbf{FL}(p_t) = -\alpha_t (1 - p_t)^\gamma \log(p_t), \tag{3}$$

where γ is a key hyperparameter to control the strength of the modulating term, and α_t is proved to increase the accuracy slightly in practice. Compared with CE function, FL adds a modulating factor as $(1 - p_t)^\gamma$. If $\gamma = 0$, FL is equivalent to CE. As γ increases, the easy examples, contributed less useful

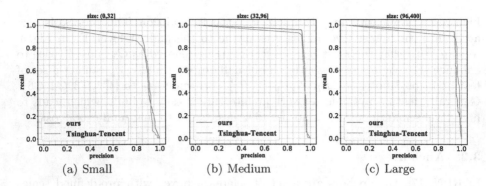

(a) Small   (b) Medium   (c) Large

Fig. 4. Precision-recall curves of compared methods tested on Tsinghua-Tencent dataset for small, medium and large sizes.

gradient, can be further discounted. During training, the extreme imbalance between foreground and background class lead the gradient dominate by easily classified negatives, which comprise the majority of the loss. Because of the modulating factor, the gradient from easy examples can be automatically filtered out. In essence, the focal loss focuses training on a sparse set of hard examples and effectively discount the effect of easy examples.

4 Experiments

4.1 Experimental Dataset and Measurement

We conduct our experiments on Tsinghua-Tencent 100K dataset [21]. The challenge of this dataset reflects in three aspects: (i) The dataset from 100000 Tencent Street View panoramas is very large, going beyond previous benchmarks. (ii) The images in the dataset are very different in illuminance and weather conditions. (iii) The original images are high resolution with the shape of 2048×2048. Moreover, some typical traffic signs occupy only a small fraction of an image. For the reasons above, few people carry out experiments on this dataset. Both precision and recall are used as the performance measures in the evaluation.

4.2 Training Details and Parameters

We first introduce some training details. The proposal network is trained on a NVIDIA 1080 Ti GPU. As Fig. 3 shows, ResNets-50, pre-trained on ImageNet1k, is used as the backbone to build the feature pyramid network (FPN) with a top-down pathway and lateral connections. Following [14], we also use the focal loss to supervise our first stage region proposal network, and standard smooth $L1$ loss is used for box regression. In Eq. 3, we set $\gamma = 2$ and $\alpha_t = 0.25$, which have been proved to be effective in practice. As to the classification network, we borrow the idea from Densenet [6], and train a fine classification network

Table 1. Detection performance of different methods for different sizes of traffic signs in Tsinghua-Tencent 100K. (R): Recall, (P): Precision (in %). Small, medium and large correspond to different areas of smaller than 32^2 pixels, between 32^2 and 96^2 pixels, and larger than 96^2 pixels.

Object size	Small	Medium	Large
Fast R-CNN [3] (R)	46	71	77
Fast R-CNN [3] (P)	74	82	80
Zhu et al. [21] (R)	87	94	88
Zhu et al. [21] (P)	82	91	91
Ours (R)	**93**	**97**	**92**
Ours (P)	**84**	**95**	**96**

Table 2. Detection performance of some commonly used classes on Tsinghua-Tencent 100K. (R): Recall, (P): Precision (in %).

Class	i2	i4	i5	il100	il60	il80	io	ip	p10	p11	p12	p19	p23	p26	p27
Fast R-CNN [3] (R)	51	74	84	44	61	10	70	73	54	71	21	42	65	63	36
Fast R-CNN [3] (P)	82	86	85	85	70	91	75	80	72	73	47	48	79	74	100
Zhu et al. [21] (R)	82	94	95	97	91	94	89	92	95	91	89	94	94	93	96
Zhu et al. [21] (P)	72	83	92	100	91	93	76	87	78	89	88	53	87	82	78
Ours (R)	87	97	96	97	98	100	94	88	92	95	95	91	94	95	98
Ours (P)	90	92	94	93	98	94	86	90	89	90	94	75	93	89	98

Class	pl120	p5	p6	pg	ph4	ph4.5	ph5	pl100	p3	p120	p130	p140	pl5	pl50	pl60
Fast R-CNN [3] (R)	39	78	8	88	32	77	18	68	50	14	18	58	69	34	41
Fast R-CNN [3] (P)	92	87	100	86	92	82	88	86	85	89	59	78	88	65	73
Zhu et al. [21] (R)	98	95	87	91	82	88	82	98	91	96	94	96	94	94	93
Zhu et al. [21] (P)	98	95	87	91	82	88	82	98	91	96	94	96	94	94	93
Ours (R)	97	98	97	98	86	90	90	100	96	98	97	97	94	97	98
Ours (P)	99	91	90	93	94	80	78	98	81	90	92	91	92	90	95

Class	pl70	pl80	pm20	pne	pm55	pn	pm30	po	pr40	w13	w32	w55	w57	w59	wo
Fast R-CNN [3] (R)	2	34	43	90	58	87	19	46	95	32	41	43	73	74	16
Fast R-CNN [3] (P)	100	84	70	87	76	85	67	66	78	40	100	57	66	64	55
Zhu et al. [21] (R)	93	95	88	93	95	91	91	67	98	65	71	72	79	82	45
Zhu et al. [21] (P)	95	94	91	93	60	92	81	84	76	65	89	86	95	75	52
Ours (R)	93	99	94	96	97	96	96	82	100	90	91	95	94	93	42
Ours (P)	98	92	98	97	86	90	97	81	97	90	95	95	90	68	50

with depth 40 and growth rate 24. The region of interest (ROI) is resized to 48×48, then fed to the classification network to give the specific category. For optimization, we train our network with stochastic gradient descent (SGD), and set weight decay to 0.0001 and momentum to 0.9.

4.3 Result and Analysis

We compare our approach with Fast R-CNN [3] and Zhu et al. [21] on Tsinghua-Tencent 100K dataset. Different from the other two methods which use CE loss, we employ focal loss to train our network. Through the experiments, we find that this modification makes a difference significantly, improving the precision about 5%. According to [21], based on the area size, the dataset is divided into three categories: small objects (area $< 32^2$), medium objects ($32^2 <$ area $< 96^2$) and large objects (area $> 96^2$). From Table 1, we can observe that our approach offers a substantial improvement in terms of average recall and precision on three subsets of different object sizes. More concretely, the increase in the small object detection is more obvious (6% and 2% in average recall and precision). For the precision-recall curves, referring to Fig. 4, this clearly demonstrates the effectiveness of our approach, outperforming Fast R-CNN [3] and Zhu et al. [21]. The number of categories of traffic sign is about one hundred and fifty, and here we exhibit precision and recall of most commonly used traffic signs in Table 2. Our method achieves the best performance in most categories, with some classes reach 100% recall, such as "il80" and "pl100". Referring to Fig. 2, we show some examples of the detection results for small objects.

5   Conclusion

In this paper, we proposed a novel two-stage network to address the challenging problem of small object detection. In region proposal stage, we use ResNet-50 as the backbone to build feature pyramids, which greatly enhance the semantic representation of small objects. Different from the existing detectors, focal loss is also used to supervise our region proposal network, which applies a modulating term to the CE loss and focuses the training phase on hard negative examples. For classification, we adopt the idea of Densenet. Although our approach is simple and easy for implementation, extensive evaluations demonstrate that we outperform the state-of-the-art methods in term of accuracy on Tsinghua-Tencent 100K dataset.

References

1. Chen, X., Kundu, K., Zhu, Y., Ma, H., Fidler, S., Urtasun, R.: 3D object proposals using stereo imagery for accurate object class detection. IEEE Trans. Pattern Anal. Mach. Intell. **40**(5), 1259–1272 (2018)
2. Dai, J., Li, Y., He, K., Sun, J.: R-FCN: object detection via region-based fully convolutional networks. In: Advances in Neural Information Processing Systems 29: Annual Conference on Neural Information Processing Systems 2016, 5–10 December 2016, Barcelona, Spain, pp. 379–387 (2016)
3. Girshick, R.B.: Fast R-CNN. In: 2015 IEEE International Conference on Computer Vision, ICCV 2015, Santiago, Chile, 7–13 December 2015, pp. 1440–1448 (2015)
4. Girshick, R.B., Donahue, J., Darrell, T., Malik, J.: Rich feature hierarchies for accurate object detection and semantic segmentation. In: 2014 IEEE Conference on Computer Vision and Pattern Recognition, CVPR 2014, Columbus, OH, USA, 23–28 June 2014, pp. 580–587 (2014)
5. He, K., Zhang, X., Ren, S., Sun, J.: Deep residual learning for image recognition. In: 2016 IEEE Conference on Computer Vision and Pattern Recognition, CVPR 2016, Las Vegas, NV, USA, 27–30 June 2016, pp. 770–778 (2016)

6. Huang, G., Liu, Z., van der Maaten, L., Weinberger, K.Q.: Densely connected convolutional networks. In: 2017 IEEE Conference on Computer Vision and Pattern Recognition, CVPR 2017, Honolulu, HI, USA, 21–26 July 2017, pp. 2261–2269 (2017)
7. Jin, J., Fu, K., Zhang, C.: Traffic sign recognition with hinge loss trained convolutional neural networks. IEEE Trans. Intell. Transp. Syst. **15**(5), 1991–2000 (2014)
8. Krizhevsky, A., Sutskever, I., Hinton, G.E.: Imagenet classification with deep convolutional neural networks. In: Advances in Neural Information Processing Systems 25: 26th Annual Conference on Neural Information Processing Systems 2012. Proceedings of a meeting held 3–6 December 2012, Lake Tahoe, Nevada, United States, pp. 1106–1114 (2012)
9. Le, T.T., Tran, S.T., Mita, S., Nguyen, T.D.: Real time traffic sign detection using color and shape-based features. In: Nguyen, N.T., Le, M.T., Świątek, J. (eds.) ACIIDS 2010. LNCS (LNAI), vol. 5991, pp. 268–278. Springer, Heidelberg (2010). https://doi.org/10.1007/978-3-642-12101-2_28
10. Li, H., Lin, Z., Shen, X., Brandt, J., Hua, G.: A convolutional neural network cascade for face detection. In: IEEE Conference on Computer Vision and Pattern Recognition, CVPR 2015, Boston, MA, USA, 7–12 June 2015, pp. 5325–5334 (2015)
11. Li, J., Liang, X., Wei, Y., Xu, T., Feng, J., Yan, S.: Perceptual generative adversarial networks for small object detection. In: 2017 IEEE Conference on Computer Vision and Pattern Recognition, CVPR 2017, Honolulu, HI, USA, 21–26 July 2017, pp. 1951–1959 (2017)
12. Li, Z., Peng, C., Yu, G., Zhang, X., Deng, Y., Sun, J.: DetNet: a backbone network for object detection (2018)
13. Lin, T., Dollár, P., Girshick, R.B., He, K., Hariharan, B., Belongie, S.J.: Feature pyramid networks for object detection. In: 2017 IEEE Conference on Computer Vision and Pattern Recognition, CVPR 2017, Honolulu, HI, USA, 21–26 July 2017, pp. 936–944 (2017)
14. Lin, T., Goyal, P., Girshick, R.B., He, K., Dollár, P.: Focal loss for dense object detection. In: IEEE International Conference on Computer Vision, ICCV 2017, Venice, Italy, 22–29 October 2017, pp. 2999–3007 (2017)
15. Liu, W., et al.: SSD: single shot multibox detector. In: Leibe, B., Matas, J., Sebe, N., Welling, M. (eds.) ECCV 2016. LNCS, vol. 9905, pp. 21–37. Springer, Cham (2016). https://doi.org/10.1007/978-3-319-46448-0_2
16. Redmon, J., Farhadi, A.: YOLO9000: better, faster, stronger. In: 2017 IEEE Conference on Computer Vision and Pattern Recognition, CVPR 2017, Honolulu, HI, USA, 21–26 July 2017, pp. 6517–6525 (2017)
17. Ren, S., He, K., Girshick, R.B., Sun, J.: Faster R-CNN: towards real-time object detection with region proposal networks. In: Advances in Neural Information Processing Systems 28: Annual Conference on Neural Information Processing Systems 2015, 7–12 December 2015, Montreal, Quebec, Canada, pp. 91–99 (2015)
18. Sermanet, P., LeCun, Y.: Traffic sign recognition with multi-scale convolutional networks. In: The 2011 International Joint Conference on Neural Networks, IJCNN 2011, San Jose, California, USA, 31 July–5 August 2011, pp. 2809–2813 (2011)
19. Yang, F., Choi, W., Lin, Y.: Exploit all the layers: fast and accurate CNN object detector with scale dependent pooling and cascaded rejection classifiers. In: 2016 IEEE Conference on Computer Vision and Pattern Recognition, CVPR 2016, Las Vegas, NV, USA, 27–30 June 2016, pp. 2129–2137 (2016)

20. Zhang, K., Zhang, Z., Li, Z., Qiao, Y.: Joint face detection and alignment using multi-task cascaded convolutional networks. IEEE Signal Process. Lett. **23**(10), 1499–1503 (2016)
21. Zhu, Z., Liang, D., Zhang, S., Huang, X., Li, B., Hu, S.: Traffic-sign detection and classification in the wild. In: 2016 IEEE Conference on Computer Vision and Pattern Recognition, CVPR 2016, Las Vegas, NV, USA, 27–30 June 2016, pp. 2110–2118 (2016)

Dataset Refinement for Convolutional Neural Networks via Active Learning

Siwen Liu[1,2], Rong Zhu[1,2(✉)], Yimin Luo[2,3], Zhongyuan Wang[1,2], and Liguo Zhou[1,2]

[1] National Engineering Research Center for Multimedia Software,
Computer School, Wuhan University, Wuhan, China
zhurong@whu.edu.cn
[2] Collaborative Innovation Center for Geospatial Information Technology,
Wuhan, China
[3] Remote Sensing Information Engineering School,
Wuhan University, Wuhan, China

Abstract. Convolutional neural networks (CNNs) have shown significant advantages in computer vision fields. For the optimizations of CNNs, most research works focus on feature extraction, which creates deeper structures and better activations, but the optimizations on dataset is rarely discussed. Due to the boom of data, most CNNs suffer from serious problems of dataset redundancy and following high computational burden. To this end, this paper brings in an informativeness ranking thought and proposes a new methodology for dataset refinement based on Active Learning. Extensive experiments prove its effectiveness to achieve a higher classification accuracy for CNNs at a less training cost. Moreover, for classification problems with a large number of class, this paper further proposes Entropy Ranking, a new Active Learning method, to enhance the optimization ability.

Keywords: Dataset optimization · CNN · Active learning · Entropy ranking

1 Introduction

Deep learning is a hot research direction in Computer Vision [1]. Convolutional Neural Networks (CNNs), a typical kind of Deep Learning algorithms, have been widely used in visual classification tasks. Compared to traditional models of computational structures, like support vector machines (SVMs) [2], CNNs can automatically learn more abstract features [3]. Despite of its strong learning ability, CNNs are subjected to considerable expenditure of computing resource and difficulties of parameter adjustment caused by the boom of data. Therefore, refinement for dataset is required, and this paper achieves this refinement based on Active Learning (AL).

AL is a machine learning technique for dataset optimization, and it achieves this optimization by selecting the most informative samples for training [4]. Therefore, sample selection strategies are the key to AL, and these strategies are all based on the thought of informativeness ranking from classification outcomes. Compared to random sampling, AL significantly improves the efficiency of classification, and its

R. Hong et al. (Eds.): PCM 2018, LNCS 11166, pp. 565–574, 2018.
https://doi.org/10.1007/978-3-030-00764-5_52

informativeness criterion has been widely applied to many machine learning tasks [5–7]. However, most applications for AL are aimed at solving the problem of sample scarcity and high labeling cost. For dataset redundancy, actually, AL can also be applied as a mean of optimization. Here, to reduce training costs of CNNs, AL can work as a filter which establishes a new dataset with the most informative samples selected from the original dataset.

From a measuring standpoint of informativeness, there are two main kinds of AL algorithms: probability-based AL and distance-based AL. Breaking Ties [8] is a typical probability-based AL algorithm, and it ranks samples' informativeness according to their two highest posterior class probabilities. As CNNs' output of the last fully-connected layer can be processed as samples' class probabilities by Softmax [1], Breaking Ties can be directly applied to CNNs as a tool of dataset optimization.

However, for classification tasks which focus on a large number of class, dealing with two highest posterior class probabilities is of little representativeness. Considering that Entropy is a classical measure of informativeness, additionally, this paper proposes a new AL algorithm based on CNNs' output, named Entropy Ranking. Compared to Breaking Ties, Entropy Ranking takes care of all the class probabilities provided by CNNs, and samples with high entropy is of high uncertainty. Accordingly, selecting samples with high uncertainty is helpful to increase the informativeness of dataset, which achieves CNNs' improvement. In addition, we propose a new measurement method for dataset redundancy and make a comprehensive discussion on how many samples should be selected based on it.

This paper mainly makes following two contributions. Firstly, CNNs are improved from a perspective of dataset refinement based on AL. Secondly, AL family is enriched by the proposed Entropy Ranking method. Experiments on MNIST [9], CIFAR-10 [10] prove the effectiveness of this methodology of refinement, as higher classification accuracy can be achieved with refined dataset for training. Moreover, it is also proved that the proposed Entropy Ranking outperforms Breaking Ties in classification tasks of which the class number is tremendous.

2 Related Works

The key to Deep Learning is back propagation training, and CNN improves it by reducing the number of parameters which obtained from spatial relationships. Therefore, CNN can be regarded as a choice of topology or architecture [11]. A typical CNN consists of three types of layers: convolutional layer, pooling layer and fully-connected layer. In most CNNs, Softmax outputs samples' class probabilities for classification. Based on it, Softmax loss function, regarded as s a supervision component for deep neural network model's training [12], dedicates to continuously increase the discernibility of sample characteristics with labels. For CNNs' optimization, most research works focus on network architecture [13, 14], activation [15, 16] and loss function [12, 17].

Traditional AL algorithms, developed on support vector machines (SVMs), focus on the interaction between the user and the classifier [7]. Based on the class probabilities provided by CNNs, Breaking Ties method, a classical AL algorithm, achieves informativeness ranking though the minimum strategy. Breaking Ties considers that samples whose two highest class probabilities are very closed are of high uncertainty. Previously, class probabilities for Breaking Ties method are calculated is assessed fitting a sigmoid function to the SVM decision function [8]. In most tasks, especially in remote sensing image classification, AL returns the classification outcome for next selection of informative unlabeled samples. For Breaking Ties' optimization, most research works focus on its cooperation with spatial information for selecting unlabeled sample [18]. Conversely, in this paper, AL is applied to reduce the redundancy of labeled samples with CNNs' classification results.

3   Dataset Optimization

In previous works, as summarized in [19], the classical flowchart of dataset optimization based on AL consists of five components: unlabeled set, labeled set, sample selective strategies, classifier and supervisors. Traditionally, informative unlabeled samples should be selected out by sample selective strategies and labeled by supervisors iteratively until the labeled set is strong enough to train out a good classifier. AL significantly relieves the difficulties of sample scarcity, therefore, it has been widely used in remote sensing image classification tasks where the human annotation cost is extremely high [8, 20, 21].

Actually, dataset optimization is not just what sample scarcity calls for. In the case of sample redundancy, refining the redundant dataset is also needed. In this data explosion era, data redundancy is a common phenomenon. Despite of the great learning ability provided by CNNs, facing a tremendous dataset for training, complexity of network architecture and waste of computing resources should be also taken into consideration. Therefore, refinement is of great significance to CNNs' performance. Here, AL provides a good idea for achieving this refinement. As depicted in Fig. 1, a small amount of samples is selected from the original dataset randomly for the establishment of initial CNN model, and then, AL's sample selective strategies can be applied to the original dataset's refinement based on the obtained CNN's outcomes. Finally, a new dataset consisting of highly informative samples is established and trained with it enables us to obtain a stronger classification model.

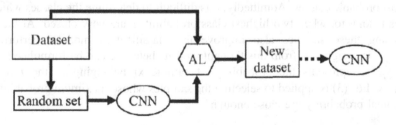

Fig. 1. AL for dataset optimization based on CNNs.

According to our flowchart in Fig. 1, the first CNN model's role in the flowchart is providing samples' class probability for AL selective strategies. Why can CNN output samples' class probability? The answer is its Softmax.

Most CNN architectures have convolutional layers, pooling layers and fully-connected layers. Softmax, which can provide class probabilities of samples lies on the last fully-connected layer of CNNs, and Softmax loss function works as a supervision component which controls back propagation training of CNNs. Softmax loss can be formulated as (1), where $p_{y_i}(x_i)$ represents the probability of sample x_i belonging to its labeled class y_i, and N represents the number of samples for training.

$$L_s = -\frac{1}{N}\sum_{i=1}^{N} log\left[p_{y_i}(x_i)\right] \tag{1}$$

This probability of which samples belong to each class is output by Softmax and is calculated by Eq. (2), where j represents class number, and W_k represents the k-th column of weight.

$$p_k(x_i) = \frac{e^{W_k^T x_i}}{\sum_{j=1}^{class} e^{W_j^T x_i}} \quad k \in \{1, 2, \ldots, class\} \tag{2}$$

In this paper, class probabilities obtained by Eq. (2) are the key to dataset refinement via AL based on CNNs.

4 Active Learning Based on CNNs

In this section, two sample selective strategies of probability-based AL (Breaking Ties and Entropy Ranking) are improved and applied to CNNs for dataset refinement. Among them, Entropy Ranking is proposed by this paper for the sake of taking care of all the class probabilities provided by CNN model.

4.1 Breaking Ties

In terms of informativeness measuring, Breaking Ties deals with the difference for the two most probable classes. Admittedly, it is difficult to determine the classes which the samples belong to, when two highest class probabilities are very closed. At the same time, using these samples also improve the classification model's performance. Breaking Ties originates from this thought, and its heuristic can be formulated as (3), where $p_j(x_i)$ represents the probability of sample xi belonging to the j-th class. Obviously, Eq. (3) is applied to selecting the samples whose maximum probability and submaximal probability are close enough.

$$\hat{x} = argmin\left[max(\{p_j(x_i)\}) - \text{submax}(\{p_j(x_i)\})\right] \tag{3}$$

Here, in order to apply Breaking Ties to CNNs, samples' class probabilities can be obtained by (2).

4.2 Entropy Ranking

Despite of Breaking Ties' effectiveness on sample selection, for classification problems with a large number of class, two highest posterior class probabilities are unable to represent samples informativeness comprehensively. Therefore, taking all the class probabilities provided by the classifier into consideration is necessary. Entropy provides us with a good idea for solving this problem, and its maximization gives a naturally multiclass heuristic.

Previously, entropy has been applied to remote sensing image classification as a selective method to reduce human annotation, named entropy query-by-bagging (EQB) [8]. However, the probability calculation of EQB is complicated, which has to set up several classifiers for voting. Considering the cost of CNNs' training, bagging is unpractical and unnecessary as consequence of its computing cost. Although, it can hardly be applied to refine sample sets in computer vision field, using entropy for informativeness ranking based on CNNs' outputs are of great significance. Accordingly, we propose a new AL method named Entropy Ranking.

Class probabilities of the proposed Entropy Ranking are calculated as (2). With samples' class probability, the entropy of each sample can be calculated as Eq. (4), where class represents class number.

$$H(x_i) = \sum_{j=1}^{class} -p_j(x_i)log_{class}p_j(x_i) \tag{4}$$

According to the obtained entropy of all samples, those satisfying (5) can be selected out for the establishment of new training set.

$$\hat{x} = arg\,max\{H(x_i)\} \tag{5}$$

5 Discussion

In terms of sample selection, selective strategies are an important part which should be taken into consideration, as it directly affects selected samples quality. On the other hand, the selected sample number is also of great significance to dataset refinement. Intuitively, to achieve a higher classification accuracy, the more informative the dataset is, the larger the selective number of samples should be chosen. Here, sample informativeness is a positive number which lies in the interval [0,1], therefore, we propose a calculation method (6) for sample redundancy.

$$R(x_i) = 1 - H(x_i) \tag{6}$$

For optimizations and application of AL, sample selective strategies are usually the core of most research works, selective number are rarely discussed. Most of time, the number of sample selection is set empirically, hence, a systematic discussion on this number is needed. Here, we establish a simple relationship between selective number and dataset informativeness, which reaches a balance between redundancy and informativeness of dataset.

$$\hat{x} = max_{\{H(x_i)\}}min_{\{R(x_i)\}}\{x_i\} \tag{7}$$

Algorithm 1 Fast method for deciding selective number

1: **Input**: Sample entropy set $\{H_i\}$;

   Output: Number of selective sample

2: $\{H_i'\} \leftarrow$ descending sort $\{H_i\}$

   $\{R_i\} \leftarrow \{1 - H_i'\}$

   $SH_0 \leftarrow 0$

   $SR_{N+1} \leftarrow 0$

   for $i = 1$ to N

       $SH_i \leftarrow SH_{i-1} + H_i'$

   end

   for $i = N$ to 1

       $SR_i \leftarrow SR_{i+1} + R_i$

   end

3: Find n which makes $SH_n \approx SR_n$

4: Choose the top n from $\{H_i'\}$.

That is to say, the selected sample sets have the largest entropy and the smallest redundancy at the same time, which matches the Algorithm 1, a fast method for finding the optimal number of sample.

6   Experiments and Results

In this section, two famous visual classification databases: MNIST [9] and CIFAR-10 [10] are used for validating the effectiveness of the proposed dataset optimization based on AL with CNNs.

6.1   Dataset

The MNIST is a handwritten digit database which has 60,000 training examples and 10,000 test examples in total. For its simple request for preprocessing and formatting, it has been widely used in experiments of pattern recognition and learning techniques. Moreover, CIFAR-10 dataset consists of 60000 32×32 color images in 10 classes, with 6000 images per class. There are 50000 training examples and 10000 test examples. We implement the CNN model using the Caffe [23] library with the proposed methodology, and details of the CNN architecture are given in Tables 1 and 2.

Table 1. The CNN architecture of the test on MNIST

Mnist					
Layer	Conv	Max pool	Conv	Max pool	FC
Num output	20	-	50	-	500
Kernel size	5×5	2×2	5×5	2×2	-
Stride	1	2	1	2	-
Pad	0	-	0	-	-

Table 2. The CNN Architecture of the Test on CIFAR-10

Cifar-10								
Layer	Conv $\times 2$	Max pool	Conv $\times 2$	Max pool	Conv $\times 3$	Max pool	FC	FC
Num Output	64	-	128	-	256	-	2048	2048
Kernel Size	3×3	2×2	3×3	2×2	3×3	2×2	-	-
Stride	1	2	1	2	1	2	-	-
Pad	1	-	1	-	1	-	-	-

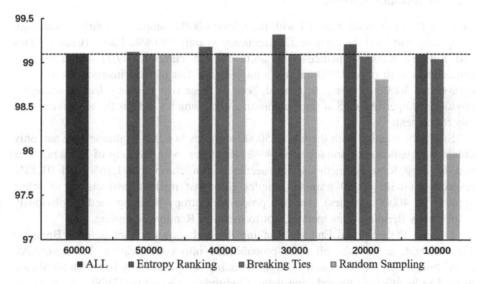

Fig. 2. Results on MNIST (using about 30000 samples selected by entropy ranking achieves the highest accuracy).

Here, 6,000 samples (600 for each class) are selected randomly from the original dataset for initial CNN models' establishment. Moreover, for fairness, we test proposed methods for 10 times for average. The classification results are compared in Figs. 2 and 3.

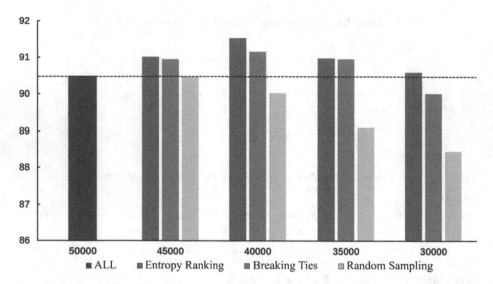

Fig. 3. Results on CIFAR-10 (using about 40000 samples selected by entropy ranking achieves the highest accuracy).

6.2 Results and Analysis

Training the classification model with the whole 60000 samples, we suffer from high computing cost, and the classification accuracy is only 99.08%. Using Breaking Ties and Entropy Ranking methods, the accuracy can climb to 99.21% and 99.29% respectively with only 30000 training samples (our fast method instructs us to select appropriate 30000 samples). In general, both of these two sample selecting strategies obviously outperforms RS at same computing cost, which validates the effectiveness of this refinement.

Similarly, training with the whole 50000 samples in Cifar-10 dataset, we can only obtain a classification accuracy of 90.49%. However, with the help of Breaking Ties and Entropy Ranking methods, the accuracy can climb to 91.16% and 91.53% respectively with 40000 training samples (our fast method instructs us to select appropriate 40000 samples). And the proposed Entropy Ranking method obviously outperforms Breaking Ties method, not to mention Random Sampling.

Overall, the proposed Entropy Ranking method shows advantage over Breaking Ties method, as its take all class probabilities into consideration. Moreover, AL algorithms which are previously used for reducing sample scarcity based on SVMs are proved to be effective for reducing dataset redundancy based on CNNs.

7   Conclusion

As the consequence of data explosion, CNNs have to face an increasingly serious problem of dataset redundancy, which may cause high computing resources, difficulty of parameter tuning and sometimes decrease of classifier performance. This paper proposes a new thought of dataset refinement using AL methods based on CNNs, moreover, we enrich AL family by proposing Entropy Ranking method and discussing how many samples should we exactly choose.

Admittedly, it's still unable for us to explain the theoretical basis of the selection strategies, therefore, we'll continue to study this problem in the future research.

Acknowledgments. The research was supported by National Natural Science Foundation of China (61671332, U1736206, 41771452, 41771454), Hubei Province Technological Innovation Major Project (2017AAA123) and the National Key Research and Development Program of China (2016YFE0202300).

References

1. Sun, Y., Chen, Y., Wang, X., Tang, X.: Deep learning face representation by joint identification-verification. In: Advances in neural information processing systems, pp. 1988–1996 (2014)
2. Huang, F.J., LeCun, Y.: Large-scale learning with SVM and convolutional for generic object categorization. In: 2006 IEEE Computer Society Conference on Computer Vision and Pattern Recognition, vol. 1, pp. 284–291. IEEE (2006)
3. Zhou, H., Huang, G.B., Lin, Z., Wang, H., Soh, Y.C.: Stacked extreme learning machines. IEEE Trans. Cybern. **45**(9), 2013–2025 (2015)
4. Zhang, L., Chen, C., Bu, J., Cai, D., He, X., Huang, T.S.: Active learning based on locally linear reconstruction. IEEE Trans. Pattern Anal. Mach. Intell. **33**(10), 2026–2038 (2011)
5. Garcia-Pedrajas, N., Perez-Rodriguez, J., de Haro-Garcia, A.: OligoIS: scalable instance selection for class-imbalanced data sets. IEEE Trans. Cybern. **43**(1), 332–346 (2013)
6. Liu, P., Zhang, H., Eom, K.B.: Active deep learning for classification of hyperspectral images. IEEE J. Sel. Top. Appl. Earth Obs. Remote. Sens. **10**(2), 712–724 (2017)
7. Du, B., et al.: Exploring representativeness and informativeness for active learning. IEEE Trans. Cybern. **47**(1), 14–26 (2017)
8. Tuia, D., Pasolli, E., Emery, W.J.: Using active learning to adapt remote sensing image classifiers. Remote Sens. Environ. **115**(9), 2232–2242 (2011)
9. Krizhevsky, A., Hinton, G.: Learning multiple layers of features from tiny images (2009)
10. Arel, I., Rose, D.C., Karnowski, T.P.: Deep machine learning-a new frontier in artificial intelligence research [research frontier]. IEEE Comput. Intell. Mag. **5**(4), 13–18 (2010)
11. Wen, Y., Zhang, K., Li, Z., Qiao, Yu.: A discriminative feature learning approach for deep face recognition. In: Leibe, B., Matas, J., Sebe, N., Welling, M. (eds.) ECCV 2016. LNCS, vol. 9911, pp. 499–515. Springer, Cham (2016). https://doi.org/10.1007/978-3-319-46478-7_31
12. He, K., Zhang, X., Ren, S., Sun, J.: Deep residual learning for image recognition. In: Proceedings of the IEEE Conference on Computer Vision and Pattern Recognition, pp. 770–778 (2016)
13. Szegedy, C., et al.: Going deeper with convolutions. In: Computer Vision and Pattern Recognition, CVPR, pp 1–9 (2015)
14. Nair, V., Hinton, G.E.: Rectified linear units improve restricted Boltzmann machines. In: Proceedings of the 27th International Conference on Machine Learning, ICML 2010, pp. 807–814 (2010)

15. He, K., Zhang, X., Ren, S., Sun, J.: Delving deep into rectifiers: surpassing human-level performance on imagenet classification. In: Proceedings of the IEEE International Conference on Computer Vision, pp. 1026–1034 (2015)
16. Liu, W., Wen, Y., Yu, Z., Yang, M.: Large-margin softmax loss for convolutional neural networks. In: International Conference on Machine Learning, ICML, pp. 507–516 (2016)
17. Pasolli, E., Melgani, F., Tuia, D., Pacifici, F., Emery, W.J.: Improving active learning methods using spatial information. In: 2011 IEEE International on Geoscience and Remote Sensing Symposium, IGARSS, pp. 3923–3926. IEEE (2011)
18. Li, M., Sethi, I.K.: Confidence-based active learning. IEEE Trans. Pattern Anal. Mach. Intell. 28(8), 1251–1261 (2006)
19. Tuia, D., Ratle, F., Pacifici, F., Kanevski, M.F., Emery, W.J.: Active learning methods for remote sensing image classification. IEEE Trans. Geosci. Remote Sens. 47(7), 2218–2232 (2009)
20. Crawford, M.M., Tuia, D., Yang, H.L.: Active learning: any value for classification of remotely sensed data? Proc. IEEE 101(3), 593–608 (2013)
21. Jia, Y., et al.: Caffe: convolutional architecture for fast feature embedding. In: Proceedings of the 22nd ACM International Conference on Multimedia, pp. 675–678. ACM (2014)

Gaze Information Channel

Lijing Ma[1], Mateu Sbert[1,2(✉)], and Miquel Feixas[2]

[1] School of Computer Science and Technology, Tianjin University, Tianjin, China
mateusbert@mac.com
[2] Institute of Informatics and Applications, University of Girona, Girona, Spain

Abstract. This paper presents the gaze information channel. Using the fact that eye tracking fixation sequences through areas of interest (AOI) have been previously modeled as a Markov chain, we go further by using the fact that a Markov chain is a special case of information, or communication, channel. Thus, the Markov chain quantities of entropy of the equilibrium distribution and conditional entropy are enriched for interpretation of the gaze sequences with the additional quantities of the information channel, such as joint entropy, mutual information and conditional entropy of each area of interest. We illustrate the gaze channel with several examples using Van Gogh paintings. Our preliminary results show that the channel quantities can be given a coherent interpretation to both classify the observers and the artworks.

Keywords: Eye-tracking · Information channel · Mutual information
Entropy

1 Introduction

Eyes are the most important sensory organ for humans, more than 80% of the information processed by humans is obtained via vision [1]. Eye tracking, which records gaze positions and eye movements, is used to understand the way of perceiving a visual stimulus [2]. Thus, eye movement metrics such as scanpaths and heatmaps are used to study viewers' perception. An observed image can be divided into several AOIs, the eye gaze trajectory can go from one AOI to another one. The classic data analytical methods have been used to define the gaze switchings between different AOIs, but they have difficulties in reflecting the dynamic features of the gaze distribution [3]. The gaze trajectories were identified as a conditional probability matrix between the AOI's first by Ellis and Stark [4] and then brought to light again as a Markov chain by Kreijtz et al. in [5,6]. In this paper we describe a new quantitative way to compare gaze, the *gaze information channel*. It uses the fact that a Markov chain can be identified as a communication, or information channel [7]. This identification of a Markov chain to an information channel has already been successfully exploited in other areas of visual computing [8,9].

© Springer Nature Switzerland AG 2018
R. Hong et al. (Eds.): PCM 2018, LNCS 11166, pp. 575–585, 2018.
https://doi.org/10.1007/978-3-030-00764-5_53

2   Background

Ellis and Stark [4] divided the cockpit display of traffic information (CDTI) into 8 AOIs, mapped the airline pilots' gaze transition to conditional probability matrices and calculated conditional entropy H_c. H_c reveals the statistical dependency in the spatial pattern of gaze represented by transition matrices and provides a way of comparing one matrix with another.

The tests of Besag and Mondal [10] verified the validity of modeling the gaze transitions as short, first-order processes. Vandeberg [11] used a multilevel Markov modeling approach to analyse gaze switching patterns. After modeling the individuals' gaze as Markov chains, Krejtz et al. [5,6], calculated the stationary entropy H_s and transition entropy H_t to interpret the overall distribution of attention over AOIs. Observe that H_c in [4] and H_t in [5,6] should be the same, as the Markov chain transition probability matrix has a dual interpretation as a conditional probability matrix. Krejtz et al. [5] to justify that $H_t \leq H_s$ used the identification of Markov chain with information channel, although they did not exploit further this concept. Raptis et al. [12] asked the participants to complete recognition tasks with various complexities, then the researchers used gaze transition entropy H_s and H_t to do eye tracking analysis, the result revealed there are quantitative differences on visual search patterns among individuals. Raptis et al. [13] stated that eye gaze, including gaze entropies, fixation duration and count, can reflect personal differences in cognitive styles.

Zhong et al. [14] modeled the relationship between the image feature and the saliency as a Markov chain, in order to predict the transition probabilities of the Markov chain, they trained a support vector regression (SVR) from true eye tracking data. At last, when given the stationary distribution of this chain, a saliency map predicting user's attention can be obtained. Krejtz et al. [15] used a entropy-based novel framework to compare gaze transition matrices, which indicated that the interactive simulation can lead to more careful visual investigation of the learning material as well as reading of the problem description.

Huang [16] used the female gaze data of browsing apparel retailers' web pages to study how the female attention was influenced by visual content composition and slot position in personalized banner ads. Gu et al. [17] used heatmap entropy (visual attention entropy, VAE) and its improved version, relative VAE (rVAE) to analyse eye tracking data of observing web pages, the result shows that VAE and rVAE have correlation with the perceived aesthetics. Hwang et al. [18] stated that it is important to notice scenes consist of objects representing not only low-level visual information but also higher-level, semantic data, and they presented transitional semantic guidance computation to estimate gaze transitions.

3   Markov Chain as an Information Channel

3.1   Markov chain

A Markov chain is a discrete-time and discrete-state sequence, which is composed by random variables $X_1, X_2, X_3, \ldots$ and defined on a countable set S. S is the state space of the chain. If X_i is a Markov chain, then the state at time n is X_n. A time-invariant Markov chain is identified by its initial state and a probability transition matrix $P = [P_{ij}]$.

If it is potential to go from any state of the Markov chain to any other state via a finite number of steps, the Markov chain is considered to be irreducible. If the maximal common factor of the length of diverse paths from a state to itself is one, the Markov chain is considered to be aperiodic [19].

If the distribution of the states at time $n + 1$ is the same as the one at time n, then the distribution is a stationary one. If the initial state of a Markov chain is drawn according to a stationary distribution, the Markov chain forms a stationary process [19].

If a finite-state Markov chain is irreducible and aperiodic, the stationary distribution of it is unique, and from any initial distribution, the distribution of X_n tends to the stationary distribution when $n \to \infty$ [19]. The stationary distribution π can be computed by solving the equation

$$\pi P = \pi \tag{1}$$

3.2 Information Channel

Shannon (1916–2001) proposed the fundamental concept of *communication channel*, or *information channel* [7]. Information channel can be used to connect any two variables sharing information. Suppose X and Y are two random variables, then we can build an information channel between X and Y. The channel includes the following main elements:

- X and Y, represent the input and output variables, respectively. $p(X)$ and $p(Y)$ are the probability distributions of X and Y.
- Probability transition matrix $p(Y|X)$, which is composed by conditional probabilities $p(y|x)$. Each row of $p(Y|X)$, represented as $p(Y|x)$, can be thought as a probability distribution.

Figure 1 shows the main elements of an information channel.

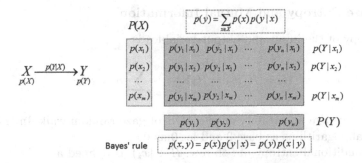

Fig. 1. Main elements of an information channel

3.3   Markov Chain as an Information Channel

From Sect. 3.1, we know that if the finite-state Markov chain is irreducible and aperiodic, the stationary distribution is unique. Next, we treat a Markov chain satisfying the above factors as an information channel.

Treating a Markov chain as an information channel demands that $P(X) = P(Y) = \pi$, which means the input and output variables of information channel are same. The matrix $P = [p_{ij}]$ in Markov chain can be seen as the conditional probabilities $p(y|x)$ of the information channel. The information channel created by a stationary Markov chain was used to study the visibility and complexity of 3D scenes by Feixas et al. [8,9].

4   Gaze Information Channel

4.1   Gaze Information Channel

Given an image, we divide it into k AOIs. A stochastic process $\{G_t\}$, $t = 1, 2, \ldots, n$ can be used to describe the gaze switching process. The Markov property and the existence of stationary distribution have been tested in [5]. We represent the transition matrix by $G = [g_{ij}]_{k \times k}$ and the stationary probability by π. Thus gaze transition can be modeled as a discrete information channel with input and output distribution, X and Y, equal to π, $\pi = (a_1, a_2, \ldots, a_k)$. The conditional probabilities in gaze information channel, g_{ij}, correspond to the probabilities $p(y|x)$ of transition matrix in Markov chain. The elements of the *gaze information channel* are:

– $p(X)$ and $p(Y)$, the marginal probabilities of input X and output Y, are both given by the stationary probability π, $\pi = (a_1, a_2, \ldots, a_k)$, which gives the frequency of visits of AOIs.
– The conditional probabilities, g_{ij}, which represent the probability that the fixations leave i-th AOI and reach to j-th AOI. Conditional probabilities fulfil $\sum_{j=1}^{k} g_{ij} = 1, \forall i \in \{1, \ldots, k\}$.

4.2   Gaze Entropy and Mutual Information

The entropy of the input and output variables is

$$H_s = H(X) = H(Y) = -\sum_{i=1}^{k} a_i \log a_i, \tag{2}$$

giving the average uncertainty on the AOIs of gaze random walk. In this paper, we treat all logarithms as base 2 and $\log 0$ as 0.

The conditional entropy of i-th row, $H(Y|x_i)$, is defined as

$$H(Y|x_i) = -\sum_{j=1}^{k} g_{ij} \log g_{ij}, \tag{3}$$

and gives the uncertainty of next fixation when current fixation is in i-th AOI.

The conditional entropy of the information channel, H_t, is the average of row entropies

$$H_t = H(Y|X) = \sum_{i=1}^{k} a_i H(Y|x_i) = -\sum_{i=1}^{k} a_i \sum_{j=1}^{k} g_{ij} \log g_{ij}. \tag{4}$$

and represents the average uncertainty of the next gaze movement.

Bayes' rule describes the relationship between $p(x, y)$ and $p(x|y)$, $p(x, y) = p(x)p(y|x)$, so in the gaze information channel, the joint distribution, $p(x, y)$, is given by $a_i g_{ij}$. The joint entropy $H(X, Y)$ of the information channel is the entropy of the joint distribution of X and Y.

$$H(X,Y) = H(X) + H(Y|X) = \sum_{i=1}^{k} \sum_{j=1}^{k} a_i g_{ij} \log a_i g_{ij}. \tag{5}$$

The mutual information of i-th row, $I(x_i; Y)$, is given by

$$I(x_i; Y) = \sum_{j=1}^{k} g_{ij} \log \frac{g_{ij}}{a_j}. \tag{6}$$

The mutual information $I(X; Y)$ is given by

$$I(X;Y) = H(X) + H(Y) - H(Y|X) = \sum_{i=1}^{k} a_i I(x_i; Y)$$

$$= \sum_{i=1}^{k} \sum_{j=1}^{k} a_i g_{ij} \log \frac{g_{ij}}{a_j}. \tag{7}$$

Figure 2 shows the relationships between different channel quantities.

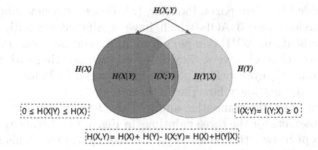

Fig. 2. Venn diagram of entropies and mutual information

4.3   Interpretation of Gaze Information Channel Quantities

The equilibrium distribution π gives the relative frequency of fixations in the AOIs. For $\pi = \pi_1, \pi_2, \ldots, \pi_n$, a higher π_i means more fixations in the i-th AOI. H_s represents the uncertainty of fixations distribution between AOIs. Higher H_s values reflect that fixations are distributed more evenly between AOIs. While lower H_s values indicate that fixations tend to be concentrated on certain AOIs [4]. H_t reflects the randomness of gaze switching. Higher H_t values are obtained when there are frequent switchings among AOIs. Lower H_t values indicate more careful viewing of AOIs [5]. $H(Y|x_i)$ measures the randomness of the gaze switching with source the i-th AOI. A lower $H(Y|x_i)$ value indicates the observer is more clear about which is the next AOI in the following observation. It may also represent that the i-th AOI gives the observer a significant clue to understand the painting. $H(X, Y)$ measures the total uncertainty. This is because, as $X \equiv Y$, then $H(X, Y) = H_s + H_t$, and $H(X, Y)$ represents the total randomness of fixations' distribution and gaze switching.

Mutual information $I(X; Y)$ is a measure of the mutual dependence or coupling between the two variables X and Y. In the gaze channel, $I(X; Y)$ measures the amount of information transferred between AOIs. $I(x_i; Y)$ measures the amount of information transferred between i-th AOI and the other AOIs of the image, and the coupling between i-th AOI and the other ones. Observe that $I(x_i; Y)$ and $H(Y|x_i)$ have opposite behaviour. A high value of $H(Y|x_i)$ represents a high indetermination about nex area of interest, while a high value of $I(x_i; Y)$ represents knowing with high probability which will be next AOI.

5   Experimental Results

The stimuli of the experiment are three paintings from Vincent van Gogh. As Fig. 3 shows, they are divided into 3 AOIs. The AOIs of painting (a) and painting (b) are equal, while painting (c) is divided into 3 non-equal AOIs because of the significant semantic differences. Painting (a) is Starry Night with no salient object but high semantic relevance between connected AOIs. Painting (b) is The Bedroom, there is also semantic relevance between connected AOIs, for example, the bed is divided into two parts. Painting (c), Beach at Scheveningen in Calm Weather, is divided into 3 AOIs which have significant semantic differences. There are two boats in AOI1, one person in AOI2 and one boat in AOI3. The participants were three observers, they observed each painting with SMI (Senso-Motoric Instruments) eye tracking glasses for 40 seconds. Before observing the paintings, they did not know how the AOIs were divided.

We show in Fig. 4 the scanpaths, and in Table 1 the transition probabilities, of the three observers for the three paintings in Fig. 3. In all transitions matrices in Table 1 except in two, the values of g_{ii} are the highest one. This is similar to the behaviour in The tempest painting example presented in [5]. It might mean that the observer gaze moves first within an AOI before moving to another one.

Table 2 shows the entropies and mutual information of the 3 gaze information channels when observing Fig. 3(a).

(a) Starry Night   (b) The Bedroom   (c) Beach

Fig. 3. The stimuli: 3 paintings by Van Gogh segmented into three AOIs each

Table 1. Transition probabilities of 3 observers for the 3 paintings in Fig. 3.

Painting	Observer N.1	Observer N.2	Observer N.3
Fig. 3(a)	0.625 0.375 0.000	0.474 0.526 0.000	0.857 0.143 0.000
	0.265 0.500 0.235	0.137 0.685 0.178	0.062 0.831 0.108
	0.000 0.348 0.652	0.000 0.325 0.675	0.000 0.250 0.750
Fig. 3(b)	0.714 0.286 0.000	0.750 0.214 0.036	0.907 0.093 0.000
	0.067 0.533 0.400	0.132 0.658 0.210	0.149 0.787 0.064
	0.000 0.316 0.684	0.056 0.194 0.750	0.000 0.600 0.400
Fig. 3(c)	0.724 0.241 0.035	0.906 0.075 0.019	0.919 0.065 0.016
	0.167 0.417 0.416	0.250 0.250 0.500	0.286 0.500 0.214
	0.068 0.119 0.813	0.026 0.066 0.908	0.016 0.049 0.934

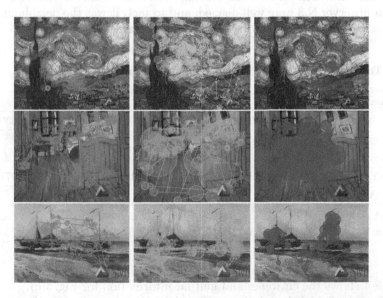

Fig. 4. Scanpaths of the three observers for the three Van Gogh paintings in Fig. 3, in orange for observer N.1, in cyan for observer N.2, and in red for observer N.3.

Table 2. Values of gaze channel' quantities for painting in Fig. 3(a).

Quantity	Observer N.1	Observer N.2	Observer N.3
π	(0.296,0.420,0.284)	(0.144,0.553,0.303)	(0.232,0.537,0.231)
$H_s = H(X)$	1.561	1.397	1.459
$H_t = H(Y|X)$	1.177	1.089	0.763
$H(Y|x)$	(0.954,1.499,0.932)	(0.998,1.210,0.910)	(0.592,0.816,0.811)
$H(X,Y)$	2.738	2.486	2.222
$I(X;Y)$	0.385	0.309	0.696
$I(x;Y)$	(0.612,0.019,0.688)	(0.776,0.065,0.531)	(1.346,0.286,0.997)

As Table 1 shows, there is no direct transition between AOI1 and AOI3 when observing Fig. 3(a). The reason might be that Starry Night painting shows a continuous scene, so switches just exist between connected AOIs. From Table 2 we know that for every observer, $H(Y|x_2)$ is the highest between $H(Y|x_i)$, and $I(x_2, Y)$ is the lowest between $I(x_i, Y)$. This means that when in AOI2, the observer can move randomly (or evenly) towards any of the neighbour AOIs, and the information shared between AOI1 and the other two AOIs is minimum.

Table 3 shows the entropies and mutual information when observing Fig. 3(b). When observing painting in Fig. 3(b), observer N.1 paid little attention to AOI1, while observer N.3 paid little attention to AOI3. The values of $I(x_1; Y)$ for observer N.1 and $I(x_3; Y)$ for observer N.3 are higher than the other values. This means that next AOI when leaving AOI1 for observer N.1 and leaving AOI3 for observer N.3 were well defined, and in fact, it was the neighbour AOI2. This behaviour can be checked in the corresponding scanpaths in Fig. 4.

Table 3. Values of gaze channel's quantities for painting in Fig. 3(b).

Quantity	Observer N.1	Observer N.2	Observer N.3
π	(0.093,0.400,0.507)	(0.275,0.372,0.353)	(0.591,0.370,0.039)
$H_s = H(X)$	1.345	1.573	1.163
$H_t = H(Y|X)$	1.046	1.085	0.648
$H(Y|x)$	(0.863,1.273,0.900)	(0.959,1.256,1.002)	(0.448,0.934,0.971)
$H(X,Y)$	2.391	2.658	1.811
$I(X;Y)$	0.300	0.488	0.515
$I(x;Y)$	(1.958,0.053,0.189)	(0.799,0.243,0.505)	(0.375,0.606,1.756)

Table 4 shows the entropies and mutual information for Fig. 3(c).

Each of the 3 AOIs in painting in Fig. 3(c) has at least one significant object, resulting in semantic differences between the 3 AOIs. Comparing the transition

Table 4. Values of gaze channel's quantities for painting in Fig. 3(c).

Quantity	Observer N.1	Observer N.2	Observer N.3
π	(0.259,0.214,0.527)	(0.376,0.085,0.539)	(0.453,0.102,0.445)
$H_s = H(X)$	1.468	1.314	1.374
$H_t = H(Y\|X)$	1.035	0.605	0.541
$H(Y\|x)$	(1.000,1.483,0.870)	(0.519,1.500,0.523)	(0.463,1.493,0.402)
$H(X,Y)$	2.503	1.918	1.915
$I(X;Y)$	0.433	0.709	0.833
$I(x;Y)$	(0.980,0.153,0.278)	(1.045,0.187,0.557)	(0.820,0.730,0.869)

matrices for the three paintings, in Table 1, we can see there are more direct transitions in painting in Fig. 3(c). For each observer, $I(X;Y)$ of painting in Fig. 3(c) is higher than for the other two paintings, owing to the smaller relationship of the 3 AOIs in painting in Fig. 3(c).

Observer N.3 has the highest $I(X;Y)$ and the lowest $H(X|Y)$ of the three observers for each painting. It might mean that Observer N.3 has a more definite strategy or more clues in exploring the paintings.

The values of $I(X;Y)$ and $H(X|Y)$ corresponding to painting in Fig. 3(c) are respectively higher and lower than for the other two paintings for every observer. It might mean that painting in Fig. 3(c) is the easiest to explore or interpret of the three. This correlates with this painting corresponding to an early period of Van Gogh life [20].

As Fig. 4 shows, the result of observer N.3 has little gaze switchings between AOIs. When observing Fig. 3(c), observer N.3 viewed it from left to right. She first observed the boats in AOI1, then the person in AOI2, and finally the boat in AOI3. In conclusion, she concentrated on a certain AOI during one duration.

The lowest $H(X,Y)$ value is obtained when observer N.3 views the painting (b). Comparing with other scanpaths in Fig. 4, the scanpaths with lowest $H(X,Y)$ has bigger fixations' diameter and less gaze switchings. In the gaze information channel, $H(X,Y) = H_t + H_s$, thus $H(X,Y)$ measures the total uncertainty of the observer.

In summary, the quantities in gaze information channel can help us understand the viewing behaviour of an observer and infer the features of the paintings.

6   Conclusions

Based on the fact that an irreducible and aperiodic Markov chain can be interpreted as an information channel, we have presented the gaze information channel, extending the previous interpretation of eye fixations as a Markov chain. Gaze information channel is a paradigm that provides us with well sounded quantitative measures and concepts to model and interpret the eye fixations between AOIs. To illustrate the channel, we have presented some preliminary examples on eye fixations on three Van Gogh paintings, showing promising results on both classifying observers and artworks. In our future work, we will on the one hand test these preliminary conclusions against data from more observers and additional paintings from the different periods of Van Gogh work, and on the other hand investigate the automatic segmentation into AOIs by maximizing the mutual information of the gaze channel.

Acknowledgements. This work has been partially funded by the National Natural Science Foundation of China (grants Nos. 61571439, 61471261 and 61771335), and by grant TIN2016-75866-C3-3-R from the Spanish Government.

References

1. Sanders, M.S., McCormick, E.J.: Human Factors in Engineering and Design. McGraw-Hill book company, New York City (1987)
2. Krueger, R., Koch, S., Ertl, T.: Saccadelenses: interactive exploratory filtering of eye tracking trajectories. In: IEEE Second Workshop on Eye Tracking and Visualization (ETVIS), pp. 31–34. IEEE (2016)
3. Cao, W., et al.: Could interaction with social robots facilitate joint attention of children with autism spectrum disorder? arXiv preprint arXiv:1803.01325 (2018)
4. Ellis, S.R., Stark, L.: Statistical dependency in visual scanning. Hum. Factors **28**(4), 421–438 (1986)
5. Krejtz, K., et al.: Gaze transition entropy. ACM Trans. Appl. Percept. (TAP) **13**(1), 4 (2015)
6. Krejtz, K., Szmidt, T., Duchowski, A.T., Krejtz, I.: Entropy-based statistical analysis of eye movement transitions. In: Proceedings of the Symposium on Eye Tracking Research and Applications, pp. 159–166. ACM (2014)
7. Shannon, C.E.: A mathematical theory of communication. Bell Syst. Tech. J. **27** (July (part I) and October (part II)) 379–423 (part I) and 623–656 (part II) (1948)
8. Feixas, M., del Acebo, E., Sbert, M.: Entropy of scene visibility. In: Proceedings of Winter School on Computer Graphics and CAD Systems (WSCG 1999), Plzen-Bory, Czech Republic, pp. 25–34, February 1999
9. Feixas, M., del Acebo, E., Bekaert, P., Sbert, M.: An information theory framework for the analysis of scene complexity. Comput. Graph. Forum **18**(3), 95–106 (1999)
10. Besag, J., Mondal, D.: Exact goodness-of-fit tests for Markov chains. Biometrics **69**(2), 488–496 (2013)
11. Vandeberg, L., Bouwmeester, S., Bocanegra, B.R., Zwaan, R.A.: Detecting cognitive interactions through eye movement transitions. J. Mem. Lang. **69**(3), 445–460 (2013)
12. Raptis, G.E., Fidas, C.A., Avouris, N.M.: On implicit elicitation of cognitive strategies using gaze transition entropies in pattern recognition tasks. In: Proceedings of the 2017 CHI Conference Extended Abstracts on Human Factors in Computing Systems, pp. 1993–2000. ACM (2017)
13. Raptis, G.E., Katsini, C., Belk, M., Fidas, C., Samaras, G., Avouris, N.: Using eye gaze data and visual activities to infer human cognitive styles: method and feasibility studies. In: Proceedings of the 25th Conference on User Modeling, Adaptation and Personalization, pp. 164–173. ACM (2017)

14. Zhong, M., Zhao, X., Zou, X.C., Wang, J.Z., Wang, W.: Markov chain based computational visual attention model that learns from eye tracking data. Pattern Recognit. Lett. **49**, 1–10 (2014)
15. Krejtz, K., Duchowski, A.T., Krejtz, I., Kopacz, A., Chrzastowski-Wachtel, P.: Gaze transitions when learning with multimedia. J. Eye Mov. Res. **9**(1), 1–17 (2016)
16. Huang, Y.T.: The female gaze: content composition and slot position in personalized banner ads, and how they influence visual attention in online shoppers. Comput. Hum. Behav. **82**, 1–15 (2017)
17. Gu, Z., Jin, C., Dong, Z., Chang, D.: Predicting webpage aesthetics with heatmap entropy. arXiv preprint arXiv:1803.01537 (2018)
18. Hwang, A.D., Wang, H.C., Pomplun, M.: Semantic guidance of eye movements in real-world scenes. Vis. Res. **51**(10), 1192–1205 (2011)
19. Cover, T.M., Thomas, J.A.: Elements of Information Theory, 2nd edn. Wiley, Hoboken (2006)
20. Rigau, J., Feixas, M., Sbert, M., Wallraven, C.: Toward Auvers period: evolution of van Gogh's style. In: Proceedings of the Sixth International Conference on Computational Aesthetics in Graphics, Visualization and Imaging. Computational Aesthetics 2010, pp. 99–106. Eurographics Association (2010)

A Multi-information Fusion Model for Shop Recommendation Based on Deep Learning

Jianwei Niu[1,2(✉)] and Yanyan Guo[1]

[1] State Key Laboratory of Virtual Reality Technology and Systems, School of Computer Science and Engineering, Beihang University, Beijing 100191, China
niujianwei@buaa.edu.cn
[2] Beijing Advanced Innovation Center for Big Data and Brain Computing (BDBC), Beihang University, Beijing 100191, China

Abstract. With the development of the e-commerce, people have become accustomed to posting their shopping experiences on websites and making consumption decisions based on reviews given by other consumers. Recommendation for shops has gradually evolved with the rise of review websites. Traditional recommendation methods are usually based on rating matrix. As abundant information is readily accessible, information of the context is gradually used by the recommendation algorithms. This paper proposes a fusion recommendation model RCFM, an integration of the Recurrent Neural Network (RNN), the Convolutional Neural Network (CNN) and the Factorization Machine (FM), which utilized the semantic information of the reviews, the feature information of the shop images and the attribute information of the customer and shop. RNN and CNN undertake tasks of text semantic extraction and image feature extraction, respectively. Feature-fusion vector is then passed into the Factorization Machine to generate the prediction rating. Finally RCFM is trained and tested with real data sets. Results show that RCFM significantly outperforms traditional baseline methods.

Keywords: Recommendation system · Deep learning
Factorization Machine · Fusion model

1 Introduction

With the fast development of e-commerce, customers are accustomed to posting comments on review sites after consumption. Recommendation systems are designed to integrate the review information from both merchants and customers to provide customers with more effective advice on consumptions. The recommendation system improves the efficiency of the trade for merchants as well. Thus, many e-commerce sites, such as Netflix and Amazon, have been designing and improving their own recommendation systems.

© Springer Nature Switzerland AG 2018
R. Hong et al. (Eds.): PCM 2018, LNCS 11166, pp. 586–595, 2018.
https://doi.org/10.1007/978-3-030-00764-5_54

Various models have been proposed for recommendation systems. For example, Collaborative Filtering (CF) [15] uses rating of the users, and Content-based recommendation model [8] relies on distribution of user preference records. However, classic recommendation models [13, 14] are not efficient at extracting features from text and images. More information in the course of online consumption, such as names of the shop and its attributes, even customer-posted reviews, photos and scores of consumption can be used by the recommendation systems to analyze the preference of users. Recently, hybrid models [1, 2, 17] are proposed based on the neural network, which is adept at extracting features from the review text. However, additional information on the consumption scenarios which can also be useful to the improvement of the recommendation model accuracy is not considered.

To address the aforementioned issue, a novel hybrid recommendation model, RCFM that uses Recurrent Neural Network (RNN) [9], Convolutional Neural Network (CNN) [7] and Factorization Machine (FM) [12], is proposed. The application scenario of the model is to recommend offline shops to customers. The main contributions of this paper are summarized as follows:

(a) A novel model, RCFM, is proposed to recommend shops to customers. RCFM innovatively introduce features of reviews, images and basic attributes to the procedure of prediction in recommendation model.
(b) The proposed model describes consumption preferences of customers by vectorizing attributes and reviews. In the meantime, RCFM creates shops description by vectorizing attributes, reviews and images of shops. The methods of vectorization utilize natural language processing techniques and deep learning techniques.
(c) The proposed model innovatively merges vectors of customers, shops, and their interaction for computation of prediction score. The experiment result based on real datasets show that RCFM outperforms baseline methods in terms of prediction accuracy.

In the rest of this paper, the related work is discussed in Sect. 2. The structure of the proposed RCFM is described in Sect. 3. In Sects. 4 and 5, establishment of vectors of customers and shops are described. And the process of predicting recommendations resulting from merged vectors are described in Sect. 6. Section 7 presents and analyzes the experimental results. Section 8 concludes this paper.

2   Related Work

The recommendation system has rapidly developed due to the easy access to the rich available information. The original recommendation model based on collaborative filtering only uses the user-item rating matrix such as the probabilistic matrix factorization (PMF) [10] and SVD++ [5]. PMF is a Gaussian-distribution-based latent factors model. SVD++ takes full advantages of neighborhood and latent factor.

Context information is gradually utilized by the recommendation algorithms due to the availability of the colossal quantity of information. TimeSVD [6] introduces time sequence to model the time drifting of user preference. The distribution of words in reviews that may contain the preference of users is considered in HFT [8]. However, the semantic information is not involved in traditional recommendation systems.

Deep learning technology performs well on extracting features from text and pictures. The DeepCoNN [17] and ConvFM [4] utilize neural network to establish semantic model of reviews. In this paper, the proposed method RCFM not only considers the reviews but also images of shops using the deep learning technology. Unlike other models, the special structure of RCFM makes more efficient use of the abstract vectors extracted from reviews and images.

3   The RCFM Model Overview

The RCFM model consists of two components: the construction of vectors for customers and shops, and the prediction with merged vectors. The structure of the model is shown in Fig. 1. RCFM constructs the vector of customers from both explicit and implicit aspects. Explicit vectors are extracted and generated from the information left by the customers on the review site. The number of reviews, the average score, and even the relationship of customers can all be utilized. On the other hand, the implicit vectors of customers are trained by the reviews. To obtain the semantic vectors that represent the consumption habits of customers, the review data are first preprocessed and then trained by the RNN [9]. Similarly, explicit vectors of shops are built from basic attributes of shops. The shop reviews and images are vectorized, and then trained by neural network to extract features for constructing implicit vectors of shops. The second part of the model is to integrate the vectors of customers and shops. The score is predicted by the FM [12], and then the recommendation degree can be obtained. Finally the loss is computed to continuously optimize RCFM through the back propagation process.

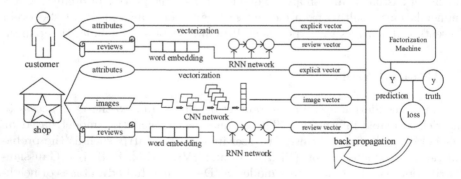

Fig. 1. The structure of the model

4   The Construction of Customer Vectors

A vector consisting of two parts is used in the RCFM to represent the overall consumption preference of a customer. The explicit part represents the behavioral characteristics of the customer in the review sites, and the implicit part represents the aforementioned semantic information of reviews.

4.1   The Construction of Customer Explicit Vector

The explicit vector is derived from the vectorization of attributes. The attributes of customers contains customer ID, customers friends on review website, and other people's interactions on reviews.

There is a friend relationship between users of the review site which represents a kind of social context of customers. Vector V_{cf} represents the relationship information of customers. Firstly, sequences of friendliness pairs are established. Then the word embedding of customers' ID $f : C_{id} \rightarrow V_{cf}^n$ is trained on sequences using the word2vec method [3], which maps customers' id C_{id} to n-dimension distributed vectors V_{cf}^n. The relative position of the vectors reveals the relationship among different customers.

Some attributes are digital quantities. Each item in these attributes is expanded to a one-hot vector v_{cp-i}. Then, an explicit vector of the customer, denoted as V_{ce}, is constructed as: $V_{ce} = V_{cf} \oplus V_{cp}$, where V_{cp} is the concatenation of v_{cp-i} and $\oplus$ is the concatenation operator.

4.2   Construction of Customer Implicit Vector

The implicit vector is derived from the features of reviews, which expresses the original consumption experience of a customer. In this paper, only the reviews corresponding to the high scores r_{chs} (above the average level) are used to establish the implicit vector because the vector of customers should describe their consumption preferences. In addition, reviews need to removing the stop words. We use $word\_list$ to represent the reprocessed result as: $word\_list = [word_1, word_2, \cdots, word_k]$, where $word_k$ is the word of the preprocessed reviews. The time sequence of wrds in $word\_list$ is maintained. After constructing the $word\_list$, the word embedding of $word_k$ is trained for all review texts using the word2vec method [3], which maps $word_k$ to m-dimension distributed vectors V_{rw}^n. This process converts $word\_list$ into a semantic matrix M_r as $M_r = [V_{rw1}^n, V_{rw2}^n \ldots \ldots, V_{rwk}^n]$, where V_{rw}^n describes the semantic information of $word_k$.

As the dimension of M_r is as large as $m \times n$ dimensions. Features in reviews need to be further extracted. Thus, the RNN [9] is used to solve this problem, the process of which are shown in Fig. 2. The hidden layer consists of k neurons which accept each word vector V_{rwi}^n in turn. Neuron i produces s_i by applying feedforward computation on word vectors V_{rwi}^n and last output s_{i-1} as follows:

$$s_i = \tanh\left(U_i \cdot V_{rwi}^n + W_{i-1} \cdot s_{i-1} + b_i\right) \tag{1}$$

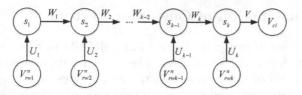

Fig. 2. The structure of RNN

Each intermediate result s_i contains the information of the sequences from V_{rw1}^n to V_{rwi}^n. The output layer activates the last hidden layer unit using the formula: $V_{ci} = \varphi(V \cdot s_k)$ where φ is the SoftMax function:

$$\varphi(x_i) = \frac{\exp(x_i)}{\sum_{i=1}^{m} \exp(x_i)} \tag{2}$$

The V_{ci} denotes the feature extracted from the semantic matrix M_r. In the meantime, V_{ci} is also used as the implicit part of the customer vector which will pass into the subsequent prediction part.

5   Construction of Shop Vectors

Vector is used in RCFM to represent the overall information of a shop. The vector of the shop is composed of three parts: vector of explicit feature consisting of shop attributes, vector of semantic feature formed by shop reviews, and vector of image feature formed by shop photos.

5.1   Construction of Shop Explicit Vector

The explicit vector is derived from the vectorization of the shop's basic information which contains shop address, average rating, number of reviews, business scope and service scope of the shop. In general, customers usually visit and compare several neighbor shops before buying the favored goods. Thus, the location of the shop has latent influence on consumption. The shop-address vector, denoted as V_{sa}, is extracted from address text through word embedding. Attributes of the shop such as the business scope represent the operating conditions of the shop. These attributes are merged as one piece of text to make up the attributes vector V_{sp} using the word2vec method. For those digital quantities, each item is expanded to a one-hot vector v_{sd-i}. Then a shop explicit vector, denoted as V_{se}, is constructed as $V_{se} = V_{sa} \oplus V_{sp} \oplus V_{sd}$, where V_{sd} is the concatenation of v_{sd-i}.

5.2   Construction of the Shop Review Vector and Image Vector

The review text is a description of the shop from the customers' perspective. Similar to the processing of user's reviews, reviews of the shop are mapped into a semantic matrix M_{sr} using word2vec method [3]. Another RNN [9] model is designed to extract features of M_{sr}, and the semantic feature vector V_{sr} of the shop is simply the output of RNN model.

Moreover, the features of photos of goods sold by the shop serve as the image part of the shop vector. A CNN [7] model is built to extract features of images. The convolution layer receives the input image m_{img} and produce feature maps by applying convolution operator with n neurons. Each neuron contains a filter K_j, and K_j in turn acts on the area of t_1 size in m_{img}. Rectified Linear Unit (ReLU) [11] is chosen as the activation function to speed the optimization convergence. The operator is defined as follows:

$$c_j = ReLU\left(K_j * m_{img} + b_j\right) \tag{3}$$

where $*$ is the convolution operator and b_j is the bias for neuron. Then max-pooling is utilized which in turn extracts the maximal feature on every feature maps. Connections between multiple convolutional layers and pooled layers are made until the output dimension reaches the desired. There are k feature maps and then each of them is reshaped to a one-dimension matrix h. In the end, the fully connect layer receives H, which is the concatenation of h, to get the final image vector V_{si} as $V_{si} = W \cdot H + b$, where W is the weight vector and b is the bias.

6 Predicting the Ratings

After the extraction process, the explicit vector V_{ce} of each customer V_{ce} and its implicit vector V_{ci} constitute the customer vector U as: $U = [V_{ce} \oplus V_{ci}]$. For each shop, an explicit vector V_{se}, an implicit vector V_{sr}, and an implicit vector V_{si} have all been extracted. They constitute the shop vector V as: $V = [V_{se} \oplus V_{sr} \oplus V_{si}]$. Then the customer vector and a to be recommended shop vector are combined to generate the predictive vector P as $P = [U \oplus V]$.

The pedictive vector P is passed into the FM [12] to predict the score. The FM introduces a weight matrix M to represent the second-order interactions of the elements in P, which yields better prediction effect than the ordinary regression algorithm. The FM prediction method is expressed as follows:

$$\hat{y}\left(P\right) = m_0 + \sum_{i=1}^{p} m_i P_i + \frac{1}{2}\sum_{i=1}^{p}\sum_{j=1}^{p} M_{ij} P_i P_j \tag{4}$$

where m_0 is the bias and m_i is the weight of P_i. And $M_{ij} = v_i^T v_j$ where v_i is the latent vector associated with P_i. The prediction of FM is first compressed to rating scales of 1–5 because the true rating is within this range. The compression function [16] is as follows:

$$C(x) = \frac{\ln(1 + \mu x)}{\ln(1 + \mu)} \tag{5}$$

where μ is the compression parameter. In addition, the final prediction score $y'_{u,i}$ also considers the grading habits of customers and the average score of shops. $y'_{u,i}$ is calculated as:

$$y'_{u,i} = C(\hat{y}) + \frac{1}{2}(d_u + d_i) \tag{6}$$

where d_c is defined as $d_c = avg - c_{avg}$, d_s is defined as $d_s = avg - s_{avg}$, avg is the global average score, c_{avg} is the average rating of the customer and s_{avg} is the average rating of shop.

The goal of the model is to minimize the square loss function as follows:

$$Loss = \sum_{y' \in O} (y' - y)^2 \tag{7}$$

where O is the set of prediction rating and y is ground truth rating. Random gradient descent method and back propagation method are used to continuously reduce the loss, and dropouts can be added to reduce the overfitting effect during actual model training.

7   Experiments

7.1   Dataset

In the experiments, the dataset is from Yelp Challenge 2018. In Yelp, there are over 142 million ratings and reviews generated from real consumption data. In addition to ratings and reviews, the dataset also contains attribute information for users and shops. The raw data set is huge and sparse, and should be filtered. After the filtering process, there are at least 20 reviews of each user, at least 20 reviews for each shop, and at least one image for each shop in the data set. The statistics of the final experimental data set are shown in Table 1.

Table 1. Description of the experimental data set

Users number	Items number	Rating scale	Reviews number	Average words per review
14219	14299	[1, 5]	613078	13

7.2   Performance Evaluation

To evaluate the performance of the proposed model RCFM, four state-of-the-art models are compared in the experiments, namely PMF [10], SVD++ [5], TimeSVD [6] and HFT [8], as introduced in Sect. 2. The experiment adopt Mean Absolute Error (MAE) and Root Mean Square Error (RMSE) for the performance evaluation. The overall rating prediction errors of the models are shown in Table 2.

Table 2. MAE and RMSE comparison with baseline

Evaluation criteria	PMF	SVD++	TimeSVD	HFT	RCFM
MAE	0.8306	0.8144	1.1556	0.8424	0.8004
RMSE	1.0511	1.0488	1.0750	1.0463	1.0382

In Table 2, the prediction result of RCFM yields the smallest error. Performance of TimeSVD is the worst due to the sparseness of dataset and the long time-spans of dataset. PMF, SVD++ and TimeSVD that use only the rating matrix perform worse than HFT and RCFM which considers the information contained in the reviews, which shows that the more information is used in model, the better result is achieved. HFT uses the distribution of the words in reviews which produces a better performance than the collaborative filtering models (e.g. PMF, SVD++, and TimeSVD). However, semantic of reviews that can provide more useful information than the distribution of words in reviews are not considered in these models. RCFM considers the semantic features of reviews and uses RNN to capture them. In addition, RCFM considers the consumption behaviors of customers and the individuality attributes of shops. The consumption is constructed by building a multi-factor fusion and high-dimension model. Thus, the prediction is the closest to real case.

7.3   Model Analysis

To identify the respective role of the three basic features, attributes, images and texts, we compare five variants of RCFM in this section: RCFM-CattSatt, RCFM-CattSimg, RCFM-CattSrev, RCFM-CrecSatt, and RCFM-CrevSRev. The vectors used in the variants of the RCFM have been processed in Sects. 4 and 5. The differences among five variants of RCFM are their respective predictive vector P as shown in Table 3.

Table 3. Differences among five variants of RCFM

Variants of RCFM	Constitution of predictive vector P
RCFM-CattSatt	$P = [V_{ce} \oplus V_{se}]$
RCFM-CattSimg	$P = [V_{ce} \oplus V_{si}]$
RCFM-CattSrev	$P = [V_{ce} \oplus V_{sr}]$
RCFM-CrevSatt	$P = [V_{cr} \oplus V_{se}]$
RCFM-CrevSrev	$P = [V_{cr} \oplus V_{sr}]$

The performance of RCFM and its five variants are shown in Fig. 3. It shows that RCFM-CattSrev and RCFM-CrecSatt delivers better performance than RCFM-CattSatt and RCFM-CattSimg. It can be concluded that the semantics of the reviews provide more contributions to the results than attribute vectors and image vectors. RCFM-CrevSrev utilizes reviews from both customers and shops. RCFM-CattSrev and RCFM-CrevSatt only model reviews from one side and are worse than RCFM-CrevSrev. The results prove that reviews of customers and shops play different roles. Finally, RCFM is lower than RCFM-CrevSrev in error, which indicates that attributes and images provides more details to the model and are indispensable.

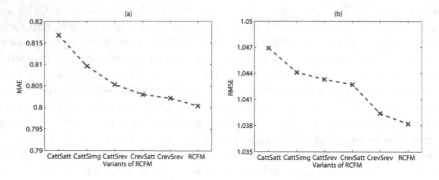

Fig. 3. (a) The MAE of RCFM and its variants, (b) The RMSE of RCFM and its variants

8 Conclusion

In this paper, a deep-learning-based fusion recommendation model, RCFM, which fuses semantic information of reviews, feature information of images, and attribute information of customers and shops, is proposed. The model generates implicit vectors for both customers and shops based on their attributes. Then RNN is used to extract features from reviews and CNN is used to extract features from images. The abstract feature vectors of reviews, images and attributes are integrated into one vector to describe the consumption and predict the rating. Experiments based on real data sets show that the proposed RCFM model has a significant improvement over other baseline methods.

Acknowledgement. This work was supported by National Key R&D Program of China (2017YFB1301100), National Natural Science Foundation of China (61572060, 61772060, 61728201) and CERNET Innovation Project (NGII20160316, NGII2017 0315).

References

1. Chen, C., Zhang, M., Liu, Y., Ma, S.: Neural attentional rating regression with review-level explanations. In: Proceedings of the 2018 World Wide Web Conference on World Wide Web, pp. 1583–1592. International World Wide Web Conferences Steering Committee (2018)
2. Chen, T., Hong, L., Shi, Y., Sun, Y.: Joint text embedding for personalized content-based recommendation (2017)
3. Goldberg, Y., Levy, O.: word2vec explained: deriving Mikolov et al'.s negative-sampling word-embedding method. arXiv preprint arXiv:1402.3722 (2014)
4. Kim, D., Park, C., Oh, J., Lee, S., Yu, H.: Convolutional matrix factorization for document context-aware recommendation. In: Proceedings of the 10th ACM Conference on Recommender Systems, pp. 233–240. ACM (2016)
5. Koren, Y.: Factorization meets the neighborhood: a multifaceted collaborative filtering model. In: Proceedings of the 14th ACM SIGKDD International Conference on Knowledge Discovery and Data Mining, pp. 426–434. ACM (2008)
6. Koren, Y.: Collaborative filtering with temporal dynamics. Commun. ACM **53**(4), 89–97 (2010)
7. Krizhevsky, A., Sutskever, I., Hinton, G.E.: Imagenet classification with deep convolutional neural networks. In: Advances in Neural Information Processing Systems, pp. 1097–1105 (2012)
8. McAuley, J., Leskovec, J.: Hidden factors and hidden topics: understanding rating dimensions with review text. In: Proceedings of the 7th ACM Conference on Recommender Systems, pp. 165–172. ACM (2013)
9. Mikolov, T., Karafiát, M., Burget, L., Černockỳ, J., Khudanpur, S.: Recurrent neural network based language model. In: Eleventh Annual Conference of the International Speech Communication Association (2010)
10. Mnih, A., Salakhutdinov, R.R.: Probabilistic matrix factorization. In: Advances in Neural Information Processing Systems, pp. 1257–1264 (2008)
11. Nair, V., Hinton, G.E.: Rectified linear units improve restricted Boltzmann machines. In: Proceedings of the 27th International Conference on Machine Learning (ICML-10), pp. 807–814 (2010)
12. Rendle, S.: Factorization machines with libFM. ACM Trans. Intell. Syst. Technol. (TIST) **3**(3), 57 (2012)
13. Rendle, S., Freudenthaler, C.: Improving pairwise learning for item recommendation from implicit feedback. In: Proceedings of the 7th ACM International Conference on Web Search and Data Mining, pp. 273–282. ACM (2014)
14. Salakhutdinov, R., Mnih, A.: Bayesian probabilistic matrix factorization using Markov chain Monte Carlo. In: International Conference on Machine Learning, pp. 880–887 (2008)
15. Sarwar, B., Karypis, G., Konstan, J., Riedl, J.: Item-based collaborative filtering recommendation algorithms. In: Proceedings of the 10th International Conference on World Wide Web, pp. 285–295. ACM (2001)
16. Vallavaraj, A., Stewart, B.G., Harrison, D.K.: An evaluation of modified μ-law companding to reduce the PAPR of OFDM systems. AEU-Int. J. Electron. Commun. **64**(9), 844–857 (2010)
17. Zheng, L., Noroozi, V., Yu, P.S.: Joint deep modeling of users and items using reviews for recommendation. In: Proceedings of the Tenth ACM International Conference on Web Search and Data Mining, pp. 425–434. ACM (2017)

An Improved SKFCM-CV Whole Heart MR Image Segmentation Algorithm

He Wang[1] , Jing Zhang[1,2(✉)] , Jie Wang[1], XiaoDong Zhang[1],
Hong Teng[3], and TianChi Zhang[1]

[1] Computer Science and Technology, Harbin Engineering University,
Harbin 150006, China
296582337@qq.com
[2] College of Engineering, Shantou University, Guangdong 515063, China
[3] Harbin Engineering University Hospital, Harbin Engineering University,
Harbin, China

Abstract. Cardiac magnetic resonance imaging (MRI) is the only effective method of inspection for some serious heart disease, such as the observation of atrial fibrillation, the evaluation of cardiac iron deposition, the diagnosis and detection of congenital heart and so on. Therefore, the whole heart segmentation of MR image is very important for medicine. In this paper: (1) A new FCM algorithm (SKFCM) is proposed based on spatial neighborhood correlation information of pixel and kernel function. Two correlation factors R_{ij}^S and R_{ij}^G are introduced to make up for the lack of original algorithm for spatial information. (2) A kernel constrained CV algorithm based on entropy and edge guidance function is proposed, which solves the problem that the fixed energy weight coefficient has poor universality for different images. And propose a simple kernel function instead of Gaussian kernel function to improve the efficiency. (3) In order to improve the accuracy and efficiency of the entropy value, the entropy weight coefficient is proposed to solve the problem that the image evolution speed is too slow due to the low entropy value of the image after SKFCM. Experiments on open data sets show that the proposed algorithm is suitable for most medical images with blur, less contrast or more noise, and the segmentation accuracy and efficiency are greatly improved.

Keywords: Whole heart segmentation · SKFCM-CV · CV · MRI

1 Introduction

Lack of blood supply caused by blockage of blood vessels at arteries is the main cause of heart disease [1]. According to the world health organization's global report, the mortality rate of cardiovascular diseases is the highest in the world. In 2012 alone, about 17.5 million people died of cardiovascular diseases worldwide, and the number of deaths is estimated to be 25 million by 2030. Therefore, the early non-invasive diagnosis and interventional therapy for cardiovascular diseases are of great significance in reducing the mortality of cardiovascular diseases.

© Springer Nature Switzerland AG 2018
R. Hong et al. (Eds.): PCM 2018, LNCS 11166, pp. 596–607, 2018.
https://doi.org/10.1007/978-3-030-00764-5_55

In recent years, the dynamic imaging technique of the heart mainly uses the cardiac MRI sequence images to sample the medical images on several spatial sequences during a heartbeat. Compared with CT images, MRI has a high soft tissue resolution and can evaluate the cardiac function by specific sequences without the influence of arrhythmia. It is the golden standard for evaluating cardiac systolic function [2]. Papers published in authoritative journals, such as the Lancet and the American heart journal, also showed that MRI is more accurate than nuclear medicine. However, the key and difficult point is how to segment the region of the heart in MR images accurately and obtain the information of the heart edge and region, so as to effectively assist clinicians in clinical diagnosis and operation navigation.

Zuluaga presented a fully automated method for the segmentation of the whole heart and the great vessels from MRI images [3]. He of Zhe Jiang University trained the average shape model of the heart by collecting the prior knowledge of the heart [4]. A robust active shape model (Robust ASM) is proposed by Zhao et al. [5]. Zhuang et al. tried to develop and evaluate a multi-atlas segmentation scheme using a new atlas ranking and selection algorithm for automatic WHS of CT data [6, 7]. Yin et al. proposed a threshold segmentation algorithm based on morphological operation and morphological gradient, to segment the whole cardiac CT [8, 9].

For the segmentation of human heart structure, many scholars have published a lot of segmentation algorithms, most of which are specific segmentation of ventricle and atrium ventricle. Since 2013, many scholars realized the importance of the whole heart segmentation, and there have been many literatures about whole heart segmentation.

According to the research results obtained from latest five years for WHS, it is clearly that the research in the field of research mainly has the following features:

(1) Most of the work was carried out with deep learning around the heart CT image. However, for MRI images, due to its high cost of time and money, it is difficult to obtain enough data sets to train the deep learning model under the current medical level.

(2) In the few studies on the segmentation of MRI heart images, it is difficult to ensure accuracy and efficiency at the same time. How to segment the region of the heart by dividing the algorithm quickly and accurately is still the key and difficult point.

In this paper, a new *SKFCM-CV* whole heart segmentation method is proposed based on the current research situation. This method can effectively segment MR images.

2 Method

On the basis of the kernel *FCM* algorithm, the space neighbourhood correlation factor R_{ij} is added to compensate for the shortage of the original algorithm in terms of spatial neighbourhood information, and the proposed method is simply called *SKFCM*. The *SKFCM* algorithm is used to rough segment the heart MR image to reduce the noise. A kernel constrained CV algorithm based on entropy and edge guidance function is proposed. In improved CV algorithm, the edge guidance function guides the evolution process according to the internal and external entropy of the evolution curve. When the evolution curve reaches the boundary of the heart contour, the internal and external entropy tends to be stable and the curve converges. The improved algorithm solves the problem that the fixed energy weights λ_1 and λ_2 have poor universality for different images and can make the evolution results of curves more accurate. When MRI is clustered by *SKFCM*, the entropy value decreases. In order to ensure the speed and accuracy of the evolution of the CV model, the initial level set is reselected according to the rough segmentation result of the *SKFCM*. And a new entropy weight coefficient ε is introduced to adjust the entropy value of the image after clustering.

2.1 Spatial Neighborhood Information and Kernel Function FCM Algorithm (SKFCM)

FCM clustering algorithm is one of the most classical classification methods. The initial *FCM* uses the characteristic space of the sample directly to cluster, but such a practice will result in the clustering result dependent on the distribution of the individual to a certain extent. James proposed a kernel clustering method to solve this problem. The Mercer kernel is used to increase the process of optimizing the characteristics of the individual, mapping a relatively low dimension space to a relatively high dimension space, and then performing the clustering operation in the feature space again. This method effectively improves the universality of *FCM* in clustering analysis of non-spherical shape distribution space samples. However, *KFCM* is sensitive to noise, because this method only considers the change of grey value in image segmentation, neglects the spatial neighbourhood information. For some images with noise, each pixel may be affected by noise, so that the images will change in the grey value, eventually lead to lose to the value of medical image segmentation [10, 11].

Based on the above problems, a new *FCM* algorithm based on spatial neighbourhood correlation information and kernel function is proposed in this chapter. By introducing neighbourhood correlation factor R_{ij} to make up for the deficiency of *FCM* on spatial neighbourhood information, so the anti-noise performance of SKFCM is improved.

Assuming that $X = \{x_1, x_2, ..., x_n\}$ is the sample set of the cluster, n is the number of samples, $\|\varphi(x_j) - \varphi(v_i)\|^2$ represents the Euclidean distance squared between the individual and the central function. The object function of algorithm can be expressed as Eq. (1), and the central function can be expressed as Eq. (2).

$$J = \sum_{i=1}^{c} \sum_{j=1}^{N} u_{ij}^m \parallel \phi(x_j) - \phi(v_i) \parallel^2 + \sum_{i=1}^{c} \sum_{j=1}^{N} u_{ij}^m \sum_{x_k \in N_j} R_{kj} u_{ij}^m \parallel \phi(x_k) - \phi(v_i) \parallel^2 \quad (1)$$

$$v_j = \left((1 - K(x_j, v_i)) + R_{ij} \sum_{r \in N_j} (1 - K(x_j, v_i))^{-1/(m-1)} \right) \bigg/ \left(\sum_{k=1}^{c} (1 - K(x_j, v_i)) + R \sum_{r \in N_j} (1 - K(x_r, v_k))^{-1/(m-1)} \right)$$

$$(2)$$

A new concept, the spatial neighbourhood correlation factor, is introduced. The correlation between spatial neighbours actually depends on the location between the two neighbourhood pixels and the grey level information between the two neighbourhood pixels, so the expression of R_{ij} can be expressed as follows (when $i \neq j$):

$$R_{ij} = R_{ij}^S \times R_{ij}^G (i \neq j) \tag{3}$$

Definition 1. If the coordinates of the given pixel i and j are respectively (x_1, y_1) and (x_2, y_2), then the spatial correlation between the two neighbourhood pixels is defined as R_{ij}.

$$R_{ij}^S = \exp\left(\sqrt{(x_1 - x_2)^2 + (y_1 - y_2)^2} \Big/ \lambda_S \right) \tag{4}$$

It can be seen in Fig. 1 that the spatial correlation increases as the distance decreases.

Definition 2. If the grey value of the given pixel i and j are respectively $g(i)$ and $g(j)$, then the grey correlation between the two neighbourhood pixels i and j is defined as R_{ij}^G.

$$R_{ij}^G = \exp\left((g(i) - g(j))^2 \Big/ \lambda_G \times \sigma_i^G\right) \tag{5}$$

In Eq. (4), λ_G represents the grey correlation factor. σ_i^G represents the grey value square deviation between i and the neighbourhood pixels of i. Under the constraint condition of $\sum u_{ij=1}$, the minimum value can be obtained by the Lagrange multiplier method, and the iterative formula between the membership degree and the central function can be obtained.

$$v_j = \sum_{i=1}^{n} u_{ij}^m K(x_i, v_j) x_i \Big/ \sum_{i=1}^{n} u_{ij}^m K(x_i, v_j),$$

$$u_{ij} = (1 - K(x_i, v_j))^{-1/(m-1)} \Big/ \sum_{j=1}^{c} (1 - K(x_i, v_j))^{-1/(m-1)} \tag{6}$$

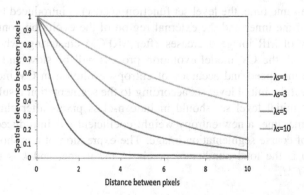

Fig. 1. The relationship between spatial relevance and spatial correlation factor.

2.2   Improved CV Algorithm Based on Entropy and Edge Guidance Function

As a simplified version of active contour model, CV model is basically consistent with that of Snake model [12]. The CV model can make full use of the global region information of the image, and it has some advantages in dealing with the uncertain medical image segmentation [13]. However, there are some common problems in original CV models. On one hand, when the energy minimization is used to solve the problem of unclear image boundaries, there will be an overflow phenomenon at the weak boundary. On the other hand, when the image with non-uniform greyscale values is segmented, over-segmentation or under-segmentation may occur.

Therefore, a kernel constraint CV model based on entropy and edge guidance function is proposed in this paper. It uses the edge guidance function to guide the evolution process according to the internal and external entropy values of the evolution curve. When the evolution curve reaches the boundary of the heart contour, the internal and external entropy tends to be stable and the curve converges. The improved algorithm solves the problem that the fixed energy weights λ_1 and λ_2 have poor universality for different images and can make the evolution results of curves more accurate.

In 1948, Shannon first proposed the concept of entropy. Suppose there are n events in a probability theory system, and the probability of occurrence of the ith event is Pi. Therefore, the entropy $E(I)$ can be expressed as Eq. (7).

$$E(I) = - \sum_{i=1}^{n} P_i \log P_i \tag{7}$$

According to the concept of entropy proposed by Shannon, it can be described in detail in the image as: can be specifically stated in the image that the closer the greyscale value in the region is (homogeneous region), the smaller the value of entropy will be; when there is more noise or in the contour of the image, the entropy is larger [14]. In this chapter, entropy is used to represent the authenticity, objectivity and similarity of the boundary information of the image region [15]. The entropy value of the image changes autonomously with the evolution of the curve, so that the proportion of the greyscale change value in the internal area and the external area is adaptively adjusted. At the same time, the level set function $\varphi(x, y)$ is introduced to represent the entropy value of the inner and the external region of the evolution contour curve C.

The entropy of MR image decreases after SKFCM clustering, when the entropy value is too small, the CV model evolution process will be slow and inaccurate. In order to ensure the speed and accuracy of entropy evolution in CV model, it is necessary to re-select the initial level set according to the segmentation results of SKFCM. The re-selected initial level set should include n-class pixels after clustering, where $n \geqslant 3$. At the same time, a new entropy weight coefficient ε is introduced to adjust the entropy value of coarse segmentation image. The expression of ε is shown in formula (8), where *sum* is the total number of pixels in the image, and k is the number of clusters in *SKFCM*.

$$\varepsilon = sum/k \tag{8}$$

The entropy values of the inner and the external region of the evolution contour C:

$$E_i = -\varepsilon \int_\Omega P_i(x|\phi) \log P_i(x|\phi)dx \tag{9}$$

$$E_o = -\varepsilon \int_\Omega P_o(x|\phi) \log P_o(x|\phi)dx \tag{10}$$

When dealing with grayscale non-uniform images, only considering the information of the inner and outer entropy values will result in over-segmentation, under-segmentation, or non-convergence of the boundaries where the greyscale distinction is not obvious [18]. Using the edge guidance function to simultaneously consider the boundary information of the contour curve will improve the segmentation accuracy of the blurred image edge. Therefore, the edge guidance function is added to the CV model, and improved CV can be expressed as formula (11).

$$E = \mu \int_\Omega \delta(\phi)|\nabla\phi|dxdy + v \int_\Omega H(\phi)dxdy +$$
$$E_i \int_\Omega g(\tau)|u - c_1|^2 H(\phi)dxdy + E_0 \int_\Omega g(\tau)|u - c_2|^2(1 - H(\phi))dxdy \tag{11}$$

Due to the inflection point and convergence of the Gaussian function, the use of a Gaussian function during the evolution of the curve will lead to a cumbersome calculation process, reduce the speed of the calculation of the function and the convergence rate of the contour curve. In order to simplify the evolution process of the curve, this paper introduces a new kernel function to constrain the level set function. The new kernel function does not need to repeatedly select the parameter μ. Its form is:

$$K(u) = \frac{1}{1+u^q}, q \in (0,1], u = x*q+y*q \tag{12}$$

It needs to point that when $q > 1$, the kernel function will degenerate into gaussian function. The segmentation steps of the FCM-CV model algorithm are as follows:

(1) Set the number of classifications k. Do pre-treatment with SKFCM, and obtain pre-segmentation results;
(2) Calculate the entropy value of the image;
(3) According to the segmentation results of SKFCM, determine the initial level set function $\varphi_0 (x, y)$ of the CV model.
(4) Set the number of termination cycle control T for the method operation;
(5) Calculate $c_1 (\varphi^n)$ and $c_2 (\varphi^n)$;
(6) Solve partial differential equation;
(7) Confirm whether the obtained solution is in a stable state. If it is in a stable state, exit the loop; If not, return (2) to continue.

3 Experiment

This chapter selects a cardiac MRI public data set from York University for testing [16], which include cardiac MR images acquired from 33 subjects. The sequence of images for each subject include 20 frames and 8–15 slices along the long axis for a total of 7980 images. In this study, 33 sets of tests were performed on the end-systolic and end-diastolic images of each patient's MR sequence images. The slicing angle of the selected MR image should be the one who can show the outline of heart most completely in all slices of this subject. The experimental data in this chapter is the average of 33 tests.

3.1 Experiment of SKFCM Algorithm

Firstly, the performance of *SKFCM* is tested by natural images, which proves that *SKFCM* do really has a better performance. Then the *SKFCM* is tested by medical images to prove that *SKFCM* is suitable for medical images.

In the performance test, select Fig. 2(a) as the test image, and test the performance of *SKFCM*, *FCM*, *FCM_S*, *KFCM* method. In this experiment, the minimum error is 0.0001, the maximum number of iterations is 200, $k = 2$ and $\sigma = 50$.

(a) From left to right: artificial image, Gaussian noise image and the result of FCM.

(b) From left to right: the result of Kernel FCM, FCM_S and improved FCM.

Fig. 2. Segment result of images with gaussian noise.

In order to verify the effectiveness of the algorithm, the segmentation accuracy (*SA*) was used as the evaluation index of the algorithm. Table 1 shows the processing time and the segmentation accuracy of each algorithm.

Table 1. Segmentation accuracy of artificial images with different levels of salt & pepper noise.

Salt & pepper noise value	FCM	Kernel FCM	FCM_S	SKFCM
0.1	92.88	93.72	96.05	99.20
0.15	90.79	91.18	94.86	98.89
0.2	88.53	89.68	93.78	98.67
Average time	0.98	0.52	0.89	0.68

According to Fig. 2 and Table 1, it is easy to know: the improvement method in presence of gaussian noise image processing effect is better than *FCM*, *KFCM* and *FCM_S* methods. In terms of speed, compared with FCM and *FCM_S*, *SKFCM* increased by about 30.6% and 30.6% respectively. Next, cardiac MR images will be processed with *FCM*, *KFCM*, *FCM_S*, and *SKFCM*, respectively, to demonstrate the applicability of the improved algorithm to cardiac MR images. Take a set of experimental end-diastolic images as an example and illustrate them with drawings in Fig. 3 (Table 2).

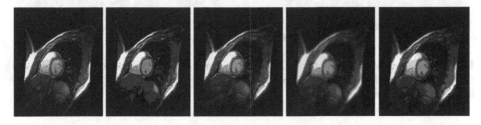

Fig. 3. Segmentation results of four algorithms for cardiac MR images. From left to right: end-diastolic image, the result of FCM, kernel FCM, FCM_S, SKFCM.

Table 2. The segmentation accuracy of four algorithms for cardiac MR images (average data)

	FCM	Kernel FCM	FCM_S	SKFCM
End-diastolic image (a)	82.54	89.70	93.60	96.62
End-systolic image (b)	82.68	89.33	93.24	96.03

It can be seen from the above experimental results that *SKFCM* can clearly preserve the boundaries of different regions in the image while reducing noise. This has important implications for dealing with complex medical images. This experiment proves that *SKFCM* method is faster and suitable for images with complex noise.

3.2   Experiment of Improved CV Model and SKFCM-CV Algorithm

Firstly, the performance of improved CV is tested by natural images, which proves the improved CV do really has a better performance. In the performance test, select Fig. 4 (a) as the test image, $\mu = 0.01$, and $\Delta t = 0.5$.

In the experimental results, the curve segmented by the original CV model can evolve to the boundary of the target. However, compared with the improved CV, the segmentation error is larger, and the speed is slow. It can be seen from the convergence curve of Fig. 5 that the improved CV model algorithm has the smallest error rate and the shortest convergence time, And the improved CV model algorithm is more stable.

Through the above experiments, it can be seen that the improved *SKFCM* and the improved CV do really improve the target boundary segmentation of the image. Next, the improved algorithm is applied to medical image segmentation to verify it is suitable for medical image segmentation. A group of end-diastolic and end-systolic images are randomly selected from 33 groups of images as an example to illustrate (as Fig. 6).

(a) Natural images   (b)CV ($\lambda 1=0.7$ $\lambda 2=0.3$)(c)CV ($\lambda 1=0.5$ $\lambda 2=0.5$)   (d) Improved CV

Fig. 4. The segmentation accuracy of CV model and improved CV model. (b) Is the result with 660 iterations, big error. (c) Is the result with 730 iterations, big error. (b) Is the result with 320 iterations, convergence.

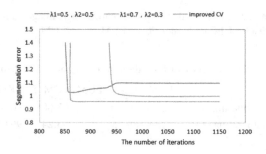

Fig. 5. The segmentation error rate of CV model and improved CV model.

Li fist mentioned to combine the original FCM and CV to segment the heart [17], so the original FCM-CV algorithm is compared as a control group in this experiment. The Precision, Recall and comprehensive evaluation index (F-Measure) are used to evaluate improved algorithm in Table 3. The number of iterations is represented by t.

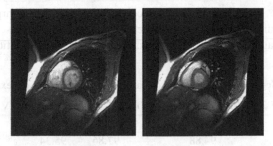

Fig. 6. MR images of end-diastolic (a) and end-systolic (b) image.

Table 3. The segmentation results of different algorithms (average data of 33 sets experiments)

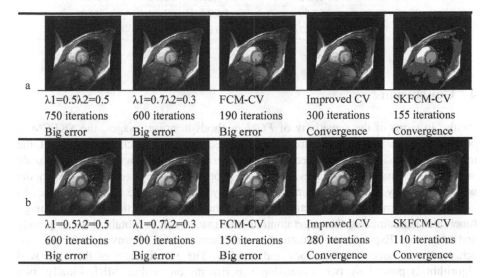

a	λ1=0.5λ2=0.5	λ1=0.7λ2=0.3	FCM-CV	Improved CV	SKFCM-CV
	750 iterations	600 iterations	190 iterations	300 iterations	155 iterations
	Big error	Big error	Big error	Convergence	Convergence
b	λ1=0.5λ2=0.5	λ1=0.7λ2=0.3	FCM-CV	Improved CV	SKFCM-CV
	600 iterations	500 iterations	150 iterations	280 iterations	110 iterations
	Big error	Big error	Big error	Convergence	Convergence

According to the data in Tables 3 and 4, the accuracy of original CV is poor. The efficiency of *FCM-CV* is about 4 times of the original CV model, but the accuracy still needs to be improved. In some images with poor greyscale, the contour curve still does not converge. The accuracy of the improved CV is higher than that of the original CV and *FCM-CV*, and the efficiency is about twice of the original CV model. However, due to the non-uniform distribution of grey value, volume effect and noise phenomenon, the boundary information and initial position are not clear, so its efficiency is lower than *FCM-CV*, and the accuracy is close to *FCM-CV*. *SKFCM-CV* eliminates the above interference information, so it can obtain relatively weak image boundary information and concave contour, and the evolution accuracy and speed are greatly improved. The efficiency of the *SKFCM-CV* is about 4–5 times than original CV, and the accuracy is about 95% and 97%, which is much higher than that of the other methods. And this method has better stability and robustness. *SKFCM-CV* method can accurately extract the contour regions corresponding to the targets in complex heterogeneous cardiac MR images. This method can achieve better segmentation effect in cardiac MR images and is an effective medical image segmentation method.

Table 4. Segmentation error rate of cardiac MR image (33 sets experiments)

Image		$\lambda1 = 0.7$ $\lambda2 = 0.3$	$\lambda1 = 0.5$ $\lambda2 = 0.5$	FCM-CV	Proposed CV	SKFCM-CV
(a)	t	1600	1100	400	500	340
	P	97.57	94.88	97.86	95.34	98.00
	R	83.34	84.79	87.52	90.11	95.37
	F	90.46	89.14	92.69	92.73	96.69
(b)	t	1100	750	285	340	255
	P	93.50	87.35	94.89	92.72	97.34
	R	85.12	87.59	86.75	89.69	94.21
	F	89.31	97.47	90.82	91.21	95.78

4 Conclusion

Firstly, in terms of the sensitivity of *FCM* methods to noise images, the SKFCM is proposed. Through two experiments, the correct and timely characteristics of the improved method are proved. Secondly, considering the inhomogeneity of grey scale and complex structure of images, this paper proposes a kernel constraint CV model based on entropy and edge guidance function. By introducing the edge guide function and using the simple kernel function instead of the gaussian function when the energy functional is minimized, the target contour information can be obtained more quickly and accurately. Experiments on natural images show that the improved CV model has achieved great success in accuracy and rapidity. The effectiveness of this improved algorithm is proved by two simulation experiments on cardiac MRI. Finally, two improved algorithms are combined to obtain *SKFCM-CV* algorithm. The *SKFCM-CV* algorithm has been improved by about 5 times in efficiency compared with the classic CV model, and the accuracy is about 95% to 97%. The SKFCM-CV has been improved obviously in three aspects: precession, recall rate and F-Measure. SKFCM-CV is more suitable for most medical images with weak contrast, high noise, and insignificant greyscale boundaries. It is a kind of effective medical image segmentation method.

Acknowledgment. This research is finance supported by 1. 2017–2020 China National Natural Science Fund (51679058); 2. 2013–2016 China Higher Specialized Research Fund (Ph.D. supervisor category) (20132304110018).

References

1. Lv, S.H.: Research on medical image segmentation based on level set. Ph.D. thesis, University of Electronic Science and Technology of China (2015)
2. Liu, Y.X.: Relationship between echocardiographic and magnetic resonance derived measures of right ventricular size and function in patients with atrial fibrillation. Chin. J. Cardiovasc. Med. **20**(2), 90–94 (2015)
3. Zuluaga, M.A., Cardoso, M.J., Modat, M., Ourselin, S.: Multi-atlas propagation whole heart segmentation from MRI and CTA using a local normalised correlation coefficient criterion. In: Ourselin, S., Rueckert, D., Smith, N. (eds.) FIMH 2013. LNCS, vol. 7945, pp. 174–181. Springer, Heidelberg (2013). https://doi.org/10.1007/978-3-642-38899-6_21
4. He, J.: Research on the whole heart segmentation algorithm based on CT image. Ph.D. thesis, Zhejiang University (2015)
5. Zhao, X., Wang, Y., Jozsef, G.: Robust shape-constrained active contour for whole heart segmentation in 3-D CT images for radiotherapy planning. In: IEEE International Conference on Image Processing, pp. 1–5. IEEE (2015)
6. Zhuang, X.: Multitask whole heart segmentation of CT data using conditional entropy for atlas ranking and selection. Med. Phys. **42**(7), 3822–3833 (2015)
7. Zhuang, X.H., Shen, J.: Multi-scale patch and multi-modality atlases for whole heart segmentation of MRI. Med. Image Anal. **31**, 77 (2016)
8. Yin, H.P., Zhang, Y.N., He, Y.: Research on whole heart segmentation method based on CT image. Mod. Comput. **32**, 62–66 (2016)
9. Yin, H.P.: New algorithm for segmentation of cavity region based on CT cardiac image. Comput. Syst. Appl. **26**(11), 292–295 (2017)
10. Yang, Y., Guo, S.X., Ren, R.Z.: Modified kernel-based fuzzy c-means algorithm with spatial information for image segmentation. J. Jilin Univ. **41**(s2), 283–287 (2011)
11. Zong, Y.S., Hu, X.H., Qu, Y.Z.: Kernel FCM image segmentation algorithm based on spatial neighboring information. Comput. Appl. Softw. **34**(4), 221–225 (2017)
12. Vese, L.A.: A multiphase level set framework for image segmentation using the mumford and shah model. Int. J. Comput. Vis. **50**(3), 271–293 (2002)
13. Li, C., Xu, C., Gui, C.: Distance regularized level set evolution and its application to image segmentation. IEEE Trans. Image Process. **19**(12), 3243 (2010)
14. Jiang, X.L., Li, B.L., Liu, J.J.: Image segmentation based on improved active contour model. Comput. Eng. **32**(4), 236–240 (2015)
15. Zheng, Q.: Research and application of medical image segmentation methods. Ph.D. thesis, Southern Medical University (2014)
16. Miao, Z.X.: Method for medical image segmentation based on active contour model. Master thesis, Southwest University (2015)
17. Andreopoulos, A., Tsotsos, J.K.: Efficient and genera-lizable statistical models of shape and appearance for analysis of cardiac MRI. Med. Image Anal. **12**(3), 335–357 (2008)
18. Li, X.P., Wang, X.: Active contour model-based medical image segmentation method collaborative with fuzzy C-means. Chin. J. Sci. Instrum. **34**(4), 860–865 (2013)

An Image Splicing Localization Algorithm Based on SLIC and Image Features

Haipeng Chen[1(✉)], Chaoran Zhao[1], Zenan Shi[1], and Fuxiang Zhu[2]

[1] College of Computer Science and Technology,
Jilin University, Changchun, China
chenhp@jlu.edu.cn
[2] Lucion Technology Corp., Ltd., Jinan, Shandong, China
zfux@163.com

Abstract. Aiming at the problem of low accuracy, high computational complexity and incomplete edge information of most image splicing localization algorithm, this paper proposes a new image splicing localization algorithm. First, the SLIC image segmentation algorithm is used to segment the image. Secondly, the noise estimation value of each super-pixel block is calculated by the FAST noise estimation algorithm. Then, weight of each image block is calculated through noise and image features. Finally, the noise value sequence is processed by clustering and statistical processing to determine the pixels of the background area and the splicing area, thus the splicing area is located. In this paper, the algorithm is tested on the color image database of Columbia, and compared with the existing image splicing localization algorithms based on block-segmentation and based on pixel. The experiment shows that the proposed algorithm can preserve the connection between image features, hold the edge of the splicing area, and effectively improve the efficiency of localization detection under the premise of ensuring the accuracy of image splicing localization.

Keywords: Image splicing · Image splicing localization · Super-pixels
Image features · Noise estimation

1 Introduction

The SLIC method has been widely used, including the following important factors: first, the edge of the pixel block generated by this method is more neat and smooth, and the intersection area between the image blocks is easy to express, which makes the method more easily used in combination with other methods; secondly, this method is not only applicable to color. The image is also suitable for grayscale images; thirdly, there are few parameters to be set in the block process, and there is no need for a large number of parameters to adjust. Fourthly, compared with other similar segmentation methods, the running speed of SLIC and the smoothness of the image block are ideal.

Image tampering technology enriches people's lives, but also leads to some adverse events. And splicing forgery is the most common and basic image tampering method. It generates some fake images by copying some parts of an image and pasting it into a

© Springer Nature Switzerland AG 2018
R. Hong et al. (Eds.): PCM 2018, LNCS 11166, pp. 608–618, 2018.
https://doi.org/10.1007/978-3-030-00764-5_56

certain area of another image [1]. No matter in daily life or in military politics, there are cases of image tampering.

At present, image splicing forgery detection methods can be divided into two categories as below.

1.1 Classification and Recognition Method Based on Feature Extraction

Ng uses higher order statistics (amplitude and phase) of dual coherence feature and its prediction error characteristics, and uses support vector machine (SVM) [2] to realize the classification and recognition of forgery images [3]. Sutthiwan uses Markov features and edge statistical features to classify image features [4]. Agarwal uses the linear filter and texture descriptor to extract image features for image forgery detection [5]. Wang takes the gray level co-occurrence matrix of the edge image as the feature to classify and detect the image [6]. Hashmi combines DCT, LBP and gobar features to identify forgery images [7]. Vaishnavi detects the authenticity of the image by extracting the feature histogram of LBP [8].

The above method extracts the image features and uses classifiers to realize the classification and recognition of natural images and forgery images. This method can only verify whether the image has been tampered with, and can not locate the splicing area in the image.

1.2 Image Forgery Detection and Localization Method

Wang proposed a demosacing method to estimate the actual corresponding object of the splicing area, so as to highlight the edge of splicing area for localization [9]. The algorithm proposed by Zampoglou can detect the change of pixels in the large data set, and regard the changed area as the splicing area [10]. Because the blur degree of the splicing area in the image is different in most cases, Bahrami trained the type frame of the fuzzy degree of the blurred image according to this characteristic [11]. Immerkær estimates the noise of image. Because the noise of splicing area and original image is different in most cases, the difference between noises is used to distinguish and locate the mosaic image [12]. Lyu proposed a method to distinguish the background image and the splicing area based on noise kurtosis [13]. Mahdian also uses noise, they segment the image and detect the change of noise [14]. Pan uses the variance of noise to distinguish the splicing area [15, 16].

At present, most splicing localization algorithms divide image into several blocks, each block is calculated, and the information is statistically classified, so as to determine the splicing area. However, the traditional mechanical block may divide the two parts from different images into one block, and also make the edge information of the splicing area incomplete, thus reducing the accuracy of localization.

Aiming at the problem of reducing the accuracy caused by the traditional mechanical block, this paper proposes to use the super pixel segmentation algorithm to block the image and locate the location by estimating the noise. SLIC super pixel segmentation is an adaptive segmentation method, which has obvious advantages in retaining the connection between image features and the boundary information of images. FAST noise estimation method is fast in operation and accurate in noise estimation. In order to make the FAST algorithm suitable for super pixel blocks, this paper adds boundary judgment to FAST noise estimation. Based on these, this paper proposes an image splicing localization algorithm based on SLIC super pixel segmentation, which can locate splicing areas effectively.

2 Basic Theory

In this paper, the SLIC super pixel segmentation method is used to block the image, and then the improved FAST algorithm is used to estimate the noise of each image block, then, computing block weights through multiple image features, finally, according to the clustering of the data sequence composed of image block, the splicing area and background are determined by clustering results.

2.1 Image Block on SLIC

Achanta proposed a super pixel block algorithm [17], it can quickly and efficiently generate color pixels as uniformly as possible, that is, super pixels. This method sets the unique parameter, that is, the number of pixels is k. Assuming that the image is based on the CIELAB color space, the original seed point is $C_i = [l_i\ a_i\ b_i\ x_i\ y_i]^T$, $i = 1$, ...,k. Assume the number of pixels in the input image is N, Super pixel block algorithm hopes to get the sizes of pixel blocks are basically the same, so the size of super pixel is N/k. And the distance of each seed point is approximately $S = \mathrm{sqrt}(N/k)$. In order to avoid the distribution of seed points at the edges of the image, the seed points are placed at the center of the 3 * 3 window and moved over the pixels and placed at the lowest gradient position.

After the super pixel segmentation of the image I, the obtained blocks are basically irregular shapes, we tag each pixel a number. The pixels with the same tag constitute a super pixel segmentation block, and then the noise of this block can be estimated.

We classify the area with the same tag of pixels as a super pixel block, denoted as I_1, I_2 ...I_k, k is the number of super pixel blocks. The number of super pixel block is decided by seed number, image size and image color complexity. The number of seed points is set to 100 by default.

2.2 Estimate Noise Variance of Super Pixel Block

FAST noise estimate is recognized as a fast and efficient method of noise estimation [12]. For the input image, the algorithm first assumes that the image has been destroyed by added to the zero mean Gauss noise [13]. The noise estimation in general steps are as follows: first compress image structure by La operator, exclude the factors that adversely affect the estimation of noise variance [14]. Then, the noise estimation formula in the literature [12] is used to calculate the mean value, and estimate the noise variance.

2.3   Compute Block Weights Through Multiple Image Features

This paper considers that the weight of image blocks is determined by three factors: the size of image block, the color similarity of image blocks, and the spatial distance between image blocks.

Because the blocks obtained by the super pixel segmentation algorithm are irregular blocks, we use the number of pixels contained in the block to determine the value of RI. This method can emphasize the importance of larger areas. In this paper, we choose the HSV color space to calculate the color similarity between different image blocks, and the calculation process is shown in Formula 1. Among them, X, Y and Z represent the values on different channels of the color space respectively.

$$D_r(r_p, r_i) = \sqrt{(X_p - X_i)^2 + (Y_p - Y_i)^2 + (Z_p - Z_i)^2} \tag{1}$$

The algorithm introduces the spatial distance of the image block to calculate the weight of the stitching area. Based on a large number of experimental studies, this paper introduces the constant and square calculation of σ_s to normalize the space distance, which makes the space distance between the sub blocks play a good role in the calculation of the weight value. The calculation method of similarity weight is shown in formula 2. w is the size of the block. S_s is a constant to control the influence of spatial distance. The final weight of the image block can be obtained by adding the weight calculated by the formula and the weight of the noise characteristics.

$$S(r_p) = \sum_{r_p \neq r_i} \left(\sqrt{\frac{D_s(r_p, r_i)}{s_s^2}} \right) w(r_p) D_r(r_p, r_i) \tag{2}$$

In most cases, the splicing area is obtained from the background image of a part of the image, so there is a certain correlation between the pieces of the splicing area. In this paper, a weight calculation method based on noise characteristics, color similarity and space distance between images is proposed. The connection of the image blocks in the splicing area is strengthened by the similarity of the features, and the weight values are used to classify the image blocks, and then the splicing area is located.

3   Purposed Method

This paper presents the procedure of the algorithm is shown in Fig. 1, the general steps are as follows: first, segment the detected image by SLIC, then estimate noise of each SLIC block; compute weight of image block; image detection.

The detailed steps of purposed method are presented as follows:

Step 1. Super pixel segmentation of input images

Super pixel segmentation is an accepted algorithm for segmentation of uniform color regions at small color distances. Because of the super pixel segmentation, each super pixel block may not be a regular rectangle, so it brings a lot of inconvenience in noise estimation of the image block. In estimating the noise of each block, it is necessary to determine the boundaries of the image blocks.

After the super pixel segmentation of image I, SLIC method adds labels to each pixel point, and the pixels with the same label constitute a block as $C_{(1,...,k)}$, k is the number of all blocks, and then do the noise estimation.

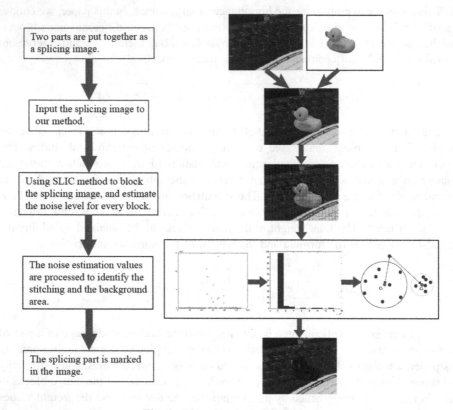

Fig. 1. Flow diagram of algorithm

Step 2. Estimates the noise variance of each block

For each pixel of C_i, do the calculation of formula (5), we can get the noise estimated value of each block, eventually form a noise estimation value sequence.

Step 3. Computing block weights through multiple image features

Compute the block of each block through noise of image block, the size of image block, the color similarity of image blocks, and the spatial distance between image blocks distance. Typically, the splicing area is less than the background area. And because the splicing area and background area come from different images, the noise of them is different. Therefore, we classify the weight sequences by K-means and classify them into two categories. In general, one is considered as splicing area, and the other as background area. The smaller one is splicing area.

Step 4. Tag and output the result image

According to the result of classification, the background area and splicing area are determined and labeled in the picture. The value of the pixel in the splicing area is set to 0, and then display image.

4 Experiment

4.1 Experiment Environment

The Columbia University image library is a benchmark for detecting splicing forgery [15]. The image has high resolution and does not carry out any compression processing. Most of the images are indoor scenes: labs, desks, books, etc. Only 27 images were taken outdoors in cloudy weather. This dataset contains two classes of images, real images and forgery images. All images are taken with 4 cameras, of which 183 are real images and 180 are spliced images, and the forgery images are obtained by splicing real images without additional post-processing.

This paper selects the Columbia University image library as the experimental data set. The [16] uses a detection algorithm based on pixel, the algorithms proposed in [13, 14] localize forgery area based on block. In this paper, [13, 14] and [16] are selected to compare with the algorithm.

The experiment is run under the MATLAB R2014b, the computer is configured as Intel Core i7-4790, clocked at 3.60 GHz, memory 4.00 GB. This algorithm uses true positive rate (TPR) and false positive rate (FPR) to evaluate the results.

4.2 Algorithm Validity

The advantage of image segmentation through SLIC method is that can preserve the boundary information of image better. As shown in Fig. 2, the Fig. 2(a) spliced an area from another image in the background area. Fig. 2(b) is to segment the image by block segmentation, and then estimate the noise variance to localize the splicing area. Fig. 2 (c) is an example diagram of this paper. It can be seen that compared with the block segmentation algorithm, the algorithm proposed has a great advantage in preserving the boundary information. Compared with the detection algorithm based on pixel, because of the super pixel segmentation, the boundary preserving ability of the algorithm is relatively good, but the computation speed will be greatly accelerated.

As shown in Fig. 3, Fig. 3(b) is detected through only noise estimation. Figure 3(c) is an example diagram of this paper. It can be seen that compared with the detection algorithm through only noise estimation, the algorithm proposed in this paper has a great advantage in reduction of leakage detection.

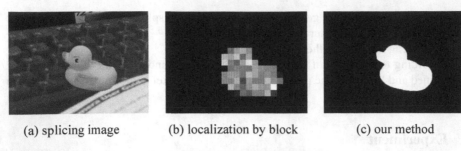

| (a) splicing image | (b) localization by block | (c) our method |

Fig. 2. Example of algorithm validity

| tamper image | (a) without image feature | (b)using image feature |

Fig. 3. Example of image feature validity

4.3 Result Analysis

4.3.1 Comparative Analysis of Detection Accuracy

As shown in Fig. 4, the leftmost column of the graph is the splicing image of Columbia University database. (d) column is the result of the proposed algorithm. Because the SLIC method is adopted, the edge of the detected splicing area is more smooth. For this reason, the accuracy of image splicing localization is improved. (a)–(c) columns are the experimental results using the algorithm proposed in [13–16], where the algorithms in [15] and [16] are the same. In these three columns, the stitching area is accurately represented by green, and red is used to indicate the region of error detection.

The image 1 corresponds to the image of the first row in Fig. 5, and the image 2 corresponds to the image of the second row, and so on. The results of image 1 show that the proposed algorithm in terms of accuracy rate and false rate are both obtained the best results; on image 2, the algorithm results are in the middle level; the experiment on image 3 has also obtained the best results compared with the [13–16] algorithm, which greatly enhances the accuracy, and reduces the false rate at the same time; from the results of image 4, the result of this algorithm is closest to the best situation of location detection.

As shown in Fig. 5, the first is the result of experiment through only noise estimation, and the second is the result of the algorithm. It can be seen that the results of this algorithm are better.

Experimental results for the selected images in Fig. 5 are given in Table 1. The best results have been shown in bold in the table. From Table 1, overall, the algorithm proposed in this paper is the best algorithm for all four images.

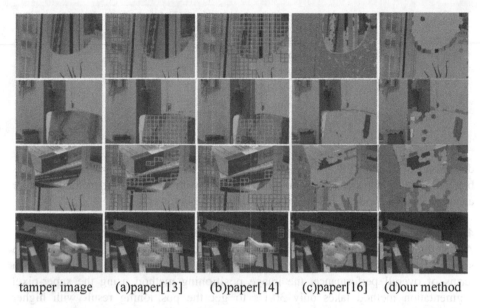

tamper image   (a)paper[13]   (b)paper[14]   (c)paper[16]   (d)our method

Fig. 4. Comparison chart of experimental results (Color figure online)

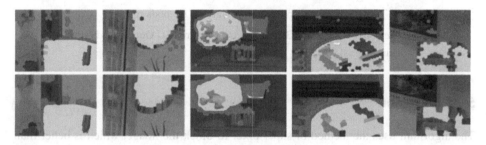

Fig. 5. Comparison chart of experimental results with image features

Table 1. Comparison of results in Fig. 3

Methods	Image 1		Image 2		Image 3		Image 4	
	TPR	FPR	TPR	FPR	TPR	FPR	TPR	FPR
Paper [13]	0.9	32.2	75.2	4.0	26.4	25.0	87.8	3.2
Paper [14]	80.7	25.1	38.4	2.2	11.5	66.3	61.6	10.0
Paper [16]	16.5	59.3	69.3	3.8	60.1	32.0	32.0	2.6
Our method	99.8	0.00	71.6	1.7	83.1	1.2	64.7	8.4

In order to further study the performance of the proposed algorithm, we use all the stitching images of the Columbia database to carry out experiments, and the experimental results are recorded in Table 2.

Table 2. Experimental results on the Columbia database

Methods	TPR	FPR
Paper [13]	30.8	21.3
Paper [14]	40.9	31.6
Paper [16]	36.8	23.0
Our method	47.6	25.8

The results of TPR show that our proposed method gets the best positioning detection result, and TPR is 41.6%. FPR produced by our method is 25.8%, which is better than that in paper [14]. Combined Table 1 with Table 2, our method is more accurate than methods in [13–16] and the false rate is lower.

4.3.2   Comparison of Computational Complexity
All of the run time for each method are counted and recorded in Table 3. Table 3 shows that our method needs to spend more time than the paper [13]. However, in terms of overall performance, the image positioning method using the super-pixel segmentation method takes only 30.1 s to get the positioning result with higher accuracy and lower false rate.

Table 3. Comparison results of various methods run time

Methods	Paper [13]	Paper [14]	Paper [16]	Our method
Run time (s)	0.42	250.5	94.2	30.1

5   Conclusion

The algorithm uses the method of super-pixel to block the image, which is better than the current mainstream method which directly blocks the image. Our method preserves the edge of the image better, and improves the precision of splicing detection.

Our experimental results show:

1. The image blocks produced by the super-pixel segmentation method contain the connection information between the image features, and using the super-pixel segmentation method retains the boundary information of the splicing area better.
2. The super-pixel segmentation method guarantees the complete information of the splicing edge, which improves the positioning detection precision compared with the current image mosaic detection algorithm.
3. On the basis of improving the accuracy of positioning detection, using the noise feature with low dimension reduces the time complexity compared with the existing localization algorithm.

In this paper, the splicing positioning detection based on the noise estimation has the generally slow operation speed, our method uses the super-pixel segmentation method combined with the fast noise estimation method, the results show that our method can ensure the accuracy rate and the pixel level detection accuracy at the same time, it also ensures the speed of operation.

References

1. Walia, S., Kumar, K.: Digital image forgery detection: a systematic scrutiny. Aust. J. Forensic Sci. **1**, 1–39 (2018). https://doi.org/10.1080/00450618.2018.1424241
2. Zhang, L., Yang, Y., Wang, M.: Detecting densely distributed graph patterns for fine-grained image categorization. IEEE Trans. Image Process. **25**(2), 553–565 (2016). https://doi.org/10.1109/TIP.2015.2502147
3. Ng, T.T., Chang, S.F., Sun, Q. (2004) Blind detection of photomontage using higher order statistics. In: Proceedings - IEEE International Symposium on Circuits and Systems, vol. 5, pp. 688–691. https://doi.org/10.1109/iscas.2004.1329901
4. Zhang, L., Gao, Y., Hong, R., Hu, Y., Ji, R., Dai, Q.: Probabilistic skimlets fusion for summarizing multiple consumer landmark videos. IEEE Trans. Multimed. **17**(1), 40–49 (2014). https://doi.org/10.1109/TMM.2014.2370257
5. Agarwal, S., Chand, S.: Image forgery detection using multi scale entropy filter and local phase quantization. Int. J. Image Graph. Sig. Process. **7**(10), 78–85 (2015). https://doi.org/10.5815/ijigsp.2015.10.08
6. Zhang, L., Hong, R., Gao, Y., Ji, R., Dai, Q., Li, X.: Image categorization by learning a propagated graphlet path. IEEE Trans. Neural Netw. Learn. Syst. **27**(3), 674–685 (2017). https://doi.org/10.1109/icip.2009.5413549. https://doi.org/10.1109/tnnls.2015.2444417
7. Hashmi, M.F., Keskar, A.G.: Image forgery authentication and classification using hybridization of HMM and SVM classifier. Int. J. Secur. Appl. **9**(4), 125–140 (2015). https://doi.org/10.14257/ijsia.2015.9.4.13
8. Vaishnavi, D., Subashini, T.S.: Recognizing image splicing forgeries using histogram features. In: MEC International Conference on Big Data and Smart City, pp. 1–4. IEEE, New York (2016). https://doi.org/10.1109/icbdsc.2016.7460342
9. Wang, B., Kong, X.: Image splicing localization based on re-demosaicing. In: Zeng, D. (ed.) Advances in Information Technology and Industry Applications. LNEE, vol. 136, pp. 725–732. Springer, Heidelberg (2012). https://doi.org/10.1007/978-3-642-26001-8_92
10. Zampoglou, M., Papadopoulos, S., Kompatsiaris, Y.: Large-scale evaluation of splicing localization algorithms for web images. Multimed. Tools Appl. **76**, 1–34 (2016). https://doi.org/10.1007/s11042-016-3795-2
11. Bahrami, K., Kot, A.C., Li, L., et al.: Blurred image splicing localization by exposing blur type inconsistency. IEEE Trans. Inf. Forensics Secur. **10**(5), 999–1009 (2015). https://doi.org/10.1109/ISCAS.2015.7168815
12. Immerkær, J.: Fast noise variance estimation. Comput. Vis. Image Underst. **64**(2), 300–302 (1996). https://doi.org/10.1109/TIFS.2015.2394231
13. Lyu, S., Pan, X., Zhang, X.: Exposing region splicing forgeries with blind local noise estimation. Int. J. Comput. Vis. **110**(2), 202–221 (2014). https://doi.org/10.1007/s11263-013-0688-y
14. Mahdian, B., Saic, S.: Using noise inconsistencies for blind image forensics. Image Vis. Comput. **27**(10), 1497–1503 (2009). https://doi.org/10.1016/j.imavis.2009.02.001
15. Pan, X., Zhang, X., Lyu, S.: Exposing image forgery with blind noise estimation. In: Thirteenth ACM Multimedia Workshop on Multimedia and Security, pp. 15–20. ACM, New York (2011). https://doi.org/10.1145/2037252.2037256

16. Pan, X., Zhang, X., Lyu, S.: Exposing image splicing with inconsistent local noise variances. In: IEEE International Conference on Computational Photography, pp 1–10. IEEE, New York (2012). https://doi.org/10.1109/iccphot.2012.6215223

17. Achanta, R., Shaji, A., Smith, K., et al.: SLIC superpixels compared to state-of-the-art superpixel methods. IEEE Trans. Pattern Anal. Mach. Intell. **234**(11), 2274–2282 (2012). https://doi.org/10.1109/TPAMI.2012.120

Effect of Checkerboard on the Accuracy of Camera Calibration

Shengju Yu, Ran Zhu, Li Yu$^{(\boxtimes)}$, and Wei Ai

School of Electronic Information and Communications, Huazhong University of
Science and Technology, Wuhan, China
{shengju_yu,hustlyu}@hust.edu.cn

Abstract. Camera calibration is an important step in many fields, and
its accuracy is influenced by many factors. In this paper, the influence of
different numbers and sizes of squares of the checkerboard on calibration
is studied, and the errors caused by this are analyzed. We propose *SBI*
algorithm to improve the calibration accuracy. In order to better judge
the quality of calibration, we map the texture map onto the depth map by
coordinate transformation to achieve the effect of registration. Besides,
we use the Canny operator to extract the depth map edges and the
mapped texture map edges, and utilize their matching results to evaluate
the calibration accuracy. Experimental results show that our registration
effect is the best and the number of mismatched pixels on the edges is
the least, which indicates our calibration accuracy is the highest.

Keywords: Camera calibration · Image registration · Edge matching

1 Introduction

The general principle of camera calibration lies in finding the correspondence
between a sufficiently large number of known 3D points in world coordinate and
their projections in 2D image [1]. A lot of researchers have proposed different
calibration methods. According to the dimensions of calibration objects, these
methods can be divided into three categories, namely 1D, 2D, 3D objects. 1D
objects, for example a wand with multiple collinear points, are the easiest to
construct. They are often used in the calibration of multiple cameras since all
cameras can see the same calibration object at the same time [2,3]. However, due
to the lower number of calibration points, these methods provide the less accurate
results. Methods based on 2D objects, for instance planar boards consisting of
checkerboards [4] or circles [5], are the most popular because of the good trade-off
between accuracy and simplicity. 3D objects, such as an object consisting of two
or three planes orthogonal to each other, can provide more accurate results. But
these approaches based on 3D objects require expensive calibration devices and
elaborate setups. Therefore, this paper will focus on 2D plane-based calibration,
more specifically the method proposed by Zhang [4].

© Springer Nature Switzerland AG 2018
R. Hong et al. (Eds.): PCM 2018, LNCS 11166, pp. 619–629, 2018.
https://doi.org/10.1007/978-3-030-00764-5_57

Zhang [4] proposed a flexible method to easily calibrate the camera, which only requires the camera to observe a planar pattern from several different directions. Not only the camera but also the planar pattern can move at an unknown scale. Although this technology has gained considerable flexibility, there is still some work to do, for instance, reducing the location error of the detected feature points, exploring the influence of the size and number of the pattern on calibration accuracy. In addition, different sets of images can lead to different calibration results even with the same pattern, and the differences between the results may be very large. So, it's essential to study how to select better images that are conducive to calibration. [4,6,7] calibrated the camera with 10, 5, 30 images respectively in their experiments. Therefore, it's also necessary to further explore the influence of the number of planar pattern images on calibration results.

We calibrate the depth camera and texture camera of kinect v1 [8] separately. And the main contents of this paper can be summarized as follows: (1) In order to select better images that are favorable for camera calibration, we propose SBI algorithm. (2) We design a series of comparative experiments to explore the influence of the size and number of black-and-white squares of the checkerboard on calibration. (3) In order to judge the quality of calibration results, we map the texture map onto the depth map by coordinate transformation to achieve the registration effect. (4) We further use the Canny operator to extract the depth map edges and the mapped texture map edges, and take advantage of [9] to match these edges. The matching results are used to evaluate the calibration accuracy.

This paper is organized as follows. Section 2 briefly describes the calibration principle. Section 3 states the proposed SBI algorithm. Section 4 describes how to register a depth camera and a texture camera. Experimental results and error analysis are reported in Sect. 5. Finally, Sect. 6 concludes the work.

2   Calibration Principle

We concentrate on the well-known pinhole camera model, which is widely used in many fields. It assumes that the camera performs a perfect perspective projection transformation from the 3D scene coordinates (x, y, z) to image plane coordinates (u, v), as shown in Fig. 1. Four coordinates are defined as follows: $O_w X_w Y_w Z_w$ is the world coordinate system. $O_c X_c Y_c Z_c$ is the camera coordinate system, and the projection center is the point O_c. $O_i xy$ is the image coordinate system in millimeters, and the intersection point of the image plane and the optical axis of the camera is defined as the origin O_i. $O_p uv$ is the pixel coordinate system and the u and v axes are parallel to the x and y axes, respectively. For modern cameras, the skew coefficient of u and v axes is often equal to zero, that is, they are completely vertical. The perspective projection transformation from the world coordinate system to the pixel coordinate system can be expressed as:

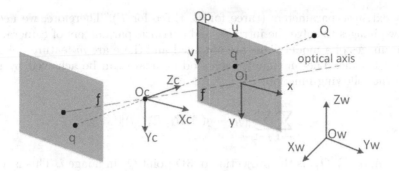

Fig. 1. Pinhole camera model

$$\lambda \begin{bmatrix} u \\ v \\ 1 \end{bmatrix} = \begin{bmatrix} \frac{1}{dx} & 0 & u_0 \\ 0 & \frac{1}{dy} & v_0 \\ 0 & 0 & 1 \end{bmatrix} \begin{bmatrix} f & 0 & 0 & 0 \\ 0 & f & 0 & 0 \\ 0 & 0 & 1 & 0 \end{bmatrix} \begin{bmatrix} R & T \\ 0^T & 1 \end{bmatrix} \begin{bmatrix} X_w \\ Y_w \\ Z_w \\ 1 \end{bmatrix} \tag{1}$$

Where $\{R, T\}$ is a rigid transformation that maps points in the world coordinate system onto the camera coordinate system. $T = [t_1, t_2, t_3]$ describes the translation between the two frames. $R = [r_{11} \ r_{12} \ r_{13}; r_{21} \ r_{22} \ r_{23}; r_{31} \ r_{32} \ r_{33}]$ is a 3 by 3 orthonormal rotation matrix and can be defined by three angles α, β, γ. If R is known, these angles can be computed using the following decomposition:

$$\alpha = \sin^{-1}(r_{31})$$
$$\beta = \mathrm{atan2}(-\frac{r_{32}}{\cos \alpha}, \frac{r_{33}}{\cos \alpha}) \tag{2}$$
$$\gamma = \mathrm{atan2}(-\frac{r_{21}}{\cos \alpha}, \frac{r_{11}}{\cos \alpha})$$

Without loss of generality, the model plane can be set on $z = 0$ of the world coordinate system. Equation (1) can be simplified as:

$$\lambda \begin{bmatrix} u \\ v \\ 1 \end{bmatrix} = A \begin{bmatrix} r_1 & r_2 & T \end{bmatrix} \begin{bmatrix} X_w \\ Y_w \\ 1 \end{bmatrix}, A = \begin{bmatrix} f_x & 0 & u_0 \\ 0 & f_y & v_0 \\ 0 & 0 & 1 \end{bmatrix} \tag{3}$$

The r_i represents the ith column of R. The parameters f_x, f_y, u_0 and v_0 are called intrinsic parameters and the $t_1, t_2, t_3, \alpha, \beta$ and γ are extrinsic parameters.

One model point Q and its image point q are related by a homography H, $H = [h_1 \ h_2 \ h_3] = A[r_1 \ r_2 \ T]$. r_1 and r_2 are orthonormal, so:

$$h_1^T A^{-T} A^{-1} h_2 = 0$$
$$h_1^T A^{-T} A^{-1} h_1 = h_2^T A^{-T} A^{-1} h_2 \tag{4}$$

Given one homography, two constraints on the intrinsic parameters can be obtained. In addition, a homography has eight degrees of freedom and there are six extrinsic parameters (three for R, three for T). Therefore, we need at least two images to solve the intrinsic and extrinsic parameters of camera.

If n images of a model plane are obtained and there are m feature points on the model plane. The maximum likelihood estimate can be achieved by minimizing the following function:

$$\sum_{i=1}^{n}\sum_{j=1}^{m}\|q_{ij} - \hat{q}(A, R_i, T_i, Q_j)\|^2 \tag{5}$$

where $\hat{q}(A, R_i, T_i, Q_j)$ is the projection of 3D point Q_j in image i. This is a nonlinear minimization problem and can be solved with the Levenberg-Marquardt Algorithm as implemented in Minpack [10].

The measurement of calibration accuracy is the reprojection error $Ereproj$.

$$Ereproj = \frac{1}{S}\sum_{i=1}^{S}\sqrt{(u_i - u_i^{reproj})^2 + (v_i - v_i^{reproj})^2} \tag{6}$$

where S is the number of feature points, (u_i, v_i) is real pixel coordinate and $(u_i^{reproj}, v_i^{reproj})$ is the reprojection coordinate.

3   Algorithm for Improving the Calibration Accuracy

Different sets of images can lead to different calibration results even with the same calibration pattern, the results are unstable and the differences between calibration results may be very large. Thus, before the camera is calibrated, it's necessary to select better images that are beneficial for calibration.

Let I denote a large set of images of the calibration pattern in different orientations, in all of which the features are correctly detected. Assuming that the size of I is N, then $I = \{image_1, image_2, \ldots, image_N\}$. Let S represent a subset of I and each element has size m. That is, $S = \{s_1, s_2, \ldots, s_n\}$, $s_i \in I^m$.

Algorithm 1. The proposed SBI algorithm for improving calibration accuracy.

1: Acquire a set I_{orig} of original images with different calibration object views.

2: Those images with successful feature detection and large difference of the orientation of calibration object are chosen as a set I, $I \in I_{orig}$ and having length N.

3: Check whether $m \leq N$, and if the expression is true, proceed.

4: Select m better images that favor camera calibration from set I:

Add a parameter $index$ for each image. The default value of $index$ is 0. Then

for i=1 to $N - m + 1$ do

  for j=i+1 to $N - m + 2$ do

    for k=j+1 to $N - m + 3$ do

- Utilize set $B = \{I_i, I_j, I_k\}$ to calibrate the camera. When E_s is the smallest, calibration results and the path of images I_i, I_j, I_k are preserved. $E_s = E_{reproj} + E_f$. $E_f = (((f_x - f_{dx})/f_{dx})^2 + ((f_y - f_{dy})/f_{dy})^2)^{1/2}$. The values of the parameter $index$ of images I_i, I_j, I_k are set to 1, which indicates that they are better images.

5: In order to explore the effect of the number of images on calibration results, a total of G images are sequentially selected.

6: Check whether $G \leq N$, if the expression is false, terminate. Otherwise:

for i=1 to $G - m$ do

for g=1 to N do

if $index_g = 0$

- Utilize set B and I_g to calibrate camera. I_g is reserved and the value of the parameter $index_g$ is set to 1 when E_s is the smallest.

When the loop of g ends, add I_g to the set B to form a new set B.

7: G better images are selected and camera calibration results from m to G images are also preserved.

---

4   Registration of Depth Camera and Texture Camera

The principle of registration is shown in Fig. 2. Q_{c_depth} is the spatial coordinate of Q point in the depth camera coordinate system. Q_{c_rgb} is the spatial coordinate of Q point in the texture camera coordinate system. $Q_{\_depth}$ and $Q_{\_rgb}$ are the projection coordinate of the point on image plane, respectively. A_{depth} is the intrinsic matrix of the depth camera. We can get:

$$Q_{\_depth} = A_{depth} \cdot Q_{c\_depth} \tag{7}$$

Depth camera and texture camera can be linked by rotation and translation matrix $(^D R_{rgb}, {}^D T_{rgb})$.

$$Q_{c\_rgb} = {}^D R_{rgb} \cdot Q_{c\_depth} + {}^D T_{rgb} \tag{8}$$

The $Q_{\_rgb}$ coordinate corresponding to the $Q_{\_depth}$ point can be obtained by the following formula:

$$Q_{\_rgb} = A_{rgb} \cdot Q_{c\_rgb} \tag{9}$$

Therefore, in order to determine the relationship between $Q_{\_depth}$ and $Q_{\_rgb}$, it's necessary to first solve $^D R_{rgb}$ and $^D T_{rgb}$. Using the extrinsic matrix, we can get:

$$Q_{c\_depth} = R_{depth} \cdot Q + T_{depth}$$
$$Q_{c\_rgb} = R_{rgb} \cdot Q + T_{rgb} \tag{10}$$

So,

$$Q_{c\_rgb} = R_{rgb} \cdot R_{depth}^{-1} \cdot Q_{c\_depth} + T_{rgb} - R_{rgb} \cdot R_{depth}^{-1} \cdot T_{depth} \tag{11}$$

Finally,

$$^D R_{rgb} = R_{rgb} \cdot R_{depth}^{-1}$$
$$^D T_{rgb} = T_{rgb} - {}^D R_{rgb} \cdot T_{depth} \tag{12}$$

For each pixel $Q_{\_depth}$ in the depth map, the corresponding coordinate $Q_{\_rgb}$ in the texture map can be obtained through the above formulas, and the color information can be assigned to the corresponding depth map to achieve the registration effect.

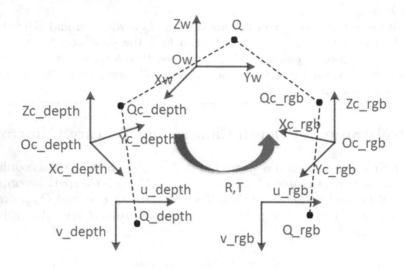

Fig. 2. Registration principle

5   Experimental Results

We use a printed flat plate with checkerboard pattern as the calibration object. The specification of checkerboard is $a \times b$. a and b indicate the number of squares in the vertical and horizontal directions, respectively. In order to explore the influence of the size of squares on calibration accuracy, we design five sets of contrast experiments, namely, 4×7, 7×10, 10×15, 15×22 and 22×27, as shown in Table 1. The checkerboard in *Full* is as big as possible. *Validation* is to validate our conjecture that the size of squares is not the bigger the better. At the same time, we also carry out 1.0 cm, 1.3 cm and 1.7 cm three groups of comparative experiments to study the effect of the number of squares on calibration accuracy. *Full* and *Validation* are also used to verify the effect of the number of images on calibration accuracy. We calibrate the depth camera and texture camera of kinect v1 separately. So, there are 50 groups of experiments in total. Theoretically, we can utilize only two images to calibrate the camera,

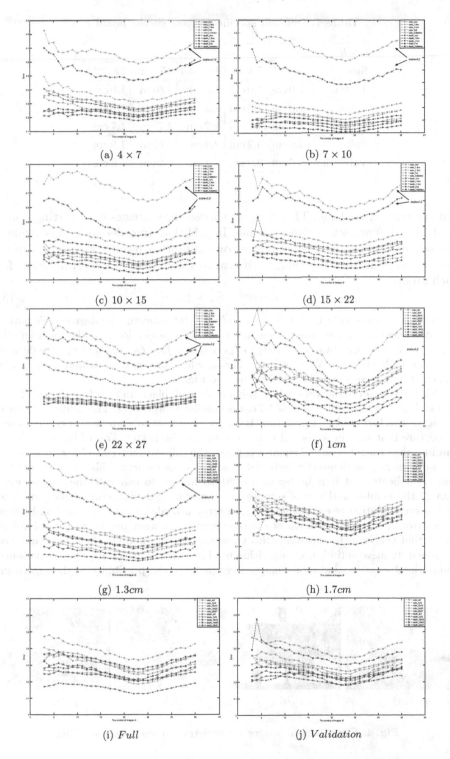

(a) 4×7

(b) 7×10

(c) 10×15

(d) 15×22

(e) 22×27

(f) $1cm$

(g) $1.3cm$

(h) $1.7cm$

(i) *Full*

(j) *Validation*

Fig. 3. Calibration *Error* of checkerboard with different specifications

Table 1. Different specifications of checkerboard

$a \times b$					
Size	4×7	7×10	10×15	15×22	22×27
1.0 cm	1.0 cm	1.0 cm	1.0 cm	1.0 cm	1.0 cm
1.3 cm	1.3 cm	1.3 cm	1.3 cm	1.3 cm	1.3 cm
1.7 cm	1.7 cm	1.7 cm	1.7 cm	1.7 cm	1.7 cm
Full	4.6 cm	3.2 cm	2.4 cm	2.0 cm	1.9 cm
Validation	5.0 cm	2.9 cm	2.1 cm	2.2 cm	1.6 cm

but the error is too large. Therefore, we choose three images as a starting point. And $N = 100$ are set as default value. In addition, in order to further explore the influence of the number of images on calibration results, we select 3 to 35 better images in turn to calibrate the camera and observe the error *Error* for each group.

$$Error = E_s + E_o \tag{13}$$

where $E_o = (((u_0 - u_{d0})/u_{d0})^2 + ((v_0 - v_{d0})/v_{d0})^2)^{1/2}$ indicates the degree of deviation of optical center. E_s contains the deviation of the reprojection and the focal length. f_x, f_y, u_0 and v_0 are obtained through experiments. f_{dx}, f_{dy}, u_{d0} and v_{d0} are derived from the FOV (Field of View) of the camera, which are used as default values. The *Error* curves of each experiment are shown in Fig. 3.

Figure 3(a)–(e) show that when the number of black-and-white squares of the checkerboard is fixed, the error of 1.7 cm is smaller than others. The calibration error of bigger checkerboard in *Validation* is larger than that in *Full*, which confirms our conjecture that the checkerboard is not the bigger the better. From Fig. 3(f)–(h), we can find that when the size of squares is fixed, the calibration result of 7×10 is better than others. So, the dense checkerboard is not good for camera calibration. Figure 3(i) and (j) indicate that it is better to use 20–26 images to calibrate the camera even though the number and size of squares are changed, and too many images will lead to the accumulation of errors and increase error probability. On the one hand, there is an error in the determination of the coordinates of corner points, and it is hard to assert that this error is consistent with Gaussian distribution. On the other hand, the nonlinear iterative optimization algorithm used in calibration can not always guarantee the optimal solution, and more images may increase the possibility of the algorithm

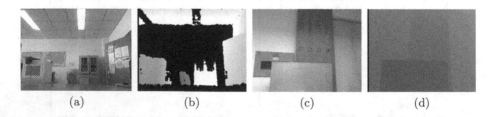

    (a)   (b)   (c)   (d)

Fig. 4. Experimental images for registration and edge matching

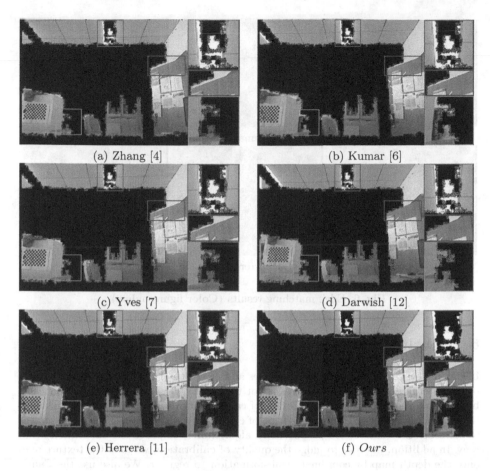

(a) Zhang [4]   (b) Kumar [6]

(c) Yves [7]   (d) Darwish [12]

(e) Herrera [11]   (f) *Ours*

Fig. 5. Registration results

falling into local optimal. Therefore, we choose 7×10 & 1.7 cm checkerboard and its better first 23 images to calibrate the kinect v1.

According to the calculated parameter values of the camera, we map the texture map onto the depth map by coordinate transformation to judge the quality of calibration results, as shown in Fig. 5. Comparing Fig. 5 with Fig. 4(a), (b), we can see that our registration result is better than others, which shows that our calibration accuracy for depth camera and texture camera is very high.

To further test the calibration results, we use the Canny operator to extract the depth map edges and the registered texture map edges, and then match these edges. The matching results are shown in Fig. 6. The blue line represents the pixels that are incorrectly matched. The red line and green line indicate the edges of the depth map and registered texture map near the mismatched pixels, respectively. One can see that the number of our mismatched pixels is the least.

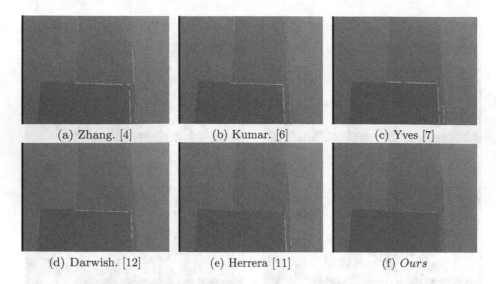

(a) Zhang. [4]   (b) Kumar. [6]   (c) Yves [7]

(d) Darwish. [12]   (e) Herrera [11]   (f) *Ours*

Fig. 6. Edge matching results (Color figure online)

6   Conclusion

In this paper, the influence of different sizes and numbers of the squares on calibration results is studied. The size of squares is not the bigger the better. The dense checkerboard can't guarantee the high accuracy of calibration results. Too many images used for calibration result in the accumulation of errors, and we suggest using 20–26 images to calibrate the camera. We propose *SBI* algorithm to improve the calibration accuracy. In addition, in order to judge the quality of calibration, we map the texture map onto the depth map by coordinate transformation to register. We also use the Canny operator to extract the depth map edges and the registered texture map edges and then match them, which is more conducive to compare the effect of registration and the accuracy of calibration.

Acknowledgment. This work was supported by the National Natural Science Foundation of China (NSFC) (No. 61231010), National High Technology Research and Development Program (No. 2015AA015901).

References

1. Carsten, S., Markus, U., Christian, W.: Machine Vision Algorithms and Applications. Trans Hsinghua University Publishing, Beijing (2008)
2. Zhang, Z.: Camera calibration with one-dimensional objects. IEEE Trans. Pattern Anal. Mach. Intell. **26**(7), 892–899 (2004)
3. Svoboda, T.: A convenient multicamera self-calibration for virtual environments. Presence **14**(4), 407–422 (2005). MIT Press
4. Zhang, Z.: A flexible new technique for camera calibration. IEEE Trans. Pattern Anal. Mach. Intell. **22**(11), 1330–1334 (2000)

5. Jiang, G., Quan, L.: Detection of concentric circles for camera calibration. In: International Conference on Computer Vision, pp. 333–340 (2005)
6. Kumar, A.: On the equivalence of moving entrance pupil and radial distortion for camera calibration. In: International Conference on Computer Vision, pp. 2345–2353 (2015)
7. Le Sant, Y.: Multi-camera calibration for 3DBOS. In: 17th International Symposium on Applications of Laser Techniques to Fluid Mechanics (2014)
8. Zhang, Z.: Microsoft kinect sensor and its effect. IEEE Multimed. **19**(2), 4–10 (2012)
9. Xiang, S., Yu, L., Chen, C.W.: No-reference depth assessment based on edge misalignment errors for T+ D images. IEEE Trans Image Process. **25**(3), 1479–1494 (2016)
10. Moré, J.J.: The Levenberg-Marquardt algorithm: implementation and theory. In: Watson, G.A. (ed.) Numerical Analysis. LNM, vol. 630, pp. 105–116. Springer, Heidelberg (1978). https://doi.org/10.1007/BFb0067700
11. Herrera, D., et al.: Joint depth and color camera calibration with distortion correction. IEEE Trans. Pattern Anal. Mach. Intell. **34**(10), 2058–2064 (2012)
12. Darwish, W., et al.: A new calibration method for commercial RGB-D sensors. Sensors **17**(6), 1204 (2017)

Weighted Multi-feature Fusion Algorithm for Fine-Grained Image Retrieval

Zhihui Wang[1], Shijie Wang[1], Hong Wang[1], Haojie Li[1(✉)],
and Chengming Li[2]

[1] Dalian University of Technology, Dalian, China
lihaojieyt@gmail.com
[2] Shenzhen Institutes of Advanced Technology,
Chinese Academy of Science, Beijing, China

Abstract. The purpose of Fine-Grained Image Retrieval is to search images that belong to the same subcategory as the query image. In this paper, we propose a Weighted Multi-Feature Fusion Algorithm (WMFFA) to improve image feature representation for fine-grained image retrieval. Firstly, we designed a new constraint to select discriminative patches, which makes use of the irregular but more accurate object region to select the discriminate patches. Secondly, based on the fact that the activation value of an object is larger in the convolution layer, a weighted max-pooling aggregation method for patch features is proposed to weaken the possible residual background information and retain effective object information as much as possible. Current methods of using multi-level features generally use a simple concatenated method, which lacks deep excavating intrinsic correlation between features. Therefore, thirdly, we introduce the Deep Belief Network to effectively fuse multi-level features, which captures the intrinsic correlation and rich complementary information in multi-level features. Experiments show our WMFFA framework achieves significantly better accuracy than existing fine-grained retrieval and general image retrieval methods.

Keywords: Convolutional Neural Network · Multi-level feature fusion
Fine-grained image retrieval

1 Introduction

Recently, fine-grained image tasks have entered the field of computer vision. Given a query image that contains a subcategory of a basic-level category, such as 'black footed albatross' of 'bird', we should return images which are in the same subcategory as the query from a database. FGIR is different from general image retrieval tasks. FGIR focuses on retrieval images that belong to the same subcategory as the query image, while general image retrieval [17–19] returns images that are similar to the query

This work was supported by the Natural Science Foundation of P.R. China (61472058, 61632019, 61771090).

R. Hong et al. (Eds.): PCM 2018, LNCS 11166, pp. 630–640, 2018.
https://doi.org/10.1007/978-3-030-00764-5_58

image's content. FGIR is very useful in product searches, search engines and biological research. However, there has not been adequate research about this topic.

We roughly divided existing fine-grained image task methods using Convolutional Neural Networks (CNN) into two groups. The first group has made great progress in locating objects precisely or increasing attention to objects, e.g., [1–4]. Lin et al. [1] designed an end-to-end network model, using two networks to complete object detection and feature extraction respectively. Wei et al. [2] located the object in an image, and then did average pooling and maximum pooling operations on corresponding depth features. The second group focuses on extracting more discriminating fine-grained features, e.g., [5–7]. Zheng et al. [5] built a multi-attention CNN based on a part learning method, which generated multiple parts by clustering and weighting activations. Yao et al. [6] designed a convolutional and normalization network to extract the object features and background features separately.

The difficulty in fine-grained image tasks is that images belonging to the same subcategory have great differences in size, posture, color, and background, while images falling into different subcategories may be very similar in these aspects, as illustrated in Fig. 1. More and more works take advantage of fine-grained features, like patch-level features of images. A bottom-up process, like selective search [8], can generate thousands of patches for one image. When aggregating features of these patches, most methods apply the sum and average operations, which treat all patches equally, without analyzing the importance of different patches. Moreover, existing methods widely use multi-level features, such as image-level, object-level and patch-level. But these methods usually just simply concatenate these features, without looking for complementary information among them.

Prothonotary Warbler

Winter Wren   Rock Wren   House Wren

Fig. 1. Large variance in the same subcategory and small variance in different subcategories (Color figure online)

This paper proposes a coarse-to-fine retrieval framework to improve the above issues, which consists of three parts: coarse retrieval, fine-grained retrieval, and query expansion. The key advantages and major contributions of our method are:

1. In order to improve the discrimination of selected patches, we designed a simple yet efficient constraint to filter patches generated by selective search, which make use of the irregular but more accurate object regions to select the discriminate patches.

2. To better aggregate huge patch features, we propose a weighted max-pooling. This method can remove useless background noise and retain effective information about an object as much as possible.
3. We introduce the Deep Belief Network (DBN) and use a joint Restricted Boltzmann Machine (RBM) layer to fuse multi-level features of an image, which could capture the intrinsic relationship and collaborative information in different level features.

The rest of this paper is organized as follows. In Sect. 2, we present our multi-feature fusion algorithm for fine-grained image retrieval. Experimental results are reported in Sect. 3, followed by the conclusion in Sect. 4.

2   Proposed Approach

This paper proposes a coarse-to-fine retrieval algorithm, as shown in Fig. 2. Given a query image q, the coarse retrieval firstly locates the object and extracts object-level CNN features to perform coarse retrieval in the entire database and return Top-K similar images to narrow the retrieval space. Then, in the fine-grained retrieval stage, we get the image patches by selective search and build constraints to select discriminate patches among them. More importantly, we propose a weighted max-pooling method to aggregate patch features, which can ignore background information while retaining more useful object-related information, thus generating a more discriminative patch feature. Next, the image-level and patch-level CNN features are inputted to the DBN network, and are combined using the joint RBM. In this way, intrinsic correlation and rich complementary information of image-level and patch-level features can be captured to obtain more powerful descriptors. We use the fusion feature to perform fine-grained retrieval in Top-K images, and receive a more accurate image sorting result. Finally, query expansion is used to further improve retrieval performance.

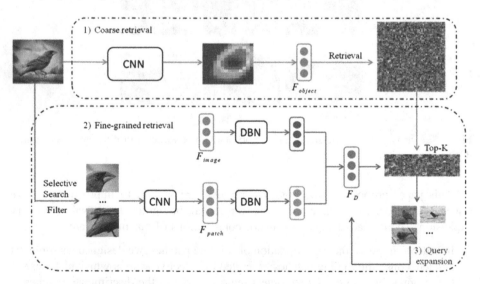

Fig. 2. The framework of WMFFA approach

2.1   Coarse Retrieval

In this paper, we adopt SCDA location methods [2] to extract object-level features with more discriminative ability. We employ global average pooling and max pooling like the original paper, and obtain feature f_{avg} and f_{max} separately. The final object-level feature F_{object} for coarse retrieval is generated by concatenating f_{avg} and f_{max}:

$$F_{object} = [f_{avg}, f_{max}] \in \mathbb{R}^{1024} \tag{1}$$

After obtaining shorter but more accurate representations for images, the Top-K images are ranked according to Euclidean distance. Besides that, image-level features are obtained and kept as input to the next stage.

2.2   Fine-Grained Retrieval

Patches Filter. A selective search algorithm is used to generate a large number of candidate patches from the image. These patches provide different perspectives and different scales from the original image, as shown in Fig. 3(a). It is necessary to remove the patches containing only background noise and retain the patches describing the object. In order to select discriminative patches, two major constraints are considered here:

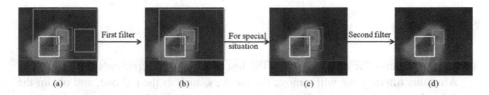

Fig. 3. Patches filter process (Color figure online)

The Constraint Between Patches and the Object. In this paper, we adopt a constraint to encourage the selected patches overlap with the object as much as possible. If directly adopting IoU to obtain the overlap between patches and the object, we need to get the minimum rectangle bounding boxes of the object area, which will introduce some background noise. So in this paper, we designed a more accurate filter condition, which uses the number of overlapping pixels as a criterion to evaluate the overlap ratio:

$$IoU_{p_i} = \frac{\sum\limits_{I \in X_{p_i} \cap X_o} 1}{\sum\limits_{I \in X_o} 1} \quad i = 1, 2, \ldots, n \tag{2}$$

where X_{p_i} represents the region of the i-th patch and X_o represents the irregularly shaped object region generated during coarse retrieval phase.

After the above filtering, we will remove the patches that only contain the background, such as the blue one in Fig. 3(a) and obtain the result demonstrated in Fig. 3(b).

However, a patch that is too large, such as the green one, causes a special situation. If a patch has a large overlap rate with the object, it can have too much background noise. So in order to ensure the selected patches have as much in-object area as possible, and less area overlapping with the background, we use an additional filter condition as follows:

$$IoU_{p_i} = \frac{\sum\limits_{I \in X_{p_i} \cap X_o} 1}{\sum\limits_{I \in X_{p_i}} 1} \tag{3}$$

After this filtering, we will remove the patches, such as the green one, and obtain the result like Fig. 3(c).

The Constraint Between Patches. The selected patches should have a small overlap with each other. Here we build a constraint to encourage the selected patches to overlap with each other as little as possible. Because the patches are all rectangular, we directly use IoU to calculate the overlap between a patch and other patches:

$$IoU_{ij} = \frac{p_i \cap p_j}{p_i \cup p_j} \quad i \neq j \tag{4}$$

where p_i represents the i-th patch after the last filter, and p_j represents the j-th patch.

After this filtering, we will remove the patches, such as the red one, and obtain the result as demonstrated in Fig. 3(d). It is clear that we will receive the discriminative patches.

Patches Feature Aggregation. Existing methods usually adopt an 'average fusion for all patches' feature to generate an available patch feature, which treats every patch feature equally. On the other hand, each selected patch represents a different object area. If directly employing the max-pooling method, much important information about an object may be missed.

In this paper, based on the property that the activation value in the object area is larger than the background area in a convolution layer, we propose a simple but effective method, named weighted max-pooling, to aggregate the patch features. While greatly reducing the redundant background information in each patch feature, it can also retain complementary information about patches, thereby generating a patch feature with stronger distinguishing ability.

The selected patches features after filtering are noted as $P=\{p_1, p_2, \ldots, p_n\}$, the final patch feature is calculated by the following formula:

$$F_{patch} = \alpha \times p_1^m + \beta \times p_2^m + (1 - \alpha - \beta) \times p_3^m, \quad \forall m \in M$$
$$st. \quad p_1^m = \max_i(p_i^m), \quad i \in 1, 2, \ldots, n$$
$$p_2^m = (second_i - \max)(p_i^m), \quad i \in 1, 2, \ldots, n \tag{5}$$
$$P_3^m = (third_i - \max)(p_i^m), \quad i \in 1, 2, \ldots, n$$

where M represents the dimension of each patch feature, and n represents the number of patch features. Weighted max-pooling method can not only weaken a certain degree of background information, but also retain more serious information in each patch. Thus, our method can generate a more distinguishing patch-level feature.

Complementary Feature Aggregation. Existing fine grained image analysis methods simply concatenate multi-level features. In this paper, we adopt a DBN network and joint RBM to fuse multi-level features. In this way, the joint distribution between features can be learned automatically. Therefore, after obtaining the image-level and patch-level features of the image, we feed them to the DBN network respectively and add a joint RBM layer to fuse two features, to capture the intrinsic correlation and rich complementary information.

The obtained image-level and patch-level CNN features are inputted to the DBN network respectively, and the probability function of network output is defined as:

$$P\left(F_{image}\right) = \sum_{h^{(1)},h^{(2)}} P\left(h^{(2)}, h^{(1)}\right) P\left(F_{image}|h^{(1)}\right) \tag{6}$$

$$P\left(F_{patch}\right) = \sum_{h^{(1)},h^{(2)}} P\left(h^{(2)}, h^{(1)}\right) P\left(F_{patch}|h^{(1)}\right) \tag{7}$$

where $h^{(1)}$ and $h^{(2)}$ represent two hidden layers in the DBN network. The output of two DBNs can preserve its high-level semantic information of image-level or patch-level features, which are denoted as Q_I and Q_p.

We then employ a joint RBM to fuse the coarse-grained image-level features and fine-grained patch-level features. The joint distribution is defined as follows:

$$P(Q_I, Q_p) = \sum_{h_1^{(1)},h_2^{(1)},h^{(2)}} P\left(h_1^{(1)}, h_2^{(1)}, h^{(2)}\right) \times \sum_{h_1^{(1)}} P\left(Q_I|h_1^{(1)}\right) \times \sum_{h_2^{(1)}} P\left(Q_p|h_2^{(1)}\right) \tag{8}$$

The resulting joint distribution is considered to be a joint representation of image-level and patch-level features, marked as F.

We perform fine-grained retrieval on the Top-K images using the fusion feature. An image sorting list that belongs to the same subcategory as the query image is returned according to Euclidean distance.

2.3   Query Expansion

Query expansion is used to further improve retrieval accuracy. Here, we simply sum and average the Top-5 image features returned by the fine-grained retrieval phase to calculate a new query descriptor. A new round of fine-grained retrieval is performed with a new query descriptor, which updates the ranking list returned in the fine-grained retrieval phase.

3   Experiments

3.1   Databases and Details

In this paper, we adopt two fine-grained image databases: CUB-200-2011 [9] (11,788 images, 200 different bird subcategories) and Oxford-Flower-102 [10] (8189 images, 102 different bird subcategories).

The experimental evaluation metric is the mean Average Precision (mAP) of the Top-5 returned, which is widely used for evaluating the retrieval accuracy.

The CNN and DBN networks are trained and tested on the Caffe platform [11]. The CNN network uses VGG [12] and the input image size is 224×224. The network is pre-trained on the ImageNet 1K database [13], and then fine-tuned on the fine-grained image database.

The image-level feature is extracted from the last layer of fully connected layers. The object-level features and patch-level features are extracted at the pool_5 layer.

The DBN networks used for fusing image-level and patch-level features have three layers. The DBN for image-level features has 4096 neurons in the input layer, while the DBN for patch-level features has 1024 neurons. The hidden layer and output layer have 2048 and 1024 neurons respectively for both DBN networks. Finally, a layer of joint RBM is put behind the two DBN networks to fuse features.

3.2   Experiment Results

Depth-adding the convolutional feature map, the part with larger activation value is considered to be the object area. The location effect is as shown in the Fig. 4.

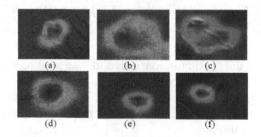

Fig. 4. Object location result

As can be seen from Fig. 4(a)–(c), the location method has achieved a better result. However, some images also show poor location. As shown in Fig. 4(d)–(f), the part with a high activation value is concentrated in the body of bird, and is missing the mouth and tail. This proves that only adopting object-level features is not enough for fine-grained image representation. Although the background area of the image has no positive effect for retrieval in theory, the image-level features are still indispensable.

In the patch filtering stage, each image finally retains five patches with a high overlap with the object and a small overlap with each other. The five patches basically guarantee coverage of the entire object region. In the weighted max-pooling, the first three maximum values in each dimension are retained, and the weights are set to 0.5, 0.3 and 0.2 respectively. We choose a section of data as a validation dataset to form the train data. We get the best value of α, β by constantly experimenting.

The experimental comparison results are shown in Table 1.

Table 1. The accuracy of different patch fusion methods

Methods	Summation and average	Max-pooling	Weighted max-pooling
mAP (%)	65.92	64.78	**66.73**

It can be seen from Table 1 that the method of summation and average can achieve a retrieval accuracy of 65.92%, and the ordinary max-pooling method reduces the retrieval accuracy. In this paper, each patch represents a different part of an object. Therefore, by only retaining the maximum value, plenty of useful information will be overlooked. The retrieval accuracy for the proposed weighted max-pooling method is 66.73%, which is 0.79% higher than summation and average method. This shows that weighted max-pooling method can weaken the residual background information, while retaining effective information as much as possible, thereby improving the accuracy of fine-grained retrieval.

In order to observe the complementarity between multi-level features, we combined multi-level features and observed their retrieval accuracy. The results are shown in Table 2.

Table 2. The accuracy of multi-level feature combination

Methods	mAP (%)
Image-level	61.29
Object-level	64.67
Patch-level	56.38
Image-level + object-level	65.51
Image-level + patch-level	65.92
Object-level + patch-level	65.25
Image-level + object-level + patch-level	**65.98**

As shown in Table 2, the CNN features of image-level, object-level and patch-level have obtained 61.29%, 64.67% and 56.38% retrieval accuracy respectively, and object-level features have greatly improved the retrieval accuracy. At the same time, patch-level features have lower accuracy than image-level features. This may be due to the fact that the patch-level features use object location results. Sometimes, the location effect is not completely accurate, which may cause a patch to contain less effective object information and have more background information.

The image-level features focus on global information, object-level features pay attention to differences in object appearance, and patch-level features reflect subtle and local differences from a more slight perspective. Therefore, fusing two level features together can further improve the retrieval accuracy. As can be seen from Table 2, the complementarity of image-level and patch-level features is stronger than other combination methods.

Finally, when we fuse the three level features, the accuracy is only 0.06% higher than the image-level and patch-level fusion. Due to image-level features and patch-level features already reflecting the object-level features to a certain extent, the effect of adding object-level features is not obvious. Considering the time complexity, only image-level and patch-level features are used for fusion in later experiments.

Linear combination methods, for example simple concatenating, are adopted in most existing methods. For this paper, we applied DBN networks and joint RBM for image-level and patch-level features to perform non-linear fusion. Next, we conducted experiments for two fusion methods, and the experimental results are shown in Table 3.

From Table 3, we can see that the non-linear fusion through joint RBM achieves better retrieval accuracy than simple concatenating fusion. It is illustrated that joint RBM can mine the intrinsic contact and rich complementary information between two features, thereby improving the ability to characterize fine-grained images.

Table 3. The accuracy of different fusion methods for features

Methods	mAP (%)
Concatenating	65.92
Joint RBM	**66.73**

In order to further test the performance of the proposed fine-grained image retrieval algorithm, it was compared with the latest image retrieval and fine-grained image retrieval methods. Table 4 summarizes the details and results on two fine-grained databases, where * indicates that the data came from the paper of SCDA [3].

The first three methods are designed for general image retrieval, which get poor performance. Compared with SCDA, the proposed WMMFA method combines multi-level information of images, and pays more attention to the effective local information. Therefore, the retrieval accuracy is further improved. In addition, the query expansion stage further improves retrieval accuracy.

Table 4. Comparison with other fine-grained retrieval methods

Methods	Dim	mAP (%)	
		CUB-200-2011	Oxford-Flower-102
CNN + CroW [14]*	256	59.69	76.16
CNN + VLAD [15]*	512	58.96	76.05
CNN + R-MAC [16]*	512	59.02	78.19
SCDA [2]	1024	64.59	76.62
SCDA+ [2]	2048	65.76	77.86
WMFFA	2048	66.73	79.58
WMFFA + QE	2048	**68.51**	**81.26**

4   Conclusion

The difficulty of fine-grained image retrieval is that the differences in images in the same subcategories can be great, and the differences in different subcategories can be small. In this paper, we propose a coarse-to-fine retrieval framework and the Weighted Multi-Feature Fusion Algorithm (WMFFA) method for alleviating these difficulties. Our work focuses on three aspects. First, we use the number of overlapping pixel between patches and irregular objects as evaluation criteria, making the filter process easier and more effective. Second, we propose a weighted max-pooling method to aggregate a lot of patch features. The proposed method can ignore possible background noise, while retaining effective object information as much as possible. Third, we introduce the DBN network and use a joint RBM layer to fuse the image-level features and patch-level features, which can capture the intrinsic relations and rich complementary information in multi-level features. Experiment results illustrate that the proposed method obtains better performance.

References

1. Lin, T.Y., Roychowdhury, A., Maji, S.: Bilinear CNN models for fine-grained visual recognition. In: IEEE International Conference on Computer Vision, Santiago, Chile, pp. 1449–1457 (2015)
2. Wei, X.S., Luo, J.H., Wu, J., et al.: Selective convolutional descriptor aggregation for fine-grained image retrieval. IEEE Trans. Image Process. **99**, 1 (2017)
3. Fu, J., Zheng, H., Mei, T.: Look closer to see better: recurrent attention convolutional neural network for fine-grained image recognition. In: IEEE Conference on Computer Vision and Pattern Recognition, Honolulu, HI, USA, pp. 4476–4484 (2017)
4. Ahmad, J., Muhammad, K., Bakshi, S., et al.: Object-oriented convolutional features for fine-grained image retrieval in large surveillance datasets. Future Gener. Comput. Syst. **81**, 314–330 (2018)
5. Zheng, H., Fu, J., Mei, T., et al.: Learning multi-attention convolutional neural network for fine-grained image recognition. In: IEEE International Conference on Computer Vision, Venice, Italy, pp. 5219–5227 (2017)

6. Yao, H., Zhang, S., Zhang, Y., et al.: One-shot fine-grained instance retrieval. In: ACM on Multimedia Conference, Mountain View, CA, USA, pp. 342–350 (2017)

7. Pang, C., Li, H., Cherian, A., Yao, H.: Part based fine-grained bird image retrieval respecting species correlation. In: IEEE International Conference on Image Processing, Beijing, China, pp. 2896–2900 (2017)

8. Uijlings, J.R.R., van de Sande, K.E., Gevers, T., Smeulders, A.W.M.: Selective search for object recognition. Int. J. Comput. Vis. **104**(2), 154–171 (2013)

9. Wah, C., Branson, S., Welinder, P., et al.: The caltech-UCSD birds 200-2011 dataset. California Institute of Technology (2011)

10. Nilsback, M.E., Zisserman, A.: Automated flower classification over a large number of classes. In: Sixth Indian Conference on Computer Vision, Graphics and Image Processing, Bhubaneswar, India, pp. 722–729 (2008)

11. Jia, Y., Shelhamer, E., et al.: Caffe: convolutional architecture for fast feature embedding. In: Proceedings of the ACM International Conference on Multimedia, Orlando, FL, USA, pp. 675–678 (2014)

12. Simonyan, K., Zisserman, A.: Very deep convolutional networks for large-scale image recognition. Computer Science (2014)

13. Deng, J., Dong, W., Socher, R., et al.: ImageNet: a large-scale hierarchical image database. In: IEEE Conference on Computer Vision and Pattern Recognition, Miami, Florida, USA, pp. 248–255 (2009)

14. Kalantidis, Y., Mellina, C., Osindero, S.: Cross-dimensional weighting for aggregated deep convolutional features. In: Hua, G., Jégou, H. (eds.) ECCV 2016. LNCS, vol. 9913, pp. 685–701. Springer, Cham (2016). https://doi.org/10.1007/978-3-319-46604-0_48

15. Ng, Y.H., Yang, F., Davis, L.S.: Exploiting local features from deep networks for image retrieval. In: IEEE Conference on Computer Vision and Pattern Recognition Workshops, Boston, MA, USA, pp. 53–61 (2015)

16. Tolias, G., Sicre, R., Jégou, H.: Particular object retrieval with integral max-pooling of CNN activations. Computer Science (2015)

17. Tang, J., Li, Z.: Weakly-supervised multimodal hashing for scalable social image retrieval. IEEE Trans. Circuits Syst. Video Technol. **PP**(99), 1 (2017). https://doi.org/10.1109/tcsvt. 2017.2715227

18. Liu, J., Zha, Z.-J., et al.: Multi-scale triplet CNN for person re-identification. In: Proceedings of the ACM Conference on Multimedia, pp. 192–196 (2016)

19. Hong, R.C., Hu, Z., Wang, R., Wang, M., Tao, D.: Multi-view object retrieval via multi-scale topic models. IEEE Trans. Image Process. **12**(25), 5814–5827 (2016)

Convolutional Neural Networks Based Soft Video Broadcast

Wenbin Yin[1(✉)], Xiaopeng Fan[1], and Yunhui Shi[2]

[1] Harbin Institute of Technology, Harbin, Heilongjiang, China
{ywb, fxp}@hit.edu.cn
[2] Beijing University of Technology, Beijing, China
syhzm@bjut.edu.cn

Abstract. Video broadcasting is becoming more and more popular in wireless networks. However, the existing digital coding and transmission approaches can hardly accommodate users with diverse channel conditions, which is called the cliff effect. Recently, a novel video broadcasting method called SoftCast has been proposed. It achieves graceful degradation with increasing noise by making the magnitude of the transmitted signal proportional to the pixel value and using a novel power allocation scheme. In this paper, we propose a novel video broadcast method that exploits deep convolutional networks and group based sparse representation. It utilizes the channel condition information generated from decoder to optimize the decoding process and reduce the various artifacts caused by source and channel coding. By utilizing soft video broadcast transmission, it achieves good broadcast performance and avoids the cliff effect. The experimental results show that the proposed scheme provides better performance compared with the traditional SoftCast with up to 1.5 dB coding gain.

Keywords: Video broadcasting · Convolutional neural networks
Soft video broadcast

1 Introduction

Wireless video broadcasting is becoming more and more popular in our daily life and its purpose is to transmit one video signal simultaneously to multiple receivers with different channel conditions. The main challenge we face is the difficulty to provide receivers with video quality that matches their channel conditions. The traditional wireless broadcasting design such as DVB-T standard [1] that combines a layered transmission scheme [2, 3] and scalable video coding (SVC) scheme [4, 5] is one of the typical wireless video broadcasting schemes. SVC encodes the video signal into one base layer (BL) and multiple enhancement layers (EL). In transmission, the hierarchical modulation (HM) [6] superimposes the multiple layer bits in one wireless symbol and allows the user to decode different numbers of layers according to their own channel condition. However, the layered scheme reduces both the compression efficiency and the transmission efficiency.

Recently, a novel solution of wireless video broadcasting called SoftCast [7] is proposed. The SoftCast consists of four steps: DCT transform, power allocation,

© Springer Nature Switzerland AG 2018
R. Hong et al. (Eds.): PCM 2018, LNCS 11166, pp. 641–650, 2018.
https://doi.org/10.1007/978-3-030-00764-5_59

Hadamard transform and direct dense modulation. DCT transform compresses the video frame by removing the spatial redundancy of a video frame. Power allocation reduces the total distortion by optimally scaling the DCT coefficients. Hadamard transform can make each packet with equal importance as a protect-coding. The most attractive difference between SoftCast and traditional approach is that SoftCast directly map the data into wireless symbols by a very dense Quadrature Amplitude Modulation (QAM). At decoder side, SoftCast uses Linear Least Square Estimator to reconstruct the video frame.

Although SoftCast achieves graceful degradation with increasing noise by making the magnitude of the transmitted signal proportional to the pixel value and using a novel power allocation scheme to against the channel noise, it still has room for improvement. Compression and transmission in its nature will introduce undesired complex artifacts, which will severely reduce the users' experience.

In recent years, several soft video broadcast schemes have been proposed [10, 11]. Meanwhile, a number of sparse coding based methods for image restoration [12, 13] have been developed and deep learning has shown impressive results on vision problems [8, 9]. In this paper, we utilize convolutional neural networks and sparse coding based representation to achieve a video multicast method with less compression and transmission artifacts. The encoder compresses the video frame by linear transformation and uses power allocation to minimize the distortion caused by channel noise. The decoder utilizes LLSE and inverse transformation to reconstruct the video frame. However, the decoded frame usually has some artifacts. The proposed scheme utilizes group based sparse representation to reduce the distortion produced by compression and CNN to reduce the distortion produced by transmission.

The rest of the paper is organized as follows. Section 2 describes the encoding and decoding process of the proposed scheme. The performance of our scheme is showed in Sect. 3, followed by concluding remarks in Sect. 4.

2 Proposed Scheme

At the encoder side, video frames are coded by block based DCT transform to compress the data and the coded components are scaled by power allocation to minimize the distortion cause of channel noise, and then transmitted to users with different channel conditions. In the traditional digital video transmission method, cliff effect affects the users' decoding experience. In our method, the scaled coefficients are directly transmitted through soft broadcast without syndrome coding over a very dense constellation that avoids the cliff effect. At decoder side, it uses LLSE to reconstruct the video frame. Since group based sparse representation model can utilize the intrinsic local sparsity and non-local similarity of nature image at same time, we exploit group based sparse representation to reduce the blocking artifacts caused by block based video compression. With initial reconstruction, we exploit convolutional networks to reduce the distortion caused by soft video transmission, since convolutional networks can extract features formed by different quality of channel noise and restore the decoded frames for different channel conditions (Fig. 1).

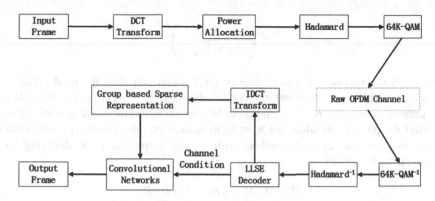

Fig. 1. Flow graph of the proposed scheme

2.1 Video Compression

Since video frames are relatively smooth and show spatial correlation. The proposed method exploits this property to compact the information in frames by taking block based DCT of pixel values. Traditional video coding scheme works with the assumption of a known channel, and encoder quantizing the DCT values as much as desired. This kind of quantization will force all receivers to the same reconstruction quality.

The proposed scheme divides the input frame into blocks, then it takes block-based DCT on this matrix to transform the frame from spatial domain into frequency domain. In general, the DCT components in the right bottom corner stand for high spatial frequencies and have low values, close or equal to zero. We can compress the video frame by discarding the zero value DCT components while these components have limited impact on the information in a frame. However, this kind of compression will cost a large amount of metadata to the decoder side to express the specific location of these discarded DCT components.

To reduce the metadata for the high frequencies DCT components, it divides the DCT values into bands and operates on bands. Specifically, we group DCT components in same position of each blocks into one band. Then we make one decision for all DCT values in a band. As we known, high frequencies components usually concentrated in same region, making one decision for a whole band can provide close performance with discarding individual DCT components. Since the proposed method has discarded few bands, it is much simpler to express the location of these bands than specific location of every discarded DCT components.

2.2 Power Allocation and Transmission

The power allocation can be treated as a protection method for each frequencies of the transmitted signal. Let P be the total power of transmission and g_{R_i} be the scaling factor of R_i that donates DCT components. According to [7], g_{R_i} is given by

$$g_{R_i} = \left(\frac{P}{\sqrt{\lambda_i} \sum_i^K \sqrt{\lambda_i}} \right)^{1/2} \tag{1}$$

where λ_i is the variance of i-th frequency DCT value and K is the total number of frequencies. We define a diagonal matrix $G = diag\{g_{R_1}, g_{R_2}, \ldots, g_{R_K}\}$, the signal M can be represent as $M = G \cdot R$. The λ_i needs to be transmitted as metadata to decoder side.

After the power allocation, we want to maximize proposed method's resilience to packet loss. We can generate packets with equally important by multiplying by a Hadamard matrix H, let, i.e.

$$U = H \cdot M = HG \cdot R = C \cdot R \tag{2}$$

In PHY layer, the metadata and DCT value are transmitted in different ways. Since the metadata needs to be transmitted without any error, we use conventional way to send the metadata. The encoder applies 8-bits scalar quantization on metadata and the quantization results are compressed by variable length coding (VLC). The compressed bit-stream is transmitted by the standard 802.11 PHY layer with FEC and modulation. To well protect the metadata, we use a 1/2 convolutional code and BPSK modulation.

Unlike the metadata, the signal consists of real values rather than binary values. In PHY layer, these real values are first mapped to complex symbol directly by 64K QAM constellation. Every two integers are quantized by an 8-bit quantizer and combined into one complex symbols as the output of the 64K QAM constellation. Given a set of complex time-domain samples, an inverse FFT is computed on each packet of symbols. The real and imaginary components are first converted to the analogue domain using D/A converters, the analogue signals are then used to modulate cosine and sine waves at the carrier frequency respectively. Then these signals are summed to give the transmission signal. With such aforementioned direct source and channel mapping method, it can let the reconstructed quality matching the channel condition in proposed method.

2.3   LLSE at Decoder

Here we define N as channel noise, and the received signal can be represented as

$$\hat{U} = U + N = HG \cdot R + N \tag{3}$$

And the received signal can be recovered by first LLSE estimator in transform domain as follows

$$\hat{R} = \Sigma_r C^T (C \Sigma_r C^T + \Sigma_N)^{-1} \hat{U} \tag{4}$$

where Σ_r and Σ_N are the covariance matrices of R and N. At high CSNR, the LLSE estimator simply inverts the encoder computation. At high CSNR, one can trust the measurement. At low CSNR, one cannot fully trust the measurements and it is better to re-adjust the estimate according to the statistics of the DCT components.

2.4 Group Based Sparse Representation for Deblocking

Due to block based DCT and power allocation, soft video broadcast usually results in visually annoying blocking artifacts in coded videos. Since the sparse representation performs well at removing the blocking artifacts and obtain visually acceptable quality for block based DCT coded videos. In the proposed method, we formulate the GSR based deblocking algorithm though maximum a posteriori (MAP) framework.

Here we define, that given first decoded video frame $\hat{x}$ and the input frame x, processed frame can be obtained by:

$$y = \arg \max \log(p(\hat{x}|x)) + \log(p(x)) \tag{5}$$

where the first term represents data-fidelity, and the second term corresponds to image priors. Inspired by the success of image group based sparse representation, the optimization problem for frame deblocking through MAP is formulated as

$$y = \arg \max \log(p(\hat{x}|x)) + \log(pGSR(x)) + \log(pQC(x)) \tag{6}$$

where $pGSR(x)$ and $pQC(x)$ stand for GSR prior and QC prior, respectively.

The decoded video frame contains transmission and compression noise. The GSR part focus on the compression noise which main causes the blocking artifacts. With the Gaussian compression noise model and compression noise variance σ_{com}^2, the first data-fidelity term can be formulated as

$$\log(p(\hat{x}|x)) = -\frac{1}{2\sigma_{com}^2}\|x - \hat{x}\|_2^2 \tag{7}$$

The group based sparse representation model [14] assumes that a few atoms of a dictionary can represent each group of image blocks. The sparse coding process of each group over dictionary is seek a sparse vector $x_{G_k} \approx D_{G_k}\alpha_{G_k}$. Then the whole image can be sparsely represented by the set of sparse codes $\{\alpha_{G_k}\}$ in the unit of group. So the second term in the Eq. (6) can be formulated as

$$\log(p_{GSR}(x)) = -\eta\|\alpha_G\|_0 \tag{8}$$

where α_G denotes the concatenation of all α_{G_k} and imposes the sparse codes vector α_G to be sparse. In order to incorporate QC prior, we define the indicator by Ω as

$$\psi(x) = \left\{ \begin{array}{ll} 0, & \text{if } x \in \Omega \\ +\infty, & \text{if } x \notin \Omega \end{array} \right\} \tag{9}$$

where Ω is the range of scaled DCT coefficients. The third term can be formulated as

$$\log(p_{QC}(x)) = -\psi(x) \tag{10}$$

Utilize the above priors, the deblocking minimization problem can be formulated as

$$(\hat{\alpha}_G, \hat{D}_G) = \arg \min_{\alpha_G, D_G} \frac{1}{2\sigma_{noise}^2} \|D_G \circ \alpha_G - \hat{x}\|_2^2 + \lambda \|\alpha_G\|_0 + \Psi(D_G \circ \alpha_G) \quad (11)$$

which can be solved by the framework of split Bergman iteration. Equation (11) is equivalently transformed into three iterative step and each separated sub-problem can acquire an efficient solution. After we get $\hat{\alpha}_G$ and $\hat{D}_G$ in hand, the de-blocked frame can be reconstructed by $y = \hat{D}_G \circ \hat{\alpha}_G$.

2.5   Convolutional Networks for Artifacts Reduction

Since the input frame is compressed by the band based coding and transmitted though the OFDM channel, the reconstructed frames usually have some compression and transmission artifacts. Since deep learning has shown impressive results on low-level vision problems, convolutional networks can extract features formed by different quality of channel noise and restore the decoded frames for different channel conditions. We adopt convolutional networks to cope with the compression and transmission artifacts. The whole convolutional networks are shown in Fig. 2.

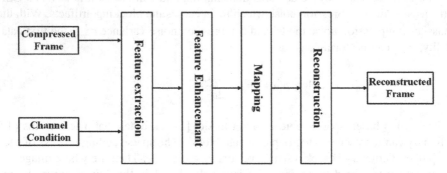

Fig. 2. Convolutional networks for video frame restoration

There are four layers in the restoration networks, each of which is responsible for a specific task. The first layer is used for patch and feature extraction, which extracts overlapping patches from the compressed frame and represents each patch as a high dimensional vector. The second layer can be seen as the feature enhancement layer which extract features from the n_1 feature maps of the first layer and form a new set of feature maps. After feature enhancement layer, at the third layer, we apply non-linear mapping layer to represents a high quality patch by a high dimensional vector. The last layer stands for reconstruction and it produce the final high resolution frames. The entire network can be express as:

$$F_i(y^{(c)}) = \max(0, W_i * y^{(c)} + B_i), i \in \{1, 2, 3\};$$
$$\hat{y}^{(c)} = W_4 * F_3(y^{(c)}) + B_4 \quad (12)$$

where W_i and B_i represent the filters and biases of the i-th layer respectively. c represent the channel condition. F_i is the output feature map and $*$ means the convolution operation. The W_i has n_i filters with size of $n_{i-1} \times f_i \times f_i$ and n_0 represent the number of channels in the input frame. Rectified Linear Unit $(ReLU, max(0, x))$ is applied on the filter responses.

Here, we define the set of un-coded frame as ground truth and represented by $\{x_i\}$. The coded frames form a set called $\{y_i\}$ and each x_i has its corresponding y_i. We choose Mean Squared Error (MSE) as the loss function:

$$L(\Omega) = \frac{1}{n} \sum_{i=1}^{n} \left\| F(y_i^{(c)}; \Omega) - x_i^c \right\|^2 \tag{13}$$

Here $\Omega = \{W_1, W_2, W_3, W_4, B_1, B_2, B_3, B_4\}$, n is the number of training samples. The loss is minimized using stochastic gradient descent with the standard back propagation.

3 Experimental Result

In experiments, we evaluate the performance of our proposed method in video unicast and multicast. We compare our scheme with SoftCast and H.264 which use standard 802.11 PHY layer with FEC and QAM modulations. The experiment method broadcasts the same video to users with different channel SNR.

We use over 400 images of size 180×180 for training. The training images are decomposed into 64×64 sub-images and then the compressed and transmitted samples are generated from the training samples with SoftCast decoder. A total of 204,800 patches are sampled with a stride of 20 on the training images. The learning rate is set as 10^{-5} in the last layer and 10^{-4} in the remaining layers. The convolutional network settings are $f_1 = 9, f_{1'} = 7, f_2 = 1, f_3 = 5, n_1 = 64, n_{1'} = 32, n_2 = 16, n_3 = 1$. A specific network is trained for each 5 dB CSNR range.

The test sequences are 'foreman_cif.yuv', 'news_cif.yuv, 'mother_cif.yuv and 'bus_cif.yuv, respectively. The video frame rate is 30 Hz. The coded signal is transmitted over OFDM channel with AWGN.

We compare the proposed method with SoftCast and the conventional frameworks based on H.264. For conventional framework, we implement 4 recommended combination of channel coding and modulation of 802.11. We calculate the corresponding bit-rate according to the bandwidth for H.264 encoder. For the proposed method, there is no bit-rate but only channel symbol rate. The video PSNR of each framework under different channel SNR is given in Fig. 3 which shows that all the four conventional transmission approaches suffer from a very serious cliff effect. In contrast, the SoftCast and the proposed method do not suffer from the cliff effect. As the channel SNR increases, the reconstruction quality increases accordingly. Figure 5 gives the performance comparison on different video sequences. Since GSR based compression artifact reduction scheme and deep convolutional networks based transmission artifact reduction scheme performs well in decoded frame restoration. Figure 6(c) and (d) shows that under similar PSNR, the proposed method not only reduce most of the artifacts, but also provides better reconstruction on both edges and textures.

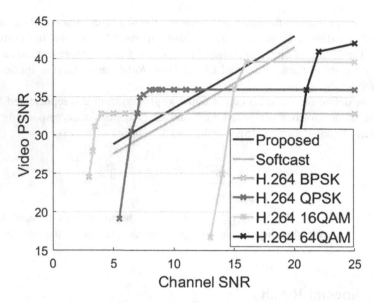

Fig. 3. Robustness comparison

We then let the frameworks serve a group of three receivers with diverse channel SNR. The channel SNR for each receiver is 5 dB, 10 dB and 20 dB. In conventional frameworks based on H.264, the server transmits the video stream by using BPSK. It cannot use higher transmission rate because otherwise the 5 dB user will not be able to decode the video. In both of SoftCast and the proposed method, the server can accommodate all the receivers simultaneously. Using our method, the 5 dB user will get slightly lower reconstruction quality than using H.264 based conventional frameworks. However, the 10 dB and 20 dB users get better reconstruction quality by using our method than conventional frameworks. The test result is given in Fig. 4.

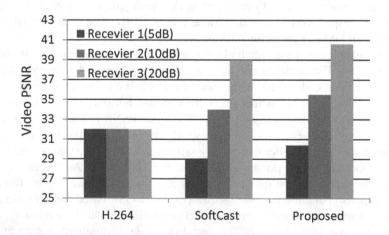

Fig. 4. Multicast comparison

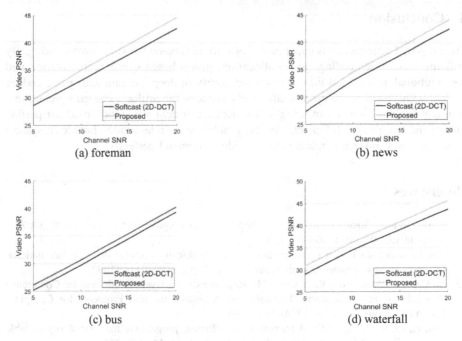

Fig. 5. Broadcast performance on different sequences

Fig. 6. Visual quality of 'foreman_cif'

4 Conclusion

The proposed scheme in this paper provides a novel method for video broadcasting. By utilizing band-based coding, power allocation, group based sparse representation and convolutional networks, it fully exploits the ability of deep learning and sparse coding to deal with vision problems and effectively reduces the artifacts caused by compression and transmission. By utilizing soft broadcast, it achieves good broadcast performance and avoids the cliff effect. Finally, it achieves wireless video broadcast system which matches modern wireless video broadcast demand perfectly.

References

1. Digital Video Broadcasting (DVB). http://www.etsi.org/deliver/etsi_en/300700_300799/300744/01.06.01_60/en_300744v010601p.pdf
2. Shacham, N.: Multipoint communication by hierarchically encoded data. In: International Conference on Computer Communications, vol. 3, pp. 2107–2114 (1992)
3. McCanne, S., Jacobson, V., Vetterli, M.: Receiver-driven layered multicast. In: Conference Proceedings on Applications, Technologies, Architectures, and Protocols for Computer Communications, pp. 117–130. ACM (1996)
4. Wu, F., Li, S., Zhang, Y.Q.: A framework for efficient progressive fine granularity scalable video coding. IEEE Trans. Circuits Syst. Video Technol. **11**, 332–344 (2001)
5. Schwarz, H., Marpe, D., Wiegand, T.: Overview of the scalable video coding extension of the H.264/AVC standard. IEEE Trans. Circuits Syst. Video Technol. **17**, 1103–1120 (2007)
6. Ramchandran, K., Ortega, A., Uz, K., Vetterli, M.: Multiresolution broadcast for digital hdtv using joint source-channel coding. In: IEEE International Conference on Communications, vol. 1, pp. 556–560 (1992)
7. Jakubczak, S., Katabi, D.: A cross-layer design for scalable mobile video. In: International Conference on Mobile Computing and Networking, pp. 289–300. ACM (2011)
8. Dong, C., Loy, C.C., He, K., Tang, X.: Image super resolution using deep convolutional networks. arXiv:1501.00092 (2014)
9. Dong, C., Loy, C.C., He, K., Tang, X.: Learning a deep convolutional network for image super-resolution. In: Fleet, D., Pajdla, T., Schiele, B., Tuytelaars, T. (eds.) ECCV 2014. LNCS, vol. 8692, pp. 184–199. Springer, Cham (2014). https://doi.org/10.1007/978-3-319-10593-2_13
10. Fan, X., Wu, F., Zhao, D.: D-cast: DSC based soft mobile video broadcast. In: International Conference on Mobile Ubiquitous Multimedia, pp. 226–235 (2012)
11. Peng, X., Xu, J., Wu, F.: Line-cast: line-based semi-analog broadcasting of satellite images. In: International Conference on Image Processing, pp. 2929–2932 (2012)
12. Jung, C., Jiao, L., Qi, H., Sun, T.: Image deblocking via sparse representation. Signal Process. Image Commun. **27**, 663–677 (2012)
13. Liu, X., Wu, X., Zhao, D.: Sparsity based soft decoding of compressed images in transfer, domain. In: International Conference on Image Processing, pp. 563–566 (2013)
14. Zhang, J., Zhao, D., Gao, W.: Group based sparse representation for image restoration. IEEE Trans. Image Process. **4**, 1–2 (2014)

Image Generation for Printed Character by Representation Learning

Kangzheng Gu[1], Jiansong Bai[2], Qichen Zhang[3], Junjie Peng[4],
and Wenqiang Zhang[1(✉)]

[1] Shanghai Key Laboratory of Intelligent Information Processing,
School of Computer Science, Fudan University, Shanghai, People's Republic of China
{kzgu17,wqzhang}@fudan.edu.cn
[2] Department of Art and Design, Fudan University,
Shanghai, People's Republic of China
baijiansong@fudan.edu.cn
[3] School of Sociology and Political Science, Shanghai University,
Shanghai, People's Republic of China
TOAD_ZQC@outlook.com
[4] School of Computer Science, Shanghai University,
Shanghai, People's Republic of China
jjie.peng@shu.edu.cn

Abstract. With the development of convolutional neural networks, generative models can synthesize really wonderful images. But most of these models are limited in generalization and extensibility. And things become difficult when generating images with multiple specified features. Therefore, this paper introduce an expandable approach to generate images with multiple features. We use our model to generate images including a single character with specified fonts and position, by learning the representations of different features from existing images, and using these representations together. Several structures are proposed to increase the training efficiency and extensibility. Finally, we arrange some experiments and show the performance of our model.

Keywords: Image generation · Represent learning · Printed character

1 Introduction

Recently, with the improvement of architecture of the Convolutional Neural Networks (CNNs) [10], methods for dealing tasks on image processing and computer vision are developing rapidly. Previously, the algorithms based on CNNs were used to solve classification [9] and segmentation problems [1,12]. Afterwards, the Auto Encoder architecture [5] and the Convolutional Neural Networks have been combined by researchers to solve image generation problems. From then on, more and more research have focused on such generation techniques.

One of the branches which discusses on generation techniques is Auto Encoder [5], followed by its variant called Variational Auto Encoder (VAE) [8],

© Springer Nature Switzerland AG 2018
R. Hong et al. (Eds.): PCM 2018, LNCS 11166, pp. 651–660, 2018.
https://doi.org/10.1007/978-3-030-00764-5_60

which achieve great success on generation problems. It holds two separate parts using neural network, which are known as encoder and decoder. The encoder extracts features of an image and the decoder generates a new image by such features. Another method is the Generative Adversarial Networks (GAN) [3]. Then, the Conditional GAN (CGAN) [13] and the Conditional VAE [7] were proposed so that researchers can synthesize images with specified styles by changing the certain values in the feature vector.

Furthermore, one work [6] used CGAN [13] to achieve image-to-image translation. However, the feature vector is in bad interpretability. That means, researchers could not determine the correspondence between features and values in feature vectors. They have to try different value in the feature vector manually and observe what these values can influence. Another shortcoming is that, they can just learn only single kind of features, like font. They are not able to deal with more than two kinds of features simultaneously, such as text and font for a printed character. In other words, they just learned a certain feature of images in network parameters, and could not extract other features at the same time.

In order to comprehend the exact meaning of such feature vectors, researchers have propose a new approach called representation learning. Representation learning usually uses methods of unsupervised learning to get more generic and robust features. One work [14] learned the relationship between partial pieces and integrated image in one sample to extract features. Another one [11] tried to learn some useful features by chronologically sorting frames in a video. Those methods indicates that finding internal relationships among data can increase the performance of feature extractors, which can be applied to many other problems.

Based on the previous discussion, we propose an approach combining the two previous methods, the generation model and representation learning, to construct a modular and extendable model to learn specific features which we want. We use the Encoder-Decoder architecture to generate images and representation learning to calculate representation of specific features. Then, we use the model to learn features of text and fonts simultaneously. Finally, we use the learned representation of features to generate new images.

2   Method

2.1   Encoder-Decoder Architecture

The basic idea of our method is the Encoder-Decoder architecture combining convolutional layers. As for encoder architecture, there are many classic network architecture used as feature extractor like VGG [15] or ResNet [4]. Indeed these architectures perform well on many problems about images, but their structures are somewhat complicated. There is no need to introduce such complicated architectures into our model. Instead, we use a more simple one, which has only four convolutional layers and two fully connected layers. The encoder takes an image containing a single character as its input and calculate a feature vector which represents information of a certain feature (Fig. 1(a)).

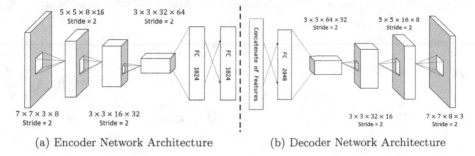

(a) Encoder Network Architecture   (b) Decoder Network Architecture

Fig. 1. (a) The encoder consists of 4 convolutional layers and 2 fully connected layers, which extracts features from the input image. (b) The decoder consists of 2 fully connected layers and 4 transpose convolutional layers, which generates images by input representation of features. Kernel shapes and sizes of fully connected layers are shown beside each module.

The decoder is like a reverse version of encoder. It takes a set of feature vectors as input and uses these inputs to generate a new image having a single character. Feature vectors firstly were transformed to a feature map by two fully connected layers. Then we use transpose convolution layers to generate a new character from small feature maps (Fig. 1(b)).

Our architecture resembles the AlexNet [9], of which we further decrease the layers and parameters. ELU [2] activation is used in our architecture, except the last layer in the encoder and decoder. The last layer of encoder uses *tanh* and the last layer of decoder uses *sigmoid*. And there is no normalization layers in our model.

Encoder and decoder can be combined in different ways. Firstly we will introduce a parallel structure, which is used directly to extract two kinds of features of printed character, their text and fonts, and generate the specified character. Then a cascade structure is proposed to deal with conditions having more than two kinds of features. In our experiment, we choose three features which are text, font and character position in image.

2.2 Parallel Structure

Parallel structure consists of several encoders and a single decoder. Each encoder learns a certain feature representation of an image containing a printed character, and generate a feature vector. Then we concatenate the output feature vectors of each encoder directly and send them into the input of the decoder, which finally generates an image having a character of certain text and font.

Section 3 will show the experiment on two cases including two kinds of features and three kinds of features of a printed character in an image.

Here we use the condition of two kinds of features to explain how the parallel structure works. For this special case, we have two encoders. One is for calculating the representation of the text of a character image, such as which character

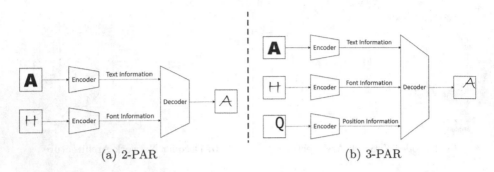

(a) 2-PAR   (b) 3-PAR

Fig. 2. (a) This figure shows how parallel structure works on two features' condition. It calculates representation of text from the first image, which contains letter 'A', while calculates representation of font from the second image, which contains font 'Bradley Hand ITC'. Then it generates a new image of 'A' with font 'Bradley Hand ITC'. (b) shows how parallel works on three features' condition. It calculates representations of text, font and position of character simultaneously, and generates a new image of 'A' with 'Bradley Hand ITC' at the top right corner.

it is, 'A' or 'H', while the other is to extract features of fonts, like 'Bradley Hand ITC' or 'Segoe UI Black'.

For example, we send an image containing a letter 'A' to the text encoder which might be in any font, and we send an image to font encoder that contains an letter in certain font like 'Bradley Hand ITC', but which might be another character such as 'H' or 'P'. Then the two encoders output two feature vectors of text and font information. That in this case, are two vectors representing 'A' and 'Bradley Hand ITC' separately. Finally the decoder generates an image having a letter 'A' with the font 'Bradley Hand ITC' according to those feature vectors (See Fig. 2(a)).

In the same way, we can construct a parallel structure which can learn more than two specified features (Fig. 2(b)). But experiment shows that it is not a good way to use parallel structure to learn more than two features because of the difficulties in neural network training. So we propose another architecture to solve that problem.

2.3   Cascade Structure

Like cascade classifiers, our cascade structure consists of many Encoder-Decoder units in series connection. Each unit is a two-feature parallel structure (Fig. 3).

If we want to learn k independent features of an image, we can connect k-1 units one by one. The first unit is used to learn and extract two features, then it generate a unified feature vector, which is used by the second unit. After that the second unit uses the unified feature vector and a learned representation vector of the third feature to generate a vector combined the previous three kinds of features, and so on (Fig. 4).

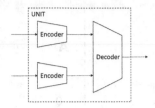

Fig. 3. The structure of a unit.

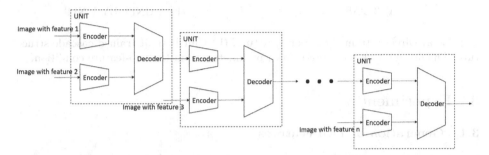

Fig. 4. The cascade structure consists of a series of units. A unit takes the previous one's output as the first input. Another input is an image having another feature different from previous ones'. And the unit provides its output to the next unit.

Similarly, we use a three-feature case shown in Fig. 5(a) to illustrate how the cascade structure is organized and works. The three features we used for experiment is text, font and position where the character stands at. We construct a cascade structure which contains two units to do with the case. In the beginning, an image including a letter 'A' in various kinds of fonts and an image including a arbitrary letter in certain font, for example 'Bradley Hand ITC', are fed into the first unit. Then, the first unit outputs an image generated by those features, for example, a letter 'A' in font 'Bradley Hand ITC'. Furthermore, the second unit takes that as one of its inputs. The other input for the second unit is an image containing an arbitrary letter in arbitrary font, but which is specified in certain position, like the coordinate (15, 15). After taking those inputs, the second unit finally generate a image which has a letter 'A' in font 'Bradley Hand ITC' at position (15, 15).

The design of cascade structure allows us to add as many as units to learn various features theoretically. However, training difficulties cannot be avoided only by substituting the cascade structure for parallel ones. The following experiments will show that, training units one by one will achieve much better performance than training them as a unified end-to-end model.

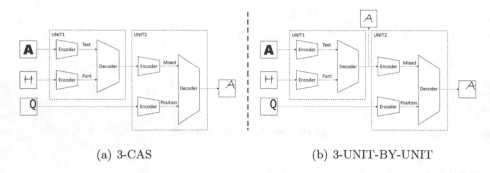

(a) 3-CAS　　　　　　　　　　　(b) 3-UNIT-BY-UNIT

Fig. 5. (a) Directly trained cascade structure. (b) Unit-by-unit trained cascade structure. The first unit uses the pre-trained parameters from the two-feature condition.

3　Experiment

3.1　Generation of Two Features

Our first experiment is to generate two-feature printed character. We use a set of 62 characters including both lowercase and uppercase English letters and digits from 0–9. The training set of fonts includes 322 Truetype files from the Windows 10 system (with several extra fonts like 'Ubuntu Mono'). Then we will test the effect of transformation on other 50 fonts, which are downloaded from the Internet.

Construct Training Data. Firstly, we choose two characters, X and Y (may be the same one), in our character set and two fonts U and V (may be the same one) in our training set of font. We generate an image A with character X and font V at the fixed position in image, and an image B with character Y and font U at the same position of A. We hope our model learn the representation of character X from image A and the representation of font U in image B, then generate an image contains a character X with font U. So we generate another image C as ground truth with character X and font U at the fixed position in image. Each image of character is in size of $(48 \times 32 \times 3)$.

Loss Function. We note the representation of character as $\phi_c(x)$, representation of font as $\phi_f(x)$ and the generator function as $g(text, font)$. In all our experiments, we use the MSE loss to evaluate the quality of the generated images, which can be formed as the following:

$$L_{two-feature}(C, A, B) = (g(\phi_c(A), \phi_f(B)) - C)^2 \tag{1}$$

Optimizer and Hyper Parameters. The optimizer we use is RMSProp [16] with learning rate 1×10^{-4} and batch size 64. And in order to decrease overfitting, we use the L2-regularization with the coefficient 1×10^{-4}.

3.2   Generation of Three Features

The second experiment introduce another variable which is the position of the character in image. So far we have three kinds of features, which are text, font and position. All the training and testing conditions are as same as the previous one. The only difference is we need to use a extra image to teach the model how to distinguish the position of character in image.

Construct Training Data. First, we choose three characters X, Y and Z, three fonts U, V and W, together with three position P, Q and R.

We then generate three image A, B and C, with X in V at R, Y in U at Q, and Z in W at P as the input of the model. Then we hope the model calculate features of the text X, the font U and the position P. Finally, it should generate a image D of a character X in font U at position P, which is the ground truth in our training data.

Loss Function. Besides the notations we used in the previous experiment, we introduce $\phi_p(x)$ as the representation of character position in image. So the loss function changes to the following one.

$$L_{three-feature}(D, A, B, C) = (g(\phi_c(A), \phi_f(B), \phi_f(C)) - D)^2 \qquad (2)$$

Architecture. In this experiment, we will try both parallel structure and cascade structure. And we will use two different approaches of training for cascade structure. One is training as a union while the other is training unit by unit. We will show all results of the three experiment in the following section.

4   Discussion

According to the training loss curves of all experiments in Fig. 6, the two-feature case is the easiest one for training. When it comes to three features, things become much more difficult. The simplest parallel structure has the worst outcome, and the cascade structure is just slightly better than the parallel one. This is because the cascade model can decrease the complexity of learning from learning three features simultaneously to two features. Furthermore, another advantage of cascade structure is useful. That is, all the units in cascade structure can be trained one by one. Firstly we train the first unit to learn to generate character images with only two features, text and font, as same as what we do in the two-feature condition. Next, we train the second unit with the pre-trained parameters of the first one. This will help to learn faster because learning features two-by-two is much more easier than learning all features at the same time. And it will also help to increase the training stability. The loss curve shows that this approach can increase the training efficiency obviously.

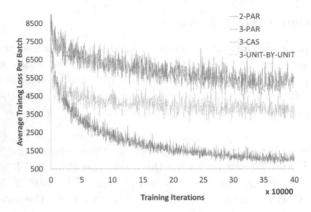

Fig. 6. This figure shows the training process of all the models above. 2-PAR represents the model generating printed character images from two features which are text and font. 3-PAR represents the parallel structure generating printed character images from three features which are text, font and position, while 3-CAS represents the cascade structure. 3-UNIT-BY-UNIT represents the unit-by-unit trained cascade structure to generate images from three features. In the remaining part we will use the same notations for models.

Figure 7 show the network output during the training process. We can see that, the network output becomes clearer and more meaningful with the process of training. When all the iterations finished, the model solved two-feature's condition very well and generated clear letter which is really close to the ground truth on the training data set, while the cascade structure generate clearer characters than the parallel structure in three-feature condition. The unit-by-unit training method for the cascade structure gets the best performance. Characters generated by it are much clearer and complete than others.

At last, we test our best model, the cascade structure trained unit by unit. The fonts used for test has never occurred during the training. We picked some good results from all the generated images. These results show that our model has a certain degree of generalization to many other kinds of fonts. Furthermore, we observe an interesting phenomenon. Sometimes, the model generates images really similar to the ground truth, like what is shown in Fig. 8(a). However sometimes, the model generates the character quiet different from the ground truth, but they still look like the same font as the provided input, which is shown in Fig. 8(b). Moreover, characters with specified position is more blurred than characters without specified position. We think this limitation is caused by the deficiency of fully connected layers that they cannot represent information of position properly. Fully convolutional architecture or some dynamic routing methods can be involved to solve this problem. Thus in future, we plan to replace the network architecture of cascade units by specially designed networks to adapt distinct features better.

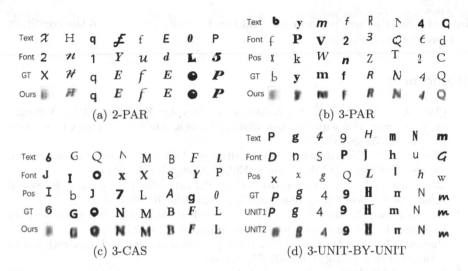

Fig. 7. All these images are selected during the training process. The first one is from the beginning of training, while the last is from the end of training. Each column contains a sample. The row marked by 'Text' contains the first input where the model learns the text representation from. The row marked by 'Font' and 'Pos' contains the second and third inputs providing representations of fonts and position. Line 'GT' represents the ground truth. Line 'Ours' includes outputs from our models. And Lines 'UNIT1' and 'UNIT2' show the outputs of each unit from the cascade structure.

(a) good results   (b) acceptable results

Fig. 8. (a) Shows results on test set which fit the ground truth well. Although results' fonts in (b) are different from the ground truth, the generated images look reasonably that have the same fonts as images shown in 'Font' lines.

5 Conclusion

This paper has presented an extendable architecture of representation learning of different components in an image of printed character, and a generator to generate images of character with specified text, fonts and position. Experiments show that both parallel structure and cascade structure can learn specified representation and generate specified image efficiently. The cascade structure trained unit by unit gains the best performance and can be easily extended to the condition of more than two features.

References

1. Chen, L.C., Papandreou, G., Kokkinos, I., Murphy, K., Yuille, A.L.: Semantic image segmentation with deep convolutional nets and fully connected CRFs. Comput. Sci. **4**, 357–361 (2014)
2. Clevert, D.A., Unterthiner, T., Hochreiter, S.: Fast and accurate deep network learning by exponential linear units (ELUs). Comput. Sci. (2015)
3. Goodfellow, I.J., et al.: Generative adversarial networks. In: Advances in Neural Information Processing Systems, vol. 3, pp. 2672–2680 (2014)
4. He, K., Zhang, X., Ren, S., Sun, J.: Deep residual learning for image recognition. In: Computer Vision and Pattern Recognition, pp. 770–778 (2016)
5. Hinton, G.E., Salakhutdinov, R.: Reducing the dimensionality of data with neural networks. Science **313**(5786), 504–507 (2006)
6. Isola, P., Zhu, J., Zhou, T., Efros, A.A.: Image-to-image translation with conditional adversarial networks. In: Computer Vision and Pattern Recognition, pp. 1125–1134 (2016)
7. Kingma, D.P., Rezende, D.J., Mohamed, S., Welling, M.: Semi-supervised learning with deep generative models. In: Advances in Neural Information Processing Systems, vol. 4, pp. 3581–3589 (2014)
8. Kingma, D.P., Welling, M.: Auto-encoding variational bayes. In: International Conference on Learning Representations (2014)
9. Krizhevsky, A., Sutskever, I., Hinton, G.E.: Imagenet classification with deep convolutional neural networks. In: International Conference on Neural Information Processing Systems, pp. 1097–1105 (2012)
10. Lecun, Y., Bottou, L., Bengio, Y., Haffner, P.: Gradient-based learning applied to document recognition. Proc. IEEE **86**(11), 2278–2324 (1998)
11. Lee, H.Y., Huang, J.B., Singh, M., Yang, M.H.: Unsupervised representation learning by sorting sequences, pp. 667–676 (2017)
12. Long, J., Shelhamer, E., Darrell, T.: Fully convolutional networks for semantic segmentation. In: IEEE Conference on Computer Vision and Pattern Recognition, pp. 3431–3440 (2015)
13. Mirza, M., Osindero, S.: Conditional generative adversarial nets. arXiv: Learning (2014)
14. Noroozi, M., Pirsiavash, H., Favaro, P.: Representation learning by learning to count, pp. 5899–5907 (2017)
15. Simonyan, K., Zisserman, A.: Very deep convolutional networks for large-scale image recognition. Comput. Sci. (2014)
16. Tieleman, T., Hinton, G.: Lecture 6.5-RMSProp: divide the gradient by a running average of its recent magnitude. COURSERA Neural Netw. Mach. Learn. **4**, 26–31 (2012)

CRNet: Classification and Regression Neural Network for Facial Beauty Prediction

Lu Xu[1], Jinhai Xiang[1(✉)], and Xiaohui Yuan[2]

[1] College of Informatics, Huazhong Agricultural University, Wuhan, China
xulu_coi@webmail.hzau.edu.cn, jimmy_xiang@mail.hzau.edu.cn
[2] Department of Computer Science and Engineering,
University of North Texas, Denton, USA
Xiaohui.Yuan@unt.edu

Abstract. Facial beauty prediction is a challenging problem in computer vision and multimedia fields, due to the variant pose and diverse conditions. In this paper, we introduce "soft label" for each annotated facial image, and propose a novel neural network–classification and regression network (CRNet) with different branches, to simultaneously process a classification and a regression task. Besides, weighted mean squared error (MSE) and cross entropy (CE) are used as the loss function, which is robust to outliers. CRNet achieves state-of-the-art performance on SCUT-FBP and ECCV HotOrNot dataset. Experimental results demonstrate the effectiveness of the proposed method and clarify the most important facial regions for facial beauty perception.

Keywords: Deep learning · Facial attractiveness analysis
Classification and regression network (CRNet)
Facial beauty prediction

1 Introduction

Facial beauty analysis [1] has been widely used in many fields such as facial image beautification APPs (e.g., MeiTu and Facetune) and social networks services [2]. In the mobile computing era, billions of images are acquired and uploaded to social networks and online platform per day. It is a new challenge for image processing and analyzing technology. Recently, with the big data and high-performance computational hardware, computational and data-driven approaches have been proposed for solving these challenges, such as face recognition [27], facial expression recognition [28], facial beauty analysis [13] and etc.

This work was primarily supported by Foundation Research Funds for the Central Universities (Program No. 2662017JC049) and State Scholarship Fund (NO. 261606765054).

© Springer Nature Switzerland AG 2018
R. Hong et al. (Eds.): PCM 2018, LNCS 11166, pp. 661–671, 2018.
https://doi.org/10.1007/978-3-030-00764-5_61

Previous methods resort to computational models to analyze facial beauty and achieve promising results [3,5,11]. The methods often extract image features (such as histogram of oriented gradients (HOG), scale-invariant feature transform (SIFT), local binary patterns (LBP), etc) firstly, and then train supervised machine learning models (such as support vector machine (SVM), k-nearest neighbors (KNN), deep neural network (DNN), logistic regression (LR), etc) to predict beauty scores [3].

In this paper, we propose a novel deep neural network, named "classification and regression neural network" (CRNet), to accurately predict facial attractiveness score from a portrait image. We test our algorithm on SCUT-FBP dataset [4] and ECCV HotOrNot dataset [5], and against state-of-the-art methods. We find that our approach achieves better performance than existing approaches.

The main contributions of this paper are as follows:

- We introduce a novel neural network named CRNet with two different branches, for a classification and a regression task respectively. These branches are trained in parallel.
- We devise a loss function that consists of cross entropy (CE) for classification and mean squared error (MSE) for regression, with weighted parameters, to enable more discriminative feature learning for improved performance.
- To enhance model's learning ability, "Soft labels" from existing annotations are developed by discretizing the corresponding continuous facial beauty values for classification task in CRNet.
- We explore the impact on facial beauty perception through detailed experimental analysis and deep feature visualization.

The rest of this paper is organized as follows. Section 2 reviews the related works of the facial descriptor and learning methods. Section 3 describes our proposed method in details. Experimental results and comparisons with baseline models are presented in Sect. 4, and Sect. 5 concludes this paper with a summary and future work.

2   Related Work

Many researchers concentrate on developing new models to achieve better classification or regression performance [4,7,8,12,13], while others focus on designing more discriminative facial descriptors [6,9]. Zhang et al. [6] introduce a method which combines both low-level and high-level representations, to form a feature vector and then perform feature selection to optimize the feature set. Huang et al. [8] propose a method to learn hierarchical representations from convolutional deep belief networks. Xie et al. [4] resort to deep convolutional neural networks (DCNN) to train a predictor and achieve state-of-the-art performance on SCUT-FBP dataset [4]. Kagian et al. [9] use rich facial features that describe facial geometry, color and texture as a representation to predict facial attractiveness.

In addition, some researchers focus on developing or improving new machine learning models. Eisenthal et al. [7] adopt KNN and SVM as classifiers to rate faces belongs to different levels. Deep convolutional neural networks (CNN) have attracted much attention in recent years, due to their extraordinary performance in image recognition tasks [15,17,19,24]. Lots of researchers turn to design better CNN architecture to precisely handle facial beauty prediction problems. Wang et al. [12] use deep auto encoders to extract features and take a low-rank fusion method to integrate scores with very promising results. Xu et al. [13] propose PI-CNN for automatically facial beauty prediction with a pearson correlation of 0.87 on SCUT-FBP dataset [4].

Despite these methods yield promising experimental results, the majority of the labeled images are under a constrained environment, which prevents these models from being used among real-life conditions.

3   Proposed Methods

In this section, we will give a detailed description about our proposed methods, which include the overall neural network architecture and the corresponding loss function.

3.1   CRNet

There has been lots of research works [14,19] indicate that jointly training can boost performance of each task. He et al. [15] propose residual networks (ResNet), which can extract more informative representation through identical mapping. Based on ResNet18's architecture, we replace the last softmax layers of ResNet18 [15] with two branched fully connected layers, for classification and regression tasks, respectively. We first reshape the feature maps into a 512-dimensional feature vector, and then feed the feature vector into the two branched layers (CBranch and RBranch).

CBranch is designed for classification and *RBranch* is designed for regression. CBranch is composed of three fully connected layers which contain 256, 64 and 3 (or 5) neurons, respectively. RBranch is also composed of three fully connected layers which contain 256, 64, and 1 neurons, respectively. The ResNet18 model used as CRNet's backbone is pretrained on ImageNet and fine-tuned on target source for fast convergence. The parameters in the backbone can be shared and these two branches are trained jointly. The output of RBranch is adopted as the predicted result. The learning procedure of CRNet is end-to-end (Fig. 1).

Mainstream deep models always treat the prediction task with continuous values as a regression task. However, [10,29] indicate that treat the regression problem as a multi-label classification task can yield more accurate and robust performance. Instead of setting the number of output neuron as 1, we discretize the output range into C parts, and the final regression value can be calculated as:

$$s_c = \frac{exp(\theta^T x)}{\sum_{c=1}^{C} exp(\theta^T x)} \tag{1}$$

$$o = \sum_{c=1}^{C} s_c \cdot c \tag{2}$$

where s_c denotes the probability output of softmax layer with the label class c, θ^T denotes learned parameters, x represents the feature vector, and o represents the final output regression value from a classification network.

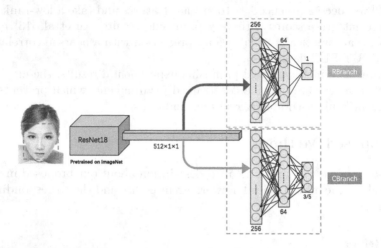

Fig. 1. Architecture of CRNet. ResNet18 [15] is adopted in our feature extraction procedure, the network is split into two branches. RBranch minimizes MSE loss, and CBranch minimize cross entropy loss.

3.2   CRLoss

Since our proposed CRNet include both classification branch and regression branch, the according cost function is illustrated as follows:

$$\mathcal{L}_c = -\frac{1}{M}\sum_{i=1}^{M} y_i \cdot log\hat{y}_i \tag{3}$$

$$\mathcal{L}_r = \frac{1}{M}\sum_{i=1}^{M}(y_i - \hat{y}_i)^2 \tag{4}$$

$$\mathcal{L} = \theta_c \cdot \mathcal{L}_c + \theta_r \cdot \mathcal{L}_r \tag{5}$$

where $\mathcal{L}_c$ denotes cross entropy loss [18] for classification task in CBranch, and $\mathcal{L}_r$ denotes MSE loss [18] for regression task in RBranch. y_i stands for the ground truth facial beauty score of image i, $\hat{y}_i$ stands for the predicted score. θ_c represents weight for CBranch loss, θ_r represents weight for RBranch loss. We set $\theta_c = 0.4$ and $\theta_r = 0.6$ in our experiments.

Since MSE loss is cost-sensitive [26], by incorporating cross entropy loss for classification and MSE loss for regression. CRLoss can improve robustness to outliers [26]. Jointly training with two branches in two different tasks can also improve the model's learning ability and performance as well [19].

4 Experiments

4.1 Datasets

SCUT-FBP. The SCUT-FBP dataset [4] is a widely used benchmark for measuring facial beauty recognition models, which contains 500 Asian females images. In order to make the prediction more reliable and reproducable, we follow the provision denoted in [4] for fair comparison. We randomly select 400 images as the training set and the rest 100 images as the test set. Finally, we average the 5 experimental results as the final performance to remove sample variances.

ECCV HotOrNot Dataset. The ECCV HotOrNot Dataset [5] is a pretty challenging benchmark dataset for facial beauty prediction, because of the variant postures, cluttered background, illumination, low resolution and unaligned faces problems, which make the facial beauty prediction more difficult. This dataset contains 2056 faces which are collected from the Internet. Each face is labeled with a score, and the dataset has already been split into 5 training and test sets. The final result is calculated by averaging the 5-fold results.

4.2 Performance Evaluation

Following the previous studies [4,5,13], the Pearson Correlation (PC) is used to evaluate the performance between the ground-truth and the predicted result.

$$PC = \frac{1}{n-1} \sum_{i=1}^{n} \left(\frac{h(x_i) - h(\overline{x})}{s_{h(x)}} \right) \left(\frac{y_i - \bar{y}}{s_y} \right) \qquad (6)$$

where n denotes the number of images, $x^{(i)}$ denotes the input feature vector of image i, $h(\bullet)$ denotes the learning algorithm, $y^{(i)}$ denotes the groundtruth attractiveness score of image i, $h(\overline{x})$ and $\bar{y}$ denote the mean of $h(x)$ and y, respectively.

PC measures the linear correlation between $h(x^{(i)})$ and $y^{(i)}$. Its value lays between 1 and -1, where 1 means absolutely positive linear correlation, 0 means no linear correlation, and -1 means absolutely negative linear correlation. In the above dataset, a larger PC value means the better performance of the model.

4.3 Experiments Settings

We implement our model with PyTorch [21] on NVIDIA Tesla K80 GPU with cuDNN acceleration. The ResNet18 used in our backbone network is pretrained on ImageNet for fast convergence [16,17]. We also adopt batch normalization [22] to accelerate training procedure. Weight decay is set as 0.0001 and dropout [23] is used with a probability of 0.5 in RBranch and CBranch to avoid overfitting. The learning rate starts at 0.001 and is divided by 10 per 20 epochs.

Table 1. Performance on SCUT-FBP and HotOrNot dataset for 5 rounds

Dataset	1	2	3	4	5	AVG
SCUT-FBP [4]	0.8723	0.8704	0.8766	0.8687	0.8735	**0.8723**
HotOrNot [5]	0.4992	0.4787	0.4715	0.5073	0.4515	**0.482**

Performance on SCUT-FBP and HotOrNot dataset for 5 rounds. Pearson Correlation (PC) is used for evaluating performance. The final result is calculated by averaging the 5 PC values. CRNet achieves a PC with 0.8723 on SCUT-FBP and a PC with 0.482 on HotOrNot dataset.

4.4 Experiments on SCUT-FBP

Since the images in SCUT-FBP are not in same size, CRNet can only support fixed squared data as input. We detect face region via [20] and crop the detected face region from the original image, then resize it to 224×224 pixels. We also normalize the input image by substracting the mean and dividing the standard variance of the pixels. Furthermore, we manually crop the central region of the image and treat it as the input for CRNet in case of failed face detection. Random rotation within a range of $-30°$ to $30°$ is also performed for data augmentation. We train CRNet with stochastic gradient descent (SGD) with momentum of 0.9 for 80 epochs.

In SCUT-FBP dataset, we take round numbers $\lceil v - \frac{1}{2} \rceil$ as the *soft labels* for classification task in CBranch (v represents groundtruth beauty value). Then we get $\mathcal{C} \in \{0, 1, 2, 3, 4\}$ for CBranch, and the number of output neurons is set to 5 accordingly. The results of 5 round experiments can be found in Table 1.

Table 2 shows performance comparison with other methods. To the best of our knowledge, our method achieves the state-of-the-art performance on SCUT-FBP [4] dataset.

Table 2. Performance comparision with baseline models on SCUT-FBP dataset

Method	PC
Gaussian Regression [4]	0.6482
CNN-based [4]	0.8187
PI-CNN [13]	0.87
Liu et al. [25]	0.6938
CRNet	**0.8723**

Performance comparison with other recently proposed methods, our method achieves the best performance. The best result is emphasized in bold style.

4.5   Experiments on ECCV HorOrNot Dataset

Since the face images in ECCV HotOrNot dataset are already in same size, and the majority of the faces are not aligned, we do not crop the face region. The portrait image is just resized to 227×227 pixels, then we randomly crop 224×224 region as input for CRNet. Besides, color jittering [21] and random rotation within a range of $-30°$ to $30°$ are also performed for data augmentation. We train CRNet with SGD with momentum of 0.9 for 30 epochs.

In ECCV HotOrNot dataset, we manually discretize the values into 3 parts based on the source training dataset:

$$c = \begin{cases} 0 ; & if \ c_i < -1 \\ 1 ; & if \ -1 \leq c_i < 1 \\ 2 ; & otherwise \end{cases}$$

where c_i represents the groundtruth beauty score, and c represents the corresponding soft label. The number of output neurons of CBranch in CRNet is set to 3 accordingly. The results of 5 round experiments can be found in Table 1.

Table 3 compares the Pearson Correlation of our proposed method with other state-of-the-art methods. Our method outperforms other methods and achieves the best performance on ECCV HorOrNot dataset without face alignment.

Table 3. Performance comparision with baseline models on ECCV HotOrNot dataset

Method	PC
Multiscale Model [5]	0.458
Wang et al. [12]	0.437
CRNet	**0.482**

Performance comparison with recent baseline models on ECCV HotOrNot dataset. To the best of our knowledge, CRNet achieves the best performance.

4.6   Analysis

From Fig. 2 we can see that the faces in second row which are not well fitted by CRNet, tend to have sharp curved forehead and pointy chinare. The organs such as eyes and mouth are more important in facial beauty perception, since CRNet tend to give higher beauty scores for those faces with large eyes and big mouth (i.e., the 9-th portrait image). People with more stereoscopic organs tend to be more attractive.

We also list both well predicted samples and imprecisely predicted samples in ECCV HotOrNot dataset in Fig. 3. It shows that the unaligned face, variant

2.380282	3.428571	2.71831	2.528571	2.6	3.056338	2.430556	2.314286
2.377618	3.416009	2.732897	2.509869	2.620086	3.088337	2.397507	2.280127

3.152778	2.802817	4.442857	4.65625	4.515625	3.028571	3.295775	3.871429
4.059937	3.55667	3.700321	3.956159	3.829313	2.420696	2.706038	4.459424

———groundtruth facial beauty score
———predicted facial beauty score by CRNet

Fig. 2. We list several predicted values given by CRNet and their ground truth scores, respectively. The ground truth beauty values are emphasized in red color, and the predicted values are emphasized in blue color. Well predicted images by CRNet in SCUT-FBP dataset are listed in the 1st row. Imprecisely predicted samples are listed in the 2nd row. (Color figure online)

Fig. 3. Well predicted by CRNet in ECCV HotOrNot dataset (up). Imprecisely predicted by CRNet in ECCV HotOrNot dataset (bottom).

pose and low-resolution images influence the performance of CRNet significantly. We believe the performance could be greatly improved through face alignment techniques. Besides, posture and facial expression may also contribute to beauty perception because our model fails to capture these samples with variant postures.

By visualizing the 512 feature activation before *average pooling layers* (see Fig. 4) as:

$$S = 0.2 \times img + 0.8 \times \frac{1}{N} \times \sum_{i=1}^{N} fm_i \tag{7}$$

where img denotes the original input image, fm_i denotes the ith feature map, N denotes the number of channels, and S denotes the output result for visualization. We can find that the *nose*, *mouth* and *face size and shape* are the most

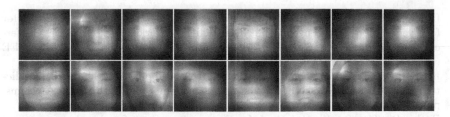

Fig. 4. Deep feature activation in face region.

important parts for facial beauty perception. The activation on attractive faces is more concentrated and symmetrical, while activation on unattractive faces is more dispersed.

5   Conclusion

In this paper, we propose a novel neural network named CRNet, guided by CRLoss for accurate face beauty prediction. CRNet outperforms other methods and achieves the state-of-the-art performance on the widely used SCUT-FBP [4] and ECCV HotOrNot dataset [5], which illustrates the effectiveness of our proposed model. We also explore the impact on facial beauty perception through detailed experimental analysis and deep feature activation visualization. However, the popular benchmark datasets such as SCUT-FBP and ECCV HotOrNot are still in a small capacity, which hinders the deeper exploration in face beauty prediction and related research fields. In our future work, a large scale labeled dataset will be released for researchers to develop better models.

References

1. Perrett, D.I., May, K.A., Yoshikawa, S.: Facial shape and judgements of female attractiveness. Nature **368**(6468), 239–42 (1994)
2. Rothe, R., Timofte, R., Van Gool, L.: Some like it hot-visual guidance for preference prediction. In: Proceedings CVPR 2016, pp. 1–9 (2016)
3. Zhang, D., Chen, F., Xu, Y.: Computer Models for Facial Beauty Analysis. Springer, Cham (2016). https://doi.org/10.1007/978-3-319-32598-9
4. Xie, D., Liang, L., Jin, L., Xu, J., Li, M.: DScut-fbp: a benchmark dataset for facial beauty perception. In: 2015 IEEE International Conference on Systems, Man, and Cybernetics (SMC), pp. 1821–1826. IEEE (2015)
5. Gray, D., Yu, K., Xu, W., Gong, Y.: Predicting Facial Beauty without Landmarks. In: Daniilidis, K., Maragos, P., Paragios, N. (eds.) ECCV 2010. LNCS, vol. 6316, pp. 434–447. Springer, Heidelberg (2010). https://doi.org/10.1007/978-3-642-15567-3_32
6. Chen, F., Xiao, X., Zhang, D.: Data-driven facial beauty analysis: prediction, retrieval and manipulation. IEEE Trans. Affect. Comput. **9**(2), 205–216 (2018)
7. Eisenthal, Y., Dror, G., Ruppin, E.: Facial attractiveness: beauty and the machine. Neural Comput. **18**(1), 119–142 (2006)

8. Huang, G.B., Lee, H., Learned-Miller, E.: Learning hierarchical representations for face verification with convolutional deep belief networks. In: 2012 IEEE Conference on Computer Vision and Pattern Recognition (CVPR), pp. 2518–2525. IEEE (2012)

9. Kagian, A., Dror, G., Leyvand, T., Cohen-Or, D., Ruppin, E.: A humanlike predictor of facial attractiveness. In: Advances in Neural Information Processing Systems, pp. 649–656 (2007)

10. Rothe, R., Timofte, R., Van Gool, L.: Deep expectation of real and apparent age from a single image without facial landmarks. Int. J. Comput. Vis. **126**(2–4), 144–157 (2018)

11. Zhang, D., Zhao, Q., Chen, F.: Quantitative analysis of human facial beautyusing geometric features. Pattern Recognit. **44**(4), 940–950 (2011)

12. Wang, S., Shao, M., Fu, Y.: Attractive or not?: beauty prediction with attractiveness-aware encoders and robust late fusion. In: Proceedings of the 22nd ACM International Conference on Multimedia, pp. 805–808. ACM (2014)

13. Xu, J., Jin, L., Liang, L., Feng, Z., Xie, D., Mao, H.: Facial attractiveness prediction using psychologically inspired convolutional neural network (PI-CNN). In: IEEE International Conference on Acoustics, Speech and Signal Processing, pp. 1657–1661 (2017)

14. Ranjan, R., Patel, V.M., Chellappa, R.: Hyperface: a deep multi-task learning framework for face detection, landmark localization, pose estimation, and gender recognition. IEEE Trans. Pattern Anal. Mach. Intell. 1 (2018)

15. He, K., Zhang, X., Ren, S., Sun, J.: Deep residual learning for image recognition. In: Proceedings of the IEEE Conference on CVPR, pp. 770–778 (2016)

16. Yosinski, J., Clune, J., Bengio, Y., Lipson, H.: How transferable are features in deep neural networks? In: Advances in Neural Information Processing Systems, pp. 3320–3328 (2014)

17. Donahue, J., et al.: Decaf: a deep convolutional activation feature for generic visual recognition. In: International Conference on Machine Learning, pp. 647–655 (2014)

18. Goodfellow, I., Bengio, Y., Courville, A., Bengio, Y.: Deep Learning, vol. 1. MIT Press, Cambridge (2016)

19. Gebru, T., Hoffman, J., Fei-Fei, L.: Fine-grained recognition in the wild: a multi-task domain adaptation approach. In: 2017 IEEE International Conference on Computer Vision (ICCV), pp. 1358–1367. IEEE (2017)

20. King, D.E.: Dlib-ml: a machine learning toolkit. JMLR.org (2009)

21. Paszke, A., Gross, S., Chintala, S., Chanan, G.: Tensors and dynamic neural networks in python with strong GPU acceleration, Pytorch (2017)

22. Ioffe, I., Szegedy, C.: Batch normalization: accelerating deep network training by reducing internal covariate shift. In: International Conference on Machine Learning, pp. 448–456 (2015)

23. Srivastava, N., Hinton, G., Krizhevsky, A., Sutskever, I., Salakhutdinov, R.: Dropout: a simple way to prevent neural networks from overfitting. J. Mach. Learn. Res. **15**(1), 1929–1958 (2014)

24. Krizhevsky, A., Sutskever, I., Hinton, G.E.: Imagenet classification with deep convolutional neural networks. In: Advances in Neural Information Processing Systems, pp. 1097–1105 (2012)

25. Liu, S., Fan, Y.-Y., Guo, Z., Samal, A., Ali, A.: A landmark-based data-driven approach on 2.5D facial attractiveness computation. Neurocomputing **238**, 168–178 (2017)

26. Khan, S.H., Hayat, M., Bennamoun, M., Sohel, F.A., Togneri, R.: Costsensitive learning of deep feature representations from imbalanced data. IEEE Trans. Neural Netw. Learn. Syst. **29**(8), 3573–3587 (2018)
27. Liu, W., Wen, Y., Yu, Z., Li, M., Raj, B., Song, L.: Sphereface: deep hypersphere embedding for face recognition. In: The IEEE Conference on Computer Vision and Pattern Recognition (CVPR), vol. 1, p. 1 (2017)
28. Yang, H., Ciftci, U., Yin, L.: Facial expression recognition by de-expression residue learning. In: Proceedings of the IEEE Conference on Computer Vision and Pattern Recognition, pp. 2168–2177 (2018)
29. Yang, T.-Y., Huang, Y.-H., Lin, Y.-Y., Hsiu, P.-C., Chuang, Y.-Y.: Ssr-net: a compact soft stagewise regression network for age estimation. In: IJCAI, pp. 1078–1084 (2018)

Stitches Generation for Random-Needle Embroidery Based on Markov Chain Model

Chen Ma, Zhengxing Sun[✉], Hao Wu, and Yuqi Guo

State Key Laboratory for Novel Software Technology, Nanjing University,
Nanjing, China
szx@nju.edu.cn

Abstract. We present a new stylization method to generate Random-needle Embroidery stitches which is a graceful Chinese embroidery art formed by intersecting stitches. First, we model the intersecting stitch and initialize stitches positions and directions for regions. The Markov chain model is used to select similar intersecting stitches for filling each region to avoid artifacts in local area. Then a hierarchical iterative stitches generation process is used to keep the characteristic of multi-layering of stitches. Finally, top layer stitches in each iteration of the generation process are slightly moved according to bottom stitches by a Lloyd's method based approach to make stitches maintain the characteristic of uniform distribution. Experiments show that our result stitches can avoid artifacts in local area and maintain the uniformity and multi-layering at the same time. Comparing with the state-of-the-art methods, our result stitches have a richer visual effect.

Keywords: Image-based artistic rendering · Random-needle embroidery
Markov chain model · Lloyd's method

1 Introduction

Image-based artistic rendering (IB-AR) focuses on technique for transforming realistic visual media to achieve various artistic styles [1]. Recently, in the area of IB-AR, a Chinese embroidery called random-needle embroidery is attracting researchers' attention [3–5]. It is a visually rich art with thousands of intersecting stitches with different lengths and angles to express texture of a reference image. In this paper, we aim to generate stitches formed by line segments that meet characteristics of this art.

To express rich texture of input images with stitches, one step is to select different stitches, i.e. style primitive in the area of IB-AR, to fill regions with different visual characteristics [1, 2]. In general, two types of methods are used to select primitives. For example, Zeng et al. [6] hard-coded the mapping relations between brushes and object categories. Another approach is to match the similarity of images and style primitives with a kind of feature such as shape, tone and texture. Kim et al. [7] selected stippling primitives based on brightness similarity. For this Chinese embroidery, Zhou et al. [3] selected intersecting stitches with different line lengths and angles according to some objective functions designed by heuristic experience. In order to use stitches fitting continuous image contents to obtain more artistic stitches, Yang et al. [4, 5] selected

© Springer Nature Switzerland AG 2018
R. Hong et al. (Eds.): PCM 2018, LNCS 11166, pp. 672–683, 2018.
https://doi.org/10.1007/978-3-030-00764-5_62

stitches by matching the image content and primitives with pixel-level similarity difference. But one limitation of previous works is that although stitches with different angles and line lengths are filled in different regions, stitches with the same angle and line length are arranged in each region which result in artifacts in local area.

Another key step is stitches optimization aims to adjust primitive attributions such as positions to keep stitches evenly distributed. In the area of IB-AR, Voronoi algorithms use efficient techniques from computational geometry to place evenly spaced primitives on a canvas. For example, Adrian et al. [8] used Voronoi diagrams by Lloyd's method for placing stippling primitives. For this Chinese embroidery art, stitches are obtained by a hierarchical iterative stitch generation process (see Fig. 4). Hence, stitches can maintain the characteristic of multi-layering of this embroidery art and artists often carefully decide the positions of top layer stitches during the iteration so that the result stitches remain evenly distributed (see Fig. 5). Although previous works [3–5] filled stitches on a canvas evenly, they arranged a sufficient number of intersecting stitches at once instead of the hierarchical iterative stitch generation process. Therefore, their results lose the multi-layered characteristic.

To avoid the artifacts in local areas generated by filling each region with the same line length and angle stitches, we model the intersecting stitches selection process as Markov chain. The chain specifies k intersecting stitches with similar angles and line lengths to fill a region with k initialized positions. To maintain the multi-layering, stitches on the canvas are generated by a hierarchical iterative process. But the result stitches will be unevenly distributed if top layer stitches is overlaid on bottom stitches naïvely during the iteration process. Therefore, we move top layer stitches slightly according to the positions of bottom stitches by a Lloyd's method based approach to keep result stitches distribution evenly at the same time.

Our contributions can be surmised as follow:

(1) A stitch selection process based on the Markov chain model is proposed, by which intersecting stitches with similar line lengths and angles are sampled for filling each region to avoid artifacts in local area.
(2) A Lloyd's method based approach is used to move top layer stitches slightly during the hierarchical iterative stitch generation process to maintain the multi-layering and uniform distribution of result stitches at the same time.
(3) The stitches have a richer visual effect because the hierarchical iterative stitch generation process is used in our framework as artists did.

2 The Proposed Method

Figure 1 shows the workflow of our method which incorporates three main modules: *Markov-based Stitch Modeling*, *Hierarchical Iterative Stitch Generation* and *Stitch Optimization based on Voronoi Diagram*. For Markov-based stitch modeling, we first define a parameterized intersecting stitch. Then positions and orientations for intersecting stitches are initialized in each region. Finally, the intersecting stitch selection process for filling each region are modeled by Markov chain. For hierarchical iterative stitch generation, the input image is segmented multiple times to obtain multi-layer regions. Then intersecting stitches are filled in each region by above method. The result stitches on the canvas are generated by an iteration process to superimpose the upper stitches on bottom stitches. For stitch optimization based on Voronoi diagram, top layer stitches are moved slightly according bottom stitches by a Lloyd's method based approach during the iteration process to keep multi-layer stitches distribution evenly.

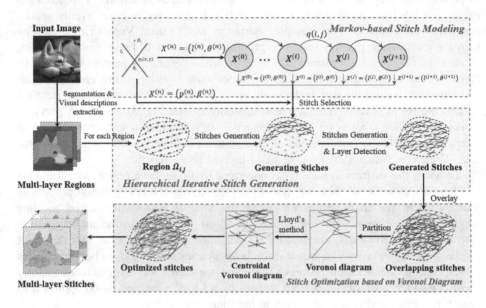

Fig. 1. The follow chart of our method.

2.1 Markov-Based Stitch Modeling

In this section we describe how we fill a region with intersecting stitches. This process can be regarded as selecting similar intersecting stitches placed on initialized positions in the region.

Intersecting Stitch. Artists always embroidery one thread to intersect with another thread as the style primitive. For this reason, two intersecting line segments with the same length are used to define the intersecting stitch. As show in Fig. 2(a), the attributes include: position, orientation, line length and angle.

Positions and Orientations Initialization. We initialize the positions and orientations of intersecting stitches in each region based on a characteristic of this embroidery: placing stitches along a fixed direction or texture direction are two common ways by artists. Two approaches are used to determine stitches positions and directions. As shown in Fig. 2(b), we extract the smallest bounding rectangle for the region, then the evenly spaced straight lines are calculated according to the direction of long side of the rectangle. Finally, we extract points at equal intervals on these straight lines as the initial positions of intersecting stitches. The orientation of stitches is the direction of the straight lines. Another approach, as shown in Fig. 2(c), evenly spaced streamline are generated by [9]. We extract points at equal intervals on these streamlines as the initial positions. The direction of an intersecting stitch is the tangent direction of the streamline at the point of its position.

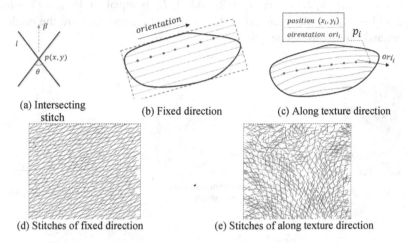

(a) Intersecting stitch

(b) Fixed direction

(c) Along texture direction

(d) Stitches of fixed direction

(e) Stitches of along texture direction

Fig. 2. Initialization of stitch modeling. (b) and (c) are the initialization of stitches positions and orientations; (d) and (e) are the results of the intersecting stitches arrangement.

Stitch Selection as Markov Chain. To avoid artifacts in local areas generated by filling a region with the same line length and angle stitches, the key is to use stitches with similar angles and line lengths to fill the region. Therefore, the selection of the next intersecting stitch should be based on the stitches in the region. Obviously, the Markov chain whose set of vertices coincides with the set of intersecting stitches is suitable for modeling the stitches selection process.

A Markov chain is a stochastic process that transitions from one state to another which only upon the current state [10]. More formally, a Markov chain is sequence of random variables $\{X^{(0)}, X^{(1)}, \ldots, X^{(n)}\}$ where $X^{(i)}$ describes the state of the process at time i. An intersecting stitch can be represented as $X^{(i)} = \left(l^{(i)}, \theta^{(i)}\right)$ where $l^{(i)}$ is the line length of stitch $X^{(i)}$ and $\theta^{(i)}$ is the angle. In order to select k intersect stitches to place on k initialized positions in the region, this process can be regarded as sampling sequence of random variables $\{X^{(0)}, X^{(1)}, \ldots, X^{(k)}\}$ where $X^{(i)} = \left(l^{(i)}, \theta^{(i)}\right)$. To sample stitches with similar angles and line lengths, we assume that angles and line lengths obey normal distribution in intervals $[l_s, l_e]$ and $[\theta_s, \theta_e]$ respectively. The two intervals serve as constraints so that the sampled stitches in a region satisfies two artistic characteristics. One is stitches with small angle are always be arranged in regions with stronger directionality of texture. Another is stitches with short line lengths are always be arranged in important regions to keep details. To calculate the two intervals for the region, we first extract two features as visual descriptions of directionality of texture and importance: *orientation dominance* [4] to indicate texture directionality and *importance value* [12] of a region. Then a contraio method [11], as shown in Fig. 3(b) and (c), is used which the analysis of histogram modes to group regions to obtain two intervals $[a_{ori}, b_{ori}]$ and $[a_{imp}, b_{imp}] \in [0, 1]$ for each region according to the orientation dominances histogram and the importance histogram respectively. Finally, $[\theta_s, \theta_e]$ is equal to $[(1 - b_{ori}) \cdot \pi/2, (1 - a_{ori}) \cdot \pi/2]$ and $[l_s, l_e]$ is equal to $[L_{min} + (1 - b_{imp}) \cdot \Delta L, L_{min} + (1 - a_{imp}) \cdot \Delta L]$ where $\Delta L = L_{max} - L_{min}$, L_{max} and L_{min} are the predefined maximum and minimum line lengths.

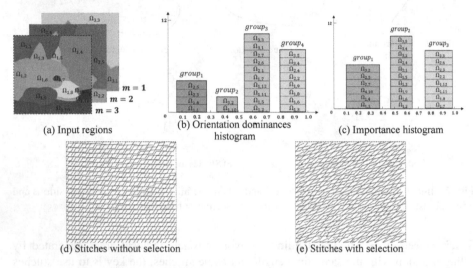

(a) Input regions

(b) Orientation dominances histogram

(c) Importance histogram

(d) Stitches without selection

(e) Stitches with selection

Fig. 3. Contraio method [11] and the comparison among stitches without and with selection. (b) and (c) are two histograms obtained by contraio method. (d) Intersecting stitches are arranged without stitches selection for filling a region. (e) Intersecting stitches are arranged with stitches selection method

Construct the Markov Chain by Gibbs Sample. Because the line lengths and angles of intersecting stitches need to be considered at the same time, we can construct the Markov chain by the Gibbs sample which is an easy simulation using Markov chain Monte Carlo methods. It is used to obtain a sequence of observations which are approximated from a specified multivariate probability distribution. Intersecting stitches are sampled from a bivariate normal distribution $X = (l, \theta)$ and $X \sim Norm(\mu, \Sigma)$ where $l \in [l_s, l_e]$ and $\theta \in [\theta_s, \theta_e]$. If k samples are needed for k positions in the region, we begin with some initial value $X^{(i)} = \left(l^{(i)}, \theta^{(i)}\right)$ where $l^{(i)} \in [l_s, l_e]$ and $\theta^{(i)} \in [\theta_s, \theta_e]$. For the next sample $X^{(i+1)} = \left(l^{(i+1)}, \theta^{(i+1)}\right)$, we update $\theta^{(i+1)}$ according to the distribution specified by $p\left(\theta^{(i+1)} | l^{(i)}\right) = Norm\left(\mu_\theta + \rho(l^{(i)} - \mu_l), \sqrt{1 - \rho^2}\right)$. And we update $l^{(i+1)}$ according to the distribution $p\left(l^{(i+1)} | \theta^{(i)}\right) = Norm(\mu_l + \rho\left(\theta^{(i)} - \mu_\theta\right), \sqrt{1 - \rho^2})$ meanwhile. Repeat the above step $k' - 1$ times. Now we obtain k' samples $\left\{ \left(l^{(1)}, \theta^{(1)} \right), \ldots, \left(l^{(k')}, \theta^{(k')} \right) \right\}$, $(k' > k)$. Note that the sample at the beginning (the so-called *burn-in period*) are common ignored. Hence, the last k samples $\left\{ \left(l^{(k'-k+1)}, \theta^{(k'-k+1)} \right), \ldots, \left(l^{(k')}, \theta^{(k')} \right) \right\}$ are obtained for k intersecting stitches line lengths and angles. These stitches with similar line lengths and angles are used to place on the k positions in the region. And the orientations of stitches are equal to the directions of correspond positions as mentioned earlier. Figure 3(e) shows the result obtained by our stitches selection method.

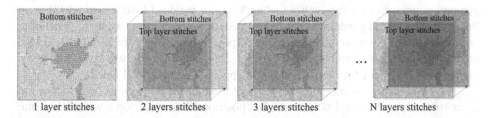

Fig. 4. The real embroidery of Random-needle is obtained by stacking single-layer stitches multiple times. To simulate this process, the result stitch is obtained by a hierarchical iterative stitch generation process, which the top layer stitches are stacked on bottom stitches in each iteration.

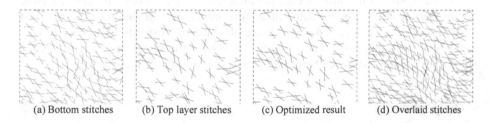

Fig. 5. Stitch optimization. (c) Top layer stitches are moved slightly according to the positions of (a) bottom stitches to keep (d) result stitches distribution evenly.

2.2   Hierarchical Iterative Stitch Generation

In order to maintain the multi-layering, as shown in Fig. 4, the result stitches are obtained by multi-layer stitches overlay as artists did (typically three layers stitches). Therefore, a hierarchical iterative stitch generation process is designed in our work.

For an input image, as shown in Fig. 3(a), we first segment it three times to obtain three layer regions. Therefore, a region can be represented as $\Omega_{m,n}$, where $m \leq 3$. m and n indicate that $\Omega_{m,n}$ is the n th region in layer m. Then we use the method described in Sect. 2.1 to fill each region with intersecting stitches from layer $m = 1$ to layer $m = 3$. Finally, stitches in layer m are overlaid on stitches in layer $m - 1$ from layer $m = 2$ to layer $m = 3$ to obtain the multi-layer stitches. But the multi-layer stitches will be unevenly distributed if top layer stitches is overlaid on bottom stitches naïvely because the positions of stitches initialized by streamlines [9] or straight lines have a certain randomness. One option would be applying Lloyd's method [8] which using the Voronoi diagram to move top layer stitches slightly to keep result stitches distribution evenly.

2.3   Stitch Optimization Based on Voronoi Diagram

In this section, we aim to move top layer stitches slightly according to bottom stitches during the hierarchical iterative stitch generation process as artists did to keep result stitches distribution evenly. A *centroidal* Voronoi diagram has the property that each generating point lies on the centroid of its Voronoi region. Lloyd's method [8] is an iterative algorithm to generate a centroidal Voronoi diagram from any set of generating points to balance point primitives. For our purpose, we use the positions of intersecting stitches as the set of points for the Voronoi diagram. Then the top layer stitches is moved (Fig. 5) by a Lloyd's method based approach as described below.

The basic idea is to compute the Voronoi diagram of all positions (including top layer stitches positions and bottom stitches positions) but move top layer stitches into the corresponding centroids. However, intersecting stitches may be moved to a remote locations. The result stitches are evenly distributed but they may not fit in these new positions (Fig. 8). To avoid this situation, our stitches optimization is designed as below. The top layer stitches on the same streamlines or straight lines are moved together in the same direction and short distance after the Voronoi diagram of all stitches positions is calculated in each iteration until they are evenly distributed. Note that intersecting stitches on the same streamline are moved together in a short distance in the direction $\delta_i \in \{\delta_1, \ldots, \delta_k\}$ which are obtained by positions of stitches to corresponding centroids. For a better result, we choose the direction $\delta_i \in \{\delta_1, \ldots, \delta_k\}$ which corresponding distance between the two points is the longest.

3   Experiments

To validate the effectiveness of our method, we first compare our stitch selection method with previous works and our method without stitch selection. Then we compare our stitch optimization method with Lloyd's method and overlaid naïvely. Finally, we compare our stitches generation method with the state-of-the-art methods [3, 5] qualitatively and quantitatively.

Evaluation of Stitches Selection. In order to evaluate if our stitches selection method can avoid artificial in local area, we conduct this experiments by comparing our stitches selection method with our method without stitches selection and the state-of-the-art methods [3, 5]. Figure 6(e) and (f) show the single-layer stitches in local area obtained by our stitches selection method. Figure 6(c) and (d) show stitches without stitches selection. Stitches in (a) and (b) are obtained by the methods of Zhou et al. [3] and Yang et al. [5] without rendering. And the numbers of intersecting stitches controlled by the density parameters of previous works are similar with numbers of stitches in other columns. Stitches with different line lengths and angles are arranged in R1, R2,

R3 and R4. We can find the previous result in artificial in local area because stitches with the same line length and angle in a region. We can obtain similar results without stitches selection (Fig. 6(d)) as previous works did. However, there are still artificial in columns (c) and (d). Comparing columns (e) and (f) with (c) and (d) respectively, stitches in (e) and (f) avoid artificial because stitches with similar instead of the same line lengths and angles are arranged in a region as artists did by the method of Markov chain model.

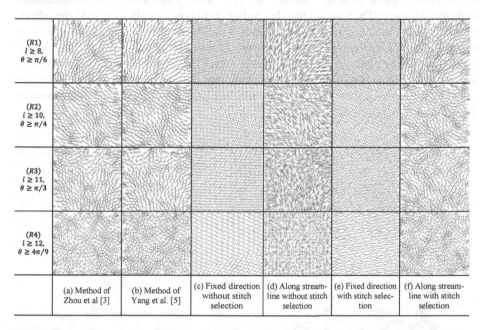

(R1) $l \geq 8,$ $\theta \geq \pi/6$						
(R2) $l \geq 10,$ $\theta \geq \pi/4$						
(R3) $l \geq 11,$ $\theta \geq \pi/3$						
(R4) $l \geq 12,$ $\theta \geq 4\pi/9$						
	(a) Method of Zhou et al [3]	(b) Method of Yang et al. [5]	(c) Fixed direction without stitch selection	(d) Along streamline without stitch selection	(e) Fixed direction with stitch selection	(f) Along streamline with stitch selection

Fig. 6. Comparison among single-layer stitches in local area by methods of (a) Zhou et al. [3], (b) Yang et al. [5] and ((c)–(f)) ours. Stitches with different line lengths and angles are arranged in rows R1, R2, R3 and R4. Comparing with previous works (a) and (b), there still artificial in the results of our method without stitch selection (c) and (d). This problem has been improved in (e) and (f) because stitches with similar instead of the same line lengths and angles are arranged in each region.

Note that, comparing with previous works (Fig. 6(a) and (b)), more visually rich stitches are obtained by our method because intersecting stitches are placed along straight lines (Fig. 6(c)) or streamlines (Fig. 6(f)). As shown in Fig. 7(R1), intersecting stitches are usually placed along a fixed direction when regions (Ω_1 and Ω_2 in (R1)) is narrow, otherwise they are arranged along the texture directions (regions except Ω_1 and Ω_2 in (R1)). Figure 7(R2) shows different result stitches are obtained by different segmentation result.

Evaluation of Stitch Optimization. To evaluation of our stitches optimization method can keep stitches distribution evenly after overlay operation, we conduct this experiments by comparing multi-layered stitches obtained by our stitches optimization

method with Lloyd's method and overlaid naïvely. Stitches in (R3) in Fig. 8(a) are
obtained by our method. Stitches in (R1) are obtained by placing top layer stitches on
bottom stitches naïvely. Stitches in (R2) are obtained by Lloyd's method to move top
layer stitches according to positions of bottom stitches. In order to get a clearer
comparison, Fig. 8(b) and (c) show two quantitative evaluations *Distribution uniformity* and *Orientation uniformity* [4] of stitches in regions of R1, R2 and R3. As
described by Yang et al. [4], the value of distribution uniformity is smaller, the stitches
distribution is more uniform. Stitches are in a mass when the value of orientation
uniformity is large. The value of blue bar in Fig. 8(b) is obviously large than the others,
and the values of red and gray bar are similar. This shows that our method has the same
effect as Lloyd's method and makes stitches distribution evenly after overlay operations. The value of the red bar in Fig. 8(c) is larger than the others, but values of blue
and gray bars are similar. This indicates that stitches are moved to far positions and in a
mass by Lloyd's method, because the stitches directions cannot be fitted texture
direction in new positions.

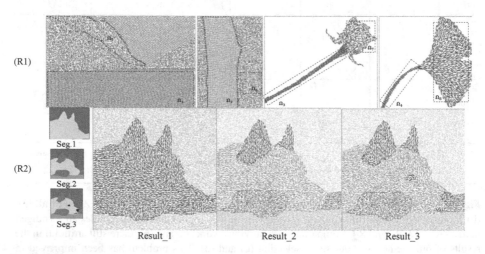

Fig. 7. (R1) Richer visual results are obtained because two intersecting stitch layout approaches
are used in our method. For example, stitches are placed along a fixed direction in narrow region
Ω_1 and Ω_2, otherwise arranged along texture directions (Ω_3 to Ω_8). Therefore, (R2) stitches
obtained by our method are related to the segmentation of the reference image. Result_1 to
Result_2 are the stitches corresponding to Seg.1 to Seg.3 respectively. Zoom 300% to view
details.

Comparison with the State-of-the-Art Methods. To evaluate if our method can
make result stitches maintain multi-layering at the same time, we conduct this experiments by comparing result stitches generated by our method and other two state-of-
the-art methods [3, 5] qualitatively and quantitatively. To clearly observe the effect of
result stitches obtained by multi-layer stitches overlay, we use the method as [3] to
render each layer stitches. Figure 9(R1) shows the rendering results of our method and
other two state-of-the-art methods. In order get a quantitative comparison, inspired by

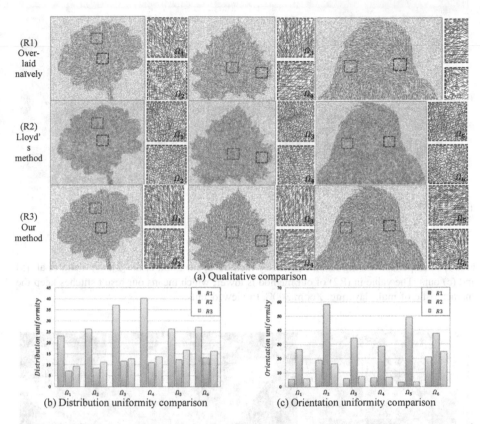

(a) Qualitative comparison

(b) Distribution uniformity comparison

(c) Orientation uniformity comparison

Fig. 8. The superposed stitches obtained by three methods. (a) The qualitative comparison of three methods. (b) and (c) are quantitative comparisons of these methods. Zoom 300% to view details. (Color figure online)

Yang et al. [4], we designed a standard deviation based evaluation indicators called *Multi-layering* to prove that our result stitches maintain the artistic characteristics of multi-layering. To calculate the *Multi-layering*, for each intersecting stitch in a region, we first record the number of intersecting stitches under it. Then the multi-layering is calculated by the standard deviation of all the numbers of stitches in a region. Hence, the multi-layering is prominent when the value is small. Figure 9(R2) shows the comparison of *Multi-layering* among the three methods. Four regions are picked from the canvas randomly, and the numbers of intersecting stitches from $Density_1$ to $Density_4$ are the same as the numbers of stitches from *2 layers* to *6 layers* respectively. We can find that we have lower values of *Multi-layering* in different regions and densities than other two methods, which means our result stitches keep the multi-layering. Figure 10 gives more result stitches of our method.

(R1)

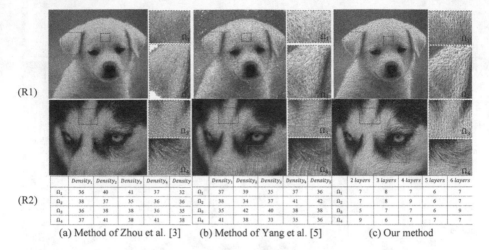

(R2)

	Density₁	Density₂	Density₃	Density₄	Density		Density₁	Density₂	Density₃	Density₄	Density₅		2 layers	3 layers	4 layers	5 layers	6 layers
Ω_1	36	40	41	37	32	Ω_1	37	39	35	37	36	Ω_1	7	8	7	6	7
Ω_2	38	37	35	36	36	Ω_2	38	34	37	41	42	Ω_2	7	8	9	6	7
Ω_3	36	38	38	36	35	Ω_3	35	42	40	38	38	Ω_3	5	7	7	6	9
Ω_4	37	41	38	41	38	Ω_4	41	38	33	35	36	Ω_4	9	6	7	7	7

(a) Method of Zhou et al. [3]   (b) Method of Yang et al. [5]   (c) Our method

Fig. 9. Comparison among result stitches by method of (a) Zhou et al. [3], (b) Yang et al. [5] and (c) ours. The value in (R2) of our method is lower, which means our result stitches keep the characteristic of multi-layering. Zoom 300% to view details.

Fig. 10. More result stitches of our method.

4 Conclusion

In this paper, we present a stylization method to generate a Chinese embroidery stitches formed by line segments. The intersecting stitches selection process is modeled by a Markov chain to sample similar stitches for each region to avoid artifacts in local area. And we use the Gibbs sample to construct the Markov chain. Result stitches are obtained by multi-layer stitches overlay. The top layer stitches are slightly moved according to bottom stitches by a Lloyd's method based approach before overlay operation to make result stitches distribution evenly. Experiments shows result stitches generated by our framework more satisfied with the artistic characteristic of this Chinese embroidery and have a richer visual effect.

Acknowledgement. This work was supported by National High Technology Re-search and Development Program of China (No. 2007AA01Z334), National Natural Science Foundation of China (Nos. 61321491 and 61272219), Program for New Century Excellent Talents in University of China (NCET-04-04605), the China Postdoctoral Science Foundation (Grant No. 2017M62 1700) and Innovation Fund of State Key Lab for Novel Software Technology (Nos. ZZKT2013 A12, ZZKT2016A11 and ZZKT2018A09).

References

1. Kyprianidis, J.E., Collomosse, J., Wang, T., Isenberg, T.: State of the "art": a taxonomy of artistic stylization techniques for images and video. TVCG **19**, 866–885 (2013)
2. Hertzmann, A.: A survey of stroke-based rendering. Comput. Graph. Appl. **23**, 70–81 (2003)
3. Zhou, J., Sun, Z., Yang, K.: A controllable stitch layout strategy for random needle embroidery. J. Zhejiang Univ. Sci. C **15**, 729–743 (2014)
4. Yang, K., Sun, Z.: Paint with stitches: a style definition and image-based rendering method for random-needle embroidery. Multimed. Tools Appl. **77**, 12259–12292 (2017)
5. Yang, K., Sun, Z., Wang, S., Li, B.: Stitch-based image stylization for thread art using sparse modeling. In: Schoeffmann, K., et al. (eds.) MMM 2018. LNCS, vol. 10704, pp. 479–492. Springer, Cham (2018). https://doi.org/10.1007/978-3-319-73603-7_39
6. Zeng, K., Zhao, M., Xiong, C., Zhou, S.C.: From image parsing to painterly rendering. ACM Trans. Graph. **29**, 2:1–2:11 (2009)
7. Kim, S., et al.: Stippling by example. In: NPAR, pp. 41–50 (2009)
8. Adrian, S.: Weighted Voronoi stippling. In: NPAR, pp. 37–43 (2002)
9. Jobard, B., Lefer, W.: Creating evenly-spaced streamlines of arbitrary density. In: Lefer, W., Grave, M. (eds.) Visualization in Scientific Computing, pp. 43–56. Springer, Vienna (1997). https://doi.org/10.1007/978-3-7091-6876-9_5
10. Pham, V.T., Roychoudhury, A.: Coverage-based greybox fuzzing as Markov chain. In: ACM SIGSAC Conference on Computer and Communications Security, pp. 1031–1043. ACM (2016)
11. Desolneux, A., Moisan, L., More, J.: A grouping principle and four applications. IEEE Trans. Pattern Anal. Mach. Intell. **25**, 508–513 (2003)
12. Huang, H., Fu, T., Li, C.: Painterly rendering with content-dependent natural paint strokes. Vis. Comput. **27**(9), 861–871 (2011)

Image Recognition with Deep Learning for Library Book Identification

Kaichen Tang[1], Hongtao Lu[1], and Xiaohua Shi[1,2]($\boxtimes$)

[1] Department of Computer Science and Engineering, Shanghai Jiao Tong University,
Shanghai, China
{tangkc,htlu,xhshi}@sjtu.edu.cn
[2] Library, Shanghai Jiao Tong University, Shanghai, China

Abstract. Book identification system is one of the core parts of a book sorting system. And the efficiency and accuracy of book identification are extremely critical to all libraries. This paper proposes a new image recognition method to identify books in libraries based on barcode decoding together with deep learning optical character recognition (OCR) and describes its application in library book identification system. The identification process relies on recognition of the images or videos of the book cover moving on a conveyor belt. Barcode is printed on or attached to the surface of each book. Deep learning OCR program is applied to improve the accuracy of recognition, especially when the barcode is blurred or faded. Book sorting system design based on this method will also be introduced. Experiment demonstrates that the accuracy of our method is high in real-time test and achieve good accuracy even when the barcode is blurred.

Keywords: Image recognition · Book cover · OCR
Library book identification system · Video analysis · Smart library

1 Introduction

With the development of image recognition technology and popularization of digital equipment, the application of computer vision and deep learning is becoming increasingly extensive. How to return books to its original place quickly, correctly, and efficiently is one of the crucial problems that every library faces. However, the accuracy and efficiency of collecting and sorting books by manpower are not high enough, because librarians have to scan barcodes from the book cover by detectors. Additionally, it is a recognized prediction that the labor cost will keep rising. This paper proposes a method of identifying the book by images or

This work was supported by NSFC (no. 61772330), the Science and Technology Commission of Shanghai Municipality (Grant No. 16Z111040011), China Next Generation Internet IPv6 project (Grant No. NGII20170609), the Arts and Science Cross Special Fund of Shanghai Jiao Tong University under Grant 15JCMY08. And the author would like to thank Sijia Hao for her help.

© Springer Nature Switzerland AG 2018
R. Hong et al. (Eds.): PCM 2018, LNCS 11166, pp. 684–696, 2018.
https://doi.org/10.1007/978-3-030-00764-5_63

videos of the book cover with barcode by real-time computer vision technology supported by deep learning OCR, to improve the identification efficiency and accuracy.

Typically, library's book sorting system can be divided into several stages. Firstly, readers return the books or librarians put the books into the identification system through conveyor belt and the system will identify the book to get the unique key information of each book, in form of collection code or International Standard Book Number (ISBN). Secondly, the unique key will be used as basic information to search the shelf or room where the book originally locates from the database. Thirdly, books will be put back to its allocating position by librarians. The first and second step can be integrated within the same book sorting machine. This paper proposes a computer vision method as a modification for the first step to increase the efficiency and accuracy of identification.

To reduce the dependency on their staff to do the identification, some libraries tended to use Radio Frequency Identification (RFID) and to replace the barcodes as an approach in library book identification and sorting system in the past. At that time, identification of barcode had to be operated by human holding a scanner on his/her hands and put it close to the barcode and there were no other alternatives like computer vision technology. The RFID tag can contain identifying information or may just be a key into a database. It can act as a kind of security device as well, being a substitute for the more traditional electromagnetic security strip [1]. It is estimated that over 30 million library items worldwide now contain RFID tags [2]. Unfortunately, RFID is not an ideal solution and still has an army of flaws and disadvantages. One of the major drawbacks of RFID technology is its tremendous cost to be implemented. RFID Readers and gate sensors used to read the information generally cost around $2,000 to $3,500 each, and the tags cost from $0.40 to $0.75 for each [3]. In addition, since the information of RFID tag can be read from a distance, the signal from one reader can interfere with the signal from another where coverage overlaps. This is called "reader collision" problem [4]. RFID tags can be read from a long distance and their contents can be read by anyone with an appropriately equipped detector, because RFID tags cannot tell the differences from one reader to another [5]. This would be inevitably harmful for the readers' privacy, if static information contained in these tags that can be relatively easily captured by readers with unauthorized identity.

Compared with RFID, for identification purposes, barcode or QR code is economical to be implemented, and is free from reader collision thus owns better privacy. According to Vinod Chachra, CEO of VTLS, a full RFID implementation, including self-checkout stations, readers, software, and tags for books, runs around $1 per book on average. Actually, due to the unaffordable conversion costs and time of a RFID system, a large proportion of libraries are still using barcode as the only or one of the identifications of their books [5]. Besides that, nearly all official published books have a barcode with ISBN on their backs. Virtually, more than a half of libraries never consider implementing RFID [6]. As we can gather the information of books from images or videos of book cover with a simple camera, the cost of implementing RFID system can be saved.

Taking advantage of the rapid growth of deep learning, hundreds of accurate image identification methods have been proposed. Tesseract [7], an optical character recognition engine which is regarded as one of the most accurate open source OCR engines, can be a perfect supplement and alternation to the recognition of book cover [8], especially when the barcode is blurred or faded. It is probable that the speed and accuracy of character and code recognition program based on images or videos will keep improving. The application of videos analysis or computer vision based recognition of books in library book sorting system, as we proposed, can be implemented to reduce the cost of labor and increase the performance of existing systems. In following parts, we will introduce the detailed algorithm we use to identify the book and the design of a system using this method to recognize books.

2   Related Work

2.1   Advanced Book Sorting System

Win et al. [13] use RFID based Intelligent shelving system to provide an efficient mechanism of books management monitoring through wireless communication between the RFID reader and the books. They put forward a prototype of a noble automatic procedure to track tagged items or books. Some techniques present in their work can be adapted to our book identification system design. We attempt to apply computer vision instead of RFID in the procedure of book identification. And we can use a camera to replace the RFID detector in their system.

Farooq et al. [14] design a book placement and book searching method for performance enhancement of existing library systems. The prototype system uses a web camera to capture the cover of the book. The image is processed to extract the title of the book which is passed to the database for getting book reference number. They use images to identify the books by English OCR. But, OCR not only is time-consuming but also cannot perform a very high accuracy recognizing intricate characters set like Chinese. Besides that, in a single book cover, there may be a variety of font types printed on it, which again incurs great difficulty to identify the book by OCR. Consequently, it would be wiser to combine other traditional computer vision method together with OCR. For this purpose, we propose to a book identification system in real-time by a combination of barcode reader and OCR.

2.2   Barcode Recognition Algorithm

ZBar barcode reader [12] is a kind of code reader that can interpret barcode and QR code, and is regarded as one of the most accurate open source barcode readers nowadays. Adopting a cue from modern processing paradigms, ZBar's design uses a layered streaming model. Processing is separated into independent layers with well-defined interfaces, which can be transplanted together or plugged individually into any other systems. After capturing an input image, image scanner makes scan passes over a two-dimensional image to create a linear stream of intensity samples. The input images may come from the video capture module, or any external image source (such as an image file output by a flatbed scanner or digital camera and so on). This module also incorporates the optional inter-frame consistency heuristic applied to a video stream. After this, linear scanner scans a stream of abstract intensity samples to produce a "bar width" stream. Intensity samples could be pixel values from the built-in image scanner, pixel values from an alternate exterior image scanner, or even raw sensor samples from a "decoder-less" wand or laser sensor. The bars are identified and measured by applying some very basic 1D signal processing to the input sample stream. Eventually, the decoder searches a stream of bar width for recognizable patterns and produces a stream of completely decoded symbol data. The current release implements decoding for EAN-13, UPC-A, UPC-E, EAN-8, Code 128 and Code 39 symbologies.

However, we find in real test that one important weakness of ZBar is that it cannot decode skewed barcode in the image. But this kind of image occurs quite commonly in practical cases. As a result, it is difficult to use pure ZBar in practice for book identification with barcode.

2.3   Existing Deep Learning and Computer Vision Application in Libraries

There has been a rapid growth in the field of computer vision and deep learning. A variety of real-time applications have emerged in various smart libraries. Recently, artificial intelligence technology, including speech recognition, OCR, text detection, and face detection, has been used broadly in different library systems.

Lyu et al. [9] develop a multilingual, multimodal digital video content management system called iVIEW for intelligent searching and access of video contents in English and Chinese. In their work, their system uses text modality and face modality to extract information from video and arrange it as a digital library.

Iwata et al. [10] put forward a novel library system using images of books by identifying book cover images and borrower's faces. They use four directional features fields for book identification and obtain information concerning the book by partial area matching. And this system has been applied to a library in the small scale, such as in a reference room of a laboratory.

Yang et al. [11] present a library inventory building and retrieval system based on scene text reading with deep learning. They specifically design a text recognition model using rich supervision to accelerate training and achieve state-of-the-art performance on several benchmark datasets.

Face recognition technology based on deep learning and computer vision has also been used in the entrance guard system in many libraries [19, 20] in recent years.

With these successful examples of application of image recognition and deep learning in libraries, we could conclude that computer vision technology can be and has a great potential to be widely used in libraries.

3   Our Approach

There are some existing libraries and packages for the detection and decoding of barcode, like ZBar and ZXing. However, they are not robust enough to handle low-resolution or skewed images. After being used for several years, the barcode on the book cover would easily be faded or blurred. To improve the accuracy and success rate of detection, we propose to combine barcode localization and detection together with deep learning OCR to achieve a real-time recognition of books with high accuracy. Our method can be divided into roughly three steps (localization of the barcode, affine transformation, and recognition) as follows. The processed results of the first and second step are shown in Fig. 1. And Fig. 2 describes the procedure of the identification process we proposed.

Fig. 1. The procedure to recognize the barcode on a book cover. First figure: the original input image. Second figure: the binary image after filtering. Third figure: localization of the barcode. Forth figure: magnified barcode depending on the localization rectangle.

3.1   Localization of the Barcode

For a given image of the book cover in the RGB color model, we convert it into gray style firstly. After the conversion, we compute the Scharr gradient magnitude representation of the image to extract the edges with large gradient, in another word, greater contrast to their background, from the image. Then, we smooth the gradient magnitude matrix by a SobelFeldman operator kernel in the shape of 3×3:

$$\begin{bmatrix} +3 & 0 & -3 \\ +10 & 0 & -10 \\ +3 & 0 & -3 \end{bmatrix} \tag{1}$$

and its transpose matrix to calculate the gradient in x-direction and y-direction. Then, at each point in the image, the resulting gradient approximations can be combined to give the gradient magnitude, using:

$$G = \sqrt{G_x{}^2 + G_y{}^2}. \tag{2}$$

After this, we apply fixed-level thresholding to the gradient matrix to remove noise and extract the edges in the original image. Since barcode consists of several isolated bars which contain a lot of edges with larger gradient, it will not be filtered by this procedure. However, since these bars were unconnected but close to each other, we could dilate the remaining edges of the image and then find the connected pixels with the largest area to locate the barcode. Finally, we just need to use a rectangle with the least area to surround these pixels, since barcodes are always in regular shape. By this step, we can find the location and expansion of barcode as well as the angle it skews in the original image.

3.2   Affine Transformation

After getting the position of the barcode and the rectangle frame surrounding it, we could rotate the image so that the bar would become vertical, thus can be recognized by existing barcode reader through counting the pixel. Compared with other approaches to get the deskewed image, like Hough transformation, our method is more robust. The purpose of Hough transformation is to find imperfect instances of objects within a certain class of shapes by a voting procedure, like lines in the image. Hough transformation is likely to be influenced by the complex patterns and lines which are quite common to be found on the book cover. As a result, Hough transformation is deficient to deskew the image with complex pattern like a book cover. Furthermore, having surround the barcode by a rectangle in the first step, we could cut out that rectangle from the original image with its vicinity, where the code of barcode in the form of readable number locates, as the input image for following steps. Smaller image will reduce decoding overhead of barcode reader, while comparatively larger characters lead to better OCR accuracy.

3.3   Recognition

After getting the image of the barcode and its vicinity, again, we convert it into gray style and smooth it to remove the noise cause by the sensor of the camera with a similar technique in the first step. With the specified location, barcode can be identified with high accuracy. This step can be accomplished by existing libraries like Zbar.

However, sometimes, when the barcode is blurred or faded, it will be hard to identify the book by barcode decoder. Vulnerability is often considered as one of the most outstanding deficiency of barcode compared with RFID. As a result, we need some supplementary methods when this happens. However, a blurred barcode typically will not affect the localization step as we introduced, if the blur merely fills the gap between bars or some bars became thinner as they faded. Tesseract is an optical character recognition engine and it is reckoned as one of the most accurate open source OCR engines. Most widely used linear barcodes also print identification readable numbers below the pattern of bars. These numbers were added originally as a supplement to be read by human when barcode is not recognizable. Therefore, Tesseract or other OCR method will be a wonderful supplement for the identification of the book to replace human labors to recognize these numbers when it is necessary. The architecture of Tesseract will be introduced in the subsequent part of this section.

In our work, we use Tesseract packet when ZBar barcode reader is not capable to decode the barcode. This procedure is described in the flowchart diagram in Fig. 2.

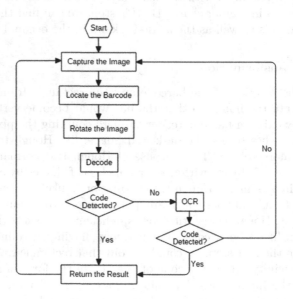

Fig. 2. The flowchart diagram of the book recognition procedure.

3.4   Architecture of Tesseract

Tesseract[1] is an open-source OCR engine with trained parameter sets available online. We adjust the parameter of the enhancer to increase the performance of

---

[1] https://github.com/tesseract-ocr/tesseract.

OCR. The function of Tesseract OCR Engine can be divided into 2 parts: line and word finding, and word recognition.

First and foremost, Tesseract takes 4 steps to locate the words in line and separate each distinct word from its background: line finding, baseline fitting, fixed pitch detection and chopping, and proportional words finding. In line finding step, algorithm is developed so that a skewed page can be recognized without having to deskew, thus saving loss of image quality. The key parts of the process are blob filtering and line construction. Blob filtering enables the use of a running average to measure the gradual vertical shift across the page. It is updated for each blob using:

$$Shift_{n+1} = \alpha Shift_n + (1 - \alpha)Newshift, \tag{3}$$

where $0.5 < \alpha < 0.7$ and increasing to the larger end with an increasing number of rows on the page [15]. Once the text lines have been observed, the baselines are fitted more precisely using a quadratic spline. This enables Tesseract to handle pages with curved baselines. Baselines are fitted by partitioning the blobs into groups with a reasonably continuous displacement for the original straight baseline. In the third step, Tesseract tests the text lines to determine whether they are fixed pitch. Finally, Tesseract finds proportional word by measuring gaps in a restricted vertical range between the baseline and mean line.

In Tesseract's newest 4.0 version, it adds a new OCR engine based on Long Short-Term Memory (LSTM) neural networks [16] to do the word recognition. LSTM, proposed by Hochreiter et al. in 1997, has many derivations as well as a variety of applications. Recently, LSTM has served in the field of OCR and yielded state-of-the-art performance [17]. Tesseract 4.00 includes a novel neural network subsystem configured as a text line recognizer. It has its roots in OCRopus' LSTM implementation. OCRopus is developed by Breuel, the author of [18]. 1D bidirectional LSTM architecture was used for recognition.

4   System Design

When it comes to the design identification system, the fact is barcode decoder and OCR can work independently after the position of the barcode has been detected. So it is worth reiterating that we have taken 2 kinds of design into consideration: firstly, barcode decoder and OCR can work simultaneously to get the result and an arbitration will be called when they have disputes; secondly, the OCR program will be called only when the barcode decoder failed to decode the image. We choose the second strategy as it is outlined in Fig. 2, because the OCR is time consuming, and the barcode decoder can get the right result successfully in most cases, even when the barcode is partly faded. In short, as we can depend on barcode to finish the recognition for the most time, call OCR will not be necessary during these cases because it will increase the overhead of our system and waste a lot of time. In addition, if we choose the first design, we will have to come up with an arbitration policy when the two modules have a different result if we use the first strategy. As a result, we choose the second design for the recognition process.

In our real test, we can put a camera with 12 million pixels 30 cm above the conveyor belt to capture the real time video of the book cover. Absolutely, the closer the camera is, less requirement we have on its resolution. As the book is moving with the belt at the speed of approximately 4 cm/s, the camera will sample a frame from the video every 0.2 s and send it as an input image to the recognition system described in Fig. 2.

We design 2 strategies for 2 distinct usage conditions for the consideration of either speed or accuracy. In the first strategy, as long as recognition system has given the same result for 3 pictures, the system will take this as the final result. And the belt will accelerate to the speed of 6 cm/s, until the recognition system realizes that the old book has been transported out of the screen. Otherwise, if the system still cannot make a decision after receiving 5 results, the system will throw an exception to signal the belt to stop and take another picture for the ultimate decision. This policy guarantees the accuracy. However, in the second policy, we will check the recognition result in the database of the library as soon as we get the returned value. If there is such a book, we will take this as the final result. This strategy makes sure the speed. The two strategies can be used in different circumstances: the first mode can be used when the library is not so busy; and the second mode can be used as the librarians are check the books in list, which means a lot of work has to be done in a short time.

5   Evaluation

5.1   Dataset

As outlined in Fig. 3, we collected more than 500 books from the main library of Shanghai Jiao Tong University (SJTU). These photos display the cover of more than 100 randomly picked books with backgrounds in different colours, each of which has about 5 different pictures captured from variety angles. On each book cover, there is a sticker with the barcode in the standard of CODE128.

5.2   Test on Images

In the experiment, we divided the dataset into 3 parts: vertical, skewed and blurred. The first part consists of images where the book cover is vertical as well as inverted, which can be recognized by existing barcode readers. The second part comprises of images where the book cover has larger incline (more than 30°). And in the third part, the barcode on the book cover is blurred, which makes it impossible to recognize the book according to the barcode. We compared 4 different kinds of method in accordance with their accuracy on our dataset, and the accuracy of these methods can be found in Table 1.

ZBar. Use only ZBar to decode the barcode. Since ZBar cannot handle the skewed image, although it performs well on the vertical images, it produces the correct result for less than one third of the skewed images.

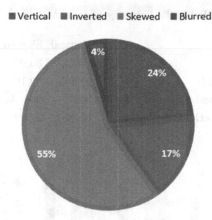

Fig. 3. The distribution of the dataset: vertical (126), inverted (89), skewed (288), blurred (21). To be noticed, in the following part, we will do not distinguish the vertical section from the inverted section, since they both can be decoded by existing libraries like ZBar.

ZBar + Hough. Use ZBar after Hough transformation, a widely-used approach to offset the incline. We use Hough transformation as a base line method to deskew the image. Specifically, we try to use the Hough transformation to identify the edge of the book cover as straight lines in the image to find the inclination of the book, and then use affine transformation to get the deskewed image. At last, ZBar is used to decode the barcode. As we can see from the Table 1, involving Hough transformation do increase the accuracy by about 20% of the skewed image. However, this method's accuracy on the vertical images is not as good as merely ZBar, because the identification of the book's edge can be interrupted easily by complex lines and pattern on the book cover.

Proposed Method Without OCR. Use our method as described in section III, but without OCR even when the barcode is blurred. As we can see from Table 1, when the image is vertical or skewed, the accuracy of this method is high. Nevertheless, similar to above two methods, when the barcode is blurred, this method cannot get a right result, which accordingly emphasizes the importance of supplementary identification method during this condition.

Proposed Method. Use our proposed method. Different the third method, this time, we use OCR, when the barcode reader cannot get any output. The experimental result proves that OCR can identify the book with an accuracy of more than 70% when the barcode is blurred, which makes our proposed method the best among these 4 methods.

All in all, the over-all accuracy of these 4 methods are 58.0%, 66.4%, 94.3%, 97.5% in turn. The result is shown in the form of bar graph in Figs. 4 and 5.

Table 1. Accuracy of four methods

Method	Vertical	Skewed	Blurred
ZBar	100%	30.9%	0%
ZBar + Hough	92.1%	52.1%	0%
Proposed method without OCR	100%	96.9%	0%
Proposed method	100%	97.6%	71.4%

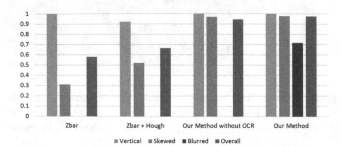

Fig. 4. Bar graph of the accuracy of the four methods on four types of images.

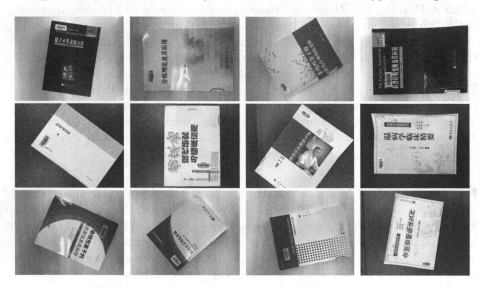

Fig. 5. Typical pictures used in our test.

5.3   Test on Videos

In our experiment, we collect dozens of videos of books moving on a conveyor belt from the main library of SJTU. By extracting frames from videos, we can use our proposed method to perform the detection. Our experiment shows that with a little pause (less than 0.5 s) of the book on the conveyor in the video, the detection can be finished successfully, which indicates that our method can be employed in real time as it is described in the System Design section.

6   Conclusion

We introduce an identification method for book images or videos for library book sorting system. The method combines traditional barcode identification algorithm with our modification to locate and deskew the image. And we involve deep learning OCR to enhance the accuracy when the barcode is blurred or partly faded. The approach we proposed is robust with high accuracy and good performance, even though input pictures are not in high resolution and the book covers are not always vertical. In our future work, we will try more advanced deep learning solution on this problem to improve the performance of our method and try to use it in smart libraries.

References

1. Butters, A.: Radio frequency identification: an introduction for library professionals. Aust. Public Libr. Inf. Serv. **19**(4), 164–174 (2006)
2. Sing, J., Brar, N., Fong, C.: The state of RFID applications in libraries. Inf. Tech. Libr. **25**(1), 4–32 (2006)
3. Shahid, S.-M.: Use of RFID technology in libraries: a new approach to circulation, tracking, inventorying, and security of library materials. Libr. Philos. Pract. **8**(1), 1–9 (2005)
4. Leong, K.S., Ng, M.L., Cole, P.H.: The reader collision problem in RFID systems. In: IEEE International Symposium on Microwave, Antenna, Propagation and EMC Technologies for Wireless Communications, vol. 1, pp. 658–661. IEEE (2005)
5. Yu, D.: Implementation of RFID Technology in Library Systems Case Study: Turku City Library (unpublished)
6. Liu, S., Du, Y., Zhang, L.: The analysis of the implementation of RFID technology in library. J. Acad. Libr. **1**(9), 23–44 (2011)
7. Smith, R.: An overview of the Tesseract OCR engine. In: International Conference on Document Analysis and Recognition, pp. 629–633. IEEE Computer Society (2007)
8. Yang, L., Shen, X.: Book Cover Recognition (unpublished)
9. Lyu, M.R., Yau, E., Sze, S.: A multilingual, multimodal digital video library system. In: ACM/IEEE-CS Joint Conference on Digital Libraries, pp. 145–153. ACM (2002)
10. Iwata, K., Yamamoto, K., Yasuda, M., Kato, K., Ishida, M., Murata, K.: Book cover identification by using four directional features field for a small-scale library system. In: Document Analysis and Recognition on Proceedings, pp. 582–586. IEEE, Washington, D.C. (2001)
11. Yang, X., He, D., Huang, W., Ororbia, A., Zhou, Z., Kifer, D.: Smart library: identifying books on library shelves using supervised deep learning for scene text reading. In: ACM/IEEE Joint Conference on Digital Libraries, pp. 1–4. IEEE Computer Society (2017)
12. ZBar Bar Code Reader. http://zbar.sourceforge.net. Accessed 15 July 2011

13. Win, T.-M., Salami, M.-J.-E., Martono, W.: RFID-Based Intelligent Books Shelving System, 1st edn. IEEE, Washington, D.C. (2007)
14. Farooq, U., Amar, M., Hasan, K.M., Asad, M.U., Iqbal, A.: Automatic book placement and searching technique for performance enhancement of library management system. Int. J. Comput. Theory Eng. **2**(4), 574–580 (2010)
15. Smith, R.: A simple and efficient skew detection algorithm via text row accumulation. In: International Conference on Document Analysis and Recognition, pp. 1145–1148. DBLP (1995)
16. Schmidhuber, J., Hochreiter, S.: Long short-term memory. Neural Comput. **9**(8), 1735–1780 (1997)
17. Ul-Hasan, A., Breuel, T. M.: Can we build language-independent OCR using LSTM networks? In: International Workshop on Multilingual, p. 9. ACM (2013)
18. Breuel, T. M., Ul-Hasan, A., Al-Azawi, M. A., Shafait, F.: High-performance OCR for printed English and Fraktur using LSTM networks. In: International Conference on Document Analysis and Recognition, pp. 683–687. IEEE (2013)
19. Wang, F., Cheng, W., Yu, B.: An entrance guard system based on face recognition. J. Changshu Inst. Technol., 4–14 (2016)
20. Li, S., Pang, N., Tang, R.: Face recognition system based on OMAP-L138and OV2460. Microcontrollers & Embedded Systems (2017)

Learning Affective Features Based on VIP for Video Affective Content Analysis

Yingying Zhu[1] , Min Tong[1], Tinglin Huang[1], Zhenkun Wen[1(⊠)], and Qi Tian[2]

[1] College of Computer Science and Software Engineering, Shenzhen University,
Shenzhen 518060, Guangdong, People's Republic of China
zhuyy@szu.edu.cn, {tongmin2016,huangtinglin}@email.szu.edu.cn,
tmtyy2005@163.com
[2] Department of Computer Science, The University of Texas at San Antonio,
San Antonio, TX 78249, USA
qitian@cs.utsa.edu

Abstract. Video affective computing aims to recognize, interpret, process, and simulate human affective of videos from visual, textual, and auditory sources. An intrinsic challenge is how to extract effective representations to analyze affection. In view of this problem, we propose a new video affective content analysis framework. In this paper, we observe the fact that only a few actors play an important role in video, leading the trend of video emotional developments. We provide a novel solution to distinguish the important one and call it the very important person (VIP). Meanwhile, we design a novel keyframes selection strategy to select the keyframes including the VIPs. Furthermore, scale invariant feature transform (SIFT) features corresponding to a set of patches are first extracted from each VIP keyframe, which forms a SIFT feature matrix. Next, the feature matrix is fed to a convolutional neural network (CNN) to learn discriminative representations, which make CNN and SIFT complement each other. Experimental results on two public audio-visual emotional datasets, including the classical LIRIS-ACCEDE and the PMSZU dataset we built, demonstrate the promising performance of the proposed method and achieve better performance than other compared methods.

Keywords: Affective analysis · Video content analysis
Very important person information · Convolutional neural network
Scale-invariant feature transform

1 Introduction

The last decades witnessed a sharp rise in the number of the video. Some scholars study the video or image content from the perspective of aesthetics or events [6,7,12,13]. However, most videos convey the amount of affective information, which brings us rich emotional experiences. Analyzing the emotion in video content not only improves the accuracy of video indexing and summarization

© Springer Nature Switzerland AG 2018
R. Hong et al. (Eds.): PCM 2018, LNCS 11166, pp. 697–707, 2018.
https://doi.org/10.1007/978-3-030-00764-5_64

but also improves the accuracy of personalized multimedia content. Therefore, video affective analysis has gained more and more attention in the research community.

Due to the facial expression is the most effective way to express emotions, some researchers regard the face as an important element in video emotion recognition [17,18,23]. However, such works are mostly ignored the fact that only a few faces in the video are important to analyze the emotion. Some faces are insignificant and play supporting roles in the video, which make less contribution to expressing the emotions. In contrast, some faces appear frequently in videos. They mostly belong to leading roles who lead the trend of emotional development of the videos. While watching the video, the mood of audiences fluctuates with the joys and sorrows of these leading roles, and the emotions of audiences are closely associated with the fates of these leading roles. Hence, a key problem is how to distinguish those leading roles.

One work [25] provides a solution, they believe that the protagonist is the leading role. Based on the prior knowledge of actor in the video, such as actor's name, they search the protagonist's face from the Internet. Nevertheless, there are some drawbacks to this solution. Firstly, it is true that we cannot get the prior knowledge of protagonist for each video. Secondly, the method of manually searching auxiliary information and face takes much time and effort. It is inconvenient to deal with the large-scale data. Thirdly, some studies show that a video clip with minimum 250 ms length can express an identifying emotion [16]. The protagonist of the entire video is not necessarily the protagonist of a video clip.

To address the above issues, in this paper, we introduce a new method. For all lengths of video, we can automatically distinguish leading roles without prior knowledge. According to the image saliency, the region of interest usually occupies a large area in the image. Meanwhile, in film-making, the director utilizes a close-up to express important content, such as characters' emotions, so that the close-up object is filled with the whole frame, occupying the largest area in the whole frame. Therefore, we consider that the face with the largest area is most important and conveys most affective information. We regard the face with the largest area as the primary face. In these different primary faces, the primary face with maximum occurrence is called as the very important person(VIP)'s face here. The VIP is a leading role and leads the change of emotion in the video.

Based on VIP, in this paper, we employ a novel audio-visual framework for video affective content analysis. We design a keyframe selection strategy to extract keyframe containing VIP. With the merit of invariance to scale, rotation and illumination changes, scale invariant feature transform (SIFT) [8] features have demonstrated promising performance in feature representation. The recent explosive research on deep neural network has contributed to the great success of the learning-based features in the vision field. With deep architecture, the convolutional neural network (CNN) has exhibited state-of-the-art performance in various tasks. The activation from the middle or top layers of deep CNN model are taken out as high-level and semantic-aware features to represent images. Witnessing the great success of low-level SIFT feature and its complementary nature to the semantic-aware CNN feature, many approaches [4, 9, 24] fuse these two features in a multi-level and complementary way. In order to bridge the semantic gap between low-level features and high-level human perception of emotion, we propose to learn a middle-level representation using CNN. The middle-level representations are learned from low-level SIFT features extracted from keyframe. The learned middle-level representations(called as visual CNN features in this paper), incorporated with hand-crafted visual and audio features, coupled with a support vector machine (SVM) or a support vector regression (SVR), are then used for emotional analysis.

The rest of this paper is organized as follows. Section 2 briefly reviews the previous work for emotion analysis. Section 3 introduces the systemic framework and the detail of our method based on VIP information. Section 4 presents experimental results. Section 5 concludes this paper with remarks and future work.

2   Related Work

So far, the relationship between the machine and human emotion is still not well understood, many researchers have made numerous efforts to gain the solutions that use the machine to recognize the human emotion.

In the earlier time, inspired by cinematic principles and concepts, most of the previous works [14, 20, 22] focus on using hand-crafted features to represent the video's affective content, such as zero crossing rate (ZCR), Mel-frequency Cepstral Coefficients (MFCC), tempo, lighting, color, etc. Zhang et al. [22] extract five audio-visual features including the short switch, ZCR, tempo, motion intensity and strength for arousal emotion recognition. Meanwhile, the authors employ five audio-visual features ranking from saturation, lighting, rhythm regularity, pitch, color energy for valence emotion recognition.

The recently emerged deep learning techniques have been quite effective in learning features representation. Many works [5, 10, 23] start using the deep learning methods, such as CNN, long short-term memory (LSTM), for video content affective analysis. For instance, Ding et al. [5] present an LSTM for audio emotion recognition, and a CNN model for extracting visual representation. Then, they integrate audio and visual modality at score level to analysis the affection.

From the above-mentioned works, we can get some conclusions. Firstly, low-level hand-crafted features are widely-used to analyze the emotion. These features build upon knowledge gained in decades of audio-visual emotion recognition research and have shown to be robust for many domains. Secondly, automatic feature learning techniques present a promising performance in high-level feature extraction. Considering these, in our paper, we use both hand-crafted feature and the feature extracted by CNN.

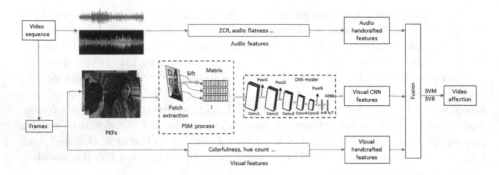

Fig. 1. The proposed framework

3 Proposed Method

3.1 System Framework

The overview of the proposed method framework is described in Fig. 1. The hybrid representation of video emotion content is composed of the audio hand-crafted features, visual CNN features and visual hand-crafted features. In the extraction of visual CNN features, we first design a new VIP-based strategy to extract keyframe, and the extracted keyframe is called PKF. Next, SIFT features extracted from patches in PKF are used to form features matrix, and we name this process as PSM process. After doing the PSM process, the PKFs are represented as matrices and fed into the CNN model to produce the middle-level representation, called as visual CNN features. Finally, after fusing three kinds of features, we use the hybrid representation for affective analysis.

3.2 Visual CNN Features

First of all, we describe how to detect the VIP. As shown in the Fig. 2, the face A occupies the largest area in the frame, so it is the primary face in this frame. Figure 3 shows the formation of the VIP's face in a sample video clip. In detail, for each frame, we run a face detection [1] to detect the primary face. Then, we transform and crop these faces. By doing face recognition and face comparison, all primary faces are binned according to the occurrence times to construct a histogram. The primary face with maximum occurrence is the VIP's face in this clip.

Next, we introduce the PKF selection strategy. In order to classify a video with a particular emotion, we need to process hundreds of frames. Considering the influence of VIP information, here, we design a selection strategy to define a set of keyframes per video, named PKFs.

We hypothesize that a video clip V contains n frames, then the video clip can be represented as $V = \{F_1, F_2, F_3, \ldots, F_n\}$, where F_n is the n-th frame in

 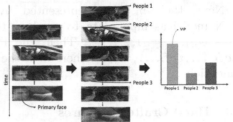

Fig. 2. The primary face example

Fig. 3. Schematic representation showing the formation of VIP face for a sample video

the video V. k PKFs, $KF = \{KF_1', KF_2', \ldots, KF_k'\}$, are extracted based on the VIP information from each video segment V by the selection strategy.

We define the RGB histogram of the i-th frame as $H(F_i)$, where $i = 1 \ldots n$. The mean RGB histogram of the entire clip is:

$$H(\overline{F}) = \frac{1}{n} \sum_{i=1}^{n} H(F_i), \tag{1}$$

Then, we calculate the Manhattan distance D_i between i-th frame and the mean RGB histogram of the whole clip. After that, we get a distance list D, $D = \{D_1, D_2, \ldots, D_n\}$.

Inspired by [21], based on the distance list D, we do a k-means clustering. Then, we rank all the frames of one class in accordance with the distance. Therefore, for each class i in k class, we can get a candidate keyframe (CKF) set $KF_i = \{KF_i^1, KF_i^2, \ldots, KF_i^m\}$. The KF_i^1 is the frame with the smallest distance to the clustering center i. So, a n-frames video clip is represented as

$$V = \sum_{i=1}^{k} \sum_{j=1}^{m_i} KF_i^j, and \sum_{i=1}^{k} m_i = n, \tag{2}$$

where k indicates the number of classes and m_i presents the number of frames in i-th class.

For a CKFs list of each clustering center i, $KF_i = \{KF_i^1, KF_i^2, \ldots, KF_i^m\}$, we start from the first frame KF_i^1 to the last frame KF_i^m to search the VIP. The first frame with the VIP in CKFs list is regarded as the PKF, KF_i'. By repeating the above steps, we can obtain k PKFs, just as $KF = \{KF_1', KF_2', \ldots, KF_k'\}$.

After introducing the PKF selection strategy, we show how to do the PSM process. For each PKF, following the work of [11,24], we extract a 24×24 patch every four pixels at five scales. So, we can obtain m patches per frame. Different from [11], we only use SIFT descriptors. We extract a set of SIFT feature vector to represent each patch, in which each SIFT feature vector is a 128-dimensional vector. Then, we put the m SIFT feature vectors together to form a feature matrix for each PKF. After applying the PSM process, the PKF is represented by a matrix.

Next, the i-th PKF represented by feature matrix are fed into the CNN model as input data to produce the 4096 dimension visual CNN features vcf_i, $i = 1, 2, \ldots, k$. Then, the PKFs list are represented as $vcf = \{vcf_1, \ldots, vcf_i, \ldots, vcf_k\}$, where k is the number of PKFs and vcf_i is 4096-dimension.

3.3 Hand-Crafted Features

Following [2, 25], for arousal space, we extract twelve features, such as harmonization energy, audio flatness envelops, slope of the power spectrum, colorfulness, lighting, normalized number of white frames. For affective analysis in the valence space, we extract seventeen audio and video features, including ZCR, colorfulness, hue count, audio asymmetry envelope, audio flatness, depth of field and so on. The extracted hand-crafted features for a video clip are corresponding defined as HF.

3.4 Fusion and Learning

Fusion at the features-level [15, 19] is the most common and straightforward way that concatenated all extracted features into a single high-dimensional feature vector. We concatenate three of the features, i.e., visual CNN features, audio hand-crafted features, visual hand-crafted features, at features-level as new features. All features are normalized by using the standard score. Then, the representation of a PKF is presented as

$$f(KF_i') = [vcf_i, HF], \tag{3}$$

Thus, a video clip can be represented as $f(V) = \{f(KF_1'), f(KF_2'), \ldots, f(KF_k')\}$, where $f(KF_i')$ is i-th PKF features, and k is the number of PKFs.

Finally, a SVM or a SVR is used to map from the features to the emotion.

4 Experiments and Results

4.1 Experiment Setting

To evaluate the performance of our audio-visual system based on VIP information, two datasets, i.e., a publicly available dataset LIRIS-ACCEDE [2] and a new annotated emotional dataset PMSZU constructed by us, are employed here. Both of the two datasets consist of different kinds of video clips. The LIRIS-ACCEDE is composed of 9800 video clips with two emotion dimensions: arousal and valence. While the PMSZU consists of 386 video clips with the arousal dimension and the genre of the movies, i.e., horror and comedy. The video clips in PMSZU dataset are extracted from 8 popular movies. The name and corresponding clips number of the movies in the dataset is shown in the Fig. 4. The video clip lasts from 16 to 280 s. The total time of all 8 movies is 12 h, 16 min and 26 s. The annotations of the video clips in this dataset are collected from

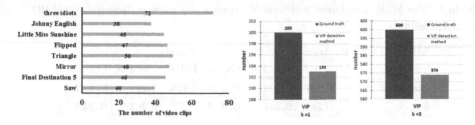

Fig. 4. The names and the corresponding number of video clips for each movie in PMSZU dataset

Fig. 5. The results of detecting VIP with ground truth and VIP detection method

twenty participants, i.e., nine females and eleven males, ranging in age from 20 to 29. For each video clip, the annotators are asked to assess the arousal values ranging from 0 to 5 and select the genre of movies.

In the experiment section, for a fair comparison, we use a subset of LIRIS-ACCEDE dataset same as in the work [25]. We implement the CNN model with the Caffe platform on a Tesla K80 GPU. The SVR and SVM we employed are from the LIBSVM toolkit [3]. To perform the assessment of the proposed method, the MSE and accuracy are adopted as the criterion in our experiments. The statistical experiment is repeated five times, and we only report the average result.

4.2 Experiment 1: Evaluate of the VIP Detection Method

The VIP information is an important factor in our method to analyze emotion. In this section, we demonstrate the performance of the VIP detection method. We manually annotate the VIP of 200 video clips as the ground truth. Figure 5 shows the results of detecting VIP with ground truth and our VIP detection method, where $k = 1$ and $k = 3$ mean the number of keyframes is 1 and 3, respectively. Compared with the ground truth, 96.5% and about 95.67% VIP's faces are distinguished correctly by using the VIP detection method, respectively. It is obvious that our VIP detection method is performed well in automatically distinguish VIP without prior knowledge.

4.3 Experiment 2: Evaluate of the VIP Effect

To present the advantages of the VIP features extracted in our method, we directly compare our performance with the reported results of previous work [25] on the LIRIS-ACCEDE subset. Because the previous work has been explored the importance of human face and further demonstrated that the protagonists' faces are more important than other faces in the video affective analysis. We only take their experimental results about protagonist information for comparison. The CNN model proposed in this experiment is the original AlexNet. Table 1 gives the performance comparisons between our method and the comparison

Table 1. The MSE of different methods in arousal and valence dimension on LIRIS-ACCEDE subset.

LIRIS-ACCEDE	Arousal			Valence		
	PWF	$PWPF$	$Ours$	PWF	$PWPF$	$Ours$
$k=1$	0.307	0.298	**0.289**	0.298	0.291	**0.282**
$k=3$	0.281	0.267	**0.253**	0.272	0.263	**0.248**

methods, PWF [25] and PWPF [25], in the arousal and valence dimension for emotion recognition.

As shown in the Table 1, in arousal dimension, when $k=1$ (it means the number of keyframes is 1), our method outperforms previous methods PWF and PWPF by more than about 0.009. As well as $k=3$, our method reduces the MSE to 0.253. For PWF, PWPF and our method, the video clips represented by three PKFs achieved better performance, i.e., reducing its MSE from 0.307 to 0.281, 0.298 to 0.267, and 0.289 to 0.253, respectively. Similarly, in the valence dimension, our method reduces the MSE at least 0.015, which achieves lowest MSE 0.248 in these three methods.

From Table 1, we can see that our method outperforms other methods in both of arousal and valence dimensions. It shows that after applying the PKFs selection strategy and the PSM process, our VIP features learned from CNN model are more discriminative than the protagonist features learned from other works. Additionally, the promising performance of our method clearly indicates that the VIP concept we described is reasonable.

4.4 Experiment 3: Evaluate on Different Fine-Tune Schemes

In this section, we fine-tune the CNN model with different labeled emotional data. Then, the performance of our proposed method is evaluated on PMSZU dataset.

We fine-tune the weights of AlexNet with different schemes. For the AlexNet, we replace its final fully connected layers with new FC layers, which correspond to the emotion labels on the different target dataset. We first use the PMSZU dataset for fine-tuning (scheme1). Due to the limit of PMSZU size, next, we consider employing both of PMSZU dataset and LIRIS-ACCEDE dataset to fine-tune the network too (scheme2). The learning rate is set to 0.01 to speed the training process. When the network is basically reached a stable state, we stop training this network and start using this fine-tuned model as a features extractor.

Table 2 shows the MSE results based on the original CNN model and two fine-tuned CNN models on the PMSZU dataset. Compared with directly using the pre-trained AlexNet, we can find two points. On the one hand, the method, which fine-tuned the pre-trained model, achieves lower MSE. On the other hand, the method fine-tuned with a larger scale dataset obtains the better performance. Our method with scheme2 achieves the best performance 0.215.

Table 2. The MSE of different fine-tuned schemes in arousal dimension on PMSZU

PMSZU	$AlexNet$	$scheme1$	$scheme2$
k = 1	0.241	0.228	**0.226**
k = 3	0.227	0.218	**0.215**

Table 3. The classification accuracy of different methods in the genre of movie on PMSZU

PMSZU	MoB_1	MoB_2	MoB_3	$Ours$
k = 1	58.7%	59.8%	59.6%	**60.8%**
k = 3	59.2%	60.1%	60.3%	**62.1%**

The experimental results clearly indicate that fine-tuned CNN model is powerful in feature learning, and it produces more discriminative features than the pre-trained CNN model. Deep models require a large amount of training data which motivates us to transfer pre-trained models to other domains. And fine-tuning allows deep models pre-trained on other domains to learn significative feature representations for emotion recognition. Accordingly, we are spurred to fine-tune the models. And above experiments show the validity of our fine-tuning strategy.

4.5 Experiment 4: Results on PMSZU Dataset

To evaluate the performance of the emotion classification accuracy of comedy or horror genre on the PMSZU dataset, we apply the CNN model with scheme2 as feature extractor, and compare our proposed audio-visual method with three methods. The results are shown in the Table 3. The method MoB_1 is the method of Baveye [2]. The method MoB_2 is an improvement of the method MoB_1, and it uses fine-tuned scheme1 to fine-tune the model. The method MoB_3 is an improvement of the method MoB_1 too, and it utilizes fine-tuned scheme2 to fine-tune the model. As can be seen, compared with the other three methods, our proposed method with scheme2 achieves the best performance 62.1%.

5 Conclusions and Future Work

In this paper, we present an audio-visual framework by combining hand-crafted visual and audio features and visual CNN features to recognize the emotions in videos. There are three advantages to this system, i.e., the detection of VIP, PKF selection strategy, and PSM process. We design a selection strategy to extract keyframes containing VIP based on face detection and selection. The experimental results indicate that the VIP we described is crucial for analyzing video affective content and the keyframes with the VIPs' face are more effective to present emotions in the video content. The proposed PSM process can combine the advantages of SIFT and CNN channels. For which, our framework is very helpful to recognize the emotion in the video.

Multimodal fusion has already become a popular trend in the field of affective computing nowadays. For future work, it is helpful to fuse other modalities to represent emotion, such as gestures, actions, captions, etc. In addition, the temporal information can be considered into the model.

Acknowledgments. This work was funded by: (i) National Natural Science Foundation of China (Grant No. 61602314); (ii) Natural Science Foundation of Guangdong Province of China (Grant No. 2016A030313043); (iii) Fundamental Research Project in the Science and Technology Plan of Shenzhen (Grant No. JCYJ20160331114551175).

References

1. Amos, B., Ludwiczuk, B., Satyanarayanan, M.: Openface: a general-purpose face recognition library with mobile applications. CMU School of Computer Science (2016)
2. Baveye, Y., Dellandrea, E., Chamaret, C., Chen, L.: LIRIS-ACCEDE: a video database for affective content analysis. IEEE Trans. Affect. Comput. **6**(1), 43–55 (2015)
3. Chang, C.C., Lin, C.J.: LIBSVM: a library for support vector machines. ACM Trans. Intell. Syst. Technol. (TIST) **2**(3), 27 (2011)
4. Connie, T., Al-Shabi, M., Cheah, W.P., Goh, M.: Facial expression recognition using a hybrid CNN–SIFT aggregator. In: Phon-Amnuaisuk, S., Ang, S.-P., Lee, S.-Y. (eds.) MIWAI 2017. LNCS (LNAI), vol. 10607, pp. 139–149. Springer, Cham (2017). https://doi.org/10.1007/978-3-319-69456-6_12
5. Ding, W., et al.: Audio and face video emotion recognition in the wild using deep neural networks and small datasets. In: Proceedings of the 18th ACM International Conference on Multimodal Interaction, pp. 506–513. ACM (2016)
6. Hong, R., Zhang, L., Tao, D.: Unified photo enhancement by discovering aesthetic communities from flickr. IEEE Trans. Image Process. **25**(3), 1124–1135 (2016)
7. Hong, R., Zhang, L., Zhang, C., Zimmermann, R.: Flickr circles: aesthetic tendency discovery by multi-view regularized topic modeling. IEEE Trans. Multimed. **18**(8), 1555–1567 (2016)
8. Lowe, D.G.: Distinctive image features from scale-invariant keypoints. Int. J. Comput. Vis. **60**(2), 91–110 (2004)
9. Lv, Y., Zhou, W., Tian, Q., Sun, S., Li, H.: Retrieval oriented deep feature learning with complementary supervision mining. IEEE Trans. Image Process. **27**, 4945–4957 (2018)
10. Noroozi, F., Marjanovic, M., Njegus, A., Escalera, S., Anbarjafari, G.: Audio-visual emotion recognition in video clips. IEEE Trans. Affect. Comput. 1 (2017). https://doi.org/10.1109/taffc.2017.2713783
11. Perronnin, F., Larlus, D.: Fisher vectors meet neural networks: a hybrid classification architecture. In: Proceedings of the IEEE Conference on Computer Vision and Pattern Recognition, pp. 3743–3752 (2015)
12. Sabirin, H., Yao, Q., Nonaka, K., Sankoh, H., Naito, S.: Toward real-time delivery of immersive sports content. IEEE MultiMedia **25**(2), 61–70 (2018). https://doi.org/10.1109/mmul.2018.112142739
13. Shi, X., Shan, Z., Zhao, N.: Learning for an aesthetic model for estimating the traffic state in the traffic video. Neurocomputing **181**, 29–37 (2016)
14. Wagner, J., Lingenfelser, F., Andr, E., Kim, J., Vogt, T.: Exploring fusion methods for multimodal emotion recognition with missing data. IEEE Trans. Affect. Comput. **2**(4), 206–218 (2011)

15. Wang, Y., Guan, L., Venetsanopoulos, A.N.: Kernel cross-modal factor analysis for information fusion with application to bimodal emotion recognition. IEEE Trans. Multimed. **14**(3), 597–607 (2012)

16. Wöllmer, M., Kaiser, M., Eyben, F., Schuller, B., Rigoll, G.: Lstm-modeling of continuous emotions in an audiovisual affect recognition framework. Image Vis. Comput. **31**(2), 153–163 (2013)

17. Yan, J., et al.: Multi-clue fusion for emotion recognition in the wild. In: Proceedings of the 18th ACM International Conference on Multimodal Interaction, pp. 458–463. ACM (2016)

18. Yao, A., Shao, J., Ma, N., Chen, Y.: Capturing au-aware facial features and their latent relations for emotion recognition in the wild. In: Proceedings of the 2015 ACM on International Conference on Multimodal Interaction, pp. 451–458. ACM (2015)

19. Zeng, Z., Pantic, M., Roisman, G.I., Huang, T.S.: A survey of affect recognition methods: audio, visual, and spontaneous expressions. IEEE Trans. Pattern Anal. Mach. Intell. **31**(1), 39–58 (2009)

20. Zeng, Z., Tu, J., Pianfetti, B.M., Huang, T.S.: Audio-visual affective expression recognition through multistream fused HMM. IEEE Trans. Multimed. **10**(4), 570–577 (2008)

21. Zhang, Q., Yu, S.P., Zhou, D.S., Wei, X.P.: An efficient method of key-frame extraction based on a cluster algorithm. J. Hum. Kinet. **39**(1), 5 (2013)

22. Zhang, S., Huang, Q., Jiang, S., Gao, W., Tian, Q.: Affective visualization and retrieval for music video. IEEE Trans. Multimed. **12**(6), 510–522 (2010)

23. Zhang, S., Zhang, S., Huang, T., Gao, W., Tian, Q.: Learning affective features with a hybrid deep model for audio-visual emotion recognition. IEEE Trans. Circuits Syst. Video Technol. 1 (2017). https://doi.org/10.1109/tcsvt.2017.2719043

24. Zhang, T., Zheng, W., Cui, Z., Zong, Y., Yan, J., Yan, K.: A deep neural network-driven feature learning method for multi-view facial expression recognition. IEEE Trans. Multimed. **18**(12), 2528–2536 (2016)

25. Zhu, Y., Jiang, Z., Peng, J., Zhong, S.: Video affective content analysis based on protagonist via convolutional neural network. In: Chen, E., Gong, Y., Tie, Y. (eds.) PCM 2016. LNCS, vol. 9916, pp. 170–180. Springer, Cham (2016). https://doi.org/10.1007/978-3-319-48890-5_17

Research on Multitask Deep Learning Network for Semantic Segmentation and Object Detection

Ting Rui[1], Feng Xiao[2(✉)], Jian Tang[1], Fukai Zhang[2],
Chengsong Yang[1], and Min Liu[3]

[1] School of Field Works, PLA Army Engineering University,
Nanjing 210007, China
1193221332@qq.com
[2] School of Graduate, PLA Army Engineering University,
Nanjing 210007, China
[3] Changsha Mystical Bow Information Technology Co, Ltd.,
Nanjing 210007, China

Abstract. After analyzing methods of object detection under the existing deep learning framework, a multitask learning model (Fully Convolution Object Detection Network, FCDN) is proposed, which can realize complete end to end semantic segmentation and object detection through deep learning, without delimiting the default boxes. First, this paper analysis the reason why the current mainstream object detection network needs the default box delineated in advance; second, an object detection network with no delimited default box needed is proposed. It uses the semantic segmentation to detect all boundaries and key points of object at the pixel level, and then obtain prediction boxes by combining the category information of the semantic segmentation map. Finally, the feasibility of the method is verified on the VOC 2007 datasets, and compared with the performance of current mainstream object detection algorithm. Results show that the semantic segmentation and object detection can be realized at the same time by the new model. Trained by the same training sample, detection precision of FCDN is superior to that of classic detection models.

Keywords: Deep learning · Object detection · Semantic segmentation
Object boundary key points · Default boxes

1 Introduction

There are two main types of object detection models at present. The first class is based on the idea of regional proposal. Its typical representatives are R-CNN [1], SPP-net [2], Fast R-CNN [3], Faster R-CNN [4], and R-FCN [5]. The second type uses the idea of regression. The typical representations of the second class are YOLO [6, 7] and SSD [8]. Both types need to define the default boxes (anchors etc.) in a certain way so as to establish the relationship of default boxes, predict boxes, and real object boxes for training. Thus, the basic reason why the default box needs to be artificially delineated is that the number of detected objects is unpredictable. Assuming that the minimum pixel

© Springer Nature Switzerland AG 2018
R. Hong et al. (Eds.): PCM 2018, LNCS 11166, pp. 708–718, 2018.
https://doi.org/10.1007/978-3-030-00764-5_65

region can be recognized on the image is A_{min}. In theory, the possibility of the number of objects in an image sized $W \times H$ is from 0 to $W \times H/A_{min}$. If the deep learning network is to output the object boxes directly, and one object box corresponds to an object, because the number of objects being detected is unknown, the dimension of the output of deep learning network cannot be determined. In order to quantify the output of the deep learning network, it is necessary to artificially define the default boxes of the model, then classifying each default box by image classification networks [26–28, 30, 31], filtering the default boxes according to the classification results to get the prediction box.

R-CNNs are carried out in two steps. The first step is to filter the default boxes (anchors) by region proposal network (RPN) to generate the candidate boxes which containing the object with high probability; the second step is to further classify the candidate boxes after the first step and do a regression of the exact location of the object. Finally, the final prediction boxes are obtained by removing the overlapped prediction box by Non-Maximum Suppression algorithm. The YOLO and SSD are only removed the first step above. It's still necessary to delimit the default boxes in a certain way in advance, and calculate the confidence of all the default boxes, then take the high confidence default boxes as the final prediction boxes in the end. Its essence is also the bottom-up default boxes exclusion method.

The object detection models based on deep learning such as R-CNNs, YOLO and SSD are called the default boxes exclusion methods. In order to ensure a high recall rate, the more and denser the default box specifications are set, but there are too many areas of repeated detection in the image, which results in too much computational cost. Therefore, the default boxes exclusion methods has the defect of low detection efficiency.

2 Our Method

The principle and basis of the object detection model are that the candidate boxes which containing the object has a greater score. The current object detect networks are expanded on the image classification network. The image classification network can only classify an image and calculate the confidence. The more complete of the object and the less of the other background, the more accurate the classification result of the image classification network output, and the prediction confidence higher.

The semantic segmentation model [10, 14–17, 24, 25, 29] is different, because its training set has the pixel level annotation to the object, so the object information extracted by the semantic segmentation network is more accurate and rich. It realizes the classification of each pixel of the image. In order to classify a pixel, it must combine the pixel points around that pixel. That is, it must be supported by region information. In fact, the semantic segmentation model extracts a lot of region information as the intermediate features in the process of semantic segmentation. Since the semantic segmentation model can judge which kind of object a pixel belong to, it can also judge whether this pixel is boundary key points.

The result of semantic segmentation of each pixel and even the intermediate features contain a lot of information which can distinguish the objects in the image. The method proposed in this paper is to make full use of the result of semantic segmentation and the intermediate features extracted by semantic segmentation network, so as to realize object recognition and location. In this paper, the information which can be used to determine which kind of object pixel belong to is called category information, and the information which can be used to judge whether pixel points are located on the boundary of objects is called boundary information. Object detection task requires category information of objects to achieve recognition of different categories of objects, the boundary information of objects is also needed to distinguish and locate objects. The boundary of the object which is tangent to the real object boxes is regarded as the object boundary key points, as shown in Fig. 1.

Fig. 1. A diagram of object boundary key points.

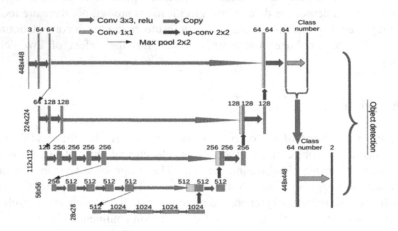

Fig. 2. Framework of full convolution detection network (FCDN).

Object boundary key points are independent of the category of object and can be regarded as the outermost edge of the object. It is a part of the edge of the object which can determine the object box.

The semantic segmentation model such as FCN can only extract the classification information of each pixel, but it cannot judge whether the pixel belong to object boundary, so it cannot identify different object in the same category from the result of semantic segmentation. A new model (Fully Convolution Detection Network, FCDN) is proposed. The model uses full convolution neural network to perform two dense prediction tasks simultaneously, and extracts the category information and boundary information of each pixel. The first task is to implement semantic segmentation, and the second task is to implement the prediction of boundary key points. Based on the VGG and other pre-trained network models, the experiment is carried out. The specific network structure is shown in Fig. 2.

3   Implementation and Training of the FCDN Model

3.1   Feature Extraction Module and Semantic Segmentation Module

Using the method of transfer learning, the full convolutional layers of these image classification models [11–13] which has good classification performance is used as feature extractor. The main function of the feature extractor is to extract the convolutional feature map of the image, which is shared by the following semantic segmentation and boundary key points prediction.

Learn from the ideas of semantic segmentation models such as FCN [10], Seg-Net [14], PSP-Net [15], Deep Lab [16], Refine-Net [17], U-Net [9]. We complete the semantic segmentation task based on feature extraction module, which are used to extract features to realize the dense classification of each pixel. In order to verify the feasibility of simultaneous semantic segmentation and boundary key points predicting, the classical and relatively simple FCN model is used to accomplish the semantic segmentation task in our experiment.

3.2   Boundary Key Points Predicting Module

The boundary key points predicting module shares the output of the semantic segmentation module and the convolution feature map of the feature extractor. The boundary key points are used to locate different objects. As shown in Fig. 3, the intersection points of the ground truth boxes and the semantic segmentation annotations in the dataset are used as the boundary key points. In the object detection model, after the model gets the semantic segmentation map of the detected image, the boundary key points are used to divide the semantic segmentation map, the semantic segmentation map provides the category information, and the boundary key points provide the location information. The model completes the object detection task by combining two kinds of information.

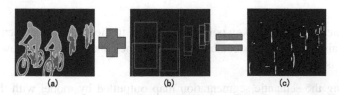

Fig. 3. Getting the tagging of the boundary key points based on VOC dataset. (a) Semantic Segmentation tagging in the dataset; (b) Tagging of the true object box in the dataset; (c) The boundary key points tagging obtained from a, b.

The training objective of the boundary key predicting module is:

$$min\left\{\sum_{(x,y)\in T} L(f_\theta(x),y) + \lambda R(\theta) : \theta \in \Theta\right\} \qquad (1)$$

In the formula, Θ is all the parameter of model, L is the loss function and R is a regular term.

The principle of the boundary key point prediction layer is the same as that of the semantic partition layer, which is a binary classification problem of whether the pixel is a boundary key point. Therefore, the loss function uses the cross entropy [18] and the loss function is usually decomposed into the sum of the loss of each pixel of the semantic segmentation map:

$$L(y,\hat{y}) = \frac{-1}{n}\left(\sum l_{y\hat{y}}\right) \qquad (2)$$

3.3   Object Detection Module

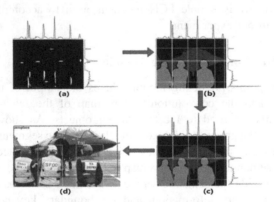

Fig. 4. Object detection by mapping boundary key points to semantic segmentation map. (a) Getting transverse and longitudinal representation of boundary key points map; (b) All possible regions of object; (c) Getting bounding box based on semantic segmentation map; (d) object detection result.

In this module, the boundary key points are mapped to the semantic segmentation map, and then obtain prediction box containing object. The concrete flow is shown in Fig. 4. The process is as follows:

1. De-noising the semantic segmentation map outputted by model with the majority filter to eliminate the semantic segmentation error, The majority filter is a nonlinear digital filtering technique used to remove noise from an image, the main idea of the

majority filter is to run through the signal entry by entry, replacing each entry with the majority of neighboring entries.

2. Obtaining the distribution of the boundary key points on the image based on the boundary key points map outputted by the model.
3. Generating the boundary lines according to the distribution of the boundary key points.
4. Obtaining all possible boxes divided by boundary lines (including the four borders of the semantic segmentation map).
5. Calculate the category and confidence for each boundary lines obtained by step 4 based on semantic segmentation map.
6. Dividing the boundary lines into groups of four based the category of boundary line, and every group boundary lines form a prediction box.
7. Determine the type and confidence of objects in the prediction box.

There are some errors in the semantic segmentation map outputted by network. The credibility of the class of objects in the prediction box is defined as:

$$C^i = \sum v_i \bigg/ \left(\sum_{1}^{1+cls\_num} \sum v_i \right) \tag{3}$$

In the expression i represent the category; cls_num is the total category of objects; $\sum v_i$ represents the number of pixels in the detection box of class i.

3.4   Small Object Re-detection Module

Through experiment, we found that the boundary key points of small objects are relatively small and the prediction error is relatively great. Some small objects may be filtered out as noise points in the process of de-noise. In order to solve this problem, a small object redetection module is designed in Fig. 5.

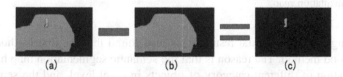

Fig. 5. Principle of the small objects re-detection module. (a) The original semantic segmentation diagram; (b) Removing noise pixel points from the detected large object; (c) Only the semantic segmentation of small object.

After each object is detected through object detect module, removing noise pixel points from each prediction box in the semantic segmentation map. Then XOR the processed semantic segmentation map with the original semantic segmentation map, the pixel points of all the objects that have been detected can be filtered, and only some small undetected objects with few pixels can be retained. For this remaining small object segmentation map, the threshold is set to retain connected domains. At last, the reserved small objects combined with all the previously detected objects are operated by maximum suppression algorithm, and the final reserved prediction boxes are obtained.

4 Experiment and Analysis

4.1 Feasibility Experiment of Boundary Key Points Predicting

In order to verify the feasibility of the boundary key points predicting, two boundary key points predict methods are designed. The first method is to use the convolutional feature map of the original image extracted by the feature extraction module to directly complete the boundary key points predicting. The semantic segmentation map and the map of boundary key points are separate outputted at the same time. The second method is to take the semantic segmentation map as the intermediate result, combining with the convolutional feature maps to predict the boundary key points. It makes full use of semantic segmentation map. The experimental results show that the convergence of the training of the first method is slow and the error of the prediction of the boundary key points is great, as shown by the orange curve in Fig. 6.

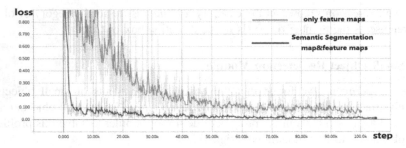

Fig. 6. Experimental comparison of training boundary key points using original map and adding semantic segmentation map.

Analyzing the experimental results, it is also found that the first method is worse than the second method. The reason is that the semantic segmentation map has realized the classification of different category of objects in pixel level, and the semantic segmentation map already contains rich boundary information of different category objects, but there is no boundary information of different individual object in the same category, the convolutional feature maps contain boundary information between different individuals of the same kind of object, the second method combines semantic segmentation map and convolutional feature maps.

4.2 FCDN Multi Task Deep Learning Experiment

One network performs two dense classify tasks simultaneously. In order to verify the feasibility of sharing all neurons in the semantic segmentation module by the boundary key points predicting module, we trained the FCDN model to complete semantic segmentation and boundary key points predict at the same time. The training results are shown in Fig. 7.

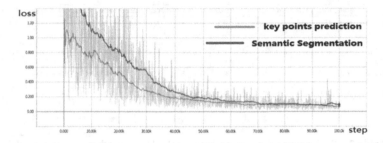

Fig. 7. Simultaneous training of semantic segmentation and boundary key point predicting task.

The experimental results show that two tasks can converge well. It can be seen that the boundary key points predicting module can share all layers of the semantic segmentation model and it doesn't affect the semantic segmentation. The two tasks can be trained at the same time. The training objectives of the model:

$$min(\alpha L_{seg} + \beta L_{kps}) \qquad (4)$$

The loss function of semantic Segmentation represented by L_{seg} [10], we adopts the same semantic segmentation loss function as FCN, L_{kps} represents the loss function of the boundary key points prediction, α, β are constant coefficients.

4.3   Contrast Test of FCDN Detection Performance

Finally, the PASCAL VOC data-set [23] is used to verify the detection performance of the FCDN model. The data set contains 20 classes of objects. Because tagging boundary key points are obtained on the basis of semantic segmentation annotation, only 3335 images with semantic segmentation annotation in VOC2007 and VOC 2012 can be used for training, and the test set of VOC 2007 can be used for testing. Mean Average Precision (mAP) is used to measure the detection accuracy of the detection model. All experiments were carried out in a single block Titan X GPU. In order to get the boundary key points annotation quickly, we use the method of Fig. 3 to obtain the sub-datasets from the VOC data set. In order to make the experimental results more comparable. Other detection models for comparison must also be trained with the 3335 object detection images with semantic segmentation annotation and the trained model is tested with test set of the VOC 2007 dataset. All the detection models used in the contrast experiment are based on VGG-16 as the backbone network. The experimental results are shown in Table 1. Because the training data set which is a sub-data set of VOC, is relatively small, compared with the model trained by complete data set, the detect performance of the comparative model is lower, it can be improved by adding training samples. The experimental results show that the performance of FCDN is better than that of the current mainstream object detection model under the same condition of training samples (Table 2).

Table 1. Test results of models trained with a subset of PASCAL VOC data sets

Method	Aero	Bike	Bird	Bottle	Boat	Bus	Car	Cat	Chair	Cow	Table	Dog	Horse	Mbike	Person	Plant	Sheep	Sofa	Train	Tv
Faster R-CNN [4]	73.5	73.6	66.9	42.1	65.5	73.1	74.7	73.4	37.2	74.9	53.7	72.8	72.6	67.5	76.7	38.8	67.6	63.9	65.3	62.6
SSD [8]	62.4	64.7	58.4	33.8	53.2	66.2	63.5	66.3	32.8	67.1	45.2	64.9	65.2	53.9	69.7	30.3	68.9	73.9	56.5	57.3
YOLO v2 [7]	74.8	66.9	60.2	31.3	56.9	66.5	57.8	60.9	30.3	57.6	50.5	65.6	57.5	60.2	62.1	34.4	54.8	62.6	64.3	58.6
FCDN	75.8	76.9	70.2	41.3	66.9	76.5	77.8	70.9	36.3	77.6	60.5	75.6	77.5	70.2	72.1	36.4	64.8	72.6	70.3	68.6

Table 2. Performance comparison between FCDN and other detection models

Method	Speed (fps)	mAP
Faster R-CNN [4]	6	64.8
SSD [8]	19	57.7
YOLO v2 [7]	21	56.7
FCDN	11	66.4

5   Conclusion

After analyzing the existing problem of the object detection model, we propose a new idea of object detection, mapping boundary key points to the semantic segmentation map to complete the object detect task, and a multi task full convolution network is proposed which complete semantic segmentation and object detection. Through the experiment, it is proved that it is feasible to use the full convolutional object detection network to output semantic segmentation and boundary key points at the same time. The task of object detection and even instance segmentation can be completed on the basis of semantic segmentation map and boundary key points.

References

1. Girshick, R., Donahue, J., Darrell. T., et al.: Rich feature hierarchies for accurate object detection and semantic segmentation. In: IEEE Conference on Computer Vision and Pattern Recognition, pp. 580–587. IEEE Computer Society (2014)
2. He, K., Zhang, X., Ren, S., et al.: Spatial pyramid pooling in deep convolutional networks for visual recognition. IEEE Trans. Pattern Anal. Mach. Intell. **37**(9), 1904–1916 (2015)
3. Girshick, R.: Fast R-CNN. Computer Science (2015)
4. Ren, S., He, K., Girshick, R., Sun, J.: Faster r-cnn: Towards real-time object detection with region proposal networks. In: Advances in Neural Information Processing Systems, pp. 91–99 (2015)
5. Li, Y., He, K., Sun, J., et al.: R-fcn: Object detection via region based fully convolutional networks. In: Advances in Neural Information Processing Systems, pp. 379–387 (2016)

6. Redmon, J., Divvala, S., Girshick, R., Farhadi, A.: Youonly look once: unified, real-time object detection. In: Proceedings of the IEEE Conference on Computer Vision and Pattern Recognition, pp. 779–788 (2016)
7. Redmon, J., Farhadi, A.: Yolo9000: better, faster, stronger (2016). arXiv:1612.08242
8. Liu, W., et al.: SSD: single shot multibox detector. In: Leibe, B., Matas, J., Sebe, N., Welling, M. (eds.) ECCV 2016. LNCS, vol. 9905, pp. 21–37. Springer, Cham (2016). https://doi.org/10.1007/978-3-319-46448-0_2
9. Ronneberger, O., Fischer, P., Brox, T.: U-Net: convolutional networks for biomedical image segmentation. In: Navab, N., Hornegger, J., Wells, W.M., Frangi, A.F. (eds.) MICCAI 2015. LNCS, vol. 9351, pp. 234–241. Springer, Cham (2015). https://doi.org/10.1007/978-3-319-24574-4_28
10. Long, J., Shelhamer, E., Darrell, T.: Fully convolutional networks for semantic segmentation. In: IEEE Conference on Computer Vision and Pattern Recognition, pp. 3431–3440. IEEE Computer Society (2015)
11. Krizhevsky, A., Sutskever, I., Hinton, G.E.: ImageNet classification with d deep convolutional encoder-decoder architecture for image segmeeep convolutional neural networks. In: International Conference on Neural Information Processing Systems, pp. 1097–1105. Curran Associates Inc. (2012)
12. Simonyan, K., Zisserman, A.: Very deep convolutional networks for large-scale image recognition. Computer Science (2014)
13. He, K., Zhang, X., Ren, S., et al.: Deep residual learning for image recognition. In: Computer Vision and Pattern Recognition, pp. 770–778. IEEE (2016)
14. Badrinarayanan, V., Kendall, A., Cipolla, R.: SegNet: antation. IEEE Trans. Pattern Anal. Mach. Intell. **PP**(99), 1 (2015)
15. Zhao, H., Shi, J., Qi, X., et al.: Pyramid scene parsing network (2016)
16. Chen, L.C., Papandreou, G., Schroff, F., et al.: DeepLab v3: rethinking atrous convolution for semantic image segmentation (2017)
17. Lin, G., Milan, A., Shen, C., et al.: RefineNet: multi-path refinement networks for high-resolution semantic segmentation (2016)
18. Bulò, S.R., Neuhold, G., Kontschieder, P.: Loss max-pooling for semantic image segmentation (2017)
19. Lin, T.-Y., Dollar, P., Girshick, R., He, K., Hariharan, B., Belongie, S.: Feature pyramid networks for object detection. arXiv preprint arXiv:1612.03144 (2016)
20. He, K., Gkioxari, G., Dollar, P., Girshick, R.: Mask R-CNN. arXiv preprint arXiv:1703.06870 (2017)
21. Ren, S., He, K., Girshick, R., Zhang, X., Sun, J.: Object detection networks on convolutional feature maps. IEEE Trans. Pattern Anal. Mach. Intell. **39**(7), 1476–1481 (2017)
22. Lin, T.-Y., Goyal, P., Girshick, R., He, K., Dollar, P.: Focal loss for dense object detection. arXiv preprint arXiv:1708.02002 (2017)
23. Everingham, M., Van Gool, L., Williams, C.K.I., Winn, J., Zisserman, A.: The Pascal visual object classes (VOC) challenge. Int. J. Comput. Vis. **88**(2), 303–338 (2010)
24. Pohlen, T., Hermans, A., Mathias, M., et al.: Full-resolution residual networks for semantic segmentation in street scenes, 3309–3318 (2017)
25. Cheng, J., Liu, S., Tsai, Y.H., et al.: Learning to segment instances in videos with spatial propagation network (2017)
26. He, K., Zhang, X., Ren, S., et al.: Deep residual learning for image recognition. In: IEEE Conference on Computer Vision and Pattern Recognition, pp. 770–778. IEEE Computer Society (2016)
27. Huang, G., Liu, Z., Laurens, V.D.M., et al.: Densely connected convolutional networks, 2261–2269 (2016)

28. Szegedy, C., Ioffe, S., Vanhoucke, V, et al.: Inception-v4, Inception-ResNet and the impact of residual connections on learning (2016)
29. Hong, R., Li, L., Cai, J., Tao, D., Wang, M., Tian, Q.: Coherent semantic-visual indexing for large-scale image retrieval in the Cloud. IEEE Trans. Image Process. **26**(9), 4128–4138 (2017)
30. Hong, R., Zhenzhen, H., Wang, R., Wang, M., Tao, D.: Multi-view object retrieval via multi-scale topic models. IEEE Trans. Image Process. **25**(12), 5814–5827 (2016)
31. Hong, R., Zhang, L., Zhang, C., Zimmermann, R.: Flickr circles: aesthetic tendency discovery by multi-view regularized topic modeling. IEEE Trans. Multimed. **18**(8), 1555–1567 (2016)

Learning to Match Using Siamese Network for Object Tracking

Chaopeng Li[1], Hong Lu[1(✉)], Jian Jiao[1], and Wenqiang Zhang[2]

[1] Shanghai Key Laboratory of Intelligent Information Processing,
School of Computer Science, Fudan University, Shanghai, China
honglu@fudan.edu.cn
[2] Shanghai Engineering Research Center for Video Technology and System,
School of Computer Science, Fudan University, Shanghai, China

Abstract. The general object tracking problem traditionally been tackled by modeling the object's appearance. In this paper we consider object tracking as a similarity measurement problem. We focus on learning a matching mechanism with great generalization ability. We present a Siamese convolutional neural network as a matching function to perform object tracking. First we simply match the exemplary target in previous frame with the candidates in a new frame using cosine similarity and return the most similar one by the learnt matching function. Then we perform bounding box regression to refine the target position given by the network as the final result. Extensive experiments on real-world benchmark datasets validate the superior performance of our approach.

Keywords: Object tracking · Siamese network
Convolutional neural network

1 Introduction

Object tracking is an important task within the fields of computer vision. It's prevalent in the scenarios of motion-based recognition, human-computer interaction, traffic monitoring, augmented reality and so on [7,28]. We have witnessed significant advances in object tracking with the development of efficient algorithms and fruitful applications in recent years. But it still remains a complex problem due to noise in images, deformation and occlusion, cluttered background, illumination variation, real-time processing requirements, etc [25]. Therefore, an efficient and accurate tracker for robust object tracking is required.

Currently, the most successful paradigm for object tracking has been to learn an appearance-based model of the object in an online fashion using examples extracted from the video itself. One group of these models performs a matching of the representation of the target model built from the previous frame. Another group takes a different view of tracking by building a model on the distinction of the target foreground against the background. It is called tracking-by-detection as in [18]. Despite the success of these methods, there is one major

© Springer Nature Switzerland AG 2018
R. Hong et al. (Eds.): PCM 2018, LNCS 11166, pp. 719–729, 2018.
https://doi.org/10.1007/978-3-030-00764-5_66

drawback that they rely on low-level hand-crafted features which are incapable to capture semantic information of targets, not robust to significant appearance changes, and only have limited discriminative power. Thus, it limits the richness of the model they can learn [21]. While other problems in computer vision, such as image classification, semantic segmentation, object detection and many others, have seen an increasingly pervasive adoption of deep convolutional neural networks (CNNs) trained from large supervised datasets. Different from hand-crafted features, features learned by CNNs from huge training datasets which cover a wide range of variations in the combination of target and background carry rich high-level semantic information. These features can be powerful in object tracking.

In this paper, we consider object tracking as a similarity measurement problem. Specifically, what we consider to solve is to find a candidate image patch that most similar to the exemplary patch in previous frame among dozens of image patches in search area. Hence, we focus on learning a matching mechanism with great generalization ability. We present a Siamese neural network based on VGG network [17] as a matching function to perform object tracking. Similar Siamese architecture [3,14,19] has shown the best performance on either accuracy or efficiency up to now. We use weights pre-trained on ImageNet dataset to initiate our proposed network, and further fine-tune the network using ALOV++ video datasets [18]. When it comes to tracking, we first simply match the exemplary target in previous frame with the candidates in a new frame using cosine similarity and return the most similar one by the learnt matching function. Then we perform bounding box regression to refine the target position given by the network as the final result. Extensive experiments on real-world benchmark datasets validate the superior performance of our approach.

The main contribution of our work are summarized as follows. First, we present a fully-convolutional Siamese network as a matching function to perform tracking. Second, we propose to use cosine similarity as the distance measurement between image patches. Third, we combine feature map obtained from the convolutional layer's output with super-pixel segmentation's result of raw image patch to perform bounding box regression to refine the given target position.

The rest part of this paper is organized as follows. We first review related work in Sect. 2, and discuss our model in detail in Sect. 3. Exhaustive experiments on various video sequences and result analysis are shown in Sect. 4, and Sect. 5 gives our conclusion.

2   Related Work

Deep Learning in Tracking. Despite tremendous success of deep CNNs on a wide range of computer vision applications, only a limited number of tracking algorithms have been proposed. In [23], a stacked denoising autoencoder is offline trained using auxiliary natural images to perform online tracking. In [13], an online tracking algorithm using a single CNN for learning effective feature representations of the target object over time is presented. [21] proposes a new approach for general object tracking with fully CNN. It utilize features extracted from both lower layer and top layer. But it only runs at 3 fps with a GPU. A few recent approaches [11,22] transfer features of CNN which pretrained for image classification task to online tracking. But the representation may not be very effective due to the fundamental difference between classification and tracking tasks.

Siamese Network Based Tracking. Once the object tracking is seen as a similarity measurement problem, it is straightforward to train a Siamese CNN model to select target from candidate image patches cropped from a searching region through a matching function learned offline on image pairs. A Siamese network is a Y-shaped neural network that shares the parameters and joins two CNN branches in final layers to produce a single output [16]. Siamese CNNs have been shown work well on face and signature verification [4,5]. In [10], they propose two convolutional branches inherited from AlexNet [12]. The tracker learns a generic relationship between object motion and appearance and can be used to track novel objects that do not appear in the training set. At run time, the object's bounding box is directly regressed by the model. SiamFC [3] also uses AlexNet as their base architecture, a novel cross-correlation layer links the last layer of two branches. The output is a scalar-valued score map with high values at pixels indicating object presence. [19] proposes two identical query and search networks inherited from AlexNet and VGGNet. Both networks are unconnected but share the same weights. The presented tracker, which is called SINT (Siamese INstance search Tracker), simply matches the initial patch of the target in the first frame with candidates in a new frame and returns the most similar patch by a learned matching function. Despite the model's simplicity, it only runs at about 2 fps on GPU. In [29], they use a Siamese region proposal network to identify potential targets across the whole frame. The model is updated with a short- and long-term strategy. Still, the tracker can run only at 3 fps on GPU. CFNet [20] adds a correlation filter and crop layers to the branch that concerns the template. This work is the first to interpret the Correlation Filter learner as a differentiable layer in a deep neural network.

3   Model

Inspired by the previous mentioned trackers, especially from SINT, we present a new fully-convolutional Siamese network tracker that is trained with a modified margin contrastive loss. The network's architecture is illustrated in Fig. 1. How the Siamese network is trained and how it works during tracking inference will be discussed in detail.

3.1   Similarity Measurement

There exists fundamental difference between tracking and image classification tasks. It is inferior to directly apply CNNs to tracking, since the power of CNNs

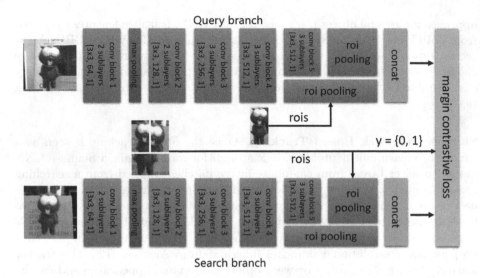

Fig. 1. The proposed fully-convolutional Siamese network. Numbers in square brackets are convolutional layers' kernel size, number of outputs and stride. The kernel size and stride of max pooling layer is 2×2 and 2, respectively. Image patch pairs cropped from search area of a frame and the corresponding box pairs as the network input.

relies on large-scale training. While dealing with tracking tasks, the only information we know is the first frame and the position of the target to be tracked, thus it's not a good idea to perform tracking in a classification or object detection way. Since deep CNNs extract high-level features, which are invariant to appearance changes or deformations, it's natural to utilize a robust matching function for matching arbitrary, generic objects that may undergo different kinds of appearance variations. Basically, the matching function indicates the similarity between exemplary target $O_1 \in \Re^{m \times n \times 3}$ and candidate patch $Z_t^i \in \Re^{m_i \times n_i \times 3}$ that cropped from a searching area in ith frame, where m, n stands for rows and columns of the image patch, respectively, and $i \in \{1, 2, \cdots, k\}$, k is the number of candidates. Our goal is to find the most similar patch by measuring the distance of their features. This can be formalized as:

$$\hat{Z}_t^i = \arg \min_i D\left(f\left(O_1\right), f\left(Z_t^i\right)\right). \tag{1}$$

where $f\left(\cdot\right)$ stands for the feature extractor, e.g. a CNN branch, while $D\left(\cdot, \cdot\right)$ denotes distance between two features. By this means, we can find the target's positions in subsequent frames.

3.2   Fully-Convolutional Siamese Architecture

As shown in Fig. 1, our Siamese network composed of two CNN branches, each branch is similar with VGG network except we remove some max-pooling layer and utilize the multi-scale region pooling strategy. The network process two inputs separately and join them in the last contrastive margin loss layer.

Fewer Max-Pooling Layer. Max-pooling makes the extracted features invariant to scale or orientation changes and more robust to deformations. But at the same time, targets in visual tracking task are typically small, the spatial resolution of the feature map is aggressively reduced (50% in the case of 2×2 stride). Too many max-pooling layers results in wider receptive fields in top layer, leading to imprecise localization of target. Therefore, we only have one max-pooling layer at the very early stage to maintain the network's robustness to local noise, meanwhile, keep the resolution of feature map from being too small.

Multi-stage Region Pooling Layer. When comparing the similarity between image patches, one can simply pass through the candidate regions independently, but this would leads to redundant computation of overlapping regions, slow down the training and tracking speed. Therefore, we employ a region pooling layer to process overlapping candidate regions. Besides, in a deep CNN, a top layer encodes more semantic features which is more invariant to appearance changes, while a lower layer carries more discriminative information and can better separate the target from distracters with similar appearance [21]. For this reason, we propose to fuse features from different stage of CNN as the final representation of image patches. Here, we perform region pooling on the last layer of conv block 4 and 5, e.g. conv4_3 and conv5_3.

Margin Contrastive Loss. In the end, the two branches are connected to the last single loss layer. Inspired by [5], we employ the margin contrastive loss:

$$\mathcal{L}(x_i, x_j, y_{ij}) = y_{ij} D^2 + (1 - y_{ij}) \max(0, \epsilon - D)^2. \tag{2}$$

where D is the distance between two features, $y_{ij} \in \{0, 1\}$ indicates whether x_i and x_j are the same target or not, and ϵ is the minimum distance margin that pairs depicting different objects should satisfy. Generally, the Euler distance is employed as distance measurement, here we propose to use *cosine distance*. As the features are flattened to one-dimension vector, it's preferable to measure how closely the two vectors in the same direction rather than compute the distance of tow "points" in high dimensional space. Moreover, the features produced by CNN are not normalized, the Euler distance between two vectors could be extremely large or small, making it difficult to set a proper value of the hyperparameter ϵ. Although we can normalize Euler distance to $[0, 1]$, it will be intuitive to use cosine distance directly, which is formalized as follow:

$$D(x_i, x_j) = 1 - \frac{x_i \cdot x_j}{\|x_i\| \|x_j\|}. \tag{3}$$

3.3   Training

Instead of training the Siamese network from scratch, we initialize the network using parameters pre-trained for ImageNet classification task, and fine-tune the Siamese network with ALOV++ video dataset. Since the lower layers capture basic visual patterns such as edges and corners, whereas higher layers get activated the most on more complex patterns, we assume that the features of lower layers for different visual tasks are similar. Thus the conv block 1 and 2 are fixed during training. The two branches of Siamese network share the same parameters, therefore, only one branch is fine-tuned actually. For one branch, named query branch, we randomly pick a frame from a video sequence, together with the ground truth bounding box as the input of this branch. For another branch, named search branch, we pick a frame from the same video sequence but doesn't need to be adjacent to the frame of query branch. For the frame of search branch, we sample boxes and the ones that overlap with the ground truth larger than 0.7 are labeled as positive and less than 0.5 are labeled as negative, together with the frame of search branch as the input. By this means, we form positive and negative box pairs for training.

3.4   Tracking

Once we have finished the matching function learning as previously described, a robust generic Siamese CNN Tracker is obtained. The network's parameters are fixed without any further adapting during tracking. There is one thing to be noticed, the training network composed of two branches, while the tracking network only have one single branch after removing the last loss layer. When the tracking is started, the first frame and its ground truth bounding box are fed to the network, the obtained feature is stored as the exemplary template. We sample boxes in the subsequent frames and select the one that matches best to the exemplary template as the result according to Eq. (1).

Candidate Box Sampling. To make the candidate box sampling procedure simple, we just randomly generate boxes in the search region. Assume that in tth frame, the bounding box BB_t given by the network is centered at (x_t, y_t) with a shape of $w \times h$. Then we define the search area in $t+1$ frame as a rectangle that is also centered at (x_t, y_t) but with a shape of $2w \times 2h$. Then we randomly generate P positive boxes and N negative boxes in three different scales $0.95, 1, 1.05$. The definition of positive and negative boxes are same as training stage. The value of P and N are both set to 128 in our experiment.

Template Updating. To cope with the target appearance changes, we update the exemplary template obtained from the first frame incrementally. Specifically, we update the exemplary template T^* every 10 frames using the best matched result within the intervening frames. It can be formalized as:

$$T^* \leftarrow (1 - \eta)T^* + \eta T_i, \tag{4}$$

where T_i stands for the best matched result in 10 frames, and η is updating rate.

Bounding Box Refinement. As employed in SINT and R-CNN [8], a refinement step of the bounding boxes can improve the tracking or localization performance significantly. In SINT, they utilize the region pooling layer's output of conv block 4, which is high-level features, to train four Ridge regressors for width, height, and coordinates of the center of the box based on the first frame. The goal is to learn a transformation that maps a proposed box to ground truth box. As presented in the superpixel tracker [24], they perform superpixel segmentation for each frame to construct a discriminative appearance model to distinguish object from cluttered background with mid-level cues. Actually, superpixel is a group of connected pixels with similar colors. Superpixel segmentation algorithms group raw pixels into perceptually meaningful atomic regions, provide a convenient primitive from which to compute image features [1]. We propose to combine region pooling layer's output with color information captured in superpixels to train regressors. Once the regressors are learned, they will be used to fine-tune the box given by the network.

4   Experiments

4.1   Dataset and Evaluation Metrics

We pick 10 sequences with substantial variations such as fast motion, deformation, background clutter and occlusion from the OTB benchmark data set [26] as the test sequences to evaluate the proposed Siamese network. The evaluation is based on two metrics: center location error and bounding box overlap ratio. The one-pass evaluation (OPE) is employed to compare our algorithm with the four trackers including CPF [15], SMS [6], Frag [2] and SBT [9].

4.2   Implementation Details

We use ALOV++ video dataset for training, the frame pairs and boxes pairs are generated as described in Sect. 3.3. All the image patches are resized to $96 \times 96 \times 3$ as network input. In total, we have sampled from ALOV dataset more than 60,000 pairs of frames and each pair of frames has 128 pairs of boxes. The minimum distance margin ϵ in Eq. (2) is set to 0.3 and the template updating rate η is set to 0.01 based on empirical study. We use Adam [27] as the learning rate schedule and the initial learning rate is set to 0.001. The learning rate of conv block 3 and 4 are set 100 and 10 times smaller than conv block 5, respectively. The number of training epochs is 6. Our algorithm is implemented in Python with MXNet library, and runs at 12 fps with NVIDIA GTX 1070.

4.3   Evaluation

Overall Comparison. Figure 2 shows the success plot and precision plot based on center location error and bounding box overlap ratio, respectively. A frame is defined as successfully tracked if the overlap between the predicated bounding box and the ground truth box is larger than a threshold. The success plot is obtained by assigning different overlap threshold in [0, 1]. While the precision plot demonstrates the percentage of frames where the distance between the predicted target location and the ground truth location is within a given threshold. The precision plot is obtained by assigning different location error threshold in [0, 50]. It can be seen from the figure that the proposed tracker outperforms all the other trackers in terms of precision score. To facilitate better analysis on the tracking performance, we further evaluate all the trackers on sequences with different attributes (illumination variation, occlusion, deformation and so on). Due to the space limit, performance plots on two attributes are shown in Fig. 3.

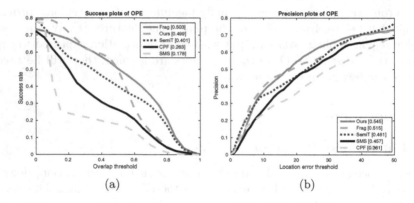

Fig. 2. Success plot and precision plot.

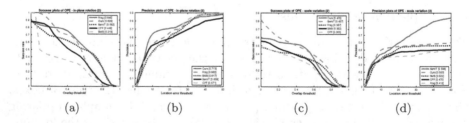

Fig. 3. The success plots and precision plots for different attributes.

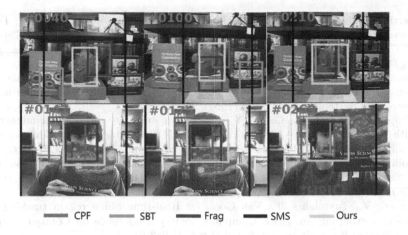

CPF   SBT   Frag   SMS   Ours

Fig. 4. Qualitative results of the proposed method on some sequences.

Result Analysis. Qualitative comparison of our approach with other trackers on some sequences are shown in Fig. 4. As can be seen, our proposed tracker are more robust to illumination changes and occlusion.

5 Conclusions

In this paper, we present a fully-convolutional Siamese network to solve the tracking problem in a similarity measurement way. The strength of tracker comes from powerful matching function. The cosine similarity used in the loss function is more suitable for measuring the distance of two feature vectors which makes the training procedure easier. The multi-stage region pooling layer combines features from both lower and top layers. It makes our tracker more robust to cluttered background and possess the ability of precise localization. The bounding box refinement helps the network to produce a better result. Experiment results on OTB dataset show that the proposed tracking algorithm performs favorably against existing methods.

Acknowledgements. This work was supported in part by National Natural Science Foundation of China (No. 81373555) and Shanghai Committee of Science and Technology (14JC1402200 and 14441904403).

References

1. Achanta, R., Shaji, A., Smith, K., Lucchi, A., Fua, P., Süsstrunk, S.: Slic superpixels compared to state-of-the-art superpixel methods. IEEE Trans. Pattern Anal. Mach. Intell. **34**(11), 2274–2282 (2012)
2. Adam, A., Rivlin, E., Shimshoni, I.: Robust fragments-based tracking using the integral histogram. In: IEEE Computer Society Conference on Computer Vision and Pattern Recognition, vol. 1, pp. 798–805. IEEE (2006)

3. Bertinetto, L., Valmadre, J., Henriques, J.F., Vedaldi, A., Torr, P.H.S.: Fully-convolutional siamese networks for object tracking. In: Hua, G., Jégou, H. (eds.) ECCV 2016. LNCS, vol. 9914, pp. 850–865. Springer, Cham (2016). https://doi.org/10.1007/978-3-319-48881-3_56

4. Bromley, J., Guyon, I., LeCun, Y., Säckinger, E., Shah, R.: Signature verification using a "siamese" time delay neural network. In: Advances in Neural Information Processing Systems, pp. 737–744 (1994)

5. Chopra, S., Hadsell, R., LeCun, Y.: Learning a similarity metric discriminatively, with application to face verification. In: IEEE Computer Society Conference on Computer Vision and Pattern Recognition, vol. 1, pp. 539–546. IEEE (2005)

6. Collins, R.T.: Mean-shift blob tracking through scale space. In: IEEE Computer Society Conference on Computer Vision and Pattern Recognition, vol. 2, pp. II-234. IEEE (2003)

7. Ferrari, V., Tuytelaars, T., Van Gool, L.: Real-time affine region tracking and coplanar grouping. In: IEEE Computer Society Conference on Computer Vision and Pattern Recognition, vol. 2, pp. II-226–II-233 (2001)

8. Girshick, R., Donahue, J., Darrell, T., Malik, J.: Rich feature hierarchies for accurate object detection and semantic segmentation. In: IEEE Computer Society Conference on Computer Vision and Pattern Recognition, pp. 580–587 (2014)

9. Grabner, H., Leistner, C., Bischof, H.: Semi-supervised on-line boosting for robust tracking. In: Forsyth, D., Torr, P., Zisserman, A. (eds.) ECCV 2008. LNCS, vol. 5302, pp. 234–247. Springer, Heidelberg (2008). https://doi.org/10.1007/978-3-540-88682-2_19

10. Held, D., Thrun, S., Savarese, S.: Learning to track at 100 FPS with deep regression networks. In: Leibe, B., Matas, J., Sebe, N., Welling, M. (eds.) ECCV 2016. LNCS, vol. 9905, pp. 749–765. Springer, Cham (2016). https://doi.org/10.1007/978-3-319-46448-0_45

11. Hong, S., You, T., Kwak, S., Han, B.: Online tracking by learning discriminative saliency map with convolutional neural network. In: International Conference on Machine Learning, pp. 597–606 (2015)

12. Krizhevsky, A., Sutskever, I., Hinton, G.E.: Imagenet classification with deep convolutional neural networks. In: Advances in Neural Information Processing Systems, pp. 1097–1105 (2012)

13. Li, H., Li, Y., Porikli, F.: Robust online visual tracking with a single convolutional neural network. In: Cremers, D., Reid, I., Saito, H., Yang, M.-H. (eds.) ACCV 2014. LNCS, vol. 9007, pp. 194–209. Springer, Cham (2015). https://doi.org/10.1007/978-3-319-16814-2_13

14. Nam, H., Han, B.: Learning multi-domain convolutional neural networks for visual tracking. In: IEEE Computer Society Conference on Computer Vision and Pattern Recognition, pp. 4293–4302. IEEE (2016)

15. Pérez, P., Hue, C., Vermaak, J., Gangnet, M.: Color-based probabilistic tracking. In: Heyden, A., Sparr, G., Nielsen, M., Johansen, P. (eds.) ECCV 2002. LNCS, vol. 2350, pp. 661–675. Springer, Heidelberg (2002). https://doi.org/10.1007/3-540-47969-4_44

16. Pflugfelder, R.P.: Siamese learning visual tracking: a survey. arXiv preprint arXiv:1707.00569 (2017)

17. Simonyan, K., Zisserman, A.: Very deep convolutional networks for large-scale image recognition. arXiv preprint arXiv:1409.1556 (2014)

18. Smeulders, A.W.M., Chu, D.M., Cucchiara, R., Calderara, S., Dehghan, A., Shah, M.: Visual tracking: an experimental survey. IEEE Trans. Pattern Anal. Mach. Intell. **36**(7), 1442–1468 (2014)

19. Tao, R., Gavves, E., Smeulders, A.W.: Siamese instance search for tracking. In: IEEE Computer Society Conference on Computer Vision and Pattern Recognition, pp. 1420–1429. IEEE (2016)
20. Valmadre, J., Bertinetto, L., Henriques, J., Vedaldi, A., Torr, P.H.: End-to-end representation learning for correlation filter based tracking. In: IEEE Computer Society Conference on Computer Vision and Pattern Recognition, pp. 5000–5008. IEEE (2017)
21. Wang, L., Ouyang, W., Wang, X., Lu, H.: Visual tracking with fully convolutional networks. In: Proceedings of the IEEE International Conference on Computer Vision, pp. 3119–3127 (2015)
22. Wang, N., Li, S., Gupta, A., Yeung, D.Y.: Transferring rich feature hierarchies for robust visual tracking. arXiv preprint arXiv:1501.04587 (2015)
23. Wang, N., Yeung, D.Y.: Learning a deep compact image representation for visual tracking. In: Advances in Neural Information Processing Systems, pp. 809–817 (2013)
24. Wang, S., Lu, H., Yang, F., Yang, M.H.: Superpixel tracking. In: 2011 IEEE International Conference on Computer Vision (ICCV), pp. 1323–1330. IEEE (2011)
25. Wang, X., O'Brien, M., Xiang, C., Xu, B., Najjaran, H.: Real-time visual tracking via robust kernelized correlation filter. In: IEEE International Conference on Robotics and Automation, pp. 4443–4448 (2017)
26. Wu, Y., Lim, J., Yang, M.H.: Online object tracking: a benchmark. In: IEEE Computer Society Conference on Computer Vision and Pattern Recognition, 2013. pp. 2411–2418. IEEE (2013)
27. Xu, L., Krzyzak, A., Oja, E.: Rival penalized competitive learning for clustering analysis, RBF net, and curve detection. IEEE Trans. Neural Netw. 4(4), 636–649 (1993)
28. Yilmaz, A., Javed, O., Shah, M.: Object tracking: a survey. ACM Comput. Surv. (CSUR) 38(4), 13 (2006)
29. Zhang, H., Ni, W., Yan, W., Wu, J., Bian, H., Xiang, D.: Visual tracking using siamese convolutional neural network with region proposal and domain specific updating. Neurocomputing 275, 2645–2655 (2018)

Macropixel Based Fast Motion Estimation for Plenoptic Video Compression

Lingjun Li, Xin Jin$^{(\boxtimes)}$, Haixu Han, and Qionghai Dai

Shenzhen Key Lab of Broadband Network and Multimedia, Graduate School at
Shenzhen, Tsinghua University, Shenzhen 518055, China
jin.xin@sz.tsinghua.edu.cn

Abstract. A generalized plenoptic video coding method is proposed to realize
fast motion estimation in integer pixel precision by exploiting the imaging
structure correlation among macropixels. Based on the imaging structure correlation, collocated blocks in the neighboring macropixels are used to predict the
current prediction unit (PU) and the 8-step neighborhood refinement improves
the performance further. The experimental results demonstrate the proposed
algorithm outperforms the state-of-the-art plenoptic compression methods, like
Test Zone Search (TZS) algorithm embedded in the HEVC Test Model (HM),
Rotation Scan Mapping (RSM) method and Light Field Multiview Video
Coding algorithm (LF-MVC) with an average of 7.11%, 69.72% and 41.14%
bitrate reduction, respectively, under the comparative computing complexity.

Keywords: Plenoptic video coding · Light field coding · Macro pixel
Inter-prediction · HEVC/H.265

1 Introduction

In recent years, the plenoptic data comprising both spatial and angular information,
which can be used for integral imaging [1], saliency detection [2] and depth map
estimation [3] are attracting great interests from both industry and academy. Among the
plenoptic data acquiring systems, the hand-held plenoptic cameras such as Lytro [4]
present their superior convenience in acquiring plenoptic data. The architecture of a
typical plenoptic camera is shown in Fig. 1, where a microlens array is inserted at the
imaging plane of the main lens. The light rays emitting from the object will be scatted
by microlens and imaged on the sensor as a group of pixels called macropixel. Based
on the optical architecture, super-high-resolution plenoptic data captured by the camera
can record both the spatial and angular information, which require big storage space or
transmission bandwidth even for one image. With the development in plenoptic
applications, plenoptic video, which consists of a sequence of plenoptic images
recording not only the spatial and angular information but also the temporal plenoptic
variations, becomes the challenging data format. Distinct spatial structure inside each
frame and complex plenoptic correlations among the frames bring great difficulties to
plenoptic video compression. Although some conventional compression methods can
be also used to encode the plenoptic videos such as multiview video coding (MVC) [7],

© Springer Nature Switzerland AG 2018
R. Hong et al. (Eds.): PCM 2018, LNCS 11166, pp. 730–739, 2018.
https://doi.org/10.1007/978-3-030-00764-5_67

the compression efficiency is still limited. Thus, efficient plenoptic video compression method is highly desired.

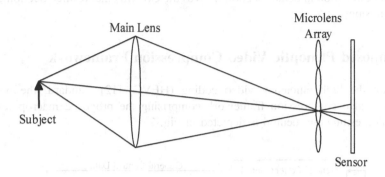

Fig. 1. Architecture of a typical plenoptic camera

However, to the best of our knowledge, there have been no methods directly compressing the plenoptic videos. The only method dedicated to plenoptic videos, light field multiview video coding (LF-MVC) [9], compresses the subaperture videos rendered from the plenoptic video, which may lead to plenoptic information loss. Based on our investigations, some existing plenoptic image compression methods may be possible to be extended to encode plenoptic videos, such as plenoptic-image-based compression, multiview-based compression and pseudo-video-based compression. Based on plenoptic image, some compression methods attempt to remove the redundancy between macropixels using conventional 2D image compression methods [5, 6]. These algorithms cannot sufficiently exploit the imaging structure correlation of the macropixels in different frames to remove the temporal redundancy. Multiview-based methods regard subaperture videos extracted from macropixels as different views and the elaborate prediction structures are designed for compression by MVC. Some existing works on MVC can be extended to the plenoptic video compression such as Central 2D Prediction [8]. However, the parameters that describe the reference relationship among the views consume a huge number of bits in MVC, while a large number of frames result in massive header bits consumption. Pseudo-video-based method regards each subaperture as a frame and the pseudo video is generated by reordering all subapertures based on the elaborate subaperture scan order. There have been several subaperture orders in the existing works, including line scan mapping [10], rotation scan mapping [10] and horizonal zigzag scanning [11]. Similarly, these methods introduce massive overhead bits, such as slice header.

Therefore, the first method that directly compresses the plenoptic video is proposed in this paper, which can exploit the imaging structure correlation among macropixels in different plenoptic frames to find the best reference block fast and accurately by searching a small number of high correlated candidates. Compared with existing coding methods, the proposed compression method brings compression efficiency improvement with an average of 69.72% and 41.14% bitrate reduction relative to RSM and LF-MVC, respectively. It also outperforms HEVC fast search algorithm with an average of 7.11% bitrate reduction under the comparative computing complexity.

The rest of this paper is organized as follows. Section 2 describes the proposed plenoptic video compression framework. Section 3 demonstrates the proposed model of motion estimation in detail. Section 4 gives the experimental results. Section 5 gives the conclusions.

2 Proposed Plenoptic Video Compression Framework

Based on the high efficiency video coding (HEVC) [12] standard, the proposed plenoptic video compression framework comprising the proposed macropixel-based fast motion estimation method is depicted in Fig. 2.

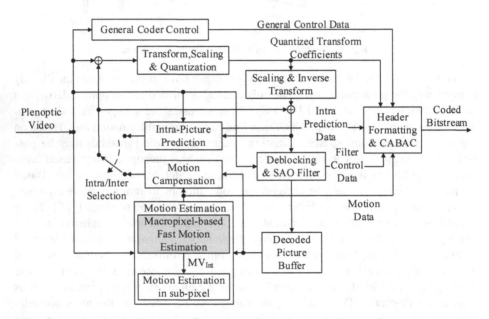

Fig. 2. Proposed plenoptic video compression framework

The input of the encoder is a plenoptic video sequence, which consists of a series of plenoptic images captured by plenoptic cameras. The plenoptic video recodes not only the spatial and angular information but also the temporal plenoptic variations. Each frame in the video is a plenoptic image which has been preprocessed by the JPEG-PLENO recommended method including demosaicing, devignetting, rotating and scaling [13]. The plenoptic images are arranged as the plenoptic video following the temporal order. Then, the plenoptic video is compressed by the video encoder comprising the proposed macropixel-based fast motion estimation method, which predicts the current prediction unit (PU) using the collocated blocks in the neighboring macropixels within the search range and complementary reference blocks determined by the 8-step neighborhood refinement. It can sufficiently exploit the structure consistency among macropixels in the neighboring frames to accelerate motion estimation by searching much less while much more correlated reference candidates. The proposed method is described in detail in Sect. 3.

3   The Proposed Fast Motion Estimation Algorithm

Motion estimation is the most computationally expensive task in video compression, which consumes 60% to 80% of the total encoding time [14]. Therefore, much less reference candidates with much higher correlation are selected by the proposed algorithm to maintain the efficiency of motion estimation and to reduce the computational complexity simultaneously. The proposed algorithm contains two parts: macropixel-based search (MS) and refinement. MS exploits the imaging structure correlation among macropixels in different frames to find the best reference block fast. Collocated blocks in the macropixels within the search range are used as the reference candidates. Refinement aims at improving the accuracy of MS by 8-step neighborhood refinement with early termination.

3.1   The Macropixel-Based Search Method

Based on Ng's optical analysis [15], the continuity of spatial pixels is reflected by the continuity of collocated pixels in neighboring macropixels. Simultaneously, the collocated block in different macropixels can be considered as recording the rays with the specific directions. Therefore, imaging structure correlation is utilized in the proposed method to reduce the number of candidates as well as to improve the accuracy of candidates. The search pattern of MS is depicted in Fig. 3. As shown in the figure, the reference block derived by the motion vector predictor (MVP) is regarded as the highly correlated block of the current PU. Setting it as the center of the search range, inter prediction candidates that are spatially collocated in macropixels within the search range are selected as prediction candidates. The candidates present high imaging correlation with the current PU. Then, block-matching is performed and the best prediction is determined by minimizing the rate-distortion (RD) cost between the current block and the prediction block.

Since the macropixels are arranged in hexagonal structure, the coordinate of the top-left corner of prediction candidates can be computed by:

$$
\begin{cases}
x = x_0 + p \cdot w + \textit{offset} \\
y = y_0 + q \cdot h
\end{cases}
$$
$$
p \in Z, q \in Z, (x, y) \in SR,
$$
$$
\textit{offset} = \begin{cases}
0, & q \in E \\
\frac{w}{2}, & q \in O
\end{cases}
$$

(1)

where (x, y) represents the coordinate of the top-left corner of prediction candidates; (x_0, y_0) represents the coordinate of the top-left corner of central prediction candidate; $h(h = 14)$ represents the pixel distance between the neighboring macropixels in vertical direction as shown in Fig. 3; $w(w = 16)$ represents the pixel distance between the neighboring macro pixels in horizontal direction as shown in Fig. 3; SR represents the

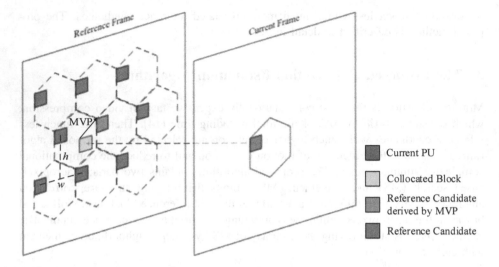

Fig. 3. Search pattern of macropixel-based search algorithm

search range of the current motion estimation process; E represents the set of even numbers; O represents the set of odd numbers.

Based on Eq. (1), the total number of candidates can be calculated as:

$$
n_1 =
\begin{cases}
(k+1) \cdot (2\lfloor \frac{W}{w} \rfloor + 1) + 2k \cdot \left\lfloor \frac{W+w/2}{w} \right\rfloor, & k \in E \\
k \cdot (2\lfloor \frac{W}{w} \rfloor + 1) + 2(k+1) \cdot \left\lfloor \frac{W+w/2}{w} \right\rfloor, & k \in O
\end{cases}
$$
$$
k = \left\lfloor \frac{H}{h} \right\rfloor
$$
(2)

where W represents the half width of the search range; H represents the half height of the search range; k represents the half number of rows of macropixels within the search range.

3.2   Refinement Search

Based on the best reference block determined by MS, the 8-step neighborhood refinement with early termination is designed to improve the performance further by exploiting the correlation in neighboring spatial pixels. In each step, the maximum of eight neighborhoods around the best candidate determined in the last step are searched and compared by RD cost. The refinement can be early terminated as the search center is still the best reference block after comparing with all the candidates in the current step.

The search process of refinement is shown in Fig. 4, using 4-step as an instance. The black points represent the best matching found in the step numbered in white. The gray points represent the rest of candidates in the step.

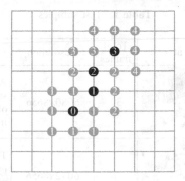

Fig. 4. Search pattern of refinement

Based on the above analysis, the total number of candidates in the refinement can be calculated by:

$$n_2 = 8 + \sum_{k=2}^{n} s_k \quad n \leq 8, \tag{3}$$

where n represents the total number of steps; s_k represents the increased number of candidates. If the search center is located at the vertex of the eight neighborhoods in the last step, the s_k equals to 5, otherwise s_k equals to 3.

4 Experiments and Analysis

4.1 Experiment Conditions

To validate the performance, the proposed algorithm is implemented into the HEVC codec software (HM-16.15) [16]. Four plenoptic videos, acquired by our Lytro Illum camera with resolution 7728 × 5368 and angular resolution 15 × 15 are tested. All videos have 30 frames acquired under the outdoor environment with different photographing methods. The detailed specification of test samples is shown in Table 1.

As described in Fig. 2, each frame in the plenoptic videos is preprocessed by Matlab LFToolbox (Version 0.4) [17] and converted from RGB color space to YCbCr4:4:4 color space before the compression. All data are encoded by HEVC range extension (RExt) with "Low Delay B" configuration [18] at four QPs 22, 27, 32 and 37. In this configuration, the search range of uni-prediction and bi-prediction is set as 64 and 4, respectively. Therefore, the total number of reference candidates in the proposed algorithm vary from 85 to 120 according to Eqs. (2) and (3). The compression efficiency among the approaches is measured by BD-bitrate [19], in which the bitrate is the average number of bits per frame of encoded video, and the PSNR is the frame-averaged PSNR which is calculated between the subaperture images rendered from the original plenoptic frame and those rendered from the reconstructed plenoptic frame [20].

Table 1. Test samples

Video name	Content	Total frames	Resolution	Description
Auto Race		30	6080×8656	A red car manually moved in straight line with the speed of 0.5 cm per frame approximately
Flowers		30	6072×8656	The static flowers captured by the camera rotating approximately 4 degrees per frame
Rubik's Cube		30	6080×8656	The Rubik's cube rotated manually at approximately 7.5 degrees per frame
Matryoshka		30	6072×8656	Still toys in line captured by camera rotating approximately 2 degrees per frame.

4.2   Experimental Results and Analysis

The efficiency of the proposed method is demonstrated by comparing with four state-of-the-art methods: HEVC Test Model (HM) test zone search (TZS), a diamond search pattern based fast search algorithm; HM full search (FS), a high-performance algorithm by searching all candidates within the search range; rotational scan mapping (RSM) [10], a state-of-the-art pseudo-video-based algorithm; light field multiview video coding (LF-MVC) [9], an advanced multiview-based algorithm. To evaluate the refinement performance, the macropixel-based search algorithm without refinement is also tested under the consideration of trade-off between extra computing complexity introduced from refinement and the extra performance gains.

Compression efficiency comparison results are shown in Table 2. As shown in the table, the proposed algorithm reduces compression bitrate by an average of 69.72% and 41.14% relative to RSM and LF-MVC, respectively, which demonstrates that the proposed method has much higher performance than pseudo-video-based and multiview-based method. The proposed algorithm also outperforms TZS with 7.11% bitrate reduction on average by using the same number of reference candidates in larger search range. Compared with FS method, much less candidates, only 1.68% relative to that of FS, result in only 1.68% performance loss on average. According to the MS result, the refinement brings 0.47% bitrate reduction on average. The RD curves for the

Table 2. Compression performance comparation with other methods

Video	Proposed vs. TZS	Proposed vs. FS	Proposed vs. RSM	Proposed vs. LF-MVC	Proposed vs. MS
Auto race	− 8.62%	1.41%	− 69.67%	− 51.63%	− 0.98%
Flowers	− 3.78%	1.93%	− 75.50%	− 51.33%	− 0.26%
Rubik's cube	− 4.92%	1.14%	− 74.12%	− 44.27%	− 0.28%
Matryoshka	− 11.12%	2.26%	− 59.59%	− 17.34%	− 0.37%
Average	**− 7.11%**	**1.68%**	**− 69.72%**	**− 41.14%**	**− 0.47%**

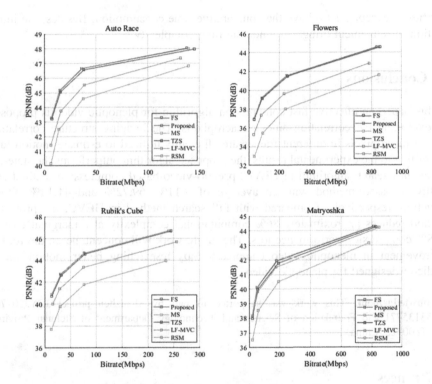

Fig. 5. Rate distortion results for the test videos

test videos are depicted in Fig. 5, which further demonstrate that the proposed algorithm outperforms the compared algorithms at all testing bitrates.

The encoding time complexity is measured by the time consumption in the complete encoding process, which represents the computational complexity. The ratios between the proposed algorithm and other reference algorithms are provided in Table 3. FS is the algorithm with the highest complexity in all algorithms due to more than 100 times larger number of candidates in the proposed method. And the

Table 3. Time consumption ratio

Video	Proposed/ TZS	Proposed/ FS	Proposed/ RSM	Proposed/ LF-MVC	Proposed/ MS
Auto race	102.30%	11.56%	143.48%	81.92%	100.43%
Flowers	101.25%	16.10%	128.45%	89.24%	100.46%
Rubik's cube	105.08%	13.01%	127.37%	81.64%	101.12%
Matryoshka	101.68%	15.80%	133.94%	84.20%	100.59%
Average	**102.58%**	**14.12%**	**133.31%**	**84.25%**	**100.65%**

algorithms except for FS have the comparative time consumption. Besides, the introduction of refinement brings an ignorable time complexity.

5 Conclusions

In this paper, an effective fast compression algorithm for plenoptic video is proposed by exploiting the correlation among macropixels. The imaging structure correlation among macropixels in different frames are efficiently utilized to estimate motion fast. According to the experimental results, the proposed algorithm outperforms the state-of-the-art fast search algorithm in HEVC, pseudo-video-based compression method and multiview-based method with an average of 7.11%, 69.72% and 41.14% bitrate reduction, respectively. Compared with full search method in HEVC, the proposed method reduces approximately 86% computational complexity at a marginal cost of 1.68% compression performance loss. The future work will concentrate on the further improvement in plenoptic video compression by researching better motion vector predictor designed for plenoptic video.

Acknowledgement. This work was supported in part by Shenzhen project JCYJ201703 07153135771 and Foundation of Science and Technology Department of Sichuan Province 2017JZ0032c, China.

References

1. Martínezcorral, M., et al.: From the plenoptic camera to the flat integral-imaging display. In: Proceedings of SPIE—The International Society for Optical Engineering, vol. 9117, pp. 91170H–91170H-6 (2014)
2. Li, N., et al.: Saliency detection on light field. In: Computer Vision and Pattern Recognition IEEE (CVPR), pp. 2806–2813 (2014)
3. Tao, M.W., et al.: Depth from combining defocus and correspondence using light-field cameras. In: 2013 IEEE International Conference on Computer Vision (ICCV), pp. 673–680. IEEE (2013)
4. Lytro Home Page. https://www.lytro.com/

5. Elharar, E., Stern, A., Hadar, O., Javidi, B.: A hybrid compression method for integral images using discrete wavelet transform and discrete cosine transform. J. Disp. Technol. **3**, 321–325 (2007)
6. Aggoun, A., Mazri, M.: Wavelet-based compression algorithm for still omnidirectional 3D integral images. Signal Image Video Process. **2**, 141–153 (2008)
7. Shi, S., Gioia, P., Madec, G.: Efficient compression method for integral images using multi-view video coding. In: 18th IEEE International Conference on Image Processing, pp. 137–140 (2011)
8. Dricot, A., Jung, J., Cagnazzo, M., Pesquet, B., Dufaux, F.: Full parallax super multi-view video coding. In: Proceedings of ICIP, Paris, France, pp. 135–139, October 2014
9. Wang, G., et al.: Light field multi-view video coding with two-directional parallel inter-view prediction. IEEE Trans. Image Process. **25**(11), 5104–5117 (2016)
10. Dai, F., et al.: Lenselet image compression scheme based on subaperture images streaming. In: IEEE International Conference on Image Processing, pp. 4733–4737. IEEE (2015)
11. Zhao, S., et al.: Light field image coding with hybrid scan order. In: Visual Communications and Image Processing, pp. 1–4. IEEE (2017)
12. Sullivan, G.J., et al.: Overview of the high efficiency video coding (HEVC) standard. IEEE Trans. Circuits Syst. Video Technol. **22**(12), 1649–1668 (2013)
13. Dansereau, D.G., Pizarro, O., Williams, S.B.: Decoding, calibration and rectification for lenselet-based plenoptic cameras. In: Computer Vision and Pattern Recognition, pp. 1027–1034. IEEE (2013)
14. Chen, Z., et al.: Fast integer-pel and fractional-pel motion estimation for H.264/AVC. J. Vis. Commun. Image Represent. **17**(2), 264–290 (2002)
15. Ng, R.: Digital light field photography. Ph.D. thesis Stanford University **115**(3), 38–39 (2006)
16. HEVC codec software HM-16.15 download website. https://hevc.hhi.fraunhofer.de/svn/svn_HEVCSoftware/tags/HM-16.15/
17. Light Field toolbox. http://www.mathworks.com/matlabcentral/fileexchange/-49683-light-fieldtoolbox-v0-4
18. Au, O.C., Zhang, X., Pang, C., Wen, X.: Suggested common test conditions and software reference configurations for screen content coding. In: Joint Collaborative Team on Video Coding (JCT-VC), Torino, JCTVC-F696, July 2011
19. Bjontegaard, G.: Calculation of average PSNR difference between RD-curves. ITU-T VCEG-M33 (2001)
20. Light field compression evaluation. http://mmspg.epfl.ch/files/content/sites/mmspl/files/shared/LF-GC/CFP.pdf

Text to Region: Visual-Word Guided Saliency Detection

Tengfei Xing[1], Zhaohui Wang[1], Jianyu Yang[2], Yi Ji[1(✉)],
and Chunping Liu[1(✉)]

[1] School of Computer Science and Technology, Soochow University,
Suzhou, Jiangsu, China
{jiyi, cpliu}@suda.edu.cn
[2] School of Rail Transportation, Soochow University, Suzhou, Jiangsu, China

Abstract. Image/video captioning based on neural network can generate accurate description. But how to convert visual information into natural language representation is a true enigma. Existing caption-guided saliency methods take the entire sentence as input to generate a saliency map, which exposes the region-to-word mapping. However, visual information is not related to every word in caption. We eliminate these meaningless stop words such as 'the', 'of' to avoid misleading. We also utilize MFB (Multi-modal Factorized Bilinear Pooling) to fuse C3D features, which could provide richer spatiotemporal information to exposure visual-word guided saliency. Such the system produces better spatiotemporal heatmaps for both predicted captions and arbitrary query sentences without introducing attentional layers. The experimental results on MSR-VTT and Flickr30K dataset surpasses the state-of-the-art by a significant margin.

Keywords: Image/video caption · Saliency map · MFB

1 Introduction

So far, top-down task-driven neural saliency methods have achieved remarkable results. They can effectively generate saliency heatmaps through given high-level semantic input. For example, top-down visual search [2, 20] can find out the important areas related to the object category without any pixel supervision at training time. Caption-guided saliency [9] can search for visual scenes through the salient elements described by natural language description, or search for relevant spatial salient regions and temporal salient regions through the description of the actions. But as in [6], stop words 'the' and 'to' do not have corresponding canonical visual signal. Non-visual words may mislead or weaken the validity of the process of generating a significant area.

Deep Image/video caption models [3, 12, 13, 19] are good at transforming visual input into language output. It can potentially discover a mapping between visual signal and language. But this is not enough to explain their inner principle and cannot provide a clear explanation about the internal mapping between images and generated words, because of its high degree of opacity.

© Springer Nature Switzerland AG 2018
R. Hong et al. (Eds.): PCM 2018, LNCS 11166, pp. 740–749, 2018.
https://doi.org/10.1007/978-3-030-00764-5_68

In our work, we proposed a visual-words guided saliency method to solve these problems. First, we used the visual-word extractor to find the visual-words in the captions, and then used our model to generate saliency for images and videos. Our method is based on an encoder-decoder model, and can generate spatial and temporal heatmaps for a given visual-word of input captions.

The main contributions of this paper:

1. We proposed a visual-word extractor that can effectively distinguish visual words from non-visual words and used visual-word to guide saliency generation.
2. We utilized MFB [18] to fuse C3D features and CNN features, which could improve the problem of discarding spatiotemporal information.

2   Related Work

2.1   Saliency

Saliency algorithms include bottom-up and top-down methods. Bottom-up methods focus on exploring low-level vision features such as color [4], background [23], and contrast [22]. Top-down saliency methods [2, 7, 17, 20, 21] are based on high-level task-specific prior knowledge. These methods do not need to define and adjust salient features manually, but directly learn the low-level features and high-level feature useful for saliency detection from the image. These methods use isolated object labels to recover the pixel importance for a given class. We apply this idea to language sentences. Our approach directly extracts the mapping between input pixels and visual-words from the encoder-decoder model. Without modifying the network and without requiring any explicit modeling of temporal or spatial attention, our method can capture the dependencies between the input and the output sequences by using memory cells and gating mechanisms.

2.2   Soft-Attention

The attention model simulates the human visual attention mechanism and solves the problems of missing semantics and information dilution. Recently, Xu et al. [15] first applied the soft-attention framework for machine translation [1] to image caption. Soft-attention assigns different weights to different areas of the image and processes all image areas selectively. Similarly, in video caption, soft-attention layer attends to specific temporal segments while generating the description [16]. Compared to our model, the soft-attention models requires an attention layer in addition to the encoder-decoder models.

2.3   Caption-Guided Saliency

The caption-guided saliency method [9] can generate temporal and spatial heatmaps for predictive captions and arbitrary query sentences, and estimates the significance of each temporal and spatial region by calculating the information gain generated by each frame and spatial region. However, Lu et al. [6] indicates that most image/video captioning methods associate visual information to every word in captions even non-visual words such as 'the' or 'to', but the decoder likely requires little visual information from image or video to predict these non-visual words such as 'the' or 'to'. Similarly, gradients from non-visual words in caption could mislead and diminish the overall effectiveness of the visual signal in guiding the saliency generation process.

3   Our Approach

3.1   Review on MFB

Multi-modal Factorized Bilinear Pooling (MFB) is a method of feature fusion. It first maps two vectors to the same higher dimension, finds the Hadamard product, and then reduces the dimensions through pooling and outputs the results. It can be expressed as follows:

$$\text{mfb}_{u,v} = MFB(u, v) \tag{1}$$

It can be extended as follows:

$$
\begin{aligned}
e_u &= W_{u,p}u + b_{u,p} \\
e_v &= W_{v,p}v + b_{v,p} \\
e_{u,v} &= e_u \odot e_v \\
\textit{mfb}_{u,v} &= sumpool(e_{u,v})
\end{aligned}
\tag{2}
$$

where $W_{u,p}$ and $W_{v,p}$ is weight matrix, $b_{u,p}$ and $b_{v,p}$ is bias. e_u and e_v represent the matrix after dimension transformation and they have the same dimensions. $e_{u,v}$ represents the result of the Hadamard product of two matrices. $\textit{mfb}_{u,v}$ indicates the result of the final fusion.

3.2   Feature Fusion Model

In case of a given video, we calculate the most salient spatiotemporal areas that corresponds to the given sentences description of an event. We divide the video into N segments and use the C3D model to extract the feature x_i of each segment, then choose one frame from each video segment to extract CNN feature y_i. After two features are obtained, concatenation or element-wise summations are most frequently used for multi-modal feature fusion. Since the distributions of two feature sets in different modalities may vary significantly, the representation capacity of the fused features may be insufficient and limit the final prediction performance, so we use MFB feature fusion methods [18] to fuse two features (see Fig. 1).

$$z_i = MFB(x_i, y_i) \forall\, i \in \{0, \ldots, n\} \tag{3}$$

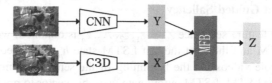

Fig. 1. Using MFB feature fusion method to fuse C3D and CNN features, improve the expression ability of fusion features, and provide rich spatiotemporal information for the entire network.

3.3 Visual-Word Extractor

Not all words in a sentence has correspondent visual information in the image and video. By analyzing captions, we know that noun phrases, verbs and adjectives normally have visual information in images or video frames, and some stop words do not have any visual information. In fact, gradients from non-visual words could mislead even diminish the overall effectiveness of visual signal in guiding the saliency generation process. Based on the above reasons, we have filtered useless words when we feed captions as input. We use NLTK's list of stop word, including 128 English stop words, to filter words for more concise input.

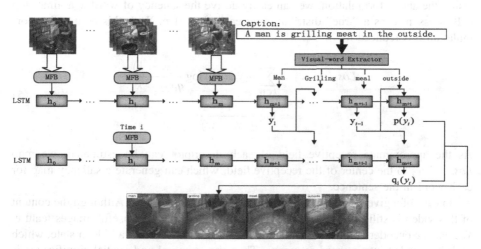

Fig. 2. Overview of our proposed visual-word guided visual saliency approach for temporal saliency in video. We introduce an encoder-decoder model to produce temporal saliency values for each frame i and visual-word t in a given input sentence. The values are computed by removing all but the ith descriptor from the input sequences, doing a forward pass, and comparing to the original word probability distribution. A similar idea can be applied to spatial image saliency.

3.4   Visual-Word Guided Saliency

After video fusion feature sequence $Z' = z_1', z_2', \ldots, z_n'$ is obtained, we feed the feature sequence into LSTM and then get the last LSTM state h_m^e. This vector represents the information of whole videos. At the beginning of the sentence generation h_m^e will be input to the decoder LSTM. LSTM output gate uses this vector to predict words at each time step t. As the words are generated, the visual concept in the decoder LSTM continues to evolve, and the evolved vectors in turn interacts with the LSTM output gate to produce the next word.

Our method uses a decoder to predict the probability distribution of words from the vocabulary at each time step. We use it as our "true" distribution. Then in order to measure how much information the measurement item i descriptor carries in the time step t, we remove the descriptors other than the i descriptors from the code. After passing forward through the encoder and decoder, we get a new probability distribution. Then we use KL-divergence to calculate the information loss of the two distributions (see Fig. 2):

$$
\begin{aligned}
p(y_t) &= P(y_t|y_{1:t-1}, z'_{1:26}) \\
q_i(y_t) &= P(y_t|y_{1:t-1}, z_i') \\
Loss(t, i) &= D_{KL}(p(y_t)\|q_i(y_t))
\end{aligned}
\tag{4}
$$

Using the above formulation, we can easily derive the saliency of word w at time step t. It is assumed as a "true" distribution at each time step. With this assumption formulation 6 is reduced to:

$$
\begin{aligned}
Loss(t, i, w) &= \sum_{k \in W} p(y_t = k) \log \frac{p(y_t = k)}{q_i(y_t = k)} \\
&= \log \frac{1}{q_i(y_t = w)}
\end{aligned}
\tag{5}
$$

As the approximate receptive field of each descriptor can be estimated, we map $Loss(t, i, w)$ to the center of the receptive field, which can generate a saliency map for each word in the sentence.

In case of a given image, we treat the image as a video frame. Although the content of this video is still, we can still use the C3D model to extract useful images feature. We use the encoder LSTM to encode this visual information into a hidden state, which is then decoded into a word sequence. Then the temporal and spatial significance is achieved through the same process as the previous section.

4   Experiments

4.1   Dataset

We evaluated our method on two datasets: Flickr30K Entities [8] and Microsoft Research Video to text (MSR-VTT) [14].

Flickr30K. Flickr30K contains 31,783 images collected from Flickr. The majority of these images depict human activities. Each image contains five captions. We use the publicly splits containing 1000 images for validation and test each. In addition we use Flickr30k Entities which is an expansion of the dataset Flickr30k that contains a manual bounding box of each caption in the dataset.

MSR-VTT. The dataset contains 10,000 video segments. A total of 41.2 h of online video covers a very comprehensive category and different visual content. It is divided into three parts: training set, verification and test set. Each video clip is labeled with about 20 sentences.

4.2 Implementation Details

Our model is based on the TensorFlow framework [17]. We pre-train InceptionV3 [11] on ImageNet [5] and C3D model was pre-trained on UCF101 [10] to extract visual descriptions. For the video we have proposed a pooling layer for video frames and video segments. For images we use the convolutional layer as the input to the encoder.

Table 1. Evaluation of the proposed method on localization all noun phrases from the groundtruth captions in the Flickr30kEntities dataset using the pointing game protocol. "Baseline random" samples the points of maximum saliency uniformly from the whole image and "Baseline center" corresponds to always pointing to the center.

	Bodyparts	Animals	People	Instruments	Vehicles	Scene	Others	Clothing	Avg per NP
Baseline random	0.100	0.240	0.318	0.179	0.275	0.524	0.246	0.151	0.268
Baseline center	0.201	0.599	**0.647**	**0.496**	0.644	0.652	0.384	**0.397**	0.492
Caption guided saliency [14]	0.194	0.690	0.601	0.458	**0.645**	**0.667**	0.427	0.360	0.501
Our model	**0.214**	**0.701**	0.613	0.445	0.643	0.653	**0.438**	0.351	**0.507**

4.3 Quantitative Analysis

To evaluate our saliency performance, we use the Point game strategy [20] to test our method. We feed ground truth captions from the test split of Flickr30k feed into our model. A hit is counted if the maximum saliency point lies on the one of the annotated instances of noun phrase in each GT caption of Flickr30kEntities, otherwise a miss is counted. We use the formula $Acc = \frac{\#Hits}{\#Hits + \#Missed}$ to calculate the accuracy of each noun phrase category in each caption. The overall performance is measured by the mean across different categories.

Table 1 shows the average accuracy of all noun phrases in Flickr30kEntities obtained by our model and the accuracy of each category. We compare them with "baseline random", where the maximum saliency point is sampled from the entire image. And compared with the "baseline center", where the baseline is assumed to be the maximum point is always in the center of the image and mimic the center bias present in consumer photos. Compared our method with Caption-guided Saliency [14], our method greatly improved the algorithm's shortcomings and has also been greatly improved performance of saliency.

4.4 Saliency Visualizations

Figure 3 show the comparison of spatial and temporal saliency map on videos from MSR-VTT dataset with caption-guided saliency. The arbitrary query comes from the ground truth descriptions. For each nounphrase, the saliency map is generated by the visual-words in the phrase and then renormalizing them. Red shows the most saliency and blue is the lowest in the map. From the figure we can see our model method almost always localizes the correct areas and generate more higher saliency in correct area (e.g., 'meat', 'outside').

Figure 4 compares our saliency heatmaps on images from Flickr30kEntities with caption-guided saliency. Most discriminative images for each word are outlined in the same color as the word. Darker gray indicates higher magnitude of saliency for words. From the figure we can see our method have better performance than caption-guided saliency. An interesting observation about the image is that our method exactly find the area of "girl", "snowsuit" and "mountain". Comparing with us, caption-guided saliency locates wrong regions.

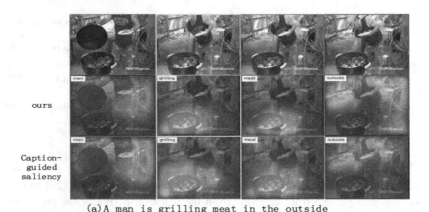

ours

Caption-
guided
saliency

(a) A man is grilling meat in the outside

Fig. 3. Spatial and temporal saliency in videos. For each word, darker grey indicates higher relative saliency of the frame. Most relevant frames for each word are shown at the bottom, highlighted with the same color. (Color figure online)

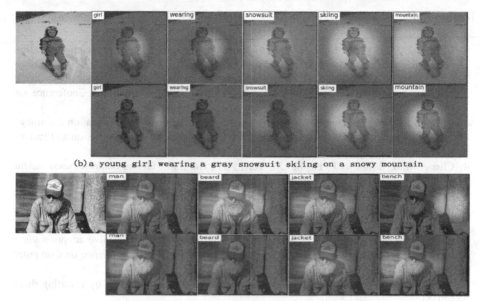

(b) a young girl wearing a gray snowsuit skiing on a snowy mountain

(a) An old man with long beard in jacket is sitting on bench

Fig. 4. Saliency maps (red to blue denotes high to low value) in Flickr30kEntities generated for an arbitrary query sentence (shown below). Each row shows saliency map for different visual-words extracted from the query. Comparing our saliency maps with caption-guided saliency. (Color figure online)

5 Conclusion

In this paper, we proposed a saliency detection approach guided by visual-words to understand the complex decision processes in image/video captioning generation. And we adopt MFB to fuse C3D features and CNN features, which provide richer spatiotemporal information for our network. We have found that our approach achieves more accurate heatmaps than existing method. The model is general and can be used to understanding a wide variety of encoder-decoder architectures.

Acknowledgements. This work was partially supports by National Natural Science Foundation of China (NSFC Grant No. 61773272, 61272258, 61301299, 61572085, 61170124, 61272005), Provincial Natural Science Foundation of Jiangsu (Grant No. BK20151254, BK20151260), Science and Education Innovation based Cloud Data fusion Foundation of Science and Technology Development Center of Education Ministry (2017B03112), Six talent peaks Project in Jiangsu Province (DZXX-027), Key Laboratory of Symbolic Computation and Knowledge Engineering of Ministry of Education, Jilin University (Grant No. 93K172016K08), and Collaborative Innovation Center of Novel Software Technology and Industrialization.

References

1. Bahdanau, D., Cho, K., Bengio, Y.: Neural machine translation by jointly learning to align and translate. arXiv preprint arXiv:1409.0473 (2014)
2. Cao, C., et al.: Look and think twice: capturing top-down visual attention with feedback convolutional neural networks. In: Proceedings of the IEEE International Conference on Computer Vision, pp. 2956–2964 (2015)
3. Chen, X., Lawrence Zitnick, C.: Mind's eye: a recurrent visual representation for image caption generation. In: Proceedings of the IEEE Conference on Computer Vision and Pattern Recognition, pp. 2422–2431 (2015)
4. Cheng, M.M., Mitra, N.J., Huang, X., Torr, P.H., Hu, S.M.: Global contrast based salient region detection. IEEE Trans. Pattern Anal. Mach. Intell. **37**(3), 569–582 (2015)
5. Deng, J., Dong, W., Socher, R., Li, L.J., Li, K., Fei-Fei, L.: Imagenet: a large-scale hierarchical image database. In: IEEE Conference on Computer Vision and Pattern Recognition, CVPR 2009, pp. 248–255. IEEE, Washington (2009)
6. Lu, J., Xiong, C., Parikh, D., Socher, R.: Knowing when to look: adaptive attention via a visual sentinel for image captioning. In: Proceedings of the IEEE Conference on Computer Vision and Pattern Recognition (CVPR), vol. 6 (2017)
7. Mahendran, A., Vedaldi, A.: Understanding deep image representations by inverting them (2015)
8. Plummer, B.A., Wang, L., Cervantes, C.M., Caicedo, J.C., Hockenmaier, J., Lazebnik, S.: Flickr30k entities: collecting region-to-phrase correspondences for richer image-to-sentence models. In: 2015 IEEE International Conference on Computer Vision (ICCV), pp. 2641–2649. IEEE (2015)
9. Ramanishka, V., Das, A., Zhang, J., Saenko, K.: Top-down visual saliency guided by captions. In: Proceedings of the IEEE Conference on Computer Vision and Pattern Recognition (CVPR), vol. 1, p. 7 (2017)
10. Soomro, K., Zamir, A.R., Shah, M.: Ucf101: a dataset of 101 human actions classes from videos in the wild. arXiv preprint arXiv:1212.0402 (2012)
11. Szegedy, C., Vanhoucke, V., Ioffe, S., Shlens, J., Wojna, Z.: Rethinking the inception architecture for computer vision. In: Proceedings of the IEEE Conference on Computer Vision and Pattern Recognition, pp. 2818–2826 (2016)
12. Venugopalan, S., Xu, H., Donahue, J., Rohrbach, M., Mooney, R., Saenko, K.: Translating videos to natural language using deep recurrent neural networks. arXiv preprint arXiv:1412.4729 (2014)
13. Vinyals, O., Toshev, A., Bengio, S., Erhan, D.: Show and tell: a neural image caption generator. In: 2015 IEEE Conference on Computer Vision and Pattern Recognition (CVPR), pp. 3156–3164. IEEE, Washington (2015)
14. Xu, J., Mei, T., Yao, T., Rui, Y.: MSR-VTT: a large video description dataset for bridging video and language. In: 2016 IEEE Conference on Computer Vision and Pattern Recognition (CVPR), pp. 5288–5296. IEEE, Washington (2016)
15. Xu, K., et al.: Show, attend and tell: neural image caption generation with visual attention. In: International Conference on Machine Learning, pp. 2048–2057 (2015)
16. Yao, L., et al.: Describing videos by exploiting temporal structure. In: Proceedings of the IEEE International Conference on Computer Vision, pp. 4507–4515 (2015)
17. Young, P., Lai, A., Hodosh, M., Hockenmaier, J.: From image descriptions to visual denotations: new similarity metrics for semantic inference over event descriptions. Trans. Assoc. Comput. Linguist. **2**, 67–78 (2014)

18. Yu, Z., Yu, J., Fan, J., Tao, D.: Multi-modal factorized bilinear pooling with co-attention learning for visual question answering. In: Proceedings of IEEE International Conference on Computer Visualization, vol. 3 (2017)
19. Zeiler, M.D., Fergus, R.: Visualizing and understanding convolutional networks. In: Fleet, D., Pajdla, T., Schiele, B., Tuytelaars, T. (eds.) ECCV 2014. LNCS, vol. 8689, pp. 818–833. Springer, Cham (2014). https://doi.org/10.1007/978-3-319-10590-1_53
20. Zhang, J., Lin, Z., Brandt, J., Shen, X., Sclaroff, S.: Top-down neural attention by excitation backprop. In: Leibe, B., Matas, J., Sebe, N., Welling, M. (eds.) ECCV 2016. LNCS, vol. 9908, pp. 543–559. Springer, Cham (2016). https://doi.org/10.1007/978-3-319-46493-0_33
21. Zhou, B., Khosla, A., Lapedriza, A., Oliva, A., Torralba, A.: Learning deep features for discriminative localization. In: 2016 IEEE Conference on Computer Vision and Pattern Recognition (CVPR), pp. 2921–2929. IEEE, Washington (2016)
22. Zhou, L., Yang, Z., Yuan, Q., Zhou, Z., Hu, D.: Salient region detection via integrating diffusion-based compactness and local contrast. IEEE Trans. Image Process. **24**(11), 3308–3320 (2015)
23. Zhu, W., Liang, S., Wei, Y., Sun, J.: Saliency optimization from robust background detection. In: Proceedings of the IEEE Conference on Computer Vision and Pattern Recognition, pp. 2814–2821 (2014)

Text-Guided Dual-Branch Attention Network for Visual Question Answering

Mengfei Li, Li Gu, Yi Ji[✉], and Chunping Liu[✉]

School of Computer Science and Technology, Soochow University,
Suzhou, Jiangsu, China
{jiyi, cpliu}@suda.edu.cn

Abstract. VQA is a multimodal joint learning task of AI-complete. The goal of our work is to present a text-guided dual-branch attention network (TDAN) for visual question answer. The focus of different attention models is different, and merging multiple models can produce better answers. So TDAN has two branches (i.e., two sub-models) and it first separately generates predictions for the answers through dual branch, and then generates weight to dual branch through question guidance. Thus, the key layers of two branches are merged into one and produce final output. We also exploit a one-dimensional gated convolutional neural network (1D-GCNN) encoding question texts and text embedding method with position information for high efficiency. In experiments, ours model is superior to the previous models, on the VQA 2.0 dataset from 61.89% to 63.94%, and on the COCOQA dataset from 62.5% to 63.98%.

Keywords: VQA · 1D-GCNN · TDAN

1 Introduction

Visual question answer (VQA) has become a very significant research issue in the modern era [2, 6, 13, 15]. Recently, VQA has explored models based on attention mechanism [16–19], with the following models: (1) Highlight areas related to the question in the image. (2) Highlight words (related to the image or not) in the question. (3) Generate text or image features highlighting key channels. Though for different models, the focus of their attention model will be non-overlapping. Notice this characteristic, we propose to fuse the key layers of different models to produce better output. Specifically, this model has two unique features:

Text-guided Dual-branch attention model: We propose a new mechanism to fuse the key layers of different models. The weight is guided by the question text, and is used to fuse the key layers of two previous models to produce the final result.

One-dimensional gated convolutional neural networks: We use one-dimensional gated CNN (1D-GCNN) to process question features. We first concatenate the word embedding and position embedding, then perform a one-dimensional convolution operation (the convolution kernels have various dimensions as 1, 3, and 5), then use the gated linear unit to activate. Overall, the main contributions of our work are:

© Springer Nature Switzerland AG 2018
R. Hong et al. (Eds.): PCM 2018, LNCS 11166, pp. 750–760, 2018.
https://doi.org/10.1007/978-3-030-00764-5_69

- We propose dual-branch attention model, which combine the key layers of two models and generate the final results.
- We use one-dimensional gated CNN to process the text. This can give full play to the GPU's parallel processing capabilities. We also proposes a word embedding method with position information, which improves the ability of the model to understand the semantics by determining the position of each word in the question.
- We use tensorflow to implement the MFH-co-attention [21] model and use it as a submodel.
- Finally, we evaluated our proposed model for two large datasets VQA [3] and COCO-QA [15]. Our model has established a new level of art in both VQA and COCO-QA. The best method is 2.05% and 1.48% higher than before. We also conducted ablation studies to quantify the role of different components in our proposed model.

2 Related Work

2.1 Multi-modal Bilinear Model of VQA

Multi-modal feature fusion plays an important and fundamental role in VQA. Fukui et al. [5] proposed a multi-modal compact bilinear pool (MCB) that uses the outer product of two eigenvectors to generate a very high-dimensional quadratic unwrapping feature. Kim et al. [8] proposed a multi-modal low-rank bilinear pooling (MLB) method based on Hadamard products of two eigenvectors. Yu et al. proposed Multi-modal Factorized Bilinear Pooling (MFB) [20]. MFB is also a Hadamard product based on two eigenvectors. The difference is that MFB will have a pooled operation, MFB can be expressed as follows:

$$z = \text{SumPooling}\left(U^T x \odot V^T y, k\right) \tag{1}$$

Where $U \in R^{m \times (o*k)}$ and $V \in R^{n \times (o*k)}$ are projection matrices, o is the dimension of the output feature, and k is the expansion coefficient. Yu et al. [21] also proposed Multi-modal Factorized High-order Pooling (MFH). The MFH uses MFB multiple times and inputs the last intermediate result to the next calculation. Finally, all MFB results are stitched together as the final output.

2.2 The Attention Mechanism in VQA

The attention mechanism is widely used in VQA, which allows the model to selectively extract useful visual or textual information. Lu et al. [11] proposed an image-question co-attention mechanism that not only focuses on the relevant image regions but also attends to the important question words. Yang et al. [19] introduced soft attention and proposed a stacking attention model. We [22] have previously propose a multi-layer cross-guided attentional network (MCAN) that utilizes multi-modal information to fully intersect.

3   Method

The ultimate goal of the VQA is to infer the best answer to the relevant question. The process of predicting the answer can be expressed as:

$$ans = \arg\max_{a \in A} p(a|v, q, \theta) \tag{2}$$

Where θ is the model parameter and A is the answer set. The overall structure of TDAN is shown in Fig. 1. In this section, we first introduce image models and language models, and then detail our network.

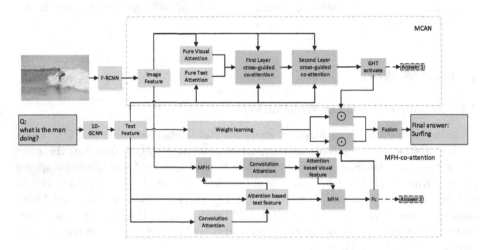

Fig. 1. The overall structure of TDAN. The two sub-modules generate answer vectors respectively, and the model weight is used to weight-add the two key layers and output the result.

3.1   Input Indication

Bounding Box Image Encoding. We used a fine-tuned, pre-trained, faster R-CNN to encode images with ResNet-101 CNN. Select the 36 highest-bounded boundary box features as ours image representation.

Question Text Encoding. We use 1D-GCNN to encode question texts. Given a question q, we first embed the question into the semantic space. At the same time, the word position information is stitched behind the word feature vector to form a new feature vector. All eigenvectors of a question are spliced into a matrix qe. Then perform a one-dimensional convolution of qe. The features of the convolution output is $2 * f$, where f represents the number of features that will eventually be output. We divide it into two parts. The number of features in each part is f. Then softmax is performed on the second part, and the result after softmax is multiplied with the first part as the output of the one-dimensional gated convolutional neural network.1D-GCNN can be expressed as follows:

$$qe = concat(qs, p) \tag{3}$$

$$c = [c_1, c_2] = convolution(qe) \tag{4}$$

$$o = c_1 \odot softmax(c_2) \tag{5}$$

Where qs represents question embedding, p represents position embedding, convolution() represents one-dimensional convolution, and $\odot$ represents Hadamard product.

3.2   Parallel Cross-Guide Co-attention (Branch 1)

This model maintains the state of the representation of each attentional layer through the three context memory vectors generated by the image and question representations. In this section, we describe in detail the attention mechanisms used by each layer.

Visual Attention. Our model implements two visual attention mechanisms. The first attention mechanism is called the Pure Visual Attention Mechanism (PVA), which only considers the original information of the image. The second type of attention mechanism is a classical question-guided visual attention (QVA) mechanism similar to most attention-focused VQA models [18, 19].

Text Attention. We also implement two textual attention mechanisms to focus on the specific words in question. The first is called the pure text attention mechanism (PTA), which aims to focus attention on the original meaningful words in the sentence. Usually, the image contains information related to the question. Inspired by this, we implement an image-guided text attention (ITA) mechanism.

Nonlinear Activation Unit. We use gated hyperbolic tangent (GHT) activation as NLA. This is a gating operation similar to recursive units such as LSTM and GRU.

MCAN for Visual Question Answering. MCAN has set up three memory units to maintain the status of images, questions, and multi-modal joint learning functions, respectively. These units are recursively updated by:

$$m_v^k = m_v^{k-1} + v_m^k \tag{6}$$

$$m_u^k = m_u^{k-1} + u_m^k \tag{7}$$

$$m_j^k = m_j^{k-1} + m_v^k \odot m_u^k \tag{8}$$

Where k is the number of cross-guided attentional layers set to at least 1, m_v^k and m_u^k are the visual and textual memory states of the kth attentional layer, v_m^k and u_m^k are weighted sums of visual and textual attention, and m_j^k is a contextual joint learning memory unit that contains important content about images and questions. We feed the last joint learning memory vector m_j^k into the NLA function and then use the single-layer softmax classifier with cross-entropy to predict the answer by Eq. (2):

$$h^k = NLA\left(m_j^k\right) \tag{9}$$

$$p_a = softmax\left(W_a h^k + b_a\right) \tag{10}$$

Where h^k is the output of NLA, p_a is the probability distribution of candidate answers, W_a is a learning weight matrix, b_a is bias.

3.3   Alternating Co-attention (Branch 2)

In this section, we will elaborate on the attentional mechanism of alternating co-attention.

Multi-modal Factorized High-Order Pooling (MFH). Multi-modal Factorized High-order Pooling (MFH) is an improved version of MFB [20] that performs MFB operations several times and inputs the last intermediate result to the next calculation, resulting in better results than MFB. MFH can be expressed as follows:

$$e_u^k = W_{u,p}^k u + b_{u,p}^k \tag{11}$$

$$e_v^k = W_{v,p}^k v + b_{v,p}^k \tag{12}$$

$$e_{u,v}^k = \begin{cases} e_u^k \odot e_v^k & (k = 1) \\ e_u^k \odot e_v^k \odot e_{u,v}^{k-1} & (k > 1) \end{cases} \tag{13}$$

$$o_{u,v}^k = sumpool\left(e_{u,v}^k\right) \tag{14}$$

$$mfh_{u,v} = concat\left(o_{u,v}^1, o_{u,v}^2, o_{u,v}^3 \ldots\right) \tag{15}$$

Where $W_{u,p}^k$ and $W_{v,p}^k$ are learned parameters, $b_{u,p}^k$ and $b_{v,p}^k$ are the bias.

MFH with Co-attention. This model first makes the text self-attention, and then merges with the image features to guide the attention of the image. After adding the attention, the image features are merged back to the previously added text features of self-attention and finally output the results.

3.4   Text-guided Dual-Branch Attention Network

Both MCAN and MFH-Co-Attention can generate VQA answers independently. Text-guided dual-branch attention network (TDAN) can give two models different attention levels and integrate them to produce the final results.

We take the output of the one-dimensional gated convolutional neural network as the input for the TDAN, and perform the pooling operation first. After pooling, the results are processed in two layers, and finally two outputs are obtained, namely the weights of the key layers of the two models. The process can be expressed as follows:

$$i = maxpool(qe) \tag{16}$$

$$hs = softmax(W_{i,h}i + b_{i,h}) \tag{17}$$

$$ia = hs \odot i \tag{18}$$

$$os = softmax(W_{h,o}ia + b_{h,o}) \tag{19}$$

$$out = os_0 \odot k1 + os_0 \odot k1 \tag{20}$$

Where $W_{i,h}$ and $W_{h,o}$ is weight matrix. $b_{i,h}$ and $b_{h,o}$ is bias. $k1$ and $k1$ represent the key layers of the first and second models, *out* indicates the final output.

4   Experiments

We performed experiments on the VQA 2.0 dataset and the COCO-QA dataset on tensorflow.

Table 1. Comparison on the VQA 2.0 (test-std) dataset.

Method	Accuracy			
	All	Yes/no	Numb	Other
per Q-type prior [3] as reported in [1]	32.06	64.42	26.95	8.76
d-LSTM Q [3] as reported in [1]	43.01	67.95	30.97	27.20
Deeper LSTM Q norm. I [10] as reported in [1]	51.61	73.06	34.41	39.85
HieCoAtt [11] as reported in [7]	51.02	70.77	34.55	42.33
NMN [2] as reported in [1]	51.62	73.38	33.23	39.93
SAN [19] as reported in [1]	52.02	68.89	34.55	43.80
MCB [5] as reported in [1]	59.71	77.91	37.47	51.76
Oracle (GVQA, SAN) [1]	61.96	**85.65**	**43.76**	48.75
TDAN (ours)	**63.94**	79.49	43.13	**55.41**

4.1   Experiment Details

The network's 1D-GCNN feature number is set to 512. We trained our network with a random gradient descent and used an Adam solver with a learning rate of 0.0005. With the batch size set to 100. For the VQA 2.0 dataset, the train + val data set is used for training and the candidate answer is set to 3000. For the COCO-QA data set, the candidate answer is set to 430.

4.2   Results and Analysis

Table 1 shows the performance of our method on the VQA 2.0 dataset and comparison with the cutting-edge methods. Oracle (GVQA, SAN) is the latest and best method.

Table 1 shows that our method achieved better overall accuracy than the Oracle (GVQA, SAN) on VQA 2.0. It can be noted that Oracle (GVQA, SAN) has shown the best performance on issues like Yes/no because it specifically selects the Yes/no questions to be handled separately. Even so, our model is still higher than Oracle (GVQA, SAN) in overall accuracy. Another point to emphasize is that the Object type question is more difficult to answer than the Yes/No type question, and our model is nearly 4% higher than the MCB on it. This also proves the superiority of our model.

Table 2. Comparison on the COCO-QA dataset, in percentage and '-' represents the result is not available.

Method	All	Object	Number	Color	Location	WUPS0.9	WUPS0.0
2-VIS + BLSTM [15]	55.0	58.2	44.8	49.5	47.3	65.3	88.6
ATT-VGG-SEG [4]	58.1	62.5	45.7	46.8	53.7	68.4	89.9
IMG-CNN [12]	58.4	-	-	-	-	68.5	89.7
DPPnet [14]	61.2	-	-	-	-	70.8	90.6
SAN [19]	61.6	65.4	48.6	57.9	54.0	71.6	90.9
QRU [9]	62.5	65.1	46.9	**60.5**	57.0	72.6	**91.6**
TDAN (ours)	**63.98**	**67.20**	**49.00**	59.45	**57.18**	**73.34**	91.35

We also evaluated our model on the COCO-QA dataset and illustrate performance in Table 2. We also compared accuracy with other methods. Our approach increased state-of-the-art QRU [9] from 62.5% to 63.98%. Specifically, our model has improved on almost categories of issues. Our model also uses WUPS to measure performance. Compared with QRU [9], our model increased by 1.02% on WUPS0.9. Though it is a bit inferior in WUPS0.0. It may be due to the fact that our model is more pursuing accuracy and ignores similarities.

4.3 Ablation Study

In this section, we conduct ablation studies to analyze the contribution of each component to our proposed model. We eliminated our complete model and evaluated the VQA test development set. Table 3 shows the performance of our ablation experiments on the VQA 2.0 test development set.

Table 3. Ablation study on the VQA 2.0 test-std

No.	Method	Acc
Model 1	Bidirectional gru based MCAN model	61.89
Model 2	1D-GCNN based MCAN model (without position embedding)	61.92
Model 3	1D-GCNN based MCAN model (with position embedding)	62.57
Model 4	Model that sums key layers of MCAN and MFH-co-attention	63.03
Model 5	TDAN model incorporating MCAN and MFH-co-attention	63.94

Comparing Model 1 and Model 2, after using 1D-GCNN instead of bidirectional GRU to encode the questions text, we get an accuracy improvement of 0.03%. By comparing Model 2 and Model 3, we found that after adding the position information to the words in the question, the accuracy increased by 0.65%. For Model 3 and Model 4, we found that combining the key layers of two independent models yielded better results. Comparing Model 4 and Model 5, we found that after using the TDAN fusion method, the accuracy increased by 0.91%.

Ablation study does demonstrate the rationality of the proposed model and that the question coding method has the least impact on performance.

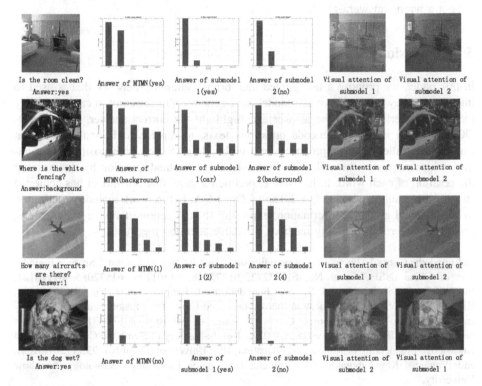

Fig. 2. Qualitative example of a VQA 2.0 test set. Submodel 1 indicates MCAN, and submodel 2 indicates MFH-co-attention. The first three lines are correct examples, and the last line is an error example.

4.4 Qualitative Evaluation

Figure 2 shows several typical cases in our model. In the first line of Fig. 2, when asked 'if the room is clean?', MCAN looks at most of the room and answers the correct answer for 'yes'. The MFH-co-attention model only focuses on a painting on the wall, and then makes a 'no' mistake. After the two sets of probabilities are merged by TDAN, we get the correct answer to 'yes'. In the second row of Fig. 2, when asked 'where the white fence is?', MCAN barely noticed the white fence on the second floor of the far end. Its answer was 'car'. The MFH-co-attention is concerned with the remote background and the near-end car and gives the correct answer. The two sets of answer probabilities are merged into one group and the correct answer is obtained. What we want to emphasize is the third row of Fig. 2. The question is: 'How many aircrafts are there?'. The correct answer is '1', while the two sub-models get the answers of '4' and '2' respectively, which are not the correct answer, but '1' has the second most probability in both sub-models, after TDAN fusion. We got the correct answer '1'. This also fully proves the advanced nature of our model. But our algorithm is not omnipotent. The fifth row in Fig. 2 is a counterexample. The question is: Is the dog wet? The two models received different answers. After being merged by TDAN, we got a wrong answer.

5   Conclusions

In this paper, we propose a text-guided dual-branch attention network that is used to integrate the key layers of two sub-models and generate a final output. It can effectively reduce the interference in the answer and highlight the correct answer. We use 1D-GCNN, which is used to encode questions texts, to replace traditional RNNs that cannot be parallelized. We propose a text embedding method with position information, which improves the ability of the model to understand question by determining the position of each word. Exhaustive experiments show that our model has achieved state-of-arts results on both VQA 2.0 and COCO-QA. Future work includes continuing to add external guidance information, exploring the extraction of image features, and providing better joint learning methods for these attention mechanisms.

Acknowledgements. This work was partially supported by National Natural Science Foundation of China (NSFC Grant No. 61773272, 61272258, 61301299, 61572085, 61170124, 61272005), Provincial Natural Science Foundation of Jiangsu (Grant No. BK20151254, BK20151260), Science and Education Innovation based Cloud Data fusion Foundation of Science and Technology Development Center of Education Ministry (2017B03112), Six talent peaks Project in Jiangsu Province (DZXX-027), Key Laboratory of Symbolic Computation and Knowledge Engineering of Ministry of Education, Jilin University (Grant No. 93K172016K08), and Provincial Key Laboratory for Computer Information Processing Technology, Soochow University.

References

1. Agrawal, A., Batra, D., Parikh, D., Kembhavi, A.: Don't just assume; look and answer: Overcoming priors for visual question answering. arXiv preprint arXiv:1712.00377 (2017)
2. Andreas, J., Rohrbach, M., Darrell, T., Klein, D.: Neural module networks. In: Proceedings of the IEEE Conference on Computer Vision and Pattern Recognition, pp. 39–48 (2016)
3. Antol, S., Agrawal, A., Lu, J., Mitchell, M., Batra, D., Lawrence Zitnick, C., Parikh, D.: VQA: visual question answering. In: Proceedings of the IEEE International Conference on Computer Vision, pp. 2425–2433 (2015)

4. Chen, K., Wang, J., Chen, L.C., Gao, H., Xu, W., Nevatia, R.: ABC-CNN: an attention based convolutional neural network for visual question answering. arXiv preprint arXiv: 1511.05960 (2015)
5. Fukui, A., Park, D.H., Yang, D., Rohrbach, A., Darrell, T., Rohrbach, M.: Multimodal compact bilinear pooling for visual question answering and visual grounding. arXiv preprint arXiv:1606.01847 (2016)
6. Gao, H., Mao, J., Zhou, J., Huang, Z., Wang, L., Xu, W.: Are you talking to a machine? Dataset and methods for multilingual image question. In: Advances in Neural Information Processing Systems, pp. 2296–2304 (2015)
7. Goyal, Y., Khot, T., Summers-Stay, D., Batra, D., Parikh, D.: Making the V in VQA matter: elevating the role of image understanding in visual question answering. In: CVPR. vol. 1, p. 9 (2017)
8. Kim, J.H., On, K.W., Lim, W., Kim, J., Ha, J.W., Zhang, B.T.: Hadamard product for low-rank bilinear pooling. arXiv preprint arXiv:1610.04325 (2016)
9. Li, R., Jia, J.: Visual question answering with question representation update (QRU). In: Advances in Neural Information Processing Systems, pp. 4655–4663 (2016)
10. Lu, J., Lin, X., Batra, D., Parikh, D.: Deeper LSTM and normalized CNN visual question answering model (2015)
11. Lu, J., Yang, J., Batra, D., Parikh, D.: Hierarchical question-image co-attention for visual question answering. In: Advances in Neural Information Processing Systems, pp. 289–297 (2016)
12. Ma, L., Lu, Z., Li, H.: Learning to answer questions from image using convolutional neural network. In: AAAI, vol. 3, p. 16 (2016)
13. Malinowski, M., Rohrbach, M., Fritz, M.: Ask your neurons: a neural-based approach to answering questions about images. In: Proceedings of the 2015 IEEE International Conference on Computer Vision (ICCV), pp. 1–9. IEEE Computer Society (2015)
14. Noh, H., Hongsuck Seo, P., Han, B.: Image question answering using convolutional neural network with dynamic parameter prediction. In: Proceedings of the IEEE Conference on Computer Vision and Pattern Recognition, pp. 30–38 (2016)
15. Ren, M., Kiros, R., Zemel, R.: Exploring models and data for image question answering. In: Advances in neural information processing systems, pp. 2953–2961 (2015)
16. Shih, K.J., Singh, S., Hoiem, D.: Where to look: focus regions for visual question answering. In: Proceedings of the IEEE Conference on Computer Vision and Pattern Recognition, pp. 4613–4621 (2016)
17. Xiong, C., Merity, S., Socher, R.: Dynamic memory networks for visual and textual question answering. In: International Conference on Machine Learning, pp. 2397–2406 (2016)
18. Xu, H., Saenko, K.: Ask, attend and answer: exploring question-guided spatial attention for visual question answering. In: Leibe, B., Matas, J., Sebe, N., Welling, M. (eds.) ECCV 2016. LNCS, vol. 9911, pp. 451–466. Springer, Cham (2016). https://doi.org/10.1007/978-3-319-46478-7_28
19. Yang, Z., He, X., Gao, J., Deng, L., Smola, A.: Stacked attention networks for image question answering. In: Proceedings of the IEEE Conference on Computer Vision and Pattern Recognition, pp. 21–29 (2016)
20. Yu, Z., Yu, J., Fan, J., Tao, D.: Multi-modal factorized bilinear pooling with co-attention learning for visual question answering. In: Proceedings of IEEE International Conference on Computer Visualization, vol. 3 (2017)

21. Yu, Z., Yu, J., Xiang, C., Fan, J., Tao, D.: Beyond bilinear: generalized multimodal factorized high-order pooling for visual question answering. IEEE Trans. Neural Netw. Learn. Syst. 1–13 (2018)
22. Haibin, L., Shengrong, G., Yi, J., Jianyu, Y., Tengfei, X., Chunping, L.: Multi-layer cross-guided attention networks for visual question answering. In: Proceedings of the 2018 International Conference on Computer Modeling, Simulation and Algorithm (in printing)

A Fast Zero-Quantized Percentage Model for Video Coding with RDO Quantization

Haoyun Yang, Haibing Yin$^{(\boxtimes)}$, and Xiaofeng Huang

School of Communication Engineering,
Hangzhou Dianzi University, Hangzhou, China
15382363770@163.com

Abstract. In video coding, the percentage of zero-quantized coefficients, denoted by ρ, is directly determined by the quantization algorithm adopted. ρ-domain rate distortion (RD) modeling is widely employed to optimize the implementation algorithm for customizable modules such as rate control and mode decision etc. How to calculate or estimate ρ according to quantization algorithm is the first step task for ρ-domain RD modeling. There are two typical quantization algorithm, soft-decision quantization such as dead-zone, and soft-decision quantization such as rate distortion optimized quantization (RDOQ). RDOQ is more frequently employed in the latest video encoders compared with deadzone quantization due to its inspiring coding performance. In HDQ based video encoder, ρ can be easily obtained by simply rounding. However, it is computation-intensive to calculate ρ in video encoder with RDOQ, in which complicated trellis search is employed. This paper focus on developing estimation model for quickly estimating ρ for RDOQ based video coding. The contribution of this article is as follows: First, this paper develops the ρ model adaptively according to an adaptive deadzone offset model, which is modeled by imitating the behavior of RDOQ. Second, an accurate ρ model is adaptively built offline as function of weighted *SATD* (sum of absolute transformed distortion) denoted by *WSATD*, quantization step size q, and average *WSATD/q* estimated from ensemble. The weight in *WSATD* is adaptively determined according to the adaptive offset to simulate the behavior pattern of RDOQ as much as possible. Experimental results verify that the proposed model can quickly and accurately predict the ρ results of RDOQ with moderate implementation complexity. The proposed ρ model can be employed to estimate the percentage of zero quantized coefficients which can be used for fast all-zero detection and ρ domain rate distortion modeling in RDOQ based HEVC video encoder.

Keywords: Video coding · RDOQ · Rate-distortion model
Percentage of zero-quantized coefficients

1 Introduction

In video coding, the rate-distortion (R-D) function model is widely employed for rate control and mode decision. Several R-D models for DCT-based video coding have been proposed in the literature [1]. Accurately calculating R and D will cost video

© Springer Nature Switzerland AG 2018
R. Hong et al. (Eds.): PCM 2018, LNCS 11166, pp. 761–771, 2018.
https://doi.org/10.1007/978-3-030-00764-5_70

encoder high computation complexity especially in the latest HEVC coders with computation-intensive RDOQ and CABAC entropy codec. Developing fast estimation model for R and D had attracted intensive research interests in the past twenty years. Quantization directly determine the coding distortion and rate consumption. Thus, some works explore rate distortion models as function of quantization parameter. Some works explore the relationship between RD models with the percentage of zero-quantized coefficients denoted by ρ. ρ-domain modeling or q-domain modeling are two typical rate distortion modelling methods. Relatively, ρ-domain modeling can more accurately describe the microcosmic characteristics of CABAC in terms of accurate R estimation. Several works had also verified that ρ has a critical effect on the coding bit rate R, especially at low bit rates [2].

On the other hand, there are great amounts of blocks are quantized to all-zero in the latest HEVC standard, especially at low bit applications. In the latest standard, rate distortion optimization is applied in quantization. In RDO based quantization, quantization consumes high computation complexity due to that the quantizer need to evaluate the distortion and rate for all possible candidate results and select an optimal result in the sense of rate distortion optimization. If we can do all zero block (AZB) detection before using RDO, we can reduce the HEVC coding burden. In the past two decades, there are several AZB detection algorithms reported in the literature [3]. From the viewpoint of target quantization algorithm, these AZB decision algorithm were usually designed for video encodes with HDQ, for example dead-zone quantization, in which RD optimized RDOQ is not supported [3]. Thus, we can establish a ρ model for RDOQ to indirectly implement all-zero block decisions before using RDO.

ρ model is used to estimate the percentage of zero-quantized coefficients, and thus it is highly related with the quantization algorithm adopted in video encoder. There are two typical quantization algorithms in prevailing video encoders, hard-decision quantization (HDQ) such as deadzone and soft-decision quantization (SDQ) such as rate distortion optimized quantization (RDOQ). In HDQ based video encoder, ρ can be easily obtained by simply rounding. However, RDOQ can achieve superior coding rate distortion performance compared with HDQ. It is computation-intensive to calculate ρ in video encoder with RDOQ, in which complicated trellis search is employed. Thus, it is meaningful to develop ρ model for RDOQ based video coding.

Hence, this paper proposes ρ model for RDOQ based video coding. Firstly, we formulate ρ as functions of quantization step size, weighted *SATD* (sum of absolute transformed difference) and the mean of *WSATD*. By accurately measuring the adaptive deadzone offset estimated from the DCT coefficient distribution parameter Λ, we define an adaptive weight model for weighted *SATD* to obtain the adaptive offset δ and apply it to the weight model. Then, the three-dimensional ρ model is constructed using statistical curve fitting method from ensemble. The ρ model is developed individually in the cases of different types of TU blocks. The proposed model can quickly and accurately predict the ρ results of RDOQ.

This paper is organized as follows. Problem formulation and motivation analysis are given in Sect. 2. The proposed ρ model is given in Sect. 3. Section 4 gives the experimental results. Section 5 concludes the whole paper.

2   Problem Formulation and Motivation

2.1   RDOQ and ρ Model

ρ model is used to estimate the percentage of zero-quantized coefficients, and thus it is highly related with the quantization algorithm adopted in video encoder. There are two typical quantization algorithm, soft-decision quantization such as dead-zone, and soft-decision quantization such as rate distortion optimized quantization (RDOQ). In HDQ based video encoder, the quantization result z_i is adjusted using a rounding deadzone offset f described as follows:

$$z_i = floor\left(\frac{|c_i|}{q} + f\right) \tag{1}$$

where floor(.) is a direct integer operation, c_i is the DCT coefficient, and q is the quantization step size. In deadzone HDQ, ρ can be easily obtained by simply rounding.

In RDOQ, several candidate quantization results are determined according to the results of HDQ. Then, rate distortion optimization is employed to further refine the optimal quantization result from three candidate results. In RDOQ, inter-coefficient correlation is taken into consideration by joint rate distortion optimization with context adaptive binary arithmetic coding (CABAC). Suppose there are N coefficient in the current transform block, and there are m candidate quantization results preselected for further refinement. For a specific coefficient c_i, its candidate results are $l_{i1}, l_{i2},..., l_{im}$ which are centered about the result of HDD obtained with fixed rounding offset $f = 0.5$. RDOQ checks all candidates to select an optimal result l_i using RDO described [4] as follows.

$$l_i = \underset{k=1\sim m}{\arg\min}\left\{D(c_i, l_{ik}) + \lambda \cdot R(l_{ik}) + \sum_{j=i+1}^{N} D(c_j, \overline{l_j}) + \lambda \cdot R(\overline{l_j})\right\} \tag{2}$$

where $D(c_i, l_{ik})$ and $R(l_{ik})$ are the coding distortion and rate when c_i is quantized to l_{ik}, and λ is the Lagrangian multiplier, and $\overline{l_j}$ is the initial center of all candidates, i.e. the HDQ quantization result of the j-th coefficient in the current block. As shown in the above equation, inter-coefficient influence is considered in RDOQ. The backward inter-coefficient rate propagation is taken into consideration to optimally determine a specific coefficient's quantization result.

In general, trellis search is employed to solve this dynamic programming problem. In HEVC video codec reference model (HM), simplified trellis search is implemented to alleviate the heavy computation of the full trellis search. Nevertheless, the computation is still relatively high. As a result, it is computation-intensive to calculate ρ in video encoder with RDOQ, in which complicated trellis search is still desired instead of simple rounding. Since RDOQ can achieve superior coding rate distortion performance compared with HDQ, thus, it is meaningful to develop ρ model for RDOQ based video coding.

2.2   Analysis on RDOQ Based ρ Model

In order to develop ρ model, we need to build a function between ρ and some characteristic parameters which characterize the block and are relatively easy to be obtained for reasonable model computation complexity.

First, the sum of absolute transformed distortion denoted by *SATD* is usually employed to describe the current block's characteristics. In addition, it is also easy to obtain due to that it is available after mode decision. *SATD* was widely employed to develop the RD models in traditional works. In this work, we also tend to explore the relationship between ρ and *SATD*. In video coding, *SATD* is defined as follows.

$$SATD = \sum_i |c_i| \tag{3}$$

where c_i is the DCT coefficient. We can conclude that *SATD* is directly proportional to the amplitudes of all DCT coefficients as shown in Eq. (3).

We cannot directly apply *SATD* into ρ model building. In terms of ρ modeling, there is nonlinear relationship between *SATD* and ρ. For example, if one coefficient is larger than the optimal zero-quantized threshold, it is quantized to non-zero coefficient. However, increasing the coefficient intensity do not change a certain coefficient's contribution to ρ of the current block. In other words, if $| c_i |$ is larger than the optimal zero-quantized threshold, no matter how large $| c_i |$ is, the contribution of the current coefficient is stable.

What if we change the *SATD* definition accounting for the above nonlinearity in terms of ρ contribution? This paper tends to adopt weighted *SATD* to alleviate the nonlinearity extent between *SATD* and ρ. We propose a weight model according to the intensity of c_i to approximate a linear relationship between weight *SATD* and ρ.

Second, there is an intrinsic relationship between ρ and the quantization step size q. In general, the larger q is, the more coefficients are quantized to zero. There is proportional relationship between ρ and q. By jointly taking *WSATD* and q into consideration, we employ *WSATD/q* denoted by χ to build a relationship among ρ, *WSATD* and q. Suppose we apply a weight factor w_i for the coefficient c_i, we can then define the composite parameter χ as follows.

$$\chi = \sum_i \frac{w_i \times c_i}{q} \tag{4}$$

Where c_i is the DCT coefficient, w_i is the weight factor, and q is the quantization step size. Here, how to adaptively determine the weight factor w_i is one important problem.

Third, in our simulation, we find that the scatter results of (ρ, χ) samples are not convergent enough to use a close-form function to formulate it. RDOQ uses complicated trellis search to deal with inter-coefficient influence. It is not enough to accurately describe ρ only according to χ. Therefore, we need to introduce another parameter to develop more accurate three-dimensional ρ-model to imitate the behavior pattern of the optimal RDOQ.

By statistical analysis on the ρ-χ samples, we found that the average of χ estimated from a sliding window in the case of the same ρ, denoted by ω, also have a regular and obvious functional relationship with ρ. χ can measure the ensemble characteristics from the viewpoint of large sample analysis. Consequently, this paper will take the ω (average χ) as the third parameter for developing the ρ model. Figure 1 below shows the framework of the proposed ρ model.

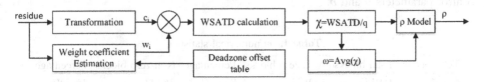

Fig. 1. Framework of the proposed ρ model

3   The Proposed ρ Model

3.1   *WSATD* with Deadzone Offset Adaptive Weight

Judging whether one DCT coefficient is quantized to zero or not accurately is the key to building the weight model for *WSATD* as analyzed in Sect. 2. The quantization determines the weight directly in terms of *SATD* weighting. In HDQ, whether or not c_i is quantized to zero can be determined by simple rounding according to deadzone offset f. However, in the RDOQ, it is computation-intensive to determine whether c_i is quantized to zero or not accurately.

In our previous work, we had made in-depth research to model an adaptive deadzone offset δ to improve the deadzone HDQ. By imitating the behavior pattern of RDOQ using statistic analysis, the offset δ mode is modeled as function of quantization parameter, quantization remainder, and the DCT coefficient distribution parameter Λ. This model is built offline, a three-dimension table is given offline. With this model, δ can be simply estimated by simple table lookup [5]. Based on Laplacian model, the distribution of DCT coefficients Λ can be estimated as follows.

$$\Lambda = \frac{1}{n} \sum_{i=1}^{n} |c_i| \qquad (5)$$

where c_i represents the DCT coefficient and n represents the number of statistical coefficients.

In general, the larger *SATD* is, the higher probability that coefficients are quantized to nonzero, and the smaller ρ is. Aiming at develop *WSATD* which is inversely proportional to ρ, this work proposes a weight model to adaptively adjust the contribution to *WSATD* according to coefficient the intensity $|c_i|$. If one coefficient c_i is quantized by RDOQ to zero, its contribution to *WSATD* is also close to zero. If one coefficient c_i is quantized to non-zero by RDOQ, its contribution to *WSATD* is identical regardless of its intensity $|c_i|$. Intuitively, the larger $|c_i|$ is, the smaller the corresponding weight w_i is.

That is, $|c_i|$ and w_i approximately comply with negative exponential function. This paper adopts the adaptive weight model shown as follows.

$$w_i = e^{-(|c_i|-b)/a} \tag{6}$$

where c_i represents the DCT coefficient, and w_i is the resulting weight. We can control the slope and centroid of the functional curve according to the $|c_i|$ by employing two control parameters a and b.

Table 1. m numerical statistics

TU type	m	Total blocks	ALL-zero blocks	Percentage	Non-zero blocks	Percentage
4×4	2	159173	132871	83.48%	26302	16.52%
8×8	3	108598	100647	92.68%	7951	0.72%
16×16	6	140682	135398	96.24%	5284	3.76%
32×32	6	26594	26445	99.44%	149	0.056%

In order to quickly determine w_i, we can pre-judge whether c_i is quantized to zero in the case of RDOQ according to c_i. On one hand, according to the principle of HDQ, coefficients whose $|c_i|$ are within the range $[0, (1 - f)q)$ are quantized by HDQ to zero, and these coefficients are also quantized to zero ones by RDOQ. On the other hand, for the coefficients whose $|c_i|$ are within the range $[(1 - f)q, q)$, the coefficients are quantized to nonzero by HDQ, they may be quantized to zero or nonzero coefficients by RDOQ. In general, if one TU only contain m possible nonzero coefficients with intensity $[(1 - f)q, q)$, the current block may be quantized to all-zero block by RDOQ in the sense of rate distortion optimization due to that quantizing all coefficients to zero will save some coding bits. We have observed some sample blocks that are quantized to all-zero by RDOQ although some coefficients are quantized to nonzero by HDQ. The typical m is given in Table 1 in the cases of different TU blocks with different block size.

As analyzed above, if c_i is quantized to non-zero by RDOQ, different $|c_i|$ intensities contribute to final ρ identically. Using this property, we propose heuristic way to determine the control parameters a and b to obtain accurate WSATD for ρ modeling. We need to deterministic control points $(|c_i|, w_i)$ for parameter selection. On one hand, suppose a coefficient with intensity $(1 - \delta)q$ have the normalized contribution 1 for WSATD estimation and ρ. As a result, $((1 - \delta)q, 1)$ is used as one control point. On the other hand, suppose that the maximal of $|c_i|$ is c_{max}, a corresponding minimal weight w_{min} is supposed to be determined. The coefficient with maximal $|c_i|$ is supposed to contribute the same degree with the coefficient with intensity $(1 - \delta)q$. As a result, another control point (c_{max}, w_{min}) is derived according to $c_{max}*w_{min} = (1 - \delta)q$. As a result, we can derive a and b according to Eq. (6) using the above two control points. a and b vary dynamically according to the TU block, resulting in accurate weight model w_i.

3.2 ρ Model

As analyzed in Sect. 2, ρ has an intrinsical functional relationship with χ and ω. This work determines a three-dimensional ρ model (denoted by ρ-ω-χ) via surface fitting. How to accurate estimate ω online is the first task here. In order to achieve fast ρ estimation, we need to build a model to estimate the averaged χ, i.e. ω.

Suppose that the percentage of zero-quantized coefficients in the case of HDQ is denoted as ρ'. Using RDO to determine the optimal quantization result, only very few coefficients with nonzero-quantized HDQ results are finally quantized to zero coefficients. As a result, ρ' is basically identical with ρ in general. Supposed that the averaged χ in the case of identical ρ' is denoted as ω'. Similarly, ω' is usually identical with ω. As a result, we can estimate ω according to ω' instantly due to that ω' can be easily obtained online.

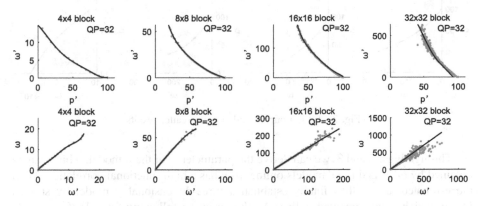

Fig. 2. Fitting curves and scatter plots between ω-ω'

In terms of ω-ω' function modeling, we need to clean the samples collected. The first case is that ω in the case of certain ρ may not exist. In this case, we need to remove the corresponding ω' sample in the case of the corresponding ρ' (whether or not the mean exists). The second case is that ω' in the case of certain ρ' may not exist. In this case, we need to determine ω in the case of corresponding ρ with ω' (if ω exists). There is high correlation between ρ' and ω', as a result we can develop modeling ω'-ρ'. Two function models ω'-ρ' and ω-ω' can be obtained by curve fitting respectively:

$$\omega' = g(\rho') \tag{7}$$

where ρ' is the percentage of zero-quantized coefficients by HDQ based pre-quantization, ω' is the mean of χ in the case of identical ρ'. There is a monotonically decreasing function of the second order between ω and ω', and this function can be formulated as follows.

$$\omega = f(\omega') \tag{8}$$

Where ω' is the mean of χ in the case of identical ρ', ω is the mean of χ in the case of identical ρ. Figure 2 shows the fit curve and scatter plot results between ω and ω'.

According to Fig. 2, we can derive the following conclusions. When the TU block type is 4×4 and 8×8, we can accurately predict ω according to Eqs. (7) and (8). For other types of TU blocks, Eqs. (7) and (8) cannot accurately predict ω sometimes. In order to ignore these outlier samples, we employ the estimated ω' to replace ω for ρ-ω modeling. Finally, the function model ρ-ω is obtained by curve fitting. Figure 3 below shows the fit curve and sample scatter results.

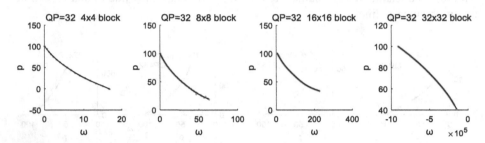

Fig. 3. Fitting curves and sample scatter results

Through Figs. 2 and 3, we can derive the parameter ω of the ρ model, which can be obtained by ω'-ρ' and ω-ω' models online. ρ-ω has a high functional correlation. With the estimated ω, we then finally establish a three-dimensional ρ model by surface fitting, which is implemented online. As shown in the following formula (9).

$$\rho = f(\omega, \chi) \tag{9}$$

The three-dimensional ρ model is a polynomial with respect to the one order of χ and the third order of ω.

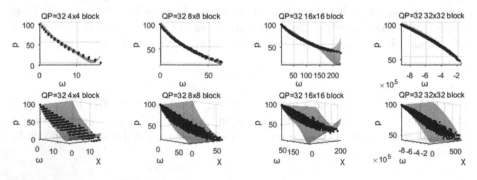

Fig. 4. Fitting surfaces and sample scatter results

Figure 4 gives the corresponding fitted surfaces and sample scatter results. Figure 4 are intensively shown from two angles of view for better understanding.

From the results in Fig. 4, we can draw the following conclusions. On one hand, the proposed ρ model can accurately predict the ρ results of RDOQ. Compared with the percentage of zero-quantized coefficients by HDQ, the prediction results obtained by the proposed model is very close to the actual ρ samples. In order to evaluate the accuracy of the proposed ρ model, we also report the estimated error ratio. The resulting estimation error ratios are given in Fig. 5, in which the histogram results of the prediction error are given.

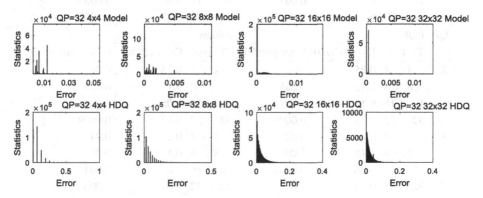

Fig. 5. Error frequency histogram

4 Experimental Result

The proposed ρ model can quickly and accurately predict the ρ results of RDOQ. In order to evaluate the model accuracy of ρ model, we take the ρ results of RDOQ as the comparison anchor and investigate the estimation error of the proposed ρ model relative to the results of RDOQ. Here, we compare the simple ρ model estimated from simple deadzone HDQ with the proposed ρ model. The estimation error results of two ρ models are given in Table 2. There simulation results in the cases of different QP and TU block size are given.

According to the results in Table 2, we can draw the following conclusions. In the cases of different TU blocks, the proposed ρ model established off-line can quickly and accurately predict the ρ results of RDOQ, and the error results are usually smaller than 0.01. Comparatively, in the case of HDQ based ρ model, a great amount of samples suffer from estimation error larger than 0.01, and some samples have estimation error ratio close to 0.2. In terms of computational complexity, only an additional curve and surface fitting are desired. The additional complexity of proposed model is moderate.

Table 2. ρ estimation error results of the two models compared with RDOQ

QP	BasketballPass			BasketballDrill		
	TU type	Error (HDQ)	Error (model)	TU type	Error (HDQ)	Error (model)
32	4 × 4	0.35	0.05	4 × 4	0.25	0.04
	8 × 8	0.2	0.01	8 × 8	0.15	0.01
	16 × 16	0.13	0.005	16 × 16	0.06	0.005
	32 × 32	0.1	0.001	32 × 32	0.06	0.001
37	4 × 4	0.25	0.05	4 × 4	0.25	0.05
	8 × 8	0.2	0.01	8 × 8	0.15	0.01
	16 × 16	0.1	0.005	16 × 16	0.06	0.005
	32 × 32	0.1	0.001	32 × 32	0.05	0.0009
QP	BQMall			Johnny		
	TU type	Error (HDQ)	Error (model)	TU type	Error (HDQ)	Error (model)
32	4 × 4	0.25	0.03	4 × 4	0.25	0.04
	8 × 8	0.15	0.006	8 × 8	0.15	0.009
	16 × 16	0.1	0.001	16 × 16	0.05	0.003
	32 × 32	0.08	0.0009	32 × 32	0.05	0.0009
37	4 × 4	0.25	0.02	4 × 4	0.2	0.04
	8 × 8	0.15	0.006	8 × 8	0.1	0.007
	16 × 16	0.06	0.0001	16 × 16	0.03	0.003
	32 × 32	0.05	0.0005	32 × 32	0.04	0.0004
QP	FourPeople			KristenAndSara		
	TU type	Error (HDQ)	Error (model)	TU type	Error (HDQ)	Error (model)
32	4 × 4	0.2	0.04	4 × 4	0.2	0.03
	8 × 8	0.15	0.01	8 × 8	0.1	0.01
	16 × 16	0.06	0.005	16 × 16	0.04	0.003
	32 × 32	0.03	0.001	32 × 32	0.03	0.0009
37	4 × 4	0.2	0.03	4 × 4	0.2	0.03
	8 × 8	0.1	0.01	8 × 8	0.1	0.01
	16 × 16	0.025	0.001	16 × 16	0.04	0.003
	32 × 32	0.022	0.001	32 × 32	0.03	0.001

5 Conclusion

In video encoder, the percentage of zero-quantized coefficients ρ is useful for rate distortion model building and all-zero block detection. Developing fast ρ estimation model plays important role in rate distortion optimization for video coding. This paper proposes a fast ρ model for RDOQ based video coding. An accurate ρ model is adaptively built offline as function of weighted $SATD$, quantization step size, and average $WSATD/q$ estimated from ensemble. Experimental results verify that the proposed model can quickly and accurately predict the ρ results of RDOQ with moderate implementation complexity. The proposed ρ model can be employed to optimized all-zero block detection and rate distortion model building.

References

1. He, Z., Kim, Y.K., Mitra, S.K.: Low-delay rate control for DCT video coding via ρ-domain source modeling. IEEE Trans. Circuits Syst. Video Technol. **11**(8), 928–940 (2001)
2. Milani, S., Celetto, L., Mian, G.A.: An accurate low-complexity rate control algorithm based on (ρ, Eq)-domain. IEEE Trans. Circuits Syst. Video Technol. **18**(2), 257–262 (2008)
3. Lee, K., et al.: A novel algorithm for zero block detection in high efficiency video coding. IEEE J. Sel. Topics Signal Process. **7**(6), 1124–1134 (2013)
4. Yang, K.H.: Methods and systems for rate-distortion optimized quantization of transform blocks in block transform video coding. US, US7957600 (2011)
5. Wang, H., Yin, H., Shen, Y.: A Novel Hard-Decision Quantization Algorithm Based on Adaptive Deadzone Offset Model. In: Chen, E., Gong, Y., Tie, Y. (eds.) PCM 2016. LNCS, vol. 9917, pp. 335–345. Springer, Cham (2016). https://doi.org/10.1007/978-3-319-48896-7_33

Partially Separated Networks for Person Search

Chuanchuan Chen, Jingbo Fan, Yuesheng Zhu$^{(\boxtimes)}$, and Guibo Luo

Communication and Information Security Lab, Shenzhen Graduate School,
Peking University, Shenzhen, China
{chenchuanchuan,fjb}@pku.edu.cn, zhuys@pkusz.edu.cn,
luoguibo@sz.pku.edu.cn

Abstract. The multi-task learning framework that considers pedestrian detection and person re-identification jointly is an effective solution for person search. However, the existing joint frameworks simply share the backbone network without considering the negative interaction between the two tasks. To alleviate this conflict and meet the different requirements in detection and re-identification, a Partially Separated Network (PSN) for person search is proposed in this paper. Unlike the traditional joint frameworks, our backbone network is partially separated for detection and identification, and feature maps with different scales are provided according to different characteristics. Our experiment results have demonstrated that on CUHK-SYSU dataset our mAP and top-1 on ResNet-50 are 5.4% and 4.4% higher, and on PRW dataset our mAP and top-1 on PVANet are 8.0% and 5.0% higher compared with the state-of-the-art methods. Specially, the improvements can be more impressive in the case of large gallery, occlusion and low resolution.

Keywords: Person search · Pedestrian detection
Person re-identification · Multi-task learning
Partially separated network

1 Introduction

As a new and challenging field, person search [29,38] has attracted increasing interest in recent years, which is more practical than person re-identification [9, 33]. The purpose of person search is to localize specific persons matching the provided query in gallery images or video frames, so it has important applications in many fields such as multi-camera person tracking [32] and person activity analysis [31]. However, it suffers the difficulties encountered in both object detection [23] and person re-identification, such as small or low-resolution instances, huge variance of appearance across multiple cameras, varying poses.

Most of person re-identification methods focus on learning discriminative features and distance metrics [4,5,19,24]. Traditional feature learning methods [2,3,19,37] mainly extract shallow features from cropped pedestrian images.

© Springer Nature Switzerland AG 2018
R. Hong et al. (Eds.): PCM 2018, LNCS 11166, pp. 772–782, 2018.
https://doi.org/10.1007/978-3-030-00764-5_71

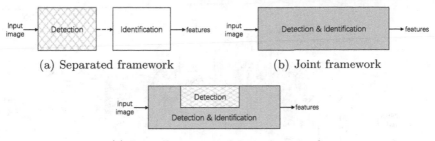

(a) Separated framework   (b) Joint framework

(c) Partially separated Joint framework

Fig. 1. Three person search frameworks that involve detection and identification. Only the core part of the backbone networks is shown. (a) Detection and identification are trained separately. (b) Detection and identification share network parameters completely. (c) Detection and identification are partially separated.

In recent years, many feature learning methods [1,18,26,27,35] based on deep learning achieve impressive performance. For pedestrian detection, traditional pedestrian detectors, such as ACF [6] and LDCF [22], exploit various filters on Integral Channel Features (ICF) [7] with sliding window strategy to localize each target. And CNN-based deep learning models [10,11,13,16,21,23] have become mainstream object detection methods recently.

At present, the combining detection and re-identification frameworks based on convolutional neural networks (CNN) [17] are popular for person search. In [38], the tasks of detection and identification (i.e. discriminative person feature extraction) are trained separately corresponding to Fig. 1(a). The two tasks do not interfere mutually during training and a combination of various detection and identification methods can be used. In [29], a joint framework is proposed to optimize detection and identification simultaneously in a single network (Fig. 1(b)). In essence, it is a multi-task learning framework that unites pedestrian detection (implemented by RPN [23], i.e. region proposal network) and classification (online instance matching loss function). We consider that two tasks share all the parameters in the backbone network (not including RPN) because the network assigns the loss functions of detection and identification together at the end. We note that this multi-task framework is just a simple combination of detection and identification, without considering the negative interference and different characteristics between the two tasks.

In this paper, we investigate the multi-task person search framework in [29] and notice that the detection accuracy increases when the backpropagation of the identification part is off. Our experiments show that this is because the backpropagation of identification loss updates the parameters of the backbone network before RPN, which makes the parameters conflict and affects the performance of RPN indirectly. Also, the appropriate feature maps for detection and identification are of different hierarchies. As for detection, deeper feature maps are better as richer semantics can be provided, while shallower feature maps layers are better for small-size object detection [16,34]. For identification, the

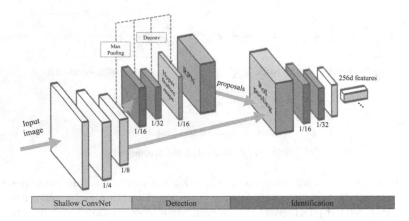

Fig. 2. Proposed partially separated network. On ResNet-50, the first 3 stages (conv1 to conv3_4) comprise the Shallow ConvNet. Layers of conv4_1 to conv5_1 are duplicated into two partially separated blocks (red). 1/4, 1/8, 1/16 and 1/32 present the ratio of the feature map size and input size. (Color figure online)

sufficient local features of the pedestrian body play an important role in distinguishing identities [35], which makes identification benefit from high-resolution maps. So it is inappropriate for person search to put detection and identification at the same feature maps.

To ease the negative interference between two tasks and feed their different requirements for feature maps, we design an end-to-end trained partially separated network framework. Our framework partially separates detection and identification in the backbone network, and the backbone network of detection does not share parameters with other parts of the network (Fig. 1(c)).

To validate the effectiveness of our general-purpose framework, we implement partially separated networks on ResNet-50 and PVANet. We test our method on CUHK-SYSU dataset [29] and PRW dataset [38]. On CUHK-SYSU dataset, our mAP and top-1 accuracy of PSN on ResNet-50 are 7.8% and 6.9% higher than those in [29] without partial separation. In the case of lager gallery size, occlusion and low resolution, our performance improvements are more obvious compared to the existing methods.

2   Partially Separated Networks

2.1   Overall Framework

We have two goals to achieve: (i) to ease negative interaction between detection and identification in a joint network, and (ii) provide different feature levels for two tasks suitably. The proposed partially separated network can satisfy these two objectives simultaneously with fine detection results and more discernible features.

As shown in Fig. 2, our proposed network consists of three sections: Shallow ConvNet, detection and identification. The front stages (conv1 to conv3_4) of the chosen backbone CNN (such as ResNet-50 [14], PVANet [15]) comprise Shallow ConvNet. The part of the chosen backbone CNN between Shallow ConvNet and last layers (for instance, layers after conv5_1 on ResNet-50) is duplicated into two blocks, called partially separated blocks. One is used for detection, and the other is used for identification.

In detection section, we properly fuse the coarse-to-fine CNN features in detection section to generate hyper feature maps which are exploited by RPN. Following Shallow ConvNet, the first partially separated block with private parameters continues to extract the deep detection features. Distinct sampling strategies are employed for different layers to combine multi-level maps of the same resolution. The output maps of Shallow ConvNet whose scale is 1/8 of input are subsampled by the following max pooling layer. On the contrary, the output (1/32 scale) of this partially separated block experiences an upsampling operation which is done by a deconvolutional (Deconv) layer. Then we choose the output of a layer (1/16 scale) from partially separated block as reference and concatenate it with the results of subsampling and upsampling to get hyper feature maps.

In identification section, an RoI pooling layer [10,13] is adopted to extract features with a fixed resolution (28×28 for ResNet-50 and 32×32 for PVANet) for proposed regions from high resolution feature maps, i.e. maps after Shallow ConvNet. The features are sequentially processed by the second partially separated block (has the same structure and initial parameters as the former one but does not share parameters during training) and last layers. Then we project the features into a L2-normalized 256 dimensional subspace as output. The OIM loss function is adopted to supervise the training in the same fashion as [29] owing to its excellent performance under the circumstance of a large number of categories with few instances in each class. At inference stage cosine similarity is directly used to measure the similarity between the query and the gallery.

3 Experiments and Analysis

3.1 Datasets and Protocol

CUHK-SYSU: CUHK-SYSU is a large-scale person search dataset composed of 18,184 images, 8,432 identities, and 96,143 pedestrians bounding boxes. The training set contains 11,206 images and 5,532 query persons. Correspondingly, the query set contains 2,900 persons and the gallery contains 6,978 images within the testing set. To test the performance in low resolution and occlusion situations, two subsets which contain 290 and 187 query persons respectively are included.

PRW: The PRW dataset contains 11,816 video frames of scenes from 6 cameras, generating 43,110 pedestrian bounding boxes. Furthermore, 34,304 pedestrians

Table 1. Comparisons of Baseline-1, Baseline-2 and PSN on CUHK-SYSU.

| | mAP (%) | top-1 (%) | All-detection |
			Recall (%)	AP (%)
Baseline-1	78.4	81.1	81.7	77.0
Baseline-1 w/o OIM	-	-	84.0	80.5
Baseline-1 PSN tail	73.9	77.0	82.9	78.6
PSN-Res	**83.0**	**83.7**	83.0	80.1
Baseline-2	78.9	82.3	81.6	75.3
Baseline-2 w/o OIM	-	-	86.9	82.0
Baseline-2 PSN-tail	78.8	82.2	81.9	76.6
PSN-PVA	**86.8**	**89.1**	86.1	82.3

are annotated with 932 IDs among the bounding boxes. The results are split into a training set and a testing set, which individually contains 5,134 frames of 482 IDs and 2057 query persons with a gallery of 6,112 images.

Evaluation Protocol: We adopt mean average precision (mAP) and top-1 matching rate as evaluation protocols, which are commonly used in the community [20,29,38]. The mAP metric reflects the accuracy of detecting the query person from the gallery images. For top-1 matching rate, the predicted top-1 box is counted as a matching if it overlaps with ground truth larger than or equal to the Intersection-over-Union (IoU) threshold 0.5. We also use detection recall and detection AP to specifically evaluate the detection performance of all pedestrians.

3.2 Effectiveness of Partially Separated Networks

In this section, we locate the problem of parameter conflict and evaluate the effectiveness of PSN on the CUHK-SYSU dataset. For a fair comparison, Baseline-1 and Baseline-2 are built on ResNet-50 and PVANet separately without the partially separated structure just as in [29], and anchors of two aspect ratios (width to height): 0.33, 0.26 and four scales: 4, 8, 16, 32 are adopted according to the statistics of CUHK-SYSU dataset. The mAP and top-1 of our two baselines are summarized in Table 1. The gallery size is 100 by default.

As shown in Table 1, when we turn off the backpropagation of OIM (Baseline w/o OIM) on both Baseline-1 and Baseline-2, i.e. turning off the identification section, the detection recall and AP of the two baselines are improved, suggesting that identification has some negative effects on detection in such joint structure. The parameter conflict may occur in backbone network after RPN or before RPN. We duplicate the backbone network after RPN into two partially separated blocks (Baseline PSN-tail): one is followed by detection loss, and the other is followed by identification loss. From the result in Table 1, we find that detection

Table 2. Comparison of PSN's performance on CUHK-SYSU with the state-of-the-arts.

Method	mAP (%)	top-1 (%)
ACF+DSIFT+Euclidean	21.7	25.9
ACF+DSIFT+KISSME	32.3	38.1
ACF+BoW+Cosine	42.4	48.4
ACF+LOMO+XQDA	55.5	63.1
ACF+IDNet	56.5	63.0
CCF+DSIFT+Euclidean	11.3	11.7
CCF+BoW+Cosine	26.9	29.3
CCF+LOMO+XQDA	41.2	46.4
CCF+IDNet	50.9	57.1
CNN+DSIFT+Euclidean	34.5	39.4
CNN+BoW+Cosine	56.9	62.3
CNN+LOMO+XQDA	68.9	74.1
CNN+IDNet	68.6	74.8
JDI-PS	75.5	78.7
IAN	77.3	80.5
NPSM	77.9	81.2
PSN-Res	83.3	85.6
PSN-PVA	**86.8**	**89.1**

performance is nearly not improved compared with baseline, and person search performance decreased, which prove that interference after RPN is not the key. Following experiments will validate the problem that mainly exists before RPN.

We partially separate the backbone network before RPN. For a better comparison with [29], the partially separated blocks in Table 1 start at conv3_3 and end at conv4_3 on ResNet-50, so the detection can be done after conv4_3 without hyper feature maps. We can see that the recall and AP of both PSN-Res and PSN-PVA approximately reach the level without OIM, which indicates that the negative interference between the two tasks mainly occurs in backbone network before RPN and the partially separated block mitigates the interaction effectively. The mAP and top-1 of our method outperform all baselines by over 6% which is a considerable improvement for person search.

3.3   Comparisons with State-of-the-Art Methods

CUHK-SYSU. We compare PSN with some state-of-the-art methods on the CUHK-SYSU dataset, including person search methods (IAN [28], JDI-PS [29], NPSM [20]) and some other methods that split the task into pedestrian detection (ACF [6], CCF [30], Faster R-CNN [23]) and person re-identification (BoW [37], DSIFT [36], KISSME [24], LOMO and XQDA [19] etc.). When the gallery size is

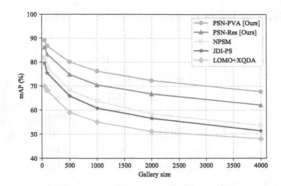

Fig. 3. Test mAPs of different approaches under different gallery sizes on CUHK-SYSU dataset.

Table 3. Comparison of PSN's performance on the occlusive subset and low-resolution subset of CUHK-SYSU with the state-of-the-arts.

Method	Occlusive		Low-resolution	
	mAP (%)	top-1 (%)	mAP (%)	top-1 (%)
JDI-PS	51.7	54.0	55.9	59.7
IAN	53.0	54.6	52.6	54.9
PSN-Res	64.3	65.8	69.7	74.5
PSN-PVA	**73.0**	**74.3**	**75.4**	**79.7**

set to 100, the results are gathered in Table 2. "CNN" represents Faster R-CNN with ResNet-50, and "IDNet" is the re-identification part in the framework of JDI-PS. Our PSN outperforms the others by large margin. Comparing with our baseline method JDI-PS implemented on ResNet-50, PSN-Res improves mAP by 7.8% and top-1 by 6.9%. On a machine with 64GB memory, NVIDIA Titan X (Pascal) GPU and i7-6700K CPU, PSN-Res takes 413 ms to detect targets and extract features from a whole image, which is only 18 ms more than JDI-PS. Compared with the results of NPSM, PSN-Res increases by 5.4% of mAP and 4.4% of top-1. PSN-PVA outperforms these approaches more under our evaluation protocols, and only takes 270 ms for detection and features extraction.

Person search becomes more challenging as the search space increases. We compare the mAP of several methods in different gallery sizes, i.e. 50, 100, 500, 1000, 2000, and 4000. As shown in Fig. 3, our method is superior to the others under all gallery sizes. Especially, it is worth noting that our mAP decreases slower than other methods as the gallery size becomes larger. When the gallery size is 4000, PSN-Res and PSN-PVA outperform the mAP of NPSM by more than 10% and 17%, respectively.

In addition, with the presence of occlusion and low resolution, the performance of PSN and several existing methods are compared in Table 3. Our method outperforms the others impressively. Compared with JDI-PS implemented on

Table 4. Comparison of PSN's performance on PRW dataset with the state-of-the-arts.

Method	#detection = 3		#detection = 5	
	mAP (%)	top-1 (%)	mAP (%)	top-1 (%)
DPM-Alex+LOMO+XQDA	13.4	34.9	13.0	34.1
DPM-Alex+IDE$_{det}$	20.2	48.2	20.3	47.4
DPM-Alex+IDE$_{det}$+CWS[38]	20.0	48.2	20.5	48.3
ACF-Alex+LOMO+XQDA	10.5	31.8	10.3	30.6
ACF-Alex+IDE$_{det}$	17.0	45.2	17.5	43.6
ACF-Alex+IDE$_{det}$+CWS	17.0	45.2	17.8	45.2
LDCF+LOMO+XQDA	11.2	31.6	11.0	31.1
LDCF+IDE$_{det}$	17.5	45.3	18.3	44.6
LDCF+IDE$_{det}$+CWS	17.5	45.5	18.3	45.5
JDI-PS	15.6	36.3	16.0	36.3
JDI-PS+CWS	15.7	36.8	16.7	37.0
PSN-Res	17.9	41.5	17.9	40.5
PSN-Res+CWS	17.9	41.6	18.8	41.6
PSN-PVA	28.2	53.2	28.9	51.6
PSN-PVA+CWS	**28.3**	**53.7**	**30.0**	**53.3**

ResNet-50, our PSN-Res improves by 12.6% and 11.8% for mAP and top-1 in the case of occlusion, while the performance of PSN-PVA overtakes JDI-PS by 21.3% and 20.3%. With the presence of low resolution, the performance of our method is still outstanding among others in the case of occlusion. The experiments in the condition of different gallery sizes, occlusion and low resolution show that our model is more effective in dealing with hard negative samples, and more robust in feature generation.

PRW. We compare PSN with JDI-PS [29] and some separated methods proposed by [38] which combine different detectors (respective R-CNN [12] detectors of DPM [8], ACF [6], LDCF [22] based on AlexNet [17]) and recognizers (LOMO and XQDA [19], IDE$_{det}$). According to [38], incorporating different recognizers for DPM and ACF, AlexNet achieves better performance than ResNet [14] and VGGNet [25]. The results on the PRW dataset with 3 and 5 detected bounding boxes per image are reported in Table 4. "IDE$_{det}$" model is built on the ImageNet pre-trained model. It is trained as a two-class recognition model using the detection data and then trained as a 482-class recognition model using the training data of PRW. Among the methods (the last six rows in Table 4) where single-model is trained end-to-end, our PSN always outperforms JDI-PS. With 3 detection boxes per image, PSN-PVA outperforms the best multi-model methods by 8.0% on mAP and 5.0% on top-1.

4   Conclusion

In this work, we focus on the negative interaction between pedestrian detection and person re-identification in person search, and the characteristics of these two tasks. Accordingly, a partially separated network is proposed which can relieve the negative interaction between two tasks, and feature maps with different depths are also used for detection and identification respectively. Experiments demonstrate that our general-propose network is suitable for the traits of person search.

Acknowledgement. This work is supported by the Shenzhen Municipal Development and Reform Commission (Disciplinary Development Program for Date Science and Intelligent Computing), and by Shenzhen International cooperative research projects GJHZ20170313150021171.

References

1. Ahmed, E., Jones, M., Marks, T.K.: An improved deep learning architecture for person re-identification. In: Computer Vision and Pattern Recognition, pp. 3908–3916 (2015)
2. Chen, D., Yuan, Z., Chen, B., Zheng, N.: Similarity learning with spatial constraints for person re-identification. In: IEEE Conference on Computer Vision and Pattern Recognition, pp. 1268–1277 (2016)
3. Chen, D., Yuan, Z., Hua, G., Zheng, N.: Similarity learning on an explicit polynomial kernel feature map for person re-identification. In: Conference on Computer Vision and Pattern Recognition, pp. 1565–1573 (2015)
4. Chen, W., Chen, X., Zhang, J., Huang, K.: Beyond triplet loss: a deep quadruplet network for person re-identification. In: IEEE Conference on Computer Vision and Pattern Recognition, pp. 1320–1329 (2017)
5. Ding, S., Lin, L., Wang, G., Chao, H.: Deep feature learning with relative distance comparison for person re-identification. Pattern Recognit. **48**(10), 2993–3003 (2015)
6. Dollar, P., Appel, R., Belongie, S., Perona, P.: Fast feature pyramids for object detection. IEEE Trans. Pattern Anal. Mach. Intell. **36**(8), 1532–1545 (2014)
7. Dollr, P., Tu, Z., Perona, P., Belongie, S.: Integral channel features. In: British Machine Vision Conference (2009)
8. Felzenszwalb, P.F., Girshick, R.B., Mcallester, D., Ramanan, D.: Object detection with discriminatively trained part-based models. IEEE Trans. Pattern Anal. Mach. Intell. **47**(2), 6–7 (2014)
9. Gheissari, N., Sebastian, T.B., Hartley, R.: Person reidentification using spatiotemporal appearance. In: IEEE Computer Society Conference on Computer Vision and Pattern Recognition, pp. 1528–1535 (2006)
10. Girshick, R.: Fast R-CNN. In: IEEE International Conference on Computer Vision (2015)
11. Girshick, R., Donahue, J., Darrell, T., Malik, J.: Rich feature hierarchies for accurate object detection and semantic segmentation. In: IEEE Conference on Computer Vision and Pattern Recognition (2014)

12. Girshick, R., Donahue, J., Darrell, T., Malik, J.: Region-based convolutional networks for accurate object detection and segmentation. IEEE Trans. Pattern Anal. Mach. Intell. **38**(1), 142 (2016)
13. He, K., Zhang, X., Ren, S., Sun, J.: Spatial pyramid pooling in deep convolutional networks for visual recognition. IEEE Trans. Pattern Anal. Mach. Intell. **37**(9), 1904–1916 (2015)
14. He, K., Zhang, X., Ren, S., Sun, J.: Deep residual learning for image recognition. In: Computer Vision and Pattern Recognition, pp. 770–778 (2016)
15. Hong, S., Roh, B., Kim, K.H., Cheon, Y., Park, M.: PVANet: lightweight deep neural networks for real-time object detection. In: International Workshop on Efficient Methods for Deep Neural Networks (2016)
16. Kong, T., Yao, A., Chen, Y., Sun, F.: Hypernet: towards accurate region proposal generation and joint object detection. In: Computer Vision and Pattern Recognition, pp. 845–853 (2016)
17. Krizhevsky, A., Sutskever, I., Hinton, G.E.: Imagenet classification with deep convolutional neural networks. In: International Conference on Neural Information Processing Systems, pp. 1097–1105 (2012)
18. Li, W., Zhao, R., Xiao, T., Wang, X.: DeepReID: deep filter pairing neural network for person re-identification. In: IEEE Conference on Computer Vision and Pattern Recognition, pp. 152–159 (2014)
19. Liao, S., Hu, Y., Zhu, X., Li, S.Z.: Person re-identification by local maximal occurrence representation and metric learning. In: Computer Vision and Pattern Recognition, pp. 2197–2206 (2015)
20. Liu, H., et al.: Neural person search machines, pp. 493–501 (2017)
21. Liu, W., et al.: SSD: single shot multibox detector. In: Leibe, B., Matas, J., Sebe, N., Welling, M. (eds.) ECCV 2016. LNCS, vol. 9905, pp. 21–37. Springer, Cham (2016). https://doi.org/10.1007/978-3-319-46448-0_2
22. Nam, W., Dollr, P., Han, J.H.: Local decorrelation for improved pedestrian detection. In: Advances in Neural Information Processing Systems (2014)
23. Ren, S., He, K., Girshick, R., Sun, J.: Faster R-CNN: towards real-time object detection with region proposal networks. In: NIPS (2015)
24. Roth, P.M., Wohlhart, P., Hirzer, M., Kostinger, M., Bischof, H.: Large scale metric learning from equivalence constraints. In: IEEE Conference on Computer Vision and Pattern Recognition, pp. 2288–2295 (2012)
25. Simonyan, K., Zisserman, A.: Very deep convolutional networks for large-scale image recognition. arXiv:1409.1556 (2014)
26. Sun, Y., Zheng, L., Deng, W., Wang, S.: SVDNet for pedestrian retrieval. In: IEEE International Conference on Computer Vision, pp. 3820–3828 (2017)
27. Varior, R.R., Haloi, M., Wang, G.: Gated siamese convolutional neural network architecture for human re-identification. In: Leibe, B., Matas, J., Sebe, N., Welling, M. (eds.) ECCV 2016. LNCS, vol. 9912, pp. 791–808. Springer, Cham (2016). https://doi.org/10.1007/978-3-319-46484-8_48
28. Xiao, J., Xie, Y., Tillo, T., Huang, K., Wei, Y., Feng, J.: IAN: the individual aggregation network for person search. arXiv: 1705.05552 (2017)
29. Xiao, T., Li, S., Wang, B., Lin, L., Wang, X.: Joint detection and identification feature learning for person search. In: CVPR (2017)
30. Yang, B., Yan, J., Lei, Z., Li, S.Z.: Convolutional channel features. In: IEEE International Conference on Computer Vision, pp. 82–90 (2015)
31. Yang, Y., Liao, S., Lei, Z., Li, S.Z.: Large scale similarity learning using similar pairs for person verification. In: Thirtieth AAAI Conference on Artificial Intelligence, pp. 3655–3661 (2016)

32. Yu, S.I., Yang, Y., Hauptmann, A.: Harry Potter's marauder's map: localizing and tracking multiple persons-of-interest by nonnegative discretization. In: Computer Vision and Pattern Recognition, pp. 3714–3720 (2013)
33. Zajdel, W., Zivkovic, Z., Krose, B.J.A.: Keeping track of humans: have I seen this person before? In: IEEE International Conference on Robotics and Automation, pp. 2081–2086 (2005)
34. Zhang, L., Lin, L., Liang, X., He, K.: Is faster R-CNN doing well for pedestrian detection? In: Leibe, B., Matas, J., Sebe, N., Welling, M. (eds.) ECCV 2016. LNCS, vol. 9906, pp. 443–457. Springer, Cham (2016). https://doi.org/10.1007/978-3-319-46475-6_28
35. Zhao, H., et al.: Spindle net: person re-identification with human body region guided feature decomposition and fusion. In: IEEE Conference on Computer Vision and Pattern Recognition, pp. 907–915 (2017)
36. Zhao, R., Ouyang, W., Wang, X.: Unsupervised salience learning for person re-identification. In: Computer Vision and Pattern Recognition, pp. 3586–3593 (2013)
37. Zheng, L., Shen, L., Tian, L., Wang, S., Wang, J., Tian, Q.: Scalable person re-identification: a benchmark. In: IEEE International Conference on Computer Vision, pp. 1116–1124 (2015)
38. Zheng, L., Zhang, H., Sun, S., Chandraker, M., Yang, Y., Tian, Q.: Person re-identification in the wild. In: CVPR (2017)

Action Tree Convolutional Networks: Skeleton-Based Human Action Recognition

Wenjie Liu$^{(\boxtimes)}$, Ziyi Zhang, Bing Han, and Chenhui Zhu

University of Electronic Science and Technology of China,
Chengdu 611731, Sichuan, China
vinjay@hit.edu.cn, {2015200106011,2015010912019,
2015200106005}@std.uestc.edu.cn

Abstract. This paper is mainly about addressing the problem of skeleton-based human activity recognition: ignoring the structure and relationship between skeleton joints and body-parts, the existence of a large amount of useless information in the activity data, and poor generalization ability. In order to solve the shortcomings of existing mainstream methods used for human action recognition, we propose a novel method named Action Tree Convolutional Networks (ATCNs). This method uses a data based auto-designed Action Tree network to dynamically generate a tree of nodes/body-parts and a semantic attention center, profoundly emphasizing the relations and semantics of nodes/body-parts. This method we introduced has a great improvement on the previous algorithm's neglect of the importance of nodes/body-parts relation, and improves the generalization ability of the algorithm. Through experiments on Kinetics and NTU-RGB+D datasets, our method achieves better performance improvements over other state-of-the-art methods.

Keywords: Action recognition · Action tree · Skeleton relation

1 Introduction

Human action recognition plays an important role in the computer's understanding of the real world environment. Therefore, people have generally paid attention to this issue in recent years. It refers to the original information obtained through the sensor, and through the algorithm to determine the specific category of human movement. It is an important topic in computer vision, and it has been widely used in many fields such as intelligent surveillance, image/video annotation, and human-computer interaction in public places.

Currently, in the field of human action recognition, there are many mainstream recognition methods, such as RGB (C3D Networks [1, 2]), Optical flow (Glimpse Clouds [3]), and Skeleton (PA-LSTM [4]). Most of the existing methods focus on body-part and individual skeleton joint, and most of them cannot express the relation between skeleton joints and body-parts. Moreover, in many human action scenes, features are not well characterized and classified from discrete body-parts or single skeleton joints. This leads to the fact that existing methods do not have very good generalization capabilities for different actions.

Z. Zhang, B. Han and C. Zhu—These authors contributed equally to this work.

© Springer Nature Switzerland AG 2018
R. Hong et al. (Eds.): PCM 2018, LNCS 11166, pp. 783–792, 2018.
https://doi.org/10.1007/978-3-030-00764-5_72

In order to solve the above problems, we recognize that the tree structure itself has an excellent description of the connection relationships and structural layers between nodes. At the same time, the graph convolution has been proved to be very effective for data in the form of spatial structure and temporal structure, such as Graph Convolutional Networks (GCNs) [5], Spatial Temporal Graph Convolutional Networks (ST-GCN) [6]. Therefore, we make some changes to the graph structure and extend it to the field of tree structure. For this project, we mainly study the tree structure [7]: generation of tree root (semantic attention), action tree convolution and node neighborhood definition, and finally propose a new method named Action Tree Convolutional Networks (ATCNs). In this method, we first generate a semantic attention in the Action Tree algorithm. The semantic attention as the root node of the Action Tree continuously absorbs connections of the skeleton joint and the semantic attention and connections of the skeleton joint and other skeleton joint (could be regarded as the trunk of the tree). Then an Action Tree will be constructed dynamically. Finally, we use the modified convolution operation and the spatial configuration partitioning strategy [6] of the node set to identify human actions. The main contributions of our work are as follows. (1) The algorithm is the first to use the body-part relation described by the tree structure for human action recognition (we present a novel Action Tree Convolutional Network). (2) We make full use of the different levels of the importance of body-parts and skeleton joints relation. By creating Action Tree in different specific movements of different skeleton joints, we can enhance the human action recognition, and this recognition method can better extend to different actions and different environments. Finally, we conducts experiments on two large databases and the final results of experiments show that compared to the existing methods such as ST-GCN and PA-LSTM, our ATCNs method has greater performance improvements, reaching the state-of-the-art level.

2   Related Work

At present, there are various sensors used in human action recognition, and are generally divided into wearable sensors and non-wearable sensors. Wearable sensors mostly use tri-axial acceleration sensors to recognize and monitor the posture of the human body so as to achieve the effect of action recognition. Non-wearable sensors include traditional RGB cameras and popular depth cameras, infrared cameras, etc., which can generate RGB video sequences, depth map sequences, and more. In addition, it can also generate skeleton sequence information through the skeleton joint estimation algorithm. Compared to 2D information, such as RGB video sequence and depth sequence and 3D point cloud, skeleton sequence information has less information redundancy, obvious features, and light invariance character, for which reason this is increasingly favored by researchers.

2.1   Spatial Temporal Graph Convolutional Networks

While the CNN has the disadvantage of being unable to process the data of Non Euclidean Structure and extract spatial features effectively, the Graph Neural Network

(GCN) has once become a research hotspot. In the field of human action recognition, there is a natural connection between different human skeleton joints. This undoubtedly provides a natural graph for the GCN in the field of action recognition. The Spatial-Temporal Graph Convolutional Network (ST-GCN) method constructs spatial temporal graph from skeleton sequences.

However, the graph used by the ST-GCN to generate spatial temporal graph is predefined and invariable. [6] This leads to the inability of this method to apply to different actions. By contrast, Our ATCNs method uses a dynamic tree for the first time. According to the specific focus of different actions and the hierarchical structure of human skeleton joints, a unique dynamic tree for each action is generated, thereby further improving the accuracy of action recognition.

2.2 Part-Aware Long Short-Term Memory

The Part-Aware Long Short-Term Memory (PA-LSTM) method is an action recognition method using 3D skeleton joints. The basic idea of the PA-LSTM method is dividing the skeleton joints into five major parts according to the body parts, namely the head, left and right arms, and left and right legs. [4] PA-LSTM divides the body into five major parts according to the physical connection of the body, and memorizes and learns separately. This kind of processing can more effectively learn the context information of individual body parts, and use the combination of the context information of these parts as output, so that the method itself acquires a certain part perception ability and improves the action recognition.

However, the connections of skeleton joints are also predefined while the context information and relations are different for different actions, so this method is also limited in generalization ability.

3 Action Tree Convolutional Networks

In this section, we first describe in detail how to generate the semantic attention of an action, and then explain how to expand the human skeleton joints and body-parts into a tree structure with the action semantic attention as the root node to describe human actions, and finally use the graph convolutional neural network to classify actions. In addition, we provide more details on debugging, training, and valuing using our method Action Tree Convolutional Networks (ATCNs).

3.1 Pipeline Overview

At present, there are many human action recognition methods based on skeleton sequences information in the field. However, most of the existing methods focus on a single skeleton joint, so that the relationships between the skeleton joints and the body-parts are not well expressed. In many scenarios, actions are often associated a semantic center with multiple body-parts and skeleton joints, such as brushing tooth. When people brush their tooth, the part of the hand (arm) should connect to the semantic center of the brushing tooth action (an area represents the toothbrush and the mouth).

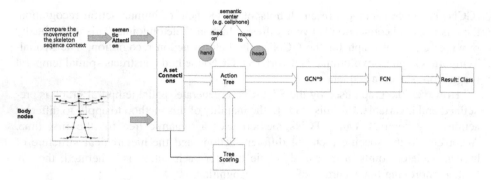

Fig. 1. Overview of pipeline

This is a high frequency, repeatedly moving relation. In addition, the head skeleton node should have "fixing" relation with the semantic center (an area represents the toothbrush and the mouth). If we can design an Action Tree, the tree selects some meaningful body-parts nodes or skeleton nodes and parses their structural hierarchy and relationships [7], which can enhance understanding of human actions and overcome the ignoring of relation. ATCNs, the newly established tree-structured action recognition method is the first of the existing methods using tree-structure until now and will be helpful for the following research on body relation (Fig. 1).

Due to the variety of human actions, the connection between the body-part or skeleton point and the attention of the body-part or skeleton point are also different, so how to find a suitable method to generate the Action Tree specific to each action is a problem key. In the case of a huge scale of data in the human activities dataset, deeper and more complex deep neural networks can often effectively extract the features and accomplish classification task. We use the ATCNs, an end-to-end, back-propagation-trained network. The input of the network is a frame of skeleton joints (hand, wrist, elbow, shoulder... they have their own fixed number) multiplied by the time (the number of frames in the skeleton sequence). The network structure starts from the recursive tree-scoring network. The middle is nine layers, and each layer is a graph convolutional network. After global pooling, add the final full-connection network, and classify the result of the action by the standard SoftMax function.

3.2   Root of Tree: Attention

The root node of the Action Tree is also the semantic attention. It is generated in the Action Tree algorithm and contains the information of the position and size of the area. The area generated by the Action Tree algorithm has two main purposes and functions: one is to be the root node of all body-parts and skeleton nodes; the other is as attention for network to train the weight of the connections of body-parts and skeleton joints. Therefore, the weight of each node connection of the Action Tree is obtained, that is, the degree of correlation with the current input action data. The weight will be used as a basis for pruning and node selection of the tree.

The Action Tree algorithm makes up for the drawbacks of the body part relation in some previous algorithms and we will elaborate on the implementation of the algorithm. First, we calculate the action amplitude weight $W_{k,t}$ for each skeleton point at each moment t relative to the previous moment $t - 1$:

$$W_{k,t} = \sqrt{\left((X_{k,t} - X_{k,t-1})^2 + (Y_{k,t} - Y_{k,t-1})^2 + (Z_{k,t} - Z_{t,t-1})^2 \right)} \tag{1}$$

k is the k-th skeleton point (in each single-frame skeleton sequence), t is the t-th frame in the skeleton sequence, and $W_{k,t}$ is the amplitude weight of the k-th skeleton point at time t. Then, for each human movement at time t, we can calculate the region S based on the magnitude of the movement and the skeleton joints with significant movement, including his position and size information.

$$S_t = \sum_{k=0}^{K} \sqrt{W_{k,t} * \left((X_{k,t} - X_{k-1,t})^2 + (Y_{k,t} - Y_{k-1,t})^2 + (Z_{k,t} - Z_{k-1,t})^2 \right)} \tag{2}$$

It is worth noting here that k is the skeleton point that exceeds the threshold of action movement amplitude (distinguishable from all skeleton joints in previous equal). $X_{k,t}$ is the X coordinate of the k-th skeleton point at time t, and $Y_{k,t}$ is the k-th time at t The Z coordinate of the skeleton point, $Z_{k,t}$ is the Z coordinate of the k-th skeleton point at time t. By analogy, the S region is a semantic attention. It sums up the results of multiplying the amplitude weight $W_{k,t}$ times the skeleton point distance of itself and skeleton point adjacent. Usually, the RGBD camera is stationary and the above algorithm can get good results. For example, the brushing tooth action, the maximum $W_{k,t}$ is concentrated to the position of the forearm and the head, and the selected skeleton point is also with the position near the head and the arm, and the S area is activated for the arm and the area near the head.

However, for certain common special situations in our true life, for example, camera is moving. The above algorithm has some problems: due to the movement of the camera's viewing angle, the human skeleton will move as a whole. Hence, we pre-correct the skeleton point to ($X_k = X_k - X_0$, $Y_k = Y_k - Y_0$, $Z_k = Z_k - Z_0$), where (X_0, Y_0, Z_0) is a fixed point on the body that is not easily shaded and masked. What is more, for the movement of the camera's viewing angle, there is another problem that the skeleton point of the human body would be shaded and masked, which will cause the skeleton data to mutate. We tried to add filter in the algorithm and leave it for future deeper work.

3.3 Tree Scoring and Construction

From the root of tree/attention generation algorithm in Sect. 3.2, the region position and size information are inputs into the scoring network. The scoring network is used to generate the optimal set of connections of skeleton joints or body-parts and the semantic attention and finally constitute an Action Tree. The recursive algorithm is: adding new connections recursively. If the score is lower, delete it. If the score is higher, add it in the set. The scoring rule is: calculate Cross-entropy-loss of output and label according to the prediction of the last network output, and then consider the number of skeleton nodes in the set and the tree branch density contribution of the tree, finally return score.

$$Score = \frac{1}{Loss_{lable}} \cdot W_{structure} \tag{3}$$

$$W_{structure} = \frac{N_{connections}}{N_{nodes}} \tag{4}$$

$Loss_{lable}$ is the CrossEntropyLoss of output and label. $N_{connections}$ is number of connectors, N_{nodes} is number of nodes. For example, the structure of a complete binary tree used to describe the action is more reasonable, and a unipolar end-to-end tree structure is considered to get a worse score.

For example, the brushing tooth action, the S region, maybe around the forearm and the neck, is a semantic attention, and then the recursive tree scoring network pre-generates a set of connectors of body-part with the semantic center S or other body-part. For example: head to semantic center, hand to head, and so on. Then, a set of connections is initialized, recursively joining the generated connections pool, and passing through a pre-trained scoring network using a pre-defined universal tree structure, and finally we can obtain a set of highest scored connections. In the subsequent training process, these connections are constructed into an Action Tree, which is inputed into the graph convolutional network, and the tree is continuously updated to adapt to different motion data to achieve excellent performance.

3.4   Action Tree Convolutions

Based on the generated Action Tree and considering the characteristics of the tree similar to the graph, we then use nine layers of GCNs. For graph convolutions within a single skeleton frame, we used the spatial graph convolution [6]:

$$f_{out}(v_{ti}) = \sum_{v_{ti} \in B(v_{ti})} \frac{1}{Z_{ti}(v_{tj})} f_{in}(p(v_{ti}, v_{tj})) w(v_{ti}, v_{tj}) \tag{5}$$

Which is proposed by ST-GCN. The most difference of ours' method is in spatial configuration partitioning strategy for nodes: the first subset is the node itself, the second is the set of neighbors whose spatial position is closer to the root node of the tree than the current node, and the third is the set of neighbors whose spatial position is farther away from the root node of the tree than the current node.

$$l_{ti(V_{tj})} = \begin{cases} 0, r_{j,root} = r_{i,root} \\ 1, r_{j,root} < r_{i,root} \\ 2, r_{j,root} > r_{i,root} \end{cases} \tag{6}$$

r is the distance between itself and the root, and there are 3 subsets.

3.5 Training and Testing

In order to reduce the common overfitting problems and slow convergence speeds in neural networks, we also introduced the method of using the dropout layer proposed by Hinton. During each training, the output of a certain node in the network is randomly removed with a fixed probability p, setting it to zero. Then continue to train this "cut out" network, repeat the discard and training operations until the convergence. Most importantly, the overall networks are built on PyTorch.

Training process: There are three steps. (1) Firstly, we need to generate the semantic attention of the Action Tree. (2) Then, recursively pre-generates the connections of skeleton joints and body-parts and the semantic attention. Meanwhile, scores the set of connections to obtain a streamlined and optimal set. (3) Select connections/nodes to generate an Action Tree for end-to-end network training to increase recognition accuracy.

Test process: During the test, all parameters are frozen, and the validation dataset is fed into the network to get the output.

We initially trained 20 epochs using a learning rate of 0.1, then reduced it to 0.1 times, trained 30 epoch, returned to the initial learning rate and repeated the previous step at a learning rate of 0.1, and finally decreased to 0.1 times the previous learning rate every 30 cycles. Hence, the learning rate is 0.1, then 0.01, then 0.1, then 0.01, then 0.1, 0.01, 0.001, 0.0001...

4 Experiment

In order to efficiently implement the ATCNs method and some existing methods, we use deep learning framework PyTorch, and use python as the language for implementing those methods. We perform tests using the ATCNs method on datasets NTU-RGB+D and Kinetics [8]. Our experimental framework mainly implements the human action recognition function and we achieves state-of-the-art. This greatly facilitates the performance comparison between the ATCNs method and existing methods.

4.1 Dataset

Kinetics
Kinetics is a large-scale, high-quality human action data set from YouTube. The Kinetics-400 contains 300,000 videos. It has 400 different human actions ranging from daily activities, covering almost every aspect of human life. Among them, there are 400–1150 video clips for each action, and the length of each video is about 10 s. Since the Kinetics dataset only provides RGB video, we used OpenPose [9–11] to generate a skeleton sequence from the video.

NTU-RGB+D
NTU-RGB+D is a dataset with 3D skeletons (body joints), Masked depth maps, Full depth maps, RGB videos, and IR videos currently used for human action recognition tasks. The dataset contains 56000 action clips of 60 action classes with three camera

views. The provided comment gives the 3D joint position (X, Y, Z) in the camera coordinate system, which is used to distinguish the positions of different skeleton joints.

4.2   Test Results and Comparison

Finally, we obtained the following results of experiment: After training with the Kinetics data set, using the benchmark test it provided, the performance of the existing methods show in Table 1 below:

Table 1. Kinetics

Method	Top-1	Top-5
Deep LSTM	16.4%	35.3%
Temporal convolution	20.3%	40.0%
ST-GCN	30.6%	52.8%
ATCN	32.3%	55.2%

Compared to several previous state-of-the-art methods, such as Temporal Convolution [12], our method improve classification accuracy.

After trained with the NTU RGB+D data set, we use the cross subject and cross view test. The performance of each existing methods show in Table 2 below.

Table 2. NTU-RGB+D

	X-Sub	X-View
Differential RNN	50.1%	52.8%
Hierarchical RNN	59.1%	64.0%
Deep LSTM	60.7%	67.3%
PA-LSTM	62.9%	70.3%
ST-LSTM+TS	69.2%	77.7%
Temporal convolution	74.3%	83.1%
C-CNN+MTLN	79.6%	84.8%
ST-GCN	80.6%	88.8%
ATCN	83.2%	91.2%

From Table 2, we can see the comparison with other methods. With the state-of-the-art method, such as Differential RNN [13], Hierarchical RNN [14] and ST-LSTM [15], our method improve classification accuracy.

Further, for cross subject, we used the ST-GCN method and the ATCN method to test the samples from different perspectives, and obtained the accuracy of different perspectives in the cross subject. The results show in Table 3:

Table 3. Performance of individual views on NTU-RGB+D X-Sub

View	0°	45°	90°	−45°	−90°
ST-GCN	82.4%	82.9%	73.0%	83.8%	79.4%
ATCN	85.4%	84.6%	78.3%	85.4%	80.4%
Performance optimization ratio	3.64%	2.05%	7.26%	1.91%	1.26%

As can be seen from the Table 3, the lowest score of ST-GCN method is 73.0% and the highest score is 83.8%. The lowest score for the ATCNs method is 78.3% and the highest score is 85.4%. Compared with the ST-GCN method, the ATCNs method improves in all aspects of the test. The performance optimization rate fluctuates from 1.26% to 7.26%, and the average performance optimization rate is 3.22%. When the view is 90°, the performance optimization ratio has a big gap compared with the other four perspectives. Based on the analysis of the environment and human behavior habits, the main reason for this phenomenon is that most people in the training group are right-handers. The right-handedness behavior determines that most of the movements related to the skeleton joints in the hand are concentrated in the right part. Observed from the right side, the performance optimization rate is the highest. At the same time, this also explains that at view = −90°, the performance optimization ratio is the smallest. Based on the comprehensive experimental results, we can conclude that the performance of the ATCNs method is significantly better than other methods.

In the cross view as Table 4 showing, we go further and divide the view into two groups, which are the training set and the test set respectively. Compared with the ST-GCN method, our ATCN method has obvious optimization of human action recognition performance.

Table 4. Performance of individual views on NTU-RGB+D X-View

X-View (1, 3 views train; 0, 2, 4 views validation)			
	Total	View 0	View 2	View 4
ST-GCN	78.7%	84.6%	68.9%	76.5%
ATCN	83.4%	89.3%	74.7%	80.0%

X-View (0, 2, 4 views train; 1, 3 views validation)		
	Total	View 1	View 3
ST-GCN	88.8%	88.0%	89.7%
ATCN	91.2%	90.4%	91.9%

5   Conclusion

This project is the relevant models and methods in the field of human action recognition, and proposes the ATCNs method. Our approach is inspired by the highly successful neural network structures: Graph Convolutional Networks (GCNs) and EM-type training methods, and constructs an Action Tree Convolutional Networks using a tree structure that can describe relationships perfectly. In our networks, the semantic tree [7] and convolutional neural network have completed efficient cooperation. Finally, we use the NTU-RGB+D and Kinetics datasets to perform experiments on existing methods and ATCNs method. The results show that our method has better performance. However, there are still some problems in our method. As the camera's view angle changing, some skeleton joints will be masked by other parts, so we will work on it in the future.

References

1. Karpathy, A., Toderici, G., Shetty, et al.: Large-scale video classification with convolutional neural networks. In: CVPR, pp. 1725–1732 (2014)
2. Tran, D., Bourdev, L., Fergus, R., Torresani, L., Paluri, M.: Learning spatiotemporal features with 3D convolutional networks. In: ICCV, pp. 4489–4497. IEEE, Washington (2015)
3. Baradel, F., Wolf, C., Mille, J., Taylor, G.W.: Glimpse clouds: human activity recognition from unstructured feature points. arXiv preprint arXiv:1802.07898 (2018)
4. Amir, S., Jun, L., Tian-Tsong, N., Gang, W.: NTU RGB+D: a large scale dataset for 3D human activity analysis. In: CVPR (2016)
5. Kipf, T.N., Welling, M.: Semi-supervised classification with graph convolutional networks. In: ICLR (2017)
6. Yan, S., Xiong, Y., Lin, D.: Spatial temporal graph convolutional networks for skeleton-based action recognition. In: AAAI (2018)
7. Lin, L., Wang, G., Zhang, R., et al.: Deep structured scene parsing by learning with image descriptions. In: Computer Vision and Pattern Recognition, pp. 2276–2284. IEEE, Washington (2016)
8. Kay, W., Carreira, J., Simonyan, K., et al.: The kinetics human action video dataset. In: arXiv preprint arXiv:1705.06950 (2017)
9. Cao, Z., Simon, T., Wei, S.E., Sheikh, Y.: Realtime multi-person 2D pose estimation using part affinity fields. In: CVPR, vol. 1, no. 2, p. 7 (2017, July)
10. Simon, T., Joo, H., Matthews, I., Sheikh, Y.: Hand keypoint detection in single images using multiview bootstrapping. In: CVPR, vol. 2, July 2017
11. Wei, S.E., Ramakrishna, V., Kanade, T., Sheikh, Y.: Convolutional pose machines. In: CVPR, pp. 4724–4732 (2016)
12. Kim, T.S., Reiter, A.: Interpretable 3D human action analysis with temporal convolutional networks. In: CVPRW, pp. 1623–1631. IEEE, Washington, July 2017
13. Veeriah, V., Zhuang, N., Qi, G.J.: Differential recurrent neural networks for action recognition. In: Computer Vision (ICCV), pp. 4041–4049. IEEE, Washington, December 2015
14. Du, Y., Wang, W., Wang, L.: Hierarchical recurrent neural network for skeleton based action recognition. In: CVPR, pp. 1110–1118 (2015)
15. Liu, J., Shahroudy, A., Xu, D., Wang, G.: Spatio-temporal LSTM with trust gates for 3D human action recognition. In: Leibe, B., Matas, J., Sebe, N., Welling, M. (eds.) ECCV 2016. LNCS, vol. 9907, pp. 816–833. Springer, Cham (2016). https://doi.org/10.1007/978-3-319-46487-9_50

Research of Secret Sharing Digital Watermarking Scheme Based on Spread Spectrum Algorithm and PCA

Xiaohong Li[✉], Dawei Niu, Shu Zhan, and Wubiao Chen

School of Computer and Information,
Hefei University of Technology, Hefei, China
jsjlxh@hfut.edu.cn

Abstract. With the extending popularization and promotion of information technology, the secret sharing scheme applied to the digital watermark domain is becoming increasingly important. In fact, secret sharing is a very useful cryptographic technique, which divides and stores secret information according to certain rules. This algorithm combines with digital watermarking to protect the copyright of the image processing system with higher efficiency, higher reliability and better security. Meanwhile, during this paper, a synthetic scheme will be researched and discussed. One is the PCA algorithm, as a data processing technique, it can reduce the dimensionality of the original complex data. In this process, feature coefficients, the most representative features of the original carrier images that were selected randomly, and extracted from them by means of PCA transform. Then, for the sake of improving capacity of resisting distortion and disturbance, applying spread spectrum to digital watermarking, it can guarantee the overall system stability and robustness. The watermark image embedded in the main component coefficients of the carrier image is generated by the secret sharing preprocessing of original image. Finally, we can summarize the overall performance of the digital watermarking system and compare the similarity between the recovered image and the original image by observing the values of NC and PSNR that were from all kinds of experiment.

Keywords: Secret sharing · Watermarking · PCA · Spread spectrum

1 Introduction

With the high-speed development of digital information, the theft of digital products and information such as pictures and videos is increasingly rampant and now it is out of control. So it is necessary to preprocess the watermarking information before embedding it into carrier data to ensure the better robustness and stability of the watermarking system [3]. All preparation procedures must be as rigorous as possible precisely. The good news is that the secret image sharing scheme precisely can meet this performance requirement. In the last few years, it was no doubt that the digital watermark technique has been become a major tool which is used for solving the copyright issue of digital multimedia and image content authentication [1]. At the same time, the secret image sharing watermarking algorithm provides a more effective

© Springer Nature Switzerland AG 2018
R. Hong et al. (Eds.): PCM 2018, LNCS 11166, pp. 793–803, 2018.
https://doi.org/10.1007/978-3-030-00764-5_73

solution [5]. Taking the characteristic of digital multimedia products and the secret sharing algorithm into consideration, some relevant protective technologies have been proposed specially. First of all, the idea of applying digital watermarking technology is to embed some digital watermarking information with certain and representative features into digital multimedia products, and then made it a part of carrier information, in the light of some digital information processing skills, extracting watermarking to check whether the carrier image has been copied, pirated and even juggled with illegal phenomena. In order to realize the protection of digital multimedia information products. In the 1970s, Shamir and Blakely alone demonstrated the conception of secret sharing by themselves. For instance (t, n) threshold scheme, The main theories and ideologies of secret sharing are that the secret S is divided up among n participants, it is t or more than t members that accomplishing the secret re-construction needs, however, it cannot realize the image recovery when less than t members. In general, the ideas of traditional secret sharing are used for resolving the security problems. The first step is to divide shared secret information into several secret shares which are called the sub keys, next distributing these secrets to a number of different participants, and restricting the rights of these participants, a part of them can reconstruct secret information together, so that the confidentiality of the whole system will be improved substantially. The idea of secret sharing is not only a very effective idea in secret key system, but also can protect and disseminate important information. In addition to this, the theory of spread spectrum communication was put forward actually in the earlier years. Nevertheless, the real study, began in the United States in the mid-1950s, and was first applied to military communications. Compared with the general communication models, the spread spectrum communication is different from them, and has the characteristics of stronger anti-interference ability and better confidentiality. With the in-depth study of digital watermarking and spread spectrum, a great number of algorithms have sprung up. The application of spread spectrum scheme to digital watermark is one of them. Spread-spectrum technology is a kind of information processing technique, using spread spectrum code which is independent of the data to be transmitted to spread the spectrum of the signal to be transmitted, and making its bandwidth far more than the necessary to transmit information. Then, the same spread spectrum code is used for the de-spreading and recovery of data in the receiver terminal. It is can greatly enhance the anti-interference, confidentiality and invisibility of digital watermarking system by applying the spread spectrum technology into watermarking field [6].

Taking all things above what we have discussed into consideration, we put the original image after dealing with the secret sharing and spread spectrum modulation as a watermark image, incorporation with the carrier image after PCA algorithm operation to extract feature data [4], then embedding and extracting the watermarking information to complete the whole digital watermarking system. Through these operations, we can further improve the robustness of the experiment, resist attack, confidentiality, stability and visual invisibility.

2 Proposed Algorithm Flow or Use

2.1 The (t, n) Secret Sharing Scheme

First, the original image needs scrambling transformation, processed by the secret sharing algorithm to generate watermark image. There are two kinds of threshold secret sharing schemes. One is the traditional threshold secret sharing scheme, called Shamir threshold secret sharing scheme [5], and the other is based on linear equations, called Simmons threshold secret sharing scheme.

There are two main stages of secret sharing, one is the stage of secret distribution, the other is the stage of secret reconstruction.

- **Scrambling Transformation**

Arnold scrambling transformation is a simple and easy periodic algorithm to implement, which is widely used in information hiding technology. This algorithm requires an iterative algorithms, and the size of the scrambled images must be $N \times N$. We supposed that (x, y) is the coordinates corresponding to a pixel cell point, and (x', y') is the coordinates of the points after the Arnold transformation. The relationship between them is as shown in the following expression.

$$\begin{pmatrix} x' \\ y' \end{pmatrix} = \begin{bmatrix} 1 & 1 \\ 1 & 2 \end{bmatrix} \cdot \begin{pmatrix} x \\ y \end{pmatrix} mod\ N \tag{1}$$

As a matter of fact, Arnold scrambling transformation is the position change of a point. For a square digital image, we can express it as the following formula

$$I = \begin{bmatrix} F_{11} & \cdots & F_{1N} \\ \vdots & \ddots & \vdots \\ F_{N1} & \cdots & F_{NN} \end{bmatrix} \tag{2}$$

N is the size of the digital image matrix. F_{xy} is the gray value corresponding to the pixel point (x, y). In digital images, the Arnold transform is adopted, that is to say, through changing the coordinate value of the pixel positions, the layout of the entire image pixel value is changed, so as to achieve the purpose of scrambling. This coordinate transformation is also iterative operation. The Arnold transform has a certain periodicity, and when the iterative operation reaches a certain number of steps, it will recover the original image.

Logistic chaotic map scrambling transformation, is a classical chaotic system algorithm. The mathematical expressions, defining the concept of Logistic chaotic mapping, are as follows.

$$x_{k+1} = \lambda \times X_k \times (1 - X_k), k = 1, 2, 3, \ldots \tag{3}$$

In this equation, λ is a constant value, when the value is between 3.569945... and 4, the system is in a chaotic state, and it is in a surjective chaotic state while $\lambda = 4.0$. Generated by the mapping relationship, Logistic chaotic sequence has the characteristics of aperiodicity, non-convergence and boundedness, which is susceptible to initial values. In a general way, the value of λ is 4, and the initial value is about 0.3.

Then in the next step, using the above definition formula to generate two chaotic sequences [7], and according to these two sequences, the row of original binary watermark image is scrambled firstly, its resolution ratio is $M \times N$, next the column scrambling is carried out, denoted as

$$I' = [s_{ij}], 1 \leq i \leq M, 1 \leq j \leq N \tag{4}$$

The I' is divided into n segments, each of which consisting of t pixels, and $I' = [s_{ij}], 1 \leq i \leq M, 1 \leq j \leq N$ can be represented as:

$$I' = \{(s_{11}, \ldots, s_{1t}), (s_{21}, \ldots, s_{2t}), \ldots, (s_{n1}, \ldots, s_{nt})\} \tag{5}$$

Constructing a nonsingular $n \times t$ matrix $W = [w_{ij}] (i = 0, 1, \ldots, n - 1, j = 0, 1, \ldots, t - 1)$, $w_{ij} \in GF_P$, $S_i = (s_{i1}, \ldots, s_{it})$ is a vector which is made up of t pixels. According to the principle of secret sharing algorithm, the value of secret D can be calculated until all pixels have been processed.

$$D = W \cdot S^T \bmod 2 \tag{6}$$

In the above formula, $D = (D_1, \ldots, D_n)$, n sub secret images, the size of them is $M \times N$, they are composed of elements of the D in proper sequence, at the end of secret sharing scheme.

- Distribution Phase

The distributor Deal, in a secret message $s \in Z_p$, randomly selects n $x_i \in Z_p^*(x_i = i, i = 1, 2, \ldots, n)$ different element values and assigns them to n participants $P_i(i = 1, 2, \ldots, n)$, these $x_i(i = 1, 2, \ldots, n)$ are the public keys. Constructing a t − 1 order polynomial $f(x)$ has to satisfy the formula $f(0) = s$. Then selecting the t − 1 constant values discretionarily, and assigning these to $a_1, a_2, \ldots, a_{t-1}$, in addition to, putting s in a_0, so that the polynomial expression is shown below

$$f(x) = a_{t-1}x^{t-1} + a_{t-2}x^{t-2} + \cdots + a_1x + a_0 \tag{7}$$

Then n resulting function values $f(x_i)$ are assigned to n participants $P_i(i = 1, 2, \ldots, n)$. $s_i = f(x_i) \bmod P$ among them.

- Reconfiguration Phase

N participants can recover secret message by owning any t sub secrets. It means that t owners $P_i(i = 1, 2, \ldots, t)$ releases $s_i(i = 1, 2, \ldots, t)$. By according to the Lagrange interpolation method, the polynomial can be reconstructed, which is shown below.

$$f(x) = \sum_{k=1}^{k} y_i \prod_{j-1, j \neq i}^{k} \frac{(x - x_i)}{(x_i - x_j)} \, mod \, P \tag{8}$$

2.2 Generation Strategy of Watermark Image Information

The spread spectrum algorithm is used to modulate the scrambled and secretly shared images to improve the robustness of the watermarking system and to generate binary watermark information embedded in the carrier images.

Spread spectrum scheme based on the concept of byte rate, mainly through extending the original watermark information and the redundancy of information, to reach the target of the implementation of spread spectrum information. According to the byte rate, the spectrum of original information is spread, and then the pseudo-random sequence is modulated with it, the final watermark information will be generated.

At the first, the watermark image is transformed into corresponding binary information, and 0 is mapped to -1, and 1 is mapped to 1, then we can get a sequence $b_j \in \{-1, 1\}, j = 1, 2, \ldots, n/R_1$, The macroblock of the luminance component of the video image with 16×16 division is n among them.

Next, expanding b_j to get a new sequence c_i, R_1 is the byte rate, $c_i = b_j, i = 1, \ldots, n, (j-1)R_1 < i < jR_1 + 1$, c_i is modulated by a pseudo random sequence m_i with zero mean. After modulation, d_i is obtained, $m_i \in \{-1, 1\}, d_i = c_i m_i, i = 1, \ldots, n$. Corresponding to 16×16 module, the sequence d_i is extended again, then generating a new sequence e_j, R_2 also is the byte rate, in this algorithm, $e_j = d_i, j = 1, \ldots, n, (j-1)R_2 < i < jR_2 + 1$. e_j is modulated by a pseudo random sequence m'_j with zero mean. After modulation, f_j is obtained, $m'_j \in \{-1, 1\}$, $f_j = e_j m'_j, j = 1, \ldots, n$.

Thus, by the above scheme steps, the whole process of spread spectrum modulation is completed, and then the processed watermarked image is generated.

2.3 Principal Component Analysis of Carrier Images

As a matter of fact, the principal component analysis method is a mathematical statistical calculation method [2]. For n sample variables $X_1, \ldots, X_n$, P dimensional vector $x = (x_1, \ldots, x_P)^T, i = 1, \ldots, n, n > P$, A mathematical sample matrix is constructed as follows

$$X = \begin{bmatrix} X_{11} & \cdots & X_{1n} \\ \vdots & \ddots & \vdots \\ X_{P1} & \cdots & X_{Pn} \end{bmatrix} \tag{9}$$

The process of principal component calculation is as follows. The most important thing is that these basic data are basically standardized, the sample matrix X is normalized,

$$Z_{ij} = \frac{x_{ij} - \overline{x}_j}{s_j}, i = 1, \ldots, n, j = 1, \ldots, P \tag{10}$$

Among them,

$$\overline{x}_j = \frac{\sum_{i=1}^n x_{ij}}{n}, \quad s_j^2 = \frac{\sum_{i=1}^n \left(x_{ij} - \overline{x}_j\right)}{n-1} \tag{11}$$

$\overline{x}_j$ is the mean value; s_j is the variance of the component of X. Solving the correlation coefficient matrix R of the normalized matrix Z.

$$R = \left[r_{ij}\right]xP = \begin{bmatrix} 1 & r_{12} & \cdots & r_{1P} \\ r_{21} & 1 & \cdots & r_{2P} \\ \vdots & \vdots & \ddots & \vdots \\ r_{P1} & r_{P2} & \cdots & 1 \end{bmatrix} \tag{11}$$

$$r_{ij} = \frac{\sum Z_{ij} Z_{ij}}{n-1}, i, j = 1, \ldots, P. \tag{12}$$

The characteristic equation of sample correlation matrix is solved, according to the following formula, P eigenvalues also would be solved.

$$|R - \lambda I_P| = 0 \tag{13}$$

Arranging them in descending order, like $\lambda_1 \geq \cdots \geq \lambda_P \geq 0$, the eigenvector $e_i (i = 1, \ldots, P)$ is evaluated according to the eigenvalue λ_i, and the feature vector matrix $U = (e_1, \ldots, e_P)^T$ is made up of the eigenvector e_i. In the principal component analysis scheme, it is critical key to determine the number of principal components, if chosen improperly, it can lead to loss of some important data.

It is the last step that transforming the real principal component, calculated according to the following formula,

$$F_J = U_j^T Z, j = 1, \ldots, m \tag{14}$$

Using the above theoretical scheme into digital image processing and supposing the original image is PI_{MN}, M and N represent the width and height of the image respectively, Dividing the image PI_{MN} into many sub blocks, rearranging the blocks, and representing an image as $L = [l_1, l_2, \ldots, l_G]$. such as, the number of sub blocks are G. Calculating the covariance matrix of L, and getting the matrix C, $C = (L - m) \times (L - m)^T$, the value of m is the mean of the sub block image in the formula. By solving the eigenvalue equation, the eigenvalue λ of C and the eigenvector φ are calculated. Based on the eigenvalue size, the orthogonal matrix $\varphi = [e_1, e_2, \ldots, e_K]$ is obtained by selecting and calculating the corresponding eigenvector, e_K is the basis function, and then principal component coefficient Y of the image are obtained. $Y = \varphi^T L = [y_1, y_2, \ldots, y_K]^T$, K is the principal component coefficient, determined by the cumulative contribution rate, could be deduced by formula $E_i = \lambda_i / \sum_{i=1}^{k} \lambda_i$. From this, we can get the carrier image after PCA processing.

After the processing of watermark image information, then the original carrier image is analyzed and optimized by PCA principal component analysis. This scheme is a feature extraction method which based on second-order statistics, which selects some of the important components and discards some unimportant information, so as to achieve the purpose of optimizing the parameter of watermarking system and improve data processing efficiency.

2.4 Embedding of Secret Watermark Image

The reliability of watermarking technology depends on the robustness and insensitivity of watermarks, all of these requirements are completed in the watermark embedding stage. Firstly, selecting a set of 128×128 gray image as testing carrier image, the carrier image is divided into blocks, and each sub block is transformed. The eigenvalue of sub block carrier images is calculated by using PCA principal component analysis algorithm. Then, the principal component coefficients of the embedded watermark image are extracted. Through the above analysis, the 64*64 original binary watermark image is used in the experiment of embedding operation, in order to visualize the effect. The watermark image, carried out scrambling transformation of the related algorithm, generates N sub secret images after secret sharing process. By using pseudo random sequence, spread spectrum modulation is performed on the corresponding sub secret graphs in proper order, a new final watermarking image is obtained. Next selecting a part of the information data bit in the watermarking image, according to the principal component obtained by the PCA algorithm to select the appropriate watermark embedding location, and then these data bits are embedded into the carrier image.

2.5 Extraction and Recovery of Watermark Image

(t, n) threshold secret sharing scheme is adopted in this paper. Therefore, the original secret watermark image can be recovered by arbitrarily selecting t secret image, the sub secret image information can be extracted, through the inverse process of improving watermarking scheme above.

PSNR, peak signal-to-noise ratio, it is an objective indicator to measure the distortion degree of images or videos, and also can be the objective evaluation to measure the visual changes degree of the image compression, video compression and image or video quality before and after processing (such as embedding), the numerical result of the evaluation is decibels. Generally, the value of PSNR has a common benchmark of 30 dB, which is considered to be seriously distorted below this value. Normalized cross correlation function, NC coefficient, is an objective indicator to measure the similarity, comparing the original watermark and the extracted watermark from the videos or images, so the closer the NC value is to 1, the better the system will be.

$$PSNR = 10lg \left[\frac{max_{\forall(m,n)} f^2(m,n)}{\frac{1}{N_f} \sum_{\forall(m,n)} [f_w(m,n) - f(m,n)]^2} \right] \tag{15}$$

In the equation, N_f is the number of pixels of the image frame, and $f(m,n)$ is the value of the original image frame pixel point, $f_w(m,n)$ is the value of the processed image frame pixel point.

$$NC = \frac{\sum_i \sum_j w^*(i,j) w(i,j)}{\sum_i \sum_j [w(i,j)]^2} \tag{16}$$

In this equation, $w^*(i,j)$ represents the extracted watermark images, and $w(i,j)$ represents the original watermark images.

3   Experiment Results

Watermarking preprocessing scheme proposed previously, is used to generate watermark information. The experiment of this improved algorithm is based on MATLAB simulation platform. Combining with the invisibility, anti-quantification attack and robustness of the watermark image, the simulation results of the experiment were analyzed. What's more, the size of the carrier test image is 128 * 128, the size of the original watermark image is 64 * 64, which is chosen as the word "watermark information", and then applying the preprocessing scheme including scrambling transformation, (t, n) threshold secret sharing and spectrum spreading, to complete the entire experiment, the parameters of embedding and extraction are optimized by PCA algorithm operation. Effect comparison is made between the original carrier images and the watermarked images, and calculating their peak signal to noise ratio (PSNR) and Normalized cross-correlation function value (NC) that can reflect the differences before and after the embedded watermark would be done in the final (Figs. 1, 2, and 3).

Image	PSNR (dB)
Cameraman. tif	36.67
Rice.png	36.83
Lena.gif	37.92
Straw.tif	36.76
Shenlin.jpg	37.72
House.tif	37.34

Next, several groups of watermark attack test experiments are carried out. The first is the attack experiment of salt and pepper noise, and the results are shown in the following table.

Fig. 1. (1) Cameraman.tif (2) Cameraman image containing watermark information.tif (3) Rice. png (4) Rice image containing watermark information.png

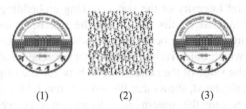

Fig. 2. (1) Original image (2) scrambling image (3) restore image

Fig. 3. (1) Original watermark image (2) spread spectrum modulated image (3) restore watermark image

Noise attack	Salt and pepper noise (0.03)	Salt and pepper noise (0.06)	Salt and pepper noise (0.09)	Salt and pepper noise (0.13)	Salt and pepper noise (0.16)
NC	0.9888	0.9787	0.9681	0.9540	0.9438

Then the scaling and clipping attack experiments are presented, as shown in the following table.

Scaling and clipping attack	Scaling (1/2)	Scaling (1/4)	Cliping (1/4)	Cliping (1/8)	Cliping (1/16)
NC	0.7220	0.8400	0.8835	0.9513	0.9848

4   Conclusion

This paper focuses on the theme of "secret sharing watermarking algorithm based on spread spectrum and principal component analysis". The theoretical analysis and simulation experiments of digital image watermarking, such as resistance signal, noise attack and geometric attack, are carried out, and the robustness of the watermarking algorithm is improved. After pixel matrix of the image is manipulated, watermark is embedded in the PCA principal component parameters of the carrier image, processed by some preprocessing methods such as pseudo random spread spectrum modulation and secret sharing etc. Combined with the principal component analysis algorithm to control the position and intensity of the watermarking embedding, the algorithm was improved. The experimental results show that the watermarking system has good robustness to the common image processing methods and attack behaviors. The size of PSNR was obtained, which objectively indicates that the scheme also has a good sense of insensibility, then the value of the similarity NC between the original images and the restored images was calculated, shows that the value is much larger than 0.6, indicating that the system can restore the watermark information very well in this improved scheme.

Acknowledgment. This work was supported by National Natural Science Foundation of China under Grant 61371156.

References

1. Jain, K., Raju, U.S.N.: A digital video watermarking algorithm based on LSB and DCT. J. Inf. Secur. Res. (2015)
2. Sinha, S., Bardhan, P., et al.: Digital video watermarking using discrete wavelet transform and principal component analysis, a treatise on electricity and magnetism. Int. J. Wisdom Based Comput. **1**(2), 7–12 (2011)

3. Chimanna, M.A., Khot, S.R.: Digital video watermarking techniques for secure multimedia creation and delivery. Int. J. Eng. Res. Appl. (IJERA) **3**(2), 839–844 (2013). ISSN 2248-9622
4. Patil, S.A., Srivastava, N.: Digital video watermarking using DWT and PCA. IOSR J. Eng. (IOSRJEN) **3**(11), 45–49 (2013). e-ISSN 2278-8719, p-ISSN 2278-8719
5. Ulutas, M., Ulutas, G., Nabiyev, V.V.: Medical image security and EPR hiding using Shmir's secret sharing scheme. J. Syst. Softw. **84**(3), 341–353 (2011)
6. Su, K., Kundur, D., Hatzinakos, D.: Statistical invisibility for collusion-resistant digital video watermarking. IEEE Trans. Multimed. **7**(1), 43–51 (2005)
7. Neri, A., Campisi, P., Carli, M., et al.: Watermarking hiding in video sequences. In: First International Workshop on Video Processing and Quality Metrics for Consumer Electronics, Scottsdale, Arizona, USA. SPIE (2005). (LTS-CONF-2005-009)

Convolutional Neural Networks Based Image Classification for Himawari-8 Stationary Satellite Imagery

Jinglin Zhang[1], Pu Liu[1], Jianwei Zheng[2], and Cong Bai[2(✉)]

[1] School of Atmospheric Science,
Nanjing University of Information Science and Technology, Nanjing, China
jinglin.zhang@nuist.edu.cn
[2] College of Computer Science, Zhejiang University of Technology, Hangzhou, China
congbai@zjut.edu.cn

Abstract. Cloud plays an extremely important role in the atmosphere, which directly influences the radiation balance and indicates the potential weather and climate change. It also builds up the strength of locally thermal and dynamic processes. A novel convolutional neural networks (CNNs) approach, namely Satellites Model (SatMod), is introduced to satellite imagery classification, which performs well under multiple contrails conditions. We take contrails into account in the satellite imagery classification, which leads to satellite imagery classification more challenging than existing satellite imagery database, as clear understanding of contrails would facilitate the study of the effects of contrails on global warming. A novel dataset based on Himawari-8 stationary satellite imagery (HSSI) is proposed to represent 5 different scenes. Extensive experiments and evaluation indicate that the proposed SatMod achieves a good performance on HSSI database.

Keywords: Satellite imagery · Image classification
Convolutional neural network · Contrails

1 Introduction

Clouds are formed during the evolution of different synoptic systems and can predict future weather climate changes. Clouds are also the important part of the atmospheric radiation system. The change of cloud coverage will affect the balance of energy budget of the whole atmosphere, which impacts the climate system [1]. It has been universally acknowledged that different levels and types of clouds have different feedback effect on the atmosphere. Moreover, different types of clouds indicate different weather changes in the future. Experimental results indicate that accurate cloud classification will have a significant impact on the numerical models of the atmosphere, the accuracy of forecasting climate models and specific business operations.

© Springer Nature Switzerland AG 2018
R. Hong et al. (Eds.): PCM 2018, LNCS 11166, pp. 804–810, 2018.
https://doi.org/10.1007/978-3-030-00764-5_74

Generally, clouds classification mainly includes satellite and ground observation, while satellite images could obtain detailed large-scale information of clouds for climate prediction. Therefore, the current research on cloud classification mainly focuses on ground observation instruments [2] and satellite observation [3,4]. With the continuous development of observational instruments, such as the whole-sky imagery (WSI), it could provide continuous and high-resolution cloud images. Based on these instruments, increasing number of research works have been done on the extraction of the texture-structure features. For instance, the features are extracted from cloud images using the image texture measurement algorithm such as the local binary pattern (LBP) [5], auto-correlation and co-occurrence matrices [6]. The automatic cloud classification algorithm extracts spectral features and some simple texture features to classify seven different classes of clouds [7]. While [8] adopts the statistical color deformation to obtain texture and color information. Integrating both color and texture information also gets good results in cloud taxonomies [9].

Contrail is a special kind of cloud, caused by the aircraft engine emissions from the exhaust gas mixed with the surrounding air and the condensation of watervapor [10,11]. Meyer et al. [3] introduces a contrail detection algorithm (CDA) to recognize linear-shaped contrails on satellite images, while Hetzheim [4] proposes a very complicated approach using mathematical methods to detect contrails based on texture or stochastic behaviors of contrails. Although these efforts have contributed to clouds even contrails classification, the issue of accuracy has not been well addressed. In recent years, convolution neural networks (CNNs) has achieved remarkable results in other fields such as face recognition and image classification. However, there are very little insight on cloud classification using deep learning. The neural network approach has the capability to learn patterns whose complexity makes them difficult to define using previous approaches [17].

In this paper, we introduce the Satellites Model (SatMod), a framework of convolutional neural network, which achieve exceeding progress compared with the conventional approach in the satellite imagery classification. We also establish one satellite-based cloud database that is convictive and discriminative, which mainly consists of the 5 category of scenes under Meteorological standards. In addition, contrails is taken into account in the database. The total number of sample images in our database is several times as much as the previously studied database. Our method not only improves the accuracy of cloud classification, but also avoids the conventional reliance on many empirical classification mistakes.

The reminder of the paper is structured as follows: Sect. 2 describes the proposed SatMod, followed by satellite-based cloud database in Sect. 3. And then the experiments are detailed in Sect. 4. Finally, conclusions are given in Sect. 5.

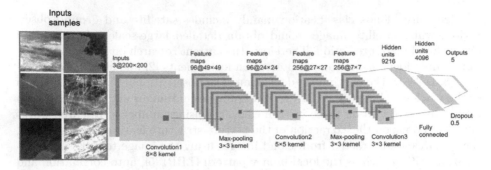

Fig. 1. An illustration of the SatMod architecture, which consists of 3 layers of convolutional neural network followed by 2 fully-connected layer.

2   Model Architecture

Here, we propose a specific architecture, termed Satellites Model (SatMod), which is partly inspired by the CNNs for the sequence-to-sequence learning task. The architecture of the SatMod is shown in the Fig. 1, which contains 5 layers and is very straightforward compared to other architectures [12]. Our model takes a series of satellite sub-images as input, and a sequence of label predictions as outputs, which represent the probability of each category.

In the SatMod architecture, Batch Normalization [13] is applied in the first and second convolutional layer and a rectified linear activation [18] is used for each layer. In the second and fourth layer, the corresponding Max Pooling operation subsamples their input. As illustrated in previous work [8], deep layers and hidden layers of CNNs could reflect the high-level and fine features of clouds respectively. Visualization of cloud features in different convolution layer [8] also demonstrates that introducing two convolutional layers promote the feature representation, especially for contrails. The third convolutional and fully connected layers are direct connected, due to the difficulty of cloud feature extraction. Meanwhile, dropout [16] is applied to the fifth fully connected layer with the nonlinearity. The final fully connected layer and softmax activation produce a distribution over the 5 output probabilities for each category. We use the Stochastic Gradient Descent optimizer with the default parameters in [14]. Finally we save the best model configuration as evaluating with the test set during the optimization process.

3   Data

In order to achieve remarkable performance, the essential factor is plenty of labeled data. For example, the ImageNet [15] collects over million labeled images for classification and recognition task. The Singapore Whole-sky Imaging Categories database (SWIMCAT) [9] contains 784 images of sky and cloud samples

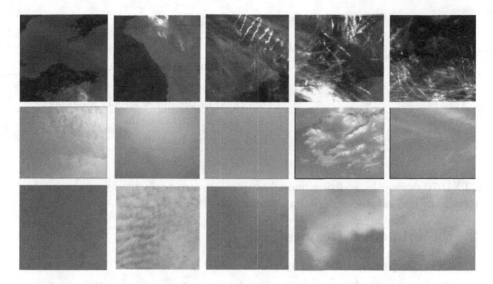

Fig. 2. Cloud samples from different databases. The first line: sample images of HSSI database; the second line: cloud images come from 6 class HUST database [8]; the third line: cloud images come from SWIMCAT [9].

with 5 distinct categories: *clear sky, patterned clouds, thick dark clouds, thick white clouds and veil clouds*, as shown in Fig. 2.

In order to get better classification results, we construct a dataset-HSSI (Himawari-8 stationary satellite imagery) to classify the different scenes in satellite imagery, especially for contrails recognition. The dataset contains 1000 unique sub-images of satellite imagery, which is divided from Himawari-8 stationary satellite imagery. The Advanced Himawari Imager (AHI) on board Himawari-8 is a 16-channel multispectral imagery to capture visible light and infrared images of the Asia-Pacifc region, which produces images with a resolution down to 500 m and provides full disk observations every 10 mins. Based on the subjective classification of meteorological experts, we define 5 different scenes in HSSI datasets: *clear sky, thick cloud, thin cloud, single contrail and cross contrails*, as shown in the first row of Fig. 2. All sample images are with 200 × 200 pixels resolution in the JPEG format.

3.1 Implementation Details

The HSSI dataset consists of a training set and a test set. The training set contains 800 sample images and the test set contains 200 sample images. We also make sure that there is no sample image overlapping between the training and test sets. Before training the SatMod, we need to subtract the mean activity over the training set from each pixel. The network is trained with stochastic gradient utilizing the distributed machine learning tools *caffe* on NVIDIA GeForce

Table 1. The description of HSSI database.

Category	Number of sample	Description
Clear sky	200	Clear sky, land, ocean without any clouds
Thick cloud	200	Thick cloud in the lower level of atmosphere
Thin cloud	200	Thin cloud in the upper level of atmosphere
Single contrail	200	Single and parallel contrails
Cross contrails	200	Multi and cross contrails

GTX1070 device. In addition, model evaluations are performed using a running average of the parameters computed over time (Table 1).

4   Experimental Results

4.1   The Experimental Setup

In order to objectively evaluate our model of performance on this multi-class classification task and the HSSI database mentioned above, we use the HSSI dataset to train SatMod. We evaluate the individual classification accuracy for each of the 5 categories. We also report the average classification accuracy, macro-precision, micro-precision, precision and recall. In such a multi-class classification task, macro-values give an indication of system performance across all the categories, whereas micro-values quantify the performance of per image classification [9].

Table 2. Performance evaluation of HSSI database

Category	F-measures	Precision	Recall	Accuracy
Clear sky	0.87	0.83	1	0.93
Thick cloud	0.91	0.93	1	0.93
Thin cloud	0.96	0.92	1	0.97
Single contrail	0.80	1	0.67	0.93
Cross contrails	0.69	0.83	0.57	0.91

4.2   Results Analysis

Table 2 shows the performance evaluation of cloud recognition rates for different categories. As far as we know, we firstly introduce and analyse the multiple scene of contrails in the HSSI dataset, which leads to a more challenging classification task of satellite imagery classification. More clear understanding of contrails will promote the study of how contrails impact on global warming. Based on the previous research [10], the result of satellite imagery classification will promote the task of contrails recognition, while some different scenarios of contrails recognition are illustrated in Fig. 3.

Fig. 3. Contrails recognition examples based on SatMod and HSSI dataset

5   Conclusion

In this paper, we propose a CNNs-based model, termed SatMod, which exceeds
the performance in satellite imagery classification. Due to large annotated HSSI
dataset that we collected and SatMod that can map a sequence of raw clouds
images to a sequence of ground truth label. As the better understanding of contrails would facilitate the study of the effects of contrails on global warming and
climate change, contrails are taken into account in the satellite imagery classification, which leads to satellite imagery classification more challenging than
exiting satellite imagery databases. Based on the HSSI database and SatMod
model, extensive experimental results show that the effectiveness of our proposed model on the satellite imagery classification.

Acknowledgements. The work of Jinglin Zhang is funded by the Scientific Research
Foundation (BK20150931) of JiangSu Province, Natural Science Foundation of China
(Grants No. 61702275 and 41775008). The work of Jianwei Zheng and Cong Bai
is funded by Natural Science Foundation of China under Grant No. 61502424 and
61602413, Zhejiang Provincial Natural Science Foundation of China under Grant
LY18F020032.

References

1. Duda, D.P., Minnis, P., Khlopenkov, K., Chee, L.: Estimation of 2006 Northern
   Hemisphere contrail coverage using MODIS data. Geophys. Res. Lett. **40**, 612–617
   (2013)
2. Liu, S., Zhang, Z.: Learning group patterns for ground-based cloud classification in
   wireless sensor networks. EURASIP J. Wirel. Commun. Netw. **2016**(1), 1–6 (2016)

3. Meyer, R., Mannstein, H., Meerkotter, R., Schumann, U., Wendling, P.: Regional radiative forcing by line-shaped contrails derived from satellite data. J. Geophys. Res.: Atmos. **107**(D10), ACL 17-1–ACL 17-15 (2002)
4. Hetzheim, H.: Automated contrail and cirrus detection using stochastic properties. Int. J. Remote Sens. **28**(9), 2033–2048 (2007)
5. Ojala, T., Pietikäinen, M., Mäenpää, T.: Gray scale and rotation invariant texture classification with local binary patterns. In: Vernon, D. (ed.) ECCV 2000. LNCS, vol. 1842, pp. 404–420. Springer, Heidelberg (2000). https://doi.org/10.1007/3-540-45054-8_27
6. Buch Jr., K.A., Sun, C.H.: Cloud classification using whole-sky imager data. In: Ninth Symposium on Meteorological Observations & Instrumentation, vol. 16, no. 3, pp. 353–358 (2005)
7. Heinle, A., Macke, A., Srivastav, A.: Automatic cloud classification of whole sky images. Atmos. Meas. Tech. Discuss. **3**(3), 557–567 (2010)
8. Zhuo, W., Cao, Z., Xiao, Y.: Cloud classification of ground-based images using texture structure features. J. Atmos. Oceanic Technol. **31**(1), 79–92 (2014)
9. Dev, S., Lee, Y.H., Winkler, S.: Categorization of cloud image patches using an improved texton-based approach. In: IEEE International Conference on Image Processing (2015)
10. Zhang, J., Shang, J., Zhang, G.: Verification for different contrail parameterizations based on integrated satellite observation and ECMWF reanalysis data. Adv. Meteorol. **1**, 1–11 (2017)
11. Burkhardt, U., Karcher, B.: Global radiative forcing from contrail cirrus. Nat. Clim. Change **1**(1), 54–58 (2011)
12. Szegedy, C., Vanhoucke, V., Ioffe, S., Shlens, J., Wojna, Z.: Rethinking the inception architecture for computer vision. In: IEEE Conference on Computer Vision and Pattern Recognition (CVPR), pp. 2818–2826 (2016)
13. Ioffe, S., Szegedy, C.: Batch Normalization: Accelerating Deep Network Training by Reducing Internal Covariate Shift. arXiv e-prints, pp. 448–456 (2015)
14. Krizhevsky, A., Sutskever, I., Hinton, G.E.: ImageNet classification with deep convolutional neural networks. Commun. ACM **60**(2) (2012)
15. Deng, J., Socher, R., Fei-Fei, L., Dong, W., Li, K., Li, L.-J.: ImageNet: a large-scale hierarchical image database. In: 2009 IEEE Computer Society Conference on Computer Vision and Pattern Recognition Workshops (CVPR Workshops), pp. 248–255 (2009)
16. Srivastava, N., Hinton, G., Krizhevsky, A., Sutskever, I., Salakhutdinov, R.: Dropout: a simple way to prevent neural networks from overfitting. J. Mach. Learn. Res. **15**, 1929–1958 (2014)
17. Lee, J., Weger, R.C., Sengupta, S.K., Welch, R.M.: A neural network approach to cloud classification. IEEE Trans. Geosci. Remote Sens. **28**(5), 846–855 (1990)
18. Glorot, X., Bordes, A., Bengio, Y.: Deep sparse rectifier neural networks. In: International Conference on Artificial Intelligence and Statistics, pp. 315–323 (2011)

Author Index

Printed in the United States
By Bookmasters

Printed in the United States
By Bookmasters